D0022544

THEORETICAL MECHANICS
OF PARTICLES AND CONTINUA

International Series in Pure and Applied Physics

Park: *Introduction to the Quantum Theory*
Present: *Kinetic Theory of Gases*
Read: *Dislocations in Crystals*
Richtmyer, Kennard, and Cooper: *Introduction to Modern Physics*
Rossi and Olbert: *Introduction to the Physics of Space*
Schiff: *Quantum Mechanics*
Schwartz: *Introduction to Special Relativity*
Schwartz: *Principles of Electrodynamics*
Slater: *Quantum Theory of Atomic Structure, Vol. I*
Slater: *Quantum Theory of Matter*
Slater: *Electronic Structure of Molecules: Quantum Theory of Molecules and Solids, Vol. 1*
Slater: *Insulators, Semiconductors, and Metals: Quantum Theory of Molecules and Solids, Vol. 3*
Slater: *The Self-consistent Field for Molecules and Solids: Quantum Theory of Molecules and Solids, Vol. 4*
Slater and Frank: *Introduction to Theoretical Physics*
Smythe: *Static and Dynamic Electricity*
Stratton: *Electromagnetic Theory*
Tinkham: *Group Theory and Quantum Mechanics*
Tinkham: *Introduction to Superconductivity*
Townes and Schawlow: *Microwave Spectroscopy*
Wang: *Solid-State Electronics*
White: *Introduction to Atomic Spectra*

The late F. K. Richtmyer was Consulting Editor of the series from its inception in 1929 to his death in 1939. Lee A. DuBridge was Consulting Editor from 1939 to 1946; and G. P. Harnwell from 1947 to 1954. Leonard I. Schiff served as consultant from 1954 until his death in 1971.

Park: *Introduction to the Quantum Theory*
Present: *Kinetic Theory of Gases*
Read: *Dislocations in Crystals*
Richtmyer, Kennard, and Cooper: *Introduction to Modern Physics*
Rossi and Olbert: *Introduction to the Physics of Space*
Schiff: *Quantum Mechanics*
Schwartz: *Introduction to Special Relativity*
Schwartz: *Principles of Electrodynamics*
Slater: *Quantum Theory of Atomic Structure, Vol. I*
Slater: *Quantum Theory of Matter*
Slater: *Electronic Structure of Molecules: Quantum Theory of Molecules and Solids, Vol. 1*
Slater: *Insulators, Semiconductors, and Metals: Quantum Theory of Molecules and Solids, Vol. 3*
Slater: *The Self-consistent Field for Molecules and Solids: Quantum Theory of Molecules and Solids, Vol. 4*
Slater and Frank: *Introduction to Theoretical Physics*
Smythe: *Static and Dynamic Electricity*
Stratton: *Electromagnetic Theory*
Tinkham: *Group Theory and Quantum Mechanics*
Tinkham: *Introduction to Superconductivity*
Townes and Schawlow: *Microwave Spectroscopy*
Wang: *Solid-State Electronics*
White: *Introduction to Atomic Spectra*

The late F. K. Richtmyer was Consulting Editor of the series from its inception in 1929 to his death in 1939. Lee A. DuBridge was Consulting Editor from 1939 to 1946; and G. P. Harnwell from 1947 to 1954. Leonard I. Schiff served as consultant from 1954 until his death in 1971.

THEORETICAL MECHANICS OF PARTICLES AND CONTINUA

Alexander L. Fetter

John Dirk Walecka

Professors of Physics
Stanford University

McGraw-Hill Book Company

New York   St. Louis   San Francisco   Auckland   Bogotá   Hamburg
Johannesburg   London   Madrid   Mexico   Montreal   New Delhi
Panama   Paris   São Paulo   Singapore   Sydney   Tokyo   Toronto

This book was set in Times Roman.
The editors were C. Robert Zappa and Madelaine Eichberg;
the production supervisor was Charles Hess.
The drawings were done by Santype International Limited.
Fairfield Graphics was printer and binder.

THEORETICAL MECHANICS OF PARTICLES AND CONTINUA

234567890 FGRFGR 89876543210

Library of Congress Cataloging in Publication Data

Fetter, Alexander L
    Theoretical mechanics of particles and continua.

    (International series in pure and applied physics)
    Includes index.
    1. Continuum mechanics.   2.   Field theory (Physics)
I.   Walecka, John Dirk, date   joint author.
II.   Title.   III.   Series.
QA808.2.F47   530.1   79-15624
ISBN 0-07-020658-9

CONTENTS

Chapter **10** Surface Waves on Fluids

Chapter **11** Heat Conduction

Chapter **12** Viscous Fluids

PREFACE

This book developed from a first-year graduate course each of us has taught at Stanford since 1965. Most first-year physics graduate students enroll, along with some advanced undergraduates in physics and many graduate students from other departments. Originally, the course treated particle mechanics and mathematical physics, but the latter portion gradually evolved into a course on the physics of classical continuous media, not only for its own intrinsic interest but also as a natural outgrowth of the earlier material. We feel that a broad and thorough training in classical physics is essential for modern students of physics, independent of their subsequent choice of career. For example, familiarity with continuum mechanics, hydrodynamics, acoustics, and wave phenomena is fundamental in understanding the world around us, yet these subjects are generally missing from the standard graduate physics curriculum. In addition, classical mechanics provides a natural framework for introducing many of the advanced mathematical concepts in physics. A student's physical intuition concerning these everyday systems helps distinguish the mathematical questions from the physical ones, in contrast to the situation in classical electrodynamics or quantum mechanics, where the less familiar physics may itself be a source of difficulty.

We intend this frankly as a *textbook* and aim to provide a lucid and self-contained account of classical mechanics, together with appropriate mathematical methods. Although two quarters suffice to teach much of the material, a full year would allow a more complete and leisurely treatment. The material divides naturally into two parts: particles and continua. The first part starts from Newton's laws of motion and systematically develops the dynamics of classical particles, with chapters on basic principles, rotating coordinate systems, lagrangian formalism, small oscillations, dynamics of rigid bodies, and hamiltonian formalism, including a brief discussion of the transition to quantum mechanics. This part of the book also considers examples of the limiting behavior of many particles, facilitating the eventual transition to a continuous medium. The second part deals with classical continua, including chapters on strings, membranes, sound waves, surface waves on nonviscous fluids, heat conduction, viscous fluids, and elastic media. Each of these latter chapters is self-contained, providing the relevant physical background and developing the appropriate mathematical techniques. Thus the text treats lagrangian field theory, eigenfunctions and Sturm-Liouville theory, variational methods, perturbation theory, Green's functions, Fourier and Laplace transforms, and asymptotic techniques like the method of stationary phase. In addition, appendixes provide brief summaries of the theory of complex variables, vector and tensor calculus in curvilinear orthogonal coordinates, separation of variables, and integral representations of special functions.

Any treatment of classical mechanics must confront the question of special relativity. We have decided to omit it entirely, for we feel that it fits more naturally into classical electrodynamics, where the Lorentz invariance facilitates the treatment of four-dimensional space-time. In contrast, the customary relativistic generalization of Newton's laws of motion strikes us as cumbersome at best.

A textbook on mechanics faces a difficult problem in selecting references. Since our aim is to teach current physics for modern applications, we have not included primary sources, which students frequently find obscure or irrelevant. Some historical perspective is valuable, however, and we end this preface with a short chronological list of significant names associated with mechanics and mathematical physics. In addition, we have listed in Appendix G several familiar basic texts and monographs. These sources suffice for most sections, but where appropriate we have added selected references that we have found particularly clear or helpful, as a guide to further study.

Every chapter contains several homework problems of varying degrees of difficulty. We consider them an integral part of the text, and all students should attempt several from each chapter. Since classical mechanics is an old subject, no effort has been made to trace the origin of our examples and problems, many of which are modified versions of those in the list of texts and monographs.

The reader is assumed to be familiar with intermediate mechanics at the level of J. B. Marion, *Classical Dynamics of Particles and Systems*, 2d ed., Academic, New York, 1970, and with the elements of linear algebra and partial differential equations. A working knowledge of the first and second law of thermodynamics, at the level of F. Reif, *Fundamentals of Statistical and Thermal Physics*, McGraw-Hill, New York, 1965, will make some of the later sections on sound waves, heat conduction, and viscous fluids more meaningful.

We are grateful to our own teachers, in particular S. D. Drell and G. F. Carrier, for introducing us to many of these beautiful topics. We would also like to thank Victoria LaBrie for her invaluable help in the preparation of this manuscript.

Alexander L. Fetter
John Dirk Walecka

SIGNIFICANT NAMES IN MECHANICS AND MATHEMATICAL PHYSICS†

Isaac Newton (1642–1727)
Daniel Bernoulli (1700–1782)
Leonhard Euler (1707–1783)
Jean Le Rond d'Alembert (1717–1783)
Joseph Louis Lagrange (1736–1813)
Pierre Simon de Laplace (1749–1827)
Adrien Marie Legendre (1752–1833)
Jean Baptiste Joseph Fourier (1768–1830)
Karl Friedrich Gauss (1777–1855)
Siméon-Denis Poisson (1781–1840)
Friedrich Wilhelm Bessel (1784–1846)
Augustin-Louis Cauchy (1789–1857)
George Green (1793–1841)
Carl Gustav Jacob Jacobi (1804–1851)
William Rowan Hamilton (1805–1865)
Joseph Liouville (1809–1882)
George Gabriel Stokes (1819–1903)
Hermann Ludwig Ferdinand Helmholtz (1821–1894)
Gustav Robert Kirchhoff (1824–1887)
William Thomson (Lord Kelvin) (1824–1907)
Georg Friedrich Bernhard Riemann (1826–1866)
John William Strutt (Lord Rayleigh) (1842–1919)

† Detailed accounts of their contributions can be found in C. C. Gillispie (ed.), "Dictionary of Scientific Biography," Scribners, New York, 1970.

THEORETICAL MECHANICS
OF PARTICLES AND CONTINUA

BASIC PRINCIPLES

Classical mechanics involves the application of Newton's laws of motion to explain and predict the dynamical motion of point particles and bulk continuous matter. As such, it concerns the behavior of familiar classical macroscopic objects—natural and artificial satellites, the atmosphere and the oceans, laboratory solids, and even the earth itself. Indeed, one principal aim in studying classical mechanics is to understand the everyday world and to learn how to describe its properties quantitatively. In addition, classical mechanics has proved basic in deriving quantum descriptions of atomic matter, far from the original realm of classical physics. Finally, the challenge of characterizing continuous media has stimulated much of the basic mathematics of modern theoretical physics. Thus the study of bulk systems provides a natural framework for introducing and illustrating these techniques.

1 NEWTON'S LAWS

Although Newton's laws of motion are easily stated, their full implications involve subtle and complicated nonlinear phenomena that remain only partially explored. Since these laws are central to all our subsequent work, we briefly review them and some of their most basic corollaries and consequences.

Statement of Newton's laws

We first define a primary inertial coordinate system that is at rest with respect to the fixed stars. Newton's first law then states:

In this primary inertial frame, every body remains at rest or in uniform motion unless acted on by a force **F**. *The condition* $\mathbf{F} = 0$ *thus implies a constant velocity* **v** *and a constant momentum* $\mathbf{p} = m\mathbf{v}$.

In effect, Newton's first law asserts that such an inertial frame exists to arbitrary accuracy. If we construct an inertial frame and eliminate the forces as accurately as we can, Newton's first law appears to hold. Note that any experimental verification of this law must be approximate, for gravitational forces are always present in the universe as we know it.

Newton's second and third laws then state:

In the primary inertial frame, application of a force alters the momentum, in an amount specified by the quantitative relation

$$\mathbf{F} = \frac{d\mathbf{p}}{dt} \equiv \dot{\mathbf{p}} \tag{1.1}$$

Here a dot denotes a time derivative.

To each action, there is an equal and opposite reaction. Thus if $\mathbf{F}_{21}$ *is the force exerted on particle 1 by particle 2, then*

$$\mathbf{F}_{12} = -\mathbf{F}_{21} \tag{1.2}$$

and these forces act along the line separating the particles.

In applying these laws, several remarks are relevant. First, if mass is conserved and constant in time, the relation $\mathbf{p} = m\mathbf{v}$ reduces Eq. (1.1) to the familiar form

$$\mathbf{F} = m\frac{d\mathbf{v}}{dt} \tag{1.3}$$

Otherwise, it is essential to retain the original expression, e.g., in studying the dynamics of an evaporating droplet. Second, Eqs. (1.2) and (1.3) serve to define a given amount of mass in terms of a fundamental unit m^* that acquires unit acceleration under the influence of a unit force. More precisely, if the standard particle 1 (mass m^*) interacts with any other particle (m_2, say), the magnitude of their relative accelerations a_{12} and a_{21} specifies m_2 through the relation $m_2|a_{12}| = m^*|a_{21}|$. These considerations are independent of the particular force law. Thus they apply to the gravitational force between two particles with masses m_1 and m_2

$$\mathbf{F}_{21} = -Gm_1m_2\frac{\mathbf{r}_1 - \mathbf{r}_2}{|\mathbf{r}_1 - \mathbf{r}_2|^3} \tag{1.4a}$$

with G the universal constant of newtonian gravitation, and equally to Coulomb's force between two electrified objects with charges Q_1 and Q_2

$$\mathbf{F}_{21} = Q_1Q_2\frac{\mathbf{r}_1 - \mathbf{r}_2}{|\mathbf{r}_1 - \mathbf{r}_2|^3} \tag{1.4b}$$

in cgs units. It is striking that both these basic forces vary as the inverse square of the separation. It is the physicist's task to classify and enumerate the forces acting on a system; Newton's laws then allow one to calculate the subsequent motion.

As a final remark, we can verify the principle of galilean relativity that any frame moving with constant velocity relative to an inertial frame is again inertial. Thus, two observers moving uniformly with respect to each other and with respect to the primary inertial coordinate system infer the same basic laws of motion, at least in the usual case that $\mathbf{F}_{21}$ depends only on the vector separation of the particles.

PROOF Let $\mathbf{r}$ and $\mathbf{r}'$ be the coordinates as seen in two different frames moving with constant relative velocity $\mathbf{V}$. Evidently $\mathbf{r}' = \mathbf{r} + \mathbf{V}t$, so that $\mathbf{r}_i - \mathbf{r}_j = \mathbf{r}'_i - \mathbf{r}'_j$ and $\mathbf{F}_{ij} = \mathbf{F}'_{ij}$. Moreover, the usual rules of calculus ensure that $d^2\mathbf{r}/dt^2 = d^2\mathbf{r}'/dt^2$, implying that both the forces and the accelerations are the same in the two frames.

Conservation Laws

It is possible to work directly with Newton's laws, but there are distinct conceptual advantages in introducing special derived quantities like linear and angular momentum and energy, which turn out to satisfy certain simple relations.

Linear momentum Equation (1.1) can be reinterpreted as the statement that the applied force determines the rate of change of $\mathbf{p}$. In particular, $\mathbf{p}$ is a constant vector whenever $\mathbf{F}$ vanishes, and this relation holds separately for each vector component.

Angular momentum Define the angular momentum

$$\mathbf{L} = \mathbf{r} \times \mathbf{p} \tag{1.5a}$$

and assume that m is constant, implying

$$\mathbf{L} = m\mathbf{r} \times \mathbf{v} \tag{1.5b}$$

The rate of change of $\mathbf{L}$ is given by

$$\dot{\mathbf{L}} = m\dot{\mathbf{r}} \times \mathbf{v} + m\mathbf{r} \times \dot{\mathbf{v}}$$

and the observation that $\dot{\mathbf{r}} = \mathbf{v}$ eliminates the first term on the right-hand side. Use of Eq. (1.3) then gives

$$\dot{\mathbf{L}} = \mathbf{r} \times \mathbf{F} \equiv \boldsymbol{\Gamma} \tag{1.6}$$

where $\boldsymbol{\Gamma}$ is the torque. Once again, we have obtained a vector conservation law, any specified component of angular momentum remaining constant whenever the corresponding component of torque vanishes. In contrast to $\mathbf{p}$, however, we note that $\mathbf{L}$ depends on the choice of coordinate frame, since a shift of the origin by $-\mathbf{r}_0$ transforms $\mathbf{r}$ into $\mathbf{r} + \mathbf{r}_0$ and $\mathbf{L}$ correspondingly becomes $m\mathbf{r} \times \mathbf{p} + m\mathbf{r}_0 \times \mathbf{p}$, where

$\dot{\mathbf{r}}_0$ is assumed zero. The even more complicated case of transformation to moving coordinates will be considered in Chap. 2.

Energy and work Consider a static force field $\mathbf{F}(\mathbf{r})$ defined throughout some region of space. If a test particle is inserted at $\mathbf{r}$ and moved a small distance $d\mathbf{s}$, the work done on the particle is $dW = \mathbf{F}(\mathbf{r}) \cdot d\mathbf{s}$. Consequently, the work in moving the test particle a finite distance from point 1 to point 2 along some particular path is just the line integral

$$W_{1\to2} = \int_1^2 d\mathbf{s} \cdot \mathbf{F}(\mathbf{r}) \tag{1.7a}$$

In general, this relation cannot be simplified. For the special case that the particle starts at $\mathbf{r}_1$ and follows a dynamical trajectory that passes through $\mathbf{r}_2$, however, the element of length $d\mathbf{s}$ is then just $\mathbf{v}\, dt$, and the dynamical principle (1.3) allows us to integrate Eq. (1.7a) directly

$$W_{1\to2} = \int_1^2 d\mathbf{s} \cdot \left(m\frac{d\mathbf{v}}{dt} \right) = \int_1^2 dt\, m\mathbf{v} \cdot \frac{d\mathbf{v}}{dt} = m \int_1^2 dt\, \frac{d}{dt}\tfrac{1}{2}v^2 = \tfrac{1}{2}mv_2^2 - \tfrac{1}{2}mv_1^2 \tag{1.7b}$$

independent of the intervening path. If $T \equiv \tfrac{1}{2}mv^2$ denotes the kinetic energy, the work done in moving a particle from 1 to 2 is precisely the increase in the kinetic energy $T_2 - T_1$.

This result can be sharpened if $\mathbf{F}(\mathbf{r})$ has the special form

$$\mathbf{F}(\mathbf{r}) = -\nabla U(\mathbf{r}) \tag{1.8}$$

where U is known as the *potential*. Such forces are called conservative; although they occur frequently, it is important to realize that they are quite restrictive, the scalar function $U(\mathbf{r})$ being specified by only one number at each point whereas a general vector field requires three. For such conservative forces, the right-hand side of Eq. (1.7a) is readily rewritten $-\int_1^2 d\mathbf{s} \cdot \nabla U(\mathbf{r})$, and the integrand is now just the differential change in U in moving from $\mathbf{r}$ to $\mathbf{r} + d\mathbf{s}$. Thus

$$-\int_1^2 d\mathbf{s} \cdot \nabla U(\mathbf{r}) = -\int_1^2 dU = -U_2 + U_1 \tag{1.9}$$

A combination with Eq. (1.7b) immediately yields the relation $T_2 - T_1 = -U_2 + U_1$, or, equivalently, the conservation law

$$T_1 + U_1 = T_2 + U_2 \tag{1.10}$$

for the total energy $E = T + U$ in the presence of conservative forces. To conclude this section, we may also recall two other equivalent criteria for conservative forces (see Prob. 1.1):†

$$\nabla \times \mathbf{F}(\mathbf{r}) = 0 \qquad \text{for all } \mathbf{r} \tag{1.11a}$$

$$\oint d\mathbf{s} \cdot \mathbf{F}(\mathbf{r}) = 0 \qquad \text{for all closed paths} \tag{1.11b}$$

† Problems will be found at the end of each chapter.

2 SYSTEMS OF PARTICLES

The preceding considerations are readily generalized to systems of interacting particles, assuming only that the individual masses are constant in time and that Newton's laws of motion remain valid. To be specific, consider N particles with masses $\{m_i\}$ located at instantaneous positions $\{r_i\}$ in an inertial coordinate system. We then define the vector center of mass $\mathbf{R}$ by the relation

$$\sum_i m_i \mathbf{r}_i = \sum_i m_i \mathbf{R} \tag{2.1a}$$

or, equivalently,

$$\mathbf{R} = M^{-1} \sum_i m_i \mathbf{r}_i \tag{2.1b}$$

where

$$M = \sum_i m_i \tag{2.2}$$

is the total mass. Here, the sums run over all particles $i = 1, \ldots, N$.

Center-of-Mass Motion

It is convenient to separate the force $\mathbf{F}_i$ acting on the ith particle into an external contribution $\mathbf{F}_i^{(e)}$ arising from outside sources and the contributions $\mathbf{F}_{ji}$ arising from the other particles $j \neq i$. Thus we write $\mathbf{F}_i = \mathbf{F}_i^{(e)} + \sum_{j \neq i} \mathbf{F}_{ji}$. Since the ith particle moves according to the total force $\mathbf{F}_i$, Newton's second law immediately yields

$$\dot{\mathbf{p}}_i = \mathbf{F}_i^{(e)} + \sum_{j \neq i} \mathbf{F}_{ji} \tag{2.3}$$

Evidently, the system typically undergoes a complicated motion, for the forces depend on the instantaneously changing positions. Despite this situation, there remain certain conserved quantities that help simplify the description. In particular, consider the acceleration of the center of mass

$$M\ddot{\mathbf{R}} = \sum_i m_i \ddot{\mathbf{r}}_i = \sum_i \dot{\mathbf{p}}_i = \sum_i \mathbf{F}_i^{(e)} + \sum_i \sum_{j \neq i} \mathbf{F}_{ji}$$

To analyze the last term, it is convenient to define $\mathbf{F}_{ii} \equiv 0$, reflecting the absence of self forces; the double sum then vanishes identically, since a change of dummy summation indices and Newton's third law (1.2) yield

$$\sum_i \sum_{j \neq i} \mathbf{F}_{ji} = \frac{1}{2} \sum_{ij} (\mathbf{F}_{ji} + \mathbf{F}_{ij}) = 0$$

We thus reach the very important conclusion that

$$M\ddot{\mathbf{R}} = M\dot{\mathbf{V}} = \sum_i \mathbf{F}_i^{(e)} \equiv \mathbf{F}^{(e)} \tag{2.4}$$

namely, the center of mass $\mathbf{R}$ moves as if the total external force $\mathbf{F}^{(e)}$ acted on the total mass M concentrated at $\mathbf{R}$. As an immediate yet powerful corollary, note that the total momentum of the system

$$\mathbf{P} = \sum_i \mathbf{p}_i \qquad (2.5)$$

is a vector constant of the motion whenever $\mathbf{F}^{(e)}$ vanishes, which follows from the various alternative representations

$$\mathbf{P} = \sum_i m_i \dot{\mathbf{r}}_i = \sum_i m_i \mathbf{v}_i = M\dot{\mathbf{R}} = M\mathbf{V} \qquad (2.6)$$

where $\mathbf{V}$ is the velocity of the center of mass.

Angular Momentum

The total angular momentum $\mathbf{L}$ is defined in an analogous manner

$$\mathbf{L} = \sum_i \mathbf{r}_i \times \mathbf{p}_i \qquad (2.7)$$

with the time derivative

$$\dot{\mathbf{L}} = \sum_i \dot{\mathbf{r}}_i \times \mathbf{p}_i + \sum_i \mathbf{r}_i \times \dot{\mathbf{p}}_i \qquad (2.8)$$

The first term on the right-hand side is zero, just as in the derivation of Eq. (1.6) for a single particle, and a combination with Eq. (2.3) yields

$$\dot{\mathbf{L}} = \sum_i \mathbf{r}_i \times \left(\mathbf{F}_i^{(e)} + \sum_j \mathbf{F}_{ji} \right)$$

The last term can be rewritten by interchanging dummy indices and then using Eq. (1.2)

$$\sum_{ij} \mathbf{r}_i \times \mathbf{F}_{ji} = \frac{1}{2} \sum_{ij} (\mathbf{r}_i \times \mathbf{F}_{ji} + \mathbf{r}_j \times \mathbf{F}_{ij})$$

$$= \frac{1}{2} \sum_{ij} (\mathbf{r}_i - \mathbf{r}_j) \times \mathbf{F}_{ji}$$

This quantity vanishes when the force $\mathbf{F}_{ji}$ lies along $\mathbf{r}_i - \mathbf{r}_j$, as in Eq. (1.2);† thus we find the dynamical law

$$\dot{\mathbf{L}} = \sum_i \mathbf{r}_i \times \mathbf{F}_i^{(e)} \equiv \mathbf{\Gamma}^{(e)} \qquad (2.9)$$

---

† If the direction of $\mathbf{F}_{ij}$ differs from that of $\mathbf{r}_i - \mathbf{r}_j$, our formulation of Newton's third law [Eq. (1.2) and following material] no longer holds. In that case, the total angular momentum $\mathbf{L}$ of the particles need not obey Eq. (2.9). Such noncollinearity generally reflects the presence of other nonmechanical degrees of freedom that contribute to the overall conservation laws. One simple example is the magnetic force between charged particles when the electromagnetic fields themselves contain intrinsic angular momentum.

relating the rate of change of total angular momentum $\mathbf{L}$ to the external torque $\mathbf{\Gamma}^{(e)}$. In particular, $\mathbf{L}$ is a constant of the motion if $\mathbf{\Gamma}^{(e)} = 0$. Note that changes in $\mathbf{L}$ arise only from external forces, similar to the behavior of the total momentum in Eq. (2.4); internal contributions from the other particles affect the individual motions, but they cancel out in evaluating the total momentum and angular momentum. It is important to emphasize the vector nature of Eq. (2.9), which holds independently for each component. As a simple example, consider a uniformly precessing spinning top supported at the origin in a uniform vertical gravitational field; the external torque always lies in the horizontal plane, so that L_z is a strict constant of the motion, but L_x and L_y vary periodically with time.

It is often preferable to transform to an internal coordinate system located at the (moving) center of mass, writing

$$\mathbf{r}_i = \mathbf{R} + \mathbf{r}'_i \tag{2.10a}$$

with $\mathbf{r}'_i$ the coordinate of the ith particle measured from $\mathbf{R}$ (see Fig. 2.1). Differentiation with respect to time yields the analogous expression for velocities

$$\mathbf{v}_i = \mathbf{V} + \mathbf{v}'_i \tag{2.10b}$$

We first derive a useful relation by comparing Eqs. (2.1a) and (2.10a)

$$\sum_i m_i \mathbf{r}_i = \sum_i m_i(\mathbf{R} + \mathbf{r}'_i) = M\mathbf{R} + \sum_i m_i \mathbf{r}'_i \tag{2.11}$$

By definition, the left-hand side is just $M\mathbf{R}$, so that

$$\sum_i m_i \mathbf{r}'_i = 0 \qquad \sum_i m_i \mathbf{v}'_i = 0 \tag{2.12}$$

where the second relation is simply the time derivative of the first. Consider now the total angular momentum expressed in these variables:

$$\mathbf{L} = \sum_i (\mathbf{R} + \mathbf{r}'_i) \times m_i(\mathbf{V} + \mathbf{v}'_i)$$

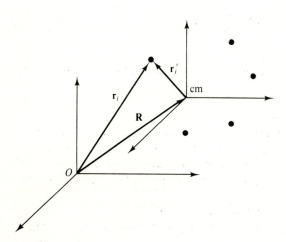

Figure 2.1 Center-of-mass coordinates for a system of particles.

Two of these terms are zero because of (2.12), and we find

$$L = \sum_i m_i \, \mathbf{R} \times \mathbf{V} + \sum_i m_i \mathbf{r}_i' \times \mathbf{v}_i' = \mathbf{L}_{cm} + \mathbf{L}' \qquad (2.13a)$$

Here

$$\mathbf{L}_{cm} \equiv M\mathbf{R} \times \mathbf{V} = \mathbf{R} \times \mathbf{P} \qquad (2.13b)$$

is the angular momentum of the center of mass relative to the fixed origin, and

$$\mathbf{L}' = \sum_i m_i \mathbf{r}_i' \times \mathbf{v}_i' \qquad (2.13c)$$

is the internal angular momentum about the center of mass, properly expressed solely in terms of the primed (internal) variables. If $\mathbf{V} = 0$, then $\mathbf{L}$ is independent of the position of $\mathbf{R}$, but this property is not true in general.

These relations apply directly to the dynamical behavior of the composite system. As in the previous cases, the time derivative of $\mathbf{L}_{cm}$ has a simple form

$$\dot{\mathbf{L}}_{cm} = \frac{d}{dt}(\mathbf{R} \times \mathbf{P}) = \mathbf{R} \times \dot{\mathbf{P}}$$

Use of Eqs. (2.4) and (2.6) leads to

$$\dot{\mathbf{L}}_{cm} = \mathbf{R} \times \sum_i \mathbf{F}_i^{(e)} = \mathbf{R} \times \mathbf{F}^{(e)} \qquad (2.14)$$

showing that the angular momentum of the center of mass changes as if the total external force $\mathbf{F}^{(e)}$ exerted a torque $\mathbf{R} \times \mathbf{F}^{(e)}$ at $\mathbf{R}$. A combination with Eq. (2.9) yields the additional and crucial relation

$$\dot{\mathbf{L}}' \equiv \dot{\mathbf{L}} - \dot{\mathbf{L}}_{cm} = \sum_i (\mathbf{r}_i - \mathbf{R}) \times \mathbf{F}_i^{(e)} = \sum_i \mathbf{r}_i' \times \mathbf{F}_i^{(e)} \qquad (2.15)$$

It states that the rate of change of the angular momentum about the center of mass is equal to the external torque about the center of mass. This relation holds for arbitrary $\mathbf{F}_i^{(e)}$ and hence for *arbitrary motion of the center-of-mass coordinate* $\mathbf{R}$ by Eq. (2.4). In particular, the dynamical relation (2.15) holds *even if the center-of-mass coordinate system in* Fig. 2.1 *is not an inertial frame*. We thus infer that $\mathbf{L}'$ is a strict constant of the motion if there are no torques about the center of mass, independent of its motion. In Chap. 5 this observation will be used to explain the wobbling motion of a spinning axisymmetric object in free fall, because the external (gravitational) forces exert no net torque. We also note that differentiation of Eq. (2.13c) provides an alternative expression

$$\dot{\mathbf{L}}' = \sum_i m_i \mathbf{r}_i' \times \dot{\mathbf{v}}_i' = \sum_i \mathbf{r}_i' \times \dot{\mathbf{p}}_i' \qquad (2.16)$$

properly expressed in terms of the primed variables.

Energy

The total kinetic energy is defined as the sum of the individual contributions

$$T = \frac{1}{2} \sum_i m_i v_i^2 \qquad (2.17)$$

Use of (2.10b) and (2.12) separates T into its two physically distinct components

$$T = T_{cm} + T' \tag{2.18a}$$

where $\qquad T_{cm} = \tfrac{1}{2}MV^2 \qquad$ and $\qquad T' = \dfrac{1}{2}\sum_i m_i v_i'^2 \tag{2.18b}$

are the energies associated with the motion of the center of mass and the internal motion about **R**. As before, it is convenient (and usually sufficient) to assume conservative forces, with

$$\mathbf{F}_i^{(e)} = -\mathbf{V}_i V^{(e)}(\mathbf{r}_i) \tag{2.19a}$$

$$\mathbf{F}_{ji} = -\mathbf{V}_i V(r_{ij}) = -\mathbf{V}_{ij} V(r_{ij}) \tag{2.19b}$$

Here we assume an isotropic interparticle potential, which holds whenever $\mathbf{F}_{ji}$ lies along $\mathbf{r}_{ij} \equiv \mathbf{r}_i - \mathbf{r}_j$, and use $\mathbf{V}_i$ to denote the gradient with respect to the coordinate $\mathbf{r}_i$. Suppose that the system is altered from configuration 1 to configuration 2 by moving each particle along prescribed trajectories; the work done on the system is given by

$$W_{1\to 2} = \sum_i \int_1^2 d\mathbf{s}_i \cdot \mathbf{F}_i = \sum_i \int_1^2 d\mathbf{s}_i \cdot \mathbf{F}_i^{(e)} + \sum_{ij} \int_1^2 d\mathbf{s}_i \cdot \mathbf{F}_{ji}$$

where $\mathbf{F}_{ii}$ is again assumed to vanish. Relabeling the dummy indices in the last term yields

$$\frac{1}{2}\sum_{ij} \int_1^2 (d\mathbf{s}_i - d\mathbf{s}_j) \cdot \mathbf{F}_{ji}$$

and we next observe that $d\mathbf{s}_i - d\mathbf{s}_j$ is just the change in relative separation $d\mathbf{r}_{ij}$. Use of (2.19b) then reduces the term to

$$\sum_{ij} \int_1^2 d\mathbf{s}_i \cdot \mathbf{F}_{ji} = -\frac{1}{2}\sum_{ij} \int_1^2 d\mathbf{r}_{ij} \cdot \mathbf{V}_{ij} V(r_{ij}) = -\frac{1}{2}\sum_{ij} [V(r_{ij})]_1^2$$

A similar calculation yields the work done by the external forces. Moreover, the total work done is the change in the total kinetic energy by Eq. (1.7). Exactly as for the one-body system, we can therefore infer the N-body conservation law

$$T + \sum_i V^{(e)}(\mathbf{r}_i) + \frac{1}{2}\sum_{ij} V(r_{ij}) = T + V = E = \text{const} \tag{2.20}$$

valid for conservative forces. Here V is the total potential energy

$$V = \sum_i V^{(e)}(\mathbf{r}_i) + \frac{1}{2}\sum_{ij} V(r_{ij}) \tag{2.21}$$

where the factor $\tfrac{1}{2}$ can be interpreted as indicating that each pair of particles contributes only once to V. Note also that $V(r_{ii}) = 0$ because of the absence of self forces, so that the internal potential energy is sometimes written $\sum_{i<j} V(r_{ij})$. In the special case of a rigid body with fixed $|r_{ij}|$, the internal contribution to V is constant, and only the external potentials affect the change in kinetic energy.

3 CENTRAL FORCES

Perhaps the most important application of the preceding formalism is the motion of a particle of mass m subject to a central force $\mathbf{F}(\mathbf{r}) = \hat{r}f(r)$ directed toward a fixed point (here taken as the origin), as illustrated in Fig. 3.1. It is easy to verify that $\nabla \times \mathbf{F}$ is zero, implying that $\mathbf{F}$ is conservative; thus it has the representation $\mathbf{F} = -\nabla V(r)$, or, taking radial components

$$f(r) = -\frac{dV(r)}{dr} \tag{3.1}$$

Moreover, the torque $\mathbf{r} \times \mathbf{F}$ vanishes by inspection, so that the angular momentum $\mathbf{l} = \mathbf{r} \times \mathbf{p}$ is a constant of the motion, with fixed *direction* and *magnitude*. As a consequence, the motion is strictly planar, since the position vector $\mathbf{r}$ of the particle measured from the center of force is always perpendicular to this fixed vector $\mathbf{l}$:

$$\mathbf{r} \cdot \mathbf{l} = \mathbf{r} \cdot (\mathbf{r} \times \mathbf{p}) = 0 \tag{3.2}$$

Note that similarly $\mathbf{p} \cdot \mathbf{l} = \mathbf{p} \cdot (\mathbf{r} \times \mathbf{p}) = 0$. The description of such motion requires only two variables, which may be taken as the plane polar coordinates (r, ϕ) with

$$x = r \cos \phi \qquad \text{and} \qquad y = r \sin \phi \tag{3.3}$$

Conservation Laws

We start by invoking the conservation of energy for a conservative force $E = \frac{1}{2}m(\dot{x}^2 + \dot{y}^2) + V(r)$. Use of Eq. (3.3) yields the relation

$$E = T + V = \tfrac{1}{2}m(\dot{r}^2 + r^2\dot{\phi}^2) + V(r) \tag{3.4}$$

known as a *first integral* because it involves only first time derivatives, in contrast to the second time derivatives in Newton's second law (see Prob. 1.6). Further-

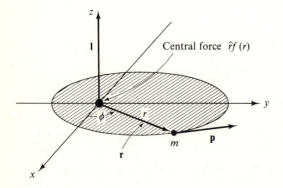

Figure 3.1 The central-force problem.

more, the angular momentum lies perpendicular to the plane of the motion with fixed magnitude $l_z = xp_y - yp_x = m(x\dot{y} - y\dot{x})$. A simple calculation with Eq. (3.3) provides a second constant of the motion

$$l = mr^2\dot{\phi} \tag{3.5}$$

Any physical motion must correlate $r(t)$ and $\phi(t)$ to ensure the constancy of E and l; thus if $\dot{\phi}$ increases, r^2 must correspondingly decrease to maintain fixed l. Moreover, $\dot{\phi}$ necessarily has the sign of l for all time because $r^2 > 0$; hence the particle in a central potential always advances, with no retrograde motion.

Equation (3.5) has one simple but important consequence. Consider a particle moving along some planar trajectory $[r(t), \phi(t)]$ as illustrated in Fig. 3.1. In a small interval of time dt, it moves through an angular interval $d\phi = \dot{\phi}\, dt$, sweeping out an element of area (Fig. 3.2)

$$dA = \tfrac{1}{2}r(r\, d\phi) = \tfrac{1}{2}r^2\dot{\phi}\, dt \tag{3.6}$$

Comparison with (3.5) shows that $dA = (l/2m)\, dt$, implying that the areal velocity has a constant value

$$\dot{A} = \frac{l}{2m} = \text{const} \tag{3.7}$$

Kepler established empirically that $\dot{A}$ is constant for planetary orbits; it constitutes the second of his three laws. The others will be considered subsequently because they depend on the detailed form of the gravitational interaction.

Effective Potential

Before studying the special case of gravitational forces, we shall consider some general features of the motion $\mathbf{r}(t)$. In general, the solution involves six constants, which may be taken, for example, as the initial position and velocity. Alternatively, it is often preferable to choose a different more physical set including E and $\mathbf{l}$, with E and l taken from Eqs. (3.4) and (3.5) and $\hat{l}$ fixed by the two constants that specify the orientation of the orbital plane, e.g., its normal. This choice leaves only two remaining constants, to be determined as follows. Equation (3.5) implies that $\dot{\phi} = l/mr^2$, and substitution into (3.4) yields an equivalent one-dimensional conservation law [compare Eq. (1.10)]

$$E = \tfrac{1}{2}m\dot{r}^2 + V_{\text{eff}}(r) \tag{3.8}$$

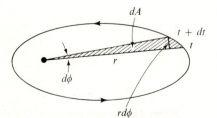

Figure 3.2 The areal velocity in a central field.

for motion with positive r in an effective one-dimensional potential

$$V_{\text{eff}}(r) = V(r) + \frac{l^2}{2mr^2} \tag{3.9}$$

In this interpretation, the angular momentum augments the original potential $V(r)$ with a centrifugal barrier $l^2/2mr^2$. For the typical case of an attractive potential $-\lambda r^{-1}$ with λ positive, V_{eff} is repulsive at short distances owing to the centrifugal barrier but attractive for large r (see Fig. 3.3). Since $E - V_{\text{eff}} = \frac{1}{2}m\dot{r}^2$ must be positive, the motion for a given E necessarily occurs in the region where $E > V_{\text{eff}}$, and the positions r_0 satisfying the relation $E = V_{\text{eff}}(r_0)$ are known as *turning points* because $\dot{r} = 0$ there. In Fig. 3.3, for example, the motion is bounded internally for positive E since there is only one turning point. Correspondingly, the trajectory approaches from infinity, reaches the turning point, and then recedes to infinity again. If E is negative, however, the motion is bounded both internally and externally with r oscillating between the two turning points. As E decreases, these turning points approach each other and coalesce at some negative minimum energy E_{min}, when only a circular orbit remains. Evidently, no motion can occur for $E < E_{\text{min}}$. Note that the turning points and E_{min} depend explicitly on l. Similar but more complicated diagrams describe other potentials, for example $V(r) = -\lambda e^{-r/a} r^{-1}$ or $-\lambda r^{-n}$; moreover, the typical orbit never closes, even for $E_{\text{min}} < E < 0$, because the angular period is generally incommensurate with the radial one.

In principle, Eq. (3.8) provides the complete solution to the trajectory, just as in a one-dimensional problem. Solving for dr, we readily find

$$t = \pm(\tfrac{1}{2}m)^{1/2} \int^r dr \, [E - V_{\text{eff}}(r)]^{-1/2} + t_0 \tag{3.10}$$

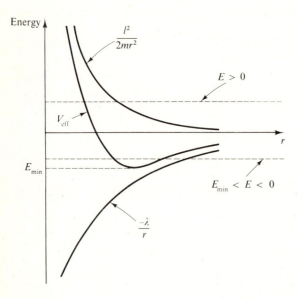

Figure 3.3 Effective potential in a central potential $V(r) = -\lambda/r$.

where t_0 is one additional constant of the motion and the sign of the square root is fixed by the direction of the radial motion at the time t (in or out). Inversion of Eq. (3.10) determines the radial position $r(t)$, and substitution into (3.5) then gives the remaining function

$$\phi = lm^{-1} \int^t dt \, [r(t)]^{-2} + \phi_0 \qquad (3.11)$$

including the final constant ϕ_0. Note that these last two constants are rather trivial since they depend on the arbitrary choice of initial time and angle. Since $r(t)$ and $\phi(t)$ completely fix the time-dependent orbit, they provide the general solution.

In practice, the above prescription rapidly becomes complicated, and it is frequently simpler to forgo any knowledge of the time dependence, seeking instead only the geometric orbit $r(\phi)$. We start from Eq. (3.5), rewritten as $\dot{r} = (dr/d\phi)\dot{\phi} = (dr/d\phi)(l/mr^2)$; substitution into Eq. (3.8) yields

$$E = \frac{l^2}{2mr^4}\left(\frac{dr}{d\phi}\right)^2 + V_{\text{eff}}(r) \qquad (3.12)$$

which can be integrated to give the formal equation of the orbit

$$\phi = \pm l(2m)^{-1/2} \int^r dr \, r^{-2}[E - V_{\text{eff}}(r)]^{-1/2} + \phi_0 \qquad (3.13)$$

At worst, the integral can be evaluated numerically, but an analytic solution is feasible in simple cases.

Inverse-Square Force: Kepler's Laws

The simplest and most important example is the motion in an attractive gravitational field (1.4a), where

$$V(r) = -m\gamma r^{-1} \qquad (3.14a)$$

Here
$$\gamma \equiv GM \qquad (3.14b)$$

and M is the mass of the fixed central body. The substitution $u = r^{-1}$ simplifies Eq. (3.13) to the form

$$\phi = \phi_0 \mp \int^u du \left(\frac{2mE}{l^2} + \frac{2m^2\gamma u}{l^2} - u^2\right)^{-1/2} \qquad (3.15a)$$

and a standard integration gives

$$\phi = \phi_0 \pm \arccos \frac{1 - ul^2/m^2\gamma}{(1 + 2El^2/m^3\gamma^2)^{1/2}} \qquad (3.15b)$$

Solving for u, we find

$$r^{-1} = C[1 - \epsilon \cos(\phi - \phi_0)] \qquad (3.16a)$$

where
$$\epsilon = \left(1 + \frac{2El^2}{m^3\gamma^2}\right)^{1/2} \quad \text{and} \quad C = \frac{m^2\gamma}{l^2} \tag{3.16b}$$

The initial angle ϕ_0 determines the orientation of the orbit; for definiteness, we set $\phi_0 = 0$ to obtain the conventional representation

$$r^{-1} = C(1 - \epsilon \cos \phi) \tag{3.17}$$

describing a conic section. Only the bound orbits are considered in this section, where E is negative and the orbit turns out to represent an ellipse. Section 5 treats other possibilities.

To verify the identification of (3.17), we recall some alternative definitions of an ellipse.

1. The simplest characterization asserts the existence of two fixed points (*foci*) located at $x = \pm f$, $y = 0$, such that the orbital point maintains a constant total distance $d + d'$ from the two foci. Consideration of Fig. 3.4 for $y = 0$ shows that

$$d + d' = 2a \tag{3.18}$$

where a is the semimajor axis, and it is conventional to define the eccentricity ϵ by the relation

$$f = \epsilon a \tag{3.19}$$

Evidently, $\epsilon < 1$ for an ellipse, the limit $\epsilon = 0$ representing a circle.

2. Put the origin at the left focus and introduce plane polar coordinates. By definition, d' is the radial distance r, and the law of cosines gives

$$d^2 = d'^2 - 4d'f \cos \phi + 4f^2 \tag{3.20a}$$

A combination with Eqs. (3.18) and (3.19) readily yields the polar equation of an ellipse with its center at the left focus:

$$r(1 - \epsilon \cos \phi) = a(1 - \epsilon^2) \tag{3.20b}$$

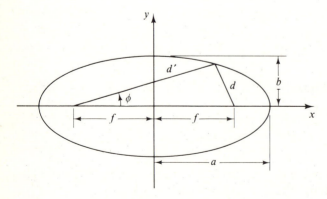

Figure 3.4 The ellipse as a conic section. Here $f = \epsilon a$.

3. Introduce plane cartesian coordinates with the origin at the center of the ellipse. Pythagoras' rule implies

$$d^2 = (x - f)^2 + y^2 \qquad \text{and} \qquad d'^2 = (x + f)^2 + y^2 \tag{3.21a}$$

or, equivalently,

$$\tfrac{1}{2}(d'^2 + d^2) = x^2 + y^2 + f^2 \tag{3.21b}$$

$$\tfrac{1}{2}(d'^2 - d^2) = 2xf \tag{3.21c}$$

and a simple manipulation with Eqs. (3.18) and (3.19) gives

$$d' - d = \frac{2xf}{a} = 2\epsilon x \tag{3.22}$$

As a final step, we note the algebraic identity

$$\tfrac{1}{2}(d^2 + d'^2) - \tfrac{1}{4}(d - d')^2 = \tfrac{1}{4}(d + d')^2 = a^2 \tag{3.23}$$

whose left-hand side can be constructed from Eqs. (3.21b) and (3.22). In this way, we find

$$\frac{x^2}{a^2} + \frac{y^2}{b^2} = 1 \tag{3.24}$$

where the length b is defined by

$$b^2 = a^2(1 - \epsilon^2) = a^2 - f^2 \tag{3.25}$$

Inspection of Fig. 3.4 for $x = 0$ shows that b is the semiminor axis.

Comparison of Eqs. (3.17) and (3.20b) verifies that the orbit of a particle with $E < 0$ in a gravitational field is indeed an ellipse with the center at one focus; this conclusion constitutes Kepler's first law of planetary motion. Moreover, Eq. (3.16b) determines the eccentricity

$$\epsilon = \left(1 - \frac{2|E|l^2}{m^3\gamma^2}\right)^{1/2} \tag{3.26a}$$

and the semimajor axis

$$a = [C(1 - \epsilon^2)]^{-1} = \frac{m\gamma}{2|E|} \tag{3.26b}$$

in terms of the physical quantities $|E|$ and l. Alternatively, Eq. (3.26b) can be rewritten

$$E = -\frac{m\gamma}{2a} = -\frac{MmG}{2a} \tag{3.27}$$

showing that the energy of the orbit depends only on the semimajor axis and not on the angular momentum. These relations hold for any inverse-square force; similar behavior occurs even in the quantum theory of the hydrogen atom, where the energy levels depend only on the principal quantum number and are independent of the angular momentum.

The last of Kepler's laws follows directly from the preceding analysis. Let τ be the period of the motion. By definition, the total area $A = \pi ab$ of the ellipse swept out in one orbit can be obtained from Eq. (3.7)

$$A = \pi ab = \int_0^\tau dt \dot{A} = \frac{l\tau}{2m}$$

Use of Eqs. (3.25) and (3.26) then leads to

$$A = \pi a^2 (1 - \epsilon^2)^{1/2} = \pi a^2 \left(\frac{2|E||l^2|}{m^3 \gamma^2} \right)^{1/2}$$

and a little manipulation with (3.27) gives the final relation

$$\tau = 2\pi a^{3/2} \gamma^{-1/2} \tag{3.28}$$

This equation constitutes Kepler's third law, namely, that the period varies as the $\frac{3}{2}$ power of the semimajor axis and expresses the numerical coefficient in terms of fundamental constants. Alternatively, the constant can be evaluated for a particle executing a circular orbit with an inverse-square force; elementary considerations give $\gamma m/a^2 = mv^2/a$, which is precisely Eq. (3.28). Note that the mass m of the particle cancels out, so that all test particles follow the same trajectory. As a corollary, a measurement of the period fixes only the mass of the heavy fixed central body M; for example, knowledge of the earth's orbital period around the sun can determine the sun's mass but not that of the earth.

The same method holds for more general potentials, with the orbits bounded by the two radii $r_{\min}$ and $r_{\max}$ that are the turning points (see the discussion of Fig. 3.3). In contrast to the keplerian motion, however, the orbit is typically open. Indeed, Eq. (3.13) shows that the particle's angular position increases by

$$\Delta\phi = l(2m)^{-1/2} \int_{r_{\min}}^{r_{\max}} dr \, r^{-2} [E - V_{\text{eff}}(r)]^{-1/2} \tag{3.29}$$

in going from its minimum radius to its maximum. If $\Delta\phi$ is a rational multiple of π, the orbit is closed; otherwise the radial and angular motions are incommensurate, and the particle never returns to its initial position, the orbit eventually sweeping out the whole region $r_{\min} < r < r_{\max}$. Two simple attractive potentials produce closed orbits: the newtonian (or Coulomb) potential $V(r) \propto -r^{-1}$, already considered, and the harmonic potential $V(r) \propto r^2$, where the orbit is again an ellipse but with the origin at the center (see Prob. 1.10).

4 TWO-BODY MOTION WITH A CENTRAL POTENTIAL

We have previously assumed a fixed central-force field, with $\mathbf{F}(r) = \hat{r} f(r)$ acting along the radial direction. In practice, however, the motion typically involves two particles with masses m_1 and m_2 at $\mathbf{r}_1$ and $\mathbf{r}_2$, interacting through a central

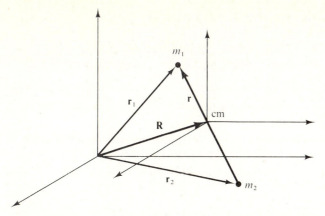

Figure 4.1 The two-body central-force problem with potential $V(|\mathbf{r}_1 - \mathbf{r}_2|)$.

potential $V(|\mathbf{r}_1 - \mathbf{r}_2|)$ that depends only on their separation (see Fig. 4.1). Newton's second law readily yields the equations of motion

$$m_1\ddot{\mathbf{r}}_1 = -\nabla_1 V(r_{12}) = -V'(r_{12})\frac{\mathbf{r}_1 - \mathbf{r}_2}{r_{12}} \tag{4.1a}$$

$$m_2\ddot{\mathbf{r}}_2 = -\nabla_2 V(r_{12}) = -V'(r_{12})\frac{\mathbf{r}_2 - \mathbf{r}_1}{r_{12}} \tag{4.1b}$$

since $(\mathbf{r}_1 - \mathbf{r}_2)r_{12}^{-1}$ is a unit vector along the line joining the two particles. Introduce the center of mass $\mathbf{R}$ and relative positions $\mathbf{r}_1'$ and $\mathbf{r}_2'$ by the familiar relations (2.10a), and define the separation of the two particles

$$\mathbf{r} = \mathbf{r}_1 - \mathbf{r}_2 = \mathbf{r}_1' - \mathbf{r}_2' \tag{4.2}$$

A little algebra then gives

$$\mathbf{r}_1 = \mathbf{R} + \frac{m_2\mathbf{r}}{M} \tag{4.3a}$$

$$\mathbf{r}_2 = \mathbf{R} - \frac{m_1\mathbf{r}}{M} \tag{4.3b}$$

where $M = m_1 + m_2$ is the total mass. Substitution into (4.1) immediately leads to the two basic equations:

$$M\ddot{\mathbf{R}} = 0 \tag{4.4}$$

showing that the center of mass moves without acceleration, and

$$\mu\ddot{\mathbf{r}} = -V'(r)\frac{\mathbf{r}}{r} = -V'(r)\hat{r} \tag{4.5}$$

where

$$\mu = \frac{m_1 m_2}{m_1 + m_2} \tag{4.6a}$$

is the reduced mass; an alternative formula is

$$\frac{1}{\mu} = \frac{1}{m_1} + \frac{1}{m_2} \tag{4.6b}$$

Equation (4.5) reduces the dynamics to an equivalent one-body problem of the sort considered in Sec. 3, with the one-body mass m replaced by the reduced mass μ.

Planetary orbits about the sun provide a particularly interesting application of this formalism. Newton's law of universal gravitation (1.4a) specifies the potential $V(r) = -m_1 \gamma/r$ in Eq. (3.14), so that Eq. (4.5) becomes

$$\ddot{\mathbf{r}} = -\gamma \frac{m_1}{\mu} \frac{\mathbf{r}}{r^3} = -\bar{\gamma} \frac{\mathbf{r}}{r^3} \tag{4.7a}$$

with

$$\bar{\gamma} = \frac{\gamma m_1}{\mu} = G(m_1 + m_2) \tag{4.7b}$$

replacing γ in all the previous expressions. For example, this change affects only the third of Kepler's laws, which now becomes

$$\tau = 2\pi a^{3/2} \bar{\gamma}^{-1/2} = 2\pi a^{3/2} [G(m_1 + m_2)]^{-1/2} \tag{4.8}$$

In principle, a planet's period depends on its own mass m_1 as well as the sun's mass m_2, but the small value of m_1/m_2 makes this correction unimportant except in the case of Jupiter, where $m_1/m_2 \approx 10^{-3}$.

As discussed in Sec. 2, the kinetic energy T and angular momentum $\mathbf{L}$ assume especially simple forms when expressed in terms of the center of mass and relative variables. In the present case, use of Eq. (4.3) in the general expression (2.18b) for the kinetic energy T' about the center of mass yields the expression

$$T' = \tfrac{1}{2}\mu \dot{\mathbf{r}}^2 = \tfrac{1}{2}\mu v^2 \tag{4.9}$$

where $\mathbf{v} = \dot{\mathbf{r}}$. Similarly, the angular momentum $\mathbf{L}'$ about the center of mass [Eq. (2.13c)] reduces to

$$\mathbf{L}' = \mu \mathbf{r} \times \dot{\mathbf{r}} = \mu \mathbf{r} \times \mathbf{v} \tag{4.10}$$

Both these quantities have the form expected from the equivalent one body equation (4.5).

5 SCATTERING

The previous sections have concentrated on the bound orbits, where the negative energy prevents the particle from moving to infinity. The situation is different for positive energy $(E > 0)$, and we now consider the unbounded orbits in a gravitational potential.

Hyperbolic Orbits in Gravitational Potential

A hyperbola can be defined in one of several distinct ways, analogous to those in Sec. 3 for an ellipse.

1. One of the simplest states that there exist two points (foci) located at $x = \pm f$, $y = 0$, such that the orbital point maintains a constant difference in distance between the foci; with the parameters indicated in Fig. 5.1, this condition becomes

$$d - d' = \pm 2a \tag{5.1}$$

where the upper and lower signs refer to the left and right branches, respectively, and $2a$ is the minimum separation of the two branches. We define the eccentricity

$$\epsilon = \frac{f}{a} > 1 \tag{5.2}$$

or $f = \epsilon a$, analogous to Eq. (3.19).

2. Set the origin at the left focus and introduce plane polar coordinates (r, ϕ). By definition, d' is just r, and the law of cosines gives

$$d^2 = d'^2 - 4d'f \cos \phi + 4f^2 \tag{5.3}$$

A combination with Eqs. (5.1) and (5.2) yields

$$\epsilon \cos \phi \pm 1 = a(\epsilon^2 - 1)r^{-1} \tag{5.4}$$

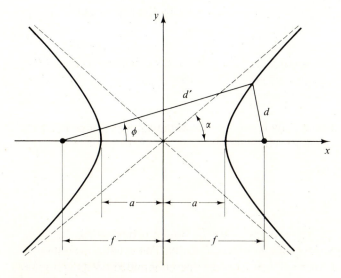

Figure 5.1 The hyperbola as a conic section.

as the polar equation of the hyperbola, wholly analogous to Eq. (3.20*b*) for an ellipse. The upper and lower signs again refer to the left and right branches; moreover, the positive nature of the right side of (5.4) identifies the asymptotes of the two branches, which are oriented at $\cos \alpha = \epsilon^{-1}$ (see Fig. 5.1).

3. Introduce cartesian coordinates with the origin at the center. Then the relations

$$\tfrac{1}{2}(d'^2 + d^2) = x^2 + y^2 + f^2 \tag{5.5a}$$

$$\tfrac{1}{2}(d'^2 - d^2) = 2xf \tag{5.5b}$$

follow just as in the case of an ellipse [see Eq. (3.21)], and comparison with Eq. (5.1) implies

$$d + d' = \mp \frac{2xf}{a} = \mp 2\epsilon x = 2\epsilon |x| \tag{5.6}$$

As a final step, we use the algebraic identity

$$\tfrac{1}{2}(d^2 + d'^2) - \tfrac{1}{4}(d + d')^2 = \tfrac{1}{4}(d - d')^2 = a^2 \tag{5.7}$$

whose left side can be constructed directly from (5.5a) and (5.6). In this way we obtain the familiar form

$$\frac{x^2}{a^2} - \frac{y^2}{b^2} = 1 \tag{5.8}$$

where the length *b* is defined by [compare (3.25)]

$$b^2 = a^2(\epsilon^2 - 1) = f^2 - a^2 \tag{5.9a}$$

or

$$a^2 + b^2 = f^2 \tag{5.9b}$$

We may note that *b* has a simple geometric interpretation as the perpendicular distance from the focus to the asymptote. This identification follows readily from Fig. 5.2, where $\alpha = \arccos \epsilon^{-1}$ is the angle between the asymptote and the *x* axis. The perpendicular distance *AF* has length $f \sin \alpha$, and simple trigonometry transforms this quantity into $(f/\epsilon)(\epsilon^2 - 1)^{1/2}$. Use of Eqs. (5.2) and (5.9a) then confirms that the length *AF* is indeed just *b*; as seen below, it also has the interpretation as the *impact parameter* in a collision.

This discussion applies directly to an encounter between two massive gravitating bodies, where the trajectory has the form [compare Eqs. (3.16) with $\phi_0 = 0$ and (4.7*b*)]

$$r^{-1} = C(1 - \epsilon \cos \phi) \tag{5.10}$$

with

$$C = \frac{\mu^2 \bar{\gamma}}{l^2} \quad \text{and} \quad \epsilon = \left(1 + \frac{2El^2}{\mu^3 \bar{\gamma}^2}\right)^{1/2} \tag{5.11}$$

A simple reidentification of angles $\phi \to \pi - \phi$ in the first of Eqs. (5.4) indicates that Eq. (5.10) represents the right branch of a hyperbola but with the origin at the right focus as illustrated in Fig. 5.2. Note that $\cos \phi$ now cannot exceed $1/\epsilon$. In

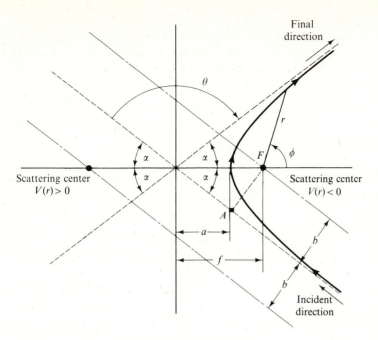

Figure 5.2 Hyperbolic orbits in a potential $V(r) = -\mu\bar{\gamma}/r$.

Eq. (5.11) E and l are the energy and angular momentum in the center-of-mass frame. Suppose that the particles approach from infinity with relative speed v_∞ and impact parameter b (the distance of closest approach if the trajectory were undeflected), as illustrated in Fig. 5.3; the corresponding energy and angular momentum are

$$E = \tfrac{1}{2}\mu v_\infty^2 \qquad \text{and} \qquad l = \mu v_\infty b \qquad (5.12)$$

so that Eq. (5.11) reduces to

$$\epsilon = \left[1 + \left(\frac{v_\infty^2 b}{\bar{\gamma}}\right)^2\right]^{1/2} \qquad (5.13a)$$

Moreover, the actual distance of closest approach is

$$r_{\min} = f - a = (\epsilon - 1)a = \left(\frac{\epsilon - 1}{\epsilon + 1}\right)^{1/2} b \qquad (5.13b)$$

where we have used Eq. (5.9a). Note that knowledge of v_∞ and b suffices to determine $r_{\min}$, which is important, for example, in aiming a rocket to pass near (but miss) a planet. As expected physically, $r_{\min}$ increases with increasing v_∞ because a fast object is less deflected, and $r_{\min} \to b$ for a distant passage ($b \to \infty$).

For some purposes, the most interesting aspect of the trajectory is the angle of total deflection θ measured from the initial direction. Figure 5.2 shows that in the

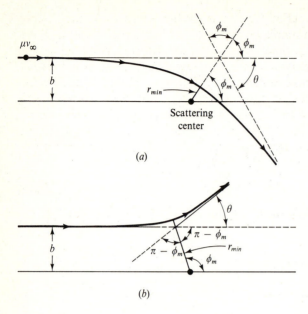

(a)

(b)

Figure 5.3 Unbounded trajectories in a central potential: (a) attractive; (b) repulsive. Note that the definition of the polar angle here differs from that in Fig. 5.2 by $\alpha - \pi$.

gravitational scattering problem $\theta = \pi - 2\alpha$ and, as noted previously, $\cos \alpha = \epsilon^{-1}$. Hence we have

$$\cot \tfrac{1}{2}\theta = \cot \left(\tfrac{1}{2}\pi - \alpha\right) = \tan \alpha = (\epsilon^2 - 1)^{1/2}$$

Use of Eq. (5.13) yields the basic relation

$$\cot \tfrac{1}{2}\theta = \frac{v_\infty^2 b}{\bar{\gamma}} \tag{5.14}$$

with the expected limits of small deflection ($\theta \to 0$) for large $v_\infty^2 b$ and large deflection ($\theta \to \pi$) for small $v_\infty^2 b$. In newtonian gravitational attraction these relations involve only the total mass $m_1 + m_2$ and are independent of the reduced mass μ, exactly as in Eq. (4.8). Thus all small test particles with $\mu \ll M$ follow essentially the same trajectories. We may note that such a simple description holds only in the center-of-mass frame ($\mathbf{R} = 0$); owing to their own motion, earth-bound observers naturally would see a much more complicated trajectory for a rocket passing near a planet.

General Scattering Orbits

The preceding analysis illustrates all the essential features of a two-body collision, and the behavior for a general central potential $V(r)$ differs only in the detailed form of the orbit. Consider a particle incident from $x = -\infty$ at an impact parameter b. Attractive and repulsive potentials differ slightly, and we first consider the attractive case, illustrated in Fig. 5.3a. Let ϕ_m be the polar angle at the distance of closest approach r_{min}. Then simple geometry shows that the deflection angle θ is related to ϕ_m by $\theta = \pi - 2\phi_m$. Since $\phi_m < \frac{1}{2}\pi$ for an attractive potential, we see that $\theta > 0$. Moreover, we shall assume that $\phi_m > 0$, which excludes the possibility of orbits that circle the attractive center more than once. Inclusion of such trajectories is not difficult, but the analysis becomes more intricate owing to the multiple-valued relation between θ and b. A repulsive potential involves similar considerations (Fig. 5.3b), with $\theta = 2\phi_m - \pi$. Since ϕ_m necessarily exceeds $\frac{1}{2}\pi$, however, the deflection angle is again positive. The two cases can be combined in the single general expression

$$\theta = |\pi - 2\phi_m| \qquad (5.15)$$

For a given $V(r)$, the angle ϕ_m follows directly from Eq. (3.13) with $\phi = \pi$ when $r = \infty$:

$$\phi_m = l \int_\infty^{r_{min}} dr\, r^{-2} \left[2\mu E - 2\mu V(r) - \frac{l^2}{r^2} \right]^{-1/2} + \pi \qquad (5.16a)$$

Here r_{min} is the root of the equation

$$2\mu[E - V(r_{min})] = \frac{l^2}{r_{min}^2} \qquad (5.16b)$$

and E and l are given in Eq. (5.12). In this way, we obtain a formal solution for ϕ_m and θ in terms of the physical parameters v_∞ and b.

Cross Section

We have introduced this discussion by calculating the positive-energy orbits of macroscopic and astronomical bodies in the gravitational potential $V(r) = -m\bar{\gamma}/r$. It is interesting that the same concepts remain valid in the *microscopic* domain for the scattering of particles by a potential $V(r)$. We shall, in fact, derive the celebrated Rutherford formula for the scattering of two charged atomic nuclei. While the distances of interest in the interaction of the earth with the sun are $\approx 10^{13}$ cm, the distances relevant to the collision of two charged nuclei are $\approx 10^{-13}$ cm. The range of validity of classical physics as presented by Newton's laws together with the gravitational and electrostatic potentials is truly awe-inspiring.

In the microscopic problem, the target is typically subject to a uniform incident beam of particles with relative initial velocity v_∞. We characterize the incident beam through its *flux*

$$F \equiv \text{incident flux}$$

$$= \text{number of particles crossing unit transverse}$$
$$\text{area per unit time} \qquad (5.17)$$

The experimental quantity of interest is the number of events of a certain type p occurring per unit time, which is best described with the concept of a *differential cross section* $d\sigma_p$. Since classical particles follow definite trajectories, those which participate in the specific process p necessarily pass through a definite element of transverse area (see Fig. 5.4). This element of area defines the differential cross section $d\sigma_p$. Figure 5.4 shows that the number of particles passing through $d\sigma_p$ per unit time is just $F\,d\sigma_p$, and we therefore have

$$\text{Number of events of type } p \text{ per unit time} \equiv F\,d\sigma_p \qquad (5.18)$$

More generally, the number of events of the particular type per unit time must be proportional to the total flux F, and the constant of proportionality may be taken to define the differential cross section.

As a concrete example, we shall study elastic scattering and define an event p to be the elastic scattering through a deflection angle lying between θ and $\theta + d\theta$. We can compute the quantity $d\sigma_{el}$ directly from the preceding study of orbits in a central potential. Consider the particles incident with impact parameter lying between b and $b + db$ (Fig. 5.5). The number of such particles passing through the associated transverse area $2\pi b\,db$ per unit time is $2\pi b\,db\,F$. The axial symmetry implies that all these particles are scattered with deflection angles lying between θ and $\theta + d\theta$, where θ is assumed to be a known function of b, as in Eq. (5.14). Since all the particles passing through the ring of area $2\pi b\,db$ are scattered through the angle θ, the differential elastic scattering cross section is, by definition,

$$F\,2\pi b\,db = F\,d\sigma_{el}(\theta) \qquad (5.19)$$

Imagine the scattering center to be surrounded with a macroscopically large sphere of radius R. The particles entering the original ring all emerge through the area $dA = 2\pi R^2 \sin\theta\,d\theta$ (see Fig. 5.5). The corresponding *solid angle* $d\Omega$ into which the particles are scattered is then given by

$$d\Omega = \frac{dA}{R^2} = 2\pi \sin\theta\,d\theta \qquad (5.20)$$

where θ is the usual polar angle measured from the incident direction. We can exhibit the proportionality to the solid angle by multiplying and dividing the right-hand side of Eq. (5.19) by $d\Omega$

$$F\,d\sigma_{el}(\theta) = F\left(\frac{d\sigma}{d\Omega}\right)_{el} d\Omega \qquad (5.21)$$

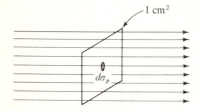

Figure 5.4 Definition of incident flux and differential cross section. Incident flux F = number of particles crossing this unit (transverse) area per unit time; $F\,d\sigma_p$ = number of events of type p per unit time.

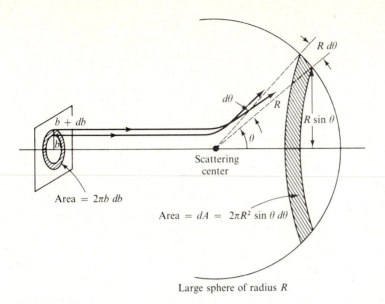

Figure 5.5 The scattering problem and relation of cross section to impact parameter.

Comparison of Eqs. (5.19) to (5.21) yields the *cross section for elastic scattering per unit final solid angle*

$$\left(\frac{d\sigma}{d\Omega}\right)_{el} = \frac{b}{\sin\theta}\left|\frac{db}{d\theta}\right| \tag{5.22}$$

which is commonly referred to simply as the *differential cross section*. The absolute-value sign is needed in Eq. (5.22) because $d\sigma/d\Omega$ is positive whereas $db/d\theta$ is typically negative (particles with large b are less deflected, as seen in Fig. 5.5). This last expression has some interesting features. First, note that $d\sigma/d\Omega$ has the dimension of an area, consistent with its name. Second, the total number of particles scattered per unit time in all directions for a unit incident flux is known as the *total cross section* σ_T; it is obtained by integrating over all solid angles

$$\sigma_T = \int d\Omega \left(\frac{d\sigma}{d\Omega}\right)_{el} \tag{5.23}$$

Conservation of particles implies that $\sigma_T F$ is also the number of particles removed from the incident beam per unit time and therefore characterizes the attenuation per scattering site.

Rutherford Scattering

The calculation of the differential cross section has now been reduced to a detailed study of the orbits for various impact parameters. As a specific case of great practical interest, we consider the scattering of particles with charge ze and mass

m from a nucleus with charge Ze and mass M. The analysis is very similar to that for the Kepler problem, because the relative orbit in the center-of-mass frame satisfies [compare Eqs. (4.5) and (4.7a)] the equation

$$\mu \ddot{\mathbf{r}} = \frac{zZe^2 \mathbf{r}}{r^3} \tag{5.24}$$

Evidently, the analysis of hyperbolic gravitational orbits remains valid if $\bar{\gamma}$ in Eq. (4.7b) is replaced by

$$\bar{\gamma} = -\frac{zZe^2}{\mu} \tag{5.25}$$

which now explicitly contains the reduced mass μ. To interpret the minus sign, we recall the general equation (5.10) for a hyperbola with $\epsilon > 1$ and $C = \mu^2\bar{\gamma}/l^2$. Evidently, C is negative for like-charged projectiles, so that Eq. (5.10) thus takes the form

$$r^{-1} = |C|(\epsilon \cos \phi - 1) \tag{5.26}$$

Comparison with Eq. (5.4) and Fig. 5.1 shows that the trajectory lies on the right branch for a scattering center located at the left focus. Thus in a potential $V(r) = -\mu\bar{\gamma}/r$ the orbit is *unchanged* when the sign of $\bar{\gamma}$ is reversed, provided only that the scattering center is placed at the other focus. Note that the impact parameter b in Fig. 5.2 is identical in these two cases. It is thus easy to see that Eq. (5.14) describes both situations if $\bar{\gamma}$ is replaced by its absolute value $|\bar{\gamma}|$. For Rutherford scattering, the desired relation between impact parameter b and scattering angle θ is therefore

$$b = \frac{|zZ|e^2}{\mu v_\infty^2} \cot \tfrac{1}{2}\theta \tag{5.27}$$

A combination with the general relation (5.22) leads to the Rutherford formula

$$\left(\frac{d\sigma}{d\Omega}\right)_{el} = \left(\frac{zZe^2}{2\mu v_\infty^2}\right)^2 \frac{1}{\sin^4 \tfrac{1}{2}\theta} = \left(\frac{zZe^2}{4E \sin^2 \tfrac{1}{2}\theta}\right)^2 \tag{5.28}$$

where $E = \tfrac{1}{2}\mu v_\infty^2$ is the incident kinetic energy in the center-of-mass frame.

This basic result has several interesting implications.

1. The differential cross section is independent of the signs of z and Z, and such scattering cannot fix the sign of the nuclear charge.
2. The Rutherford cross section varies as θ^{-4} for small angles (large impact parameters). *All* particles are scattered, at least slightly.†
3. Correspondingly, the total cross section diverges because the nucleus affects every particle.†
4. This rather elementary theory is extremely important, not only for its historical role in developing the planetary model of the atom but also even today in studying heavy-ion reactions.

---

† In an atom, the Coulomb charge on the nucleus is shielded by the atomic electrons at large distances, and the total cross section is finite.

Scattering by a Hard Sphere

As seen in Eq. (5.22), the calculation of the differential cross section becomes straightforward as soon as the relation between b and θ is known. Unfortunately, determining that relation is usually quite difficult, but it sometimes follows from elementary considerations. One particularly simple example is the scattering by a hard sphere of radius a, where the projectile rebounds elastically (Fig. 5.6). In this case the differential cross section is unexpectedly simple, although perhaps not intuitively so. To analyze the behavior, consider a particle with impact parameter $b < a$. Inspection of Fig. 5.6 shows that $b = a \sin \phi_m$ and that $\theta = 2\phi_m - \pi$, as in the general relation (5.15). Consequently, we find

$$b = a \cos \tfrac{1}{2}\theta \qquad \text{for} \qquad b < a \qquad (5.29)$$

and, clearly, $\theta = 0$ for $b > a$. Substitution into (5.22) then yields the desired result

$$\left(\frac{d\sigma}{d\Omega}\right)_{el} = \tfrac{1}{4}a^2 \qquad (5.30)$$

implying an *isotropic* differential cross section. Thus particles emerge uniformly in all directions after scattering by a hard sphere. The total cross section is just the geometrical area

$$\sigma_T = \int d\Omega \left(\frac{d\sigma}{d\Omega}\right)_{el} = \pi a^2 \qquad (5.31)$$

because all particles incident on that region are removed from the initial beam.

It is sometimes useful to note an equivalent relation for the total cross section obtained directly from Eq. (5.19)

$$\sigma_T = 2\pi \int_0^{b_{max}} b \, db = \pi b_{max}^2 \qquad (5.32)$$

where b_{max} is the limiting impact parameter for zero deflection. This expression immediately reproduces (5.31) for a hard sphere because $\theta = 0$ for $b > a$, and it also confirms that σ_T is infinite for Rutherford scattering. More generally, the classical cross section σ_T is finite and equal to πa^2 whenever the potential $V(r)$ is strictly zero for $r > a$.

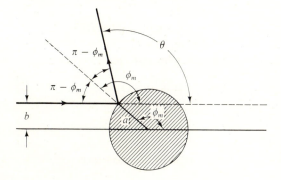

Figure 5.6 Elastic scattering by a hard sphere.

PROBLEMS

1.1 Verify that the three alternative definitions of a conservative force in Eqs. (1.8), (1.11a), and (1.11b) are equivalent.

1.2 A uniform spool of mass M and diameter d rests on end on a frictionless table. A massless string wrapped around the spool is attached to a weight m which hangs over the edge of the table. If the spool is released from rest when its center of mass is a distance l from the edge of the table, what is the velocity of the weight m when the center of mass of the spool reaches the edge of the table?

1.3 A massless string is placed over a massless pulley, and each end is wound around and fastened to a vertical hoop. The hoops have masses $M_{1,2}$ and radii $R_{1,2}$. The apparatus is placed in a uniform gravitational field g and released with each end of the string aligned along the field.
    (a) Show that the tension in the string is $\tau = gM_1 M_2(M_1 + M_2)^{-1}$.
    (b) Show that the acceleration of the center of mass of hoop 1 is $a_1 = M_1(M_1 - M_2)^{-1}a_1^{\text{point}}$, where a_1^{point} is the acceleration when both hoops are replaced with point masses.

1.4 A rocket with initial mass m_0 emits exhaust gases at a constant rate m_0/τ with constant speed v_0 relative to the rocket. It starts from rest at the earth's surface and rises vertically. Treating the earth's gravitational field as uniform, find the height h as a function of the time. Discuss the behavior for $t \ll \tau$ and for $t \to \tau$. Sketch carefully the resulting speed and displacement.

1.5 A particle of mass m moves in a one-dimensional force field with potentials (a) $V(x) = A|x|^n$, (b) $V(x) = -V_0 \operatorname{sech}^2 \alpha x$, and (c) $V(x) = V_0 \tan^2 \alpha x$. For each case, sketch the potential and state the range of energy and position within which bounded oscillations can occur. Determine the period of oscillation as a function of the energy E. Discuss the limiting behavior in case (a) for $n = 2$.

1.6 (a) Starting from Newton's second law $m\ddot{\mathbf{r}} = \mathbf{F}$ for a central force $\mathbf{F} = f(r)\hat{r}$, show that the equations of motion in two-dimensional polar coordinates become

$$m(\ddot{r} - r\dot{\phi}^2) = f(r) \qquad \text{and} \qquad m(2\dot{r}\dot{\phi} + r\ddot{\phi}) = 0$$

    (b) Integrate these equations explicitly; hence relate them to the treatment of central-force motion discussed in Sec. 3.

1.7 A rocket of mass m is in circular orbit around the earth at a distance R from the center.
    (a) What tangential impulse, that is, $m\,\Delta v$, must be given to the body so that it just escapes to infinity?
    (b) Describe the resulting orbit.
    (c) Compare with the radial impulse that must be given to a particle initially at rest at R if it is to acquire sufficient velocity to escape to infinity.

1.8 A rocket of mass m is in a circular orbit with radius R about a planet of radius R_0.
    (a) What tangential impulse will cause the rocket to graze the back side of the planet?
    (b) Describe the resulting orbit.

1.9 A particle of mass m moves in an attractive central potential $V(r) = -\lambda r^{-1}e^{-r/a}$, where λ and a are both positive. Reduce the equations of motion to an equivalent one-dimensional problem. Use the effective potential to discuss the qualitative nature of the orbits for different values of the energy and angular momentum.

1.10 A particle of mass m moves in a central harmonic potential $V(r) = \frac{1}{2}kr^2$ with a positive spring constant k.
    (a) Use the effective potential to show that all orbits are bound and that E must exceed $E_{\min} = (kl^2/m)^{1/2}$.
    (b) Verify that the orbit is a closed ellipse with origin at the center. If the relation $E/E_{\min} = \cosh \xi$ defines the quantity ξ, verify that the orbital parameters are given by $a^2 = e^{\xi}l(mk)^{-1/2}$, $b^2 = e^{-\xi}l(mk)^{-1/2}$, and $\epsilon^2 = 1 - e^{-2\xi}$. Discuss the limiting cases $E \to E_{\min}$ and $E \gg E_{\min}$.
    (c) Prove that the period is $2\pi(m/k)^{1/2}$, independent of E and l. Discuss this elementary result.

1.11 A particle of mass m moves in a singular central potential with $V(r) = -\lambda r^{-n}$ with $n > 2$. Reduce the equations of motion to an equivalent one-dimensional problem and discuss the qualitative nature of the orbit for different values of the energy. For the bound orbits, show that the particle takes a *finite* length of time to spiral into the center of force, passing through a *finite* number of revolutions. Can you say anything about the subsequent motion?

1.12 The orbit of the planet mercury has an eccentricity 0.206 and a period 0.241 year; moreover, the perihelion advances slowly at the rate of 43 seconds of arc per century. One possible explanation of this effect is that the potential energy around the sun has the form $V = -(mMG/r)(1 + \alpha GM/rc^2)$, where α is a dimensionless constant and $MG/c^2 \approx 1.475$ km characterizes the sun's gravitational field. Demonstrate that the resulting orbit indeed represents a precessing ellipse. Find the magnitude and sign of α needed to fit the observed data.

1.13 A rocket with velocity v_∞ and impact parameter b approaches a planet of radius R_0 and mass M. What is the condition that the rocket will strike the planet? If it just misses, what is its angle of deflection?

1.14 The cross section to strike the nuclear surface is of interest in discussing nuclear reactions during heavy-ion scattering. By integrating over the appropriate impact parameters, show that the cross section to strike a nucleus of radius R in Rutherford scattering is given by $\sigma_r = \pi R^2(1 - V_c/E)$, where $V_c = zZe^2/R$ is the repulsive Coulomb barrier at the nuclear surface and it is assumed that $E \geqslant V_c > 0$.

1.15 (*a*) Obtain that the following general relation between the scattering angle θ and impact parameter b for a repulsive central potential

$$\theta(b) = \pi - 2b \int_0^{u_0} \frac{du}{[(1 - V/E) - b^2 u^2]^{1/2}}$$

where $u = 1/r$ and u_0 is the classical turning point.
  (*b*) Rederive the results obtained in the text for Rutherford scattering from this expression.
  (*c*) What is the corresponding general expression for the differential cross section?
  (*d*) What is the expression corresponding to part (*a*) for attractive potentials?

1.16 A uniform beam of particles with energy E is scattered by a repulsive central potential $V(r) = \gamma/r^2$. Derive the differential elastic cross section

$$\left(\frac{d\sigma}{d\Omega}\right)_{el} = \frac{\gamma\pi^2}{E \sin\theta} \frac{\pi - \theta}{\theta^2(2\pi - \theta)^2}$$

Sketch carefully the angular dependence. Discuss the total cross section. What happens if the potential is attractive, that is, $\gamma < 0$?

1.17 A uniform beam of particles with energy E is scattered by an attractive central potential

$$V(r) = \begin{cases} 0 & r > a \\ -V_0 & r < a \end{cases}$$

Show that the orbit of a particle is identical with that of a light ray refracted by a sphere of radius a and index of refraction $n = [(E + V_0)/E]^{1/2}$. Prove that the differential elastic cross section for $\cos\frac{1}{2}\theta > n^{-1}$ is

$$\left(\frac{d\sigma}{d\Omega}\right)_{el} = \frac{n^2 a^2}{4 \cos\frac{1}{2}\theta} \frac{[n\cos(\frac{1}{2}\theta) - 1](n - \cos\frac{1}{2}\theta)}{(1 + n^2 - 2n\cos\frac{1}{2}\theta)^2}$$

What is the total cross section?

1.18 A particle with large impact parameter b is slightly deflected from a uniform trajectory by a central potential $V(r)$.
  (*a*) In the impulse approximation, the (small) integrated deflecting force is evaluated along the original straight-line trajectory. Use this approxmation to derive the expression

$$\theta \approx \frac{2b}{mv_\infty^2} \left| \int_b^\infty \frac{dr}{(r^2 - b^2)^{1/2}} \frac{dV}{dr} \right|$$

for the (small) deflection angle.

(b) If $V(r) = \gamma r^{-n}$ with positive n and γ, find the differential cross section for small-angle scattering and discuss its behavior as $\theta \to 0$. Show that the answer reproduces the known results for $n = 1$ (Sec. 5) and 2 (Prob. 1.16). Is σ_T well defined for any n?

(c) If $V(r) = \gamma e^{-\lambda r}$, show that b varies approximately like $\lambda^{-1} \ln (1/\theta)$. Hence obtain the approximate form of the differential cross section. Is σ_T well defined?

(d) In quantum mechanics, the small-angle part of σ_T is finite whenever $r^2 V(r) \to 0$ as $r \to \infty$. Discuss briefly why the classical behavior is different.

TWO

ACCELERATED COORDINATE SYSTEMS

In the preceding chapter we noted that the analysis of two interacting particles becomes particularly simple in a frame that moves with the center of mass because the motion then reduces to that of an equivalent single body. Other choices of coordinate frames naturally lead to different (usually more complicated) descriptions, e.g., the moons of Jupiter as seen by an observer on the surface of the earth. In this case, of course, the motion is easily analyzed by transforming to the center-of-mass frame of the Jupiter system, but it is sometimes preferable to work directly in the moving frame of the observer, the most important example being earthbound motion seen from a laboratory fixed to the rotating earth. The same techniques will also turn out to simplify considerably the description of rigid-body motion in Chap. 5. Thus we now turn to the theory of moving coordinate systems, starting with pure rotations.

6 ROTATING COORDINATE SYSTEMS

Consider an inertial frame with a fixed set of orthonormal coordinate axes $\hat{e}_i^0$ and an origin O; attach a second set of (moving) orthonormal coordinate axes $\hat{e}_i$ with the same origin O (Fig. 6.1). As seen in the inertial frame, $\hat{e}_i^0$ are fixed but $\hat{e}_i$ vary with time; conversely, in the moving "body-fixed" frame, $\hat{e}_i^0$ appear time-dependent, but $\hat{e}_i$ are stationary. For definiteness, the moving frame has been identified with one fixed in some moving body, but the transformation is completely general.

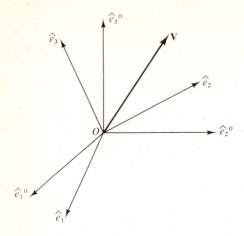

Figure 6.1 Transformation to a rotating coordinate system.

Let **V** be a general vector, e.g., the position of a particle. This vector can be characterized by its components with respect to either orthonormal triad. Thus we can write

$$\mathbf{V} = \sum_{i=1}^{3} V_i^0 \hat{e}_i^0 \tag{6.1a}$$

$$\mathbf{V} = \sum_{i=1}^{3} V_i \hat{e}_i \tag{6.1b}$$

where $V_i^0 = \mathbf{V} \cdot \hat{e}_i^0$ and $V_i = \mathbf{V} \cdot \hat{e}_i$ are just the projections onto the appropriate axes. Suppose now that an observer in the inertial frame sees **V** change with time. Since the unit vectors $\hat{e}_i^0$ are fixed in that frame, the rate of change of **V** seen in the inertial frame can arise only from the time dependence of the components in Eq. (6.1a)

$$\left(\frac{d\mathbf{V}}{dt}\right)_{\text{inertial}} = \sum_{i=1}^{3} \frac{dV_i^0}{dt} \hat{e}_i^0 \tag{6.2}$$

On the other hand, the time derivative of (6.1b) yields

$$\left(\frac{d\mathbf{V}}{dt}\right)_{\text{inertial}} = \sum_{i=1}^{3} \frac{dV_i}{dt} \hat{e}_i + \sum_{i=1}^{3} V_i \frac{d\hat{e}_i}{dt} \tag{6.3}$$

and the first term on the right-hand side is just the rate of change of **V** as seen by an observer in the moving (body-fixed) frame

$$\left(\frac{d\mathbf{V}}{dt}\right)_{\text{body}} \equiv \sum_{i=1}^{3} \frac{dV_i}{dt} \hat{e}_i \tag{6.4}$$

We infer from Eqs. (6.2) to (6.4) that

$$\left(\frac{d\mathbf{V}}{dt}\right)_{\text{inertial}} = \left(\frac{d\mathbf{V}}{dt}\right)_{\text{body}} + \sum_{i=1}^{3} V_i \frac{d\hat{e}_i}{dt} \tag{6.5}$$

7 INFINITESIMAL ROTATIONS

Equation (6.5), which holds for any vector **V**, involves the rate of change of the body-fixed unit vectors $\hat{e}_i$ as seen from the inertial frame. In analyzing these quantities, it is convenient to consider two neighboring times t and $t + dt$, when the configurations $\hat{e}_i(t)$ and $\hat{e}_i(t + dt) = \hat{e}_i(t) + d\hat{e}_i$ differ only by an infinitesimal amount $d\hat{e}_i$. Here $d\hat{e}_i$ is the infinitesimal vector that must be added to $\hat{e}_i(t)$ to produce $\hat{e}_i(t + dt)$. As a first step, we recall that $\hat{e}_i$ form an orthonormal triad at any time t, so that

$$\hat{e}_i \cdot \hat{e}_j = \delta_{ij} \tag{7.1}$$

where δ_{ij} denotes the *Kronecker delta*, defined to be 1 for $i = j$ and 0 otherwise. In particular, $\hat{e}_i \cdot \hat{e}_i = 1$, and a first-order expansion at $t + dt$ yields

$$\hat{e}_i \cdot d\hat{e}_i = 0 \tag{7.2}$$

Consequently, the vector $d\hat{e}_i$ representing the infinitesimal change in the unit vector $\hat{e}_i$ necessarily lies perpendicular to $\hat{e}_i$ itself. Furthermore, the vectors $\hat{e}_i$ constitute a complete set at any time, so that $d\hat{e}_i$ can be expanded in this basis set

$$d\hat{e}_i = \sum_{j=1}^{3} d\Omega_{ij} \hat{e}_j \tag{7.3}$$

with some infinitesimal coefficients $d\Omega_{ij}$. The relation (7.2) shows that $d\Omega_{ii} = 0$ for $i = 1, 2,$ or 3; more generally, the coefficients $d\Omega_{ij}$ can be identified as the scalar products

$$d\hat{e}_i \cdot \hat{e}_j = d\Omega_{ij} \tag{7.4}$$

Although there initially appear to be six such quantities, the differential of Eq. (7.1) implies that $d\hat{e}_i \cdot \hat{e}_j + \hat{e}_i \cdot d\hat{e}_j = 0$, so that $d\Omega_{ij} = -d\Omega_{ji}$. Consequently, only three of these infinitesimal coefficients are independent, and it is convenient to relabel them as

$$d\Omega_{12} \equiv d\Omega_3 \qquad d\Omega_{23} \equiv d\Omega_1 \qquad d\Omega_{31} \equiv d\Omega_2 \tag{7.5}$$

With this new notation, we have, for example,

$$d\hat{e}_1 = d\Omega_3 \hat{e}_2 - d\Omega_2 \hat{e}_3$$

which can be written more compactly as a vector product

$$d\hat{e}_1 = d\mathbf{\Omega} \times \hat{e}_1$$

where $$d\mathbf{\Omega} \equiv d\Omega_1 \hat{e}_1 + d\Omega_2 \hat{e}_2 + d\Omega_3 \hat{e}_3 \tag{7.6}$$

A similar relation holds for the other infinitesimal vectors $d\hat{e}_2$ and $d\hat{e}_3$, yielding the general result

$$d\hat{e}_i = d\mathbf{\Omega} \times \hat{e}_i \tag{7.7}$$

The vector $d\mathbf{\Omega} = \sum_i d\mathbf{\Omega}_i = \sum_i d\Omega_i \hat{e}_i$ has a simple interpretation as an infinitesimal rotation obtained by combining separate infinitesimal rotations $d\mathbf{\Omega}_i = d\Omega_i \hat{e}_i$ about each of the three coordinate axes. These elementary infinitesimal rotations can be assigned a direction along the axis of rotation and a magnitude equal to the (infinitesimal) amount of rotation about this axis. Here and henceforth we use the right-hand convention for positive rotations. Consider the simple case of a single rotation $d\mathbf{\Omega}_1 = d\Omega_1 \hat{e}_1$ about $\hat{e}_1$. Inspection of Fig. 7.1 shows that for this rotation

$$d\hat{e}_1 = 0 = d\mathbf{\Omega}_1 \times \hat{e}_1$$

$$d\hat{e}_2 = d\Omega_1 \hat{e}_3 = d\mathbf{\Omega}_1 \times \hat{e}_2 \tag{7.8}$$

$$d\hat{e}_3 = -d\Omega_1 \hat{e}_2 = d\mathbf{\Omega}_1 \times \hat{e}_3$$

Thus $d\hat{e}_1$, $d\hat{e}_2$, and $d\hat{e}_3$ are indeed given correctly by Eq. (7.7), and similar considerations hold for rotations $d\mathbf{\Omega}_2$ and $d\mathbf{\Omega}_3$. To first order, the total change $d\hat{e}_i$ is obtained by adding the separate components, precisely reproducing (7.7).

These preliminary ideas now permit a straightforward analysis of Eq. (6.5). In an element of time dt, the change in $\hat{e}_i$ is given by

$$d\hat{e}_i = \frac{d\hat{e}_i}{dt} dt = \frac{d\mathbf{\Omega}}{dt} \times \hat{e}_i \, dt = \mathbf{\omega} \times \hat{e}_i \, dt \tag{7.9}$$

where $\mathbf{\omega}$ is the instantaneous angular velocity vector of the rotating frame as seen in the inertial frame

$$\mathbf{\omega} \equiv \frac{d\mathbf{\Omega}}{dt} \tag{7.10}$$

Its direction is along the axis of rotation $d\mathbf{\Omega}$, and its magnitude is just the angular speed $|d\mathbf{\Omega}/dt|$. Substitution into the last term of Eq. (6.5) leads to

$$\sum_{i=1}^{3} V_i \frac{d\hat{e}_i}{dt} = \sum_{i=1}^{3} V_i \mathbf{\omega} \times \hat{e}_i = \mathbf{\omega} \times \mathbf{V}$$

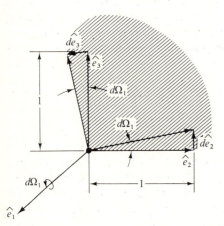

Figure 7.1 Infinitesimal rotation $d\mathbf{\Omega}_1 = d\Omega_1 \hat{e}_1$ about the 1 axis.

where the last expression follows from Eq. (6.1*b*). Consequently, Eq. (6.5) acquires the final form

$$\left(\frac{d\mathbf{V}}{dt}\right)_{\text{inertial}} = \left(\frac{d\mathbf{V}}{dt}\right)_{\text{body}} + \boldsymbol{\omega} \times \mathbf{V} \qquad (7.11)$$

where $\boldsymbol{\omega}$ is the instantaneous angular velocity of the rotating (body-fixed) coordinate axes as seen in the inertial frame.

Equation (7.11) is very powerful, for it applies to an arbitrary moving vector **V**. To clarify this rather formal result, we shall rederive the same expressions from a more physical point of view. Consider a coordinate system instantaneously rotating about the $\hat{z}$ axis with angular speed ω (see Fig. 7.2) and some vector **r** fixed in the moving system. In a time dt, the rotation angle ϕ increases by $d\phi = \omega\, dt$, and the following simple argument gives the change in **r** as seen by an inertial observer. Evidently, $d\mathbf{r}$ is perpendicular both to $\hat{r}$ and to $\hat{z}$, lying along the direction $\hat{z} \times \hat{r}$. In addition, its magnitude is $r \sin \theta\, d\phi$, where θ is the (spherical polar) angle between $\hat{r}$ and $\hat{z}$. Hence $d\mathbf{r}$ can be written as a vector product

$$d\mathbf{r} = d\phi\, \hat{z} \times \mathbf{r} = d\boldsymbol{\Omega} \times \mathbf{r} \qquad (7.12)$$

where

$$d\boldsymbol{\Omega} \equiv d\phi\, \hat{z} \qquad (7.13)$$

is indeed the infinitesimal positive rotation about the z axis as defined in Eq. (7.6). More generally, we would have

$$d\boldsymbol{\Omega} = d\phi\, \hat{n} \qquad (7.14)$$

for an infinitesimal rotation $d\phi$ about an arbitrary axis $\hat{n}$.

Note that Eqs. (7.12) and (7.14) completely characterize the infinitesimal rotation of the body-fixed coordinate system in terms of the *vector displacements* $d\mathbf{r}$ of every point in that system. The same three parameters $d\boldsymbol{\Omega} = d\phi\, \hat{n}$ describe the displacement of every point. Conversely, if the vector displacement $d\mathbf{r}$ of each point has the form of Eqs. (7.12) and (7.14), the transformation necessarily represents an infinitesimal rigid rotation about $\hat{n}$ through an angle $d\phi$.

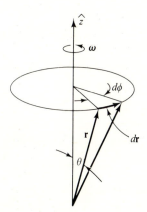

Figure 7.2 Rotation about $\hat{z}$ axis with angular speed ω.

Consider next two successive infinitesimal rotations $d\Omega_1$ and $d\Omega_2$, which need not be collinear. Under these operations, a vector $\mathbf{r}$ fixed in the rotated frame changes successively first to $\mathbf{r}_1 = \mathbf{r} + d\Omega_1 \times \mathbf{r}$ and then to

$$
\begin{aligned}
\mathbf{r}_2 &= \mathbf{r}_1 + d\Omega_2 \times \mathbf{r}_1 \\
&= \mathbf{r} + d\Omega_1 \times \mathbf{r} + d\Omega_2 \times (\mathbf{r} + d\Omega_1 \times \mathbf{r}) \\
&\approx \mathbf{r} + (d\Omega_1 + d\Omega_2) \times \mathbf{r}
\end{aligned}
\tag{7.15}
$$

Here we have used the known addition law for vector displacements and have neglected the second-order corrections. This important relation shows that $\mathbf{r}_2$ differs from $\mathbf{r}$ by a *single infinitesimal rotation* $d\Omega \equiv d\Omega_1 + d\Omega_2$ *obtained as the vector sum of* $d\Omega_1$ *and* $d\Omega_2$. Note that infinitesimal rotations commute, since Eq. (7.15) also holds to first order if the rotations are performed in reverse order. Thus we have explicitly established that infinitesimal rotations can indeed be added as vectors, as in Eq. (7.6). In contrast, finite rotations do not commute and hence cannot be represented by vectors, even though they can also be described by a rotation of finite magnitude $\Delta\phi$ about the axis $\hat{n}$.

We can now relate this approach to our previous one. Divide the relation (7.12) by dt to find

$$
\left(\frac{d\mathbf{r}}{dt}\right)_{\text{inertial}} = \frac{d\Omega}{dt} \times \mathbf{r} = \omega \times \mathbf{r}
\tag{7.16}
$$

where ω defined by (7.10) is the instantaneous angular velocity. On the other hand, we can apply the general expression (7.11) to a vector $\mathbf{r}$ fixed in the body frame; since $(d\mathbf{r}/dt)_{\text{body}} = 0$, we immediately reproduce Eq. (7.16).

The quantity ω is the instantaneous angular velocity of the rotating frame as seen in the inertial frame. Since the arguments in (7.15) have demonstrated that $d\Omega$ is a vector, we conclude that $\omega = d\Omega/dt$ is also a vector. Like any other vector, ω can then be decomposed in any frame, as in Eq. (6.1). Moreover, Eq. (7.11) may be applied to ω, in which case we find

$$
\left(\frac{d\omega}{dt}\right)_{\text{inertial}} = \left(\frac{d\omega}{dt}\right)_{\text{body}}
\tag{7.17}
$$

because $\omega \times \omega = 0$. Thus two observers, one fixed in the inertial frame and one fixed in the rotating frame, agree on the *rate of change* of ω.

8 ACCELERATIONS

As emphasized in the preceding section, the general expression (7.11) can be applied to the coordinate vector $\mathbf{r}$ of a moving particle, relating the velocity seen by observers fixed in the inertial frame and in the rotating frame

$$
\left(\frac{d\mathbf{r}}{dt}\right)_{\text{inertial}} = \left(\frac{d\mathbf{r}}{dt}\right)_{\text{body}} + \omega \times \mathbf{r}
\tag{8.1}
$$

To study the dynamical motion viewed from a moving coordinate system, however, it is necessary to consider the corresponding *accelerations* seen by the two observers. Since Eq. (8.1) in fact holds true for *any* vector, we see that the time derivatives formally satisfy the operator equation

$$\left(\frac{d}{dt}\right)_{\text{inertial}} = \left(\frac{d}{dt}\right)_{\text{body}} + \boldsymbol{\omega} \times \tag{8.2}$$

Applying this operation twice, we find

$$\left(\frac{d^2\mathbf{r}}{dt^2}\right)_{\text{inertial}} = \left[\left(\frac{d}{dt}\right)_{\text{body}} + \boldsymbol{\omega} \times\right]\left[\left(\frac{d\mathbf{r}}{dt}\right)_{\text{body}} + \boldsymbol{\omega} \times \mathbf{r}\right]$$

$$= \left(\frac{d^2\mathbf{r}}{dt^2}\right)_{\text{body}} + 2\boldsymbol{\omega} \times \left(\frac{d\mathbf{r}}{dt}\right)_{\text{body}} + \frac{d\boldsymbol{\omega}}{dt} \times \mathbf{r} + \boldsymbol{\omega} \times (\boldsymbol{\omega} \times \mathbf{r}) \tag{8.3}$$

where $d\boldsymbol{\omega}/dt$ is the same in both frames [see Eq. (7.17)]. Hence the acceleration seen by an observer in the rotating frame differs from that seen by an inertial observer by a variety of terms that depend on $\boldsymbol{\omega}$ and on $d\boldsymbol{\omega}/dt$.

9 TRANSLATIONS AND ROTATIONS

This discussion has been restricted to the case of two coordinate systems with a common origin, one an inertial frame and the other in arbitrary rotation. It is often necessary to generalize the description to include translation of the origin, which is easily done as follows. Consider an inertial coordinate system with axes (x_0, y_0, z_0) and a moving (in general rotating) coordinate system with axes (x, y, z) and origin O located at the instantaneous point $\mathbf{a}$ (see Fig. 9.1). If a

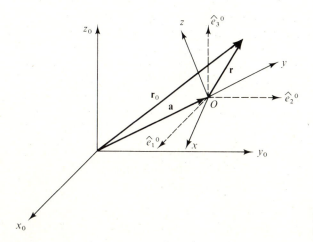

Figure 9.1 Translated and rotated coordinate system.

particle has the coordinates $\mathbf{r}_0$ in the inertial frame and $\mathbf{r}$ in the body-fixed frame, these coordinates are related by

$$\mathbf{r}_0 = \mathbf{a} + \mathbf{r} \tag{9.1}$$

Consequently, the translational acceleration of the moving axes leads to the generalized expression

$$\left(\frac{d^2\mathbf{r}_0}{dt^2}\right)_{inertial} = \left(\frac{d^2\mathbf{a}}{dt^2}\right)_{inertial} + \left(\frac{d^2\mathbf{r}}{dt^2}\right)_{inertial} \tag{9.2}$$

Now the second term on the right-hand side is just that calculated in Eq. (8.3). This is true since the change in the vector $\mathbf{r}$ as observed in the inertial and body-fixed frames differs only due to the rotation of the body-fixed frame about the origin O, even though this point is itself accelerating. (Readers are urged to convince themselves of this fact.) Thus we find

$$\left(\frac{d^2\mathbf{r}_0}{dt^2}\right)_{inertial} = \left(\frac{d^2\mathbf{a}}{dt^2}\right)_{inertial} + \left(\frac{d^2\mathbf{r}}{dt^2}\right)_{body}$$

$$+ 2\boldsymbol{\omega} \times \left(\frac{d\mathbf{r}}{dt}\right)_{body} + \frac{d\boldsymbol{\omega}}{dt} \times \mathbf{r} + \boldsymbol{\omega} \times (\boldsymbol{\omega} \times \mathbf{r}) \tag{9.3}$$

It is important to note that $\boldsymbol{\omega}$ is the instantaneous angular velocity of the body-fixed frame relative to the frame whose origin is at $\mathbf{a}$ and whose axes are parallel to those of the inertial frame (x_0, y_0, z_0). The origin at $\mathbf{a}$ may now be undergoing arbitrary acceleration.

10 NEWTON'S LAWS IN ACCELERATED COORDINATE SYSTEMS

Application of Newton's second law for a single particle in the inertial frame gives

$$m\left(\frac{d^2\mathbf{r}_0}{dt^2}\right)_{inertial} = \mathbf{F}^{(e)} \tag{10.1}$$

where $\mathbf{F}^{(e)}$ is the external force. A combination with Eq. (9.3) then yields the equivalent dynamical equation as seen by an observer fixed in the translating and rotating frame

$$m\left(\frac{d^2\mathbf{r}}{dt^2}\right)_{body} = \mathbf{F}^{(e)} - m\left(\frac{d^2\mathbf{a}}{dt^2}\right)_{inertial} - 2m\boldsymbol{\omega} \times \left(\frac{d\mathbf{r}}{dt}\right)_{body}$$

$$- m\boldsymbol{\omega} \times (\boldsymbol{\omega} \times \mathbf{r}) - m\frac{d\boldsymbol{\omega}}{dt} \times \mathbf{r} \tag{10.2}$$

suitably expressed in the body-fixed time derivatives. Evidently, the rotation produces several extra terms that act like forces. The term $-m(d^2\mathbf{a}/dt^2)_{inertial}$ arises from the acceleration of the origin of the body system as seen in the original inertial frame. The term $-2m\boldsymbol{\omega} \times (d\mathbf{r}/dt)_{body}$, called the *Coriolis force*, vanishes unless the particle moves in the rotating frame along a direction different from $\hat{\omega}$. In contrast, the *centrifugal force* $-m\boldsymbol{\omega} \times (\boldsymbol{\omega} \times \mathbf{r})$ acts even on a stationary particle. Finally, the contribution $-m(d\boldsymbol{\omega}/dt) \times \mathbf{r}$ occurs only for a coordinate system with angular acceleration.

11 MOTION ON THE SURFACE OF THE EARTH

The previous analysis has important applications to the dynamics of terrestrial objects. The earth's motion has two principal components: an annual nearly circular trajectory about the sun with radius $R_{se} = 1$ astronomical unit (AU) $\approx$ 1.50×10^{13} cm, period $\tau_{se} \approx 3.16 \times 10^7$ s, and angular frequency $\omega_{se} = 2\pi/\tau_{se} \approx$ 1.99×10^{-7} s^{-1}, and a daily rotation about the direction of the north pole star with equatorial radius $R_e \approx 6.38 \times 10^8$ cm, period $\tau_e \approx 8.62 \times 10^4$ s,† and angular frequency

$$\omega_e = \frac{2\pi}{\tau_e} \approx 7.29 \times 10^{-5} \text{ s}^{-1} \tag{11.1}$$

Apart from the very slow motion of the earth's rotation axis (precession of the equinoxes with a period of about 25,000 years), the angular velocity vectors $\boldsymbol{\omega}_{se}$ and $\boldsymbol{\omega}_e$ are constant, allowing us to neglect $d\boldsymbol{\omega}/dt$ in Eq. (10.2).

Consider an inertial frame fixed at the sun's center and a noninertial frame fixed in the rotating earth with origin at the earth's center. This point follows a nearly circular trajectory $\mathbf{a}(t)$ relative to the sun with radius 1 AU, and the associated earth-fixed frame has an angular velocity $\boldsymbol{\omega}_e + \boldsymbol{\omega}_{se}$. In this case, Eq. (10.2) becomes

$$m\left(\frac{d^2\mathbf{r}}{dt^2}\right)_e = \mathbf{F}^{(e)} - m\left(\frac{d^2\mathbf{a}}{dt^2}\right)_{inertial} - 2m(\boldsymbol{\omega}_e + \boldsymbol{\omega}_{se}) \times \left(\frac{d\mathbf{r}}{dt}\right)_e$$
$$- m(\boldsymbol{\omega}_e + \boldsymbol{\omega}_{se}) \times [(\boldsymbol{\omega}_e + \boldsymbol{\omega}_{se}) \times \mathbf{r}] \tag{11.2}$$

where $\mathbf{F}^{(e)}$ includes the sun's gravitational field, the earth's gravitational field, and any other forces that may act. Since $M_e \approx 5.98 \times 10^{27}$ g and $M_s \approx 1.99 \times 10^{33}$ g, we note that the earth's gravitational field at the earth's surface exceeds that of the sun by a factor of about $(M_e/M_s)(R_{se}/R_e)^2 \approx 1.66 \times 10^3$. Moreover, at the center of the earth, the sun's field precisely cancels the contribution $-m(d^2\mathbf{a}/dt^2)_{inertial}$ in Eq. (11.2) that arises from the earth's annual motion about the sun. Neglecting smaller corrections of relative order $R_e/R_{se} \approx 4.25 \times 10^{-5}$ arising from gradients

---

† Note that τ_e is less than 24 h = 8.64×10^4 s by a factor of ≈ 0.997 because of the daily orbital angular displacement ($\approx 1°$) about the sun. French [1], chap. 8, has an interesting historical account of these measurements. Numbers refer to Selected Additional Readings at the end of the chapter; more general references are listed in Appendix G and indicated by author and date.

in the sun's field and of order $\omega_{se}/\omega_e \approx 2.73 \times 10^{-3}$, we therefore obtain the approximate equation of motion for a particle on the rotating earth

$$m\left(\frac{d^2\mathbf{r}}{dt^2}\right)_e = \mathbf{F}_g + \mathbf{F}' - 2m\boldsymbol{\omega} \times \left(\frac{d\mathbf{r}}{dt}\right)_e - m\boldsymbol{\omega} \times (\boldsymbol{\omega} \times \mathbf{r}) \qquad (11.3)$$

Here, $\boldsymbol{\omega} \approx \boldsymbol{\omega}_e$, $\mathbf{r}$ is measured from the center of the earth, $\mathbf{F}'$ represents additional terrestrial forces (a supporting string, for example), and $\mathbf{F}_g$ is the force of the earth's gravity. For simplicity, we treat the earth as an essentially uniform sphere, with the familiar force

$$\mathbf{F}_g = -GM_e m \frac{\mathbf{r}}{r^3} \qquad (11.4)$$

but the generalization to include the earth's inhomogeneous and aspherical form is not difficult.

Particle on a Scale

The simplest application of (11.3) is a stationary particle on a scale at the earth's surface. The corresponding velocity and acceleration vanish identically, and a combination with (11.4) yields the condition of "static" equilibrium

$$0 = -GM_e m \frac{\mathbf{r}}{r^3} + \mathbf{F}' - m\boldsymbol{\omega} \times (\boldsymbol{\omega} \times \mathbf{r})$$

or

$$\mathbf{F}' + m\mathbf{g} = 0 \qquad (11.5a)$$

where, by definition

$$\mathbf{g} \equiv -GM_e \frac{\mathbf{r}}{r^3} - \boldsymbol{\omega} \times (\boldsymbol{\omega} \times \mathbf{r}) \qquad (11.5b)$$

is the net gravitational and centrifugal force on a stationary object with unit mass near the surface of the earth. Note that m disappears from this definition, exemplifying Einstein's equivalence principle that all test particles follow the same trajectory in a gravitational field, independent of mass. In addition, $\mathbf{g}$ is not strictly radially inward but instead acquires an outward equatorial component perpendicular to $\boldsymbol{\omega}$ (see Fig. 11.1). At the poles, $\mathbf{g}$ is indeed radial with magnitude $-g_0 = -GM_e R_e^{-2}$, but at the equator Eq. (11.5b) becomes $\mathbf{g} = (-g_0 + \omega^2 R_e)\hat{r}$. Numerical evaluation yields $g_0 \approx 980$ cm s^{-2} and $\omega^2 R_e \approx 3.39$ cm s^{-2}, implying that the centrifugal terms have a small but significant effect. If the particle is to remain in static equilibrium, we must supply an additional force $\mathbf{F}' = -m\mathbf{g}$, such as tension in a string suspended from a crossbar. As a corollary, a plumb line necessarily points along the direction $\hat{g}$, which in general differs from $-\hat{r}$. By definition, however, $\hat{g}$ is "vertical," automatically including the centrifugal contribution.

To conclude this discussion, we can project $\mathbf{g}$ into its spherical polar coordinates

$$\mathbf{g} = -(GM_e R_e^{-2} - \omega^2 R_e \sin^2 \theta)\hat{r} + \tfrac{1}{2}\sin 2\theta \, \omega^2 R_e \, \hat{\theta} \qquad (11.6)$$

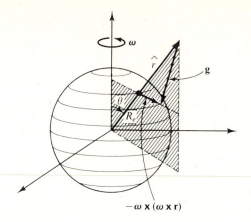

$$-\boldsymbol{\omega} \times (\boldsymbol{\omega} \times \mathbf{r})$$

Figure 11.1 Newton's laws in an earth-fixed system.

where $\hat{\omega} \cdot \hat{r} = \cos \theta$, with θ the colatitude measured from the north pole. Equation (11.6) displays the net tangential component of $\mathbf{g}$ that acts toward the equator. If the earth were an ideal fluid, matter would indeed flow from the poles, producing an equatorial bulge and ensuring that the surface would everywhere be perpendicular to $\mathbf{g}$ (see Fig. 11.1). In fact, the earth's early solidification produced just such a small quadrupole mass deformation, and Eq. (11.4) must be modified to include the altered gravitational attraction. The resulting equipotential surface, known as the *geoid*,† is conventionally defined by the shape of a thin co-rotating fluid layer. The corresponding value of g at sea level increases by about 5.2 cm s^{-2} in moving from the equator to the pole, somewhat exceeding the simple estimate $\omega^2 R_e \approx 3.39$ cm s^{-2} evaluated on a spherical surface for the newtonian potential $-m\gamma/r$ (see Prob. 2.7 for a calculation of this effect).

Falling Particle

We now return to the more general case of a particle moving near the surface of the rotating earth with velocity $\mathbf{v}$ as seen by a terrestrial observer. A combination of Eqs. (11.3) to (11.5) provides the dynamical equation

$$m\dot{\mathbf{v}} = \mathbf{F}' - 2m\boldsymbol{\omega} \times \mathbf{v} + m\mathbf{g} \tag{11.7}$$

where $\dot{\mathbf{v}}$ is the corresponding acceleration seen from the earth. Consider a particle released from rest a (small) height $h \ll R_e$ above the earth's surface. If the earth were stationary, the particle would fall vertically along a plumb line, but the Coriolis force displaces its landing site by an amount proportional to ω. What is the direction of the displacement? (East? West?) To analyze this situation, we rewrite Eq. (11.7) in a form more appropriate for free fall ($\mathbf{F}' = 0$) as

$$m\ddot{\mathbf{r}} = m\mathbf{g} - 2m\boldsymbol{\omega} \times \dot{\mathbf{r}} \tag{11.8}$$

† Stacey [2], sec. 4.1.

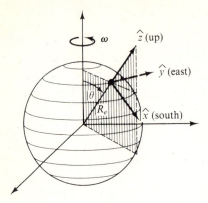

Figure 11.2 Earth-fixed frame.

where **g** is again the net gravitational and centrifugal acceleration at the earth's surface. Since this vector equation depends only on $\dot{\mathbf{r}}$ and $\ddot{\mathbf{r}}$, it can be resolved in *any* convenient body-fixed coordinate basis. We choose a local frame on the earth's surface, with $\hat{x}$ southward, $\hat{y}$ eastward, and $\hat{z}$ vertically upward (Fig. 11.2). In fact, $\hat{z}$ is not strictly along the radial direction $\hat{r}$, owing to the centrifugal term in (11.5), but this effect is second order in ω [Eq. (11.1)] and negligible for the present considerations.† Thus we take $\mathbf{g} = -g\hat{z}$ and retain only terms linear in ω.

Let $\mathbf{r}(0) = h\hat{z}$ be the particle's initial position and let $\mathbf{v}(0) = 0$. It is convenient to use a perturbation analysis; let $\mathbf{r}(t) = \mathbf{r}_0(t) + \mathbf{r}_1(t)$, where $\mathbf{r}_0(t)$ describes the trajectory on a nonrotating earth and $\mathbf{r}_1(t)$ incorporates the small corrections proportional to ω. Evidently, $\mathbf{r}_0(t)$ satisfies the equation

$$\ddot{\mathbf{r}}_0 = \mathbf{g} = -g\hat{z} \tag{11.9}$$

A simple integration gives

$$\mathbf{r}_0(t) = \mathbf{r}(0) + \tfrac{1}{2}\mathbf{g}t^2 = \mathbf{r}(0) - \tfrac{1}{2}gt^2\hat{z} \tag{11.10}$$

Substituting the more general form $\mathbf{r}_0 + \mathbf{r}_1$ into Eq. (11.8) and expanding to first order in ω, we find

$$\ddot{\mathbf{r}}_0 + \ddot{\mathbf{r}}_1 = \mathbf{g} - 2\boldsymbol{\omega} \times \dot{\mathbf{r}}_0 \tag{11.11}$$

Comparison with Eq. (11.9) shows that the zero-order terms cancel identically, leaving the simple inhomogeneous equation

$$\ddot{\mathbf{r}}_1 = -2\boldsymbol{\omega} \times \mathbf{g}t \tag{11.12}$$

Use of the initial condition $\mathbf{r}_1(0) = \dot{\mathbf{r}}_1(0) = 0$ readily yields the solution

$$\mathbf{r}_1(t) = -\tfrac{1}{3}\boldsymbol{\omega} \times \mathbf{g}t^3 = \tfrac{1}{3}\omega gt^3 \sin\theta\, \hat{y} \tag{11.13}$$

---

† The relevant dimensionless parameter is ωt_0, where $t_0 \approx (2h/g)^{1/2}$ is the time of free fall from a height h. For $h \approx 100$ m, $\omega t_0 \approx 3.3 \times 10^{-4}$.

and the total trajectory becomes

$$\mathbf{r}(t) = \hat{z}(h - \tfrac{1}{2}gt^2) + \tfrac{1}{3}\omega gt^3 \sin \theta \, \hat{y} \qquad (11.14)$$

It has several interesting features:

1. The vertical motion is independent of ω to first order, with the simple time of fall $t_0 = (2h/g)^{1/2}$.
2. The particle is deflected *eastward*, with magnitude $\tfrac{1}{3}\omega g \sin \theta \, t_0^3 = \tfrac{1}{3}\omega g \sin \theta \, (2h/g)^{3/2}$. This effect is the same in the Northern and the Southern Hemisphere; it vanishes at the poles, where $\sin \theta = 0$, and is maximum at the equator. If $h = 100$ m, then $t \approx 4.5$ s and the equatorial deflection is a measurable 2.2 cm.
3. At first sight, the eastward deflection may be surprising, for the earth itself turns to the east. Recall, however, that the particle initially starts from rest in the rotating frame and then falls toward the earth's surface. In an inertial frame, the initial motion is eastward, and conservation of angular momentum in that frame [compare Eq. (3.5)] demands that it speed up as it falls. The detailed analysis of this same motion as seen from an inertial frame is an interesting exercise in orbital dynamics (see Prob. 2.3).

Horizontal Motion

The vector nature of the Coriolis force in Eq. (11.8) implies quite different behavior for horizontal and vertical motion on the rotating earth. This difference is most striking at the north or south pole, where we have seen that a falling particle follows a plumb line with no transverse deflection. In contrast, inspection of the vector cross product in Eq. (11.8) shows that a projectile fired horizontally near the north pole deflects to the right when viewed along its trajectory. Such an effect is obvious when considered from an inertial frame, for the projectile's orbit remains rectilinear while the earth turns eastward. (In this case, conservation of angular momentum plays a smaller role, because the initial motion is perpendicular to the gravitational force.)

The analysis becomes only slightly more complicated for general horizontal motion. Consider a particle at colatitude (polar angle) θ, moving with horizontal velocity $\mathbf{v}$ directed at an angle ϕ in a positive (counterclockwise) sense from the southerly direction. Using the same coordinates as in Fig. 11.2, we have $\hat{\omega} = \hat{z} \cos \theta - \hat{x} \sin \theta$ and $\mathbf{v} = v(\hat{x} \cos \phi + \hat{y} \sin \phi)$. A simple calculation gives the Coriolis force

$$\mathbf{F}_c = -2m\boldsymbol{\omega} \times \mathbf{v} = 2m\omega v(\cos \theta \sin \phi \, \hat{x} - \cos \theta \cos \phi \, \hat{y} + \sin \theta \sin \phi \, \hat{z}) \quad (11.15)$$

For example, a north-moving particle ($\phi = \pi$) in the Northern Hemisphere ($\cos \theta > 0$) experiences an easterly force, whereas a south-moving particle ($\phi = 0$) in the Northern Hemisphere experiences a westerly force. This effect is proportional to ω and therefore small, but it must be included in accurate calculations of ballistic trajectories (see Prob. 2.5). It also accounts for the sense of the major

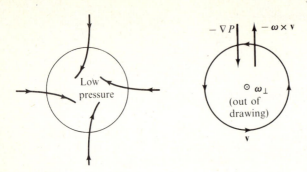

Figure 11.3 Cyclones and hurricanes in the Northern Hemisphere.

ocean currents like the Gulf Stream, which flow clockwise in the Northern Hemisphere when viewed from above. In addition, the Coriolis force explains the slow displacement in the Northern Hemisphere of rivers toward the right when facing downstream. Finally, we may consider a low-pressure atmospheric region in the Northern Hemisphere. The net inward mass flow experiences a Coriolis force, impelling it into counterclockwise motion as seen from above. The resulting circular flow (Fig. 11.3) reaches a dynamical equilibrium with the negative pressure gradient $-\nabla P$ equal and opposite to the Coriolis force, directed along $-\boldsymbol{\omega} \times \mathbf{v}$. In the Southern Hemisphere, the normal component of the angular velocity $\omega_\perp$ points *into* the ground, which reverses the sense of these effects.

12 FOUCAULT PENDULUM

One of the simplest yet most interesting terrestrial demonstrations of the earth's rotation is the Foucault pendulum; its plane of oscillation rotates slowly with a period comparable with 1 day. Indeed, at the north pole, the point of support is fixed in space and exerts no transverse forces on the string. Consequently, an inertial observer merely finds that the point mass oscillates in a fixed plane; an earthbound observer, however, sees the plane of oscillation rotate clockwise viewed from above, once per day, because of the earth's rotation.

The general case is considerably more complicated and illustrates clearly the usefulness of the formalism developed in Sec. 11. On the earth's surface, the point of support is fixed, and the only motion involves the point mass m, for which the equation of motion becomes [see Eq. (11.7)]

$$m\ddot{\mathbf{r}} = m\mathbf{g} + \mathbf{F}' - 2m\boldsymbol{\omega} \times \dot{\mathbf{r}} \qquad (12.1)$$

Here the velocity and acceleration are those seen by the terrestrial observer. It is again convenient to use the local coordinate system of Fig. 11.2, centered at the position of the unperturbed particle, with $\hat{x}$ south, $\hat{y}$ east, and $\hat{z}$ up, neglecting corrections of order ω^2. The acceleration then becomes $(\ddot{x}, \ddot{y}, \ddot{z})$, and the Coriolis forces has components

$$-2m\boldsymbol{\omega} \times \dot{\mathbf{r}} = 2m\omega(\dot{y} \cos \theta, -\dot{x} \cos \theta - \dot{z} \sin \theta, \dot{y} \sin \theta) \qquad (12.2)$$

where, as previously, θ is the polar angle on the earth's surface (namely, the colatitude). The only remaining unknown quantity in Eq. (12.1) is the force $\mathbf{F}'$ arising from the tension in the string. Figure 12.1 illustrates the situation, with ψ the deflection angle measured from the downward direction and ϕ the angle between the x axis and the plane of oscillation. If l denotes the length of the string and $r = (x^2 + y^2)^{1/2}$ is the transverse displacement, we have the components

$$m\mathbf{g} + \mathbf{F}' = (-T \sin \psi \cos \phi, -T \sin \psi \sin \phi, T \cos \psi - mg) \qquad (12.3)$$

where $\sin \psi = r/l$.

These various relations provide all the information needed to construct the trajectory. Consider first the vertical motion

$$m\ddot{z} = T \cos \psi - mg + 2m\omega\dot{y} \sin \theta \qquad (12.4)$$

In a typical case, ωv is much smaller than g (recall that $\omega v \approx 7 \times 10^{-5}$ cm s^{-2} for $v = 1$ cm s^{-1}), so that the last term of Eq. (12.4) is negligible relative to the other contributions. Moreover, the vertical displacement is given by $z = l(1 - \cos \psi)$, and this also is small for small displacements from equilibrium ($r \ll l$ implies $\psi \approx r/l$ and $z \approx r^2/2l \ll r$). Consequently, the left-hand side of Eq. (12.4) is also negligible, leading to the familiar conclusion that

$$T \cos \psi \approx T \approx mg \qquad (12.5)$$

We now consider the horizontal motion

$$m\ddot{x} = -T \sin \psi \cos \phi + 2m\omega\dot{y} \cos \theta \qquad (12.6a)$$

$$m\ddot{y} = -T \sin \psi \sin \phi - 2m\omega(\dot{x} \cos \theta + \dot{z} \sin \theta) \qquad (12.6b)$$

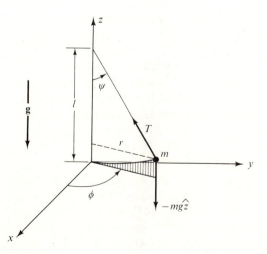

Figure 12.1 Foucault pendulum.

and the preceding discussion shows that the last term on the right-hand side of (12.6b) can be neglected. Since $\sin \psi \cos \phi \approx x/l$ and $\sin \psi \sin \phi \approx y/l$ (Fig. 12.1), a combination of Eqs. (12.5) and (12.6) yields the simple linear equations

$$\ddot{x} = -\frac{g}{l} x + 2\omega \dot{y} \cos \theta \tag{12.7a}$$

and

$$\ddot{y} = -\frac{g}{l} y - 2\omega \dot{x} \cos \theta \tag{12.7b}$$

where $\omega \cos \theta = \boldsymbol{\omega} \cdot \hat{z} \equiv \omega_\perp$ is the vertical projection of the earth's angular velocity.

These coupled differential equations can be solved in several ways. The most straightforward one is to notice that the time-independent coefficients permit a solution in the exponential form $x = x_0 e^{-i\sigma t}$ and $y = y_0 e^{-i\sigma t}$, which will be used frequently in subsequent chapters. Here, however, we prefer to use a trick that applies whenever the velocity-dependent terms have the special structure seen in Eqs. (12.7). Introduce the complex linear combination

$$\zeta(t) = x(t) + iy(t) \tag{12.8}$$

The corresponding equation of motion follows by adding i times (12.7b) to (12.7a)

$$\ddot{\zeta} = -\frac{g}{l} \zeta - 2i\omega \cos \theta \, \dot{\zeta} \tag{12.9}$$

This equation again has constant coefficients, and it therefore has an exponential solution $\zeta(t) = \zeta_0 e^{-i\sigma t}$. Substitution into (12.9) yields the consistency condition $\sigma^2 - 2\omega \cos \theta \, \sigma - g/l = 0$, with the two roots

$$\sigma_\pm = p \pm q \tag{12.10a}$$

$$p = \omega \cos \theta \equiv \omega_\perp \qquad q = \left(\omega_\perp^2 + \frac{g}{l}\right)^{1/2} \tag{12.10b}$$

Consequently, the general solution is a linear combination of two terms

$$\zeta(t) = A e^{-i(p+q)t} + B e^{-i(p-q)t} \tag{12.11}$$

with the complex constants A and B chosen to fit the initial conditions.

For definiteness, suppose the pendulum bob is displaced southward a small distance a and released from rest. Then $\zeta(0) = x(0) = a$ and $\dot{\zeta}(0) = 0$, and some elementary algebra gives the corresponding time-dependent solution

$$\zeta(t) = a e^{-ipt}(\cos qt + ipq^{-1} \sin qt) \tag{12.12}$$

Note that $\zeta(t)$ never vanishes exactly, implying that the particle always remains a finite distance from the origin. This behavior arises from the Coriolis force (Sec. 11), which acts to the right in the Northern Hemisphere. Although the exact solution is readily analyzed, we can simplify the treatment by noticing that p is of order $10^{-4} \, \text{s}^{-1}$, much smaller than $q = (g/l + \omega_\perp^2)^{1/2}$, which is essentially the free-pendulum angular frequency and is typically of the order of $1 \, \text{s}^{-1}$. To leading order, the second term of (12.12) is negligible compared to the first, and the motion is given approximately by

$$\zeta(t) \approx a(\cos qt)e^{-ipt} \tag{12.13}$$

with
$$|\zeta(t)|^2 = |x(t) + iy(t)|^2 = x(t)^2 + y(t)^2 \approx a^2 \cos^2 qt \tag{12.14}$$

Evidently, this approximation omits the small transverse motion, for the displacement undergoes simple harmonic motion with amplitude a and frequency $q \approx (g/l)^{1/2}$, as expected for a pendulum on a stationary earth. The phase factor e^{-ipt} in (12.13) still has an important effect, which is most easily studied by taking real and imaginary parts:

$$x(t) = \text{Re } \zeta(t) \approx a \cos pt \cos qt \approx a \cos \omega_\perp t \cos \left[\left(\frac{g}{l}\right)^{1/2} t \right]$$

$$\tag{12.15}$$

$$y(t) = \text{Im } \zeta(t) \approx -a \sin pt \cos qt \approx -a \sin \omega_\perp t \cos \left[\left(\frac{g}{l}\right)^{1/2} t \right]$$

Since $\omega_\perp = \omega \cos \theta \ll (g/l)^{1/2}$, these equations represent a superposition of two perpendicular oscillatory motions proportional to $\cos [(g/l)^{1/2}t]$ but with slowly varying amplitudes $a \cos \omega_\perp t$ and $-a \sin \omega_\perp t$. If $\omega_\perp = 0$ (either because $\omega = 0$, or at the equator where $\cos \theta = 0$), the motion remains simple harmonic in the north-south plane $(y = 0)$ with angular frequency $(g/l)^{1/2}$. More generally, however, $\omega_\perp$ is nonzero, and the pendulum bob acquires an east-west component. In the present approximation, Eq. (12.14) shows that the motion is plane-polarized at an angle ϕ relative to the x axis, where ϕ is given by the relation

$$\tan \phi = \frac{y(t)}{x(t)} = -\tan \omega_\perp t \tag{12.16a}$$

or, equivalently,
$$\phi = -\omega_\perp t \tag{12.16b}$$

Thus the plane of oscillation rotates uniformly at an angular rate $-\omega_\perp = -\omega \cos \theta$ or with a period $|\sec \theta|$ days. In the Northern Hemisphere, $\omega_\perp = \omega \cos \theta > 0$ and $\sec \theta \geq 1$, and we indeed find a clockwise rotation when viewed from above, consistent with the rightward sense of the Coriolis force. At the north pole, this relation reproduces the simple daily period, and it verifies that there is no precession to order ω at the equator. In the Southern Hemisphere, the perpendicular component of $\boldsymbol{\omega}$ reverses sign and points into the ground; as a result, $\omega_\perp = \omega \cos \theta$ becomes negative, and the plane of oscillation of the Foucault pendulum rotates in the opposite direction (counterclockwise when viewed from above).

PROBLEMS

2.1 Larmor's theorem

(a) The Lorentz force implies the equation of motion $m\ddot{\mathbf{r}} = e(\mathbf{E} + c^{-1}\mathbf{v} \times \mathbf{B})$. Prove that the effect of a weak uniform magnetic field $\mathbf{B}$ on the motion of a charged particle in a central electric field $\mathbf{E} = E(|\mathbf{r}|)\hat{\mathbf{r}}$ can be removed by transforming to a coordinate system rotating with an angular frequency $\boldsymbol{\omega}_L = -(e/2mc)\mathbf{B}$ (state precisely what "weak" means).

(b) Extend this result to a system of particles of given ratio e/m interacting through potentials $V_{ij}(|\mathbf{r}_i - \mathbf{r}_j|)$.

2.2 Assume that over the time interval of interest, the center of mass of the earth moves with approximately constant velocity with respect to the fixed stars and that ω, the angular velocity of the earth, is a constant. Rederive the terrestrial equations of particle motion (11.8) and (11.6) by writing Newton's law in a body-fixed frame with origin at the *surface* of the earth (Fig. 11.2).

2.3 An observer that rotates with the earth drops a particle from a height h above the earth's surface. Analyze the motion from an inertial frame of reference and rederive the net eastward deflection of $\approx \frac{1}{3}\omega g \sin\theta\,(2h/g)^{3/2}$, where θ is the observer's polar angle.

2.4 A particle is projected vertically upward from the surface of the rotating earth at colatitude (polar angle) θ. It rises to a height $h\,(\ll R_e)$ and then falls back to the surface. Show that it strikes the ground at a distance $\frac{8}{3}\omega \sin\theta\,(2h^3/g)^{1/2}$ to the west of the initial position. What is its westward displacement at its maximum height?

2.5 A cannon is placed on the surface of the earth at colatitude (polar angle) θ and pointed due east.

(a) If the cannon barrel makes an angle α with the horizontal, show that the lateral deflection of a projectile when it strikes the earth is $(4V_0^3/g^2)\omega \cos\theta \sin^2\alpha \cos\alpha$, where V_0 is the initial speed of the projectile and ω is the earth's angular-rotation speed. What is the direction of this deflection?

(b) If R is the range of the projectile for the case $\omega = 0$, show that the change in the range is given by $(2R^3/g)^{1/2}\omega \sin\theta\,[(\cot\alpha)^{1/2} - \frac{1}{3}(\tan\alpha)^{3/2}]$. Neglect terms of order ω^2 throughout.

2.6 Extend the calculation of the falling particle on the rotating earth by solving Eq. (11.8) to second order in ω. Show that the time to fall a vertical distance h is increased by $\frac{1}{6}\omega^2(2h/g)^{3/2}\sin^2\theta$ and that the particle is deflected toward the equator by an amount $(\omega^2 h^2/3g)\sin 2\theta$.

2.7 (a) Show that the centrifugal force $-m\boldsymbol{\omega}\times(\boldsymbol{\omega}\times\mathbf{r})$ at a colatitude (polar angle) θ can be written $-m\nabla\Phi_c$, where

$$\Phi_c = -\tfrac{1}{2}|\boldsymbol{\omega}\times\mathbf{r}|^2 = -\tfrac{1}{2}\omega^2 r^2 \sin^2\theta$$

(b) If Φ_g is the earth's exterior gravitational potential, prove that a thin surface layer of water (the "ocean") would assume the shape of an equipotential surface $\Phi_g + \Phi_c = $ const (known as the *geoid*).

(c) Suppose that the earth has a small quadrupole deformation, with

$$\Phi_g = -M_e G r^{-1}\left[1 - J_2\left(\frac{R_e}{r}\right)^2 P_2(\cos\theta)\right]$$

Here $P_2(\cos\theta) = \frac{1}{2}[3(\cos^2\theta) - 1]$ and $J_2 \approx 1.083 \times 10^{-3}$ characterizes the earth's deviation from sphericity (see Prob. 5.7). Obtain the approximate relation

$$\frac{\Delta R}{R_e} = \tfrac{3}{2}J_2 + \frac{1}{2}\frac{\omega^2 R_e^3}{M_e G}$$

where ΔR is the difference between the geoid's equatorial and polar radii. Compare the resulting value of $\Delta R/R_e$ with $(R_e - R_p)/R_e \approx 3.35 \times 10^{-3}$ determined from the measured value of the earth's equatorial and polar radii.

(d) Explain why $-\nabla(\Phi_g + \Phi_c)$ on the geoid is the local acceleration $\mathbf{g}$ due to gravity at sea level, and show that its magnitude is given approximately by $g \approx \partial(\Phi_g + \Phi_c)/\partial r$. Hence derive Clairaut's formula

$$g = g_e\left[1 + (\cos^2\theta)\left(\frac{5}{2}\frac{\omega^2 R_e^3}{M_e G} - \frac{\Delta R}{R_e}\right)\right]$$

Use the measured values $g_e \approx 978.03$ cm s^{-2} and $g_p \approx 983.20$ cm s^{-2} to estimate $\Delta R/R_e$, and compare with the previous value.

SELECTED ADDITIONAL READINGS

1. A. P. French, "Newtonian Mechanics," Norton, New York, 1971.
2. F. D. Stacey, "Physics of the Earth," 2d ed., Wiley, New York, 1977.

THREE

LAGRANGIAN DYNAMICS

13 CONSTRAINED MOTION AND GENERALIZED COORDINATES

In many cases, the motion of a particle is restricted by one or more constraints that reduce the number of independent degrees of freedom. It is advantageous to incorporate the constraints explicitly and work with a set of *independent coordinates*. In addition, the constraints exert forces on the particle to maintain its restricted motion. These *forces of constraint* are complicated, since they depend on the actual trajectory of the particle. It is also advantageous to formulate classical mechanics to eliminate these forces of constraint entirely. We shall see that both goals can be achieved simultaneously. Let us start by characterizing the types of constraints that can occur.

Constraints

Consider the motion of N particles. This system has $n\,(=3N)$ degrees of freedom. Label the coordinates x_i and order the labels so that i runs over the values $i = (1, 2, 3)$, $(4, 5, 6)$, ..., $(n - 2,\ n - 1,\ n)$, where the first triplet refers to particle 1, the second triplet to particle 2, and so on. Suppose there are k equations

$$f_j(x_1, \ldots, x_n, t) = c_j \qquad j = 1, 2, \ldots, k \tag{13.1}$$

that relate the coordinates and possibly the time. We refer to these k relations as *holonomic constraints*. Figure 13.1 illustrates three examples of such constraints. In the first, the particle is constrained to move on a surface $z(x, y)$. Thus the system in

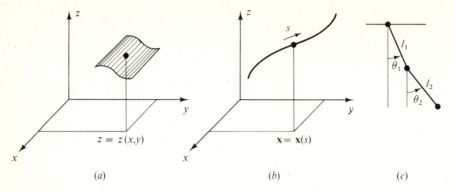

Figure 13.1 Three examples of holonomic constraints: (a) motion on a surface, (b) motion on a curve, (c) double planar pendulum.

fact possesses only two degrees of freedom. In the second, the particle is constrained to move along a specified curve $\mathbf{x} = \mathbf{x}(s)$, where s is the distance measured along the curve (a frictionless wire, for example). Here the system possesses just one degree of freedom. The third example is a double planar pendulum where the strings connecting the two particles and the first particle with the support have constant length. These relations constitute two holonomic constraints, leaving two degrees of freedom that can be specified, e.g., by the angles between each string and the vertical.

Constraints can be either time-independent or time-dependent. Figure 13.1b provides a simple example of the latter type if the wire itself executes prescribed time-dependent motion in space.

Physical systems can also involve *nonholonomic constraints*, which have no simple equation relating the coordinates and the time. For example, consider a mass point sliding without friction on a sphere of radius R, where the radial coordinate r of the particle necessarily satisfies the constraint $r \geqslant R$. In the earth's gravitational field, a particle sliding on such a sphere will eventually leave its surface and execute free fall, with no further constraint on its motion. Thus, no analytic equation can represent the geometric constraint on the coordinates of the particle for all times.

We may also encounter *nonintegrable* constraints. This special subclass of nonholonomic constraints can only be written in differential form $\sum_i h_i \, dx_i = 0$, where h_i may depend on the particle coordinates, the particle velocities, and the time. In fact, as we shall see, we can readily incorporate such nonintegrable constraints into our discussion of lagrangian dynamics.

Generalized Coordinates

Consider N particles with $n = 3N$ degrees of freedom moving under k holonomic constraints [Eq. (13.1)]. There are $n - k$ independent degrees of freedom, and we choose $q_1, q_2, \ldots, q_{n-k}$ to be any set of independent coordinates that completely specify the configuration of the system. We refer to these as *generalized coordinates*. In Fig. 13.1*b*, for example, the distance *s* along the curve provides a suitable coordinate, and the two angles θ_1 and θ_2 are appropriate in Fig. 13.1*c*. By assumption, these generalized coordinates completely determine the three-dimensional configuration of every particle. In addition, we may also have to specify the time if the constraints themselves depend on time. For example, if the wire in Fig. 13.1*b* is moving, the three-dimensional configuration of a bead on the wire is specified not only by the distance *s* along the wire but also by the time. In each of these cases, the fundamental relations between the cartesian coordinates of all the particles and the generalized coordinates can be written as

$$x_1 = x_1(q_1, q_2, \ldots, q_{n-k}, t)$$
$$\cdots\cdots\cdots\cdots\cdots\cdots\cdots\cdots\cdots\cdots\cdots\cdots\cdots \tag{13.2}$$
$$x_n = x_n(q_1, q_2, \ldots, q_{n-k}, t)$$

Suppose that all the generalized coordinates and the time are changed slightly by amounts $\{dq_i\}$ and dt. These differentials $\{dq_i\}$ may, for example, represent the dynamical change in the actual configuration when *t* increases to $t + dt$, but they also describe more general cases. The fundamental rule for partial differentiation gives the change induced in the variable x_1 as

$$dx_1 = \frac{\partial x_1}{\partial q_1}\, dq_1 + \cdots + \frac{\partial x_1}{\partial q_{n-k}}\, dq_{n-k} + \frac{\partial x_1}{\partial t}\, dt \tag{13.3}$$

where the partial derivatives imply that all the other independent variables in Eq. (13.2) are kept constant. This relation can be summarized for all the differentials as

$$dx_i = \sum_{\sigma=1}^{n-k} \frac{\partial x_i}{\partial q_\sigma}\, dq_\sigma + \frac{\partial x_i}{\partial t}\, dt \qquad i = 1, \ldots, n \tag{13.4}$$

Virtual Displacements

We now introduce the concept of a *virtual displacement*, which will be essential in reexpressing the basic principles of mechanics. A virtual displacement δx_i ($i = 1, \ldots, n$) is defined as an *infinitesimal, instantaneous* displacement of the coordinates $x_1, \ldots, x_n$ *consistent with the constraints*. Virtual displacements thus have three characteristics: they are infinitesimal, they occur at a *given* instant in time, and they are to be carried out in a manner consistent with the (here holonomic) constraints in the problem. In effect, we imagine freezing the system at some instant and then performing infinitesimal displacements δx_i permitted by the constraints. Since Eq. (13.4) holds for a general displacement of coordinates satisfying the constraints, it follows that the virtual displacements are given by

$$\delta x_i = \sum_{\sigma=1}^{n-k} \frac{\partial x_i}{\partial q_\sigma}\, \delta q_\sigma \tag{13.5}$$

This equation differs from Eq. (13.4) only in that there is no change in the time, since, by definition, virtual displacements are instantaneous.

14 D'ALEMBERT'S PRINCIPLE

The basic observation leading to d'Alembert's principle is that

> The reaction forces, or forces of constraint, do no work under a virtual displacement.

This statement contains no new principle of mechanics, for we are still dealing with Newton's law. Instead, it simply characterizes how the forces of constraint act during the motion of the particles. (Throughout the discussion in this section, it will be assumed that no frictional forces are present.) We shall not give a general proof of this assertion but merely illustrate it for the three examples shown in Fig. 13.1. In the absence of friction, it is clear in cases a and b that the forces of constraint are perpendicular to any motion allowed by a virtual displacement. Recalling that the work done by a force is the vector dot product of the force and the distance through which it moves, we see that these forces of constraint do no work when the particles move in a manner consistent with the constraints. It does not matter that the constraints themselves may be moving as a function of time since virtual displacements have been defined to be *instantaneous*; thus the constraints are frozen during the virtual displacement. The situation in Fig. 13.1c is more complex than in the first two, and we shall analyze it in some detail (see Fig. 14.1). Here the constraint has no dynamical properties itself, serving merely to transmit the reaction force between the particles through the tension τ in the string. In accordance with Newton's third law, we see that the reaction forces between the particles are equal and opposite. We now write the virtual displacements of the first particle as δs_1 and of the second particle as $\delta s_2 = \delta s_1 + \delta s_r$ and observe that δs_r must correspond to a rotation about point 1 since the length l_2 is constant. The virtual work done by the constraints is then given by

$$\delta W = \tau \cdot \delta s_1 - \tau \cdot \delta s_2 = -\tau \cdot \delta s_r = 0 \tag{14.1}$$

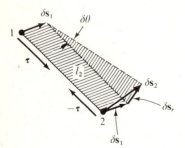

Figure 14.1 Virtual displacement for Fig. 13.1c.

because $\delta \mathbf{s}_r$ is perpendicular to the tension τ. Although the reaction force does work on each of the separate particles, the net work done by the reaction force on the entire system vanishes.

The observation on the nature of the work done by reaction forces under virtual displacement will now be used to reformulate the basic equations of mechanics. Newton's second law for the system of N particles can be written

$$\dot{p}_i = F_i^{(a)} + R_i \qquad i = 1, \ldots, n \tag{14.2}$$

where $p_i = i$th component of momentum
$F_i^{(a)} = i$th component of applied force
$R_i = i$th component of reaction force

(recall our convention that the first triplet of coordinates refers to particle 1, the second triplet to particle 2, etc.). Multiply this equation by the ith virtual displacement δx_i and sum over all the coordinates

$$\sum_i (F_i^{(a)} + R_i - \dot{p}_i)\, \delta x_i = 0 \tag{14.3}$$

Since the forces of constraint do no virtual work, it follows that

$$\sum_i R_i\, \delta x_i = 0 \tag{14.4}$$

This notation is somewhat compact, for it contains the vector dot product of the reaction forces and the virtual displacement for all the N particles of the system. A combination of Eqs. (14.4) and (14.3) yields the basic statement of *d'Alembert's principle*

$$\sum_i (F_i^{(a)} - \dot{p}_i)\, \delta x_i = 0 \tag{14.5}$$

Suppose first that there are no constraints, so that the δx_i are independent. Then all of them except one can be set equal to zero. Since this nonzero variation is itself arbitrary, its coefficient must vanish. This procedure can be repeated for each of the variations in turn, and we therefore just recover Newton's second law for unconstrained motion.

The situation is more complicated for constrained motion, because the δx_i *are no longer independent* and we cannot conclude that the coefficients in (14.5) vanish. In fact, they do not, as is evident from the forces of constraint R_i in Eq. (14.2). Nevertheless, d'Alembert's principle as stated in Eq. (14.5) has the great advantage that *the forces of constraint have now disappeared from the problem*. This accomplishes one of the two major objectives mentioned at the start of Sec. 13. From now on, the superscript on $F_i^{(a)}$ will be suppressed, F_i denoting the applied forces.

15 LAGRANGE'S EQUATIONS

It is difficult to use d'Alembert's principle (14.5) directly for constrained motion because not all the virtual displacements are independent. To eliminate this complication, we shall now rewrite Eq. (14.5) in terms of the generalized coordinates q_σ, using the usual rules of calculus to change variables.

Consider first the virtual work done by the applied forces under a virtual displacement

$$\delta W = \sum_i F_i \, \delta x_i \tag{15.1}$$

Substitution of Eq. (13.5) gives

$$\delta W = \sum_{\sigma=1}^{n-k} \left(\sum_{i=1}^{n} F_i \frac{\partial x_i}{\partial q_\sigma} \right) \delta q_\sigma \tag{15.2}$$

where the order of summations has been reversed, which is always possible for finite sums. The quantity in parentheses can be taken to define a *generalized force*

$$Q_\sigma \equiv \sum_{i=1}^{n} F_i \frac{\partial x_i}{\partial q_\sigma} \tag{15.3}$$

and Eq. (15.2) can then be rewritten as

$$\delta W = \sum_{\sigma=1}^{n-k} Q_\sigma \, \delta q_\sigma \tag{15.4}$$

This generalized force can be computed analytically from Eq. (15.3), given the external forces and the basic equations (13.2) relating the cartesian components to the generalized coordinates. It is often simpler, however, to note that the virtual displacements of the generalized coordinates δq_σ are independent by definition. Thus the generalized forces can be determined by setting all except one of the δq_σ equal to zero and then computing the virtual work done by the applied forces. Comparison with Eq. (15.4) immediately identifies the generalized force Q_σ.

The second term in Eq. (14.5) requires a more intricate analysis. By definition, particles with constant mass obey the relations

$$\dot{p}_i = m_i \ddot{x}_i \qquad i = 1, 2, \ldots, n \tag{15.5}$$

where we recall that $i = (1, 2, 3)$ means particle 1, and so on. Again introducing the definition (13.5) of the virtual displacement and interchanging orders of summation, we have

$$\sum_i \dot{p}_i \, \delta x_i = \sum_i m_i \ddot{x}_i \, \delta x_i = \sum_\sigma \left(\sum_i m_i \ddot{x}_i \frac{\partial x_i}{\partial q_\sigma} \right) \delta q_\sigma \tag{15.6}$$

It is now convenient to express the particle accelerations in terms of the particle velocities. This can be done by rewriting the quantity in parentheses in Eq. (15.6) as a total time derivative

$$\sum_i m_i \ddot{x}_i \frac{\partial x_i}{\partial q_\sigma} \equiv \sum_i m_i \frac{d\dot{x}_i}{dt} \frac{\partial x_i}{\partial q_\sigma}$$

$$= \sum_i m_i \left[\frac{d}{dt} \left(\dot{x}_i \frac{\partial x_i}{\partial q_\sigma} \right) - \dot{x}_i \frac{d}{dt} \frac{\partial x_i}{\partial q_\sigma} \right] \tag{15.7}$$

where the second term cancels the additional contribution.

To proceed further, we need an explicit expression for the particle velocities. It can readily be obtained by dividing the general expression (13.4) for the particle displacements by the time interval dt

$$\frac{dx_i}{dt} \equiv \dot{x}_i = \sum_{\sigma} \left(\frac{\partial x_i}{\partial q_{\sigma}}\right) \dot{q}_{\sigma} + \left(\frac{\partial x_i}{\partial t}\right) \qquad i = 1, \ldots, n \qquad (15.8)$$

Recall now that the partial derivatives in these expressions were obtained by holding constant all the other independent variables in Eq. (13.2). Thus the partial derivatives in this expression are themselves functions of the generalized coordinates q_{σ} and the time. As a result, the particle velocities have the following functional form

$$\dot{x}_i = \dot{x}_i(q_1, \ldots, q_{n-k}; \dot{q}_1, \ldots, \dot{q}_{n-k}; t) \qquad i = 1, \ldots, n \qquad (15.9)$$

Equation (15.8) provides an explicit function of the indicated variables and shows that $\dot{x}_i$ in fact depends linearly on the generalized velocities $\{\dot{q}_{\sigma}\}$. Thus we can readily evaluate the partial derivative $\partial \dot{x}_i / \partial \dot{q}_{\sigma}$ to obtain

$$\frac{\partial \dot{x}_i}{\partial \dot{q}_{\sigma}} = \frac{\partial x_i}{\partial q_{\sigma}} \qquad (15.10)$$

where the partial derivative on the left-hand side of this relation implies that all the other variables appearing in parentheses in Eqs. (15.9) *are to be kept constant.* Note that the independent variables now appearing in Eq. (15.9) are also *physically independent*, in the sense that each can be *specified independently* at a given instant of time. The subsequent motion of the system is then, of course, determined by the equations of motion, which will in general be second-order differential equations in time. The expression (15.10) can now be substituted into the first term on the right-hand side of Eq. (15.7), with the result that

$$\sum_i m_i \frac{d}{dt}\left(\dot{x}_i \frac{\partial x_i}{\partial q_{\sigma}}\right) = \sum_i m_i \frac{d}{dt}\left(\dot{x}_i \frac{\partial \dot{x}_i}{\partial \dot{q}_{\sigma}}\right)$$

$$= \frac{d}{dt}\left[\frac{\partial}{\partial \dot{q}_{\sigma}}\left(\frac{1}{2}\sum_i m_i \dot{x}_i^2\right)\right] \qquad (15.11)$$

Note that the final expression in parentheses in Eq. (15.11) is just the total kinetic energy T of the system.

It only remains to rewrite the second term on the right-hand side of Eq. (15.7). We claim that

$$\frac{d}{dt}\left(\frac{\partial x_i}{\partial q_{\sigma}}\right) = \frac{\partial}{\partial q_{\sigma}}\left(\frac{dx_i}{dt}\right) \qquad (15.12)$$

This is not an obvious relation, because different variables are kept constant in carrying out the indicated partial differentiations. Nevertheless, it can be proved by simply evaluating each side. Since the quantity appearing in parentheses on the left-hand side of Eq. (15.12) is only a function of the generalized coordinates $\{q_{\lambda}\}$

and the time, its total differential can be evaluated as in Eq. (13.4). If the result is divided by dt to give the total time derivative, it leads to

$$\text{lhs} = \sum_\lambda \left(\frac{\partial}{\partial q_\lambda} \frac{\partial x_i}{\partial q_\sigma} \right) \dot{q}_\lambda + \frac{\partial}{\partial t} \frac{\partial x_i}{\partial q_\sigma} \tag{15.13}$$

which is the direct analog of Eq. (15.8) with x_i replaced by $\partial x_i/\partial q_\sigma$ and a change of dummy summation variables.

In evaluating the right-hand side of Eq. (15.12), we note that the partial derivative here means: *keep all the other variables in Eq. (15.9) fixed*. In particular, q_σ and $\dot{q}_\sigma$ are treated as independent. Since the variables $\{q_\sigma\}$ appear only in the quantities in parentheses in the right-hand side of Eq. (15.8), this differentiation can be immediately carried out [after first relabeling the summation variable in Eq. (15.8) as λ]

$$\text{rhs} = \sum_\lambda \left(\frac{\partial}{\partial q_\sigma} \frac{\partial x_i}{\partial q_\lambda} \right) \dot{q}_\lambda + \frac{\partial}{\partial q_\sigma} \frac{\partial x_i}{\partial t} \tag{15.14}$$

But now observe that Eqs. (15.13) and (15.14) are identical, for it is always permissible to interchange the order of partial derivatives referring to the same sets of variables. This observation establishes the desired identity (15.12), which permits us to rewrite the second term on the right-hand side of Eq. (15.7) as

$$\sum_i m_i \dot{x}_i \frac{d}{dt} \frac{\partial x_i}{\partial q_\sigma} = \sum_i m_i \dot{x}_i \frac{\partial}{\partial q_\sigma} \frac{dx_i}{dt}$$

$$= \frac{\partial}{\partial q_\sigma} \left(\frac{1}{2} \sum_i m_i \dot{x}_i^2 \right) \tag{15.15}$$

Note that the final quantity appearing in parentheses is again just the total kinetic energy T of the system.

It is now possible to rewrite d'Alembert's principle (14.5), using Eqs. (15.3) and (15.4) for the virtual work done by the applied forces and Eqs. (15.6), (15.7), (15.11), and (15.15) for the acceleration term. A simple analysis yields

$$\sum_\sigma \left(\frac{d}{dt} \frac{\partial T}{\partial \dot{q}_\sigma} - \frac{\partial T}{\partial q_\sigma} - Q_\sigma \right) \delta q_\sigma = 0 \tag{15.16}$$

which is the principal result. Since the variations of the generalized coordinates δq_σ ($\sigma = 1, \ldots, n - k$) are indeed *independent*, each of the coefficients must separately vanish.† We therefore obtain *Lagrange's equations*

$$\frac{d}{dt} \frac{\partial T}{\partial \dot{q}_\sigma} - \frac{\partial T}{\partial q_\sigma} = Q_\sigma \qquad \sigma = 1, \ldots, n - k \tag{15.17}$$

† Just set $\delta q_{\sigma \neq \lambda} = 0$ and $\delta q_\lambda \neq 0$, where $\lambda = 1, \ldots, n - k$ in turn.

where T is the total kinetic energy

$$T = \frac{1}{2} \sum_i m_i \dot{x}_i^2 = T(q_1, \ldots, q_{n-k}; \dot{q}_1, \ldots, \dot{q}_{n-k}; t) \tag{15.18}$$

The partial derivatives appearing in Eqs. (15.16) and (15.17) imply that all the other independent variables *appearing in Eq. (15.18) are kept fixed during this differentiation.* Equations (15.17) provide $n - k$ equations, one for each generalized coordinate, and the constraint forces have disappeared. We have thus accomplished all the objectives set out at the start of Sec. 13. Note that the equations of motion can be obtained without ever using the x_i; all that is needed is the kinetic energy T expressed in terms of the generalized coordinates and the generalized velocities according to Eq. (15.18). The most straightforward and unambiguous way to obtain this expression, however, is to start from the cartesian coordinates (13.2) themselves. On the other hand, the generalized force Q_σ can be obtained directly from Eq. (15.4) by considering the virtual work done by the applied forces in carrying out any particular independent virtual displacement δq_σ.

Although these equations have been derived for constrained motion, d'Alembert's principle (15.16) and Lagrange's equations (15.17) hold equally well for unconstrained motion. We can, in fact, interpret the above discussion as the general transformation of Newton's laws from a cartesian basis into *any set of generalized coordinates.*

These results take a particularly simple form for *conservative forces.* In this case, there is a potential energy V that depends only on the cartesian coordinates of all the N particles. If V is rewritten with the general coordinate transformation (13.2), it takes the form

$$V(x_1, \ldots, x_n) = V(q_1, \ldots, q_{n-k}, t) \tag{15.19}$$

and the left-hand side must be independent of time because the forces are conservative. Since the applied forces are simply the negative gradients of this expression, the generalized force in Eq. (15.3) becomes

$$Q_\sigma \equiv \sum_i F_i \frac{\partial x_i}{\partial q_\sigma} = -\sum_i \left[\frac{\partial}{\partial x_i} V(x_1, \ldots, x_n) \right] \frac{\partial x_i}{\partial q_\sigma}$$

$$= -\frac{\partial}{\partial q_\sigma} V(q_1, \ldots, q_{n-k}, t) \tag{15.20}$$

where the last equality follows from the general rule for differentiating an implicit function. As before, the partial derivative in the last term implies that all the other independent variables on the right-hand side of Eq. (15.19) are to be kept constant. For the final step, we note that the generalized velocities do not appear in Eq. (15.19), so that

$$\frac{\partial V}{\partial \dot{q}_\sigma} = 0 \tag{15.21}$$

Thus Lagrange's equation (15.17) for conservative forces can be rewritten as

$$\frac{d}{dt}\frac{\partial L}{\partial \dot{q}_\sigma} - \frac{\partial L}{\partial q_\sigma} = 0 \qquad \sigma = 1, \ldots, n-k \tag{15.22}$$

where the *lagrangian* has been defined according to

$$L \equiv T - V \tag{15.23}$$

By construction, the lagrangian is a function of the generalized coordinates, the generalized velocities, and the time, as on the right-hand side of Eq. (15.18). We again emphasize that all the other independent variables are to be kept fixed in carrying out the partial differentiations in Eq. (15.22). In contrast, the final derivative with respect to time is then a *total* derivative, to be evaluated in the usual fashion by including the variation of all quantities that depend on time.

16 EXAMPLES

We analyze two examples that illustrate the utility of Lagrange's equations for constrained motion.

Pendulum

Consider a simple planar pendulum of length *l*, as illustrated in Fig. 16.1. The kinetic energy is [compare Eq. (3.4)]

$$T = \tfrac{1}{2}m(l\dot{\theta})^2 \tag{16.1}$$

and the potential energy due to the uniform gravitational field has the form

$$V = -mgl \cos \theta + \text{const} \tag{16.2}$$

The lagrangian thus becomes

$$L = \tfrac{1}{2}ml^2\dot{\theta}^2 + mgl \cos \theta - \text{const} \tag{16.3}$$

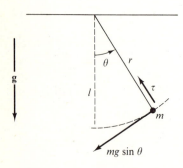

Figure 16.1 Simple planar pendulum in uniform gravitational field. The tension in the string and the component of the gravitational force in the direction of motion are also indicated.

where the angle θ has been taken as the generalized coordinate. Lagrange's equation (15.22) then reads (note that the additive constant in V plays no role in the equation of motion)

$$\frac{d}{dt} ml^2\dot\theta - (-mgl \sin \theta) = 0 \tag{16.4}$$

The final differentiation with respect to time yields the exact *pendulum equation*

$$\ddot\theta = -\frac{g}{l} \sin \theta \tag{16.5}$$

which has the familiar small-amplitude solutions ($\theta \ll 1$) that oscillate at an angular frequency $\omega = (g/l)^{1/2}$. As anticipated, the force of constraint, here the tension τ in the string, never enters this derivation.

Bead on a Rotating Wire Hoop

A bead slides without friction on a hoop that rotates with *constant* angular velocity ω about an axis perpendicular to the plane of the hoop and passing through the edge of the hoop (see Fig. 16.2). Note that this problem ignores both friction and gravity. The angle θ, which measures the displacement of the bead from the diameter of the hoop, serves as a generalized coordinate (see Fig. 16.2). The cartesian coordinates of the particle [compare Eq. (13.2)] are given by

$$x = a \cos \omega t + a \cos (\omega t + \theta) \qquad y = a \sin \omega t + a \sin (\omega t + \theta) \tag{16.6}$$

and the cartesian components of the velocity [compare Eq. (15.9)] then follow by differentiation

$$\dot x = -a\omega \sin \omega t - a(\omega + \dot\theta) \sin (\omega t + \theta)$$
$$\dot y = a\omega \cos \omega t + a(\omega + \dot\theta) \cos (\omega t + \theta) \tag{16.7}$$

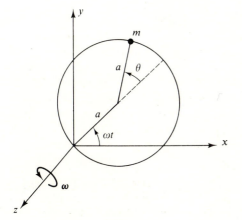

Figure 16.2 Bead moving without friction on circular hoop that rotates with constant angular velocity ω about an axis perpendicular to the hoop and passing through the edge of the hoop (the origin above). There is no gravity in this problem.

where the angular velocity of the hoop ω is assumed constant. The kinetic energy has the general form

$$T = \tfrac{1}{2}m(\dot{x}^2 + \dot{y}^2) \tag{16.8}$$

when expressed in terms of the cartesian components of the velocities. This familiar cartesian relation is always the best and safest starting point in any lagrangian problem. Now use the trigonometric identity

$$\cos \omega t \cos (\omega t + \theta) + \sin \omega t \sin (\omega t + \theta) = \cos (\omega t + \theta - \omega t) = \cos \theta \tag{16.9}$$

The lagrangian, which in this problem is just the kinetic energy, can then be written as

$$T = L = \tfrac{1}{2}ma^2[\omega^2 + (\omega + \dot{\theta})^2 + 2\omega(\omega + \dot{\theta}) \cos \theta] \tag{16.10}$$

Despite the explicit time dependence in Eq. (16.6), this lagrangian, expressed in terms of θ and $\dot{\theta}$, has no explicit dependence on the time. Direct evaluation of lagrange's equation for the generalized coordinate θ yields

$$\frac{d}{dt}\{\tfrac{1}{2}ma^2[2(\omega + \dot{\theta}) + 2\omega \cos \theta]\} - \tfrac{1}{2}ma^2[-2\omega(\omega + \dot{\theta}) \sin \theta] = 0 \tag{16.11}$$

and a straightforward reduction gives the final result

$$\ddot{\theta} + \omega^2 \sin \theta = 0 \tag{16.12}$$

Once again, we find the exact pendulum equation, with an angular frequency ω that is simply the angular frequency of the hoop itself. This similarity to Eq. (16.5) illustrates the general equivalence between gravitational and centrifugal forces (compare Sec. 11), for g is merely replaced by $\omega^2 l$. Indeed, the forces in the present problem can be nicely interpreted by transforming to a co-rotating frame (see Prob. 3.15). Note, however, that the lagrangians (16.3) and (16.10) have a very different structure, so that equivalent dynamical equations need not reflect equivalent physical situations (see the end of Sec. 20).

17 CALCULUS OF VARIATIONS

We now turn to the calculus of variations, which initially seems unrelated to our previous considerations. As seen subsequently, this new formalism provides still another and more powerful formulation of Newton's laws, in which a single variational principle implies the whole set of Lagrange's equations. As an introduction, consider first the following basic problem. Let x be an independent variable defined over the closed interval $[x_1, x_2]$, and let $y(x)$ be some differentiable function of x defined on this interval, as indicated in Fig. 17.1. The derivative of $y(x)$ will be denoted $y'(x) \equiv dy/dx$. Suppose one also has a known relation

$$\phi \equiv \phi[y(x), y'(x), x] \tag{17.1}$$

that yields the number ϕ for given values of the independent variable x, the

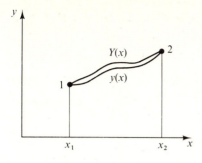

Figure 17.1 Quantities appearing in discussion of basic problem in the calculus of variations.

function $y(x)$, and its derivative $y'(x)$. Such a function of a function is called a *functional*; it is assumed to retain a fixed form throughout the discussion. The basic problem of the calculus of variations is the following:

Find the function $y(x)$ that makes

$$I \equiv \int_{x_1}^{x_2} \phi(y, y', x)\, dx \tag{17.2}$$

an *extremum* (usually a minimum).

To illustrate the idea, consider two examples. First, given two points 1 and 2 in a plane, what function $y(x)$ minimizes the distance $\int_1^2 ds$ between 1 and 2? Since the infinitesimal element of distance is given by

$$ds = [(dx)^2 + (dy)^2]^{1/2} = [1 + (y')^2]^{1/2}\, dx \tag{17.3}$$

the functional in this problem takes the form

$$\phi = [1 + (y')^2]^{1/2} \tag{17.4}$$

In this case, the integral I in (17.2) is just the total distance between points 1 and 2.

A more intricate example is given by the problem that originally stimulated the calculus of variations.† Suppose an initially stationary bead travels without friction along a wire from points 1 and 2 in a uniform gravitational field, as illustrated in Fig. 17.2. What shape of the wire minimizes *the time of travel from points 1 to 2*? Since the time to travel an infinitesimal distance ds is just ds/v, the total transit time is given by

$$t_{12} = \int_1^2 \frac{ds}{v} \tag{17.5}$$

Energy conservation can now be used to relate the velocity to the vertical displacement of the particle

$$\tfrac{1}{2}mv^2 = mgy \tag{17.6}$$

† For a good historical survey, see Bliss [1], chap. 1.

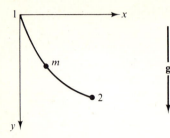

Figure 17.2 The *brachistochrone* problem. A bead falls without friction on a wire in a uniform gravitational field. What shape of the wire minimizes the transit time from point 1 to point 2?

A combination of Eqs. (17.3), (17.5), and (17.6) yields the following expression for the transit time

$$t_{12} = \int_1^2 \frac{[1 + (y')^2]^{1/2}}{(2gy)^{1/2}} \, dx \qquad (17.7)$$

which implies that this problem has the functional

$$\phi = \left[\frac{1 + (y')^2}{2gy} \right]^{1/2} \qquad (17.8)$$

The resulting curve (see Prob. 3.10) is called a *brachistochrone* ("shortest time").

To solve the basic problem in the calculus of variations as stated above, consider two infinitesimally different paths with common endpoints, as indicated in Fig. 17.1. Let

$$y(x) = \text{the path that solves the problem} \qquad (17.9)$$

and let the neighboring path be

$$Y(x) = y(x) + \epsilon\eta(x) \qquad (17.10)$$

where ϵ is an infinitesimal quantity and $\eta(x)$ is an *arbitrary* function of x satisfying

$$\eta(x_1) = \eta(x_2) = 0 \qquad (17.11)$$

By construction, all admissible curves pass through the same fixed endpoints. Compute the value of the integral I for the function $Y(x)$:

$$I(\epsilon) = \int_{x_1}^{x_2} \phi[Y(x), \, Y'(x), \, x] \, dx$$

$$= \int_{x_1}^{x_2} \phi[y(x) + \epsilon\eta(x), \, y'(x) + \epsilon\eta'(x), \, x] \, dx \qquad (17.12)$$

Since $Y(x)$ lies infinitesimally close to the function $y(x)$ that actually makes I an extremum and thus solves the problem, it is natural to expand the integrand in a Taylor series about $\epsilon = 0$

$$I(\epsilon) = \int_{x_1}^{x_2} \phi(y, \, y', \, x) + \epsilon \int_{x_1}^{x_2} \left[\frac{\partial\phi}{\partial y}\eta(x) + \frac{\partial\phi}{\partial y'}\eta'(x) \right] dx + O(\epsilon^2) \qquad (17.13)$$

Note that the partial derivative here means that all other variables in $\phi(y, y', x)$ are kept fixed.

If I is an extremum, e.g., minimum or maximum, for the function $y(x)$, the integral can only either increase or decrease when any other function is added to $y(x)$. This observation means that $I(\epsilon)$, considered as a function of ϵ, must have a vanishing derivative at the point $\epsilon = 0$

$$\left[\frac{dI(\epsilon)}{d\epsilon}\right]_{\epsilon=0} = 0 \qquad (17.14)$$

Figure 17.3 illustrates the situation. Unfortunately, demanding that the first derivative vanish only ensures that the function $I(\epsilon)$ has an extremum (either a minimum, maximum, or a point of inflection); to prove that $\epsilon = 0$ represents a true minimum or a maximum requires examination of the second derivative $[d^2I(\epsilon)/d\epsilon^2]_{\epsilon=0}$. In general, this added step is complicated. In most cases, however, the physics of the problem indicates whether we have a minimum or a maximum. For example, if we require that the distance between two points in a plane be an extremum and then find a unique solution, it is clear that this solution must represent a minimum, for one is always free to make the distance between the two points arbitrarily large.

Application of (17.14) to the explicit function of ϵ given in Eq. (17.13) yields the condition for an extremum

$$\int_{x_1}^{x_2} \left[\frac{\partial \phi}{\partial y}\eta(x) + \frac{\partial \phi}{\partial y'}\frac{d}{dx}\eta(x)\right] dx = 0 \qquad (17.15)$$

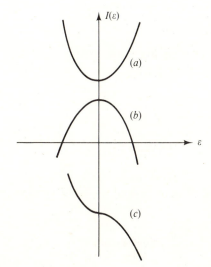

(a)

(b)

(c)

Figure 17.3 The three situations where the function $I(\epsilon)$ has an extremum at $\epsilon = 0$: (a) minimum, (b) maximum, (c) point of inflection. We shall generally be interested in case (a) in the calculus of variations.

The second term in this expression can be partially integrated by rewriting the integrand as a total derivative and subtracting off the extra contribution

$$\frac{\partial \phi}{\partial y'} \frac{d}{dx} \eta(x) = \frac{d}{dx} \left[\frac{\partial \phi}{\partial y'} \eta(x) \right] - \eta(x) \frac{d}{dx} \frac{\partial \phi}{\partial y'} \tag{17.16}$$

Furthermore, the integral of the first term on the right-hand side of Eq. (17.16) vanishes because of the condition of fixed endpoints (17.11):

$$\int_{x_1}^{x_2} \frac{d}{dx} \left[\frac{\partial \phi}{\partial y'} \eta(x) \right] dx = \left[\eta(x) \frac{\partial \phi}{\partial y'} \right]_{x_1}^{x_2} = 0 \tag{17.17}$$

Thus Eq. (17.15) can be rewritten as

$$\int_{x_1}^{x_2} \eta(x) \left(\frac{\partial \phi}{\partial y} - \frac{d}{dx} \frac{\partial \phi}{\partial y'} \right) dx = 0 \tag{17.18}$$

If this equation is to hold for *arbitrary* $\eta(x)$, the coefficient of $\eta(x)$ in this expression must vanish for every x in the interval $[x_1, x_2]$†

$$\frac{d}{dx} \frac{\partial \phi}{\partial y'} - \frac{\partial \phi}{\partial y} = 0 \tag{17.19}$$

This result is known as the *Euler-Lagrange equation* for the variational problem (17.2). If I is to be an extremum, the function $y(x)$ must satisfy this second-order differential equation (17.19). Conversely if $y(x)$ satisfies this differential equation, then I is an extremum.

Let us return to the example of finding the shortest distance between two points, where the functional is given by Eq. (17.4)

$$\phi = [1 + (y')^2]^{1/2} \tag{17.20}$$

Since this ϕ has no explicit dependence on y, the Euler-Lagrange equation takes the form

$$\frac{d}{dx} \frac{y'}{[1 + (y')^2]^{1/2}} = 0 \tag{17.21}$$

The differentiation can be performed, to give

$$y'' \left\{ \frac{1}{[1 + (y')^2]^{1/2}} - \frac{(y')^2}{[1 + (y')^2]^{3/2}} \right\} = \frac{y''}{[1 + (y')^2]^{3/2}} = 0 \tag{17.22}$$

and the positive-definite character of the denominator implies that

$$y'' = 0 \tag{17.23}$$

---

† The proof is analogous to that in obtaining Eq. (15.17), for $\eta(x)$ can be chosen to vanish except in a small interval about some particular x.

Thus the Euler-Lagrange equation for this particular variational problem has a simple form. A trivial integration yields the equation for a straight line

$$y(x) = ax + b \qquad (17.24)$$

where the arbitrary constants a and b can be chosen to pass the curve through the fixed points 1 and 2. It is clear from the previous discussion that this curve minimizes the distance between the points.

It is frequently helpful to reformulate the preceding analysis in a slightly different language, which will allow us to carry out variational calculations very concisely. Define the difference between the function $Y(x)$ and the true solution $y(x)$ as the variation of y, written as $\delta y(x)$

$$Y(x) - y(x) = \epsilon\eta(x) \equiv \delta y(x) \qquad (17.25)$$

The derivative of this equation with respect to x yields

$$Y'(x) - y'(x) = \epsilon\eta'(x) \equiv \delta y'(x) \qquad (17.26)$$

which defines the variation of the derivative $\delta y'(x)$. It is obvious by comparing Eqs. (17.25) and (17.26) that

$$\delta y'(x) = \frac{d}{dx}\,\delta y(x) \qquad (17.27)$$

In addition, the difference between the functional ϕ computed with $Y(x)$ and its value computed with the true solution $y(x)$ is called the *variation of the functional*

$$\phi[Y(x),\ Y'(x),\ x] - \phi[y(x),\ y'(x),\ x] \equiv \delta\phi \qquad (17.28)$$

The Taylor-series expansion used in Eq. (17.13) makes it evident that $\delta\phi$ takes the form

$$\delta\phi = \frac{\partial\phi}{\partial y}\,\delta y + \frac{\partial\phi}{\partial y'}\,\delta y' \qquad (17.29)$$

correct to first order. Equation (17.13) also gives the variation of I from its value computed with the true solution

$$\delta I = \int_{x_1}^{x_2} \delta\phi\, dx = \int_{x_1}^{x_2} \left(\frac{\partial\phi}{\partial y}\,\delta y + \frac{\partial\phi}{\partial y'}\,\delta y'\right) dx$$

$$= \int_{x_1}^{x_2} \left(\frac{\partial\phi}{\partial y}\,\delta y + \frac{\partial\phi}{\partial y'}\frac{d}{dx}\,\delta y\right) dx \qquad (17.30)$$

where the last line follows from Eq. (17.27). The second term of this equation can be partially integrated just as in (17.15) to (17.18), and the condition of fixed endpoints

$$\delta y(x_1) = \delta y(x_2) = 0 \qquad (17.31)$$

again eliminates the contribution of the total derivative. As a result, the variation of I can be written as

$$\delta I = \int_{x_1}^{x_2} \delta y \left(\frac{\partial \phi}{\partial y} - \frac{d}{dx} \frac{\partial \phi}{\partial y'} \right) dx \tag{17.32}$$

For an extremum, we demand that the variation of I vanish

$$\delta I = 0 \tag{17.33}$$

which simply restates the condition given in Eq. (17.14). If this result is to hold for an arbitrary variation $\delta y(x)$, the coefficient of $\delta y(x)$ in Eq. (17.32) must vanish. Hence we again obtain the Euler-Lagrange equation for the variational problem posed in Eq. (17.2)

$$\frac{d}{dx} \frac{\partial \phi}{\partial y'} - \frac{\partial \phi}{\partial y} = 0 \tag{17.34}$$

Note that this modified derivation relies on precisely the same logic used in obtaining Eq. (17.19) and differs only in the more compact notation.

18 HAMILTON'S PRINCIPLE

The Euler-Lagrange equation (17.19) or (17.34) for the calculus of variations closely resembles Lagrange's equations for mechanics [Eqs. (15.22) and (15.23)]. This similarity suggests that Lagrange's equations have a variational basis, and the resulting variational statement of mechanics is known as *Hamilton's principle*. We shall first state Hamilton's principle for a system with a single generalized coordinate q subject to conservative forces, where Lagrange's equation can be constructed by direct analogy with Eq. (17.34). We shall then immediately generalize the result to many degrees of freedom.

For a single degree of freedom, Hamilton's principle has the concise form

$$\delta \int_{t_1}^{t_2} L[q(t), \dot{q}(t), t] \, dt = 0 \tag{18.1}$$

$$\delta q(t_1) = \delta q(t_2) = 0 \tag{18.2}$$

where the time integral of the lagrangian (15.23) is called the *action*. In words, Hamilton's principle asserts that of all possible paths between the fixed endpoints $q(t_1)$ and $q(t_2)$ (illustrated in Fig. 18.1), the actual dynamical trajectory makes the action an extremum or stationary. Note that the variation $\delta q(t)$ considered here is precisely the virtual displacement introduced in the discussion of d'Alembert's principle in Secs. 13 and 14.

The Euler-Lagrange equation for the variational problem stated in Eqs. (18.1)

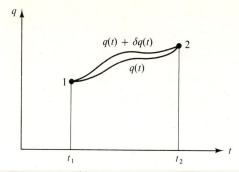

Figure 18.1 Quantities appearing in Hamilton's principle.

and (18.2) follows immediately from the treatment in Sec. 17 by making the obvious substitutions

$$x \to t \qquad y(x) \to q(t) \qquad y'(x) \to \frac{dq}{dt} = \dot{q} \tag{18.3}$$

In particular, comparison with Eq. (17.34) yields the Euler-Lagrange equation for the variational problem stated in Hamilton's principle

$$\frac{d}{dt}\frac{\partial L}{\partial \dot{q}} - \frac{\partial L}{\partial q} = 0 \tag{18.4}$$

which is precisely Lagrange's equation. Since Lagrange's equations have been proved equivalent to Newton's laws for a conservative system, *we can take Hamilton's principle as a fundamental statement of mechanics*:

$$\delta \int_{t_1}^{t_2} L \, dt = 0 \qquad \text{fixed endpoints} \tag{18.5}$$

We proceed to summarize and discuss this crucial result.

As illustrated in Fig. 18.1, the value of the action depends on the path considered for the possible motion. The actual path is the one that "minimizes" the action (or, in general, makes it an extremum).

Since the lagrangian is given by the difference of the kinetic and potential energies, $L = T - V$, Hamilton's principle says that the particle takes the path that minimizes the integrated difference between these quantities.

This result can immediately be extended to a system with n arbitrary degrees of freedom. We use our variational notation to write

$$\delta q_\sigma(t) = \epsilon \eta_\sigma(t) \qquad \sigma = 1, \ldots, n \tag{18.6}$$

Hamilton's principle then states that

$$0 = \delta \int_{t_1}^{t_2} L(q_1, \ldots, q_n; \dot{q}_1, \ldots, \dot{q}_n; t) \, dt$$

$$= \int_{t_1}^{t_2} \sum_{\sigma=1}^{n} \left(\frac{\partial L}{\partial q_\sigma} \delta q_\sigma + \frac{\partial L}{\partial \dot{q}_\sigma} \delta \dot{q}_\sigma \right) dt \tag{18.7}$$

Just as in Eq. (17.27), one has

$$\delta \dot{q}_\sigma = \frac{d}{dt}\, \delta q_\sigma \qquad (18.8)$$

and the time derivative in the last term of (18.7) can again be partially integrated using the condition of fixed endpoints, which is an essential part of Hamilton's principle,

$$\delta q_\sigma(t_1) = \delta q_\sigma(t_2) = 0 \qquad \sigma = 1, \ldots, n \qquad (18.9)$$

This manipulation leads to the condition

$$\int_{t_1}^{t_2} \left[\sum_\sigma \delta q_\sigma \left(\frac{\partial L}{\partial q_\sigma} - \frac{d}{dt}\frac{\partial L}{\partial \dot{q}_\sigma} \right) \right] dt = 0 \qquad (18.10)$$

Now consider two possibilities.

Possibility 1 If all the q_σ are independent, they are suitable generalized coordinates and the δq_σ are *independent variations*. Thus the coefficients of each variation must vanish to give

$$\frac{d}{dt}\frac{\partial L}{\partial \dot{q}_\sigma} - \frac{\partial L}{\partial q_\sigma} = 0 \qquad \sigma = 1, \ldots, n \qquad (18.11)$$

yielding Lagrange's equations for this mechanical system.

Possibility 2 Suppose there are k holonomic constraints

$$f_j(q_1, \ldots, q_n, t) = c_j \qquad j = 1, \ldots, k \qquad (18.12)$$

where c_j are constants. We are then free to choose any $n - k$ independent coordinates, which we have referred to as the generalized coordinates for the problem, and reexpress L in terms of these variables. For variations of these independent coordinates, the above argument remains exactly as stated, leading to $n - k$ equations of the form (18.11) for the generalized coordinates labeled $\sigma = 1, \ldots, n - k$.

This formulation of mechanics is independent of the choice of coordinates and only depends on the kinetic and potential energies. One is free to express T and V in *any* convenient set of coordinates and then simply substitute these expressions into Hamilton's principle and carry out the variation.

19 FORCES OF CONSTRAINT

It is often preferable to incorporate the k holonomic constraints of Eq. (18.12) into Hamilton's principle (18.10) in another fashion using the method of *Lagrange multipliers*.† We first notice that Eq. (18.10) holds even when the set $\{\delta q_\sigma\}$ are not

---

† All the arguments in this section can alternatively be based on d'Alembert's principle, since the use of Lagrange multipliers in no way requires Hamilton's principle.

all independent variations. Now take the variation of Eq. (18.12)

$$\delta f_j = \sum_{\sigma=1}^{n} \frac{\partial f_j}{\partial q_\sigma} \delta q_\sigma = 0 \qquad j = 1, \ldots, k \tag{19.1}$$

These equations provide k constraining relations between the variations δq_σ. Multiply each of these equations by a Lagrange multiplier λ_j, which is an as yet unspecified function of the q's, sum over all j, and integrate over time from t_1 to t_2. Since none of these operations changes the right-hand side of Eq. (19.1), the result still vanishes. If this expression is added to Eq. (18.10), Hamilton's principle then takes the form

$$\int_{t_1}^{t_2} \sum_{\sigma=1}^{n} \delta q_\sigma \left(\frac{\partial L}{\partial q_\sigma} - \frac{d}{dt} \frac{\partial L}{\partial \dot{q}_\sigma} + \sum_j \lambda_j \frac{\partial f_j}{\partial q_\sigma} \right) dt = 0 \tag{19.2}$$

There are now $n - k$ independent variations δq_σ, and we choose (arbitrarily) to label them $\sigma = 1, \ldots, n - k$. As a result, the remaining variations $\delta q_{n-k+1}, \ldots, \delta q_n$ are no longer independent. We can, however, *choose* the k independent functions $\lambda_1, \ldots, \lambda_k$ so that the coefficients of $\delta q_{n-k+1}, \ldots, \delta q_n$ in Eq. (19.2) vanish identically. By assumption, the remaining $n - k$ variations $\delta q_1, \ldots, \delta q_{n-k}$ are independent, and the coefficient of each one in Eq. (19.2) must also vanish. In this way, Hamilton's principle, together with the constraints incorporated in Eq. (19.2), leads to the result

$$\frac{d}{dt} \frac{\partial L}{\partial \dot{q}_\sigma} - \frac{\partial L}{\partial q_\sigma} = \sum_{j=1}^{k} \lambda_j \frac{\partial f_j}{\partial q_\sigma} \qquad \sigma = 1, \ldots, n \tag{19.3}$$

$$f_j(q_1, \ldots, q_n, t) = c_j \qquad j = 1, \ldots, k \tag{19.4}$$

We now have $n + k$ equations and $n + k$ unknowns $(q_1, \ldots, q_n; \lambda_1, \ldots, \lambda_k)$. The reader may understandably wonder why we complicate the problem, which could just as well have been described with $n - k$ independent generalized coordinates. In fact, the Lagrange multipliers will turn out to determine the reaction forces, which provides new and important information about the system, in addition to the description of the motion. To derive this result, we return to the original form of Lagrange's equations (15.17) for n independent degrees of freedom

$$\frac{d}{dt} \frac{\partial T}{\partial \dot{q}_\sigma} - \frac{\partial T}{\partial q_\sigma} = Q_\sigma \qquad \sigma = 1, \ldots, n \tag{19.5}$$

Here Q_σ is the generalized force, defined in terms of the work done (15.1) by the applied forces through the relation

$$\delta W = \sum_\sigma Q_\sigma \, \delta q_\sigma \tag{19.6}$$

Since $L = T - V$ with $V = V(q_1, \ldots, q_n, t)$, Eq. (19.3) can be written in the similar form

$$\frac{d}{dt} \frac{\partial T}{\partial \dot{q}_\sigma} - \frac{\partial T}{\partial q_\sigma} = -\frac{\partial V}{\partial q_\sigma} + \sum_{j=1}^{k} \lambda_j \frac{\partial f_j}{\partial q_\sigma} \qquad \sigma = 1, \ldots, n \tag{19.7}$$

Comparison with Eq. (19.5) shows that the generalized forces for n degrees of freedom subject to k constraints contain two contributions

$$Q_\sigma = -\frac{\partial V}{\partial q_\sigma} + Q_\sigma^r \qquad (19.8a)$$

The first term is just the applied force from the potential V; the second term

$$Q_\sigma^r \equiv \sum_{j=1}^{k} \lambda_j \frac{\partial f_j}{\partial q_\sigma} \qquad (19.8b)$$

which is the right-hand side of Eq. (19.3), arises from the forces of constraint and depends explicitly on the Lagrange multipliers. It is the part Q_σ^r of the generalized force that is not included in the lagrangian L for the unconstrained motion. Equivalently, the set Q_σ^r are just the generalized forces that must be exerted by the constraints to force the particles to follow the constrained trajectories.

To confirm this identification, we shall temporarily forget the constrained problem and consider a new one for n degrees of freedom subject to the same lagrangian L but with all the constraints replaced by an equivalent set of extra applied forces Q_σ^{app} that produce exactly the same motion as in the constrained case. Lagrange's equations then take the form

$$\frac{d}{dt}\frac{\partial L}{\partial \dot{q}_\sigma} - \frac{\partial L}{\partial q_\sigma} = Q_\sigma^{app} \qquad \sigma = 1, \ldots, n \qquad (19.9)$$

where all the n coordinates q_σ ($\sigma = 1, \ldots, n$) are now independent. The quantities Q_σ^{app} are readily obtained from the virtual work done by the extra applied forces

$$\delta W^{app} = \sum_\sigma Q_\sigma^{app} \, \delta q_\sigma \qquad (19.10)$$

where the superscript on the left-hand side emphasizes the restriction to that part of the applied force not already contained in the potential V.

The solutions to Eq. (19.9) automatically satisfy the constraint conditions

$$f_j(q_1, \ldots, q_n, t) = c_j \qquad j = 1, \ldots, k \qquad (19.11)$$

Furthermore, as discussed above, the physical motion arising from (19.9) is identical with that of the constrained problem (19.3). Consequently, the equations of motion must also be identical, and comparison of the right-hand sides of Eqs. (19.9) and (19.3) yields the identification

$$Q_\sigma^{app} = Q_\sigma^r \equiv \sum_{j=1}^{k} \lambda_j \frac{\partial f_j}{\partial q_\sigma} \qquad \sigma = 1, \ldots, n \qquad (19.12)$$

showing that the right-hand side of Eq. (19.3) indeed gives the generalized reaction forces for the constrained problem. Note that the Lagrange multipliers λ_j, and hence the reaction forces Q_σ^r, can be obtained only after solving Eq. (19.3). In addition, the interpretation of Q_σ^r in general requires consideration of independent variations δq_σ that *violate the constraints*. This point is evident from the fictitious unconstrained problem (19.9), where all the constraint forces are treated as applied forces, interpreted through Eq. (19.10) by performing a virtual displacement and computing the corresponding virtual work.

This discussion provides a basis for incorporating as many constraining forces as is desirable and convenient. At one extreme, we can omit all the constraining forces; the resulting $n - k$ equations involve $n - k$ independent generalized coordinates, and the forces of constraint never appear, just as in our original motivation of Lagrange's equations. On the other hand, there may be one (or more) particular constraining force that we would like to find; we then leave the corresponding generalized coordinate(s) free and incorporate the constraint(s) with the method of Lagrange multipliers, as discussed in Eqs. (19.3) and (19.4).† To fix these ideas, we study three specific examples, omitting the superscript r on Q_σ for simplicity.

Pendulum

Consider the planar pendulum (Fig. 16.1) with r and θ as generalized coordinates. The lagrangian takes the form [compare Eq. (3.4)]

$$L = \tfrac{1}{2}m(\dot{r}^2 + r^2\dot{\theta}^2) + mgr \cos \theta + \text{const} \qquad (19.13)$$

---

† Although this treatment has assumed holonomic constraints (19.4), it is readily extended to include certain nonholonomic constraints of the form

$$\sum_{\sigma=1}^{n} \alpha_{j\sigma} \, dq_\sigma + \alpha_{jt} \, dt = 0 \qquad j = 1, \dots, k$$

where $\alpha_{j\sigma}$ and α_{jt} can depend on the q_σ and t. If these coefficients satisfy the conditions

$$\frac{\partial \alpha_{j\sigma}}{\partial q_\lambda} = \frac{\partial \alpha_{j\lambda}}{\partial q_\sigma} \qquad \frac{\partial \alpha_{jt}}{\partial q_\sigma} = \frac{\partial \alpha_{j\sigma}}{\partial t}$$

they merely represent the partial derivatives of holonomic constraints

$$\alpha_{j\sigma} = \frac{\partial f_j}{\partial q_\sigma} \qquad \alpha_{jt} = \frac{\partial f_j}{\partial t}$$

Otherwise, the constraints are called *nonintegrable*. In this latter case, we note that the virtual displacements δq_σ still satisfy linear relations

$$\sum_{\sigma=1}^{n} \alpha_{j\sigma} \, \delta q_\sigma = 0$$

obtained from the above at fixed time. This equation replaces (19.1), and the previous derivation immediately yields

$$\frac{d}{dt}\frac{\partial L}{\partial \dot{q}_\sigma} - \frac{\partial L}{\partial q_\sigma} = \sum_{j=1}^{k} \lambda_j \alpha_{j\sigma} \qquad \sigma = 1, \dots, n$$

instead of Eq. (19.3). The remaining k constraint equations follow directly on dividing by dt

$$\sum_{\sigma=1}^{n} \alpha_{j\sigma} \dot{q}_\sigma + \alpha_{jt} = 0 \qquad j = 1, \dots, k$$

These $n + k$ equations in the $n + k$ unknowns constitute the generalization of (19.3) and (19.4) to nonintegrable constraints (note that this argument can be further generalized to the case where the $\alpha_{j\sigma}$ depend on $\dot{q}_\lambda$).

There is one holonomic equation of constraint, ensuring that the radial coordinate has fixed length

$$r = l \tag{19.14}$$

Variation of this equation yields

$$\delta r = 0 \tag{19.15}$$

Multiply this constraint equation by λ, integrate over the time, and add the result to Hamilton's principle, as in Eq. (19.2). This alteration affects only the coefficient of δr (namely the radial equation), and the modification simply consists in adding λ to its right-hand side. Lagrange's equations (19.3) thus read

$$m(\ddot{r} - r\dot{\theta}^2) - mg \cos \theta = \lambda \equiv Q_r \tag{19.16}$$

$$\frac{d}{dt} mr^2\dot{\theta} + mgr \sin \theta = 0 \tag{19.17}$$

which shows that there is no generalized reaction force Q_θ associated with the angular motion since Eq. (19.14) involves only the coordinate r.

To interpret the reaction force Q_r, carry out a virtual displacement of the coordinate r and compute the virtual work done by the equivalent set of forces that must be applied to make the particle follow the constrained trajectory of Eq. (19.14). It is clear from Fig. 16.1 that this constraint force is just the tension in the string. Thus the virtual work done under a virtual displacement δr is

$$\delta W = Q_r \, \delta r = -\tau \, \delta r \tag{19.18}$$

where the minus sign arises because positive δr represents an outward displacement. We immediately identify

$$Q_r = \lambda = -\tau \tag{19.19}$$

so that the generalized reaction force, in this case λ, is the negative of the tension in the string. Note that the interpretation of this reaction force requires us to perform a variation δr that violates the constraint in Eq. (19.14).

The three equations (19.16), (19.17), and (19.14) in three unknowns (r, θ, λ) are readily solved by noting that Eq. (19.14) implies

$$\dot{r} = \ddot{r} = 0 \tag{19.20}$$

Substitution into (19.16) and (19.17) yields the relations

$$-\lambda = \tau = ml\dot{\theta}^2 + mg \cos \theta \tag{19.21}$$

$$\ddot{\theta} = -\frac{g}{l} \sin \theta \tag{19.22}$$

Equation (19.22) is precisely the pendulum equation (16.5) that would have been obtained by using θ as the only generalized coordinate; in that case, the force of constraint would never have appeared. In addition, however, we have also calculated the tension in the string through Eq. (19.21). Its two terms have an immediate physical interpretation as the centrifugal force and the appropriate component of gravity. Note that the tension can be computed only after solving the equations of motion because τ and λ depend on time through $\theta(t)$ and $\dot{\theta}(t)$.

Atwood's Machine

As a second example, consider the freshman-physics problem of Atwood's machine, as illustrated in Fig. 19.1. Here we leave the lengths l_1 and l_2 independent and include the holonomic constraint

$$l_1 + l_2 = l \tag{19.23}$$

If l_1 and l_2 are independent, this system has the lagrangian

$$L = \tfrac{1}{2}m_1 \dot{l}_1^2 + \tfrac{1}{2}m_2 \dot{l}_2^2 + m_1 g l_1 + m_2 g l_2 + \text{const} \tag{19.24}$$

Varying the constraint equation and multiplying by λ gives

$$\lambda(\delta l_1 + \delta l_2) = 0 \tag{19.25}$$

If this expression is integrated over time and added to Hamilton's principle, the net effect will be to add a term λ to the right-hand side of Lagrange's equation for both l_1 and l_2. Thus this system has Lagrange's equations

$$m_1 \ddot{l}_1 - m_1 g = \lambda \equiv Q_1 \tag{19.26}$$

$$m_2 \ddot{l}_2 - m_2 g = \lambda \equiv Q_2 \tag{19.27}$$

where Q_1 and Q_2 are the reaction forces associated with the corresponding degrees of freedom l_1 and l_2. These equations differ in structure from those for a pendulum (19.16) and (19.17) in that the single constraint (19.23) here involves both coordinates.

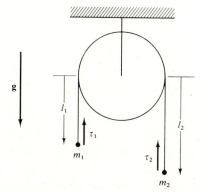

Figure 19.1 Atwood's machine.

These reaction forces can be identified from Eq. (19.10) by computing the work done in a virtual displacement of each coordinate

$$\delta W = Q_1 \, \delta l_1 + Q_2 \, \delta l_2 \qquad (19.28)$$

If l_2 is kept fixed, inspection of Fig. 19.1 shows that

$$\delta W = Q_1 \, \delta l_1 = -\tau_1 \delta l_1 \qquad (19.29)$$

implying that $-Q_1$ is just the tension τ_1 in string 1. A similar argument for fixed l_1 gives $-Q_2 = \tau_2$, and comparison with the right-hand sides of (19.26) and (19.27) yields the relation

$$-\lambda = \tau_1 = \tau_2 \qquad (19.30)$$

Without further calculation, we see that the tensions τ_1 and τ_2 are equal. As in the previous case, this interpretation of the reaction forces requires consideration of virtual displacements (19.29) that violate the constraint (19.23).

The three equations (19.23), (19.26), and (19.27) in three unknowns (l_1, l_2, λ) can be solved by noting that Eq. (19.23) implies

$$\ddot{l}_1 = -\ddot{l}_2 \qquad (19.31)$$

Subtraction of Eq. (19.27) from (19.26) gives

$$\ddot{l}_1 = \frac{m_1 - m_2}{m_1 + m_2} g \qquad (19.32)$$

and the result of multiplying Eq. (19.26) by m_2 and Eq. (19.27) by m_1 and adding is

$$Q_1 = Q_2 = \lambda = -\frac{2m_1 m_2}{m_1 + m_2} g \qquad (19.33)$$

Note that Eq. (19.32) is precisely the result that would arise from treating the coordinate l_1 as the only independent variable. In addition, however, our formalism also yields Eq. (19.33), which shows that the equal tensions τ_1 and τ_2 [compare Eq. (19.30)] are also time-independent, with a value proportional to the reduced mass (4.6a) and to g. As in the previous example of the pendulum, this last conclusion follows only by examining the detailed equations of motion.

One Cylinder Rolling on Another

Our last example is the problem of one uniform cylinder of radius R_2 and mass m rolling without slipping on another fixed cylinder of radius R_1, as illustrated in Fig. 19.2. While the cylinders remain in contact, there is only one degree of freedom θ_1, but it will be convenient to introduce two additional coordinates, characterizing the location of the moving cylinder by the polar coordinates (r, θ_1) of its center of mass and the rotation angle θ_2 about its axis, measured from the vertical (see Fig. 19.2). As a result, we need two equations of constraint to enforce

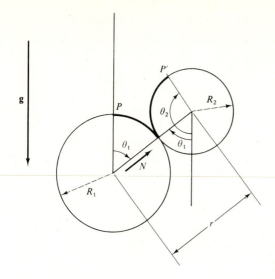

Figure 19.2 One cylinder rolling on another.

the conditions of contact between the cylinders and pure rolling. Equations (2.18) and (3.4) show that the rolling cylinder has kinetic energy

$$T = \tfrac{1}{2}m(\dot{r}^2 + r^2\dot{\theta}_1^2) + \tfrac{1}{2}(\tfrac{1}{2}mR_2^2)\dot{\theta}_2^2 \tag{19.34}$$

obtained as the sum of the kinetic energy of a point mass m located at the center of mass and the rotational energy about the center of mass. From elementary considerations, the latter takes the form $\tfrac{1}{2}I\omega^2$, where the moment of inertia of a cylinder about its axis of symmetry is

$$I_{\text{cyl}} = \tfrac{1}{2}mR^2 \tag{19.35}$$

The potential energy of the moving cylinder is that of a point particle of mass m located at the position of the center of mass

$$V = mgr \cos \theta_1 + \text{const} \tag{19.36}$$

We now assume that the two cylinders remain in contact and introduce two holonomic constraints. The first explicitly ensures this contact, requiring that

$$r = R_1 + R_2 \tag{19.37}$$

The constraint of pure rolling demands that the distance along the first cylinder's surface from the point of contact to P (see Fig. 19.2) equal the distance along the second cylinder's surface from the point of contact to P', where the points P and P' coincide at the start of the motion. Thus

$$R_1\theta_1 = R_2(\theta_2 - \theta_1) \tag{19.38}$$

The presence of *two* constraints now necessitates introducing two Lagrange multipliers. Take the variation of the first constraint and multiply by λ_1 and the variation of the second constraint and multiply by λ_2

$$\lambda_1 \, \delta r = 0 \tag{19.39}$$

$$\lambda_2[(R_1 + R_2) \, \delta\theta_1 - R_2 \, \delta\theta_2] = 0 \tag{19.40}$$

Integrate these equations over time and add them to Hamilton's principle; this procedure adds a term λ_1 to the right-hand side of the radial equation, a term $\lambda_2(R_1 + R_2)$ to the right-hand side of the θ_1 equation, and a term $-\lambda_2 R_2$ to the right-hand side of the θ_2 equation. Use of the lagrangian

$$L = \tfrac{1}{2}m[(\dot{r}^2 + r^2\dot{\theta}_1^2) + \tfrac{1}{2}R_2^2\dot{\theta}_2^2] - mgr \cos\theta_1 - \text{const} \tag{19.41}$$

leads to Lagrange's equations in the form

$$m(\ddot{r} - r\dot{\theta}_1^2 + g \cos\theta_1) = \lambda_1 \equiv Q_r \tag{19.42}$$

$$\frac{d}{dt} mr^2\dot{\theta}_1 - mgr \sin\theta_1 = (R_1 + R_2)\lambda_2 \equiv Q_1 \tag{19.43}$$

$$\tfrac{1}{2}mR_2^2\ddot{\theta}_2 = -R_2\lambda_2 \equiv Q_2 \tag{19.44}$$

A combination with the equations of constraint (19.37) and (19.38) provides five equations in five unknowns $(r, \theta_1, \theta_2, \lambda_1, \lambda_2)$.

As in the previous examples, it is helpful first to identify the Lagrange multipliers by performing virtual displacements of the coordinates. If only $\delta r \neq 0$, then the relevant force is the normal reaction N exerted by the first cylinder on the second, maintaining the condition of contact. It does work

$$\delta W = N \, \delta r \qquad \delta\theta_1 = \delta\theta_2 = 0 \tag{19.45}$$

and comparison with the general expression

$$\delta W = Q_r \, \delta r + Q_1 \, \delta\theta_1 + Q_2 \, \delta\theta_2 \tag{19.46}$$

and with Eq. (19.42) provides the identification

$$Q_r = \lambda_1 = N \tag{19.47}$$

A similar analysis for $\delta\theta_1$ and $\delta\theta_2$ relates Q_1 and Q_2 to the tangential component of the frictional force that acts at the point of contact to ensure the condition (19.38) of pure rolling. For our purposes, however, it is sufficient to consider only Eq. (19.47).

The next step is the solution of the coupled Eqs. (19.37), (19.38), and (19.42) to (19.44). The constraint equation (19.37) implies $\dot{r} = \ddot{r} = 0$, and Eq. (19.42) then becomes

$$-m(R_1 + R_2)\dot{\theta}_1^2 + mg \cos\theta_1 = \lambda_1 \tag{19.48}$$

which relates the Lagrange multiplier λ_1 to the time-dependent variable $\theta_1(t)$.

Similarly, a combination of Eqs. (19.38) and (19.44) gives an analogous relation for λ_2

$$-\tfrac{1}{2}mR_2\ddot{\theta}_2 = -\tfrac{1}{2}m(R_1 + R_2)\ddot{\theta}_1 = \lambda_2 \tag{19.49}$$

Finally, Eqs. (19.43) and (19.37) yield

$$m(R_1 + R_2)^2\ddot{\theta}_1 - mg(R_1 + R_2)\sin\theta_1 = (R_1 + R_2)\lambda_2 = -\tfrac{1}{2}m(R_1 + R_2)^2\ddot{\theta}_1 \tag{19.50}$$

where the last equality incorporates Eq. (19.49). Equation (19.50) is just the equation of motion that would have been obtained by treating θ_1 as the only independent generalized coordinate

$$\tfrac{3}{2}(R_1 + R_2)\ddot{\theta}_1 - g\sin\theta_1 = 0 \tag{19.51}$$

We now observe that the first cylinder is only capable of exerting a *positive* normal force on the second cylinder. Thus physical considerations require

$$N \geqslant 0 \tag{19.52}$$

and the second cylinder will leave the first at the point where N first vanishes. In this way, the present formalism provides a mechanism for answering the question: When and where will the rolling cylinder fall off the fixed one? It is evident from Eqs. (19.47) and (19.48) that this event occurs when the centrifugal force just balances the normal component of the gravitational force. As in our previous examples, however, determining this point requires an explicit integration of (19.51), since the condition $N = 0$ involves θ_1 and $\dot{\theta}_1$.

The remaining steps are straightforward. Multiply Eq. (19.51) by $\dot{\theta}_1$ and integrate with respect to time; this procedure yields the first integral

$$(R_1 + R_2)\dot{\theta}_1^2 = \tfrac{4}{3}g(1 - \cos\theta_1) \tag{19.53}$$

where the constant has been chosen to incorporate the initial condition that the cylinder starts from rest at $\theta_1 = 0$. In principle, Eq. (19.53) could be integrated to find $\theta_1(t)$ [compare Eq. (3.13)], but we shall be content to combine Eqs. (19.47), (19.48), and (19.53) to obtain

$$N(\theta_1) = \tfrac{1}{3}mg(7\cos\theta_1 - 4) \tag{19.54}$$

Evidently, $N > 0$ for $7\cos\theta_1 > 4$, and the rolling cylinder first leaves the fixed one at an angle $\arccos\tfrac{4}{7} \approx 55.15°$, *independent of the ratio* R_2/R_1. Beyond this point, the constraints (19.37) and (19.38) cease to act unless the two cylinders are held in contact by rigid rods that exert an attractive reaction force $N < 0$. Otherwise, the subsequent motion acquires two new degrees of freedom r and θ_2 in addition to θ_1; the corresponding equations of motion are just those found previously with $\lambda_1 = \lambda_2 = 0$, subject to the initial conditions appropriate to the instant of separation.

20 GENERALIZED MOMENTA AND THE HAMILTONIAN

For a conservative holonomic system, Lagrange's equations state that

$$\frac{d}{dt}\frac{\partial L}{\partial \dot{q}_\sigma} - \frac{\partial L}{\partial q_\sigma} = 0 \tag{20.1}$$

where throughout this section we work with $n - k$ independent generalized coordinates q_σ. Define the quantity

$$p_\sigma \equiv \frac{\partial L}{\partial \dot{q}_\sigma} \tag{20.2}$$

known as the *generalized momentum* or *canonical momentum*. Lagrange's equations can evidently be rewritten as

$$\dot{p}_\sigma = \frac{\partial L}{\partial q_\sigma} \tag{20.3}$$

If some particular generalized coordinate q_σ does not appear explicitly in the lagrangian, then $\partial L/\partial q_\sigma = 0$, and the corresponding generalized momentum p_σ is constant. Such coordinates are said to be *cyclic*; they play a very special role in mechanics. It is evident from Lagrange's equations that

> Generalized momenta p_σ corresponding to cyclic coordinates are constants of the motion.

In addition, the generalized momenta p_σ are functions of $(q_\sigma, \dot{q}_\sigma, t)$ by construction, and each cyclic coordinate therefore provides an explicit first integral.

Symmetry Principles and Conserved Quantities

The existence of conserved quantities has an important relationship to the symmetry of the problem. If the system is invariant under some continuous transformation, then T and V are unchanged by alterations in the corresponding generalized coordinate q_σ. As a result, $\partial L/\partial q_\sigma = 0$ for this particular q_σ, and Eq. (20.3) shows that the associated p_σ is constant. Thus the existence of a continuous symmetry operation implies the presence of a conserved generalized momentum.

We shall illustrate these ideas with two simple examples.

Three-dimensional motion in a one-dimensional potential If the potential depends only on one cartesian coordinate z, the lagrangian becomes

$$L = \tfrac{1}{2}m(\dot{x}^2 + \dot{y}^2 + \dot{z}^2) - V(z) \tag{20.4}$$

By assumption, the system is invariant under a shift of the coordinate axes in the xy plane; equivalently,

$$\frac{\partial L}{\partial x} = \frac{\partial L}{\partial y} = 0 \tag{20.5}$$

so that the corresponding momenta p_x and p_y are constants. Use of Eq. (20.2) verifies the expected form

$$p_x = \frac{\partial L}{\partial \dot{x}} = m\dot{x} \qquad p_y = \frac{\partial L}{\partial \dot{y}} = m\dot{y} \tag{20.6}$$

and the relations

$$\dot{p}_x = \dot{p}_y = 0 \tag{20.7}$$

merely express the conservation of linear momentum.

Planar motion in a central potential We return to the problem of a point mass m in a central potential (compare Sec. 3), as illustrated in Fig. 3.1. This system has the lagrangian

$$L = \tfrac{1}{2}m(\dot{r}^2 + r^2\dot{\phi}^2) - V(r) \tag{20.8}$$

which is obviously invariant under a rotation of the coordinate axes about the z axis. Correspondingly, the variable ϕ is cyclic, so that p_ϕ is a constant of the motion. The general definition (20.2) yields the explicit form

$$p_\phi = \frac{\partial L}{\partial \dot{\phi}} = mr^2\dot{\phi} \tag{20.9}$$

which is just the magnitude of the angular momentum [compare Eq. (3.5)]. In this case, the relation

$$\dot{p}_\phi = \frac{d}{dt}mr^2\dot{\phi} = 0 \tag{20.10}$$

expresses the conservation of angular momentum, with the direction fixed by the normal to the xy plane. The corresponding Lagrange equation for the radial coordinate is

$$\frac{d}{dt}m\dot{r} - mr\dot{\phi}^2 + \frac{dV}{dr} = 0 \tag{20.11}$$

These two equations (20.10) and (20.11) are just the results of Prob. 1.6, there obtained directly from Newton's second law in polar coordinates.

The Hamiltonian

The previous discussion has concentrated on invariance under spatial transformations, and we now turn to the analogous situation for the time dependence. It is first convenient to introduce the general definition of the *hamiltonian*

$$H \equiv \sum_\sigma p_\sigma \dot{q}_\sigma - L \tag{20.12}$$

It plays a central role in theoretical mechanics, in large part because of two very important properties.

Property 1 If the lagrangian does not depend explicitly on the time, then the hamiltonian is a constant of the motion:

$$\frac{\partial L}{\partial t} = 0 \Rightarrow \frac{dH}{dt} = 0 \tag{20.13}$$

This relation exemplifies our general treatment of symmetries, for the vanishing of $\partial L/\partial t$ implies that the system is unchanged by a shift in the initial time $t = 0$. As in the previous cases, this invariance will now be shown to imply the existence of a conserved quantity, which here is just the hamiltonian. Since H can be considered a function of all the coordinates and velocities, we again obtain a first integral $H(q_\sigma, \dot{q}_\sigma) = $ const.

The proof is direct. In a small time interval dt each of the variables in Eq. (20.12) undergoes small changes, and the total differential dH becomes

$$dH = \sum_\sigma \left(p_\sigma \, d\dot{q}_\sigma + \dot{q}_\sigma \, dp_\sigma - \frac{\partial L}{\partial q_\sigma} \, dq_\sigma - \frac{\partial L}{\partial \dot{q}_\sigma} \, d\dot{q}_\sigma \right) - \frac{\partial L}{\partial t} \, dt \tag{20.14}$$

where, as usual, L depends on the variables $(q_\sigma, \dot{q}_\sigma, t)$. The first and last terms in parentheses cancel owing to the definition in Eq. (20.2), and the term $\partial L/\partial t$ vanishes by hypothesis. Division of the remaining terms by dt yields

$$\frac{dH}{dt} = \sum_\sigma \dot{q}_\sigma \left(\frac{dp_\sigma}{dt} - \frac{\partial L}{\partial q_\sigma} \right) = \sum_\sigma \dot{q}_\sigma \left(\frac{d}{dt} \frac{\partial L}{\partial \dot{q}_\sigma} - \frac{\partial L}{\partial q_\sigma} \right) = 0 \tag{20.15}$$

where we have used Eq. (20.2) and Lagrange's equation (20.1). Thus H is indeed a constant of the motion if the lagrangian does not contain the time explicitly ($\partial L/\partial t = 0$).

One simple way to ensure that $\partial L/\partial t = 0$ is to consider systems with time-independent potentials and time-independent constraints, when the basic transformations (13.2) never involve the time explicitly. In this case, we can derive the following second important property.

Property 2 If there are only time-independent potentials $V(x_1, \ldots, x_n)$ with time-independent constraints, i.e., the time does not appear explicitly in Eqs. (13.2), then the hamiltonian is not only a constant of the motion [Eq. (20.13)] but also the total energy

$$\frac{\partial L}{\partial t} = 0 \quad \text{and} \quad \frac{\partial x_i}{\partial t} = 0 \Rightarrow H = T + V = E = \text{const} \tag{20.16}$$

To prove this result, use the general expression for the components of the velocities in cartesian coordinates [Eq. (15.8)]

$$\dot{x}_i = \sum_\sigma \frac{\partial x_i}{\partial q_\sigma} \dot{q}_\sigma + \frac{\partial x_i}{\partial t}$$

$$= \sum_\sigma \frac{\partial x_i}{\partial q_\sigma} \dot{q}_\sigma \tag{20.17}$$

where the term containing $\partial x_i / \partial t$ vanishes by assumption. The kinetic energy for the system can thus be written as

$$T = \frac{1}{2} \sum_i m_i \dot{x}_i^2 = \frac{1}{2} \sum_\sigma \sum_\lambda \left(\sum_i m_i \frac{\partial x_i}{\partial q_\sigma} \frac{\partial x_i}{\partial q_\lambda} \right) \dot{q}_\sigma \dot{q}_\lambda \tag{20.18}$$

It is a quadratic form in the generalized velocities $\dot{q}_\sigma$ with coefficients

$$\sum_i m_i \frac{\partial x_i}{\partial q_\sigma} \frac{\partial x_i}{\partial q_\lambda} \equiv m_{\sigma\lambda} \tag{20.19}$$

This quantity is evidently a real, symmetric matrix in the indices σ and λ

$$m_{\sigma\lambda} = m_{\lambda\sigma} = m_{\sigma\lambda}^* = m_{\lambda\sigma}^* \tag{20.20}$$

and depends on all the generalized coordinates through the partial derivatives in Eq. (20.19). Consequently, the kinetic energy takes the form

$$T = \frac{1}{2} \sum_\sigma \sum_\lambda m_{\sigma\lambda}(q) \dot{q}_\sigma \dot{q}_\lambda \tag{20.21}$$

Moreover, the first term appearing in the definition of the hamiltonian [Eq. (20.12)] can be rewritten

$$\sum_\sigma p_\sigma \dot{q}_\sigma = \sum_\sigma \frac{\partial L}{\partial \dot{q}_\sigma} \dot{q}_\sigma = \sum_\sigma \frac{\partial T}{\partial \dot{q}_\sigma} \dot{q}_\sigma \tag{20.22}$$

where the definition $L = T - V$ has been used, along with the observation that the potential energy is independent of the generalized velocities for conservative forces [see Eq. (15.21)]. The required derivative follows directly from Eq. (20.21)

$$p_\sigma = \frac{\partial T}{\partial \dot{q}_\sigma} = \frac{1}{2} \sum_\lambda m_{\sigma\lambda} \dot{q}_\lambda + \frac{1}{2} \sum_\lambda m_{\lambda\sigma} \dot{q}_\lambda = \sum_\lambda m_{\sigma\lambda} \dot{q}_\lambda \tag{20.23}$$

and the first term in the hamiltonian is just twice the kinetic energy

$$\sum_\sigma p_\sigma \dot{q}_\sigma = 2T \tag{20.24}$$

Finally, the hamiltonian itself becomes

$$H = 2T - (T - V) = T + V = E \tag{20.25}$$

verifying the claim made in Eq. (20.16).

The preceding derivation shows that if T is a quadratic form in the generalized velocities, the hamiltonian is the total energy of the system. Although this situation occurs frequently, it is by no means universal; to illustrate one possibility, we consider the previous example of a bead on a rotating wire hoop (Sec. 16), where [see Eq. (16.10)]

$$L = T = \tfrac{1}{2}ma^2[\omega^2 + (\omega + \dot{\theta})^2 + 2\omega(\omega + \dot{\theta}) \cos \theta] \qquad (20.26)$$

By inspection, $\partial L/\partial t = 0$ even though the transformation to generalized coordinates [Eq. (16.6)] indeed involves the time explicitly. Furthermore, the generalized momentum has an unexpectedly complicated form

$$p_\theta = \frac{\partial L}{\partial \dot{\theta}} = ma^2(\omega + \dot{\theta}) + ma^2\omega \cos \theta \qquad (20.27)$$

owing to the transformation to a rotating coordinate system [compare Eq. (20.9)]. In contrast, the hamiltonian has a simpler structure

$$H = p_\theta \dot{\theta} - L = ma^2[\tfrac{1}{2}\dot{\theta}^2 - \omega^2(1 + \cos \theta)] \qquad (20.28)$$

The general theorem (20.13) ensures that H is a *constant of the motion*, so that Eq. (20.28) could be used to express the trajectory as a definite integral. Nevertheless, H is *not* the total energy E, which is here given by T in Eq. (20.26) (recall that $V = 0$). Indeed, the transformation (16.6) involves the time explicitly, and the conditions of the general theorem (20.16) are thus violated. Furthermore, E is *not* constant in this problem because the uniformly rotating wire exerts a time-dependent force. This enforced motion also affects the structure of T in Eq. (20.26), which is not a quadratic form in $\dot{\theta}$. If $\omega = 0$, however, the extra terms vanish identically, and we then recover the expected result $H = E$.

PROBLEMS

3.1 A point mass m slides without friction along a wire bent into a vertical circle of radius a. The wire rotates with constant angular velocity Ω about the vertical diameter, and the apparatus is placed in a uniform gravitational field **g** parallel to the axis of rotation.

    (a) Construct the lagrangian for the point mass using the angle θ (angular displacement measured from the downward vertical) as generalized coordinate.

    (b) Show that the condition for an equilibrium circular orbit at angle θ_0 is $\cos \theta_0 = g/a\Omega^2$. Explain this result by balancing forces in a co-rotating coordinate system.

    (c) Demonstrate that this orbit is stable against small displacements along the wire and show that the angular frequency of oscillation about the equilibrium orbit is given by $\omega^2 = \Omega^2 \sin^2 \theta_0$. *Hint:* Write $\theta = \theta_0 + \eta(t)$, where $\eta(t)$ is a small quantity.

    (d) What happens if $a\Omega^2 < g$?

3.2 Repeat Prob. 3.1 for a straight wire rotating about the vertical direction at a fixed polar angle θ_0. Construct the lagrangian for the point mass using the distance l along the wire as a generalized coordinate. Show that the condition for an equilibrium circular orbit is $l_0 = g \cos \theta_0/(\Omega \sin \theta_0)^2$. Discuss the stability of this orbit against small displacements along the wire. Compare with the results obtained in Prob. 3.1.

3.3 A simple pendulum of mass m_2 and length l is constrained to move in a single plane. The point of support is attached to a mass m_1 which can move on a horizontal line in the same plane. Find the lagrangian of the system in terms of suitable generalized coordinates. Derive the equations of motion. Find the frequency of small oscillations of the pendulum.

3.4 A massless flat plate rests on a frictionless plane that is inclined to the horizontal at an angle α. A massless rigid rod of length h is fixed at right angles to the plate. From the top of the rod hangs a simple pendulum of mass m and length l ($<h$). Introduce suitable generalized coordinates, and find the lagrangian of the system. Derive the equations of motion. Discuss the limiting case $\alpha \to 0$.

3.5 A massless inextensible string passes over a pulley which is a fixed distance above the floor. A bunch of bananas of mass m is attached to one end A of the string. A monkey of mass M is initially at the other end B. The monkey climbs the string, and his displacement $d(t)$ with respect to the end B is a *given* function of time. The system is initially at rest, so that the initial conditions are $d(0) = \dot{d}(0) = 0$. Introduce suitable generalized coordinates and calculate the lagrangian of the system in terms of these coordinates. Show that the equation of motion governing the height Z of the monkey above the floor is

$$(m + M)\ddot{Z} - m\ddot{d} = (m - M)g$$

Integrate the equation to find the subsequent motion. In the special case that $m = M$, show that the bananas and the monkey rise through equal distances so that the vertical separation between them is constant.

3.6 An inextensible massless string of length l passes through a hole in a frictionless table. A point mass m at one end moves on the table and a point mass m hangs from the other end.

  (*a*) Write the lagrangian for this system.

  (*b*) Under what condition will the hanging mass remain stationary?

  (*c*) Starting from the situation in part (*b*), the hanging mass is pulled down slightly and released. State clearly what is conserved during this process.

  (*d*) Compute the subsequent motion of the hanging mass.

3.7 The point of suspension of a pendulum m is allowed to move in the horizontal direction. Two springs of force constant $\frac{1}{2}k$ exert a net restoring force $-kx$ on the point of suspension.

  (*a*) Use the coordinates x (the displacement of the point of support) and θ (the angular displacement of m from the vertical) to write the lagrangian and the equations of motion.

  (*b*) Assuming small oscillations, show that this system is equivalent to a simple pendulum of length $l + mg/k$.

3.8 A point mass m slides without friction inside a surface of revolution $z = \alpha \sin (r/R)$ whose symmetry axis lies along the direction of a uniform gravitational field $\mathbf{g}$. Consider $0 < r/R < \frac{1}{2}\pi$.

  (*a*) Construct the lagrangian $L(r, \phi; \dot{r}, \dot{\phi})$ and compute the equations of motion for the generalized coordinates r and ϕ.

  (*b*) Are there stationary horizontal circular orbits?

  (*c*) Which of these orbits is stable under small impulses along the surface transverse to the direction of motion?

  (*d*) If the orbit is stable, what is the frequency of oscillation about the equilibrium orbit?

3.9 Repeat Prob. 3.8 for a surface of revolution $z = \alpha(1 - e^{-r/R})$ whose symmetry axis lies along the direction of a uniform gravitational field $\mathbf{g}$.

3.10 The curve traced in space by a point on a rolling wheel is called a *cycloid*.

  (*a*) Show that the equation of the cycloid passing through the origin and generated by a wheel of radius a rolling underneath the x axis is

$$x = a(\theta - \sin \theta) \quad \text{and} \quad y = a(1 - \cos \theta)$$

where the y axis is chosen as in Fig. 17.2. Show that the distance s measured along the cycloid from the first minimum is given by

$$s = -4a \cos \tfrac{1}{2}\theta$$

  (*b*) Consider a point mass m in a uniform gravitational field $\mathbf{g} = g\hat{y}$ starting at rest from the origin and sliding without friction on a wire bent in the shape of this cycloid. Show that such a curve solves the brachistochrone problem formulated in Sec. 17.

(c) Consider oscillations of the mass m about a minimum of the cycloid. Use s as the generalized coordinate and prove that the system forms an *isochronous oscillator*, i.e., an oscillator whose period is independent of the amplitude, with a period $2\pi(4a/g)^{1/2}$.

3.11 (a) Consider the functional $I = \int_{x_1}^{x_2} dx \, \phi(y, y')$, where ϕ is independent of x. Show that the Euler-Lagrange equation has a first integral of the form $\phi - y'(\partial\phi/\partial y') = $ const.

(b) A uniform flexible rope of length $l \, (>2x_0)$ hangs between two supports at the same height a distance $2x_0$ apart. Minimize the potential energy, incorporating the constraint of fixed l with a Lagrange multiplier. Using the result of part (a), or otherwise, integrate to find the explicit equation for the rope's shape. If $l = 4x_0$, how far is the center of the rope below the line joining the supports?

3.12 In geometric optics, the trajectory of a light ray is given by Fermat's principle, which states that a ray travels between two fixed points in such a way that the time of transit is stationary with respect to small variations in the path. Consider the propagation of light rays in the earth's atmosphere, assuming the refractive index n is a function only of the distance from the center of the earth. Express Fermat's principle in integral form. Derive the following equation of a ray in the atmosphere:

$$\frac{dr}{d\phi} = r(kn^2r^2 - 1)^{1/2}$$

where k is a constant and (r, ϕ) are the two-dimensional polar coordinates of a point on the ray with the center of the earth as origin. If n is proportional to r^m, find the value of m such that a ray lying initially along a tangential direction (parallel to $\hat{\phi}$) remains a constant distance from the center of the earth for any value of r.

3.13 (a) Express the lagrangian of a simple pendulum in terms of the cartesian variables $x = l \cos \theta$ and $y = l \sin \theta$.

(b) Use the holonomic constraint $x^2 + y^2 = l^2$ to rewrite L explicitly in terms of x and $\dot{x}$. Obtain the corresponding dynamical equation and reconcile it with Eq. (16.5).

(c) Write out Eq. (18.10) in terms of the variations δx and δy. Eliminate δy in terms of δx with the holonomic constraint and hence obtain the same dynamical equation for the variable $x(t)$.

(d) As an alternative procedure, use a lagrange multiplier λ to incorporate the constraint. Hence obtain a set of equations for $x(t)$, $y(t)$, and λ. Compare with the treatment of Eq. (19.13) in two-dimensional polar coordinates.

3.14 A particle of mass m suspended by a massless string of length l is stationary in a gravitational field **g**. It is struck an impulsive horizontal blow, giving it an initial angular velocity ω.

(a) Introduce a lagrange multiplier and prove the following statements:

1. If $l\omega^2 < 2g$, the tension τ does not vanish and the particle does not reach the horizontal.
2. If $2g < l\omega^2 < 5g$, the particle passes the horizontal and the string becomes slack before the particle comes to rest.
3. If $5g < l\omega^2$, the string always remains taut and the particle executes periodic circular motion.

(b) Discuss the role of the tension in the string by showing how these results are changed if the string is replaced by a rigid massless rod.

3.15 Consider the problem discussed in Sec. 16 of a bead of mass m moving without friction on a horizontal circular hoop that rotates with constant angular velocity ω about an axis passing through the wire and perpendicular to its plane (see Fig. 16.2).

(a) Compute the reaction force of the wire on the bead by means of a Lagrange multiplier and Lagrange's equations.

(b) Interpret the result by going to a co-rotating coordinate system whose origin lies at the *center* of the circular hoop.

3.16 Generalize the treatment in Sec. 19 of a cylinder rolling on a fixed cylinder to the case that θ_1 has the initial value α. Find the position at which the cylinders separate and express the elapsed time as a definite integral. Discuss the limiting cases $\alpha \to 90°$, $\alpha \to 0°$.

3.17 (*a*) A point mass is placed on the top of a fixed smooth sphere in a gravitational field. Show that it leaves the fixed sphere at a polar angle acrccos $\frac{2}{3} \approx 48.19°$.

(*b*) The point mass is replaced by a roughened sphere that rolls without sliding on the fixed sphere. Show that it leaves the fixed sphere at a polar angle arccos $\frac{10}{17} \approx 53.97°$. Compare these two values with that for a rolling cylinder [see Eq. (19.54)].

3.18 A uniform ladder of length L and mass M has one end on a smooth horizontal floor and the other end against a smooth vertical wall. The ladder is initially at rest in a vertical plane perpendicular to the wall and makes an angle θ_0 with the horizontal. Make a convenient choice of generalized coordinates and find the lagrangian. Derive the corresponding equations of motion. Prove that the ladder leaves the wall when its upper end has fallen to a height $\frac{2}{3}L \sin \theta_0$. Show how the subsequent motion can be reduced to explicit integrals. Does the ladder ever lose contact with the floor?

3.19 A point mass m is constrained to move without friction on an arbitrary, fixed two-dimensional surface in the absence of external forces. The surface is described by a set of generalized coordinates (q^1, q^2) such that the square of the distance $d\mathbf{s} \cdot d\mathbf{s}$ between two infinitesimally close points on the surface is given by

$$d\mathbf{s} \cdot d\mathbf{s} = ds^2 = \sum_{i=1}^{2} \sum_{j=1}^{2} g_{ij}(q^1, q^2) \, dq^i \, dq^j$$

where the *metric tensor* $g_{ij}(q^1, q^2) = g_{ji}(q^1, q^2)$ is symmetric and depends on position.

(*a*) Construct the lagrangian. Show that the equations of motion for the particle are given by

$$\sum_j g_{ij} \frac{d^2 q^j}{dt^2} + \frac{1}{2} \sum_{j,k} \left(\frac{\partial g_{ij}}{\partial q^k} + \frac{\partial g_{ik}}{\partial q^j} - \frac{\partial g_{jk}}{\partial q^i} \right) \frac{dq^j}{dt} \frac{dq^k}{dt} = 0 \qquad i = 1, 2$$

Introduce the inverse $(g^{-1})_{ij} \equiv g^{ij}$ of the metric tensor, and derive the equivalent equations of motion

$$\frac{d^2 q^i}{dt^2} + \sum_{j,k} \Gamma^i_{jk} \frac{dq^j}{dt} \frac{dq^k}{dt} = 0 \qquad i = 1, 2$$

where the *affine connection* is defined by

$$\Gamma^i_{jk} \equiv \frac{1}{2} \sum_m g^{im} \left(\frac{\partial g_{mj}}{\partial q^k} + \frac{\partial g_{mk}}{\partial q^j} - \frac{\partial g_{jk}}{\partial q^m} \right)$$

(*b*) Show that the curves of minimum distance between two points on the surface, the *geodesics*, are given by

$$\frac{d^2 q^i}{d\tau^2} + \sum_{j,k} \Gamma^i_{jk} \frac{dq^j}{d\tau} \frac{dq^k}{d\tau} = 0 \qquad i = 1, 2$$

where $0 \leqslant \tau \leqslant 1$ is a uniform parametrization of the distance along the curve.

(*c*) Show that $\mathbf{v}^2 \equiv v^2 = \text{const}$ for any motion on this surface. For a given trajectory of the particle passing through two points on the surface a distance l apart, prove that we can take $\tau = (v/l)t = \text{const} \times t$. Hence conclude that the equations and curves in *a* and *b* are identical. *Note:* These observations form one of the starting points for the theory of general relativity.

SELECTED ADDITIONAL READING

1. G. A. Bliss, "Calculus of Variations," Open Court, La Salle, Ill., 1925.

FOUR

SMALL OSCILLATIONS

This chapter considers small-amplitude oscillations of mechanical systems about static equilibrium, e.g., coupled pendulums (see Fig. 23.1) and longitudinal or transverse oscillations of particles coupled with springs (see Figs. 24.1 and 24.2). Other applications include vibrations of molecules or crystals about their equilibrium configuration. We first develop the general normal-mode analysis of such systems and then apply the formalism to several specific examples. The final section illustrates the transition to continuum mechanics and introduces the concepts of continuum field theory by taking the limit where the phenomena of interest have a wavelength large compared with the interparticle spacing.

21 FORMULATION

Consider a set of n independent generalized coordinates q_σ labeled $\sigma = 1, \ldots, n$. Our fundamental assumptions are that the system is conservative with a time-independent potential

$$V = V(q_1, \ldots, q_n) \tag{21.1}$$

and that there are no time-varying constraints. Hence the cartesian coordinates are related to the generalized coordinates by

$$x_i = x_i(q_1, \ldots, q_n) \tag{21.2}$$

where the subscript i runs over all the original degrees of freedom of the system.

The kinetic energy for such a system was derived in Eqs. (20.17) to (20.21). Thus the lagrangian takes the form

$$L = L(q_1, \ldots, q_n, \dot{q}_1, \ldots, \dot{q}_n) \tag{21.3}$$

Lagrange's equations have the basic structure [see Eq. (15.17)]

$$\frac{d}{dt}\frac{\partial T}{\partial \dot{q}_\sigma} - \frac{\partial T}{\partial q_\sigma} = Q_\sigma = -\frac{\partial V}{\partial q_\sigma} \qquad \sigma = 1, \ldots, n \tag{21.4}$$

where the potential determines the generalized force as in Eq. (15.20). Static equilibrium is characterized by the conditions

$$\dot{q}_\sigma = \ddot{q}_\sigma = 0 \tag{21.5}$$

$$\sigma = 1, \ldots, n$$

$$q_\sigma = q_\sigma^0 \tag{21.6}$$

Substitution of Eq. (20.21) into the left-hand side of Eq. (21.4) and use of (21.5) shows that static equilibrium is equivalent to the vanishing of all the generalized forces

$$Q_\sigma = -\left(\frac{\partial V}{\partial q_\sigma}\right)_{q^0} = 0 \qquad \sigma = 1, \ldots, n \tag{21.7}$$

This conclusion is intuitively obvious, for if there is no net generalized force, the corresponding generalized coordinate remains fixed, retaining its constant static value (21.6).

Suppose that the system now undergoes a *small displacement* from equilibrium. The qualitative behavior can be understood by referring to Fig. 21.1, which displays the potential energy obtained by allowing one of the generalized coordinates to vary from its equilibrium value. If the potential energy has a minimum, the equilibrium will be stable, as in Fig. 21.1a. If the potential has a maximum or an inflection point (Fig. 21.1b and c), the equilibrium will be unstable because the system can lower its potential energy by moving away from the equilibrium configuration, leading to a runaway solution.

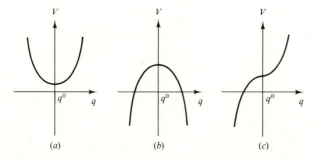

Figure 21.1 Qualitative behavior of potential energy as a function of the generalized coordinate q for system undergoing small displacements about equilibrium q^0: (a) stable; (b) and (c) unstable.

To obtain an analytic description of these small displacements from equilibrium we write

$$q_\sigma = q_\sigma^0 + \eta_\sigma \qquad \sigma = 1, \ldots, n \tag{21.8}$$

where η_σ is assumed small. Differentiation of Eq. (21.8) with respect to time leads to

$$\dot{q}_\sigma = \dot{\eta}_\sigma \qquad \sigma = 1, \ldots, n \tag{21.9}$$

The kinetic energy of the system can be immediately constructed through second order in $\dot{\eta}_\sigma$ by substituting Eq. (21.9) into Eq. (20.21)

$$T = \frac{1}{2} \sum_\sigma \sum_\lambda m_{\sigma\lambda} \dot{\eta}_\sigma \dot{\eta}_\lambda \tag{21.10a}$$

$$m_{\sigma\lambda} \equiv \sum_i m_i \left(\frac{\partial x_i}{\partial q_\sigma}\right)_{q^0} \left(\frac{\partial x_i}{\partial q_\lambda}\right)_{q^0} \tag{21.10b}$$

where the mass matrix in Eq. (21.10b) is evaluated at the equilibrium configuration. As a result $m_{\sigma\lambda}$ becomes a *constant*, real, symmetric matrix [compare Eq. (20.20)]

$$m_{\sigma\lambda} = m_{\lambda\sigma} = m_{\sigma\lambda}^* = m_{\lambda\sigma}^* \tag{21.11}$$

when considering small oscillations about equilibrium.

We next examine the behavior of the potential energy under small displacements about equilibrium

$$V(q_1, \ldots, q_n) = V(q_1^0 + \eta_1, \ldots, q_n^0 + \eta_n) \tag{21.12}$$

A Taylor-series expansion of this expression through terms of second order in η_σ yields

$$V(q_1, \ldots, q_n) = V(q_1^0, \ldots, q_n^0) + \sum_\sigma \eta_\sigma \left(\frac{\partial V}{\partial q_\sigma}\right)_{q^0}$$
$$+ \frac{1}{2} \sum_\sigma \sum_\lambda \eta_\sigma \eta_\lambda \left(\frac{\partial^2 V}{\partial q_\sigma \, \partial q_\lambda}\right)_{q^0} \tag{21.13}$$

Comparison with Eq. (21.7) shows that *the linear term in this expansion vanishes* because static equilibrium is defined by the absence of all generalized forces. This relation is just the condition that all the curves in Fig. 21.1 have zero slope at $q_\sigma = q_\sigma^0$. The potential energy in Eq. (21.13) thus takes the form

$$V = V_0 + \frac{1}{2} \sum_\sigma \sum_\lambda v_{\sigma\lambda} \eta_\sigma \eta_\lambda \tag{21.14}$$

where the coefficients $v_{\sigma\lambda}$ are given by

$$v_{\sigma\lambda} \equiv \left(\frac{\partial^2 V}{\partial q_\sigma \, \partial q_\lambda}\right)_{q^0} \tag{21.15}$$

and have the following obvious properties

$$v_{\sigma\lambda} = v_{\lambda\sigma} = v_{\sigma\lambda}^* = v_{\lambda\sigma}^* \tag{21.16}$$

These coefficients depend on the second derivative of the potential energy evaluated at the equilibrium configuration; they again constitute a constant, real, symmetric matrix.

In the case of small displacements about equilibrium, the lagrangian can be obtained from Eqs. (21.10) and (21.14) according to

$$L = T - V = \frac{1}{2} \sum_{\sigma} \sum_{\lambda} (m_{\lambda\sigma} \dot{\eta}_\lambda \dot{\eta}_\sigma - v_{\lambda\sigma} \eta_\lambda \eta_\sigma) - V_0 \tag{21.17}$$

We are now free to take the η_σ with $\sigma = 1, \ldots, n$ as the n generalized coordinates since the specification of the η_σ completely fixes the configuration of the many-particle system according to (21.8). Hamilton's principle then immediately leads to the set of Lagrange's equations

$$\frac{d}{dt} \frac{\partial L}{\partial \dot{\eta}_\sigma} - \frac{\partial L}{\partial \eta_\sigma} = 0 \qquad \sigma = 1, \ldots, n \tag{21.18}$$

Differentiation of the lagrangian in Eq. (21.17) and use of the symmetry of both $m_{\sigma\lambda}$ and $v_{\sigma\lambda}$ leads to the equations of motion

$$\sum_{\lambda=1}^{n} (m_{\sigma\lambda} \ddot{\eta}_\lambda + v_{\sigma\lambda} \eta_\lambda) = 0 \qquad \sigma = 1, \ldots, n \tag{21.19}$$

Note that the lagrangian is a quadratic function of the small displacements from equilibrium and small velocities [Eqs. (21.8) and (21.9)], whereas the equations of motion are linear in these same small quantities. Thus *linearization* of the equations of motion about static equilibrium is equivalent to constructing the lagrangian through terms quadratic in the small displacements and small velocities.

22 NORMAL MODES

Equations (21.19) form a set of n linear, homogeneous, coupled differential equations with real, constant coefficients. We proceed to discuss the solution to such equations.

Simplest Case

Consider first the case of one-dimensional motion where

$$m_{\sigma\lambda} \equiv m \tag{22.1}$$

$$v_{\sigma\lambda} \equiv k \tag{22.2}$$

are simply real numbers. The equations of motion (21.19) then take the familiar form (here $\eta_1 \equiv \eta$)

$$\ddot{\eta} = -\frac{k}{m}\eta \tag{22.3}$$

We assume stable motion about equilibrium, with $k > 0$ and a potential energy of the form shown in Fig. 21.1a. Thus the problem of small oscillations here reduces to that of simple harmonic motion, as is evident from Eq. (22.3). Although this problem is elementary, we shall discuss the solution in detail, for similar methods also apply in the more complicated case of coupled systems.

To solve Eq. (22.3), it is first convenient to introduce a complex variable z that satisfies the linear equation

$$\ddot{z} = -\frac{k}{m}z \tag{22.4}$$

Since this equation contains only real coefficients, it is evident that *any* solution to (22.4) provides an equivalent solution to Eq. (22.3) with

$$\eta = \text{Re } z \tag{22.5}$$

solving the original problem. We seek a solution to Eq. (22.4) that varies harmonically in time

$$z = z^0 e^{i\omega t} \tag{22.6}$$

Substitution into Eq. (22.4) yields

$$\left(\omega^2 - \frac{k}{m}\right)z^0 = 0 \tag{22.7}$$

which is now an *eigenvalue equation* for the squared frequency ω^2. Equation (22.7) can have nontrivial solutions with $z^0 \neq 0$ only if the coefficient of z^0 vanishes. This observation implies that there are two allowed frequencies

$$\omega = \pm\omega_1 \tag{22.8}$$

where

$$\omega_1 \equiv \left(\frac{k}{m}\right)^{1/2} \tag{22.9}$$

Since Eq. (22.4) is *linear*, its general solution can be constructed as a linear superposition of all the specific solutions. Thus we write

$$z(t) = z_+^{(1)}e^{i\omega_1 t} + z_-^{(1)}e^{-i\omega_1 t} \tag{22.10}$$

where $z_+^{(1)}$ and $z_-^{(1)}$ are complex constants, equivalent to four real constants. This solution contains enough freedom to match an arbitrary set of initial conditions, since a complete specification of the solution to Eq. (22.4) requires the initial value of z (two real quantities) and the initial value of $\dot{z}$ (two more real quantities). As a result, Eq. (22.10) is indeed the most general solution to Eq. (22.4).

The solution to the original problem (22.3) can now be obtained directly from (22.5):

$$\eta = \operatorname{Re} z = \tfrac{1}{2}(z + z^*) = \tfrac{1}{2}\{[z_+^{(1)} + (z_-^{(1)})^*]e^{i\omega_1 t} + \text{complex conjugate}\} \quad (22.11)$$

The complex number $z_+^{(1)} + (z_-^{(1)})^*$ will now be written in the form

$$z_+^{(1)} + (z_-^{(1)})^* \equiv z^{(1)} \equiv \rho^{(1)}e^{i\phi_1} \quad (22.12)$$

which defines the two real numbers $\rho^{(1)}$ and ϕ_1. Thus the solution (22.11) becomes

$$\eta = \rho^{(1)} \cos{(\omega_1 t + \phi_1)} = \operatorname{Re}{(z^{(1)}e^{i\omega_1 t})} = \operatorname{Re}{(\rho^{(1)}e^{i(\omega_1 t + \phi_1)})} \quad (22.13)$$

It is readily verified that this result is a solution to Eq. (22.3); moreover, it contains two arbitrary real constants $\rho^{(1)}$ and ϕ_1 that permit us to match any two real initial conditions in Eq. (22.3). Thus Eq. (22.13) is the most general solution to Eq. (22.3), which certainly is no surprise. It is somewhat less obvious, however, that this general real solution requires only the *positive eigenvalue* from Eq. (22.8), as illustrated in detail by the above discussion. This simplification arises from the use of the real part in (22.13).

Coupled Problem: Formulation

With this introduction, consider the full set of coupled equations (21.19). It is again convenient to introduce complex quantities z_σ, with $\sigma = 1, \ldots, n$ denoting the n generalized coordinates; the z_σ are assumed to satisfy the equations

$$\sum_{\lambda=1}^{n} (m_{\sigma\lambda}\ddot{z}_\lambda + v_{\sigma\lambda}z_\lambda) = 0 \qquad \sigma = 1, \ldots, n \quad (22.14)$$

Since $m_{\sigma\lambda}$ and $v_{\sigma\lambda}$ are real, it is again evident that the solution to Eqs. (21.19) can be obtained as

$$\eta_\sigma = \operatorname{Re} z_\sigma \qquad \sigma = 1, \ldots, n \quad (22.15)$$

We first attempt to determine the time dependence by looking for *normal modes, a type of motion in which all the coordinates oscillate with the same frequency.* Thus we seek normal-mode solutions of the form [compare Eq. (22.6)]

$$z_\sigma = z_\sigma^0 e^{i\omega t} \qquad \sigma = 1, \ldots, n \quad (22.16)$$

where the frequency ω is independent of σ. Substitution into the set of equations (22.14) yields

$$\sum_{\lambda=1}^{n} (v_{\sigma\lambda} - \omega^2 m_{\sigma\lambda})z_\lambda^0 = 0 \quad (22.17)$$

In this way, the original problem (22.14) of n linear homogeneous coupled *differential* equations with real, constant coefficients has now been reduced to one of n linear homogeneous coupled *algebraic* equations for the amplitudes z_λ^0. To clarify the subsequent discussion, we digress briefly to review the theory of linear algebraic equations.

Linear Equations: A Review†

Consider the following set of linear inhomogeneous algebraic equations for the quantities $(x_1, \ldots, x_n)$:

$$
\begin{aligned}
a_{11}x_1 + a_{12}x_2 + \cdots + a_{1n}x_n &= y_1 \\
a_{21}x_1 + a_{22}x_2 + \cdots + a_{2n}x_n &= y_2 \\
&\cdots\cdots\cdots\cdots\cdots\cdots\cdots\cdots\cdots \\
a_{n1}x_1 + a_{n2}x_2 + \cdots + a_{nn}x_n &= y_n
\end{aligned}
\tag{22.18}
$$

Here a_{ij} are real, constant coefficients, and the set of values $(y_1, \ldots, y_n)$ is assumed given. The equations (22.18) can be written compactly either with a summation symbol

$$
\sum_{j=1}^{n} a_{ij}x_j = y_i \qquad i = 1, \ldots, n
\tag{22.19}
$$

or in matrix notation

$$
\underline{a}\,\underline{x} = \underline{y}
\tag{22.20}
$$

where

$$
\underline{x} \equiv \begin{bmatrix} x_1 \\ x_2 \\ \vdots \\ x_n \end{bmatrix} \qquad
\underline{y} \equiv \begin{bmatrix} y_1 \\ y_2 \\ \vdots \\ y_n \end{bmatrix} \qquad
\underline{a} \equiv \begin{bmatrix} a_{11} & a_{12} & \cdots & a_{1n} \\ a_{21} & a_{22} & \cdots & a_{2n} \\ \multicolumn{4}{c}{\cdots\cdots\cdots\cdots\cdots\cdots} \\ a_{n1} & a_{n2} & \cdots & a_{nn} \end{bmatrix}
\tag{22.21}
$$

Cramer's rule gives the solution to this set of linear equations. If det $\underline{a} \neq 0$ (here det $\underline{a}$ denotes the determinant of the matrix $\underline{a}$), then x_i is the ratio of two determinants

$$
x_i = \frac{\det \begin{vmatrix} a_{11} & \cdots & \overset{\text{ith column}}{y_1} & \cdots & a_{1n} \\ \multicolumn{5}{c}{\cdots\cdots\cdots\cdots\cdots\cdots} \\ a_{n1} & \cdots & y_n & \cdots & a_{nn} \end{vmatrix}}{\det \begin{vmatrix} a_{11} & \cdots & a_{1n} \\ \multicolumn{3}{c}{\cdots\cdots\cdots\cdots} \\ a_{n1} & \cdots & a_{nn} \end{vmatrix}}
\tag{22.22}
$$

In the numerator, the elements in the ith column of the matrix $\underline{a}$ have been replaced by the quantities on the right-hand side of Eqs. (22.18). The numerator and denominator in this expression can be evaluated with the standard rules for determinants, and we therefore obtain the complete solution for linear algebraic inhomogeneous equations.

Suppose, however, that $y_i = 0$ for all i, when the set of linear algebraic equations (22.18) becomes *homogeneous*. There are two distinct possibilities.

† See, for example, Hildebrand [1], chap. 1.

Case 1: det $\underline{a} \neq 0$ Since one column in the numerator of Eq. (22.22) vanishes, Cramer's rule shows that Eqs. (22.18) have only the trivial solution $x_i = 0$ for all i.

Case 2: det $\underline{a} = 0$ In this case, at least one of Eqs. (22.18), say the last one, is linearly dependent. This condition means that it is some linear combination of all the preceding equations; correspondingly, the last equation contains no new information and can be discarded as redundant. We then assume without loss of generality that the solution of the resulting $n - 1$ equations has at least one nonzero component, say x_n. Divide the remaining equations through by this quantity x_n, casting Eqs. (22.18) in the following form:

$$a_{11} \frac{x_1}{x_n} + a_{12} \frac{x_2}{x_n} + \cdots + a_{1,n-1} \frac{x_{n-1}}{x_n} = -a_{1n}$$

$$\cdots\cdots\cdots\cdots\cdots\cdots\cdots\cdots\cdots\cdots\cdots\cdots\cdots\cdots\cdots\cdots\cdots\cdots \qquad (22.23)$$

$$a_{n-1,1} \frac{x_1}{x_n} + \cdots + a_{n-1,n-1} \frac{x_{n-1}}{x_n} = -a_{n-1,n}$$

We now have $n - 1$ *inhomogeneous* equations in $n - 1$ unknowns. Cramer's rule again provides a solution, with explicit values for the $n - 1$ ratios (x_i/x_n).† Observe that if all the elements a_{ij} are real, this construction necessarily gives *real values* for the $n - 1$ ratios x_i/x_n with $i = 1, 2, \ldots, n - 1$.

Coupled Problem: Eigenvectors and Eigenvalues

The preceding discussion shows that the set of linear homogeneous algebraic equations (22.17) has a nontrivial solution only if the determinant of the coefficients vanishes

$$\det |v_{\sigma\lambda} - \omega^2 m_{\sigma\lambda}| = 0 \qquad (22.24)$$

Since the $v_{\sigma\lambda}$ and $m_{\sigma\lambda}$ are real specified constants, the only freedom left in this relation is the choice of the frequencies ω^2. Evaluation of the determinant of the $n \times n$ matrix in Eq. (22.24) evidently leads to an nth-order polynomial in ω^2. Any nth-order polynomial always possesses n roots, and they here constitute the n eigenvalues, labeled

$$\omega^2 = \omega_s^2 \qquad s = 1, 2, \ldots, n \qquad (22.25)$$

We proceed to show that all the roots (22.25) of the eigenvalue equation (22.24) are *real*. First substitute the eigenvalue ω_s^2 in Eqs. (22.17) and let $z_\lambda^{(s)}$ with $\lambda = 1, \ldots, n$ denote the corresponding nontrivial solution. To be explicit, $z_\lambda^{(s)}$ satisfies the set of equations

$$\sum_{\lambda=1}^{n} (v_{\sigma\lambda} - \omega_s^2 m_{\sigma\lambda})z_\lambda^{(s)} = 0 \qquad \sigma = 1, \ldots, n \qquad (22.26)$$

---

† It is assumed here that the determinant of the coefficients on the left-hand side of Eq. (22.23) is nonzero and that the mechanical problem indeed has a solution.

Perform the operation $\sum_\sigma z_\sigma^{(s)*}$ on Eq. (22.26). This means to multiply each equation by the quantity $z_\sigma^{(s)*}$ and then sum over all σ. The resulting expression can now be solved for the quantity ω_s^2, leading to the equation

$$\omega_s^2 = \frac{\sum_\sigma \sum_\lambda z_\sigma^{(s)*} v_{\sigma\lambda} z_\lambda^{(s)}}{\sum_\sigma \sum_\lambda z_\sigma^{(s)*} m_{\sigma\lambda} z_\lambda^{(s)}} \tag{22.27}$$

Since $v_{\sigma\lambda}^* = v_{\lambda\sigma}$ and $m_{\sigma\lambda}^* = m_{\lambda\sigma}$ are real symmetric matrices [Eqs. (21.11) and (21.16)], the complex conjugate of Eq. (22.27) can be rewritten by interchanging the dummy summation indices σ and λ. In this way, we immediately conclude that all the eigenvalues in Eqs. (22.25) are *real*:

$$(\omega_s^2)^* = \omega_s^2 \qquad s = 1, \ldots, n \tag{22.28}$$

A stable system clearly requires $\omega_s^2 \geq 0$. Otherwise, if $\omega_s^2 < 0$ for any s, the corresponding imaginary frequency leads to runaway solutions in Eqs. (22.16) and (22.15). As seen in Fig. 21.1a, the equilibrium will be stable with positive eigenvalues in Eq. (22.25) if the potential energy is a minimum in all directions. This conclusion is evident from Eq. (22.27) since the denominator is a positive-definite quadratic form [compare Eqs. (20.18) and (20.21)], and the numerator will also be positive definite if the potential is a minimum at equilibrium [Eqs. (21.12) to (21.15)].

Since all the coefficients in Eq. (22.26) are real, it is evident from the discussion following Eq. (22.23) that the ratios of the components $z_\sigma^{(s)}/z_n^{(s)}$ ($\sigma = 1, \ldots, n-1$) for a given solution s are real. Thus a complex constant can appear only as an overall multiplicative factor, independent of σ. In this way, the solution to Eq. (22.26) can be written

$$z_\sigma^{(s)} = e^{i\phi_s} \rho_\sigma^{(s)} \qquad \sigma = 1, \ldots, n \tag{22.29}$$

where ϕ_s is real and one real component ($\rho_n^{(s)}$, say) can be specified arbitrarily. Substitution of Eq. (22.29) into (22.26) yields the equation for the sth eigenvalue and eigenvector

$$\sum_\lambda v_{\sigma\lambda} \rho_\lambda^{(s)} = \omega_s^2 \sum_\lambda m_{\sigma\lambda} \rho_\lambda^{(s)} \qquad \sigma = 1, \ldots, n \tag{22.30}$$

where the common phase $e^{i\phi_s}$ has been canceled from both sides. Equations (22.30) now contain only real quantities. A similar equation can be written for the tth eigenvalue and eigenvector

$$\sum_\lambda v_{\lambda\sigma} \rho_\lambda^{(t)} = \omega_t^2 \sum_\lambda m_{\lambda\sigma} \rho_\lambda^{(t)} \qquad \sigma = 1, \ldots, n \tag{22.31}$$

where the symmetry conditions of Eqs. (21.11) and (21.16) have been employed. We now perform the operation $\sum_\sigma \rho_\sigma^{(t)}$ on Eq. (22.30) (this means to multiply by $\rho_\sigma^{(t)}$ and then sum over all σ) and a similar operation $\sum_\sigma \rho_\sigma^{(s)}$ on Eq. (22.31) and then take the difference of the two resulting expressions. Since $v_{\sigma\lambda}$ is a symmetric matrix, the left-hand side of these expressions cancels identically; since $m_{\sigma\lambda}$ is a symmetric matrix, an interchange of dummy summation variables on the right-hand side yields

$$(\omega_s^2 - \omega_t^2) \sum_\lambda \sum_\sigma \rho_\sigma^{(t)} m_{\sigma\lambda} \rho_\lambda^{(s)} = 0 \tag{22.32}$$

To appreciate this important relation, suppose first that all the eigenvalues are distinct, implying that $\omega_s^2 \neq \omega_t^2$ for $s \neq t$. Equation (22.32) then shows that the eigenvectors $\rho_\sigma^{(s)}$ and $\rho_\sigma^{(t)}$ in Eqs. (22.29) to (22.31) are *orthogonal* according to

$$\sum_\lambda \sum_\sigma \rho_\sigma^{(t)} m_{\sigma\lambda} \rho_\lambda^{(s)} = 0 \qquad s \neq t \tag{22.33}$$

In addition, recall that the real homogeneous equations (22.30) and (22.31) can only determine $n - 1$ real ratios $\rho_\sigma^{(s)}/\rho_n^{(s)}$ with $\sigma = 1, \ldots, n - 1$. Multiplying all the components in Eq. (22.29) by a common factor leaves the solution unchanged. We use this freedom to *normalize* the eigenvectors, requiring that Eq. (22.33) have the value unity when $t = s$. Thus we can choose the solutions to Eqs. (22.30) to be *orthonormal*, according to the relation

$$\sum_\lambda \sum_\sigma \rho_\sigma^{(t)} m_{\sigma\lambda} \rho_\lambda^{(s)} = \delta_{st} \tag{22.34}$$

The general solution in Eq. (22.29) corresponding to the eigenvalue ω_s^2 then takes the form

$$z_\sigma^{(s)} = C^{(s)} e^{i\phi_s} \rho_\sigma^{(s)} \qquad \sigma = 1, \ldots, n \tag{22.35}$$

the *real* constants $C^{(s)}$ and ϕ_s being the only parameters that can still be specified arbitrarily. Here and henceforth the real quantities $\rho_\sigma^{(s)}$ are assumed to be completely determined by Eq. (22.34).

The situation is formally more complicated if the eigenvalue equation (22.24) has a multiple root. Even in this *degenerate* case, however, one can always orthogonalize the solutions according to Eq. (22.33) by using the Gram-Schmidt orthogonalization procedure (see Prob. 4.10). Thus Eq. (22.34) can always be assumed valid.

Coupled Problem: General Solution

We can now construct the general solution to the original problem in Eqs. (22.14) by superposing the solutions obtained in Eqs. (22.16), (22.25), and (22.35) to give [compare Eq. (22.10)]

$$z_\sigma(t) = \sum_{s=1}^n [(z_+^{(s)})_\sigma e^{i\omega_s t} + (z_-^{(s)})_\sigma e^{-i\omega_s t}] \qquad \sigma = 1, \ldots, n \tag{22.36}$$

where we have defined

$$(z_+^{(s)})_\sigma \equiv C_+^{(s)} e^{i\phi_s^+} \rho_\sigma^{(s)} \tag{22.37}$$

and

$$(z_-^{(s)})_\sigma \equiv C_-^{(s)} e^{i\phi_s^-} \rho_\sigma^{(s)} \tag{22.38}$$

Each term in Eq. (22.36) is a solution to the set of equations (22.14); since those equations are linear, any sum of solutions, as in Eq. (22.36), is again a solution. In particular, the sum in Eq. (22.36) includes *all the normal modes* (labeled by s). In contrast, the subscript σ tells which generalized coordinate is being considered. The solution in Eq. (22.36) contains $4n$ independent real constants $\{C_\pm^{(s)},\ \phi_s^\pm;\ s = 1, \ldots, n\}$, which provides enough freedom to match an arbitrary set of complex initial conditions $\{z_\sigma(t_0), \dot{z}_\sigma(t_0); \sigma = 1, \ldots, n\}$. As a result, we see that Eq. (22.36) constitutes *the most general solution* to Eq. (22.14).

The solution to the original mechanical problem Eq. (21.19) then follows by taking the real part of Eq. (22.14) and using the definition (22.15). This procedure is valid because Eq. (22.14) has only real coefficients. With the redefinition [compare Eq. (22.12)]

$$(z_+^{(s)})_\sigma + (z_-^{(s)})_\sigma^* \equiv z_\sigma^{(s)} \equiv C^{(s)}\rho_\sigma^{(s)}e^{i\phi_s} \qquad \sigma = 1, \ldots, n \tag{22.39}$$

we arrive at the final result

$$\eta_\sigma(t) = \sum_{s=1}^n \mathrm{Re}\,(C^{(s)}\rho_\sigma^{(s)}e^{i(\omega_s t + \phi_s)})$$

$$= \sum_{s=1}^n \rho_\sigma^{(s)}C^{(s)} \cos\,(\omega_s t + \phi_s) \tag{22.40}$$

This important equation requires a regrettably large number of superscripts and subscripts, and it is helpful to summarize its essential content. Equation (22.40) is the most general solution to Lagrange's equations (21.19) for small oscillations of a mechanical system about equilibrium. Equation (21.8) shows that η_σ is the small displacement of the σth generalized coordinate, and the sum in Eq. (22.40) includes *all normal modes* (labeled by s). The frequencies ω_s of the normal modes are the solutions to the eigenvalue equation (22.24). The corresponding normal-mode amplitudes $\rho_\sigma^{(s)}$ with $\sigma = 1, \ldots, n$ are given by the $n-1$ real ratios obtained by solving the set of linear equations (22.30), scaled with an overall real normalization constant chosen to satisfy the normalization condition in Eq. (22.34). The final solution (22.40) has $2n$ real constants $\{C^{(s)}, \phi_s; s = 1, \ldots, n\}$, which is just sufficient to match the set of initial conditions $\{\eta_\sigma(t_0), \dot{\eta}_\sigma(t_0); \sigma = 1, \ldots, n\}$ that must be specified to determine the solution to the set of second-order differential equations (21.19). Thus we have indeed constructed the general small-amplitude solution to Lagrange's equations.

Matrix Notation

Matrix notation permits a concise summary of the preceding results. Define the square matrix $\underline{a}$ as

$$\underline{a} = \begin{bmatrix} a_{11} & a_{12} & \cdots & a_{1n} \\ \cdots\cdots\cdots\cdots\cdots \\ a_{n1} & a_{n2} & \cdots & a_{nn} \end{bmatrix} \tag{22.41}$$

The n-dimensional column vector $\underline{x}$ and its transpose $\underline{x}^T$ are defined according to

$$\underline{x} = \begin{bmatrix} x_1 \\ x_2 \\ \vdots \\ x_n \end{bmatrix} \qquad \underline{x}^T = [x_1 \quad x_2 \quad \cdots \quad x_n] \qquad (22.42)$$

More generally, the transpose of a matrix is defined as the matrix obtained by interchanging rows and columns. We note the following two properties of matrix multiplication. First, the product of the square matrix $\underline{a}$ and the column vector $\underline{x}$ is again an n-dimensional column vector defined according to

$$\underline{a}\underline{x} = \begin{bmatrix} (\underline{a}\underline{x})_1 \\ (\underline{a}\underline{x})_2 \\ \vdots \\ (\underline{a}\underline{x})_n \end{bmatrix} = \begin{bmatrix} \sum_j a_{1j} x_j \\ \sum_j a_{2j} x_j \\ \vdots \\ \sum_j a_{nj} x_j \end{bmatrix} \qquad (22.43)$$

Second, the product of the transpose of the column vector with the quantity in Eq. (22.43) is a scalar

$$\underline{x}^T \underline{a}\underline{x} \equiv \sum_i \sum_j x_i a_{ij} x_j \qquad (22.44)$$

The equations in the preceding sections can now be recast in the following manner. The eigenvalue equation (22.24) becomes

$$\det |\underline{v} - \omega^2 \underline{m}| = 0 \qquad (22.45)$$

and the eigenvector equation (22.30) for the sth eigenvalue takes the form

$$(\underline{v} - \omega_s^2 \underline{m})\underline{\rho}^{(s)} = 0 \qquad (22.46)$$

where $\underline{v}$ and $\underline{m}$ are the real symmetric matrices defined in Eqs. (21.10) and (21.15). Note that Eq. (22.46) is just a compact way of representing a set of n linear homogeneous algebraic equations for the column vectors $\underline{\rho}^{(s)}$. The orthonormality condition Eq. (22.34) can be written

$$\underline{\rho}^{(s)T} \underline{m} \underline{\rho}^{(t)} = \delta_{st} \qquad (22.47)$$

where the superscripts s and t again denote the eigenvectors corresponding to particular eigenvalues. As a result, the general solution (22.40) to the original problem of coupled oscillations [Eq. (21.19)] is given by

$$\underline{\eta}(t) = \sum_s C^{(s)} \underline{\rho}^{(s)} \cos\left(\omega_s t + \phi_s\right) \qquad (22.48)$$

where the sum is over the *normal modes* of the system. The σth component of Eq. (22.48) describes the displacement of the σth generalized coordinate from its equilibrium value in terms of the corresponding component of the eigenvectors, as well as the $2n$ real constants $C^{(s)}$ and ϕ_s.

These remaining constants can be chosen to fit some specified set of initial conditions $\underline{\eta}(0)$ and $\underline{\dot{\eta}}(0)$ at $t = 0$. Equation (22.48) and its time derivative yield

$$\underline{\eta}(0) = \sum_s C^{(s)} \underline{\rho}^{(s)} \cos \phi_s$$
$$\underline{\dot{\eta}}(0) = -\sum_s C^{(s)} \underline{\rho}^{(s)} \omega_s \sin \phi_s$$
(22.49)

To determine the constants $C^{(s)}$ and ϕ_s, we multiply from the left with $\underline{\rho}^{(t)T} \underline{m}$. Use of the orthonormality (22.47) on the right-hand side leads to the desired results

$$\underline{\rho}^{(t)T} \underline{m} \underline{\eta}(0) = \sum_s C^{(s)} \underline{\rho}^{(t)T} \underline{m} \underline{\rho}^{(s)} \cos \phi_s = C^{(t)} \cos \phi_t$$
$$\underline{\rho}^{(t)T} \underline{m} \underline{\dot{\eta}}(0) = -\sum_s C^{(s)} \underline{\rho}^{(t)T} \underline{m} \underline{\rho}^{(s)} \omega_s \sin \phi_s = -C^{(t)} \omega_t \sin \phi_t$$
(22.50)

and it is now straightforward to separate the quantities C and ϕ. In this way, we have constructed an explicit solution to the original dynamical equations (21.19) that satisfies prescribed initial conditions.

Modal Matrix

The key to the discussion of small oscillations is the *modal matrix* $\mathscr{A}$, defined as the square matrix whose columns are the eigenvectors obtained from the ortho-normal solutions to Eq. (22.46) that obey Eq. (22.47). This very important quantity thus has the form

$$\mathscr{A} \equiv \begin{bmatrix} \rho_1^{(1)} & \rho_1^{(2)} & \cdots & \rho_1^{(n)} \\ \cdots\cdots\cdots\cdots\cdots \\ \rho_n^{(1)} & \rho_n^{(2)} & \cdots & \rho_n^{(n)} \end{bmatrix} \equiv \begin{bmatrix} \rho^{(1)} & \rho^{(2)} & \cdots & \rho^{(n)} \\ \downarrow & \downarrow & \cdots & \downarrow \end{bmatrix}$$
(22.51)

or, equivalently, in detailed matrix notation

$$\mathscr{A}_{\lambda\sigma} \equiv \rho_\lambda^{(\sigma)}$$
(22.52)

where the first, or row, index indicates the component of the eigenvector and the second, or column, index indicates the particular normal-mode solution. Note that the elements of the modal matrix in Eq. (22.51) are a set of real constants that characterize the eigenvectors, independent of the particular initial conditions.

Consider the matrix product $\mathscr{A}^T \underline{m} \mathscr{A}$, which is still an $n \times n$ matrix. By the general definitions of the transpose and of the matrix product, its $\mu\nu$th component is

$$(\mathscr{A}^T \underline{m} \mathscr{A})_{\mu\nu} \equiv \sum_\lambda \sum_\sigma (\mathscr{A}^T)_{\mu\lambda} m_{\lambda\sigma} \mathscr{A}_{\sigma\nu} = \sum_\lambda \sum_\sigma \mathscr{A}_{\lambda\mu} m_{\lambda\sigma} \mathscr{A}_{\sigma\nu}$$
(22.53)

Use of Eq. (22.52) expressing the modal matrix in terms of the eigenvectors reduces Eq. (22.53) to the simple form

$$(\mathscr{A}^T \underline{m} \mathscr{A})_{\mu\nu} = \sum_\lambda \sum_\sigma \rho_\lambda^{(\mu)} m_{\lambda\sigma} \rho_\sigma^{(\nu)} = \delta_{\mu\nu}$$
(22.54)

where the orthonormality relation (22.47) leads to the last form. Since the quantity on the right-hand side is just the $\mu\nu$th element of the unit matrix, Eqs. (22.53) and (22.54) constitute the compact matrix relation

$$\underline{\mathscr{A}}^T \underline{m}\underline{\mathscr{A}} = \underline{1} \tag{22.55}$$

Thus the modal matrix has the important feature that it *diagonalizes the mass matrix* $\underline{m}$.

We can similarly investigate the matrix product $\underline{\mathscr{A}}^T \underline{v}\underline{\mathscr{A}}$. The set of linear equations (22.30) [or, equivalently, (22.46)] satisfied by the eigenvectors allows us to rewrite the $\mu\nu$th component of this matrix relation as

$$(\underline{\mathscr{A}}^T \underline{v}\underline{\mathscr{A}})_{\mu\nu} = \sum_{\lambda} \rho_{\lambda}^{(\mu)} \left[\sum_{\sigma} v_{\lambda\sigma} \rho_{\sigma}^{(\nu)} \right] = \sum_{\lambda} \rho_{\lambda}^{(\mu)} \left[\omega_{\nu}^2 \sum_{\sigma} m_{\lambda\sigma} \rho_{\sigma}^{(\nu)} \right] \tag{22.56}$$

Since the subscript on the eigenvalue ω_{ν}^2 does not appear in the matrix summation, this quantity can be removed from the sum to give

$$(\underline{\mathscr{A}}^T \underline{v}\underline{\mathscr{A}})_{\mu\nu} = \omega_{\nu}^2 \sum_{\lambda} \sum_{\sigma} \rho_{\lambda}^{(\mu)} m_{\lambda\sigma} \rho_{\sigma}^{(\nu)} = \omega_{\nu}^2 \delta_{\mu\nu} \tag{22.57}$$

where the orthonormality condition (22.47) has again been used. The quantity on the right-hand side of Eq. (22.57) is again the $\mu\nu$th element of a diagonal matrix, whose components are the eigenvalues

$$\omega_D^2 \equiv \begin{bmatrix} \omega_1^2 & & & \\ & \omega_2^2 & & \\ & & \ddots & \\ & & & \omega_n^2 \end{bmatrix} \tag{22.58}$$

Equations (22.56) and (22.57) together make up the matrix relation

$$\underline{\mathscr{A}}^T \underline{v}\underline{\mathscr{A}} = \omega_D^2 \tag{22.59}$$

We thus obtain the remarkable result that the modal matrix *simultaneously diagonalizes* the mass matrix $\underline{m}$ and the potential matrix $\underline{v}$. In fact, the procedure summarized by Eqs. (22.45) to (22.47) is the general method for simultaneously diagonalizing two real symmetric matrices, one of which must be positive definite.

Normal Coordinates

Define a new set of generalized coordinates $\zeta_1, \ldots, \zeta_n$, linearly related to the original generalized coordinates $\eta_1, \ldots, \eta_n$ [recall Eq. (21.8)] by the modal matrix

$$\underline{\eta}(t) \equiv \underline{\mathscr{A}}\underline{\zeta}(t) \tag{22.60}$$

This relation can be inverted by multiplying from the left with $\underline{\mathscr{A}}^T \underline{m}$ and using Eq. (22.55)

$$\underline{\mathscr{A}}^T \underline{m}\underline{\eta}(t) = \underline{\zeta}(t) \tag{22.61}$$

We emphasize again that the matrices $\underline{\mathscr{A}}$ and $\underline{\mathscr{A}}^T \underline{m}$ have constant real elements, so that Eqs. (22.60) and (22.61) simply define a new linear combination of the original generalized coordinates.

To demonstrate the usefulness of these new coordinates, we shall rewrite the lagrangian (21.17) for small-amplitude motion in terms of the new variables. First, note that the lagrangian can be written in matrix notation [recall Eq. (22.44)] as

$$2L = \dot{\eta}^T \underline{m}\dot{\eta} - \eta^T \underline{v}\eta \tag{22.62}$$

where the additive constant in Eq. (21.17) has been suppressed. Insertion of the definition (22.60) into this relation then yields [recall $(\underline{a}\underline{b})^T \equiv \underline{b}^T \underline{a}^T$]

$$2L = \dot{\zeta}^T \mathscr{A}^T \underline{m}\mathscr{A}\dot{\zeta} - \zeta^T \mathscr{A}^T \underline{v}\mathscr{A}\zeta \tag{22.63}$$

Since the modal matrix $\mathscr{A}$ diagonalizes both $\underline{m}$ and $\underline{v}$ [Eqs. (22.55) and (22.59)], the lagrangian has the final simple form, expressed either as a matrix product

$$L = \tfrac{1}{2}(\dot{\zeta}^T \dot{\zeta} - \zeta^T \omega_D^2 \zeta) \tag{22.64}$$

or in terms of the matrix components

$$L = \frac{1}{2} \sum_{\sigma=1}^{n} (\dot{\zeta}_\sigma^2 - \omega_\sigma^2 \zeta_\sigma^2) \tag{22.65}$$

where ω_σ^2 with $\sigma = 1, \ldots, n$ are the normal-mode frequencies of the complete mechanical system [see Eqs. (22.25) and (22.58)]. Thus we find the second remarkable property of the modal matrix: the new coordinates ζ related to the original generalized coordinates η through the modal matrix in Eqs. (22.60) and (22.61) *diagonalize the lagrangian*. Furthermore, specification of these new coordinates ζ_σ certainly specifies the η_σ [Eqs. (22.60)], which in turn determine the original set of generalized coordinates through Eqs. (21.8) and thus, by assumption, the configuration of the entire system. As a result, we are now free to take the ζ_σ with $\sigma = 1, \ldots, n$ as a *new set of generalized coordinates*, known as *normal coordinates*.

Application of Hamilton's principle to the lagrangian (22.65), with ζ_σ as generalized coordinates, immediately yields the corresponding Lagrange's equations for the ζ_σ [compare Eq. (22.3)]

$$\ddot{\zeta}_\sigma = -\omega_\sigma^2 \zeta_\sigma \qquad \sigma = 1, \ldots, n \tag{22.66}$$

Diagonalizing the lagrangian therefore *decouples* the equations of motion, reducing the problem to a set of n independent uncoupled simple harmonic oscillators. *Each normal coordinate ζ_σ oscillates independently, with an angular frequency given by the normal-mode frequency ω_σ*. Note that Eqs. (22.65) and (22.66) are *very general*. They hold for any mechanical system undergoing small-amplitude oscillations about static equilibrium, since the introduction of normal coordinates always puts a system into normal form.

The general solution to the uncoupled simple-harmonic-oscillator equation (22.66) was given in Eq. (22.13)

$$\zeta_\sigma = C^{(\sigma)} \cos (\omega_\sigma t + \phi_\sigma) \qquad \sigma = 1, \ldots, n \tag{22.67}$$

This set of n relations can be summarized in matrix notation by writing

$$\zeta = \begin{bmatrix} C^{(1)} \cos (\omega_1 t + \phi_1) \\ \vdots \\ C^{(n)} \cos (\omega_n t + \phi_n) \end{bmatrix} \tag{22.68}$$

The definition of the modal matrix in Eq. (22.52) can then be combined with the defining relation (22.60) to express the components of the eigenvector $\underline{\eta} = \mathscr{A}\underline{\zeta}$ in the form [compare Eq. (22.40)]

$$\eta_\lambda(t) = \sum_{s=1}^{n} \rho_\lambda^{(s)} C^{(s)} \cos (\omega_s t + \phi_s) \tag{22.69}$$

Observe that the normal coordinates from Eq. (22.67) are simply the *coefficients or amplitudes of the eigenvectors* $\rho^{(s)}$ in the expansion of η. Each normal coordinate in Eqs. (22.67) undergoes uncoupled simple harmonic motion, at the corresponding normal-mode frequency of the coupled system.

23 EXAMPLE: COUPLED PENDULUMS

As an example of the analysis into normal modes, consider two identical pendulums moving in a common plane and coupled by a spring of length d (Fig. 23.1). At equilibrium, assume that the pendulums hang vertically with the spring at its natural length d_0. Let the system undergo small oscillations about this equilibrium configuration. The transverse displacements of the two pendulum bobs η_1 and η_2 serve as a suitable set of generalized coordinates.

For small displacements, inspection of Fig. 23.1 shows that

$$\frac{\eta}{l} = \sin \theta \approx \theta \tag{23.1}$$

The gravitational potential energy of the system involves the height through which each bob is raised. For small displacements, this quantity is evidently given by

$$\text{Height raised} = l(1 - \cos \theta) \approx \tfrac{1}{2}l\theta^2 = \frac{\eta^2}{2l} \tag{23.2}$$

In addition, the spring has a potential energy arising from its compression or extension against the spring constant k. For small displacements, the change in the length of the spring given by

$$d - d_0 = \eta_2 - \eta_1 + O(\eta^4) \tag{23.3}$$

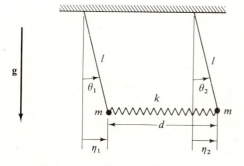

Figure 23.1 Two identical pendulums coupled with a spring and moving in a plane as example of coupled system undergoing small oscillations about equilibrium.

where the reader can easily verify that the neglected term is indeed of order η^4 by writing out the exact length of the spring in cartesian coordinates. The potential energy in the spring is $\frac{1}{2}k(d - d_0)^2$, and the total potential energy now follows directly from Eqs. (23.2) and (23.3)

$$V_{\text{spring}} + V_{\text{gravity}} \approx \frac{k}{2}(\eta_1 - \eta_2)^2 + \frac{mg}{2l}(\eta_1^2 + \eta_2^2) \tag{23.4}$$

The kinetic energy of the bobs is

$$T = \tfrac{1}{2}m[(l\dot{\theta}_1)^2 + (l\dot{\theta}_2)^2] \approx \tfrac{1}{2}m(\dot{\eta}_1^2 + \dot{\eta}_2^2) \tag{23.5}$$

where the condition (23.1) of small displacements has again been used. Finally, we assume that the spring is massless. Thus the lagrangian for the system of coupled pendulums undergoing small oscillations takes the form

$$L = \tfrac{1}{2}m(\dot{\eta}_1^2 + \dot{\eta}_2^2) - \frac{mg}{2l}(\eta_1^2 + \eta_2^2) - \tfrac{1}{2}k(\eta_1^2 + \eta_2^2 - 2\eta_1\eta_2) \tag{23.6}$$

and Lagrange's equations for the variables η_1 and η_2 read

$$m\ddot{\eta}_1 + \left(k + \frac{mg}{l}\right)\eta_1 - k\eta_2 = 0 \tag{23.7}$$

and

$$m\ddot{\eta}_2 + \left(k + \frac{mg}{l}\right)\eta_2 - k\eta_1 = 0 \tag{23.8}$$

The last term in each equation reflects the presence of the spring coupling the motion of the two pendulums.

We shall solve these equations by looking for *normal modes*. It is here sufficient to work directly with the real form of the solution and seek a solution of the type

$$\eta_\sigma = C\rho_\sigma \cos(\omega t + \phi) \qquad \sigma = 1, 2 \tag{23.9}$$

In a given normal mode, we emphasize again that both coordinates *oscillate with the same frequency* ω. Substitution of (23.9) into the equations of motion (23.7) and (23.8) and cancellation of common factors yield the resulting eigenvalue equations

$$\left(\frac{mg}{l} + k - m\omega^2\right)\rho_1 - k\rho_2 = 0 \tag{23.10}$$

$$-k\rho_1 + \left(\frac{mg}{l} + k - m\omega^2\right)\rho_2 = 0 \tag{23.11}$$

These relations can be expressed compactly in matrix form

$$(\underline{v} - m\omega^2)\underline{\rho} = 0 \tag{23.12}$$

where $\underline{\rho}$ is a two-component eigenvector

$$\underline{\rho} = \begin{bmatrix} \rho_1 \\ \rho_2 \end{bmatrix} \tag{23.13}$$

Inspection of Eqs. (23.10) and (23.11) then identifies the mass matrix $\underline{m}$ and potential matrix $\underline{v}$

$$\underline{m} = m \begin{bmatrix} 1 & 0 \\ 0 & 1 \end{bmatrix} \qquad \underline{v} = \begin{bmatrix} \dfrac{mg}{l} + k & -k \\[2ex] -k & \dfrac{mg}{l} + k \end{bmatrix} \qquad (23.14)$$

Equations (23.10) and (23.11) are linear homogeneous equations; they possess nontrivial solutions for ρ only if the determinant of the coefficients vanishes

$$\begin{vmatrix} \dfrac{mg}{l} + k - m\omega^2 & -k \\[2ex] -k & \dfrac{mg}{l} + k - m\omega^2 \end{vmatrix} = 0 \qquad (23.15)$$

Evaluation of this determinant gives a quadratic equation in ω^2

$$m^2\omega^4 - 2m\omega^2\left(k + \frac{mg}{l}\right) + \left(k + \frac{mg}{l}\right)^2 - k^2 = 0 \qquad (23.16)$$

It is particularly simple since it factorizes

$$\left(m\omega^2 - \frac{mg}{l}\right)\left[m\omega^2 - \left(\frac{mg}{l} + 2k\right)\right] = 0 \qquad (23.17)$$

and the eigenvalues, or *normal-mode frequencies*, are given by

$$\omega_1 = \left(\frac{g}{l}\right)^{1/2} \qquad (23.18)$$

$$\omega_2 = \left(\frac{g}{l} + 2\frac{k}{m}\right)^{1/2} \qquad (23.19)$$

The first frequency is that of a single free pendulum [compare Eq. (16.5)], and the second frequency is definitely higher for any positive k.

We proceed to determine the normal-mode *eigenvectors* corresponding to the normal-mode frequencies (23.18) and (23.19). It is easy to verify that Eqs. (23.10) and (23.11) become identical when ω^2 is equal to either of the eigenvalues (23.18) and (23.19). Since the equations are then linearly dependent, only one of them need be retained. Substitution of the first eigenvalue (23.18) into Eq. (23.10) gives

$$k\rho_1^{(1)} - k\rho_2^{(1)} = 0 \qquad (23.20)$$

where the superscript (1) indicates the eigenvector corresponding to the first eigenvalue. This equation implies

$$\rho_1^{(1)} = +\rho_2^{(1)} \qquad (23.21)$$

so that the displacements of the two bobs are equal in mode 1. Figure 23.2 illustrates this motion, which oscillates with the frequency of the free pendulum

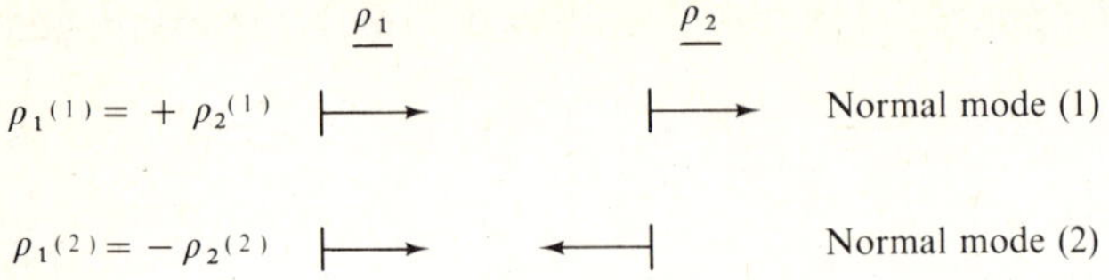

$$\rho_1 \qquad\qquad\qquad \rho_2$$

$\rho_1^{(1)} = + \rho_2^{(1)}$ ⊢───▶   ⊢───▶   Normal mode (1)

$\rho_1^{(2)} = -\rho_2^{(2)}$ ⊢───▶   ◀───┤   Normal mode (2)

Figure 23.2 Displacements of the bobs in the normal modes of the coupled pendulums.

because the spring has fixed length and plays no role whatsoever. Similarly, substitution of the second eigenvalue (23.19) into Eq. (23.10) gives

$$-k\rho_1^{(2)} - k\rho_2^{(2)} = 0 \tag{23.22}$$

or, equivalently,

$$\rho_1^{(2)} = -\rho_2^{(2)} \tag{23.23}$$

In mode 2, the displacements of the bobs are opposite but equal in magnitude (see Fig. 23.2), with the spring alternately compressed and extended, raising the normal-mode frequency above that of the free pendulum, as seen in Eq. (23.19).

Since Eqs. (23.10) and (23.11) are homogeneous, multiplication of the components of ρ by an arbitrary overall factor leaves these relations unchanged. We choose to scale the components in (23.21) and (23.23) to satisfy the normalization condition (22.47). It is readily verified that the necessary constant is $(2m)^{-1/2}$, and the normalized eigenvectors become

$$\underline{\rho}^{(1)} = \frac{1}{(2m)^{1/2}} \begin{bmatrix} 1 \\ 1 \end{bmatrix} \qquad \underline{\rho}^{(2)} = \frac{1}{(2m)^{1/2}} \begin{bmatrix} 1 \\ -1 \end{bmatrix} \tag{23.24}$$

It is instructive to verify explicitly that these eigenvectors satisfy the correct normalization conditions

$$\underline{\rho}^{(1)T} \underline{m} \underline{\rho}^{(1)} = \frac{m}{2m} \begin{bmatrix} 1 & 1 \end{bmatrix} \begin{bmatrix} 1 & 0 \\ 0 & 1 \end{bmatrix} \begin{bmatrix} 1 \\ 1 \end{bmatrix} = \tfrac{1}{2}(2) = 1 = \underline{\rho}^{(2)T} \underline{m} \underline{\rho}^{(2)} \tag{23.25}$$

and orthogonality relations

$$\underline{\rho}^{(1)T} \underline{m} \underline{\rho}^{(2)} = \tfrac{1}{2} \begin{bmatrix} 1 & 1 \end{bmatrix} \begin{bmatrix} 1 & 0 \\ 0 & 1 \end{bmatrix} \begin{bmatrix} 1 \\ -1 \end{bmatrix} = 0 = \underline{\rho}^{(2)T} \underline{m} \underline{\rho}^{(1)} \tag{23.26}$$

that were combined into the single equation (22.47). Recall that these relations involve the mass matrix $\underline{m}$, which in the present simple example happens to be proportional to the unit matrix [Eq. (23.14)].

We next construct the *modal matrix*, whose columns are made from the eigenvectors in Eq. (23.24)

$$\mathscr{A} \equiv \frac{1}{(2m)^{1/2}} \begin{bmatrix} 1 & 1 \\ 1 & -1 \end{bmatrix} \tag{23.27}$$

The reader is now urged to verify explicitly the following general relations with Eqs. (23.14) and (23.27):

$$\mathscr{A}^T m \mathscr{A} = 1 \tag{23.28}$$

$$\mathscr{A}^T v \mathscr{A} = \begin{bmatrix} \omega_1^2 & 0 \\ 0 & \omega_2^2 \end{bmatrix} \equiv \omega_D^2 \tag{23.29}$$

where ω_1^2 and ω_2^2 are the eigenvalues (23.18) and (23.19).

The *normal coordinates* ζ are defined by the implicit relation

$$\eta \equiv \mathscr{A}\zeta \tag{23.30}$$

which can be inverted with the help of Eq. (23.28) to give

$$\zeta = \mathscr{A}^T m \eta = \frac{m}{(2m)^{1/2}} \begin{bmatrix} 1 & 1 \\ 1 & -1 \end{bmatrix} \begin{bmatrix} 1 & 0 \\ 0 & 1 \end{bmatrix} \begin{bmatrix} \eta_1 \\ \eta_2 \end{bmatrix} = \left(\frac{m}{2}\right)^{1/2} \begin{bmatrix} \eta_1 + \eta_2 \\ \eta_1 - \eta_2 \end{bmatrix} \tag{23.31}$$

This matrix equation has the explicit components

$$\zeta_1 = (\tfrac{1}{2}m)^{1/2}(\eta_1 + \eta_2) \tag{23.32}$$

and
$$\zeta_2 = (\tfrac{1}{2}m)^{1/2}(\eta_1 - \eta_2) \tag{23.33}$$

and the normal coordinates here are simply the (normalized) sum and difference of the displacements of the bobs. The reader can readily verify that substitution of the normal coordinates (23.32) and (23.33) reduces the lagrangian (23.6) to the expected diagonal form [compare Eq. (22.65)]

$$L = \frac{1}{2} \sum_{\sigma=1}^{2} (\dot{\zeta}_\sigma^2 - \omega_\sigma^2 \zeta_\sigma^2) \tag{23.34}$$

with ω_σ^2 the normal-mode frequencies (23.18) and (23.19). We can now take the normal coordinates ζ_σ with $\sigma = 1, 2$ as new generalized coordinates, and Lagrange's equations then describe uncoupled simple harmonic motion

$$\ddot{\zeta}_\sigma = -\omega_\sigma^2 \zeta_\sigma \qquad \sigma = 1, 2 \tag{23.35}$$

They have the general solution

$$\zeta = \begin{bmatrix} \zeta_1 \\ \zeta_2 \end{bmatrix} = \begin{bmatrix} C^{(1)} \cos(\omega_1 t + \phi_1) \\ C^{(2)} \cos(\omega_2 t + \phi_2) \end{bmatrix} \tag{23.36}$$

where $\{C^{(\sigma)}, \phi_\sigma; \sigma = 1, 2\}$ are four constants that must be fixed by the initial conditions.† To determine these constants, it is most convenient to combine Eq. (23.31) and its first time derivative, both evaluated at $t = 0$ with Eq. (23.36). This procedure yields the following four relations [compare Eqs. (22.50)]:

$$\zeta(0) = \begin{bmatrix} C^{(1)} \cos \phi_1 \\ C^{(2)} \cos \phi_2 \end{bmatrix} = (\tfrac{1}{2}m)^{1/2} \begin{bmatrix} \eta_1(0) + \eta_2(0) \\ \eta_1(0) - \eta_2(0) \end{bmatrix} \tag{23.37}$$

$$\dot{\zeta}(0) = \begin{bmatrix} -\omega_1 C^{(1)} \sin \phi_1 \\ -\omega_2 C^{(2)} \sin \phi_2 \end{bmatrix} = (\tfrac{1}{2}m)^{1/2} \begin{bmatrix} \dot{\eta}_1(0) + \dot{\eta}_2(0) \\ \dot{\eta}_1(0) - \dot{\eta}_2(0) \end{bmatrix} \tag{23.38}$$

† We generally choose to specify the initial time as $t = 0$.

between the four constants in the normal coordinates $\{C^{(\sigma)}, \phi_\sigma; \sigma = 1, 2\}$ and the four *initial conditions* for the displacement and velocity of each bob. Thus the *initial conditions* of the mechanical problem completely determine the *normal coordinates.*

The normal coordinates, in turn, yield the general solution to the mechanical problem through the modal matrix [compare Eq. (23.30)]

$$\underline{\eta}(t) = \mathscr{A}\underline{\zeta}(t) \tag{23.39}$$

Explicit evaluation of the matrix multiplication on the right-hand side of Eq. (23.39) follows from the normal coordinates (23.36) and the modal matrix [Eqs. (23.27) and (23.24)]; it yields the components

$$\eta_\sigma(t) = \sum_{s=1}^{2} C^{(s)}\rho_\sigma^{(s)} \cos\left(\omega_s t + \phi_s\right) \tag{23.40}$$

As in Eq. (22.69), we note that the normal coordinates, which are completely determined by the initial conditions [see Eqs. (23.37) and (23.38)], are just the amplitudes of the normal-mode eigenvectors $\rho_\sigma^{(s)}$ in the general solution for the original generalized coordinates $\eta_\sigma(t)$.

As a specific example, consider a particular set of initial conditions with the left bob displaced a distance α to the right and released from rest. The generalized coordinates have the initial values

$$\eta_1(0) = \alpha \tag{23.41}$$

$$\eta_2(0) = \dot{\eta}_1(0) = \dot{\eta}_2(0) = 0 \tag{23.42}$$

and substitution into Eqs. (23.37) and (23.38) immediately yields the relations

$$\phi_1 = \phi_2 = 0 \tag{23.43}$$

$$C^{(1)} = C^{(2)} = \alpha(\tfrac{1}{2}m)^{1/2} \tag{23.44}$$

Thus the normal coordinates (23.36) have the particular form

$$\underline{\zeta} = \alpha(\tfrac{1}{2}m)^{1/2} \begin{bmatrix} \cos\omega_1 t \\ \cos\omega_2 t \end{bmatrix} \tag{23.45}$$

and the corresponding time-dependent solution to the mechanical problem is given by Eq. (23.39)

$$\underline{\eta} = \mathscr{A}\underline{\zeta} = \frac{\alpha}{2} \begin{bmatrix} 1 & 1 \\ 1 & -1 \end{bmatrix} \begin{bmatrix} \cos\omega_1 t \\ \cos\omega_2 t \end{bmatrix}$$

$$= \frac{\alpha}{2} \begin{bmatrix} \cos\omega_1 t + \cos\omega_2 t \\ \cos\omega_1 t - \cos\omega_2 t \end{bmatrix} = \begin{bmatrix} \eta_1 \\ \eta_2 \end{bmatrix} \tag{23.46}$$

Each individual displacement in (23.46) is a linear combination of two amplitudes that oscillate at the normal-mode frequencies. Since the particular motions (23.46) satisfy the initial conditions

$$\underline{\eta}(0) = \begin{bmatrix} \alpha \\ 0 \end{bmatrix} \quad \text{and} \quad \underline{\dot{\eta}}(0) = \begin{bmatrix} 0 \\ 0 \end{bmatrix} \tag{23.47}$$

it is clear that we have constructed the appropriate solution to Lagrange's equations (23.7) and (23.8). Use of the standard trigonometric identities

$$\cos a \pm \cos b = \cos\left[\tfrac{1}{2}(a+b) + \tfrac{1}{2}(a-b)\right] \pm \cos\left[\tfrac{1}{2}(a+b) - \tfrac{1}{2}(a-b)\right]$$

$$= \begin{cases} 2\cos\tfrac{1}{2}(a+b)\cos\tfrac{1}{2}(a-b) \\ -2\sin\tfrac{1}{2}(a+b)\sin\tfrac{1}{2}(a-b) \end{cases} \tag{23.48}$$

yields the final solution

$$\underline{\eta} = \begin{bmatrix} \eta_1(t) \\ \eta_2(t) \end{bmatrix} = \begin{bmatrix} [\alpha \cos\tfrac{1}{2}(\omega_2 - \omega_1)t]\cos\tfrac{1}{2}(\omega_1 + \omega_2)t \\ [\alpha \sin\tfrac{1}{2}(\omega_2 - \omega_1)t]\sin\tfrac{1}{2}(\omega_1 + \omega_2)t \end{bmatrix} \tag{23.49}$$

These functions explicitly satisfy the initial conditions and determine the motion of the bobs for all subsequent times. To clarify the behavior, consider the limit of a weak spring ($k \ll mg/l$) when ω_2 only slightly exceeds ω_1. The final factors in Eq. (23.49), which depend on the *sum* of the frequencies, oscillate rapidly, whereas their amplitudes [in square brackets in Eq. (23.49)] depend on the *difference* of the frequencies and produce only a *slow amplitude modulation*. Under these conditions the behavior of Eq. (23.49) is sketched in Fig. 23.3. After an initial displacement α, the first bob oscillates with decreasing amplitude, and the second bob oscillates with increasing amplitude until the first bob appears to become stationary just when the second bob attains a maximum amplitude. This behavior occurs after a time $\tau = \pi/(\omega_2 - \omega_1)$. The situation then reverses, and the amplitudes of the oscillations continue to vary periodically. The reader is urged to demonstrate this phenomenon explicitly because the motion can appear quite complex, even though it involves only two fundamental frequencies. If the ratio ω_2/ω_1 is rational, Eq. (23.46) shows that the motion is strictly periodic and

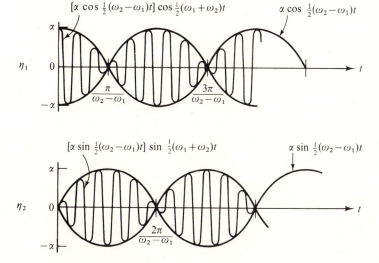

Figure 23.3 Behavior of coupled pendulums in Fig. 23.1 for the initial conditions in (23.41) and (23.42) [see Eq. (23.49)]. Here the coupling is assumed weak, so that ω_2 slightly exceeds ω_1.

repeats after some (typically long) interval. If ω_2/ω_1 is irrational, however, the system never returns exactly to its initial configuration, despite the apparent regularity discussed above.

24 EXAMPLE: MANY DEGREES OF FREEDOM

As seen in the preceding section, it is straightforward to determine the oscillation frequencies and normal modes for the small-amplitude motion of a mechanical system with few degrees of freedom. Although a system with many degrees of freedom is no more difficult in principle, evaluating a large determinant is generally prohibitive. Certain exceptions occur, however, when the system exhibits special symmetries that permit a complete solution. In this section, we formulate and solve two such N-body problems, where N may be arbitrarily large. Specifically, we consider longitudinal oscillations of particles connected by massless springs (see Fig. 24.1) and transverse planar oscillations of particles on a stretched, massless string (see Fig. 24.2). In each case, the system consists of identical elementary units, and the translational periodicity enables us to find the vibration frequencies and normal modes explicitly. These systems are important not only as examples of the analysis into normal modes but also as models for the propagation of waves in crystals or other discrete periodic media.

Two N-Body Problems

The first problem is the longitudinal oscillation of mass points connected with unstretched springs, illustrated in Fig. 24.1. Assume that all the masses m are equal with equal equilibrium separation a and that all the springs are massless with identical constants k. Let η_i with $i = 1, \ldots, N$ denote the displacement from equilibrium of the ith mass along the axis of the system; this set of variables can serve as the generalized coordinates for the longitudinal motion of the coupled system. This example provides a model for a one-dimensional crystal lattice with only nearest-neighbor interactions.

The kinetic energy arises from the one-dimensional motion of each particle, and the potential energy is just that of the extension or compression of the springs. Thus the lagrangian for the system in Fig. 24.1 can be immediately written

$$L = T - V = \tfrac{1}{2}m \sum_{i=1}^{N} \dot{\eta}_i^2 - \tfrac{1}{2}k \sum_{i=0}^{N} (\eta_{i+1} - \eta_i)^2 \tag{24.1}$$

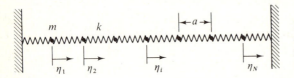

Figure 24.1 Longitudinal oscillations of mass points connected by massless springs. All subunits are taken to be identical.

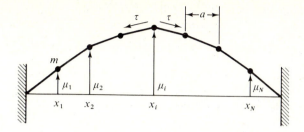

Figure 24.2 Transverse planar oscillations of mass points on a stretched massless string. All subunits are taken to be identical.

As illustrated in Fig. 24.1, the ends of the left- and right-hand springs are assumed fixed. We incorporate this condition of fixed endpoints with the convention

$$\eta_0 = \eta_{N+1} \equiv 0 \tag{24.2}$$

Consequently, the potential energy in (24.1) has $N + 1$ terms, whereas the kinetic energy has only N terms because the springs are assumed massless.

Lagrange's equations for the generalized coordinates η_i with $i = 1, \dots, N$ follow immediately from Eq. (24.1)

$$m\ddot{\eta}_i - k(\eta_{i+1} - \eta_i) + k(\eta_i - \eta_{i-1}) = 0 \tag{24.3}$$

or, equivalently

$$m\ddot{\eta}_i + 2k\eta_i - k(\eta_{i+1} + \eta_{i-1}) = 0 \qquad i = 1, \dots, N \tag{24.4}$$

where the convention (24.2) incorporates the condition of fixed endpoints. Note that the coupling to the nearest neighbors in Eq. (24.4) is completely determined by the positions of the particles since the force on an individual mass depends on the compression or extension of the adjacent springs.

The second problem is the transverse oscillation of N identical mass points m, equally spaced on a stretched massless string, illustrated in Fig. 24.2. At equilibrium, the string has a constant uniform tension τ, with adjacent particles separated a distance a. We shall treat only small transverse motion in the plane,† where the displacements μ_i with $i = 1, \dots, N$ serve as generalized coordinates.

Figure 24.3 shows the forces on the ith particle of this system. Newton's second law provides the exact equation of motion

$$m\ddot{\mu}_i = F_i = \tau(\sin \phi - \sin \theta) \tag{24.5}$$

where the right-hand side is the transverse component of the force in the positive direction. For small displacements, this quantity reduces to

$$m\ddot{\mu}_i \approx \tau(\phi - \theta) \tag{24.6}$$

and inspection of Fig. 24.3 then gives

$$m\ddot{\mu}_i \approx \frac{\tau}{a}\left[(\mu_{i+1} - \mu_i) - (\mu_i - \mu_{i-1})\right] \tag{24.7}$$

---

† The system of mass points connected by *stretched* springs is also capable of *transverse* oscillations. For small amplitudes, the longitudinal and transverse modes decouple (see Prob. 4.11).

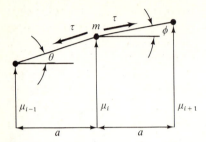

Figure 24.3 Force diagram on the ith particle for system illustrated in Fig. 24.2.

Note that the quantity in brackets in (24.7) is already linear in small quantities so that it is permissible to use the equilibrium tension τ in this expression. Equation (24.7) can be rewritten as

$$m\ddot{\mu}_i + \frac{2\tau}{a}\mu_i - \frac{\tau}{a}(\mu_{i+1} + \mu_{i-1}) = 0 \qquad i = 1, \ldots, N \tag{24.8}$$

and it is evident from Fig. 24.2 that the condition of fixed endpoints simply requires

$$\mu_0 = \mu_{N+1} \equiv 0 \tag{24.9}$$

We now observe that the equations of motion (24.4) and (24.2) for longitudinal oscillations of mass points joined by massless springs have *exactly the same form* as the equations of motion (24.8) and (24.9) for transverse oscillations of mass points on a taut, massless string. Indeed, the two problems become identical with the following identification

$$\eta_i \leftrightarrow \mu_i \qquad \text{and} \qquad k \leftrightarrow \frac{\tau}{a} \tag{24.10}$$

Substitution in Eq. (24.1) immediately provides the lagrangian for the second problem

$$L = T - V = \tfrac{1}{2}m\sum_{i=1}^{N} \dot{\mu}_i^2 - \frac{\tau}{2a}\sum_{i=0}^{N}(\mu_{i+1} - \mu_i)^2 \tag{24.11}$$

where Eq. (24.9) again ensures the condition of fixed endpoints. Since the second system has no dissipative forces, it, like the first, is conservative, allowing us to define a potential energy V. In fact, this potential energy represents the work done against the tension τ in stretching the strings from their equilibrium straight-line configuration (see Fig. 24.2). We shall subsequently calculate this work explicitly. At this point, however, it is sufficient that the lagrangian (24.11) immediately yields (24.8) as Lagrange's equations, which are the correct equations of motion as derived from Newton's laws.

Normal Modes

Equation (24.8) plays a central role in our subsequent development, and we shall solve it with two different methods. The first method seeks the normal modes for fixed endpoints and then manipulates the resulting determinantal eigenvalue equation to provide an explicit solution for the N eigenvalues. Since these techniques are independent of the order N of the determinant, the argument is sufficiently elegant and powerful to merit consideration. The second method seeks normal modes in the form of propagating plane waves. In this latter approach, the condition of fixed endpoints is satisfied by superposing the plane-wave solutions. The second method is the one used in solid-state physics to study propagating modes in crystals; it is also closer in spirit to those introduced subsequently in this text to solve problems in continuum mechanics.

We first seek normal-mode solutions to Eq. (24.8) of the form [compare Eq. (22.40)]

$$\mu_i = C\rho_i \cos(\omega t + \phi) \tag{24.12}$$

where all the displacements oscillate at the same frequency. Substitution of (24.12) into (24.8) yields

$$\left(\frac{2\tau}{a} - m\omega^2\right)\rho_i - \frac{\tau}{a}(\rho_{i+1} + \rho_{i-1}) = 0 \qquad i = 1, \ldots, N \tag{24.13}$$

and the condition of fixed endpoints becomes

$$\rho_0 = \rho_{N+1} = 0 \tag{24.14}$$

It is convenient to introduce the abbreviation

$$2 - \frac{m\omega^2 a}{\tau} \equiv \lambda \tag{24.15}$$

in which case the eigenvalue equation (24.13) takes the form

$$\lambda\rho_i - (\rho_{i+1} + \rho_{i-1}) = 0 \tag{24.16}$$

Write out these relations explicitly

$$\begin{aligned}
\lambda\rho_1 - \rho_2 \quad\quad\quad\quad\quad &= 0 \\
-\rho_1 + \lambda\rho_2 - \rho_3 \quad\quad\quad &= 0 \\
-\rho_2 + \lambda\rho_3 - \rho_4 \quad &= 0 \\
\cdots\cdots\cdots\cdots\cdots\cdots\cdots\cdots\cdots& \\
-\rho_{N-1} + \lambda\rho_N &= 0
\end{aligned} \tag{24.17}$$

They form a set of N linear homogeneous equations for the displacements ρ_i in the normal modes; a nontrivial solution exists only if the determinant of the coefficients vanishes. Thus we conclude that the $N \times N$ determinant

$$D_N \equiv \begin{vmatrix}
\lambda & -1 & 0 & 0 & 0 & \cdots & 0 \\
-1 & \lambda & -1 & 0 & 0 & \cdots & 0 \\
0 & -1 & \lambda & -1 & 0 & \cdots & 0 \\
\cdots\cdots\cdots\cdots\cdots\cdots\cdots\cdots\cdots & & & & & & \\
0 & & \cdots & & & \lambda & -1 \\
0 & & \cdots & & & -1 & \lambda
\end{vmatrix} \tag{24.18}$$

must vanish

$$D_N = 0 \tag{24.19}$$

Evidently this determinant is an Nth-order polynomial in λ and hence possesses N roots.

Before finding the roots of the eigenvalue equation (24.19), we first investigate the general structure of the determinant D_N defined in Eq. (24.18). This determinant has a simple enough form for its expansion in minors† to yield a useful recursion relation. Standard rules show that the determinant is equal to the sum of the elements in the first column, each multiplied by the determinant of the minor obtained by striking the appropriate row and column. The sign of each term is -1 raised to the sum of the row and column indices of the element in the first column. In this expansion, the determinant that is the coefficient of λ is evidently just D_{N-1}. The determinant that is the coefficient of -1 (obtained by striking the first column and second row of the original determinant) takes the form

$$\begin{vmatrix} -1 & 0 & 0 & 0 & \cdots & 0 \\ -1 & \lambda & -1 & 0 & \cdots & 0 \\ \hdotsfor{6} \\ 0 & & \cdots & & \lambda & -1 \\ 0 & & \cdots & & -1 & \lambda \end{vmatrix}$$

This determinant can be evaluated immediately by expanding in minors along the first *row*, in which case there is only one term, with the coefficient D_{N-2}. Thus the expansion in minors of the original determinant takes the form

$$D_N = \lambda D_{N-1} - (-1)(-1)D_{N-2} \tag{24.20}$$

and a rearrangement gives

$$D_N + D_{N-2} = \lambda D_{N-1} \tag{24.21}$$

This is now a two-term recursion relation for D_N; given D_1 and D_2, one can determine D_N by iteration. One simply computes D_3 in terms of D_1 and D_2, then computes D_4 in terms of D_2 and D_3, and so on. For any specified values of D_1, D_2 (and λ) this iteration procedure clearly generates a unique D_N for all N.

To solve Eq. (24.21) explicitly, it is first convenient to observe from Eq. (24.15) that $|\lambda|$ is certainly less than 2 for small ω^2. In addition, Fig. 24.3 shows that the string will be concave downward for small oscillations, implying that $\rho_{i+1} + \rho_{i-1} < 2\rho_i$. Since

$$\lambda = \frac{\rho_{i+1} + \rho_{i-1}}{\rho_i} \tag{24.22}$$

from Eq. (24.16), these normal-mode solutions again suggest that $|\lambda| < 2$. Thus we are motivated to introduce the following definition

$$\lambda \equiv 2 \cos \psi \tag{24.23}$$

† Hildebrand [1], sec. 1.4.

In fact, this definition involves no loss of generality, for all real values of λ can be represented in this form with a complex ψ. Introduction of Eq. (24.23) into Eq. (24.21) leads to the second-order finite-difference equation

$$D_N + D_{N-2} = 2 \cos \psi \, D_{N-1} \qquad (24.24)$$

Equation (24.24) has the special property that its coefficients are independent of N, which is analogous to the occurrence of constant coefficients in the second-order differential equations (22.14). There, the general solution consisted of exponentials with linear time dependence, and we here seek a solution of the similar form

$$D_N = A e^{iBN} \qquad (24.25)$$

where A and B are independent of N. Substitution into (24.24) and cancellation of the common factor A yields the condition

$$e^{iBN} + e^{iB(N-2)} = 2 \cos \psi \, e^{iB(N-1)}$$

Multiplication by $e^{-iB(N-1)}$ eliminates N entirely, giving the result

$$e^{iB} + e^{-iB} \equiv 2 \cos B = 2 \cos \psi \qquad (24.26)$$

Evidently, B has two possible values

$$B = \pm \psi \qquad (24.27)$$

The general solution of (24.24) consists of a linear combination of both possibilities

$$D_N = A_+ e^{iN\psi} + A_- e^{-iN\psi} \qquad (24.28)$$

where the constants A_+ and A_- can be determined by comparing (24.28) and (24.18) for $N = 1$ and $N = 2$.

$$D_1 = A_+ e^{i\psi} + A_- e^{-i\psi} = \lambda \equiv 2 \cos \psi \qquad (24.29)$$

$$D_2 = A_+ e^{2i\psi} + A_- e^{-2i\psi} = \lambda^2 - 1 = 4 \cos^2 \psi - 1 \qquad (24.30)$$

These linear equations for A_+ and A_- are readily solved (with Cramer's rule, for example) to give

$$A_+ = \frac{e^{i\psi}}{2i \sin \psi} \qquad (24.31)$$

and

$$A_- = A_+^* = \frac{-e^{-i\psi}}{2i \sin \psi} \qquad (24.32)$$

Thus the general expression for D_N in Eq. (24.18) is obtained from Eqs. (24.28), (24.31), and (24.32)

$$D_N = \frac{\sin (N+1)\psi}{\sin \psi} \qquad (24.33)$$

This remarkable result provides an *explicit evaluation of the determinant D_N for all integral values of N*, no matter how large or small.

With this general expression for the determinant D_N, the eigenvalue condition (24.19) now reads simply

$$\frac{\sin (N + 1)\psi}{\sin \psi} = 0 \tag{24.34}$$

It has the following N distinct solutions

$$(N + 1)\psi = n\pi \qquad n = 1, 2, \ldots, N \tag{24.35}$$

To obtain the normal-mode frequencies, combine Eqs. (24.15) and (24.23) to give

$$\lambda = 2 - \frac{m\omega^2 a}{\tau} = 2 \cos \psi \tag{24.36}$$

and then solve for ω^2

$$\omega^2 = \frac{2\tau}{ma} (1 - \cos \psi) = \frac{4\tau}{ma} \sin^2 \tfrac{1}{2}\psi \tag{24.37}$$

Substitution of (24.35) finally yields

$$\omega_n^2 = \frac{4\tau}{ma} \sin^2 \frac{n\pi}{2(N + 1)} \qquad n = 1, 2, \ldots, N \tag{24.38}$$

which are the normal-mode frequencies of the problem posed by Eqs. (24.8) and (24.9).

We have now found N distinct roots of the eigenvalue condition (24.34). On the other hand, the discussion following Eq. (24.19) assures us that this eigenvalue equation has *precisely N roots* because it is an Nth-order polynomial in λ. Since $m\omega_n^2 a/\tau < 4$, we evidently have $|\lambda| < 2$ for all N roots, justifying the assumption (24.23) with real ψ for all the normal modes. The reader may wonder about the other integral values in Eqs. (24.35). It is clear that $n = 0$ and $n = N + 1$ are *not* solutions because the denominator of Eq. (24.33), $\sin \psi$, also vanishes for these values. In addition, it is easy to verify that values of the integers $n \geq N + 2$ simply reproduce one of the eigenvalue frequencies given in Eq. (24.38) and therefore do not produce new solutions.

In principle, the N allowed eigenfrequencies (24.38) can be substituted into (24.13) to find the corresponding normal-mode amplitudes $\rho_i^{(n)}$, but this involves another finite-difference equation in the index i

$$2\left(\cos \frac{n\pi}{N + 1}\right)\rho_i^{(n)} = \rho_{i+1}^{(n)} + \rho_{i-1}^{(n)} \tag{24.39}$$

Although it can be solved by a similar procedure, we shall instead return to the original equation (24.8) and develop another technique that leads to the same results.

First, it is convenient to label the displacements in Fig. 24.2 by the position along the string

$$\mu_j \equiv \mu(x_j) \tag{24.40a}$$

$$x_j \equiv ja \tag{24.40b}$$

where j is an integer denoting the equilibrium location of each mass. Now seek a solution to Eq. (24.8) of the form

$$\mu(x_j, t) = Ae^{i(kx_j - \omega t)} \tag{24.41}$$

where, as always, the true displacement is just the real part. This solution has a simple physical interpretation as a plane-wave normal mode with wave number k, angular frequency ω, and speed ω/k propagating along the chain; when Eq. (24.41) is substituted into Eq. (24.8), the periodicity in the position of the masses will turn out to allow us to cancel the overall dependence on position, along with the time dependence. Alternatively, we may remark that the coefficients in Eq. (24.8) are *independent of both t and i*, suggesting an exponential solution that is linear in both t and i [compare the discussion preceding Eq. (24.25)]; Eq. (24.41) has just such a structure. To verify the correctness of this assumed form, substitute Eq. (24.41) into Eq. (24.8) and cancel the common factors to produce

$$-m\omega^2 + \frac{2\tau}{a} - \frac{\tau}{a}(e^{ika} + e^{-ika}) = 0 \tag{24.42}$$

This result can be rearranged to give the eigenvalue relation

$$\omega^2 = \frac{2\tau}{ma}(1 - \cos ka) = \frac{4\tau}{ma}\sin^2\left(\tfrac{1}{2}ka\right) \tag{24.43}$$

Such an expression of the form $\omega(k)$ is known as a *dispersion relation*.

To this point in the discussion, k and ω are considered continuous variables. We now render them discrete by imposing either of two boundary conditions.

Periodic boundary conditions This situation can be achieved physically for large N simply by connecting the ends of the stretched string around a cylinder and assuming that the masses move without friction along the surface of the cylinder in a direction perpendicular to the string (see Fig. 24.4). Periodic boundary conditions also apply, for example, to the transverse (and longitudinal) vibrations of benzene ($N = 6$). In the case of the string, translation through a distance Na along the string returns us to the starting point; analytically, this condition reads

$$\mu(x_i) = \mu(x_{N+i}) = \mu(x_i + Na) \tag{24.44}$$

Applying this condition to the solution in (24.41) implies

$$e^{ikNa} = 1 \tag{24.45}$$

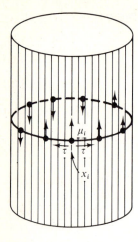

Figure 24.4 Physical situation requiring *periodic boundary conditions* for the mechanical system in Fig. 24.2. Here it is assumed that the masses can move without friction on the surface of the cylinder in a direction perpendicular to the string and that N is so large that any curvature in the string as one goes around the cylinder can be neglected.

which restricts the allowed values of the wave number k. The solutions to Eq. (24.45) take the form

$$k = \frac{2\pi n}{Na} \begin{cases} n = 0, \pm 1, \ldots, \pm\frac{1}{2}(N-1) & N \text{ odd} \\ n = 0, \pm 1, \ldots, \pm(\frac{1}{2}N - 1), +\frac{1}{2}N & N \text{ even} \end{cases} \qquad (24.46)$$

where we have again invoked the condition that there must be precisely N normal modes. Increasing the integer n by N evidently does not give a new solution in Eq. (24.41), since the replacement

$$n \to n + N \qquad (24.47)$$

implies that kNa increases by $2N\pi$ or that

$$ka \to ka + 2\pi \qquad (24.48)$$

Substitution of this relation into Eq. (24.41) and use of Eq. (24.40) show that this transformation leaves the solution (24.41) unchanged.

Fixed ends The solutions obtained in Eqs. (24.41) and (24.46) for periodic boundary conditions represent propagating waves moving to the right and left along an infinite lattice. Analogous solutions for fixed endpoints, illustrated in Fig. 24.2 and examined in our previous analysis, will now be constructed by *superposing* the two degenerate solutions with $\pm k$ in Eq. (24.41). It is clear from Eq. (24.43) that the normal-mode frequency is unaffected by a change in the sign of k. Thus we assume normal modes of the form

$$\mu(x_j, t) = A\left(e^{ikx_j - i\omega t} - e^{-ikx_j - i\omega t}\right) \qquad (24.49)$$

where the equal amplitudes have been chosen to satisfy the fixed-endpoint condition at the left end

$$\mu(x_0) = \mu(0) = 0 \qquad (24.50)$$

Similarly, the fixed-endpoint condition at the right end requires

$$\mu(x_{N+1}) = \mu[(N+1)a] = 0 \tag{24.51}$$

Imposing this condition on Eq. (24.49) yields

$$e^{ik(N+1)a} - e^{-ik(N+1)a} = 0 \tag{24.52}$$

or, equivalently,

$$\sin k(N+1)a = 0 \tag{24.53}$$

Note that the fixed-endpoint conditions (24.50) and (24.51) yield a different restriction (24.53) on the allowed values of the wave number k from that for periodic boundary conditions [Eq. (24.45)]. The solutions to Eq. (24.53) evidently take the form

$$k(N+1)a = n\pi \qquad n = 1, 2, \ldots, N \tag{24.54}$$

or

$$k = \frac{n\pi}{a(N+1)} \qquad n = 1, 2, \ldots, N \tag{24.55}$$

Equations (24.43) and (24.55) are now identical with the previous result in Eq. (24.38).

The present approach offers the further advantage of providing an explicit solution for the displacements of the mass points in a given normal mode in Eq. (24.49). Substitution of Eq. (24.55) into Eq. (24.49) yields the explicit solution

$$\mu(x_j, t) = 2iA_n \left[\sin \frac{n\pi x_j}{a(N+1)} \right] e^{-i\omega_n t} \tag{24.56}$$

where ω_n is given in Eq. (24.38); recalling the implicit real part and Eq. (24.40b), we therefore obtain normal-mode solutions (24.12) in the form

$$\mu(x_j, t) = \mu_j(t) = 2A_n \sin \frac{\pi n j}{N+1} \sin \omega_n t \tag{24.57}$$

where A_n is assumed real. Apart from normalization, the normal-mode amplitude $\rho_j^{(n)}$ for the jth particle in the nth normal mode is just that of a sine wave with n half wavelengths, *evaluated at the discrete position* $x_j = ja$ (see Fig. 24.5). The determination of the appropriate normalization constant A_n is left for Prob. 4.14, along with the verification of the proper orthogonality [Eq. (22.34)].

The N transverse normal-mode frequencies in Eq. (24.38) of the physical system in Fig. 24.2 are shown in Fig. 24.6; with the replacement (24.10) they also represent the longitudinal normal-mode frequencies of the system in Fig. 24.1. To simplify this graph, Eq. (24.38) has been rewritten as

$$\frac{\omega_n}{c} = \frac{2}{a} \sin \left(\frac{n\pi}{l} \frac{a}{2} \right) \qquad n = 1, 2, \ldots, N \tag{24.58}$$

where

$$l \equiv (N+1)a \tag{24.59}$$

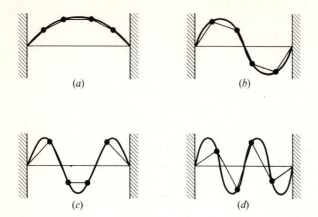

(a)   (b)

(c)   (d)

Figure 24.5 Particle displacements (with exaggerated overall normalization) for the four normal modes in Eq. (24.54) of the mechanical system in Fig. 24.2 with $N = 4$: (a) $n = 1$, $\omega_1 = 0.618(\tau/ma)^{1/2}$; (b) $n = 2$, $\omega_2 = 1.176(\tau/ma)^{1/2}$; (c) $n = 3$, $\omega_3 = 1.618(\tau/ma)^{1/2}$; (d) $n = 4$, $\omega_4 = 1.902(\tau/ma)^{1/2}$.

defines the total length of the string and

$$c \equiv \left(\frac{\tau}{m/a}\right)^{1/2} \tag{24.60}$$

represents a characteristic velocity. For small n, the hence low frequencies, it is evident that the normal-mode frequencies take the form

$$\frac{\omega_n}{c} \xrightarrow[n \to 0]{} \frac{n\pi}{l} \tag{24.61}$$

The wavelength of these modes follows from Eqs. (24.49) and (24.54) by recalling that the wave number k is related to the wavelength λ of the disturbance by $k = 2\pi/\lambda$. Equation (24.54) then implies that the wavelength is given by

$$\frac{2\pi}{\lambda}(N + 1)a = n\pi \tag{24.62}$$

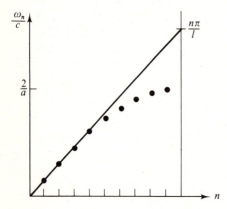

Figure 24.6 Normal-mode frequencies given by Eqs. (24.58) to (24.60) of the physical systems in Figs. 24.1 and 24.2 (with $N = 9$).

or, introducing the definition in Eq. (24.59),

$$\lambda = \frac{2l}{n} \qquad n = 1, 2, \ldots, N \qquad (24.63)$$

Low-frequency normal modes correspond to long wavelengths, where neighboring particles all move in phase. The discrete nature of the system is clearly irrelevant for such modes and, as discussed in detail in the following section, the same results can be obtained directly by treating the system as continuous.

As n increases, the normal-mode frequencies evidently fall below the linear function (24.61), as illustrated in Fig. 24.6. In addition, Eq. (24.38) implies the existence of *a maximum normal-mode frequency* since $0 < n \le N$. For large N, Eqs. (24.58) and (24.59) show that this maximum normal-mode frequency is

$$\left(\frac{\omega_n}{c}\right)_{max} \approx \frac{2}{a} \qquad (24.64)$$

Correspondingly, the disturbance has a minimum wavelength λ_{min}, which can be obtained from Eqs. (24.63) and (24.59); for large N it becomes

$$\lambda_{min} \approx 2a \qquad (24.65)$$

These results indicate that neither of the physical systems in Figs. 24.1 or 24.2 can oscillate with a higher frequency or a shorter wavelength than those given in (24.64) and (24.65), where a is the interparticle spacing. The discrete nature of the lattice provides a simple physical interpretation of these restrictions. Equations (24.41) and (24.49) show that the normal-mode amplitudes are propagating waveforms generated by the envelope of the displacements of the particles. It is evident from Fig. 24.5 that a sensible waveform for a discrete lattice requires the approximate relation

$$\lambda \gtrsim 2a \qquad (24.66)$$

which is equivalent to the condition (24.65) of minimum wavelength for the normal modes.

Three-dimensional crystals can support both types of motion: longitudinal oscillations (illustrated in Fig. 24.1) and transverse oscillations (illustrated in Fig. 24.2). In fact, there are two distinct transverse oscillations, corresponding to the two orthogonal transverse directions. In real materials, the transverse and longitudinal modes typically have different restoring forces and hence different velocities. Longitudinal and transverse modes also differ in that density variations arise predominantly from the longitudinal modes of oscillation, as is evident from Figs. 24.1 and 24.2.

25 TRANSITION FROM DISCRETE TO CONTINUOUS SYSTEMS

In this section, we examine the behavior of the preceding discrete systems when the characteristic scale of the phenomena is large compared with the interparticle spacing. The resulting limiting equations describe a continuous string, and we shall also derive the same equations directly. Finally, the same dynamics will be rederived from a lagrangian, illustrating the application of Hamilton's principle to a continuous system.

Passage to the Continuum Limit

Consider the limit of the previous analysis when the number of mass points on the string becomes infinite, keeping the length of the string and its mass density constant. This limit has the analytic form

$$N \to \infty \qquad a \to 0 \tag{25.1}$$

$$\text{Length of string} = (N + 1)a \equiv l = \text{const} \tag{25.2}$$

$$\text{Mass density of string} = \frac{m}{a} \equiv \sigma = \text{const} \tag{25.3}$$

More precisely, Eq. (25.1) requires $N \gg n$ for any fixed normal mode n in Eq. (24.38). Under conditions (25.1) to (25.3), the normal-mode frequencies in Eq. (24.38) take the form

$$\omega_n^2 \to \frac{4\tau}{ma} \left[\frac{n\pi}{2(N + 1)} \right]^2 = \frac{\tau}{\sigma} \left(\frac{n\pi}{l} \right)^2 \tag{25.4}$$

or, equivalently,

$$\omega_n = \left(\frac{n\pi}{l} \right) c \qquad n = 1, 2, \ldots, \infty \tag{25.5}$$

where

$$c \equiv \left(\frac{\tau}{\sigma} \right)^{1/2} \tag{25.6}$$

is the characteristic velocity introduced in Eq. (24.60). These are the normal-mode frequencies of the uniform string. In the present limit of infinitely many mass points, the system can oscillate in any normal mode n; the previous solutions (24.41) become plane waves propagating along the string

$$\mu(x, t) = A e^{i(kx - \omega t)} \tag{25.7}$$

where the wave number obtained from Eq. (24.46) reduces to

$$k = \frac{2\pi}{l} n \qquad n = 0, \pm 1, \pm 2, \ldots, \pm \infty \tag{25.8}$$

Direct Treatment of a Continuous String

The preceding analysis showed how the continuous displacement of a string (25.7) could be obtained from the solution $\mu_i(t)$ with $i = 1, \ldots, N$ for discrete mass points on a stretched massless string. Although such a limiting process is straightforward in principle, the solution for the discrete case can become cumbersome. Thus it is

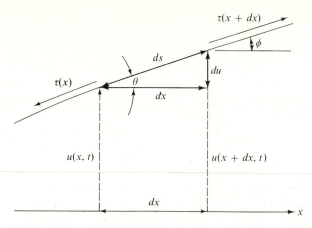

Figure 25.1 Force diagram on a small element of string with mass $\sigma(x)\,dx$.

often preferable to reformulate the problem entirely, dealing only with macro-scopic variables that characterize the continuous system. To illustrate these techniques, we consider the continuous string shown in Fig. 25.1, where $u(x, t)$ is the transverse displacement and x is a *continuous* variable labeling the position along the undisplaced string. Evidently, this description corresponds to the substitution

$$\mu_i(t) \rightarrow u(x, t) \tag{25.9}$$

Consider the motion of a mass element $\sigma\,dx$ at the point x, whose transverse displacement is given by $u(x, t)$. In the approximation of small displacements

$$\sin\theta \approx \theta \approx \tan\theta = \frac{\partial}{\partial x} u(x, t) \tag{25.10}$$

where the last quantity is the instantaneous slope of the string at time t. Newton's second law for the element dx of string then implies

$$[\sigma(x)\,dx]\frac{\partial^2 u(x, t)}{\partial t^2} = \tau(x + dx)\sin\phi - \tau(x)\sin\theta$$

$$\approx \tau(x + dx)\frac{\partial u(x + dx, t)}{\partial x} - \tau(x)\frac{\partial u(x, t)}{\partial x} \tag{25.11}$$

because the quantity on the left is the mass times the transverse acceleration and the quantity on the right is the transverse component of the force acting on the element dx of string in the approximation (25.10) of small displacements. Here the partial derivatives indicate that the other of the variables (x, t) is held fixed. The right-hand side of Eq. (25.11) can be expanded in a Taylor series to first order in the infinitesimal length dx

$$\text{rhs} = \tau(x)\frac{\partial u(x, t)}{\partial x} + dx\frac{\partial}{\partial x}\left[\tau(x)\frac{\partial u(x, t)}{\partial x}\right] + \cdots - \tau(x)\frac{\partial u(x, t)}{\partial x} \tag{25.12}$$

and the first and last terms cancel. Division by the remaining infinitesimal dx in Eq. (25.11) yields the *general one-dimensional string equation*

$$\sigma(x)\frac{\partial^2 u(x, t)}{\partial t^2} = \frac{\partial}{\partial x}\left[\tau(x)\frac{\partial u(x, t)}{\partial x}\right] \tag{25.13}$$

It is a second-order partial differential equation for the displacement of the string as a function of the position along the string and the time.

Equation (25.13) simplifies if the tension in the string and its mass density are assumed to be constant

$$\tau(x) = \tau = \text{const} \tag{25.14}$$

$$\sigma(x) = \sigma = \text{const} \tag{25.15}$$

In this case Eq. (25.13) becomes the *one-dimensional wave equation*

$$\frac{1}{c^2}\frac{\partial^2 u(x, t)}{\partial t^2} = \frac{\partial^2 u(x, t)}{\partial x^2} \tag{25.16}$$

where c is the velocity of propagation introduced in Eq. (25.6).

It is interesting to compare this direct treatment with our previous discussion of Newton's laws for the discrete case. We first rewrite Eq. (24.8) in the form

$$\ddot{\mu}_i = \frac{\tau}{m/a}\left[\frac{1}{a}\left(\frac{\mu_{i+1} - \mu_i}{a} - \frac{\mu_i - \mu_{i-1}}{a}\right)\right] \tag{25.17}$$

and note the following correspondence between a finite difference and a derivative as $a \to 0$:

$$\frac{\mu_{i+1} - \mu_i}{a} \equiv \frac{\mu_{i+1} - \mu_i}{x_{i+1} - x_i} \xrightarrow[a \to 0]{} \frac{\partial u}{\partial x} \tag{25.18}$$

The identification (25.9) and the definition (25.3) and (25.6) immediately show that Eq. (25.17) reduces to (25.16) in the limit $a \to 0$, for the quantity in square brackets in (25.17) is just the finite-difference approximation to $\partial^2 u(x, t)/\partial x^2$.

General Solution to the Wave Equation with Specified Initial Conditions

Following our general development from Sec. 22, we now look for *normal-mode* solutions to the wave equation (25.16) in the form

$$u(x, t) = C\rho(x) \cos(\omega t + \phi) \tag{25.19}$$

Substitution into Eq. (25.16) leads to the ordinary differential equation

$$\frac{d^2\rho(x)}{dx^2} + k^2\rho(x) = 0 \tag{25.20}$$

where $$k \equiv \frac{\omega}{c} \tag{25.21}$$

Equation (25.20) and a given set of boundary conditions constitute an *eigenvalue* problem since, in general, solutions exist only for definite values of k^2. These solutions to Eq. (25.20) generate the *eigenfunctions* of the problem. Note that this procedure is precisely that used in the discrete case; there, the assumption of normal modes (24.12) reduced the problem to the eigenvalue equation (24.13), which can now be recast in the form

$$\frac{\tau}{m/a}\left[\frac{1}{a}\left(\frac{\rho_{i+1}-\rho_i}{a}-\frac{\rho_i-\rho_{i-1}}{a}\right)\right]+\omega^2\rho_i=0 \qquad (25.22)$$

The same limit used in the analysis of Eq. (25.17) leads precisely to Eq. (25.20).

We next consider the boundary conditions and assume that the string has fixed endpoints

$$\rho(0)=\rho(l)=0 \qquad (25.23)$$

The ordinary differential equation (25.20) evidently has the following solution that satisfies the first of the boundary conditions (25.23)

$$\rho(x)=\left(\frac{2}{l\sigma}\right)^{1/2}\sin kx \qquad (25.24)$$

where the choice of normalization will be discussed shortly. The second boundary condition in Eq. (25.23) can be satisfied only for certain special values of k. These eigenvalues are given by

$$k_n=\frac{n\pi}{l} \qquad n=1,2,\ldots,\infty \qquad (25.25)$$

and the normal-mode frequencies are related to these values through Eq. (25.21)

$$\omega_n=\frac{n\pi}{l}c \qquad (25.26)$$

These relations are precisely those obtained in Eq. (25.5). They have the very simple pictorial interpretation indicated in Fig. 25.2. In order to match the boundary conditions of fixed endpoints, the string must contain an integral number of half wavelengths of the wave disturbance, with the allowed values

$$\lambda=\frac{2l}{n} \qquad n=1,2,\ldots,\infty \qquad (25.27)$$

Figure 25.2 First few normal modes of a uniform string of length l with fixed endpoints. λ is the wavelength of the oscillation (compare Fig. 24.5).

The elementary relation between wave number and wavelength yields the equivalent condition

$$k = \frac{2\pi}{\lambda} = \frac{n\pi}{l} \qquad n = 1, 2, \ldots, \infty \tag{25.28}$$

which is precisely Eq. (25.25).

The *general solution* to the one-dimensional wave equation in (25.16) can be constructed by taking a linear superposition of the normal modes obtained in Eqs. (25.19) and (25.24) to (25.26)

$$u(x, t) = \sum_{n=1}^{\infty} C_n \rho^{(n)}(x) \cos (\omega_n t + \phi_n) \tag{25.29}$$

or, equivalently,

$$u(x, t) = \sum_{n=1}^{\infty} \left(\frac{2}{l\sigma}\right)^{1/2} \sin k_n x \, (a_n \cos \omega_n t + b_n \sin \omega_n t) \tag{25.30}$$

where

$$a_n \equiv C_n \cos \phi_n \qquad \text{and} \qquad b_n \equiv -C_n \sin \phi_n \tag{25.31}$$

Equation (25.29) is simply the generalization to the continuous case of Eq. (22.40) for normal modes of discrete systems [recall Eq. (25.9)]

$$\mu_i(t) = \sum_{n=1}^{N} C^{(n)} \rho_i^{(n)} \cos (\omega_n t + \phi_n) \tag{25.32}$$

Note that this equation for a finite system contains only a *finite* number of terms, whereas Eq. (25.29) is an *infinite* series. This difference leads to profound mathematical complications, and the second half of this book examines the mechanics of continuous media, developing some of the necessary mathematical techniques.

Each term in Eq. (25.30) satisfies the wave equation and boundary conditions of the problem. Since we are describing a mechanical system subject to Newton's second law, a complete solution to the problem for all subsequent times still requires a specification of the initial values of the displacements and velocities of all elements of the string. Equation (25.30) shows that the *initial-value problem* takes the form

$$u(x, 0) \equiv f(x) = \sum_{n=1}^{\infty} a_n \rho^{(n)}(x) = \left(\frac{2}{l\sigma}\right)^{1/2} \sum_{n=1}^{\infty} a_n \sin k_n x \tag{25.33a}$$

$$\dot{u}(x, 0) \equiv g(x) = \sum_{n=1}^{\infty} \omega_n b_n \rho^{(n)}(x)$$

$$= \left(\frac{2}{l\sigma}\right)^{1/2} \sum_{n=1}^{\infty} \omega_n b_n \sin k_n x \tag{25.33b}$$

The expressions on the right-hand sides of these equations are just *Fourier series*. In Sec. 41 we shall prove that the eigenfunctions $\rho^{(n)}(x)$ in Eqs. (25.24) and (25.25) form a *complete set* of functions on the interval $0 \leq x \leq l$. Since the coefficients a_n and b_n are arbitrary, there is just enough freedom in the solution (25.30) to match both of the required initial conditions (25.33). It is readily verified (and also a general result proved in Sec. 40) that the eigenfunctions in Eqs. (25.24) and (25.25) are *orthonormal* [compare Eq. (22.34); recall also the present assumption (25.15)]

$$\int_0^l \rho^{(n)}(x)\sigma\rho^{(m)}(x)\,dx = \frac{2}{l\sigma}\int_0^l \sin k_n x \sin k_m x\,\sigma\,dx$$

$$= \frac{2}{l\sigma}\int_0^l \sin \frac{n\pi x}{l} \sin \frac{m\pi x}{l}\,\sigma\,dx = \delta_{nm} \qquad (25.34)$$

Thus Eqs. (25.33) can be inverted with the aid of Eq. (25.34) to yield explicit values for the coefficients a_n and b_n

$$a_n = \int_0^l \rho^{(n)}(x)f(x)\sigma\,dx = \left(\frac{2}{l\sigma}\right)^{1/2}\int_0^l \sin k_n x\, f(x)\sigma\,dx \qquad (25.35a)$$

$$\omega_n b_n = \int_0^l \rho^{(n)}(x)g(x)\sigma\,dx = \left(\frac{2}{l\sigma}\right)^{1/2}\int_0^l \sin k_n x\, g(x)\sigma\,dx \qquad (25.35b)$$

Note that this procedure is virtually identical with that in (22.49) and (22.50) for the discrete case.

In summary, each term in Eq. (25.30) is a solution to the wave equation and satisfies the boundary conditions. Moreover, Eq. (25.30) has enough freedom to match an arbitrary set of initial conditions according to Eqs. (25.33) to (25.35), so that we have indeed constructed the *general solution* to the one-dimensional wave equation with specified initial conditions. The mathematical complexities associated with the infinite series in Eq. (25.30) and the completeness of the solutions to the eigenvalue equation (25.20) for matching an arbitrary set of initial conditions (25.33) will be examined in Chap. 7.

Lagrangian for a Continuous String

In this section we shall construct the lagrangian for a continuous string, first by taking the limit of the discrete system, and then directly. For simplicity, only the case of constant mass density σ and constant tension τ is considered here. The lagrangian (24.11) for the discrete problem correctly gives the equations of motion (24.8). A slight rearrangement yields

$$L = \frac{m}{2a}\sum_{i=1}^N a\dot{\mu}_i^2 - \frac{\tau}{2}\sum_{i=1}^N a\left(\frac{\mu_{i+1}-\mu_i}{a}\right)^2 \qquad (25.36)$$

We can now pass to the continuum limit with the replacements

$$a \to 0 \qquad (25.37a)$$

$$\mu_i(t) \to u(x,t) \qquad (25.37b)$$

$$\frac{m}{a} \to \sigma \qquad (25.37c)$$

$$\sum_i a = \sum_i \Delta x_i \to \int_0^l dx \qquad (25.37d)$$

$$\frac{\mu_{i+1}-\mu_i}{a} \to \frac{\partial u(x,t)}{\partial x} \qquad (25.37e)$$

By now, only Eq. (25.37d) is unfamiliar, but it follows from the very definition of an integral. The lagrangian in Eq. (25.36) thus takes the form

$$L = \frac{\sigma}{2} \int_0^l \left[\frac{\partial u(x, t)}{\partial t}\right]^2 dx - \frac{\tau}{2} \int_0^l \left[\frac{\partial u(x, t)}{\partial x}\right]^2 dx \qquad (25.38)$$

This lagrangian for the continuum problem can also be derived directly, as illustrated in Fig. 25.1. The kinetic energy of the small string element dx is $\frac{1}{2}$(velocity)^2 × (mass of small element), and the total kinetic energy of the string is obtained by summing over all the elements according to the relation

$$T = \frac{1}{2} \int_0^l \left[\frac{\partial u(x, t)}{\partial t}\right]^2 (\sigma\, dx) \qquad (25.39)$$

The total potential energy V is the work done in deforming the string from its equilibrium configuration. Since the element of arc length in Fig. 25.1 is

$$ds^2 = dx^2 + du^2 \qquad (25.40)$$

the work done in stretching the element against the tension τ is given by

$$dW = \tau(ds - dx) = \tau\, dx\left(\left\{1 + \left[\frac{\partial u(x, t)}{\partial x}\right]^2\right\}^{1/2} - 1\right)$$

$$\approx \frac{\tau}{2} \left[\frac{\partial u(x, t)}{\partial x}\right]^2 dx \qquad (25.41)$$

where the last line assumes small displacements. The total potential energy of the string is obtained by summing this work over all elements of the string

$$V = \frac{\tau}{2} \int_0^l \left[\frac{\partial u(x, t)}{\partial x}\right]^2 dx \qquad (25.42)$$

Clearly, the lagrangian $L = T - V$ precisely reproduces Eq. (25.38).

Normal Coordinates

The linear superposition of normal modes (25.29) provides the general solution to the string problem

$$u(x, t) = \sum_{n=1}^{\infty} \rho_n(x) C_n \cos\,(\omega_n t + \phi_n) \qquad (25.43)$$

where the normal-mode eigenfunctions are defined according to Eqs. (25.24) and (25.25)†

$$\rho_n(x) = \left(\frac{2}{l\sigma}\right)^{1/2} \sin\frac{n\pi x}{l} \qquad (25.44)$$

---

† Since no confusion can now arise, the normal modes will henceforth be labeled with a subscript.

They satisfy the orthonormality relation (25.34) [compare Eq. (22.34)]

$$\int_0^l \rho_n(x)\rho_m(x)\sigma \, dx = \delta_{nm} \tag{25.45}$$

and the set of values $\{C_n, \phi_n\}$ can be obtained from the initial conditions according to Eqs. (25.35) and (25.31). Just as before in Eq. (22.67), the coefficients of the normal-mode amplitudes in Eqs. (25.43) are the *normal coordinates that diagonalize the lagrangian.* Define

$$\zeta_n(t) \equiv C_n \cos (\omega_n t + \phi_n) \qquad n = 1, 2, \ldots, \infty \tag{25.46}$$

In this way, the expansion (25.43) takes the form

$$u(x, t) = \sum_{n=1}^{\infty} \rho_n(x)\zeta_n(t) \tag{25.47}$$

Now insert this result in the lagrangian (25.38). In the kinetic-energy term, the time derivative only acts on the normal coordinates, and the orthonormality (25.45) of the eigenfunctions reduces the double sum to a single sum. To evaluate the potential energy in Eq. (25.38) it is first convenient to perform a partial integration, using the conditions of fixed endpoints to rewrite the potential energy as

$$V = -\frac{\tau}{2} \int_0^l u(x, t) \frac{\partial^2 u(x, t)}{\partial x^2} \, dx \tag{25.48}$$

Insertion of the expansion (25.47) and use of the eigenfunction equation

$$\frac{d^2 \rho_n(x)}{dx^2} = -k_n^2 \rho_n(x) \tag{25.49}$$

again permits us to reduce the double sum to a single sum with the orthonormality relation (25.45). Definitions (25.25), (25.26), and (25.6) then express the lagrangian in the form

$$L = \frac{1}{2} \sum_{n=1}^{\infty} (\dot{\zeta}_n^2 - \omega_n^2 \zeta_n^2) \tag{25.50}$$

This is indeed a remarkable result. The normal coordinates (25.46), which are the coefficients of the normal-mode amplitudes in the expansion (25.47) of the solution to the wave equation, reduce the lagrangian to a *discrete set of uncoupled simple harmonic oscillators.* The only difference from Eq. (22.65) for a finite system is that we now have an *infinite set* of normal coordinates ζ_n with $n = 1, 2, \ldots, \infty$, but they are still labeled with a *discrete index.* Just as before, we can now take the *normal coordinates as a new set of generalized coordinates.* Furthermore, it is clear that the specification of all of the normal coordinates completely determines the configuration of the string according to Eq. (25.47). Substitution of the lagrangian (25.50) into Hamilton's principle (18.5) immediately yields Lagrange's equations for the normal coordinates

$$\ddot{\zeta}_n = -\omega_n^2 \zeta_n \qquad n = 1, 2, \ldots, \infty \tag{25.51}$$

Each normal coordinate executes simple harmonic motion at the normal-mode frequency ω_n, and Eqs. (25.46) are simply the general solution to these equations of motion.

Hamilton's Principle for Continuous Systems

In the preceding section, the introduction of normal coordinates reduced the lagrangian for the continuous system to a countably infinite sum (25.50), allowing a direct application of Hamilton's principle. For most purposes, however, it is preferable to generalize Hamilton's principle itself to allow for continuous systems, where the generalized coordinates are continuous functions of both the spatial position and the time [see Eq. (25.38) where $u(x, t)$ is the generalized coordinate for the string problem]. In this case, Hamilton's principle takes the form

$$\delta \int_{t_1}^{t_2} L \, dt = \delta \int_{t_1}^{t_2} dt \int_0^l dx \, \mathscr{L}\left(u, \frac{\partial u}{\partial x}, \frac{\partial u}{\partial t}; x, t\right) = 0 \qquad (25.52)$$

where $\mathscr{L}$ is the *lagrangian density* and (x, t) are the independent variables that define the integration region as illustrated in Fig. 25.3. Note that $\mathscr{L}$ has the dimensions of energy per unit length.

The generalized coordinate that specifies the configuration is $u(x, t)$. It is this quantity which must be varied in Hamilton's principle. Thus we consider the variations

$$u(x, t) \rightarrow u(x, t) + \delta u(x, t) \qquad (25.53)$$

subject, as always in Hamilton's principle, to the condition of fixed endpoints *in time*

$$\delta u(x, t_1) = \delta u(x, t_2) = 0 \qquad \text{all } x \qquad (25.54)$$

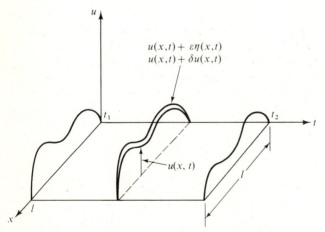

Figure 25.3 Quantities appearing in Hamilton's principle for a continuous system in one spatial dimension.

Furthermore, a string satisfying the *spatial boundary conditions* of fixed endpoints requires

$$\delta u(x = 0, t) = \delta u(x = l, t) = 0 \qquad \text{all } t \qquad (25.55)$$

In subsequent chapters, we shall extend this analysis to include more complicated spatial boundary conditions and nonuniform $\sigma(x)$ and $\tau(x)$.

Substitute the variation (25.53) into (25.52) and expand to first order in the small quantities. Hamilton's principle (25.52) then takes the form

$$\delta \int_{t_1}^{t_2} L \, dt = \int_{t_1}^{t_2} dt \int_0^l dx \left[\frac{\partial \mathscr{L}}{\partial u} \delta u + \frac{\partial \mathscr{L}}{\partial(\partial u/\partial x)} \delta \frac{\partial u}{\partial x} + \frac{\partial \mathscr{L}}{\partial(\partial u/\partial t)} \delta \frac{\partial u}{\partial t} \right] = 0 \qquad (25.56)$$

Here, the partial derivatives with respect to u or its derivatives mean that *all other variables in the lagrangian density in Eq. (25.52) are to be held fixed.* It is evident from the structure of Eq. (25.53) that we again have

$$\delta \frac{\partial u}{\partial x} = \frac{\partial}{\partial x} \delta u(x, t) \qquad (25.57a)$$

$$\delta \frac{\partial u}{\partial t} = \frac{\partial}{\partial t} \delta u(x, t) \qquad (25.57b)$$

The partial derivatives with respect to x or t in Eq. (25.57) now mean that *the other independent variable in the set (x, t) is to be held fixed.* The second and third terms in Eq. (25.56) can be partially integrated, using the condition of fixed endpoints (25.54) and (25.55) to omit the boundary terms. As a result, Eq. (25.56) can be written

$$\int_{t_1}^{t_2} dt \int_0^l dx \left[\frac{\partial \mathscr{L}}{\partial u} - \frac{\partial}{\partial x} \frac{\partial \mathscr{L}}{\partial(\partial u/\partial x)} - \frac{\partial}{\partial t} \frac{\partial \mathscr{L}}{\partial(\partial u/\partial t)} \right] \delta u(x, t) = 0 \qquad (25.58)$$

Since the variation $\delta u(x, t)$ is completely arbitrary, its coefficient in Eq. (25.58) must vanish, providing the Euler-Lagrange equation for the variational problem (25.52)

$$\frac{\partial}{\partial t} \frac{\partial \mathscr{L}}{\partial(\partial u/\partial t)} + \frac{\partial}{\partial x} \frac{\partial \mathscr{L}}{\partial(\partial u/\partial x)} - \frac{\partial \mathscr{L}}{\partial u} = 0 \qquad (25.59)$$

This is the general form of *Lagrange's equation for a continuous system.* We emphasize that the various partial derivatives mean different things. Partial derivatives of $\mathscr{L}$ are to be carried out keeping fixed all the other variables that appear in $\mathscr{L}$ in Eq. (25.52). The final partial derivatives with respect to t and x are to be carried out keeping only the other independent variable in the set (x, t) fixed. In contrast to the previous situation, where each discrete coordinate μ_i had a separate equation of motion, we now have a single equation of motion that holds for all x.

As an example, consider the lagrangian density (25.38) of the one-dimensional string

$$\mathscr{L} = \frac{\sigma}{2} \left(\frac{\partial u}{\partial t} \right)^2 - \frac{\tau}{2} \left(\frac{\partial u}{\partial x} \right)^2 \qquad (25.60)$$

For constant σ and τ, Lagrange's equation (25.59) implies

$$\sigma \frac{\partial^2 u}{\partial t^2} - \tau \frac{\partial^2 u}{\partial x^2} = 0 \tag{25.61}$$

which is precisely the one-dimensional wave equation

$$\frac{1}{c^2} \frac{\partial^2 u}{\partial t^2} = \frac{\partial^2 u}{\partial x^2} \tag{25.62}$$

with velocity of propagation defined in Eq. (25.6).

PROBLEMS

4.1 A thin hoop of radius R and mass M oscillates in its own plane with one point of the hoop fixed. Attached to the hoop is a point mass M constrained to move without friction along the hoop. The system is in a uniform gravitational field $\mathbf{g}$. Consider only small oscillations.

(a) Show that the normal-mode frequencies are

$$\omega_1 = \frac{1}{2} \left(\frac{2g}{R} \right)^{1/2} \qquad \omega_2 = \left(\frac{2g}{R} \right)^{1/2}$$

(b) Find the normal-mode eigenvectors. Sketch the motion.

(c) Construct the modal matrix.

(d) Find the normal coordinates and show that they diagonalize the lagrangian.

4.2 Consider the longitudinal oscillations, i.e., along the axis, of the mechanical system in Fig. 24.1 assuming only two masses and three springs. Work from first principles.

(a) Find the lagrangian and Lagrange's equations.

(b) What are the normal-mode frequencies and eigenvectors? Describe the motions.

(c) Construct the modal matrix and normal coordinates, and write the lagrangian in diagonal form.

(d) Suppose the mass on the left is initially displaced from equilibrium a distance α to the right. Compute the subsequent motion.

(e) Compare your treatment with the general results in Sec. 24 for $N = 2$.

4.3 A double pendulum with equal lengths and different masses m_1 and m_2 performs small oscillations in a plane. Introduce the transverse displacements of the first particle from the vertical η_1, and of the second particle from the first particle η_2 (compare Fig. 13.1c).

(a) Show that the lagrangian is given by

$$L = \tfrac{1}{2} m_1 \dot\eta_1^2 + \tfrac{1}{2} m_2 (\dot\eta_1 + \dot\eta_2)^2 - \frac{g}{2l} [(m_1 + m_2)\eta_1^2 + m_2 \eta_2^2]$$

(b) Derive the normal-mode frequencies $\omega^2 = (g/l)(1 \pm \gamma)^{-1}$, where $\gamma = [m_2(m_1 + m_2)^{-1}]^{1/2}$.

(c) Construct the normal-mode eigenvectors and describe the motions. Show that these reproduce the expected behavior for large and small values of m_1/m_2.

(d) Verify that the modal matrix has the form

$$\mathscr{A} = (2m_1)^{-1/2} \begin{bmatrix} (1-\gamma)^{1/2} & -(1+\gamma)^{1/2} \\ \gamma^{-1}(1-\gamma)^{1/2} & \gamma^{-1}(1+\gamma)^{1/2} \end{bmatrix}$$

and demonstrate explicitly that $\mathscr{A}$ diagonalizes the matrices $\underline{m}$ and $\underline{v}$.

(e) Construct the normal coordinates.

(f) Assume that $m_2 \ll m_1$. If the upper mass is displaced slightly from the vertical and released from rest, show that the subsequent motion is such that at regular intervals one pendulum is stationary and the other oscillates with maximum amplitude.

4.4 A particle with mass m slides without friction around the circumference of a circular wire hoop of radius a. The hoop is placed upright in a uniform gravitational field and rotates about a vertical diameter with angular velocity Ω (compare Prob. 3.1).

(a) Construct the lagrangian, using as generalized coordinate the angular displacement θ along the hoop measured from the downward vertical. Derive the differential equation for the motion and construct the corresponding first integral.

(b) Using the equation of motion, obtain *all* positions of dynamical equilibrium and classify them as stable or unstable. Verify these conclusions by considering forces in the co-rotating frame. For those configurations which are stable, determine the frequency of small oscillations about that position. Discuss the limiting cases $\Omega^2 \ll g/a$ and $\Omega^2 \gg g/a$.

(c) Find the hamiltonian for the system. Is it a constant of the motion? Is it the total energy? Compare with the first integral in part (a).

4.5 A point mass moves without friction on the inside of a surface of revolution $z = f(r)$, whose symmetry axis lies along a uniform gravitational field $-g\hat{z}$ (compare Probs. 3.8 and 3.9).

(a) Find the condition for a steady circular orbit of radius r_0 and show that it is stable or unstable under small impulses along the surface transverse to the direction of motion according as $3f'(r_0) + r_0 f''(r_0)$ is positive or negative. For the stable orbits, find the frequency ω of small oscillations about the equilibrium configuration.

(b) Apply the preceding theory to each of the profiles (i) $z = -(R^2 - r^2)^{1/2}$, (ii) $z = \alpha R$, and (iii) $z = \alpha[1 - \cos(\pi r/R)]$, sketching the surface of revolution for $r < R$. Relate the angular velocity Ω to r_0 and determine the ratio ω^2/Ω^2. Indicate on the sketch the region within which the motion is stable.

4.6 A particle moves in a circular orbit under the influence of an attractive central potential $V(r)$.

(a) Expand the lagrangian to second order about the equilibrium coordinates $r = r_0$, $\phi = \Omega t$, where r_0 is the radius of the circular orbit and Ω is the equilibrium angular frequency. Find the condition for stability.

(b) Show how the same criterion can be derived directly from the equivalent one-dimensional potential.

(c) If $V(r) = -\lambda r^{-n}$, show that the oscillations are stable for $n < 2$.

(d) What is the criterion for stability if the potential is $-(\lambda/r)e^{-r/a}$?

4.7 Suppose that the gravitational potential of the sun has the form $-mMGr^{-1} + \delta V(r)$, where δV is a small perturbation of the newtonian potential. Show that the perihelion of a nearly circular orbit with mean radius r_0 precesses through an angle $-(\pi r_0/mMG)[2r_0\,\delta V'(r_0) + r_0^2\,\delta V''(r_0)]$ per cycle, omitting terms of order $(\delta V)^2$ and higher. Evaluate this quantity for the specific choice (see Prob. 1.12) $\delta V = -\alpha m(MG/rc)^2$.

4.8 Four massless rods of length L are hinged together at their ends to form a rhombus. A particle of mass M is attached at each joint. The opposite corners of the rhombus are joined by springs, each with a spring constant k. In the equilibrium (square) configuration, the springs are unstretched. The motion is confined to a plane, and the particles move only along the diagonals of the rhombus. Introduce suitable generalized coordinates and find the lagrangian of the system. Deduce the equations of motion and find the frequency of small oscillations about the equilibrium configuration.

4.9 (a) A molecule consists of three identical atoms located at the vertices of a 45° right triangle, with equal spring constants k between each pair of atoms. Derive the eigenvalue equation for planar motion and show that it has three degenerate modes at $\omega^2 = 0$. What is their physical interpretation? Find the three remaining nonzero eigenfrequencies.

(b) In this connection, show that the normal-mode solutions (24.41) with finite frequency for the system illustrated in Fig. 24.4 leave the center of mass at rest.

4.10 To discuss the case where the eigenvalue equation (22.24) has multiple roots and the corresponding eigenvalues are *degenerate* consider two degrees of freedom with

$$\underline{v} \equiv \begin{bmatrix} v & v_{12} \\ v_{12} & v \end{bmatrix} \qquad \underline{m} \equiv \begin{bmatrix} m & m_{12} \\ m_{12} & m \end{bmatrix}$$

(a) Solve for the eigenvalues and eigenvectors. Show that in the limits $(m_{12}, v_{12}) \to 0$ or $(m, v) \to 0$ the eigenvalues become degenerate with $\omega_1^2 = \omega_2^2$.

(b) Show that in these limits information is *lost* concerning the corresponding eigenvectors and that one can always find two *linearly independent* solutions of the form $z_\sigma^{(s)} = e^{i\phi_s}\tilde{\rho}_\sigma^{(s)}$ with $s = 1$, 2 and $\tilde{\rho}_\sigma^{(s)}$ real.

(c) Show that these solutions can be made orthonormal according to Eq. (22.34) by choosing

$$\rho_\sigma^{(1)} \equiv C_1 \tilde{\rho}_\sigma^{(1)} \qquad \text{and} \qquad \rho_\sigma^{(2)} \equiv C_2(\tilde{\rho}_\sigma^{(2)} - \alpha\tilde{\rho}_\sigma^{(1)})$$

with

$$\alpha \equiv \frac{\sum_\lambda \sum_\sigma \tilde{\rho}_\sigma^{(2)} m_{\sigma\lambda} \tilde{\rho}_\lambda^{(1)}}{\sum_\lambda \sum_\sigma \tilde{\rho}_\sigma^{(1)} m_{\sigma\lambda} \tilde{\rho}_\lambda^{(1)}}$$

What are C_1 and C_2? This is an example of the Gram-Schmidt orthogonalization procedure.†

4.11 N identical particles with masses m are connected by $N + 1$ identical massless springs with force constant k and fixed endpoints as shown in Fig. 24.1. In the original static configuration, each spring is stretched from its equilibrium length a_0 to a new length a. Construct the lagrangian for general small oscillations in a plane and show that the longitudinal and transverse modes *decouple*. Discuss how the frequencies depend on $a - a_0$. Discuss and interpret the behavior for $a < a_0$.

4.12 A linear chain consists of $2N$ equally spaced identical particles with mass m and separation a, coupled by springs with alternating constants k_1 and k_2.

(a) Use the periodicity of the system to find the dispersion relation and sketch the resulting function $\omega(q)$. (*Hint*: Assume $\eta_{2j} = \alpha e^{i(2jqa)}$ and $\eta_{2j+1} = \beta e^{i(2j+1)qa}$.) Consider the various limiting cases $k_1 \to 0$, $k_2 \to \infty$, $k_1 \to k_2$.

(b) Apply periodic boundary conditions to find the allowed frequencies.

4.13 Consider a long chain of identical simple pendulums of mass m and length l, coupled with springs of spring constant k. In equilibrium with the pendulums all vertical, the springs are unstretched and the separation between adjacent pendulums is a. Find the dispersion relation for the propagation of waves of small amplitude, and sketch the resulting function $\omega(q)$. If there are N pendulums, apply periodic boundary conditions to determine the allowed vibration frequencies. What is the frequency of the lowest mode?

4.14 (a) Use the finite-difference techniques developed for Eq. (24.21) to solve Eq. (24.39) subject to the fixed-end boundary conditions $\rho_0^{(n)} = \rho_{N+1}^{(n)} = 0$. Hence verify the form given in Eq. (24.57).

(b) Let $z_j \equiv \exp[i\pi j/(N + 1)]$ with $j = 1, \ldots, 2N + 2$ be the $2N + 2$ roots of 1 in the complex plane. If p is an integer with $|p| < 2N + 2$, prove that

$$\sum_{j=1}^{2N+2} (z_j)^p = (2N + 2)\delta_{p,0}$$

(c) Hence, or otherwise, demonstrate that the normal-mode eigenvectors

$$\rho_j^{(n)} = \left[\frac{2}{m(N + 1)}\right]^{1/2} \sin\frac{\pi nj}{N + 1}$$

satisfy the orthonormality relations (22.34) for N identical point masses equally spaced on a string with fixed ends.

4.15 Investigate the small-amplitude longitudinal oscillations of $N + 1$ identical point masses m joined by N identical springs with spring constant κ.

(a) Use Newton's laws of motion to find the appropriate boundary conditions on the end particles $j = 0$ and N.

(b) Generalize the treatment of Eq. (24.49) to determine the allowed eigenfrequencies and eigenfunctions.

† A general discussion of multiple eigenvalues is contained in Hildebrand [1], secs 1.13 and 1.18.

(c) Show that these solutions are either even (+) or odd (−) under the transformation $\eta_j(t) \leftrightarrow \eta_{N-j}(t)$ and that the corresponding allowed wave numbers satisfy the equations

$$\tan \tfrac{1}{2}ka = \pm (\cot \tfrac{1}{2}kaN)^{\pm 1}$$

(d) Discuss the graphical solution to these equations. Show that $ka = 0$ is one root and that each of the remaining $N - 1$ roots lies in the successive intervals $(n - 1)\pi/Na < k < n\pi/Na$ with $n = 1, \ldots, N$.

(e) Compare with the case of fixed ends.

4.16 Consider a two-dimensional square array of $2N$ massless strings of unperturbed length $(N + 1)a$ with fixed endpoints. The strings are stretched to a tension τ and have point masses m located at each intersection (ia, ja) where $1 \leq i, j \leq N$.

(a) Construct the equations of motion and lagrangian for the small transverse displacements η_{ij} of the masses. Find the exact dispersion relation $\omega(\mathbf{k})$ for traveling waves with wave vector $\mathbf{k} = k_x \hat{x} + k_y \hat{y}$.

(b) In the limit of a continuous mass distribution, show that this system obeys the two-dimensional wave equation. What is the corresponding dispersion relation for traveling waves?

(c) Demonstrate explicitly that the dispersion relation for the discrete case is anisotropic in the xy plane whereas that for the continuum is isotropic. Discuss this result.

(d) For the finite system $(N < \infty)$, find the exact normal-mode frequencies and compare with the continuum limit.

(e) For the continuous system, derive the lagrangian and equations of motion directly from $L = T - V$.

4.17 A uniform string with fixed ends has length l, mass density σ, and uniform tension τ. A point mass m is attached at its center.

(a) Assuming small transverse displacements in one plane, find the equations of motion for the two halves of the string and for the mass m.

(b) Show that the modes in which m moves have frequencies that satisfy the equation

$$\frac{2c}{\omega l} \cot \frac{\omega l}{2c} = \frac{m}{\sigma l}$$

where $c^2 = \tau/\sigma$. Discuss the solutions in the limiting cases $m \to 0$ and $m \to \infty$.

(c) Prove that the eigenfunctions satisfy the orthonormality relations

$$\int_0^l dx\, \rho_p(x) m(x) \rho_q(x) = \delta_{pq}$$

where $m(x) = \sigma + m\delta(x - \tfrac{1}{2}l)$ is the mass density.

SELECTED ADDITIONAL READING

1. F. B. Hildebrand, "Methods of Applied Mathematics," 2d ed., Prentice-Hall, Englewood Cliffs, N.J., 1965.

RIGID BODIES

The dynamics of rigid bodies constitutes one of the most beautiful aspects of classical mechanics, for the complicated and unexpected behavior of tops and gyroscopes all follows from Newton's laws of motion presented in Secs. 1 and 2. To extract this behavior, however, it is necessary to introduce some additional formalism. We start by reviewing the kinematic description for rotating coordinate axes (Sec. 6).

26 GENERAL THEORY

Consider N particles with all their relative separations fixed in magnitude. This system constitutes a rigid body; it has three translational degrees of freedom, three rotational degrees of freedom, and $3N - 6$ holonomic constraints.

Motion with One Arbitrary Fixed Point

Initially, it is convenient to eliminate the translational degrees of freedom by considering a rigid body with one arbitrary fixed point, which may be taken as the origin (Fig. 26.1). As in Sec. 6, we introduce two sets of coordinates axes, one set $\{\hat{e}_i^0\}$ fixed in an inertial frame and the other $\{\hat{e}_i\}$ fixed in the body. Let $\mathbf{r}$ be the coordinate vector of some (in general, moving) point. If $\{x_i\} \equiv \{\mathbf{r} \cdot \hat{e}_i\}$ denotes the set of associated coordinate markers in the body frame, the rate of change of $\mathbf{r}$ as seen by an observer in the body frame is

$$\left(\frac{d\mathbf{r}}{dt}\right)_{\text{body}} = \sum_{i=1}^{3} \hat{e}_i \frac{dx_i}{dt} \tag{26.1}$$

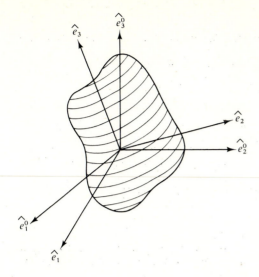

Figure 26.1 Transformation from an inertial frame $\{\hat{e}_i^0\}$ into a coordinate system $\{\hat{e}_i\}$ fixed in a rigid body.

Equation (7.11) relates this quantity to the rate of change seen in the inertial frame

$$\left(\frac{d\mathbf{r}}{dt}\right)_{\text{inertial}} = \left(\frac{d\mathbf{r}}{dt}\right)_{\text{body}} + \boldsymbol{\omega} \times \mathbf{r} \tag{26.2}$$

where $\boldsymbol{\omega}$ is the instantaneous angular velocity of the body-fixed axes as seen in the inertial frame. In particular, if $\mathbf{r}$ is fixed in the body, $(d\mathbf{r}/dt)_{\text{body}}$ vanishes and we have

$$\left(\frac{d\mathbf{r}}{dt}\right)_{\text{inertial}} = \boldsymbol{\omega} \times \mathbf{r} \tag{26.3}$$

In the inertial frame, the total kinetic energy of the rigid body is just the sum of the kinetic energy of each particle [compare Eq. (2.17)]

$$T = \frac{1}{2} \sum_p m_p v_p^2 = \frac{1}{2} \sum_p m_p (\boldsymbol{\omega} \times \mathbf{r}_p) \cdot (\boldsymbol{\omega} \times \mathbf{r}_p) \tag{26.4}$$

where we have used (26.3) and the index p is summed over all particles from 1 to N. Simple vector manipulations reduce this equation to the form

$$T = \frac{1}{2} \sum_p m_p [(\boldsymbol{\omega} \times \mathbf{r}_p) \times \boldsymbol{\omega}] \cdot \mathbf{r}_p = \frac{1}{2} \sum_p m_p [\omega^2 \mathbf{r}_p - (\boldsymbol{\omega} \cdot \mathbf{r}_p)\boldsymbol{\omega}] \cdot \mathbf{r}_p$$

$$= \frac{1}{2} \sum_p m_p [\omega^2 r_p^2 - (\boldsymbol{\omega} \cdot \mathbf{r}_p)^2] \tag{26.5a}$$

This relation describes an arbitrary set of mass points. In the limit of a continuous distribution with density $\rho(\mathbf{r})$, a volume element d^3r has mass $\rho(\mathbf{r})\,d^3r$, and Eq. (26.5a) becomes an integral

$$T = \tfrac{1}{2} \int d^3r \rho(\mathbf{r})[\omega^2 r^2 - (\boldsymbol{\omega} \cdot \mathbf{r})^2] \tag{26.5b}$$

In either form, the kinetic energy is a quadratic scalar function of the angular velocity $\boldsymbol{\omega}$. It can be written

$$T = \frac{1}{2} \sum_{i=1}^{3} \sum_{j=1}^{3} I_{ij} \omega_i \omega_j \tag{26.6}$$

where the *inertia tensor* is defined by the appropriate expressions obtained from Eqs. (26.5)

$$I_{ij} \equiv \sum_p m_p (\delta_{ij} r_p^2 - x_{pi} x_{pj}) \tag{26.7a}$$

$$I_{ij} \equiv \int d^3 r \, \rho(\mathbf{r})(\delta_{ij} r^2 - x_i x_j) \tag{26.7b}$$

Since we can always include the special case of point masses by allowing $\rho(\mathbf{r})$ to have delta-function contributions, discrete systems do not require a separate treatment.

It is important to emphasize that the kinetic energy in Eq. (26.4) is a scalar quantity. Thus the numerical value of Eqs. (26.5) and (26.6) is *independent of the choice of coordinate axes* used in constructing the components ω_i and I_{ij}. One possible set is the inertial frame, where the unit vectors $\{\hat{e}_i^0\}$ are fixed; this choice has the disadvantage that coordinates $x_{pi}^0 \equiv \mathbf{r}_p \cdot \hat{e}_i^0$ of the given particle and the associated inertia tensor [Eq. (26.7)] then *change with time* as the rigid body rotates. For this reason, the instantaneous body-fixed frame is much more convenient, despite the time dependence of its unit vectors $\{\hat{e}_i\}$, because in this frame Eq. (26.7) shows that the I_{ij} are time-independent constants that depend only on the *intrinsic mass distribution of the rigid body*.

The other important physical quantity is the angular-momentum vector

$$\mathbf{L} = \sum_p m\mathbf{r}_p \times \mathbf{v}_p \tag{26.8a}$$

Substitution of (26.3) and use of standard identities gives

$$\mathbf{L} = \sum_p m_p[\mathbf{r}_p \times (\boldsymbol{\omega} \times \mathbf{r}_p)] = \sum_p m_p[r_p^2 \boldsymbol{\omega} - (\boldsymbol{\omega} \cdot \mathbf{r}_p)\mathbf{r}_p] \tag{26.8b}$$

or
$$\mathbf{L} = \int d^3 r \, \rho(\mathbf{r})[r^2 \boldsymbol{\omega} - (\boldsymbol{\omega} \cdot \mathbf{r})\mathbf{r}] \tag{26.8c}$$

where the last form describes a continuous distribution. By inspection, $\mathbf{L}$ is a linear vector function of $\boldsymbol{\omega}$ with components

$$L_i = \sum_{j=1}^{3} I_{ij} \omega_j \tag{26.9}$$

Moreover, T is related to $\mathbf{L}$ according to

$$T = \frac{1}{2} \sum_{i=1}^{3} L_i \omega_i = \tfrac{1}{2} \mathbf{L} \cdot \boldsymbol{\omega} \tag{26.10}$$

The vectors and tensors in these equations can be projected onto any convenient coordinate axes at any instant of time, and we again use the instantaneous body-fixed system for constructing the quantities L_i and ω_i. By definition, this choice yields constant values for I_{ij}, but the components $L_i \equiv \mathbf{L} \cdot \hat{e}_i$ generally vary with time, not only because $\mathbf{L}$ may have time dependence when seen in the inertial frame but also because the rigid body and its associated body-fixed axes $\{\hat{e}_i\}$ tumble in space.

General Motion with No Fixed Point

The preceding analysis is easily generalized to the case of a rigid body in arbitrary motion. Let $\mathbf{R}$ be the coordinate vector of the center of mass as seen in the inertial frame $\{\hat{e}_i^0\}$. We first construct a second parallel set of coordinate axes $\{\hat{e}_i^0\}$ centered at $\mathbf{R}$ (Fig. 26.2). We shall refer to this as the center-of-mass (cm) frame. As in Sec. 2, we shall temporarily use primes to denote the coordinates of a vector measured from the center of mass, with $\mathbf{r} = \mathbf{R} + \mathbf{r}'$, and correspondingly for the time derivatives

$$\left(\frac{d\mathbf{r}}{dt}\right)_{\text{inertial}} = \dot{\mathbf{R}} + \left(\frac{d\mathbf{r}'}{dt}\right)_{\text{inertial}} \tag{26.11a}$$

Since the coordinate axes $\{\hat{e}_i^0\}$ describe both the inertial frame and the center-of-mass frame, the last term has the equivalent forms

$$\left(\frac{d\mathbf{r}'}{dt}\right)_{\text{inertial}} = \left(\frac{d\mathbf{r}'}{dt}\right)_{\text{cm}} = \sum_{i=1}^{3} \hat{e}_i^0 \frac{d}{dt}(\mathbf{r}' \cdot \hat{e}_i^0) \tag{26.11b}$$

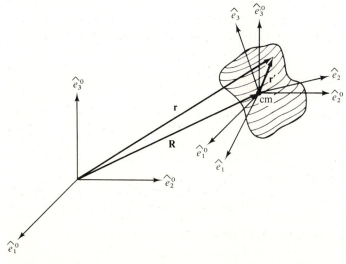

Figure 26.2 Transformation from inertial frame to center-of-mass frame for rigid body that is translating as well as rotating.

It is crucial to remember that the center-of-mass frame is not in general an inertial frame, since it may undergo *arbitrary* translational motion characterized by the vector $\mathbf{R}(t)$. As seen in Eq. (26.11b), however, this translation has no effect on $\mathbf{r}'$; if $\mathbf{r}'$ is fixed in the center-of-mass frame, for example, then $(d\mathbf{r}'/dt)_{\text{cm}}$ and $(d\mathbf{r}'/dt)_{\text{inertial}}$ both vanish.

Next, introduce another set of axes $\{\hat{e}_i\}$ with origin at $\mathbf{R}$ but fixed in the rigid body. If $(d\mathbf{r}'/dt)_{\text{body}}$ represents the time derivative as seen from this latter moving body frame, the arguments of Sec. 7 show that

$$\left(\frac{d\mathbf{r}'}{dt}\right)_{\text{cm}} = \left(\frac{d\mathbf{r}'}{dt}\right)_{\text{body}} + \boldsymbol{\omega} \times \mathbf{r}' \tag{26.12a}$$

where $\boldsymbol{\omega}$ is the instantaneous angular velocity of the body frame as seen in the center-of-mass frame at $\mathbf{R}$. Consequently, the velocity in the original inertial frame (26.11a) separates into its physically distinct components

$$\left(\frac{d\mathbf{r}}{dt}\right)_{\text{inertial}} = \dot{\mathbf{R}} + \left(\frac{d\mathbf{r}'}{dt}\right)_{\text{body}} + \boldsymbol{\omega} \times \mathbf{r}' \tag{26.12b}$$

In Sec. 2, we expressed the kinetic energy and angular momentum as a sum of contributions from the center-of-mass motion and the internal motion [Eqs. (2.18) and (2.13)]

$$T = \tfrac{1}{2}M\dot{\mathbf{R}}^2 + T' \tag{26.13a}$$

$$\mathbf{L} = \mathbf{R} \times (M\dot{\mathbf{R}}) + \mathbf{L}' \tag{26.13b}$$

Since $\mathbf{R}$ moves like the coordinate of a mass point M subject to the total external force $\mathbf{F}^{(e)}$

$$M\ddot{\mathbf{R}} = \sum_p \mathbf{F}_p^{(e)} = \mathbf{F}^{(e)} \tag{26.14}$$

the center-of-mass contributions to T and $\mathbf{L}$ are easily determined in any given situation. Moreover, the kinematic analysis of the second terms T' and $\mathbf{L}'$ is now identical with that leading to Eqs. (26.5) and (26.8). As a result, only the internal dynamics remains to be determined, and it is here sufficient to recall that torques $\boldsymbol{\Gamma}^{(e)'}$ about the center of mass affect

$$\mathbf{L}' \equiv \sum_p m_p \mathbf{r}'_p \times \left(\frac{d\mathbf{r}'_p}{dt}\right)_{\text{inertial}} = \sum_p m_p \mathbf{r}'_p \times \left(\frac{d\mathbf{r}'_p}{dt}\right)_{\text{cm}} \tag{26.15a}$$

according to the relation (2.15)

$$\left(\frac{d\mathbf{L}'}{dt}\right)_{\text{inertial}} = \left(\frac{d\mathbf{L}'}{dt}\right)_{\text{cm}} = \sum_p \mathbf{r}'_p \times \mathbf{F}_p^{(e)} \equiv \boldsymbol{\Gamma}^{(e)'} \tag{26.15b}$$

As in Eq. (26.11b), the left-hand side of this expression can be computed from the time rate of change of the components of $\mathbf{L}'$ in the center-of-mass frame. The right-hand side is the net torque about $\mathbf{R}$ arising from the external forces at a given instant in time. If $\boldsymbol{\Gamma}^{(e)'} = 0$, for example, then $\mathbf{L}'$ is an invariant vector in both

the inertial frame and the center-of-mass frame, although its components $\mathbf{L}' \cdot \hat{e}_i$ in the body frame usually vary with time. Equations (26.14) and (26.15) specify the general dynamics of rigid-body motion.

Inertia Tensor

We have seen that the inertia tensor of a rigid body is most readily expressed in the body-fixed frame, where its components are invariant constants that depend only on the internal distribution of matter. It is simplest first to consider the special body-fixed frame with its origin at the center of mass; letting $\bar{I}_{ij}$ denote the corresponding elements of the inertia tensor, we have

$$\bar{I}_{ij} = \sum_p m_p(\delta_{ij} x_p^2 - x_{pi} x_{pj}) \tag{26.16a}$$

$$\bar{I}_{ij} = \int d^3x \, \rho(\mathbf{x})(\delta_{ij} x^2 - x_i x_j) \tag{26.16b}$$

where $\mathbf{x}$ is the coordinate in this frame and the primes are henceforth omitted for notational simplicity. The constants $\bar{I}_{ij}$ completely characterize the internal dynamics of the rigid body, just as the mass does for translational motion; two different objects with the same inertia tensor will execute the same rotational motion. Furthermore, it is clear from Eq. (26.16) that $\bar{I}$ is a real symmetric matrix:

$$\bar{I}_{ij} = \bar{I}_{ji} = (\bar{I}_{ij})^* \tag{26.17}$$

In analyzing the motion of rigid bodies, it is frequently more convenient to use other sets of body-fixed axes. If the object has one point fixed in space, for example, that point becomes the natural origin. Fortunately, the *parallel-axis theorem* makes it easy to relate the inertia tensor in various frames. Consider two parallel sets of body-fixed coordinate axes, one with origin at the center of mass and one with origin displaced an arbitrary distance $\mathbf{a}$ (Fig. 26.3). Let $\mathbf{x}$ denote the coordinates of a point in the original body-fixed frame located at the center of mass, and let

$$\mathbf{y} = \mathbf{x} - \mathbf{a} \tag{26.18a}$$

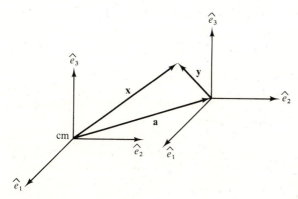

Figure 26.3 Basis for proof of parallel-axis theorem.

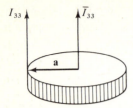

Figure 26.4 Application of parallel-axis theorem.

denote the coordinates relative to the new axes. It is slightly simpler to deal with discrete point masses, in which case the inertia tensor in the new frame has the components (along the same set of parallel axes)

$$I_{ij} = \sum_p m_p(\delta_{ij}y_p^2 - y_{pi}y_{pj}) \tag{26.18b}$$

Substitution of (26.18a) and use of Eq. (2.12) immediately give the desired *parallel-axis theorem*

$$I_{ij} = \bar{I}_{ij} + M(a^2\delta_{ij} - a_i a_j) \tag{26.19}$$

relating the inertia tensor in the two aligned coordinate frames where the second has its origin separated from the center of mass by a vector $\mathbf{a}$.

The parallel-axis theorem has great practical value. As a simple example, consider a uniform disk of radius a (Fig. 26.4). If $\bar{I}_{33}$ denotes its moment of inertia about a perpendicular axis through its center, Eq. (26.19) shows that the corresponding moment of inertia about a perpendicular axis through its edge is just

$$I_{33} = \bar{I}_{33} + Ma^2 \tag{26.20}$$

because $\mathbf{a} \cdot \hat{e}_3 \equiv \mathbf{a} \cdot \hat{e}_3^0 = 0$. Evaluation of $\bar{I}_{33}$ is elementary in cylindrical polar coordinates, whereas direct computation of I_{33} would be quite tedious.

Principal Axes

As noted previously, the elements of I_{ij} constitute a real symmetric matrix when evaluated in *any* body-fixed coordinate frame. For most purposes, it is convenient to restrict the choice further, specifying a special set (*principal axes*) in which I_{ij} is *diagonal*, with

$$I_{ij} = I_i\delta_{ij} \tag{26.21}$$

These principal axes have the following advantage. Let ξ_i temporarily denote the components of the instantaneous angular velocity $\boldsymbol{\omega}$ in this particular body-fixed frame. Comparison with Eqs. (26.6) and (26.9) shows that the corresponding kinetic energy and angular momentum have the form

$$T = \frac{1}{2}\sum_{ij=1}^{3} \xi_i I_i \delta_{ij} \xi_j = \frac{1}{2}\sum_{i=1}^{3} I_i \xi_i^2 \tag{26.22a}$$

$$L_i = \sum_{j=1}^{3} I_i \delta_{ij} \xi_j = I_i \xi_i \tag{26.22b}$$

with T a sum of squares and L_i a simple product. Since T cannot be negative, we see that $\underline{I}$ is positive definite with $I_i \geq 0$.

The existence of such a coordinate system is easily proved from the analysis of Chap. 4. In particular, we need diagonalize only the single real symmetric matrix $\underline{I}$. Consider a rigid body and an arbitrary body-fixed coordinate basis, with ω_i the components of the instantaneous angular velocity and I_{ij} the corresponding elements of the inertia tensor, which can always be evaluated in any given situation. We then seek a new set of eigenvectors with components e_i that satisfy the eigenvalue problem

$$\sum_{j=1}^{3} I_{ij}e_j = \lambda e_i \quad \text{or} \quad \sum_{j=1}^{3} (I_{ij} - \lambda\delta_{ij})e_j = 0 \tag{26.23}$$

This problem is identical with that for the normal modes of a system with three degrees of freedom, merely replacing the matrices $\underline{v}$ and $\underline{m}$ by the inertia tensor $\underline{I}$ and the unit tensor. As in that case, Eq. (26.23) has solutions only for certain values of λ, obtained from the algebraic secular equation

$$\det|I_{ij} - \lambda\delta_{ij}| = 0 \tag{26.24}$$

Let $\lambda_s(s = 1, 2, 3)$ be the three roots, and let $e_i^{(s)}$ be the components of the corresponding eigenvectors; note that the superscript (s) here labels the choice of eigenvector and the subscript i labels the components obtained by projecting $\underline{e}^{(s)}$ onto the original arbitrary orthogonal body-fixed coordinate axes.

We next introduce the modal matrix

$$\mathscr{A} = \begin{bmatrix} \underline{e}^{(1)} & \underline{e}^{(2)} & \underline{e}^{(3)} \\ \downarrow & \downarrow & \downarrow \end{bmatrix} \tag{26.25}$$

Since the unit matrix here replaces the mass matrix $\underline{m}$ from Chap. 4, Eq. (22.55) shows that $\mathscr{A}$ is an *orthogonal matrix*

$$\mathscr{A}^T\mathscr{A} = \mathscr{A}^{-1}\mathscr{A} = \underline{1} \tag{26.26}$$

Furthermore, the orthogonality relation (22.34) for the eigenvectors now takes the simple form

$$\sum_{i=1}^{3} e_i^{(s)}e_i^{(s')} = \underline{e}^{(s)T}\underline{e}^{(s')} = \mathbf{e}^{(s)} \cdot \mathbf{e}^{(s')} = \delta_{ss'} \tag{26.27}$$

demonstrating that the set $\{\mathbf{e}^{(s)}\}$ constitutes an orthonormal basis called the *principal axes*. Hence we can now denote this set of basis vectors $\{\hat{e}^{(s)}\}$, in accordance with our previous notation for unit vectors. The orthogonality of $\mathscr{A}$ ensures that the *principal axes differ from the original ones by a pure rotation*, the elements of $\mathscr{A}$ being the direction cosines between the two sets of axes.† In addition, the matrix $\mathscr{A}$ explicitly diagonalizes $\underline{I}$ through a *similarity transformation* [compare Eq. (22.59)]

$$\mathscr{A}^{-1}\underline{I}\mathscr{A} = \mathscr{A}^T\underline{I}\mathscr{A} = \lambda_D = \begin{bmatrix} I_1 & & \\ & I_2 & \\ & & I_3 \end{bmatrix} \tag{26.28}$$

† Goldstein (1950), chap. 4.

which incorporates the effect on $\underline{I}$ of the coordinate-axis rotation. In fact, given $\mathscr{A}$, it is not difficult to construct explicitly the axis and angle of the single rotation that brings the original axes onto the principal axes (Goldstein, 1950, chap. 4), but such detailed information is generally unnecessary and will be omitted here.

In the original arbitrary body axes, the instantaneous angular velocity has the components ω_i, which we denote by a three-component column vector

$$\underline{\omega} = \begin{bmatrix} \omega_1 \\ \omega_2 \\ \omega_3 \end{bmatrix} \tag{26.29}$$

Define a new vector $\underline{\xi}$ by the relation

$$\underline{\omega} = \mathscr{A}\,\underline{\xi} \tag{26.30a}$$

or its inverse

$$\underline{\xi} = \mathscr{A}^{-1}\underline{\omega} = \mathscr{A}^T\underline{\omega} \tag{26.30b}$$

Direct evaluation of this matrix product with Eq. (26.25) gives the simple result

$$\underline{\xi} = \begin{bmatrix} \xi_1 \\ \xi_2 \\ \xi_3 \end{bmatrix} = \begin{bmatrix} \hat{e}^{(1)} \to \\ \hat{e}^{(2)} \to \\ \hat{e}^{(3)} \to \end{bmatrix}\begin{bmatrix} \omega_1 \\ \omega_2 \\ \omega_3 \end{bmatrix} = \begin{bmatrix} \hat{e}^{(1)} \cdot \boldsymbol{\omega} \\ \hat{e}^{(2)} \cdot \boldsymbol{\omega} \\ \hat{e}^{(3)} \cdot \boldsymbol{\omega} \end{bmatrix} \tag{26.31}$$

Hence the element ξ_s is just the projection of $\boldsymbol{\omega}$ along the sth principal axis $\hat{e}^{(s)}$. Correspondingly, the kinetic energy (26.6) becomes

$$T = \tfrac{1}{2}\underline{\omega}^T\underline{I}\underline{\omega} = \tfrac{1}{2}(\underline{\xi}^T\mathscr{A}^T\underline{I}\mathscr{A}\,\underline{\xi}) = \tfrac{1}{2}\underline{\xi}^T\underline{\lambda}_D\,\underline{\xi} = \frac{1}{2}\sum_{s=1}^{3}\xi_s^2 I_s \tag{26.32}$$

where the last form follows from (26.28).

The same linear transformation (26.30) can be applied to any vector. For example, suppose that the angular momentum $\mathbf{L}$ has components L_i on the original body-fixed coordinate axes. Under the transformation (26.30b), it becomes

$$\underline{L}_{\text{new}} = \mathscr{A}^T\underline{L} = \begin{bmatrix} \hat{e}^{(1)} \cdot \mathbf{L} \\ \hat{e}^{(2)} \cdot \mathbf{L} \\ \hat{e}^{(3)} \cdot \mathbf{L} \end{bmatrix} \tag{26.33}$$

These three elements are precisely the projections of $\mathbf{L}$ along the principal axes. Alternatively, use of the relation $\underline{L} = \underline{I}\underline{\omega}$ [Eq. (26.9)] reproduces Eq. (26.22b)

$$\underline{L}_{\text{new}} = \mathscr{A}^T\underline{I}\underline{\omega} = \mathscr{A}^T\underline{I}\mathscr{A}\,\underline{\xi} = \underline{\lambda}_D\,\underline{\xi} = \begin{bmatrix} I_1\xi_1 \\ I_2\xi_2 \\ I_3\xi_3 \end{bmatrix} \tag{26.34}$$

It also follows from Eqs. (26.7), (26.25), and (26.28) that the principal moments of inertia I_s with $s = 1, 2, 3$ can be written in terms of the principal axes as

$$I_s = (\mathscr{A}^T\underline{I}\mathscr{A})_{ss} = \int d^3r\,\rho(\mathbf{r})[r^2 - (\mathbf{r}\cdot\hat{e}^{(s)})^2] \tag{26.35}$$

In studying bodies with a high degree of symmetry, it is often useful to introduce *body-associated coordinate systems*, rather than strictly body-fixed ones. Such a choice is useful whenever the body-associated axes remain principal axes for all times [see the discussion following Eq. (28.19) and Probs. 5.2 to 5.4].

27 EULER'S EQUATIONS

The preceding kinematic analysis introduced several basic concepts of rigid-body motion. We now turn to the central task of studying the detailed dynamics, which follows from Newton's second law. For this purpose, recall that the degrees of freedom separate into a set of three translational coordinates and a set of three rotational coordinates. Equation (26.14) reduces the first set to an equivalent one-body problem for the center of mass **R**, leaving only the rotational variables to be determined. In principle, any three parameters that fix the orientation of the rigid body could serve as the rotational variables, and the solution would then consist in determining their explicit time dependence. Sections 30 and 31 will illustrate such a procedure in studying the motion of a symmetric top. Here, however, it is more convenient to use a less direct approach based on the rate of change of the angular momentum. This latter method has great formal simplicity, but Sec. 28 will show that its interpretation can become quite intricate.

The fundamental equation (2.9)

$$\left(\frac{d\mathbf{L}}{dt}\right)_{\text{inertial}} = \mathbf{\Gamma}^{(e)} \tag{27.1}$$

describes the rate of change of the total angular momentum **L** as seen in any inertial frame. It is valid for two distinct cases:

1. If the origin of coordinates is *fixed* in some inertial frame, then Eq. (27.1) holds with **L** and $\mathbf{\Gamma}^{(e)}$ computed about that point [compare Eq. (2.9)].
2. In general, the rate of change of the angular momentum is not given by the torques taken about an accelerated origin; nevertheless, Eq. (27.1) remains valid if the origin is located at the center of mass, independent of its translational motion [compare Eq. (26.15b)]. This special but important case accounts for the preeminent role of the internal coordinate variables $\{\mathbf{r}_i\}$ measured relative to the center of mass.

Assume that one of these conditions holds, and consider a body-fixed coordinate frame oriented along the principal axes $\hat{e}^{(s)}$. Substitution of the familiar relation (7.11) into Eq. (27.1) yields

$$\left(\frac{d\mathbf{L}}{dt}\right)_{\text{inertial}} = \left(\frac{d\mathbf{L}}{dt}\right)_{\text{body}} + \mathbf{\omega} \times \mathbf{L} = \mathbf{\Gamma}^{(e)} \tag{27.2}$$

This vector relation will now be projected onto the principal body axes. Let $\Gamma_s^{(e)} = \mathbf{\Gamma}^{(e)} \cdot \hat{e}^{(s)}$, $L_s = \mathbf{L} \cdot \hat{e}^{(s)}$, and $\omega_s = \mathbf{\omega} \cdot \hat{e}^{(s)}$ be the components of the external torque, angular momentum, and angular velocity along the (typically moving) principal axes.† Equation (26.34) shows that $L_s = I_s \omega_s$ and its rate of change $(dL/dt)_{\text{body}}$ seen in the body frame has the corresponding components $(dL_s/dt)_{\text{body}} = I_s \, d\omega_s/dt$ because I_s are constants characterizing the fixed distribution of matter in the rigid body. Finally, the components of $\boldsymbol{\omega} \times \mathbf{L}$ are readily determined. Along $\hat{e}^{(1)}$, for example, we find $\omega_2 L_3 - \omega_3 L_2 = \omega_2 \omega_3 (I_3 - I_2)$. A combination of these relations immediately yields *Euler's equations*

$$I_1 \frac{d\omega_1}{dt} = \omega_2 \omega_3 (I_2 - I_3) + \Gamma_1^{(e)} \qquad (27.3a)$$

$$I_2 \frac{d\omega_2}{dt} = \omega_3 \omega_1 (I_3 - I_1) + \Gamma_2^{(e)} \qquad (27.3b)$$

$$I_3 \frac{d\omega_3}{dt} = \omega_1 \omega_2 (I_1 - I_2) + \Gamma_3^{(e)} \qquad (27.3c)$$

where the second and third follow by cyclic permutation of the indices in the first.

Although elegant and formally compact, these equations involve an unusual set of dynamical variables ω_s and driving torques $\Gamma_s^{(e)}$ because they are projected onto the *time-dependent body-fixed principal axes*. Thus integration of Euler's equations to find $\omega_s(t)$ only describes the motion seen by a body-fixed observer, and the solution for an inertial observer would need substantially more work. Indeed, the mere specification of $\Gamma_s^{(e)}$ requires this additional information, so that the general problem of rotational motion driven by external torques must be solved self-consistently. As a result of this complication, Euler's equations are most useful when the principal axes are partially constrained, reducing the number of degrees of freedom, or for torque-free motion. In the latter case, $\mathbf{L}$ is a strict constant of the motion, with fixed components in the inertial frame, although the motion of the principal axes would lead a body-fixed observer to assert that the components $L_s \equiv \mathbf{L} \cdot \hat{e}^{(s)}$ generally vary with time.

28 APPLICATIONS

The formalism of rigid-body dynamics has numerous applications, ranging from gyroscopic navigation to the coupled motion of the earth, moon, and sun. For the present purposes, however, a few illustrative examples must suffice.

Compound Pendulum: Kater's Pendulum and the Center of Percussion

A compound pendulum is a rigid body constrained to rotate about a fixed axis, which we denote $\hat{e}_3^0 = \hat{e}_3$; this axis is stationary both in the inertial frame and in

---

† Since no confusion can now arise, we shall henceforth use $\boldsymbol{\omega} = (\omega_1, \omega_2, \omega_3)$ for the angular velocity even when decomposed with respect to the body-fixed principal axes. We also use the notation of Eq. (27.1) although the analysis applies to both cases (1) and (2) discussed following that result.

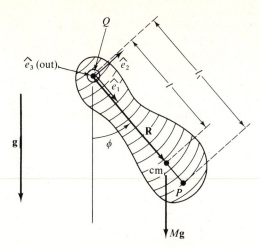

Figure 28.1 Compound pendulum.

the body frame. Construct the plane perpendicular to $\hat{e}_3$ that contains the center of mass, and let Q be the point in this plane lying on the fixed axis. The point Q serves as a natural origin of coordinates (see Fig. 28.1). Let the center of mass lie a distance l from the origin, with $\hat{e}_1$ a unit vector oriented toward the center of mass $\mathbf{R} \equiv l\hat{e}_1$. We also introduce another body-fixed unit vector $\hat{e}_2 = \hat{e}_3 \times \hat{e}_1$ perpendicular to $\mathbf{R}$ and to the rotation axis; this set $\{\hat{e}_i\}$ constitutes an orthonormal body-fixed frame.

The system has one degree of freedom, wholly determined by the angle ϕ, whose time derivative is $\dot{\phi} = \omega_3$. The first two components of Eq. (27.1) may involve torques exerted in the 1 and 2 directions by the support axis (through Q, along $\hat{e}_3$). If there is no friction in the bearings, the support axis exerts no torque in the third direction. Thus the torque in this direction comes entirely from the additional external forces, in this case gravity. The third component of Eq. (27.1) is therefore the useful one and reads

$$\frac{dL_3}{dt} = \Gamma_3^{(e)} \tag{28.1}$$

where the time derivative can be interpreted as describing either the inertial or body-fixed frame, because $\hat{e}_3^0 = \hat{e}_3$. Equation (26.9) shows that

$$L_3 = \sum_{j=1}^{3} I_{3j}\omega_j = I_{33}\dot{\phi} \tag{28.2}$$

where the unit vectors $\hat{e}_i$ are determined solely by the geometrical configuration and need not be principal axes. We next introduce the abbreviation

$$I \equiv I_{33} = \int d^3x\, \rho(\mathbf{x})(x^2 - x_3^2)$$

$$= \int d^3x\, \rho(\mathbf{x})(x_1^2 + x_2^2) \equiv \int d^3x\, \rho(\mathbf{x})r_\perp^2 \tag{28.3}$$

which is a constant characterizing the matter in the body; it involves only the perpendicular distance from the 3 axis, $r_\perp^2 = x_1^2 + x_2^2$, and is therefore independent of the particular choice of $\hat{e}_1$ and $\hat{e}_2$. The required torque arises from gravity

$$\boldsymbol{\Gamma}^{(e)} = \sum_p m_p \mathbf{r}_p \times \mathbf{g} = M\mathbf{R} \times \mathbf{g} \tag{28.4a}$$

where Eq. (2.1) has been used to identify the center of mass. Evaluating the third component of this vector product gives

$$\Gamma_3^{(e)} = -Mgl \sin \phi \tag{28.4b}$$

and substitution into Eq. (28.1) yields the equation of motion of an arbitrary compound pendulum

$$I\ddot{\phi} = -Mgl \sin \phi \tag{28.5}$$

For small oscillations ($\phi \ll 1$), the motion becomes simple harmonic

$$\ddot{\phi} = -\Omega^2 \phi \tag{28.6}$$

with an angular frequency

$$\Omega = \left(\frac{Mgl}{I}\right)^{1/2} \tag{28.7}$$

and period $\tau = 2\pi/\Omega$. To study this quantity in more detail, we use the parallel-axis theorem (26.19) to write

$$I \equiv I_{33} = \bar{I}_{33} + Ml^2 \tag{28.8}$$

where $\bar{I}_{33}$ is the usual moment of inertia about an axis along $\hat{e}_3$ through the center of mass. It is conventional to rewrite $\bar{I}_{33}$ by introducing the *radius of gyration $\bar{k}$* about the center of mass

$$\bar{I}_{33} \equiv M\bar{k}^2 \tag{28.9}$$

Evidently, $\bar{k}$ is the equivalent radial distance of a point mass M with moment of inertia $\bar{I}_{33}$. In this way, Eq. (28.8) takes the form

$$I = M(\bar{k}^2 + l^2) = Mk^2 \tag{28.10a}$$

where

$$k \equiv (\bar{k}^2 + l^2)^{1/2} \tag{28.10b}$$

is the radius of gyration about the axis of support through Q. A combination of these relations simplifies Eq. (28.7) to

$$\Omega_Q^2 = \frac{gl}{\bar{k}^2 + l^2} = \frac{g}{l_1} \tag{28.11a}$$

where

$$l_1 \equiv l + \frac{\bar{k}^2}{l} \tag{28.11b}$$

is the length of the equivalent simple pendulum. Note that $l_1 > l$ because $\bar{k}^2$ is necessarily positive.

Equation (28.11*b*) suggests considering a point P, located a length l_1 along $\hat{e}_1$ (Fig. 28.1). Note that the location of P depends on that of Q and alterations in Q in turn change P. We shall see that P has a simple physical interpretation as the *center of percussion*. First, however, we study Kater's reversible pendulum, which is a compound pendulum that can be inverted and suspended from an axis along $\hat{e}_3$ through the point P as well as from that through Q. In the reversed configuration, Eq. (28.7) gives the corresponding oscillation frequency

$$\Omega_P^2 = \frac{Mg(l_1 - l)}{M[(l_1 - l)^2 + \bar{k}^2]}$$

and a little algebra with Eq. (28.11*b*) shows that the two modes of oscillation have *identical frequencies* and *periods*:

$$\Omega_P^2 = \frac{Mg\bar{k}^2/l}{M[(\bar{k}^2/l)^2 + \bar{k}^2]} = \frac{g}{l + \bar{k}^2/l} = \Omega_Q^2 \tag{28.12}$$

In practice, an experiment consists in determining Ω_Q and then seeking the point P that satisfies Eq. (28.12). The measured period and separation l_1 between P and Q provide an absolute value for the earth's gravitational acceleration. This device was used extensively in the nineteenth century to determine local variations in g. Accurate studies require numerous corrections, including those for finite-amplitude motion, air drag, elasticity of the support, and others. In honor of Galileo, g is measured in units of 1 cm s^{-2} = 1 gal. Modern *absolute* determinations of g use either a reversible pendulum or a falling body. They have an uncertainty of ≈ 1 mgal $\equiv 10^{-3}$ gal, which should be compared with the latitudinal variation of ≈ 5 gal. In contrast, *relative* measurements with calibrated springs can be as accurate as $\approx 10^{-2}$ mgal.†

These ideas have an interesting and unexpected application in the response of a compound pendulum to a transverse force, e.g., the impulsive transverse force imparted when hitting a baseball with a bat. Consider the configuration in Fig. 28.2, where the object suspended from an axis along $\hat{e}_3$ through Q as in Fig. 28.1 is subjected to a transverse force $\mathbf{f} = f\hat{e}_2$ at a distance l' from Q. Here we neglect gravity, which is irrelevant for these considerations. In general, the applied force generates an additional reaction force at the point of support Q, and we now investigate how the transverse component $-f_r\hat{e}_2$ of this reaction force varies with the magnitude and location of the applied transverse force $\mathbf{f}$.‡

Newton's second law determines the motion of the center of mass $\mathbf{R}$. In the presence of the applied force $\mathbf{f} = f\hat{e}_2$ (see Fig. 28.2), the component of the equation of motion for $\mathbf{R}$ in the transverse direction $\hat{e}_2$ perpendicular to $\hat{e}_1$ and to the axis of rotation $\hat{e}_3$ is

$$M\ddot{\mathbf{R}} \cdot \hat{e}_2 = f - f_r \tag{28.13}$$

† See, for example, Garland [1], chap. 2, and Heiskanen and Vening Meinesz [2], chap. 4.
‡ With a bat, an additional longitudinal force of constraint keeps the center of mass in circular orbit.

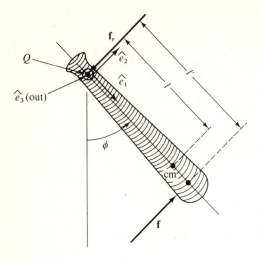

Figure 28.2 Response of a compound pendulum to a transverse force **f**, with f_r the transverse part of the reaction force.

Furthermore, the constraints restrict **R** to change only in the instantaneous direction $\hat{e}_2$, with $\ddot{\mathbf{R}} \cdot \hat{e}_2 = l\ddot{\phi}$. Thus Eq. (28.13) can be rewritten

$$Ml\ddot{\phi} = f - f_r \tag{28.14}$$

The rate of change of angular momentum about Q [Eq. (28.1)] provides a second equation. We assume here that the $\{\hat{e}_i\}$ are principal axes so that there is only angular momentum and torque in the 3 direction. In this case Eqs. (27.1) and (26.22b) give

$$\frac{dL_3}{dt} = I_3\ddot{\phi} = \Gamma_3^{(e)} = fl' \tag{28.15a}$$

because f_r exerts no torque about Q. Comparison of Eqs. (26.35) and (28.3) shows that $I_3 \equiv I$ as defined in the preceding discussion and hence Eq. (28.10) yields

$$I\ddot{\phi} \equiv M(l^2 + \bar{k}^2)\ddot{\phi} = fl' \tag{28.15b}$$

The ratio of the two equations (28.14) and (28.15b) is easily simplified to the final form

$$\frac{f_r}{f} = \frac{l_1 - l'}{l_1} \tag{28.16}$$

where l_1 is the length of the equivalent simple pendulum given in (28.11b), indicated by P in Fig. 28.1. In particular, if the force **f** is applied at P (namely $l' = l_1$), then f_r *vanishes*, and there is no transverse reaction force at the point of support; this observation explains why P is often called the *center of percussion*. If the object is struck beyond the center of percussion ($l' > l_1$), the reaction force is negative and acts in the same direction as **f**, and conversely for $l' < l_1$. Such behavior is readily demonstrated by tapping a loosely suspended baseball bat at various positions along its length; it also accounts for the "sweet spot" on a tennis racket.

Rolling and Sliding Billiard Ball

The game of billiards or pool illustrates many of these same ideas. Although a precise description would be very complicated, we can idealize the situation to allow two basic types of motion, sliding and rolling. In the first case, the ball experiences an opposing force $\mathbf{F}_f$ of sliding friction; in the second, the friction is much smaller and will here be neglected entirely.

As a first example, consider a homogeneous ball of mass M and radius a, struck impulsively dead center with a horizontal force. How long will it take to acquire pure rolling motion? Let x denote the horizontal coordinate (direction $\hat{e}_1^0$) of the center of mass (Fig. 28.3a); after the impulsive force has ceased to act, only the friction remains, and x obeys the simple equation

$$\ddot{x} = -\mu g \tag{28.17}$$

where the force of sliding friction is assumed independent of speed and proportional to the gravitational force

$$\mathbf{F}_f = -\mu g M \hat{e}_1^0 \tag{28.18}$$

with μ the coefficient of sliding friction. The absence of vertical acceleration implies an upward reaction force Mg, but this component is irrelevant here.

The rate of change of angular momentum about the center of mass provides a second dynamical relation. It is simplest to consider the behavior after the impulse has ceased. Taking moments about the center of mass in the direction $\hat{e}_3^0$ (see Fig. 28.3a), we have from the second case of Eq. (27.1)

$$\frac{dL_3}{dt} = I_3\ddot{\phi} = \Gamma_3^{(e)} = F_f a \tag{28.19}$$

since any system of orthogonal axes through the center of a sphere is a set of principal axes. The moment of inertia of a sphere through its center is given by Eq. (26.35)

$$I_3 \equiv I = \int_{x<a} d^3x\, \rho(x_1^2 + x_2^2) \tag{28.20a}$$

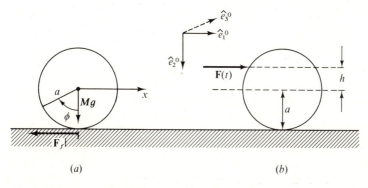

$$
\begin{array}{cc}
(a) & (b)
\end{array}
$$

Figure 28.3 Motion of a billiard ball (a) subject to sliding friction $\mathbf{F}_f$ and (b) struck impulsively a distance h above its median plane.

where $\rho = 3M/4\pi a^3$ is the mass density. The homogeneity and isotropy permit us to evaluate I in spherical polar coordinates, giving the familiar result

$$I = \frac{2}{3} \int_{x<a} d^3x \; \rho(x_1^2 + x_2^2 + x_3^2) = \tfrac{2}{3}(4\pi\rho) \int_0^a x^4 \, dx = \tfrac{2}{5}Ma^2 \qquad (28.20b)$$

In this way, we obtain the second dynamical equation

$$a\ddot{\phi} = \tfrac{5}{2}\mu g \qquad (28.21)$$

which must be solved simultaneously with Eq. (28.17).

Immediately after the initial dead-center impulsive blow, the ball satisfies the conditions $x = 0$, $\phi = 0$, $\dot{x} = v_0$, $\dot{\phi} = 0$, where v_0 is the initial velocity and the last assumption of pure initial sliding follows from the symmetry of the impulse (this assertion is proved in more detail below). Integrating Eqs. (28.17) and (28.21) once and incorporating the initial behavior, we find

$$\dot{x} = v_0 - \mu g t$$
$$a\dot{\phi} = \tfrac{5}{2}\mu g \qquad (28.22)$$

Note that friction acts immediately, starting to slow the ball down and bring it into rotation. On the other hand, pure rolling without sliding can occur only for $\dot{x} = a\dot{\phi}$, and this condition is violated until a time

$$t_1 = \frac{2v_0}{7\mu g} \qquad (28.23)$$

when sliding friction abruptly ceases and the ball rolls freely. The transition to pure rolling occurs when the ball has traveled a distance $x_1 = 12v_0^2/49\mu g$ and has an instantaneous horizontal speed $\dot{x}(t_1) = v_1 = \tfrac{5}{7}v_0$. These relations make clear the essential role of friction, for t_1 and x_1 become infinite as $\mu \to 0$; v_1, in contrast, is independent of μ, but the ball would never slow down to that speed for $\mu = 0$.

Similar treatments describe more general situations, but we shall consider only one other case, a horizontal impulse delivered to a stationary ball at a distance h above the center. Equations (28.17) and (28.21) again apply after the impulse has ceased, but determining the initial conditions is now more subtle. Let $F_x(t)$ denote the short time-dependent force. The time integral of Newton's second law relates the impulse to the net change of linear momentum

$$\text{Impulse} \equiv \int F_x(t) \, dt = \Delta p_x = M v_0 \qquad (28.24)$$

where v_0 is the horizontal speed immediately after the blow. Similarly, the time integral of Eq. (27.1) relates the impulsive torque about the center of mass to the change in angular momentum

$$\int h F_x(t) \, dt = h \cdot \text{impulse} = \Delta L = I\omega_0 \qquad (28.25)$$

where ω_0 is the initial angular velocity. A combination of Eqs. (28.20), (28.24), and (28.25) gives the appropriate initial condition

$$hv_0 = \tfrac{2}{5}a^2\omega_0 \qquad (28.26)$$

which is independent of the impulse. If $h = \tfrac{2}{5}a$, then $v_0 = a\omega_0$, and the ball immediately starts to roll without sliding; in that case, sliding friction never acts. If $h < \tfrac{2}{5}a$, the ball slides faster than it rolls ($v_0 > a\omega_0$), and sliding friction retards the horizontal motion while providing an angular torque to increase the angular velocity (as shown in Fig. 28.3a for $h = 0$). Even more interesting is the behavior for $h > \tfrac{2}{5}a$, when the ball rolls too fast for its translational motion; sliding friction then acts forward, increasing the linear velocity and decreasing the angular velocity. Thus the ball actually speeds up, as is readily demonstrated with a billiard ball on a felt surface. In all these cases, friction has the net effect of bringing the ball into pure rolling motion.†

Torque-free Motion: Symmetric Top

Euler's equations have their most important application in the *torque-free* motion of rigid bodies, when Eqs. (27.3) have the simpler form

$$I_1\dot{\omega}_1 = \omega_2\omega_3(I_2 - I_3) \qquad (28.27a)$$

$$I_2\dot{\omega}_2 = \omega_3\omega_1(I_3 - I_1) \qquad (28.27b)$$

$$I_3\dot{\omega}_3 = \omega_1\omega_2(I_1 - I_2) \qquad (28.27c)$$

Integration of these *nonlinear* differential equations determines the time development of the ω_i, which are the components of the angular velocity along the instantaneous body-fixed principal axes. In general, the ω_i bear no obvious relation to the components of angular velocity as seen in an inertial frame.

For a first example, we initially consider a *symmetric top* with

$$I_1 = I_2 \neq I_3 \qquad \text{symmetric top} \qquad (28.28)$$

and assume, for definiteness, a disklike flattened shape ($I_3 > I_1$), such as occurs for the earth. Equation (28.27c) shows that ω_3 is then *constant*, and the remaining equations become

$$\dot{\omega}_1 = -\Omega\omega_2 \qquad (28.29a)$$

$$\dot{\omega}_2 = \Omega\omega_1 \qquad (28.29b)$$

where we have introduced the abbreviation

$$\Omega \equiv \omega_3 \frac{I_3 - I_1}{I_1} \qquad (28.30)$$

Here $\Omega > 0$. This pair of equations (28.29) has a structure similar to Eq. (12.7) found in studying the Foucault pendulum, and an analogous solution shows that

---

† Other interesting examples are discussed in Sommerfeld [1952], pp. 158–161, 250–251, and 270–273.

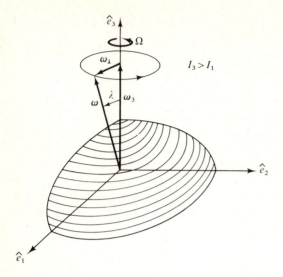

Figure 28.4 Motion of symmetric top in body-fixed frame.

both ω_1 and ω_2 vary harmonically with angular frequency Ω. Suppose that $\boldsymbol{\omega}$ initially lies in the 1-3 plane at a polar angle λ from the 3 axis (see Fig. 28.4). Incorporating the initial conditions $\omega_1 = \omega \sin \lambda, \omega_2 = 0, \omega_3 = \omega \cos \lambda$, we easily find the subsequent solution

$$\omega_1(t) = \omega \sin \lambda \cos \Omega t$$

$$\omega_2(t) = \omega \sin \lambda \sin \Omega t \qquad (28.31)$$

$$\omega_3 = \omega \cos \lambda$$

Note that the perpendicular projection of the angular velocity has constant magnitude $\omega_\perp = (\omega_1^2 + \omega_2^2)^{1/2} = \omega \sin \lambda$, so that the vector $\boldsymbol{\omega}$ maintains the fixed inclination λ with the symmetry axis. As time progresses, the magnitude of $\boldsymbol{\omega}$ also remains constant, but the axis of rotation $\hat{\omega}$ precesses about the symmetry axis in the positive (counterclockwise) sense with angular velocity Ω, sweeping out a cone with apex angle λ.

For the earth the asymmetry is characterized by $(I_3 - I_1)/I_1 \approx \frac{1}{305}$, and λ happens to be very small, so that $\omega_3 \approx \omega$ and $\omega_\perp \ll \omega$. Equation (28.30) then yields $\Omega \approx \omega/305$, implying a precession period of about 305 days for one complete cycle of the earth's rotation axis about the symmetry axis, as seen by a terrestrial observer. In fact, astronomical observations indeed verify such motion (Chandler's period), but it is somewhat irregular, with a mean period of ≈ 14 months and a mean displacement of the rotation axis on the surface of the earth at the north pole of ≈ 4 m (namely $\lambda \approx 6 \times 10^{-7}$ rad). The discrepancies are attributed variously to the earth's elasticity, its internal fluid structure, and annual meteorological variations.†

† See, for example, Stacey [3], secs. 3.3 and 3.4.

We again emphasize that these relatively simple solutions (28.31) hold only in the body frame. Determining the actual trajectory as seen in an inertial frame is considerably more difficult; this question will be deferred until we have studied the Euler angles, which constitute a more appropriate set of variables.

Torque-free Motion: Asymmetric Top

The general torque-free motion of a rigid body with unequal moments of inertia requires the integration of Eqs. (28.27). Just as in the study of central forces in Sec. 3, the existence of conserved quantities facilitates this procedure considerably. First, the kinetic energy is a strict constant; when expressed in terms of the angular velocity and inertia tensor projected onto the instantaneous body-fixed principal axes, it has the form [compare Eq. (26.32)]

$$E = T = \tfrac{1}{2}(I_1\omega_1^2 + I_2\omega_2^2 + I_3\omega_3^2) \tag{28.32}$$

and any time-dependence of ω_i must keep this first integral constant. The angular-momentum vector $\mathbf{L}$ is more difficult, for its components along the body-fixed axes change in time as the body rotates, even though $\mathbf{L}$ has constant components in the inertial frame. We avoid this complication by considering the quantity $L^2 = \mathbf{L} \cdot \mathbf{L}$, which is constant in all coordinate frames. In this way, Eq. (26.34) immediately gives another first integral

$$L^2 = I_1^2\omega_1^2 + I_2^2\omega_2^2 + I_3^2\omega_3^2 \tag{28.33}$$

which implies an additional constraint on the allowed variation of ω_i. Although these first integrals are here obtained from physical considerations, they also follow by suitable manipulations of the nonlinear differential equations (28.27).

Given these two constants of the motion, we can solve for ω_1 and ω_2 (say) in terms of E, L, and ω_3. Euler's equation (28.27c) then yields a definite integral for $\omega_3(t)$, analogous to Eqs. (3.10) and (3.11) for $r(t)$ and $\phi(t)$ in the analysis of central forces. The detailed expressions involve quite complicated elliptic integrals, however, and will be omitted here. Nevertheless, the explicit solutions for $\omega_i(t)$ in the body-fixed frame do have one important virtue, permitting a complete analysis of the motion as seen in the inertial frame.[†] In Sec. 30 the Euler angles will provide an equivalent but more transparent description of the same phenomenon for the special case of a symmetric top.

The analysis of Eqs. (28.27) becomes simpler if the rotation axis is near one of the principal axes. This situation is particularly interesting because it uses the formalism developed in Chap. 4 to study small oscillations. Suppose that the body initially rotates about an axis $\hat{\omega}$ near one of the principal axes $\hat{e}_3$ (say). Exactly as in Chap. 4, we write

$$\omega_1 = \eta_1(t) \qquad \omega_2 = \eta_2(t) \qquad \omega_3 = \omega_0 + \eta_3(t) \tag{28.34}$$

[†] Landau and Lifshitz (1960), sec. 37.

and linearize the small quantities $\eta_i(t)$. A straightforward expansion gives the linear equations

$$I_1 \dot{\eta}_1 = (I_2 - I_3)\omega_0 \eta_2$$

$$I_2 \dot{\eta}_2 = (I_3 - I_1)\omega_0 \eta_1 \tag{28.35}$$

$$I_3 \dot{\eta}_3 = O(\eta_1 \eta_2)$$

The time derivative of the first can be combined with the second to yield

$$\ddot{\eta}_1 = -\Omega_0^2 \eta_1 \tag{28.36a}$$

and similarly
$$\ddot{\eta}_2 = -\Omega_0^2 \eta_2 \tag{28.36b}$$

where
$$\Omega_0^2 = \omega_0^2 \frac{(I_3 - I_1)(I_3 - I_2)}{I_1 I_2} \tag{28.37}$$

The linearized motion is simple harmonic with frequency Ω_0. This behavior is stable only for *positive* Ω_0^2 and hence real Ω_0. Inspection of Eq. (28.37) shows that I_3 must be either the largest or smallest of the three principal moments of inertia. Thus *an asymmetric body can rotate stably about two of its principal axes, those with maximum or minimum moments of inertia.* In contrast, *motion about the intermediate principal axis is unstable* because $\eta_1(t)$ and $\eta_2(t)$ would then initially grow exponentially with time. Such behavior violates the original assumption of small displacements, and more exact solutions are needed. Nevertheless, the condition $\Omega_0^2 < 0$ correctly indicates the presence of an instability. These conclusions are easily verified experimentally with a wooden block or a tennis racket.

29 EULER ANGLES

Three parameters suffice to specify the orientation of a rigid body. Among various possible choices, the Euler angles turn out to be most convenient, and we shall analyze them in detail. Regrettably, authors in different fields of physics tend to use different conventions; here we follow the common quantum-mechanical definition, which differs slightly from that now standard in classical mechanics.†

Consider two orthonormal triads with common origin, a fixed inertial triad $\{\hat{e}_1^0, \hat{e}_2^0, \hat{e}_3^0\}$ and a movable body-fixed triad $\{\hat{e}_1, \hat{e}_2, \hat{e}_3\}$ initially coincident with the first. Perform the following sequence of operations on the body-fixed triad.

1. Rotate about $\hat{e}_3^0$ through an angle α, bringing $\hat{e}_2$ to the orientation denoted "line of nodes" in Fig. 29.1.
2. Rotate about the line of nodes through an angle β, bringing $\hat{e}_3$ to the final orientation shown in Fig. 29.1.

† Compare Goldstein (1950), pp. 107–109, with, for example, Edmonds [4], pp. 6–8, or Rose [5], pp. 48–52. In accordance with most quantum-mechanics texts, we measure the line of nodes from y instead of x, which has the advantage of yielding real representations for the quantum-mechanical rotation matrices [see also item (ii) below].

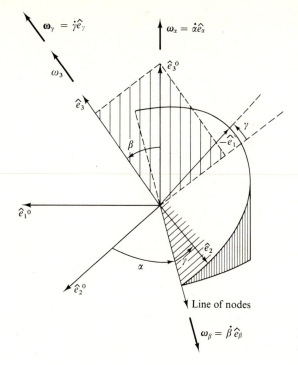

$\boldsymbol{\omega}_\gamma = \dot{\gamma}\hat{e}_\gamma$

$\boldsymbol{\omega}_\alpha = \dot{\alpha}\hat{e}_\alpha$

ω_3

$\hat{e}_3^0$

$\hat{e}_3$

γ

β

$-\hat{e}_1$

$\hat{e}_1^{\,0}$

$\hat{e}_2$

α

γ

$\hat{e}_2^{\,0}$

Line of nodes

$\boldsymbol{\omega}_\beta = \dot{\beta}\hat{e}_\beta$

Figure 29.1 Definition of Euler's angles α, β, γ.

3. Rotate about the new $\hat{e}_3$ through an angle γ. This prescription completely determines the orientation of the body axes $\hat{e}_i$ relative to the fixed ones $\hat{e}_i^0$. The corresponding *Euler angles* α, β, γ will serve as generalized coordinates that change in time as the body rotates.

The following observations are helpful in understanding the Euler angles.

 (i) The line of nodes is perpendicular to the plane determined by $\hat{e}_3^0$ and $\hat{e}_3$.
(ii) Specification of the axis $\hat{e}_3$ relative to the inertial triad $\{\hat{e}_i^0\}$ requires the two angles β and α. These have a simple interpretation as the familiar polar angle and azimuthal angle of $\hat{e}_3$ in spherical polar coordinates (readers should convince themselves of this).
(iii) The angle γ then specifies the additional rotation about the new $\hat{e}_3$ axis.
(iv) Each of the angles α, β, and γ can be varied independently. Hence they are appropriate coordinates for a lagrangian description. The usefulness of the Euler angles will become clearer after we have used them to construct the lagrangian for a symmetric top.

The first step in introducing the Euler angles is to express the kinetic energy in terms of α, β, γ and $\dot{\alpha}$, $\dot{\beta}$, $\dot{\gamma}$. As in Sec. 28, we choose to decompose $\boldsymbol{\omega}$ along the instantaneous orthonormal body-fixed principal axes $\hat{e}_1$, $\hat{e}_2$, and $\hat{e}_3$

$$T = \tfrac{1}{2}(I_1\omega_1^2 + I_2\omega_2^2 + I_3\omega_3^2) \tag{29.1}$$

where I_i are the principal moments of inertia and

$$\omega_i = \boldsymbol{\omega} \cdot \hat{e}_i \qquad (29.2)$$

The subsequent analysis is slightly intricate because the natural rotational variables

$$\boldsymbol{\omega}_\alpha = \dot{\alpha}\hat{e}_\alpha \qquad \boldsymbol{\omega}_\beta = \dot{\beta}\hat{e}_\beta \qquad \boldsymbol{\omega}_\gamma = \dot{\gamma}\hat{e}_\gamma \qquad (29.3)$$

are *not orthogonal*. Instead, Fig. 29.1 shows that they have the orientations

$$\hat{e}_\alpha = \hat{e}_3^0 \qquad \hat{e}_\beta = \text{line of nodes} \qquad \hat{e}_\gamma = \hat{e}_3 \qquad (29.4)$$

Nevertheless, as seen in Sec. 7, it is still true that angular velocities add vectorially, and *the total angular velocity is the vector sum*

$$\boldsymbol{\omega} = \boldsymbol{\omega}_\alpha + \boldsymbol{\omega}_\beta + \boldsymbol{\omega}_\gamma = \dot{\alpha}\hat{e}_\alpha + \dot{\beta}\hat{e}_\beta + \dot{\gamma}\hat{e}_\gamma \qquad (29.5)$$

To evaluate the scalar products in (29.2), it is first convenient to relate each of the three unit vectors $\hat{e}_\alpha$, $\hat{e}_\beta$, $\hat{e}_\gamma$ to those in the body frame $\hat{e}_1$, $\hat{e}_2$, $\hat{e}_3$. The last of Eqs. (29.4) immediately identifies

$$\hat{e}_\gamma = \hat{e}_3 \qquad (29.6a)$$

The vector $\hat{e}_\beta$ lies in the plane of the 1-2 axes, at an azimuth angle $\frac{1}{2}\pi - \gamma$ from $\hat{e}_1$, so that we have

$$\hat{e}_\beta = \hat{e}_1 \cos\left(\tfrac{1}{2}\pi - \gamma\right) + \hat{e}_2 \sin\left(\tfrac{1}{2}\pi - \gamma\right)$$
$$= \hat{e}_1 \sin\gamma + \hat{e}_2 \cos\gamma \qquad (29.6b)$$

Finally, the vector $\hat{e}_\alpha = \hat{e}_3^0$ has polar angle β and azimuthal angle $\pi - \gamma$ relative to the body axes, leading to

$$\hat{e}_\alpha = \hat{e}_1 \sin\beta \cos(\pi - \gamma) + \hat{e}_2 \sin\beta \sin(\pi - \gamma) + \hat{e}_3 \cos\beta$$
$$= -\hat{e}_1 \sin\beta \cos\gamma + \hat{e}_2 \sin\beta \sin\gamma + \hat{e}_3 \cos\beta \qquad (29.6c)$$

A combination of Eqs. (29.5) and (29.6) yields the desired scalar products in Eq. (29.2)

$$\omega_1 = -\dot{\alpha} \sin\beta \cos\gamma + \dot{\beta} \sin\gamma \qquad (29.7a)$$

$$\omega_2 = \dot{\alpha} \sin\beta \sin\gamma + \dot{\beta} \cos\gamma \qquad (29.7b)$$

$$\omega_3 = \dot{\alpha} \cos\beta + \dot{\gamma} \qquad (29.7c)$$

and substitution into Eq. (29.1) then expresses the kinetic energy in the proper generalized coordinates α, β, γ and velocities $\dot{\alpha}$, $\dot{\beta}$, $\dot{\gamma}$.

30 SYMMETRIC TOP: TORQUE-FREE MOTION

The Euler angles provide a flexible parametrization of rigid-body motion, but the resulting dynamical equations are complicated for an asymmetric top with three distinct principal moments. Consequently, we shall consider only a symmetric top with $I_1 = I_2 \neq I_3$, where an exact solution exists not only for the torque-free motion but also for the interesting case of motion with one fixed point in a gravitational field (Sec. 31).

Equations of Motion and First Integrals

Throw an axisymmetric homogeneous rigid object in the air. Gravity impels the center of mass into a parabolic trajectory, yet it exerts no torque about the center of mass because of the symmetric matter distribution [compare Eq. (2.15)]. Thus the object executes torque-free motion about the center of mass, and the lagrangian for the internal dynamics contains only the kinetic energy. In the present case $(I_1 = I_2)$, Eqs. (29.1) and (29.7) simplify considerably to give

$$L(\alpha, \beta, \gamma; \dot{\alpha}, \dot{\beta}, \dot{\gamma}) = T = \tfrac{1}{2} I_1 (\dot{\alpha}^2 \sin^2 \beta + \dot{\beta}^2) + \tfrac{1}{2} I_3 (\dot{\alpha} \cos \beta + \dot{\gamma})^2 \quad (30.1)$$

where α, β, and γ are the generalized coordinates; they obey the usual equations

$$\frac{d}{dt} \frac{\partial L}{\partial \dot{\alpha}} = \frac{\partial L}{\partial \alpha} \quad (30.2)$$

with similar ones for β and γ. Note that the lagrangian for the symmetric top contains neither the Euler angle α nor the angle γ. Thus both of these variables are *cyclic*, and they execute particularly simple motions, which accounts for great usefulness of the Euler angles as generalized coordinates for rigid-body motions.[†] As noted in Sec. 20, the corresponding canonical momenta p_α and p_γ are then manifest constants of the motion

$$p_\alpha = \frac{\partial L}{\partial \dot{\alpha}} = I_1 \dot{\alpha} \sin^2 \beta + I_3 (\dot{\alpha} \cos \beta + \dot{\gamma}) \cos \beta = \text{const} \quad (30.3)$$

$$p_\gamma = \frac{\partial L}{\partial \dot{\gamma}} = I_3 (\dot{\alpha} \cos \beta + \dot{\gamma}) = \text{const} \quad (30.4a)$$

These relations provide two first integrals that will enable us to construct the final solution. Comparison with Eq. (29.7c) gives the alternative expression

$$p_\gamma = I_3 \omega_3 = L_3 = \text{const} \quad (30.4b)$$

confirming that the projection of $\boldsymbol{\omega}$ along the symmetry axis $\hat{e}_3$ remains constant as the body moves [compare Eqs. (28.31)]. The remaining canonical momentum

$$p_\beta = \frac{\partial L}{\partial \dot{\beta}} = I_1 \dot{\beta} \quad (30.5)$$

obeys the equation

$$\dot{p}_\beta = I_1 \ddot{\beta} = \dot{\alpha} \sin \beta [I_1 \dot{\alpha} \cos \beta - I_3 (\dot{\alpha} \cos \beta + \dot{\gamma})] \quad (30.6a)$$

$$= \dot{\alpha} \sin \beta (I_1 \dot{\alpha} \cos \beta - I_3 \omega_3) \quad (30.6b)$$

[†] If $I_1 \neq I_2$, only α is cyclic and the analysis becomes much harder.

We first notice that p_α, p_γ, and p_β have direct physical interpretations as the projections of the angular momentum **L** along the axes $\hat{e}_3^0$, $\hat{e}_3$, and the line of nodes (equivalently, along $\hat{e}_\alpha$, $\hat{e}_\gamma$, and $\hat{e}_\beta$, as shown in Fig. 29.1). This assertion is readily proved. Recall that the vector **L** can be expressed in terms of its projections $L_i = I_i \omega_i$ ($i = 1, 2, 3$) along the body-fixed principal axes. The axisymmetry here reduces this relation to

$$\mathbf{L} = I_1(\omega_1 \hat{e}_1 + \omega_2 \hat{e}_2) + I_3 \omega_3 \hat{e}_3 \tag{30.7}$$

and the scalar product with $\hat{e}_\gamma = \hat{e}_3$ [see Eq. (29.6a)] immediately verifies that $\mathbf{L} \cdot \hat{e}_\gamma = p_\gamma$ [compare Eq. (30.4b)]. The other relations require both Eqs. (29.6) and (29.7). Straightforward manipulations eventually yield

$$\mathbf{L} \cdot \hat{e}_\alpha = I_1 \dot{\alpha} \sin^2 \beta + I_3 \cos \beta \, (\dot{\alpha} \cos \beta + \dot{\gamma}) = p_\alpha \tag{30.8a}$$

$$\mathbf{L} \cdot \hat{e}_\beta = I_1 \dot{\beta} = p_\beta \tag{30.8b}$$

where the last equalities follow from (30.3) and (30.5). In this way, the constancy of p_α and p_γ reflects the invariant projection of **L** on the space-fixed axis $\hat{e}_3^0$ as well as on the body-fixed symmetry axis $\hat{e}_3$.

Description of Motion in Inertial Frame

These results enable us to describe in detail the torque-free motion of a symmetric rigid body (the earth or a football,† for example) as seen by an inertial observer. Since **L** is a constant vector fixed in the inertial frame, we can orient the original space-fixed axes with $\hat{e}_3^0$ along **L**. For this choice, the constant $p_\alpha = \mathbf{L} \cdot \hat{e}_3^0$ has the physical interpretation as the magnitude L of the total angular momentum

$$p_\alpha = |\mathbf{L}| = \text{const} \tag{30.9a}$$

Moreover, the constant $p_\gamma = \mathbf{L} \cdot \hat{e}_3$ then has the simple form $L \cos \beta$ (see Fig. 29.1), implying that the angle β also remains constant

$$p_\gamma = p_\alpha \cos \beta = |\mathbf{L}| \cos \beta = \text{const} \tag{30.9b}$$

Since β is constant, $\dot{\beta} = 0$ and comparison with Eq. (30.5) shows that p_β vanishes identically

$$p_\beta = \mathbf{L} \cdot \hat{e}_\beta = 0 \tag{30.9c}$$

Hence **L** remains in the plane of the vectors $\hat{e}_3^0$ and $\hat{e}_3$ (shown shaded in Fig. 29.1). This result also follows directly because $\hat{e}_\beta$ is perpendicular to $\hat{e}_\alpha = \hat{e}_3^0$.

We next examine the explicit form of p_α in Eq. (30.3), which can be rewritten as

$$p_\alpha = I_1 \dot{\alpha} \sin^2 \beta + I_3 \omega_3 \cos \beta = I_1 \dot{\alpha} \sin^2 \beta + p_\gamma \cos \beta \tag{30.10}$$

---

† Throughout this work we refer to an American football, which is a symmetric top with $I_1 = I_2 > I_3$. The earth is a symmetric top with $I_1 = I_2 < I_3$.

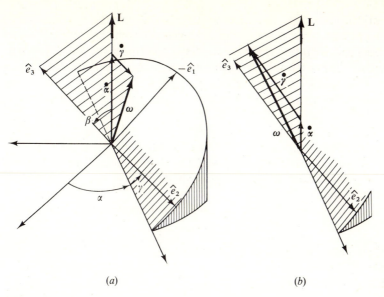

Figure 30.1 Torque-free motion of a symmetric top seen from an inertial frame: (a) $I_3 > I_1$; $\dot{\gamma} < 0$; (b) $I_3 < I_1$; $\dot{\gamma} > 0$.

The constancy of β, p_α, and p_γ implies that $\dot{\alpha}$ is constant, and comparison with Eqs. (30.4a) shows that $\dot{\gamma}$ also must be constant. Thus the angles α and γ increase linearly with time at fixed β. Since $\dot{\beta} = \ddot{\beta} = 0$, the equation of motion (30.6) expresses $\dot{\alpha}$ in terms of the constant "spin" ω_3 about the symmetry axis and the geometrical moments of inertia

$$\dot{\alpha} \cos \beta = \frac{p_\gamma}{I_1} = \frac{I_3 \omega_3}{I_1} \tag{30.11}$$

We have assumed here that $\beta \neq 0$, in which case Eqs. (30.9b) and (30.10) imply $\dot{\alpha} \neq 0$. Equation (29.7c) then leads to the relation

$$\dot{\gamma} = \omega_3 \frac{I_1 - I_3}{I_1} = -\Omega \tag{30.12}$$

between $\dot{\gamma}$ and the frequency Ω introduced previously in Eq. (28.30).

The angular velocity follows directly from Eq. (29.5) with $\dot{\beta} = 0$:

$$\boldsymbol{\omega} = \dot{\alpha}\hat{e}_3^0 + \dot{\gamma}\hat{e}_3 \tag{30.13}$$

It also lies in the plane determined by $\hat{e}_3^0$ and $\hat{e}_3$; hence we conclude that $\boldsymbol{\omega}$, **L**, and the symmetry axis $\hat{e}_3$ are coplanar. Figure 30.1 illustrates the two possible cases $I_3 > I_1$ and $I_3 < I_1$, assuming positive ω_3 and $\dot{\alpha}$.

Suppose, first, that $I_3 > I_1$, which is the situation for the earth (Sec. 28). Since $\dot{\gamma}$ in (30.12) is negative, the vector addition in (30.13) requires that **L** lie between $\hat{e}_3$ and $\boldsymbol{\omega}$, each being perpendicular to the line of nodes $\hat{e}_\beta$. The angle $\alpha(t) = \dot{\alpha}t$

increases linearly at a rate proportional to the spin ω_3 and to the ratio I_3/I_1 [Eq. (30.11)], so that the axis of symmetry precesses uniformly about $\mathbf{L} = L\hat{e}_3^0$ in the positive sense with angular velocity $\dot{\alpha}$ at a fixed polar angle β. The remaining motion of the rigid body is a uniform rotation with angular velocity $\dot{\gamma}$ about the precessing symmetry axis $\hat{e}_3$. In the present situation of negative $\dot{\gamma}$, an inertial observer sees a backward (retrograde) motion about $\hat{e}_3$, with speed Ω proportional to the rate of spin ω_3 and to the asymmetry parameter $(I_3 - I_1)/I_1$ [Eq. (30.12)]. Conversely, a body-fixed observer sees $\boldsymbol{\omega}$ precess about $\hat{e}_3$ in the positive sense with speed Ω, precisely the result found previously in Sec. 28 from Euler's equations. Finally, we may note that the magnitude ω^2 can be written $\omega_3^2 + \omega_\perp^2$, where $\boldsymbol{\omega}_\perp$ is the projection of $\boldsymbol{\omega}$ onto the plane of $\hat{e}_1$ and $\hat{e}_2$. Inspection of Fig. 30.1a shows that

$$\omega_\perp = \dot{\alpha} \sin \beta \tag{30.14}$$

because the contribution $\dot{\gamma}\hat{e}_\gamma$ lies along $\hat{e}_3$. If λ is the angle between $\boldsymbol{\omega}$ and $\hat{e}_3$ (as in Fig. 28.4), then Eq. (30.11) yields

$$\frac{\omega_\perp}{\omega_3} = \tan \lambda = \frac{I_3}{I_1} \tan \beta \tag{30.15}$$

In the present case $(I_3 > I_1)$, it is clear that $\lambda > \beta$ and $\mathbf{L}$ lies between $\hat{e}_3$ and $\boldsymbol{\omega}$, as shown in Fig. 30.1a.

This discussion applies directly to the nearly spherical earth, where $(I_3 - I_1)/I_1 \approx 1/305$ is small and positive. Moreover, the angle β also happens to be small $(\beta \approx \lambda \approx 6 \times 10^{-7})$, indicating that ω_3 is very close to the earth's angular velocity ω $(\approx 2\pi/\text{day})$ because $\omega_\perp$ is proportional to $\sin \beta$ (we here ignore the slow annual motion about the sun). As a result, Eqs. (30.11) and (30.12) show that the symmetry axis precesses rapidly (approximately daily) in the inertial frame about the space-fixed vector $\mathbf{L}$ at a rate $\dot{\alpha} \approx (306/305)\omega$ and that the body-fixed axes execute a slow retrograde motion about the symmetry axis $\hat{e}_3$ at a rate $\dot{\gamma} \approx -\omega/305$. If $\dot{\gamma}$ vanished (as for a sphere), the body would rotate once in space (seen by an inertial observer) each time α increases by 2π. In the present case, however, α must increase by $\approx 306(2\pi)$ rad for the body to complete approximately 305 revolutions as seen by an inertial observer. Finally, the condition $\dot{\alpha} > \omega$ means that the earth's line of nodes advances just faster than once per day.

The same general analysis describes an elongated body $(I_1 > I_3)$ such as a football, but there is one important distinction between the two situations. Equation (26.16) shows that the trace of the inertia tensor has the form

$$\mathrm{Tr}\underline{I} = \sum_{i=1}^{3} I_{ii} = 2 \int d^3r \, \rho(\mathbf{r})r^2 = I_1 + I_2 + I_3 \tag{30.16}$$

where the last equality follows from the invariance of the trace under a similarity transformation (26.28). Furthermore, I_3 for a symmetric object $(I_1 = I_2)$ becomes

$$I_3 = \int d^3r \, \rho(\mathbf{r})(r^2 - z^2) = \int d^3r \, \rho(\mathbf{r})r^2 \sin^2 \theta \tag{30.17}$$

where $\cos \theta = z/r$ characterizes the polar angle measured from the symmetry axis. Hence the ratio I_3/I_1 for a symmetric top can be written

$$\frac{I_3}{I_1} = \frac{2 \int d^3r\, \rho(\mathbf{r})r^2 \sin^2 \theta}{\int d^3r\, \rho(\mathbf{r})r^2(1 + \cos^2 \theta)} \tag{30.18}$$

necessarily lying between ≈ 0 (for a long cylinder) and 2 (for a flat disk or ring). As a result, the parameter $(\dot\alpha \cos \beta)/\omega_3$ can never exceed 2, but it can become small for an elongated body; similarly $\dot\gamma/\omega_3$ lies between -1 and $+1$.

For definiteness, we consider a symmetric body with $I_3 \ll I_1$ and angular velocity $\boldsymbol{\omega}$ and have sketched this case in Fig. 30.1b. Equation (30.11) shows that $\dot\alpha \ll \omega_3$ (unless $\cos \beta$ is small and $\beta \approx \frac{1}{2}\pi$), and (30.14) then implies that $\omega_3 \approx \omega$ with λ small. Consequently, the symmetry axis $\hat{e}_3$ executes a slow positive rotation in the inertial frame about the fixed vector $\mathbf{L}$, at a speed $\dot\alpha$ proportional to the total angular velocity $\omega \approx \omega_3$. On the other hand, the body axes rotate rapidly about $\hat{e}_3$ at a rate $\dot\gamma \approx \omega$. This latter behavior is difficult to observe without special equipment like a stroboscope. In contrast, the proportionality between $\dot\alpha$ and ω is readily demonstrated with a football, throwing it so that it rotates rapidly about an axis slightly different from the axis of symmetry. The relation between the precession rate $\dot\alpha$ of the symmetry axis and the spin rate ω_3 about the symmetry axis is easily observed; in particular, $\dot\alpha$ reverses sign if ω_3 does so.

31 SYMMETRIC TOP: ONE FIXED POINT IN A GRAVITATIONAL FIELD

We conclude this chapter with a study of the motion of a symmetric top placed in a uniform gravitational field and fixed at one point (Fig. 31.1). The gravitational force acts downward and affects both the vertical motion of the center of mass and the total angular momentum $\mathbf{L}$. The overall motion is quite complicated, and it is again helpful to use the Euler angles to characterize the motion seen by an inertial observer.

Dynamical Equations

The effect of gravity is readily incorporated by augmenting the kinetic energy (30.1) with the potential energy $V = Mgl \cos \beta$, equivalent to that of a particle of mass M at the center of mass, to obtain the total lagrangian

$$L(\alpha, \beta, \gamma; \dot\alpha, \dot\beta, \dot\gamma) = T - V$$

$$= \tfrac{1}{2}I_1(\dot\alpha^2 \sin^2 \beta + \dot\beta^2) + \tfrac{1}{2}I_3(\dot\alpha \cos \beta + \dot\gamma)^2 - Mgl \cos \beta \tag{31.1}$$

Inspection of Fig. 31.1 shows that the gravitational torque is along the line of nodes $\hat{e}_\beta$, so that the components p_α and p_γ of $\mathbf{L}$ in the plane perpendicular to $\hat{e}_\beta$

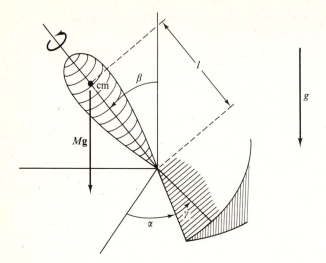

Figure 31.1 Symmetric top with one fixed point in a uniform gravitational field.

necessarily remain constant. Equivalently, we note that V depends only on β, and the *Euler angles α and γ are again cyclic*, giving two constants of the motion

$$p_\alpha = I_1 \dot\alpha \sin^2 \beta + I_3 \cos \beta \, (\dot\alpha \cos \beta + \dot\gamma) = \text{const} \qquad (31.2a)$$

$$p_\gamma = I_3(\dot\alpha \cos \beta + \dot\gamma) = I_3 \omega_3 = \text{const} \qquad (31.2b)$$

just as in Eqs. (30.3) and (30.4). It is simplest to invert these last two equations, expressing $\dot\alpha$ and $\dot\gamma$ in terms of the constants p_α and p_γ, which are here assumed positive†

$$\dot\alpha = \frac{p_\alpha - p_\gamma \cos \beta}{I_1 \sin^2 \beta} \qquad (31.3a)$$

$$\dot\gamma = p_\gamma \left(\frac{1}{I_3} + \frac{\cot^2 \beta}{I_1} \right) - \frac{p_\alpha \cos \beta}{I_1 \sin^2 \beta} \qquad (31.3b)$$

The corresponding dynamical relation for β is Lagrange's equation obtained from $L(\alpha, \beta, \gamma; \dot\alpha, \dot\beta, \dot\gamma)$ with $\{\alpha, \beta, \gamma\}$ as generalized coordinates

$$\frac{d}{dt} \frac{\partial L}{\partial \dot\beta} = I_1 \ddot\beta_1 = \frac{\partial L}{\partial \beta}$$

$$= I_1 \dot\alpha^2 \sin \beta \cos \beta - I_3(\dot\alpha \cos \beta + \dot\gamma)\dot\alpha \sin \beta + Mgl \sin \beta \qquad (31.4)$$

Fortunately, this complicated expression can be simplified considerably, because the parameters α, β, and γ must change in such a way that p_α and p_γ remain fixed; a

---

† This assumption represents a particular choice for the positive direction of the symmetry axis of the symmetric top.

similar situation occurred in the study of central forces. Use of (31.3) and (31.2b) yields the equivalent one-dimensional equation

$$I_1\ddot{\beta} = I_1 \sin \beta \cos \beta \left(\frac{p_\alpha - p_\gamma \cos \beta}{I_1 \sin^2 \beta} \right)^2 - p_\gamma \sin \beta \frac{p_\alpha - p_\gamma \cos \beta}{I_1 \sin^2 \beta} + Mgl \sin \beta$$

$$(31.5a)$$

or, upon simplification,

$$I_1\ddot{\beta} = \frac{\cos \beta}{I_1 \sin^3 \beta} (p_\alpha^2 - 2p_\alpha p_\gamma \cos \beta + p_\gamma^2) - \frac{|p_\alpha p_\gamma}{I_1 \sin \beta} + Mgl \sin \beta \quad (31.5b)$$

This is now a one-dimensional differential equation for β, although it is still highly nonlinear. Given the time-dependent solution $\beta(t)$, the corresponding behavior of the Euler angles α and γ can be inferred from Eqs. (31.3). Note the importance of maintaining the proper independent variables in this derivation. Direct substitution of (31.3) into the lagrangian, followed by differentiation at fixed p_α and p_γ would yield a different and *incorrect* equation.

Effective Potential

The form of Eq. (31.5) is reminiscent of the radial equation (20.11) in a central force, and we again anticipate the existence of a first integral for the energy. It can be constructed either directly or from the general formalism, and we shall follow the latter procedure. Since the kinetic energy is a quadratic form in the generalized velocities $\dot{\alpha}$, $\dot{\beta}$, and $\dot{\gamma}$ and the potential energy is independent of the velocities, the hamiltonian is both conserved and equal to the total energy $T + V$ (see the discussion in Sec. 20)

$$H = E = T + V = \text{const} \tag{31.6a}$$

It follows from Eqs. (31.1), (31.2b), and (31.3a) that

$$E = \tfrac{1}{2}I_1 \dot{\beta}^2 + V_{\text{eff}}(\beta) \tag{31.6b}$$

where we have defined

$$V_{\text{eff}}(\beta) \equiv \frac{(p_\alpha - p_\gamma \cos \beta)^2}{2I_1 \sin^2 \beta} + \frac{p_\gamma^2}{2I_3} + Mgl \cos \beta \tag{31.7}$$

This is the effective one-dimensional potential for the variable $\beta(t)$. Differentiating Eq. (31.6) with respect to t gives

$$I_1 \ddot{\beta} = -\frac{\partial V_{\text{eff}}}{\partial \beta} \tag{31.8}$$

which is easily seen to reproduce the differential equation (31.5).

Given Eq. (31.6), the solution reduces to a single indefinite integral analogous to that in Sec. 3 for $r(t)$. The integration is intricate and unwieldy, however, and we here treat only the special case of small oscillations about a steady configuration of dynamical equilibrium. To analyze this situation, it is helpful first

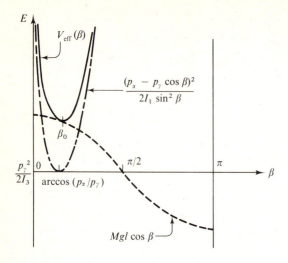

Figure 31.2 Effective potential $V_{eff}(\beta)$ in Eq. (31.7) for system illustrated in Fig. 31.1.

to examine V_{eff} in more detail. The second term in (31.7) is a constant that merely shifts the origin of energy. The first term is positive and vanishes at the value $\cos \beta = p_\alpha/p_\gamma$, which represents a physical angle only if $p_\alpha < p_\gamma$; we assume this to be the case, for it represents the usual initial condition of rapid spin about the symmetry axis. In addition, the first term of V_{eff} has the factor $\sin^2 \beta$ in the denominator, which diverges at $\beta = 0$ or π. This behavior reflects the angular-momentum barrier, similar to the term $l^2/2mr^2$ in Eq. (3.9). Consequently, the first term has a minimum at $\beta = \arccos (p_\alpha/p_\gamma) < \frac{1}{2}\pi$, and it rises steeply near the end points 0 and π (see Fig. 31.2). The remaining contribution $Mgl \cos \beta$ arises from the gravitational energy. It is antisymmetric about the midpoint $\frac{1}{2}\pi$ (see Fig. 31.2) and shifts the minimum of V_{eff} to the right of $\arccos (p_\alpha/p_\gamma)$ but maintains the same qualitative form. For given E ($> p_\gamma^2/2I_3$), the allowed motion is confined to the region $E > V_{eff}(\beta)$ with turning points $\beta_\pm$ that satisfy the equation $E = V_{eff}(\beta_\pm)$.

To analyze the motion, we first seek a stationary solution for which $\beta = \beta_0 = $ const. Because $\ddot{\beta} = 0$, the value β_0 must minimize V_{eff} by Eq. (31.8)

$$\left(\frac{\partial V_{eff}}{\partial \beta}\right)_0 = 0 \tag{31.9}$$

In addition, the condition $\dot{\beta} = 0$ requires coalescence of the turning points in Fig. 31.2 and $E = V_{eff}(\beta_0)$. Performing the indicated differentiation in Eq. (31.9) leads to the relation [compare Eq. (31.5b)]

$$p_\gamma^2 - 2p_\gamma p_\alpha \cos \beta_0 + p_\alpha^2 = \frac{p_\gamma p_\alpha \sin^2 \beta_0}{\cos \beta_0} - \frac{I_1 Mgl \sin^4 \beta_0}{\cos \beta_0} \tag{31.10}$$

which determines β_0 for given p_α, p_γ, and geometrical parameters M, l, I_1, and I_3. Correspondingly, Eqs. (31.3) imply constant values $(\dot{\alpha})_0$ and $(\dot{\gamma})_0$, so that a top in such dynamical equilibrium executes steady motion at fixed β_0 with $\alpha(t) = (\dot{\alpha})_0 t$ and $\gamma(t) = (\dot{\gamma})_0 t$.

Small Oscillations about Steady Motion

Further study can proceed in one of two ways. One possibility is to solve Eq. (31.10) explicitly for β_0, but its quartic structure in $\cos \beta_0$ greatly complicates this task. We shall subsequently perform such an analysis for a fast top $(p_y \gg p_\alpha)$. The other possibility is to consider the more general question of motion in the vicinity of β_0, keeping Eq. (31.10) in its present implicit form. This problem is similar to the small oscillations studied in Chap. 4 (see, especially, Sec. 22 and Prob. 4.6), and we therefore assume

$$\beta(t) = \beta_0 + \eta(t) \tag{31.11}$$

with $\eta(t)$ an infinitesimal. Substitute (31.11) into (31.7) for V_{eff} and expand in a Taylor series

$$V_{\text{eff}}(\beta) = V_{\text{eff}}(\beta_0) + \eta \left(\frac{\partial V_{\text{eff}}}{\partial \beta} \right)_0 + \tfrac{1}{2}\eta^2 \left(\frac{\partial^2 V_{\text{eff}}}{\partial \beta^2} \right)_0 + \cdots \tag{31.12a}$$

By construction, the linear term vanishes identically [Eq. (31.9)]. The quadratic term follows by differentiating (minus) the right-hand side of Eq. (31.5b)

$$\frac{\partial^2 V_{\text{eff}}}{\partial \beta^2} = -Mgl \cos \beta - \frac{3 p_\alpha p_y \cos \beta}{I_1 \sin^2 \beta}$$

$$+ (p_\alpha^2 - 2 p_\alpha p_y \cos \beta + p_y^2) \frac{3 - 2 \sin^2 \beta}{I_1 \sin^4 \beta} \tag{31.12b}$$

and use of Eq. (31.10) in the last term gives the result

$$\left(\frac{\partial^2 V_{\text{eff}}}{\partial \beta^2} \right)_0 = \frac{p_\alpha p_y - I_1 Mgl(4 - 3 \sin^2 \beta_0)}{I_1 \cos \beta_0} \tag{31.13}$$

To the same order of approximation, the energy integral (31.6b) becomes

$$E = \tfrac{1}{2} I_1 \dot{\eta}^2 + V_{\text{eff}}(\beta_0) + \tfrac{1}{2}\eta^2 \left(\frac{\partial^2 V_{\text{eff}}}{\partial \beta^2} \right)_0 \tag{31.14}$$

equivalent to a harmonic oscillator with dynamical equation (just differentiate with respect to time)

$$\ddot{\eta} = -\Omega^2 \eta \tag{31.15}$$

and squared angular frequency

$$\Omega^2 = \frac{1}{I_1} \left(\frac{\partial^2 V_{\text{eff}}}{\partial \beta^2} \right)_0 = \frac{p_\alpha p_y - I_1 Mgl(4 - 3 \sin^2 \beta_0)}{I_1^2 \cos \beta_0} \tag{31.16}$$

The corresponding motion

$$\beta(t) = \beta_0 + \eta_0 \cos (\Omega t + \phi_0) \tag{31.17}$$

is stable whenever $\Omega^2 > 0$, evidently requiring sufficiently large values for the product $p_\alpha p_y$. This formalism has several applications.

Sleeping top Consider a rapidly spinning top with a vertical symmetry axis; in this case, $\cos \beta_0 = 1$, $\sin \beta_0 = 0$, $\hat{e}_3 = \hat{e}_3^0$ (see Fig. 29.1), and $p_\alpha = p_\gamma$. Is such a configuration stable? Equation (31.16) indicates that Ω^2 is positive for $\beta_0 = 0$ if

$$p_\gamma^2 > 4I_1 Mgl \qquad (31.18)$$

which constitutes the criterion for stability; otherwise the symmetry axis of the top will depart from the vertical.† In practice, even though Eq. (31.18) may be satisfied initially, friction at the base gradually decreases the spin angular momentum p_γ, and the sleeping top suddenly becomes unstable when p_γ reaches the geometrical value $2(I_1 Mgl)^{1/2}$.

Precession and nutation Given the two constants p_α and p_γ, the complete solution for the top's motion follows from $\beta(t)$ and the two remaining equations (31.3) for α and γ. To illustrate this procedure, we again consider motion in the vicinity of β_0. Such behavior represents a restricted set of initial conditions, but it exemplifies most features of the general situation. Substitution of the explicit solution (31.17) for $\beta(t)$ into Eq. (31.3) and expansion in powers of $\eta(t)$ yields the approximate dynamical equations for the angles $\alpha(t)$ and $\gamma(t)$

$$\dot{\alpha}(t) \approx \left(\frac{p_\alpha - p_\gamma \cos \beta}{I_1 \sin^2 \beta}\right)_0 + \eta(t)\left(\frac{\partial}{\partial \beta} \frac{p_\alpha - p_\gamma \cos \beta}{I_1 \sin^2 \beta}\right)_0 + \cdots$$

$$\equiv (\dot{\alpha})_0 + \eta(t)(\dot{\alpha})_1 \qquad (31.19a)$$

$$\dot{\gamma}(t) \approx (\dot{\gamma})_0 + \eta(t)(\dot{\gamma})_1 + \cdots \qquad (31.19b)$$

where $(\dot{\alpha})_0$, $(\dot{\alpha})_1$, $(\dot{\gamma})_0$, and $(\dot{\gamma})_1$ are definite constants that depend on p_α, p_γ, and E through β_0. For definiteness, assume that p_α and p_γ have values that make all these four constants positive. The detailed motion has the following interesting features:

1. The rate of precession $\dot{\alpha}(t)$ of the symmetry axis $\hat{e}_3$ about $\hat{e}_3^0$ and the rate of spin $\dot{\gamma}(t)$ about the symmetry axis $\hat{e}_3$ both vary sinusoidally about the mean values $(\dot{\alpha})_0$ and $(\dot{\gamma})_0$. Neither angle increases strictly linearly with time.
2. These small oscillatory terms have *zero time average*, however, and the *mean* angles indeed have a linear form $(\dot{\alpha})_0 t$ and $(\dot{\gamma})_0 t$.
3. If $(\dot{\alpha})_1 > 0$, then $\dot{\alpha}$ is maximum for maximum β and η, which occurs when the symmetry axis is farthest from the vertical; conversely, $\dot{\alpha}$ is smallest when the symmetry axis approaches the vertical. In this way, the precession of the spin axis acquires a small ripple, known as *nutation*, superimposed on the essentially regular precession of the symmetry axis about the vertical.
4. Several distinct cases can occur, depending on the initial values η_0, $(\dot{\alpha})_0$, and $(\dot{\alpha})_1$. We emphasize that the present discussion merely requires a small value for the correction term η in Eq. (31.11). No assumption is necessary about the

---

† This set of initial conditions constitutes a singular limit of Eqs. (31.10) and (31.12b) and a more careful analysis must actually be carried out in this case; see Prob. 5.9.

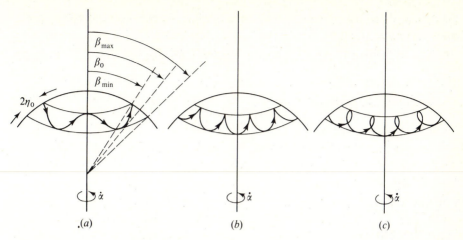

Figure 31.3 Nutation of a symmetric top: (a) $(\dot\alpha)_0 > \eta_0(\dot\alpha)_1$; (b) $(\dot\alpha)_0 = \eta_0(\dot\alpha)_1$; (c) $(\dot\alpha)_0 < \eta_0(\dot\alpha)_1$.

relative size of the terms in Eq. (31.19). In particular, the present solution is still valid even if $(\dot\alpha)_0 \approx 0$. Figure 31.3 shows the associated behavior for $(\dot\alpha)_0/\eta_0(\dot\alpha)_1 > 1$, $= 1$, and < 1, respectively. A rapidly spinning gyroscope readily illustrates these three possibilities. In principle, the spin $\gamma(t)$ also displays an oscillatory component, but observation typically requires special equipment like a stroboscope.

5. In general, $(\dot\alpha)_0$ and Ω are incommensurate, in which case the top never returns precisely to its initial configuration [compare the discussion below Eq. (3.9) for the analogous situation in a central potential].

Fast top $(p_\gamma \gg p_\alpha)$ The typical elementary discussion of gyroscopes assumes a rapid spin $(p_\gamma \to \infty)$ about an axis in the equatorial plane (see Fig. 31.1 with $\beta \approx \tfrac12\pi$). The gravitational torque is purely azimuthal, with magnitude Mgl, and the rate of change of angular momentum

$$\frac{d\mathbf{L}}{dt} = Mgl\hat\phi = \frac{Mgl}{L}\,\hat z \times \mathbf{L} \tag{31.20}$$

implies a uniform precession about $\hat z$ with angular velocity Mgl/L [compare Eq. (7.16)]. It is instructive to obtain the same expression from our more sophisticated analysis.

We start from Eq. (31.19a) for $(\dot\alpha)_0$. It can be positive only if $p_\alpha > p_\gamma \cos\beta_0$, implying that $\cos\beta_0$ becomes small for large p_γ/p_α. This behavior also follows from (31.10), which can be rewritten for given p_α

$$(p_\gamma \cos\beta_0)^2 - (p_\gamma \cos\beta_0)p_\alpha(1 + \cos^2\beta_0)$$
$$+ p_\alpha^2 \cos^2\beta_0 + I_1 Mgl \sin^4\beta_0 \cos\beta_0 = 0 \tag{31.21}$$

The product $p_\gamma \cos\beta_0$ must remain of order p_α as $p_\gamma \to \infty$, and we therefore set

$\beta_0 \approx \frac{1}{2}\pi$ in the small correction terms. In this way, we obtain the simple approximate quadratic equation

$$(p_\gamma \cos \beta_0)^2 - (p_\gamma \cos \beta_0)p_\alpha + I_1 Mgl \cos \beta_0 = 0 \qquad (31.22)$$

It is easily solved to give

$$\cos \beta_0 \approx \frac{p_\alpha}{p_\gamma} - \frac{I_1 Mgl}{p_\gamma^2} + \cdots \qquad (31.23)$$

where we choose the root that reduces to $\cos \beta_0 = p_\alpha/p_\gamma$ for $g = 0$ (see the discussion accompanying Fig. 31.2). This relation suffices to analyze the motion in considerable detail.

1. The symmetry axis lies just above the horizontal at an angle $\beta_0 \approx \frac{1}{2}\pi - p_\alpha/p_\gamma$ (see Fig. 31.1).
2. The identifications $p_\gamma = \mathbf{L} \cdot \hat{e}_\gamma = \mathbf{L} \cdot \hat{e}_3$ and $p_\alpha = \mathbf{L} \cdot \hat{e}_\alpha = \mathbf{L} \cdot \hat{e}_3^0$ imply that $\mathbf{L}$ lies essentially along the symmetry axis $\hat{e}_3$ with magnitude $L = p_\gamma$.
3. Equation (31.3b) shows that the top spins about its symmetry axis at the expected mean rate

$$(\dot{\gamma})_0 = \frac{p_\gamma}{I_3} \approx \frac{L}{I_3} \qquad (31.24)$$

4. The top's axis of symmetry precesses about the vertical at a mean rate

$$(\dot{\alpha})_0 = \frac{p_\alpha - p_\gamma \cos \beta_0}{I_1 \sin^2 \beta_0} \approx \frac{Mgl}{p_\gamma} \approx \frac{Mgl}{L} \qquad (31.25)$$

confirming the elementary estimate obtained from (31.20). Note that the angular velocity of precession varies inversely with the spin angular momentum $p_\gamma \, (\approx L)$.

5. The small-oscillation frequency (31.16) of nutation has the approximate value

$$\Omega \approx \frac{p_\gamma}{I_1} \approx \frac{L}{I_1} \qquad (31.26)$$

showing that the nutation frequency *increases* linearly with L. In particular, the product of the precession frequency $(\dot{\alpha})_0$ and the nutation frequency Ω involves only the geometric parameter Mgl/I_1 and is independent of L. This combination of rapid nutation and slow precession is sometimes called *pseudoregular precession* because the nutation is rarely observable for a fast top.

PROBLEMS

5.1 Consider the compound pendulum in Fig. 28.1 (mass M, moments of inertia $\bar{I}_{ij}$ relative to center of mass, which is a distance l from the point of support Q) but with Q attached to the bottom of a vertical spring (force constant k) and constrained to move vertically. The top of the spring is fixed to a rigid support.

(a) Construct the lagrangian for this system in terms of the generalized coordinates η (the downward displacement of Q) and ϕ. Derive Lagrange's equations.

(b) Obtain the same equations from the general principles of rigid-body dynamics concerning the motion of the center of mass and rotations about it (don't forget the force of constraint).

(c) For small oscillations, show that the motion uncouples into that of a spring and a pendulum.

5.2 Foucault gyrocompass A gyroscope in the form of a symmetric top is mounted with no gravitational torque, and its symmetry axis is constrained to move only in the horizontal plane parallel to the earth's surface. The gyroscope is set spinning about its symmetry axis with an angular velocity Ω.

(a) If $\Omega \gg \omega$, where ω is the angular velocity of the earth's rotation, show that the Coriolis force exerts the following torque on the gyroscope about its center of mass:

$$\Gamma = -2 \sum_p m_p \{ \mathbf{r}_p \times [\omega \times (\Omega \times \mathbf{r}_p)] \}$$

where the sum runs over all the particles in the gyroscope with position $\mathbf{r}_p$ relative to the center of mass.

(b) The gyroscope is located at a polar angle (colatitude) θ. Consider its angular motion about the vertical 1 (1^0) axis. Show that the symmetry axis (3) of the gyroscope will oscillate about the northerly direction according to

$$\ddot{\phi} = - \left(\frac{I_3}{I_1} \Omega\omega \sin\theta \right) \sin\phi$$

and can thus be used as a compass.

5.3 A tilted coin (a sharp-edged uniform disk) of radius a and mass M rolls without slipping on a horizontal plane in a circle of radius b. A set of orthogonal coordinate axes has its origin at the center of mass, with $\hat{e}_3$ perpendicular to the face of the coin, $\hat{e}_2$ in the plane of the coin passing through the point of contact, and $\hat{e}_1$ parallel to the horizontal plane and tangent to the trajectory. Introduce the angles θ, ϕ, and γ that specify the orientation of the coin as indicated in Fig. P5.3. The particular motion of interest (neglecting rolling friction that eventually slows down the coin) is characterized by

$$\dot{\gamma} = \text{const} \qquad \dot{\phi} = \text{const} \qquad \dot{\theta} = 0 \qquad \text{hence } \theta = \text{const}$$

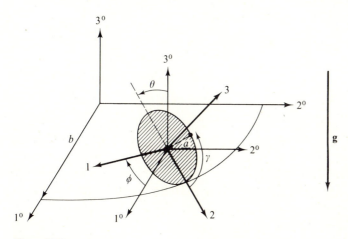

Figure P5.3

(a) Use the equations of rigid-body motion to eliminate the reaction force at the point of contact to obtain

$$\frac{d}{dt} \mathbf{L}_{cm} = Ma[g \sin\theta - \dot{\phi}^2 \cos\theta \, (b - a \sin\theta)]\hat{e}_1$$

(b) Show, in general, that

$$\boldsymbol{\Omega} = \dot{\theta}\hat{e}_1 + \dot{\phi}\cos\theta\,\hat{e}_2 - \dot{\phi}\sin\theta\,\hat{e}_3$$

$$\mathbf{L}_{cm} = I_1\dot{\theta}\,\hat{e}_1 + I_1\dot{\phi}\cos\theta\,\hat{e}_2 + I_3(\dot{\gamma} - \dot{\phi}\sin\theta)\hat{e}_3$$

where $\boldsymbol{\Omega}$ is the instantaneous angular velocity of the $(1, 2, 3)$ body-associated frame as seen in the 1^0, 2^0, 3^0 center-of-mass frame.

(c) Use the relation between $(d\mathbf{L}/dt)_{cm}$ and $(d\mathbf{L}/dt)_b$, where the subscript b now refers to the $(1, 2, 3)$ frame, to show that the period τ for motion around the circle and the angle of inclination must satisfy the equation

$$\left(\frac{\tau}{2\pi}\right)^2 = \dot{\phi}^{-2} = \frac{\cos\theta}{4g\sin\theta}(6b - 5a\sin\theta)$$

Note: For a flat disk $I_1 = I_2 = \tfrac{1}{4}Ma^2$, $I_3 = \tfrac{1}{2}Ma^2$.

5.4 A rigid body in the shape of a thumbtack formed from a thin disk of mass M and radius a and a massless stem is placed on an inclined plane that makes an angle α with the horizontal. The point of the tack remains stationary at the point P, and the head rolls along a circle of radius b. Introduce a set of laboratory coordinates whose 3^0 axis is perpendicular to the inclined plane and whose 2^0 axis points down the plane, as well as a set of body-associated principal axes with origin at the center of mass, whose 3 axis is perpendicular to the head of the tack pointing outward, whose 2 axis passes through the point of contact with the plane, and whose 1 axis is parallel to the surface and tangent to the circle. Introduce also the set of angles (θ, ϕ, γ) that specify the orientation of the tack, as indicated in Fig. P5.4.

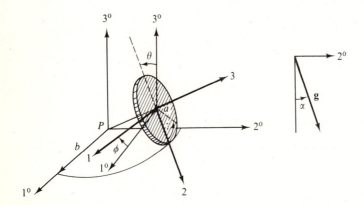

Figure P5.4

(a) Show that in general the angular velocity of the rolling tack is given by

$$\boldsymbol{\omega} = \dot{\theta}\hat{e}_1 + \dot{\phi}\cos\theta\,\hat{e}_2 + (ba^{-1} - \sin\theta)\dot{\phi}\hat{e}_3$$

(b) Show that the kinetic energy of the tack is given by

$$T = \tfrac{1}{2}I_1\dot{\theta}^2 + \tfrac{1}{2}I_1\dot{\phi}^2\cos^2\theta + (2a^2)^{-1}I_3(b - a\sin\theta)^2\dot{\phi}^2 + \tfrac{1}{2}M(b - a\sin\theta)^2\dot{\phi}^2 + \tfrac{1}{2}Ma^2\dot{\theta}^2$$

(c) Show that the potential energy of the tack is given by

$$V = -M\mathbf{g}\cdot\mathbf{R} = -Mg[(b - a\sin\theta)\sin\alpha\cos\phi - a\cos\alpha\cos\theta]$$

(d) Construct the lagrangian $L(\phi, \theta; \dot{\phi}, \dot{\theta})$ and write Lagrange's equations for ϕ and θ incorporating the constraint $\theta = \theta_0 = \arcsin(a/b)$ with a Lagrange multiplier λ_θ.

(e) Show that ϕ satisfies the pendulum equation with angular frequency given by

$$\Omega^2 = \frac{g}{a}\frac{\sin \alpha}{\cot \theta_0 \cos \theta_0}\frac{4}{6 + \tan^2 \theta_0}$$

(f) Interpret λ_θ. For small oscillations of the pendulum, show that the condition that the thumbtack not tip over is simply

$$\tan \alpha < \tan \theta_0$$

Note: For a flat disk $I_1 = I_2 = \frac{1}{4}Ma^2$; $I_3 = \frac{1}{2}Ma^2$.

5.5 (a) Verify that Eqs. (28.32) and (28.33) are indeed first integrals of Euler's equations (28.27) for torque-free motion.

(b) Eliminate ω_1 and ω_2 in terms of E, L, and ω_3, and hence reduce the resulting motion for $\omega_3(t)$ to a definite integral. Obtain the formal solution for $\boldsymbol{\omega}(t)$.

5.6 An asymmetric top $(I_1 < I_2 < I_3)$ executes torque-free motion with $2EI_2 = L^2$. If $\boldsymbol{\omega}$ initially lies in the plane of $\hat{e}_1$ and $\hat{e}_3$ (as in Fig. 28.4), integrate Euler's equations to obtain the solution

$$\omega_1(t) = \omega_\infty \left[\frac{I_2(I_3 - I_2)}{I_1(I_3 - I_1)}\right]^{1/2} \operatorname{sech} \frac{t}{\tau}$$

$$\omega_2(t) = \omega_\infty \tanh \frac{t}{\tau}$$

$$\omega_3(t) = \omega_\infty \left[\frac{I_2(I_2 - I_1)}{I_3(I_3 - I_1)}\right]^{1/2} \operatorname{sech} \frac{t}{\tau}$$

where $\omega_\infty = 2E/L$ and $\tau^{-1} = \omega_\infty[(I_3 - I_2)(I_2 - I_1)/I_3 I_1]^{1/2}$. Discuss the time dependence of ω^2 and sketch the motion of $\hat{\omega}$ as seen in the body-fixed frame.

5.7 The earth may be considered a rigid axisymmetric body with a small quadrupole deformation.

(a) If the exterior gravitational potential is written

$$\Phi_g(\mathbf{r}) = -M_e G r^{-1}\left[1 - J_2\left(\frac{R_e}{r}\right)^2 P_2(\cos \theta)\right]$$

where R_e is the equatorial radius and θ is the colatitude, show that $J_2 = (I_3 - I_1)/M_e R_e^2$ (see Prob. 2.7).

(b) Using the observed values $I_3(I_3 - I_1)^{-1} \approx 305.3$ and $J_2 \approx 1.083 \times 10^{-3}$, show that the ratio $I_3/MR_e^2 \approx 0.33$ implies an accumulation of mass toward the earth's center.

5.8 To illustrate the usefulness of the effective potential, consider the following problem. A spherical pendulum consists of a particle of mass m in a gravitational field constrained to move on the surface of a sphere of radius l.

(a) Obtain the equations of motion and reduce them to an effective one-dimensional problem. Sketch V_{eff} and discuss the qualitative features of the motion.

(b) As a special case, find the period of a conical pendulum (one whose orbit is a horizontal circle) and show that your result agrees with the expression obtained from elementary considerations.

(c) Discuss the small oscillations about this dynamical equilibrium.

5.9 A symmetric top with one fixed point in a gravitational field moves with its axis nearly vertical $(\beta \ll 1)$ and $p_\alpha = p_\gamma$.

(a) Expand the effective potential in Eq. (31.7) through terms of order β^4.

(b) If $p_\gamma^2 > 4I_1 Mgl$, show that V_{eff} has a minimum at $\beta = 0$. Sketch V_{eff} for small β. Prove that the frequency of small oscillations about this configuration is given by $\Omega^2 = (4I_1^2)^{-1}(p_\gamma^2 - 4I_1 Mgl)$. Why does this differ from Eq. (31.16) for $\beta_0 = 0$?

(c) If p_γ^2 is slightly smaller than $4I_1 Mgl$, show that V_{eff} has a maximum at $\beta = 0$ and a minimum at an angle $\beta_0 = (2 - p_\gamma^2/2I_1 Mgl)^{1/2}$. Sketch V_{eff} for small β. Prove that the frequency of small oscillations about this configuration is given by $\Omega^2 = (2I_1^2)^{-1}(4I_1 Mgl - p_\gamma^2)$.

5.10 A symmetric top with one fixed point in a gravitational field starts with the initial conditions

$$\dot{\alpha} = 2\left(\frac{Mgl}{3I_1}\right)^{1/2} \qquad\qquad \beta = 60°$$

$$\dot{\gamma} = (3I_1 - I_3)\left(\frac{Mgl}{3I_1 I_3^2}\right)^{1/2} \qquad \dot{\beta} = 0$$

(a) Find the conserved momenta p_α and p_γ and the effective potential. Use a sketch of V_{eff} to discuss the qualitative form of the solution $\beta(t)$.

(b) Show that the equation for β can be written

$$\dot{u}^2 = \frac{Mgl}{I_1}(1 - u)^2(2u - 1)$$

where $u = \cos\beta$. Hence derive the explicit solution

$$\sec\beta = 1 + \text{sech}\left[\left(\frac{Mgl}{I_1}\right)^{1/2} t\right]$$

and verify the preceding qualitative analysis in part (a). What is the corresponding behavior of $\dot{\alpha}$ and $\dot{\gamma}$?

SELECTED ADDITIONAL READINGS

1. G. D. Garland, "The Earth's Shape and Gravity," Pergamon, Oxford, 1965.
2. W. A. Heiskanen and F. A. Vening Meinesz, "The Earth and Its Gravity Field," McGraw-Hill, New York, 1958.
3. A. Sommerfeld, "Mechanics," Academic, New York, 1952.
4. F. D. Stacey, "Physics of the Earth," 2d ed., Wiley, New York, 1977.
5. A. R. Edmonds, "Angular Momentum in Quantum Mechanics," Princeton University Press, Princeton, N.J., 1957.
6. M. E. Rose, "Elementary Theory of Angular Momentum," Wiley, New York, 1961.

HAMILTONIAN DYNAMICS

Our aim in this chapter is to reformulate classical mechanics yet again. By exploiting the symmetry between generalized coordinates and generalized momenta, we shall find a broader and still more powerful description of the dynamics of classical systems. This discussion not only provides a very general basis but also serves as the most direct link between classical physics and quantum mechanics.

32 HAMILTON'S EQUATIONS

To aid our subsequent development, we first briefly review lagrangian dynamics.

Review of Lagrangian Dynamics

Hamilton's principle constitutes one of the basic statements of classical mechanics. For a system described by the generalized coordinates q_σ with $\sigma = 1$, ..., n,[†] it demands that the action be stationary under small virtual displacements about the actual motion of the system

$$\delta \int_{t_1}^{t_2} L(q_1, \ldots, q_n; \dot{q}_1, \ldots, \dot{q}_n; t) \, dt = 0 \tag{32.1}$$

---

† Throughout this section we assume for notational convenience that there are n independent generalized coordinates. If there are k holonomic constraints, the starting number of degrees of freedom of the system was $n + k$.

subject to the condition of fixed initial and final configurations

$$\delta q_\sigma(t_1) = \delta q_\sigma(t_2) = 0 \qquad \sigma = 1, \ldots, n \tag{32.2}$$

This variational problem has the Euler-Lagrange equations

$$\frac{d}{dt}\frac{\partial L}{\partial \dot{q}_\sigma} - \frac{\partial L}{\partial q_\sigma} = 0 \qquad \sigma = 1, \ldots, n \tag{32.3}$$

These are Lagrange's equations of classical mechanics. After performing the differentiations in (32.3), we find n second-order differential equations in the time. A complete solution to these equations requires the specification of all the generalized coordinates q_σ and all the generalized velocities $\dot{q}_\sigma$ at a given instant in time. The subsequent motion of the system is then fully determined.

As shown in Chap. 3, we are free to make *point transformations* to a new set of generalized coordinates

$$q_\sigma = q_\sigma(Q_1, \ldots, Q_n, t) \qquad \sigma = 1, \ldots, n \tag{32.4a}$$

where the transformations are assumed nonsingular and invertible to express Q_σ in terms of the original coordinates

$$Q_\sigma = Q_\sigma(q_1, \ldots, q_n, t) \qquad \sigma = 1, \ldots, n \tag{32.4b}$$

The lagrangian can be rewritten with the new coordinates Q_σ and the new generalized velocities $\dot{Q}_\sigma$

$$L(q, \dot{q}, t) = L[q(Q, t), \dot{q}(Q, \dot{Q}, t), t] \equiv \tilde{L}(Q, \dot{Q}, t) \tag{32.5}$$

and Hamilton's principle correspondingly takes the form

$$\delta \int_{t_1}^{t_2} \tilde{L}(Q_1, \ldots, Q_n; \dot{Q}_1, \ldots, \dot{Q}_n; t)\, dt = 0 \tag{32.6}$$

Variation of (32.4b) makes it evident that if all the virtual displacements of the original coordinates vanish at the initial and final times according to (32.2), then the variations of the new coordinates will also do so

$$\delta Q_\sigma(t_1) = \delta Q_\sigma(t_2) = 0 \qquad \sigma = 1, \ldots, n \tag{32.7}$$

Since the new variational problem is now identical with the one previously discussed in Sec. 18, the Euler-Lagrange equations for this transformed problem necessarily become

$$\frac{d}{dt}\frac{\partial \tilde{L}}{\partial \dot{Q}_\sigma} - \frac{\partial \tilde{L}}{\partial Q_\sigma} = 0 \qquad \sigma = 1, \ldots, n \tag{32.8}$$

These are just Lagrange's equations in terms of the new variables. We can state this conclusion more formally by saying that the form of Lagrange's equations remains invariant under the point transformations (32.4). In more practical terms, given the lagrangian expressed in terms of *any* set of generalized coordinates and velocities, Lagrange's equations (32.8) correctly describe the dynamics in terms of this new set of variables. Note that the variational principle (32.1) provides a much more compact proof of this invariance than the original coordinate transformation of d'Alembert's principle in Sec. 15. Many examples have demonstrated the power and usefulness of our freedom to choose appropriate generalized coordinates.

Hamiltonian Dynamics

The preceding treatment dealt only with point transformations, with the generalized velocities derived from the original coordinate transformations (32.4b) by differentiating with respect to t. We now extend the discussion by introducing the generalized momenta, treating the coordinates and momenta as equivalent variables. In this new perspective, the equations of classical mechanics will turn out to remain invariant under a much broader class of transformation, known as *canonical transformations*.

We first recall the definition of the generalized momenta in Sec. 20

$$p_\sigma \equiv \frac{\partial L}{\partial \dot{q}_\sigma} \qquad \sigma = 1, \ldots, n \tag{32.9}$$

These equations provide n relations between the generalized momenta p_σ and the original generalized coordinates q_σ and generalized velocities $\dot{q}_\sigma$ and perhaps also the time. Again these relations are assumed nonsingular and invertible to give the generalized velocities in terms of the momenta and coordinates

$$\dot{q}_\sigma = \dot{q}_\sigma(p_1, \ldots, p_n; q_1, \ldots, q_n; t) \qquad \sigma = 1, \ldots, n \tag{32.10}$$

The hamiltonian has been defined as

$$H \equiv \sum_\sigma p_\sigma \dot{q}_\sigma - L(q, \dot{q}, t) \tag{32.11}$$

and we discussed two of its important properties in Sec. 20. The total differential of Eq. (32.11) is

$$dH = \sum_\sigma \left(p_\sigma \, d\dot{q}_\sigma + \dot{q}_\sigma \, dp_\sigma - \frac{\partial L}{\partial q_\sigma} dq_\sigma - \frac{\partial L}{\partial \dot{q}_\sigma} d\dot{q}_\sigma \right) - \frac{\partial L}{\partial t} dt \tag{32.12a}$$

$$= \sum_\sigma \left(\dot{q}_\sigma \, dp_\sigma - \frac{\partial L}{\partial q_\sigma} dq_\sigma \right) - \frac{\partial L}{\partial t} dt \tag{32.12b}$$

where the defining relation (32.9) has been used to cancel the first and last terms in parentheses in Eq. (32.12a). Note that the total differential dH depends only on the differentials dp_σ, dq_σ, and dt. As a result, *the hamiltonian should be considered a function of the generalized momenta, the generalized coordinates, and the time*

$$H = H(p_1, \ldots, p_n; q_1, \ldots, q_n; t) \tag{32.13}$$

obtained by substituting Eq. (32.10) into (32.11).

In this context, the definition (32.11) can be interpreted as a *Legendre transformation*† from the original variables $(q, \dot{q}, t)$ appropriate for the lagrangian to the new set (p, q, t). Moreover, the differential of Eq. (32.13) becomes

$$dH = \sum_{\sigma} \left(\frac{\partial H}{\partial p_\sigma} dp_\sigma + \frac{\partial H}{\partial q_\sigma} dq_\sigma \right) + \frac{\partial H}{\partial t} dt \qquad (32.14)$$

where the partial derivatives of H are taken with all the other variables in parentheses in (32.13) fixed. Comparison of Eqs. (32.12b) and (32.14) immediately identifies

$$\frac{\partial H}{\partial p_\sigma} = \dot{q}_\sigma \qquad (32.15a)$$

$$\sigma = 1, \ldots, n$$

$$\frac{\partial H}{\partial q_\sigma} = -\frac{\partial L}{\partial q_\sigma} = -\dot{p}_\sigma \qquad (32.15b)$$

where Lagrange's equations (32.3) and (32.9) have been used to reexpress the right-hand side of Eq. (32.15b). Equations (32.15a) and (32.15b) are known as *Hamilton's equations*. They provide $2n$ coupled first-order differential equations in the time, and our discussion shows that they are completely equivalent to Lagrange's equations. To solve these equations one must specify values of the generalized coordinates q_σ and the generalized momenta p_σ at a given instant in time. The subsequent motion of the system is then determined by the differential equations (32.15).

Equations (32.12) and (32.14) hold for arbitrary infinitesimal changes in the variables (p_σ, q_σ, t). In particular, if p_σ and q_σ are held fixed, we obtain the additional relation

$$\frac{\partial H}{\partial t} = -\frac{\partial L}{\partial t} \qquad (32.16)$$

where the partial derivative on the left-hand side is to be evaluated at fixed $\{p_\sigma, q_\sigma\}$ and that on the right-hand side at fixed $\{q_\sigma, \dot{q}_\sigma\}$. A second important application of Eq. (32.14) is to let the system follow its actual dynamical motion from t to $t + dt$. In this case the variables p_σ and q_σ change according to $dp_\sigma = \dot{p}_\sigma \, dt$ and $dq_\sigma = \dot{q}_\sigma \, dt$. If Eq. (32.14) is divided by dt, the result is

$$\frac{dH}{dt} = \sum_{\sigma} \left(\frac{\partial H}{\partial p_\sigma} \dot{p}_\sigma + \frac{\partial H}{\partial q_\sigma} \dot{q}_\sigma \right) + \frac{\partial H}{\partial t}$$

$$= \sum_{\sigma} (\dot{q}_\sigma \dot{p}_\sigma - \dot{p}_\sigma \dot{q}_\sigma) + \frac{\partial H}{\partial t} = \frac{\partial H}{\partial t} \qquad (32.17)$$

---

† Such transformations occur frequently in thermodynamics, where the differential form of the first and second laws implies that $dE = T \, dS - p \, dV$ for a system undergoing a reversible process. Thus the internal energy $E = E(S, V)$ must be considered a function of the entropy S and volume V, and the temperature and pressure can be obtained as partial derivatives $T = (\partial E/\partial S)_V$ and $p = -(\partial E/\partial V)_S$. It is often more convenient to consider other pairs of independent variables, for example T and V. The Legendre transformation to the Helmholtz free energy $F \equiv E - TS$ achieves this aim since it follows directly that $dF = -S \, dT - p \, dV$. We see that F is indeed a function of T and V, with $S = -(\partial F/\partial T)_V$ and $p = -(\partial F/\partial V)_T$. The analogy with Eqs. (32.9) to (32.15) is evident.

where we have used the right-hand sides of Eq. (32.15) to obtain the second line. A combination with (32.16) yields

$$\frac{dH}{dt} = \frac{\partial H}{\partial t} = -\frac{\partial L}{\partial t} \qquad (32.18)$$

which shows that H varies dynamically with time only through its explicit time dependence, independent of the variations in p_σ and q_σ. In particular, we recover the previous result (Sec. 20) that the hamiltonian is a constant of the motion if the lagrangian has no explicit dependence on the time [and hence neither does the hamiltonian by (32.16)].

For a conservative system with holonomic time-independent constraints and a potential of the form $V(q_1, \ldots, q_n)$, the first term in the hamiltonian was shown in Sec. 20 to be twice the kinetic energy

$$\sum_\sigma p_\sigma \dot{q}_\sigma = 2T \qquad (32.19)$$

For such systems the hamiltonian is not only a constant of the motion but also the total energy

$$H = T + V = E = \text{const} \qquad (32.20)$$

Hamilton's equations (32.15) are fully equivalent to Newton's laws for the conservative systems under discussion. Indeed, we can choose to consider Hamilton's equations as the fundamental equations of mechanics for such systems, just as we previously viewed Lagrange's equations. In that case, a variational formulation (Hamilton's principle) provided an economical derivation of the dynamical equations that was manifestly independent of the choice of coordinates. We now investigate whether a similar type of variational principle can lead to Hamilton's equations.

Derivation of Hamilton's Equations from a Modified Hamilton's Principle

Hamilton's principle (32.1) can be rewritten in terms of the hamiltonian simply by substituting the definition (32.11)

$$\delta \int_{t_1}^{t_2} L \, dt = \delta \int_{t_1}^{t_2} \left[\sum_\sigma p_\sigma \dot{q}_\sigma - H(p, q, t) \right] dt = 0 \qquad (32.21)$$

If the generalized coordinates undergo virtual displacements δq_σ, inducing variations in the generalized velocities $\delta \dot{q}_\sigma = d(\delta q_\sigma)/dt$, then we merely rederive Lagrange's equations. Since the basic aim of hamiltonian dynamics is to maintain symmetry between the generalized coordinates and momenta, we instead introduce *a new type of variation, treating p_σ and q_σ with $\sigma = 1, \ldots, n$ as independent dynamical variables, subject to independent variations δp_σ and δq_σ.* This independence is in accordance with Hamilton's equations (32.15), whose complete solution required a specification of the variables $\{p_\sigma, q_\sigma\}$ at some initial time. In terms of this new generalized variation, we assert that Eq. (32.21) provides a variational basis for Hamilton's equations and therefore represents a modified Hamilton's principle for classical dynamics.

To be precise, we demand that the action be stationary under independent variations of the generalized coordinates and momenta

$$p_\sigma \to p_\sigma + \epsilon_1 \eta_\sigma(t) \tag{32.22}$$

$$q_\sigma \to q_\sigma + \epsilon_2 \chi_\sigma(t) \tag{32.23}$$

where the functions $\eta_\sigma(t)$ and $\chi_\sigma(t)$ are independent and arbitrary, apart from the requirement of *fixed endpoints*.

$$\delta p_\sigma(t_1) = \delta p_\sigma(t_2) = 0$$
$$\delta q_\sigma(t_1) = \delta q_\sigma(t_2) = 0 \tag{32.24}$$

The proof follows directly by carrying out the indicated variation in Eq. (32.21)

$$\int_{t_1}^{t_2} dt \left[\sum_\sigma \left(\dot q_\sigma \, \delta p_\sigma + p_\sigma \, \delta \dot q_\sigma - \frac{\partial H}{\partial p_\sigma} \delta p_\sigma - \frac{\partial H}{\partial q_\sigma} \delta q_\sigma \right) \right] = 0 \tag{32.25}$$

Since it is clear from Eq. (32.23) that

$$\delta \dot q_\sigma = \frac{d}{dt} \delta q_\sigma \tag{32.26}$$

the second term in parentheses in Eq. (32.25) can be written as

$$p_\sigma \, \delta \dot q_\sigma = p_\sigma \frac{d}{dt} \delta q_\sigma = \frac{d}{dt}(p_\sigma \, \delta q_\sigma) - \dot p_\sigma \, \delta q_\sigma \tag{32.27}$$

The total time derivative on the right-hand side of Eq. (32.27) makes no contribution to the integral in (32.21) because of the fixed endpoints (32.24). Thus Eq. (32.25) becomes

$$\int_{t_1}^{t_2} dt \left[\sum_\sigma \left(\dot q_\sigma \, \delta p_\sigma - \dot p_\sigma \, \delta q_\sigma - \frac{\partial H}{\partial p_\sigma} \delta p_\sigma - \frac{\partial H}{\partial q_\sigma} \delta q_\sigma \right) \right] = 0 \tag{32.28}$$

As is evident from Eqs. (32.22) and (32.23), our prescription is to consider the variations $\{\delta p_\sigma, \delta q_\sigma\}$ in Eq. (32.28) as independent. The coefficients of each of these terms must therefore vanish, which immediately leads to Hamilton's equations

$$\dot q_\sigma = \frac{\partial H}{\partial p_\sigma} \tag{32.29}$$

$$\sigma = 1, \ldots, n$$

$$\dot p_\sigma = -\frac{\partial H}{\partial q_\sigma} \tag{32.30}$$

This completes the proof.

In this way, the modified Hamilton's principle may be taken to be the basic statement of mechanics. The action as expressed in Eq. (32.21) must be stationary under arbitrary independent variations of the generalized coordinates q_σ and the generalized momenta p_σ subject only to the condition of fixed endpoints expressed by Eq. (32.24). This statement is equivalent to Hamilton's equations, and our previous logical development then shows it to be equivalent not only to Lagrange's equations but ultimately to Newton's laws.

33 EXAMPLE: CHARGED PARTICLE IN AN ELECTROMAGNETIC FIELD

As an example of hamiltonian dynamics we consider a particle of mass m and charge e moving in a *specified* electric field $\mathbf{E}(\mathbf{r}, t)$ and magnetic field $\mathbf{B}(\mathbf{r}, t)$. This problem is somewhat more complex than those studied so far because the electromagnetic fields can have arbitrary space-time dependence. Nevertheless, it plays a central role in physics, in particular the development of nonrelativistic quantum mechanics.

We assert that the dynamical equations of motion of the particle are derivable from a lagrangian of the form

$$L(\mathbf{r}, \dot{\mathbf{r}}, t) = \tfrac{1}{2}m\dot{\mathbf{r}}^2 - e\Phi(\mathbf{r}, t) + \frac{e}{c}\dot{\mathbf{r}} \cdot \mathbf{A}(\mathbf{r}, t) \tag{33.1}$$

where $\mathbf{A}$ and Φ are the usual vector and scalar potentials† in cgs units related to the fields by

$$\mathbf{B}(\mathbf{r}, t) = \nabla \times \mathbf{A}(\mathbf{r}, t) \tag{33.2a}$$

$$\mathbf{E}(\mathbf{r}, t) = -\nabla\Phi(\mathbf{r}, t) - \frac{1}{c}\frac{\partial}{\partial t}\mathbf{A}(\mathbf{r}, t) \tag{33.2b}$$

and c is the speed of light. Since $\mathbf{A}$ and Φ are assumed known, L indeed depends on the proper variables $L(\mathbf{r}, \dot{\mathbf{r}}, t)$. The proof of Eq. (33.1) follows through Lagrange's equations for the generalized coordinates $\mathbf{r} = (x_1, x_2, x_3)$

$$\frac{d}{dt}\frac{\partial L}{\partial \dot{x}_i} - \frac{\partial L}{\partial x_i} = 0 \qquad i = 1, 2, 3 \tag{33.3}$$

Evidently

$$\frac{\partial L}{\partial \dot{x}_i} = m\dot{x}_i + \frac{e}{c}A_i(\mathbf{r}, t) \tag{33.4a}$$

$$\frac{\partial L}{\partial x_i} = -e\frac{\partial}{\partial x_i}\Phi(\mathbf{r}, t) + \frac{e}{c}\sum_{j=1}^{3}\dot{x}_j\frac{\partial}{\partial x_i}A_j(\mathbf{r}, t) \tag{33.4b}$$

These relations are to be evaluated at the (moving) location of the particle. In evaluating the total time derivative of Eq. (33.4a), we must therefore use the relation

$$\frac{d}{dt}A_i(\mathbf{r}, t) = \sum_{j=1}^{3}\left[\frac{\partial}{\partial x_j}A_i(\mathbf{r}, t)\right]\dot{x}_j + \frac{\partial}{\partial t}A_i(\mathbf{r}, t) \tag{33.5}$$

† See, for example, Jackson [1], chap. 6.

which follows from writing the total differential and dividing by dt [compare Eq. (13.3)]. Thus Lagrange's equations (33.3) here take the form

$$m\ddot{x}_i = -e\left(\frac{\partial \Phi}{\partial x_i} + \frac{1}{c}\frac{\partial A_i}{\partial t}\right) + \frac{e}{c}\sum_{j=1}^{3}\dot{x}_j\left(\frac{\partial A_j}{\partial x_i} - \frac{\partial A_i}{\partial x_j}\right) \tag{33.6}$$

The last sum can be written

$$\sum_{j=1}^{3}\dot{x}_j\left(\frac{\partial A_j}{\partial x_i} - \frac{\partial A_i}{\partial x_j}\right) = [\dot{\mathbf{r}} \times (\nabla \times \mathbf{A})]_i \tag{33.7}$$

as is readily established by writing out the components of this vector expression. Identification of the fields through Eqs. (33.2) then leads to the vector form of Newton's second law for a particle subject to the Lorentz force

$$m\ddot{\mathbf{r}} = e\left[\mathbf{E}(\mathbf{r}, t) + \frac{1}{c}\dot{\mathbf{r}} \times \mathbf{B}(\mathbf{r}, t)\right] \tag{33.8}$$

It is indeed the correct equation of motion for a particle in a specified electromagnetic field.

To discuss this same problem from the viewpoint of hamiltonian dynamics, we first need the canonical momentum, defined by Eq. (33.4a) and given in vector form by

$$\mathbf{p} \equiv \frac{\partial L}{\partial \dot{\mathbf{r}}} = m\dot{\mathbf{r}} + \frac{e}{c}\mathbf{A}(\mathbf{r}, t) \tag{33.9}$$

It contains not only the usual mechanical momentum $m\mathbf{v} = m\dot{\mathbf{r}}$ but also an electromagnetic contribution $e\mathbf{A}/c$. In connection with the discussion of symmetries in Sec. 20, we note that if $\mathbf{A}$ and Φ are independent of a particular variable (z, say), then $\partial L/\partial z = 0$ and Eq. (33.3) then ensures that the corresponding generalized momentum p_z will be a constant of the motion. This relation is not the usual conservation of momentum, however, for Eq. (33.9) shows that the sum $mv_z + eA_z/c$ is the constant quantity, instead of mv_z alone. This alteration reflects the presence of the electromagnetic forces.

The hamiltonian for the system is defined by

$$H = \sum_{i=1}^{3}p_i\dot{x}_i - L \tag{33.10}$$

Substitution of Eqs. (33.1) and (33.9) leads to

$$H = \tfrac{1}{2}m\dot{\mathbf{r}}^2 + e\Phi(\mathbf{r}, t) \tag{33.11}$$

In this form it is evident that the hamiltonian is the sum of the kinetic and potential energies of the particle. In general, H is not a constant of the motion since

$$\frac{dH}{dt} = \frac{\partial H}{\partial t} = -\frac{\partial L}{\partial t} \neq 0 \tag{33.12}$$

if $\mathbf{E}$ and $\mathbf{B}$ vary with time. In that case, the particle can exchange energy with the fields.

It remains to express the hamiltonian in Eq. (33.11) in terms of the proper variables, and this is immediately done with the help of Eq. (33.9)

$$H(\mathbf{p}, \mathbf{r}, t) = \frac{1}{2m} \left| \mathbf{p} - \frac{e}{c} \mathbf{A}(\mathbf{r}, t) \right|^2 + e\Phi(\mathbf{r}, t) \tag{33.13}$$

which is our principal result. We now use this hamiltonian to demonstrate explicitly that Hamilton's equations

$$\frac{\partial H}{\partial p_i} = \dot{x}_i \tag{33.14a}$$

$$i = 1, 2, 3$$

$$\frac{\partial H}{\partial x_i} = -\dot{p}_i \tag{33.14b}$$

also provide the correct equations of motion for the particle. Equation (33.14a) defines the velocity in hamiltonian dynamics, and it follows from Eq. (33.13) that

$$m\dot{x}_i = p_i - \frac{e}{c} A_i(\mathbf{r}, t) \tag{33.15}$$

which is simply the component form of (33.9). The second of Hamilton's equations (33.14b) gives

$$\dot{p}_i = -e \frac{\partial}{\partial x_i} \Phi + \frac{e}{mc} \sum_{j=1}^{3} \left(p_j - \frac{e}{c} A_j \right) \left(\frac{\partial}{\partial x_i} A_j \right) \tag{33.16a}$$

because the spatial dependence of H enters only through the fields. Substitution of (33.15) simplifies this result to

$$\dot{p}_i = -e \frac{\partial}{\partial x_i} \Phi + \frac{e}{c} \sum_{j=1}^{3} \dot{x}_j \left(\frac{\partial}{\partial x_i} A_j \right) \tag{33.16b}$$

We now take the total time derivative of Eq. (33.15)

$$m\ddot{x}_i = \dot{p}_i - \frac{e}{c} \frac{d}{dt} A_i(\mathbf{r}, t) \tag{33.17}$$

and use Eqs. (33.5) and (33.16b) to rederive Eq. (33.6) and hence the correct Lorentz-force equation (33.8). It is important to note that the $\dot{\mathbf{r}}$ appearing in this expression is defined in terms of the proper variables $(\mathbf{p}, \mathbf{r}, t)$ for hamiltonian dynamics through Eq. (33.15) [or equivalently (33.9)].

34 CANONICAL TRANSFORMATIONS

Consider a transformation to a new set of generalized momenta P_σ and generalized coordinates Q_σ defined by a set of equations of the form

$$p_\sigma = p_\sigma(P_1, \ldots, P_n; Q_1, \ldots, Q_n; t) \tag{34.1a}$$

$$\sigma = 1, \ldots, n$$

$$q_\sigma = q_\sigma(P_1, \ldots, P_n; Q_1, \ldots, Q_n; t) \tag{34.1b}$$

These equations are assumed to be nonsingular, so that they can be inverted to express the new coordinates Q_σ and the new momenta P_σ in terms of the original coordinates q_σ and momenta p_σ. In analogy with our previous discussion of Lagrange's equations in Sec. 32, it is natural to ask under what conditions this transformation will preserve the form of Hamilton's equations. One obvious possibility is the set of point transformations (32.4a), with p_σ and P_σ obtained from the corresponding lagrangians according to Eq. (32.9). To consider more general cases, suppose first that the transformation (34.1) leads to the following relation when substituted into the integrand in Eq. (32.21) [cf. Eq. (32.5)]:

$$\sum_\sigma p_\sigma \dot{q}_\sigma - H(p, q, t) = \sum_\sigma P_\sigma \dot{Q}_\sigma - \tilde{H}(P, Q, t) \qquad (34.2)$$

Here $\tilde{H}$ is any function of the indicated variables. It is evident from the derivation of the modified Hamilton's principle that $\tilde{H}$ serves as the new hamiltonian, with P_σ and Q_σ obeying the corresponding Hamilton's equations. The condition (34.2) turns out to be too restrictive, but we can generalize it by adding the total time derivative of an arbitrary function F of the variables (q, Q, t). (We shall see that we have considerable freedom in choosing just which variables occur in the function F, and we start here with the pair of variables q and Q.) Thus we assume that the transformations (34.1) yield

$$\sum_\sigma p_\sigma \dot{q}_\sigma - H(p, q, t) = \sum_\sigma P_\sigma \dot{Q}_\sigma - \tilde{H}(P, Q, t) + \frac{d}{dt} F(q, Q, t) \qquad (34.3)$$

Since the original variations all vanish at the endpoints according to Eq. (32.24), the inversions of Eqs. (34.1) imply that the new variations will likewise vanish there

$$\delta Q_\sigma(t_1) = \delta P_\sigma(t_1) = 0 \quad \text{and} \quad \delta Q_\sigma(t_2) = \delta P_\sigma(t_2) = 0 \qquad (34.4)$$

Thus the total time derivative of F in Eq. (34.3) will not contribute to the modified Hamilton's principle (32.21) because the integral of the total time derivative is just the function evaluated at the endpoints, where the variations of all the variables vanish

$$\delta \int_{t_1}^{t_2} dt \, \frac{d}{dt} F(q, Q, t) = \delta[F(t_2) - F(t_1)] = 0 \qquad (34.5)$$

Furthermore, the first two terms on the right-hand side of Eq. (34.3) are again identical in form with those examined in the discussion of Eqs. (32.21), and we immediately conclude that the variational principle yields Hamilton's equations for the new variables and the new hamiltonian $\tilde{H}$

$$\dot{Q}_\sigma = \frac{\partial \tilde{H}}{\partial P_\sigma} \qquad (34.6)$$

$$\sigma = 1, \ldots, n$$

$$-\dot{P}_\sigma = \frac{\partial \tilde{H}}{\partial Q_\sigma} \qquad (34.7)$$

Whenever the transformation (34.1) is such that Eq. (34.3) is satisfied for some $\tilde{H}$

and some F, Hamilton's equations retain their form [as indicated in Eqs. (34.6) and (34.7)] and the transformation (34.1) is said to be *canonical*.

Now Eq. (34.3) is rather peculiar, and one may naturally wonder how a transformation can ever lead to such a result. We shall, in fact, reverse the process and *require* that the transformation actually yield this expression. Note first that the total time derivative of the function $F(q, Q, t)$ has the (by now) familiar form

$$\frac{d}{dt} F(q, Q, t) = \sum_\sigma \left(\frac{\partial F}{\partial q_\sigma} \dot{q}_\sigma + \frac{\partial F}{\partial Q_\sigma} \dot{Q}_\sigma \right) + \frac{\partial F}{\partial t} \qquad (34.8a)$$

so that Eq. (34.3) can be rewritten

$$\sum_\sigma \left(p_\sigma - \frac{\partial F}{\partial q_\sigma} \right) \dot{q}_\sigma - H(p, q, t) = \sum_\sigma \left(P_\sigma + \frac{\partial F}{\partial Q_\sigma} \right) \dot{Q}_\sigma - \tilde{H}(P, Q, t) + \frac{\partial F}{\partial t} \quad (34.8b)$$

Evidently, we can *guarantee* the validity of Eq. (34.3) by setting the coefficients of the generalized velocities to zero (this is the reason for considering F as a function of q and Q) and then equating the remaining terms. Thus Eq. (34.3) will hold provided that

$$p_\sigma = \frac{\partial}{\partial q_\sigma} F(q, Q, t) \qquad (34.9)$$

$$\sigma = 1, \ldots, n$$

$$-P_\sigma = \frac{\partial}{\partial Q_\sigma} F(q, Q, t) \qquad (34.10)$$

and $$\tilde{H}(P, Q, t) = H(p, q, t) + \frac{\partial}{\partial t} F(q, Q, t) \qquad (34.11)$$

Let us discuss these basic results. Equation (34.10), after inversion, simply re-expresses the original transformation Eq. (34.1b). Substituting this result in Eq. (34.9) then reproduces the original transformation Eq. (34.1a). As a final step, the new hamiltonian $\tilde{H}$ [Eq. (34.11)] is obtained by rewriting the right-hand side of Eq. (34.11) in terms of the appropriate variables P and Q. We therefore reach the following important conclusion. Whenever the original transformations in Eqs. (34.1) can be written in terms of *some* F according to Eqs. (34.9) and (34.10), then Hamilton's equations (34.6) and (34.7) necessarily hold for the new coordinates Q_σ, and new momenta P_σ, with the new hamiltonian $\tilde{H}$ given by Eq. (34.11). The function F is known as the *generator* of the canonical transformation because it determines the detailed form of Eqs. (34.9) and (34.10) and hence, after inversion the original transformations (34.1a) and (34.1b). Unfortunately, it is not always easy to determine whether such a function F actually exists for a given transformation (34.1a) and (34.1b).

On the other hand, we can turn the argument around and observe that *any* $F(q, Q, t)$ generates *some* canonical transformation, assuming only that Eqs. (34.9) and (34.10) can be inverted to yield the explicit expressions (34.1). Since Eqs. (34.9) to (34.11) are then guaranteed, the fundamental form (34.3) becomes an *identity* and the modified Hamilton's principle ensures that Hamilton's equations remain correct for the new variables and the new hamiltonian according to Eqs. (34.6) and (34.7). This is a truly remarkable result, for it implies a freedom to rewrite Hamilton's equations, and hence Newton's laws, in an arbitrary fashion by introducing an arbitrary $F(q, Q, t)$!

Given this extraordinary flexibility, can we use it to simplify the equations of motion (34.6) and (34.7)? For example, we already know that if a coordinate is cyclic, the corresponding momentum is a constant of the motion; it also follows trivially that if a momentum variable is cyclic, the corresponding coordinate is a constant of the motion. Thus we might try to construct a transformation to a new $\tilde{H}$ that contains one or more cyclic coordinates Q and momenta P. In fact, why not try to make *all the coordinates and momenta cyclic*, in which case they all become constants of the motion? This transformation indeed exists and can be constructed with the Hamilton-Jacobi theory.

35 HAMILTON-JACOBI THEORY

Instead of working with the function F in the basic relation (34.3) that characterizes a canonical transformation, it is conventional to introduce another function S defined in terms of F according to

$$F = -\sum_\sigma P_\sigma Q_\sigma + S \tag{35.1}$$

As we shall see, the function S will turn out to be the classical action. The differential of Eq. (35.1) gives [see Eq. (34.8a)]

$$dF \equiv \sum_\sigma \left(\frac{\partial F}{\partial q_\sigma} dq_\sigma + \frac{\partial F}{\partial Q_\sigma} dQ_\sigma \right) + \frac{\partial F}{\partial t} dt$$

$$= -\sum_\sigma (P_\sigma \, dQ_\sigma + Q_\sigma \, dP_\sigma) + dS \tag{35.2}$$

and Eqs. (34.9) and (34.10) allow us to solve for dS as

$$dS = \sum_\sigma (p_\sigma \, dq_\sigma + Q_\sigma \, dP_\sigma) + \frac{\partial F}{\partial t} dt \tag{35.3}$$

This expression shows that S must be considered as a function of the variables (q, P, t), and Eq. (35.1) can therefore be interpreted as a Legendre transformation from (q, Q, t) to (q, P, t). Correspondingly, the coefficients in parentheses in Eq. (35.3) are just the appropriate partial derivatives

$$p_\sigma = \frac{\partial}{\partial q_\sigma} S(q, P, t) \tag{35.4a}$$

$$\sigma = 1, \ldots, n$$

$$Q_\sigma = \frac{\partial}{\partial P_\sigma} S(q, P, t) \tag{35.4b}$$

and the remaining relation involving the time takes the form

$$\frac{\partial}{\partial t} S(q, P, t) = \frac{\partial}{\partial t} F(q, Q, t) \qquad (35.5)$$

The basic relation (34.3) written in terms of S will now hold as an *identity* provided Eqs. (35.4) are satisfied and that the new hamiltonian is identified according to [compare Eqs. (34.11) and (35.5)]

$$\tilde{H}(P, Q, t) = H(p, q, t) + \frac{\partial}{\partial t} S(q, P, t) \qquad (35.6)$$

The function $S(q, P, t)$ itself can now be taken as the generator of the canonical transformation since, after inversion, Eqs. (35.4b) give relations of the type (34.1b) and substitution into Eqs. (35.4a) leads to equations of the form (34.1a). In this way, knowledge of $S(q, P, t)$ is equivalent to knowledge of the transformation laws (34.1).

We next observe that *any* $S(q, P, t)$ will generate some canonical transformation, since Eq. (34.3) has been arranged to hold as an identity. We shall *use this freedom to transform and simplify classical mechanics* as follows:

$$\text{Choose } S \text{ so that } \tilde{H} \equiv 0 \qquad (35.7)$$

Imposing this condition transforms the mechanical problem into a lovely simple form, for $\tilde{H}$ is now manifestly independent of Q_σ and P_σ. Since *all these variables are cyclic*, Hamilton's equations (34.6) and (34.7) immediately imply that

$$\dot{Q}_\sigma = 0 \qquad \qquad (35.8)$$
$$\qquad \qquad \sigma = 1, \ldots, n$$
$$\dot{P}_\sigma = 0 \qquad \qquad (35.9)$$

Consequently, all the Q_σ and P_σ are *constants of the motion*

$$Q_\sigma = \text{const} \qquad \qquad (35.10)$$
$$\qquad \qquad \sigma = 1, \ldots, n$$
$$P_\sigma = \text{const} \qquad \qquad (35.11)$$

To guarantee condition (35.7), which simplifies the mechanical problem, the generator S must satisfy the following equation obtained by combining Eqs. (35.7), (35.6), and (35.4a)

$$H\left(\frac{\partial S}{\partial q_1}, \ldots, \frac{\partial S}{\partial q_n} ; q_1, \ldots, q_n; t\right) + \frac{\partial S}{\partial t} = 0 \qquad (35.12)$$

It is called the *Hamilton-Jacobi equation*. Thus S satisfies a first-order partial differential equation in $n + 1$ variables $q_1, \ldots, q_n, t$. We can imagine integrating this equation one variable at a time, keeping all the remaining variables fixed. Each of these integrations introduces one constant of integration, and it is evident that the complete solution to the Hamilton-Jacobi equation for the generator S will contain $n + 1$ constants of integration. If S_0 is a constant independent of q_1, $\ldots, q_n$, t we observe that the replacement $S \rightarrow S + S_0$ leaves the Hamilton-Jacobi equation unchanged, since it depends only on the partial derivatives of S with respect to the variables. Therefore, the final overall additive constant of integration S_0 is immaterial, and we choose to write S in the form

$$S = S(q_1, \ldots, q_n; \alpha_1, \ldots, \alpha_n; t) + S_0 \tag{35.13}$$

where, here and henceforth, $\alpha_1, \ldots, \alpha_n$ are *any n independent, nonadditive integration constants*.

Assume that the Hamilton-Jacobi equation has been solved in the form (35.13). In the first term on the right-hand side of Eq. (35.13) replace each of the α_σ by P_σ for $\sigma = 1, \ldots, n$. This particular solution $S(q_1, \ldots, q_n; P_1, \ldots, P_n; t)$ is called *Hamilton's principal function*; it is a well-defined solution to a certain partial differential equation, the Hamilton-Jacobi equation, written in the form of Eq. (35.13).

Our previous analysis shows that any S generates some canonical transformation, and we now study the particular *canonical transformation generated by Hamilton's principal function*. We can first generate a specific transformation from the coordinates $\{p, q\}$ to the new set of coordinates $\{P, Q\}$ according to Eqs. (35.4a) and (35.4b)

$$p_\sigma = \frac{\partial}{\partial q_\sigma} S(q_1, \ldots, q_n; P_1, \ldots, P_n; t) \tag{35.14}$$

$$\sigma = 1, \ldots, n$$

$$Q_\sigma = \frac{\partial}{\partial P_\sigma} S(q_1, \ldots, q_n; P_1, \ldots, P_n; t) \tag{35.15}$$

By comparing Eqs. (35.14), (35.15), and (35.12) with Eqs. (35.4a), (35.4b), and (35.6) we immediately verify that the transformation is canonical and that

$$\tilde{H}(P, Q, t) = 0 \tag{35.16}$$

Hamilton's equations, which are preserved under a canonical transformation, become

$$-\dot{P}_\sigma = \frac{\partial \tilde{H}}{\partial Q_\sigma} = 0 \tag{35.17}$$

$$\sigma = 1, \ldots, n$$

$$\dot{Q}_\sigma = \frac{\partial \tilde{H}}{\partial P_\sigma} = 0 \tag{35.18}$$

implying that all the Q_σ and P_σ are constants of the motion

$$P_\sigma = \text{const} \equiv \alpha_\sigma \tag{35.19}$$

$$\sigma = 1, \ldots, n$$

$$Q_\sigma = \text{const} \equiv \beta_\sigma \tag{35.20}$$

Since the α_σ are just the independent nonadditive integration constants in the solution (35.13) to the Hamilton-Jacobi equation, Eq. (35.19) immediately shows that *any* such separation constants in the Hamilton-Jacobi equation are constants of the motion. Moreover, the partial derivative (35.15) of S with respect to each of the n nonadditive constants α_σ yields n additional independent constants β_σ, as seen in Eq. (35.20).

The reader may wonder why we picked the particular set of variables (q, P, t) in Eq. (35.1). One reason becomes clear if we imagine inverting Eqs. (35.15) to express q_σ in terms of the new variables. Since the Q_σ and P_σ are constants of the motion [Eqs. (35.19) and (35.20)], this inversion immediately leads to a *solution to the mechanical problem* in the form

$$q_\sigma = q_\sigma(\alpha_1, \ldots, \alpha_n ; \beta_1, \ldots, \beta_n ; t) \qquad \sigma = 1, \ldots, n \qquad (35.21)$$

Thus Hamilton's principal function S, obtained by solving the Hamilton-Jacobi equation, immediately provides a solution (35.21) to the mechanical problem simply by inverting the generating equations (35.15). It is evident that Eqs. (35.21) and their time derivatives can be used to determine the $2n$ constants of the motion $\{\alpha, \beta\}$ in terms of the $2n$ initial conditions of the mechanical problem $\{q, \dot{q}\}$. Furthermore, the corresponding time-dependent momenta $p_\sigma = p_\sigma(\alpha, \beta, t)$ follow directly from Eqs. (35.14).

Hamilton's principal function S has the following interesting physical interpretation. Equations (35.21) determine the trajectory of the dynamical system. *Along this particular path*, the total time derivative of Eq. (35.13) becomes

$$\frac{dS}{dt} = \sum_\sigma \frac{\partial S}{\partial q_\sigma} \dot{q}_\sigma + \frac{\partial S}{\partial t} \qquad (35.22)$$

because the α_σ are constants. Equation (35.14) shows that the first term on the right-hand side is just $\sum_\sigma p_\sigma \dot{q}_\sigma$, and the Hamilton-Jacobi equation (35.12) itself identifies the last term as $-H$. In this way, we find

$$\frac{dS}{dt} = \sum_\sigma p_\sigma \dot{q}_\sigma - H = L \qquad (35.23)$$

where Eq. (32.11) has been used to obtain the last form. Thus, dS/dt is just the lagrangian, and its time integral

$$S(t) \equiv S[q_1(t), \ldots, q_n(t); \alpha_1, \ldots, \alpha_n; t]$$
$$= \int_{t_0}^{t} L(t') \, dt' + S(t_0) \qquad (35.24)$$

shows that $S(t)$ is the action evaluated along the dynamical trajectory. Unfortunately, this relation is not helpful in solving for S because it requires prior knowledge of the trajectory.

If the hamiltonian H has no explicit time dependence, H is a constant of the motion. We shall call this constant E, although it need not be the actual energy [compare Eq. (20.28)]. In this case, it is possible to perform a first separation of variables by assuming that the solution (35.13) to the Hamilton-Jacobi equation has the form

$$S(q_1, \ldots, q_n; \alpha_1, \ldots, \alpha_n; t) = W(q_1, \ldots, q_n, \alpha_1, \ldots, \alpha_n) - \alpha_1 t \qquad (35.25)$$

Substitution of this relation into the Hamilton-Jacobi equation (35.12) leads to the *time-independent* first-order partial differential equation

$$H\left(\frac{\partial W}{\partial q_1}, \ldots, \frac{\partial W}{\partial q_n}, q_1, \ldots, q_n\right) = \alpha_1 = E \tag{35.26}$$

in n variables $q_1 \ldots, q_n$. Its solution W is called *Hamilton's characteristic function*. This approach is most useful in cases where the solution W can be chosen to satisfy a *general separability condition*

$$W(q_1, \ldots, q_n) = W_1(q_1) + W_2(q_2) + \cdots + W_n(q_n) \tag{35.27}$$

so that W becomes a sum of independent additive functions, one for each independent variable.

Since the reader may be unfamiliar with this particular form of partial differential equation and its associated additive separability, we recall a simple example from elementary quantum mechanics to provide some physical insight into the condition (35.27).† Consider the time-dependent Schrödinger equation‡ for the wave function $\psi(q_1, \ldots, q_n, t)$

$$H(p_1, \ldots, p_n; q_1, \ldots, q_n; t)\psi = i\hbar \frac{\partial \psi}{\partial t} \tag{35.28}$$

where Planck's constant $h \equiv 2\pi\hbar$ and the momenta satisfy the elementary quantization condition

$$p_\sigma = \frac{\hbar}{i}\frac{\partial}{\partial q_\sigma} \qquad \sigma = 1, \ldots, n, \tag{35.29}$$

We expect classical mechanics to represent the *short-wavelength* limit of quantum mechanics, just as the rays of geometric optics are the short-wavelength limit of physical optics when diffraction and other wave effects become negligible. To extract this behavior, recall the *de Broglie* relation $\lambda = h/p$ between the wavelength and momentum; it can be rewritten $k = p/\hbar$, where $k = 2\pi/\lambda$ is the wave number. In the semiclassical or high-energy limit ($\hbar \to 0$), we therefore seek a wavelike solution in the *eikonal* form

$$\psi(q_1, \ldots, q_n, t) \propto \exp\left[\frac{i}{\hbar} S(q_1, \ldots, q_n, t)\right] \tag{35.30}$$

with S real. To leading order as $\hbar \to 0$, we need only differentiate the exponential, so that the Schrödinger equation then reads

$$\left[H\left(\frac{\partial S}{\partial q_1}, \ldots, \frac{\partial S}{\partial q_n}; q_1, \ldots, q_n; t\right) + \frac{\partial S}{\partial t}\right]\psi = 0 \tag{35.31}$$

---

† This example breaks the logic and continuity of presentation, but our experience has been that most students at this stage are more familiar with elementary quantum mechanics than with Hamilton-Jacobi theory.

‡ See, for example, Schiff [2], chap. II.

Equivalently, since the wave function in Eq. (35.31) does not vanish, we obtain

$$H\left(\frac{\partial S}{\partial q_1}, \dots, \frac{\partial S}{\partial q_n}; q_1, \dots, q_n; t\right) + \frac{\partial S}{\partial t} = 0 \tag{35.32}$$

which is *precisely the Hamilton-Jacobi equation*. As is evident from this simple illustration, Hamilton-Jacobi theory gives the most direct transition between wave mechanics and classical mechanics.

It is now clear from Eq. (35.30) that S can be thought of as the *phase of the semiclassical wave function,* and Eq. (35.24) shows that this phase is also the classical action evaluated along the path of motion. The first separation of the time dependence in Eq. (35.25) is equivalent to looking for stationary states in quantum mechanics, i.e., states with a well-defined energy E. In addition, soluble problems in quantum mechanics generally allow a separation of variables in which the wave function *factors*

$$\psi(q_1, \dots, q_n, t) = \psi_1(q_1)\psi_2(q_2) \cdots \psi_n(q_n)e^{-iEt/\hbar} \tag{35.33}$$

Evidently, this condition is equivalent to an additive separability for the phase of the wave function as in Eqs. (35.25) and (35.27).

As a simple example of the Hamilton-Jacobi method, consider the motion in a one-dimensional potential $V(q)$. This system has the lagrangian

$$L = \tfrac{1}{2}m\dot{q}^2 - V(q) \tag{35.34}$$

with the corresponding canonical momentum

$$p = m\dot{q} \tag{35.35}$$

and hamiltonian

$$H = \frac{p^2}{2m} + V(q) \tag{35.36}$$

The Hamilton-Jacobi equation (35.12) takes the form

$$\left[\frac{1}{2m}\left(\frac{\partial S}{\partial q}\right)^2 + V(q)\right] + \frac{\partial S}{\partial t} = 0 \tag{35.37}$$

Since this hamiltonian is independent of time, we can separate off the time variable according to Eq. (35.25) and look for solutions of the form

$$S(q, \alpha, t) = W(q, \alpha) - \alpha t \tag{35.38}$$

The Hamilton-Jacobi equation for Hamilton's characteristic function W then becomes

$$\frac{1}{2m}\left(\frac{dW}{dq}\right)^2 + V(q) = \alpha = E \tag{35.39}$$

where we have written $\alpha_1 \equiv \alpha$ to simplify the notation. This equation for W has the form of a first integral; it can be solved by writing

$$\frac{dW}{dq} = \pm (2m)^{1/2}[\alpha - V(q)]^{1/2} \tag{35.40}$$

and then integrating. Observe that only the indefinite integral is required since an overall additive constant of integration can be absorbed into S_0 according to Eq. (35.13) and is therefore irrelevant. Thus we find

$$W(q, \alpha) = \pm \int^q dq \, (2m)^{1/2}[\alpha - V(q)]^{1/2} \tag{35.41}$$

and the explicit solution to the Hamilton-Jacobi equation takes the form

$$S(q, \alpha, t) = W(q, \alpha) - \alpha t \tag{35.42}$$

At this point S is still a function of the three variables (q, α, t), which are as yet unrelated. To determine the trajectory $q(t)$ we recall the general discussion of Eqs. (35.20) and (35.15), which ensures that $\partial S/\partial \alpha$ is a second constant of the motion

$$\beta = \frac{\partial S}{\partial \alpha} = \pm (\tfrac{1}{2}m)^{1/2} \int^q \frac{dq}{[\alpha - V(q)]^{1/2}} - t = \text{const} \tag{35.43}$$

Note that this equation is just that obtained from the conservation of energy; it provides an implicit relation between q and t, with the two constants α and β determined from the initial conditions.†

As a specific simple example, consider a one-dimensional harmonic oscillator with

$$V(q) = \tfrac{1}{2}kq^2 \tag{35.44}$$

In this case, Eq. (35.43) becomes

$$\beta = \pm \left(\frac{m}{2}\right)^{1/2} \int^q \frac{dq}{(\alpha - \tfrac{1}{2}kq^2)^{1/2}} - t = \text{const} \tag{35.45}$$

A direct integration gives

$$\beta + t = \pm \left(\frac{m}{2}\right)^{1/2} \left\{ -\left(\frac{2}{k}\right)^{1/2} \arccos \left[\left(\frac{k}{2\alpha}\right)^{1/2} q\right] \right\} \tag{35.46}$$

and an inversion yields the desired trajectory

$$q(t) = \left(\frac{2\alpha}{k}\right)^{1/2} \cos \omega(t + \beta) \tag{35.47}$$

with

$$\omega \equiv \left(\frac{k}{m}\right)^{1/2} \tag{35.48}$$

---

† As noted previously, $S(q, \alpha, t)$ given by Eqs. (35.41) and (35.42) is the semiclassical approximation to the phase of the Schrödinger wave function, and indeed it is just the well-known WKB approximation (Schiff [2], sec. 34). In this context, Eq. (35.43) for the classical trajectory represents the condition for *constructive* interference of the de Broglie waves. Although we shall not pursue this matter here, analogous classical interference effects will be studied for sound waves and surface waves in Chaps. 9 and 10.

This result is precisely the familiar solution to the harmonic oscillator. The constants α and β can be obtained from the initial values q and $\dot{q}$, as indicated previously. In addition, differentiation of (35.42) with respect to q gives the corresponding time-dependent momentum [see Eq. (35.14)]; in the special case of Eq. (35.44), the explicit result is easily reconciled with (35.35).

36 ACTION-ANGLE VARIABLES

Hamilton-Jacobi theory has an important application to conservative systems whose motion is both *separable* and *periodic*. The first concept is familiar, since we recall from Eq. (35.27) the general separability condition

$$W(q_1, \ldots, q_n, \alpha_1, \ldots, \alpha_n) = W_1(q_1, \alpha_1, \ldots, \alpha_n) + \cdots + W_n(q_n, \alpha_1, \ldots, \alpha_n) \quad (36.1)$$

It defines a class of problems that may be readily solved with the Hamilton-Jacobi approach, describing physical systems where each variable moves independently of the others. Such motion is also called *periodic* for each coordinate and corresponding canonical momentum, either if the pair of variables (p, q) returns repeatedly to the same set of values or if the variable p is simply periodic in the variable q.

To make these ideas more concrete, consider a plot in *phase space* (Fig. 36.1) of the values p and q assumed by each of the independent degrees of freedom in Eq. (36.1) during the course of the motion. If the motion is truly periodic, the pair of variables (p, q) returns to its original value, as in Fig. 36.1a, and we say the system undergoes a *libration*. An example of such behavior is a one-dimensional simple harmonic oscillator, where Eqs. (35.36) and (35.44) show that the trajectory is an ellipse in phase space. On the other hand, if p is periodic in q, as

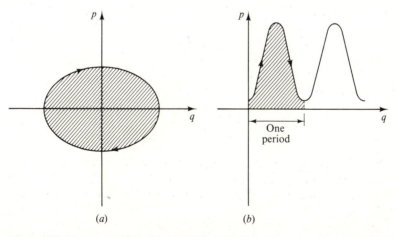

(a) (b)

Figure 36.1 Phase-space diagrams illustrating behavior of one pair of variables (p, q) in separable, periodic motion. The shaded regions correspond to the action integral. (a) Libration; (b) rotation.

indicated in Fig. 36.1*b*, we say the system undergoes a *rotation*. An example of such a system would be a simple pendulum in a gravitational field, started with sufficient energy to "go over the top." Here we take the coordinate q to be the angle θ, which increases without limit. Another interesting example is the heavy symmetrical top (Sec. 31), where the coordinates α and γ undergo rotation whereas the coordinate β undergoes libration.

Having specified the physical system under consideration, we motivate the subsequent discussion by asserting that it will permit us to compute the frequencies, or periods, of the individual independent motions without actually solving the complete multidimensional mechanical problem. This technique is very useful, for example, in astronomy, where the periods for various motions may be more interesting than the detailed dynamical behavior. Alternatively, the transition from classical to quantum mechanics focuses on the frequencies of various motions, which often bear a simple relation to the energy of light quanta emitted and absorbed.

We first define the *action variables* of the system according to

$$J_\sigma \equiv \oint p_\sigma \, dq_\sigma \qquad \sigma = 1, \ldots, n \tag{36.2}$$

Here the integral is the *area in phase space* taken over one period of the motion (see Fig. 36.1). The system has one action variable for each independent degree of freedom. According to Eqs. (35.14), (35.25), and (36.1), the canonical momenta in the Hamilton-Jacobi theory are obtained as

$$p_\sigma = \frac{\partial W}{\partial q_\sigma} = \frac{\partial W_\sigma(q_\sigma, \alpha_1, \ldots, \alpha_n)}{\partial q_\sigma} \qquad \sigma = 1, \ldots, n \tag{36.3}$$

where the last equality has used the separability condition (36.1) that the variable q_σ appear only in the function W_σ. As anticipated, each p_σ depends only on the conjugate q_σ and the various integration constants.

Substitution of Eqs. (36.3) into Eqs. (36.2) yields a sequence of relations of the form

$$J_\sigma = J_\sigma(\alpha_1, \ldots, \alpha_n) \qquad \sigma = 1, \ldots, n \tag{36.4}$$

Since the α_σ are constants of the motion, it is evident that the action integrals (36.4) are also constants of the motion. We assume that these relations can be inverted to write

$$\alpha_\sigma = \alpha_\sigma(J_1, \ldots, J_n) \qquad \sigma = 1, \ldots, n \tag{36.5}$$

In particular, Eq. (35.26) identifies the first integration constant α_1 in the Hamilton-Jacobi theory as the constant E, which we shall henceforth assume to be the total energy of the system

$$\alpha_1 = \alpha_1(J_1, \ldots, J_n) = H(J_1, \ldots, J_n) = E \tag{36.6}$$

here expressed in terms of the action variables J_σ.

We now recall that the set $\{\alpha_\sigma\}$ in Eq. (35.13) represents an arbitrary set of n independent nonadditive integration constants, which allows us great freedom to

choose a convenient parametrization. Suppose, in fact, that we take the action variables $\{J_\sigma\}$ themselves as the integration constants. For clarity, the functions obtained with the new integration constants will be denoted explicitly by a bar:

$$W[q_1, \ldots, q_n, \alpha_1(J), \ldots, \alpha_n(J)] \equiv \bar{W}(q_1, \ldots, q_n, J_1, \ldots, J_n) \qquad (36.7)$$

$$S[q_1, \ldots, q_n; \alpha_1(J), \ldots, \alpha_n(J); t] \equiv \bar{S}(q_1, \ldots, q_n; J_1, \ldots, J_n; t) = \bar{W} - \alpha_1(J)t \qquad (36.8)$$

All the arguments of Sec. 35 now apply to the function $\bar{S}(q_1, \ldots, q_n; J_1, \ldots, J_n; t)$, with the new canonical momenta $\bar{P}_\sigma \equiv J_\sigma$. In particular, the transformations (35.14) and (35.15) become

$$p_\sigma = \frac{\partial \bar{S}}{\partial q_\sigma} \qquad (36.9)$$

$$\sigma = 1, \ldots, n$$

$$\bar{Q}_\sigma = \frac{\partial \bar{S}}{\partial J_\sigma} \qquad (36.10)$$

and the quantities $\{p_\sigma\}$ are unaffected because of the equivalence of differentiation at constant $\{\alpha_\sigma\}$ and constant $\{J_\sigma\}$. In addition, $\bar{S}$ satisfies the Hamilton-Jacobi equation (35.12), and the transformation remains canonical with $\tilde{H} = 0$. As a result, the new coordinates are again constants of the motion

$$\bar{P}_\sigma \equiv J_\sigma = \text{const} \qquad (36.11)$$

$$\sigma = 1, \ldots, n$$

$$\bar{Q}_\sigma \equiv \bar{\beta}_\sigma = \text{const} \qquad (36.12)$$

We next define the *angle variables* (note they are *not* canonically conjugate to the action variables J_σ)

$$w_\sigma \equiv \frac{\partial}{\partial J_\sigma} \bar{W}(q_1, \ldots, q_n, J_1, \ldots, J_n) \qquad \sigma = 1, \ldots, n \qquad (36.13)$$

Since the general theory ensures that $\bar{\beta}_\sigma$ [given by Eqs. (36.10) and (36.12)] is a constant of the motion, Eq. (36.8) yields the relation

$$\bar{\beta}_\sigma = \frac{\partial \bar{S}}{\partial J_\sigma} = w_\sigma - \left[\frac{\partial}{\partial J_\sigma} \alpha_1(J_1, \ldots, J_n)\right] t = \text{const} \qquad (36.14)$$

Consequently, the angle variable w_σ, defined in Eq. (36.13), has the form

$$w_\sigma = v_\sigma t + \bar{\beta}_\sigma \qquad (36.15)$$

where the "frequency" v_σ is defined as

$$v_\sigma \equiv \frac{\partial}{\partial J_\sigma} \alpha_1(J_1, \ldots, J_n) = \frac{\partial}{\partial J_\sigma} H(J_1, \ldots, J_n) \qquad (36.16)$$

Furthermore, Hamilton-Jacobi theory guarantees that each angle variable in (36.15) will increase linearly with time because $\bar{\beta}_\sigma$ and v_σ in Eqs. (36.12) and (36.16) are true constants of the motion.

We now assume that *the motion of the entire system under consideration is truly periodic in time*, returning to its initial configuration after some specific time. How long this time may be is unimportant, although it is likely to be extremely long for a complicated multidimensional system. Such strict periodicity requires that the ratios of all of the frequencies in the system be rational numbers, and this, in effect, is the content of our assumption. If it is not true for the actual system under consideration, we can always make it true for a model system that differs arbitrarily little from the actual one, merely by making infinitesimal changes in the parameters.

Assume, then, that *after this (in general long) time, all the individual degrees of freedom have executed some integral number of periods.* What is the corresponding change in the angle variables? Since $\{J_\sigma\}$ are constants of the motion, the angle variables in Eqs. (36.13) change only because the coordinates $\{q_\sigma\}$ change, which is described by the differential relation

$$dw_\lambda = \sum_\sigma \frac{\partial w_\lambda}{\partial q_\sigma} dq_\sigma \qquad (36.17)$$

Inserting Eq. (36.13) and interchanging the order of partial differentiation, we find

$$dw_\lambda = \frac{\partial}{\partial J_\lambda} \sum_\sigma \frac{\partial \bar{W}}{\partial q_\sigma} dq_\sigma \qquad (36.18)$$

The separability of the motion [see Eq. (36.1)] and Eq. (36.3) simplify this relation to

$$dw_\lambda = \frac{\partial}{\partial J_\lambda} \sum_\sigma \frac{\partial \bar{W}_\sigma}{\partial q_\sigma} dq_\sigma \qquad (36.19)$$

$$dw_\lambda = \frac{\partial}{\partial J_\lambda} \sum_\sigma p_\sigma \, dq_\sigma \qquad (36.20)$$

Now *integrate this relation over one period Δt of the entire system, during which time each degree of freedom q_λ undergoes an integral number n_λ of periods*

$$\Delta t = n_\lambda \tau_\lambda \qquad \lambda = 1, \ldots, n \qquad (36.21)$$

Here τ_λ is the fundamental period of the λth degree of freedom and n_λ is the number of times that the λth periodic motion has been executed. Since the angle variables are known to be linear functions of the time [see Eq. (36.15)], the total change in the λth angle variable over this time Δt can be written as

$$\Delta w_\lambda = v_\lambda \, \Delta t = v_\lambda n_\lambda \tau_\lambda \qquad (36.22)$$

On the other hand, the total change in the same angle variable also can be obtained by integrating Eq. (36.20) (recall that $\{J_\sigma\}$ are constants of the motion)

$$\Delta w_\lambda = \frac{\partial}{\partial J_\lambda} \sum_\sigma \oint p_\sigma \, dq_\sigma \qquad (36.23)$$

Each degree of freedom has executed precisely an integral number of periods, and use of the fundamental definition (36.2) of the action variable gives

$$\Delta w_\lambda = \frac{\partial}{\partial J_\lambda} \sum_\sigma n_\sigma J_\sigma = n_\lambda \tag{36.24}$$

Comparison of Eqs. (36.22) and (36.24) leads to the basic conclusion

$$\nu_\lambda \tau_\lambda = 1 \qquad \lambda = 1, \ldots, n \tag{36.25}$$

Thus the ν_λ are the fundamental frequencies of the system; Eq. (36.16) shows that they are just partial derivatives of the energy with respect to the action variables

$$\nu_\sigma = \frac{\partial}{\partial J_\sigma} H(J_1, \ldots, J_n) = \frac{\partial}{\partial J_\sigma} E(J_1, \ldots, J_n) \tag{36.26}$$

Although the derivation of Eq. (36.26) is somewhat intricate, the final result is simple and elegant. For any multidimensional system that is truly *separable* and *periodic*, the fundamental frequencies and periods of the individual coordinates follow directly from the hamiltonian, expressed as a function of all the action variables. These integrals over one cycle in phase space are defined in Eq. (36.2); they must be calculated for each independent degree of freedom.

As an example of the usefulness of action-angle variables, consider a simple harmonic oscillator in two dimensions with two different spring constants. The hamiltonian takes the form [compare Eqs. (35.36) and (35.44)]

$$H = \frac{1}{2m} p_1^2 + \frac{1}{2m} p_2^2 + \tfrac{1}{2}k_1 q_1^2 + \tfrac{1}{2}k_2 q_2^2 = \alpha \tag{36.27}$$

Here we change the notation slightly, calling this basic separation constant α, so that the subscripts 1 and 2 can be used for the separate degrees of freedom. The Hamilton-Jacobi equation (35.26) for Hamilton's characteristic function becomes

$$\frac{1}{2m}\left[\left(\frac{\partial W}{\partial q_1}\right)^2 + \left(\frac{\partial W}{\partial q_2}\right)^2\right] + \tfrac{1}{2}k_1 q_1^2 + \tfrac{1}{2}k_2 q_2^2 = \alpha \tag{36.28}$$

Separation of variables according to (35.27)

$$W(q_1, q_2) = W_1(q_1) + W_2(q_2) \tag{36.29}$$

leads to

$$\left[\frac{1}{2m}\left(\frac{dW_1}{dq_1}\right)^2 + \tfrac{1}{2}k_1 q_1^2\right] + \left[\frac{1}{2m}\left(\frac{dW_2}{dq_2}\right)^2 + \tfrac{1}{2}k_2 q_2^2\right] = \alpha \tag{36.30}$$

The first term in brackets is a function only of the variable q_1, and the second term in brackets is a function only of the variable q_2; the only way this relation can hold is for each bracketed term in fact to be constant [compare Eq. (35.39)]

$$\frac{1}{2m}\left(\frac{dW_1}{dq_1}\right)^2 + \tfrac{1}{2}k_1 q_1^2 = \alpha_1 \tag{36.31}$$

$$\frac{1}{2m}\left(\frac{dW_2}{dq_2}\right)^2 + \tfrac{1}{2}k_2 q_2^2 = \alpha_2 \tag{36.32}$$

The sum of the separation constants is evidently given by

$$\alpha_1 + \alpha_2 = \alpha \qquad (36.33)$$

which is also just the total energy. Equations (36.31) and (36.32) can be solved to give

$$p_\sigma = \frac{dW_\sigma}{dq_\sigma} = \pm [2m(\alpha_\sigma - \tfrac{1}{2}k_\sigma q^2)]^{1/2} \qquad \sigma = 1, 2 \qquad (36.34)$$

where we have used Eq. (36.3). Introduce a new variable θ related to q by

$$q_\sigma \equiv \left(\frac{2\alpha_\sigma}{k_\sigma}\right)^{1/2} \sin \theta \qquad (36.35)$$

$$dq_\sigma = \left(\frac{2\alpha_\sigma}{k_\sigma}\right)^{1/2} \cos \theta \, d\theta \qquad (36.36)$$

Note that as one integrates around one cycle in phase space, the quantity $p \, dq$ is always positive, as indicated in Fig. 36.1a. Thus the action integral defined by Eq. (36.2) can be written

$$\oint p_\sigma \, dq_\sigma = 2\alpha_\sigma \left(\frac{m}{k_\sigma}\right)^{1/2} \int_0^{2\pi} \cos^2 \theta \, d\theta = 2\pi\alpha_\sigma \left(\frac{m}{k_\sigma}\right)^{1/2} \equiv J_\sigma \qquad (36.37)$$

This relation can be inverted to give the separation constant α_σ in terms of the action variable J_σ according to

$$\alpha_\sigma = \frac{1}{2\pi} \left(\frac{k_\sigma}{m}\right)^{1/2} J_\sigma \qquad (36.38)$$

There is one such relation for each independent degree of freedom. Thus the total hamiltonian defined by Eqs. (36.33) and (36.27) becomes

$$H(J_1, J_2) = \frac{1}{2\pi} \left[J_1 \left(\frac{k_1}{m}\right)^{1/2} + J_2 \left(\frac{k_2}{m}\right)^{1/2} \right] \qquad (36.39)$$

The fundamental frequencies of the independent modes of motion now follow from Eq. (36.26)

$$\nu_1 = \frac{\partial H}{\partial J_1} = \frac{1}{2\pi} \left(\frac{k_1}{m}\right)^{1/2} \qquad (36.40)$$

$$\nu_2 = \frac{\partial H}{\partial J_2} = \frac{1}{2\pi} \left(\frac{k_2}{m}\right)^{1/2} \qquad (36.41)$$

which are precisely the fundamental frequencies of oscillation in the 1 and 2 directions. Note that if the ratio ν_1/ν_2 turns out to be irrational, a slight change in the ratio k_1/k_2 can always make the frequencies commensurate.

37 POISSON BRACKETS

In this section we introduce yet another formulation of classical mechanics, one that provides the most direct transition between the classical mechanics and quantum mechanics.

Basic Formulation

Consider a function $F(q_1, \ldots, q_n; p_1 \ldots, p_n; t)$ of the generalized coordinates, momenta, and the time. The *Poisson bracket* of two such functions F and G is defined as

$$[F, G]_{PB} \equiv \sum_\sigma \left(\frac{\partial F}{\partial q_\sigma} \frac{\partial G}{\partial p_\sigma} - \frac{\partial F}{\partial p_\sigma} \frac{\partial G}{\partial q_\sigma} \right) \qquad (37.1)$$

where, as usual, the partial derivative means that all the other variables in F and G are to be held fixed when carrying out the indicated differentiation. By inspection, the Poisson bracket is evidently antisymmetric

$$[F, G]_{PB} = -[G, F]_{PB} \qquad (37.2)$$

Poisson brackets with the hamiltonian play a special role

$$[H, F]_{PB} = \sum_\sigma \left(\frac{\partial H}{\partial q_\sigma} \frac{\partial F}{\partial p_\sigma} - \frac{\partial H}{\partial p_\sigma} \frac{\partial F}{\partial q_\sigma} \right) \qquad (37.3)$$

Insertion of Hamilton's equations (32.29) and (32.30) into this relation gives

$$[H, F]_{PB} = -\sum_\sigma \left(\frac{\partial F}{\partial p_\sigma} \dot{p}_\sigma + \frac{\partial F}{\partial q_\sigma} \dot{q}_\sigma \right) = -\left(\frac{dF}{dt} - \frac{\partial F}{\partial t} \right) \qquad (37.4)$$

where the last form has been rewritten with the total time derivative of the function $F(q_1, \ldots, q_n; p_1, \ldots, p_n; t)$. Equivalently, Hamilton's equations relate the total time derivative of F to the Poisson bracket of the hamiltonian with F

$$\frac{dF}{dt} = -[H, F]_{PB} + \frac{\partial F}{\partial t} \qquad (37.5)$$

Conversely, Eq. (37.5) *implies* Hamilton's equations, for one can simply let $F = q_\sigma$ and $F = p_\sigma$ in turn and use the defining relation (37.3) for the Poisson bracket of any quantity with the hamiltonian. Since all the other variables of the set $(q_1, \ldots, q_n; p_1, \ldots, p_n; t)$ are to be kept fixed when carrying out the indicated partial differentiation, only a single term contributes and it immediately follows that

$$\dot{q}_\sigma = -[H, q_\sigma]_{PB} = \frac{\partial H}{\partial p_\sigma} \qquad (37.6)$$

$$\dot{p}_\sigma = -[H, p_\sigma]_{PB} = -\frac{\partial H}{\partial q_\sigma} \qquad (37.7)$$

Equations (37.6) and (37.7) are precisely Hamilton's equations (32.29) and (32.30). The last of Hamilton's equations (32.18) also follows from Eqs. (37.5) with the observation that $[H, H]_{PB} = 0$ by Eq. (37.2). Since the Poisson-bracket formulation of classical mechanics [Eqs. (37.1) and (37.5)] is fully equivalent to Hamilton's equations, each provides a complete description of the dynamics with all the content of Newton's laws.

One particular set of Poisson brackets is central, both in the classical theory and in the transition to quantum mechanics

$$[p_\alpha, q_\beta]_{PB} = \sum_\sigma \left(\frac{\partial p_\alpha}{\partial q_\sigma} \frac{\partial q_\beta}{\partial p_\sigma} - \frac{\partial p_\alpha}{\partial p_\sigma} \frac{\partial q_\beta}{\partial q_\sigma} \right)$$

$$= -\sum_\sigma \delta_{\alpha\sigma} \delta_{\beta\sigma} = -\delta_{\alpha\beta} \tag{37.8}$$

Thus the Poisson bracket of a coordinate and its canonical momentum has a particularly simple form

$$[p_\alpha, q_\beta]_{PB} = -\delta_{\alpha\beta} \tag{37.9}$$

and a similar calculation yields

$$[p_\alpha, p_\beta]_{PB} = [q_\alpha, q_\beta]_{PB} = 0 \tag{37.10}$$

Suppose one makes a canonical transformation from a set of variables $\{p, q\}$ to a new set of variables $\{P, Q\}$. Any canonical transformation preserves the form of Hamilton's equations, so that Eqs. (37.5) to (37.7) still hold for the new variables, as do Eqs. (37.9) and (37.10). Thus the Poisson-bracket description of mechanics is preserved under a canonical transformation. Equivalently,† a canonical transformation can be defined as one that preserves the Poisson-bracket description of mechanics. Indeed, Eqs. (37.9) and (37.10) provide the most convenient way to decide whether a given transformation of the form (34.1) is, in fact, canonical.

Transition to Quantum Mechanics

Hamiltonian dynamics is one of the more formal aspects of classical mechanics, and its main interest today is in the transition from classical to quantum mechanics. Here, we shall briefly discuss the role of the various reformulations that have been studied in this chapter.

The earliest prescription for quantizing a classical system was suggested separately by Sommerfeld and Wilson.‡ They proposed that the action variable in a separable, periodic system should be quantized according to

$$J_\sigma = \oint p_\sigma \, dq_\sigma = n_\sigma h \qquad n_\sigma = 1, 2, \ldots \tag{37.11}$$

where n_σ is an integer and h is Planck's constant. For example, consider a one-

---

† We shall not pursue this in detail. See Goldstein (1950), sec. 8-4, for a more extended discussion.
‡ Schiff [2], sec. 2.

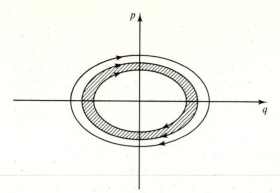

Figure 37.1 Sommerfeld-Wilson prescription of quantized action for the simple harmonic oscillator.

dimensional simple harmonic oscillator, whose phase-space diagram is sketched in Fig. 37.1. Evidently, the Sommerfeld-Wilson condition quantizes the *area* of quantum-mechanically allowed orbits in phase space. Since the action integral for a simple harmonic oscillator was evaluated in Eq. (36.37), we see that the Sommerfeld-Wilson quantization prescription requires

$$J = \oint p \, dq = 2\pi\alpha\left(\frac{m}{k}\right)^{1/2} = nh \tag{37.12}$$

Inverting this relation and identifying α with the energy [see Eq. (36.6)], one finds

$$\alpha = E = n\frac{h}{2\pi}\omega \equiv nh\omega \tag{37.13}$$

where [see Eq. (35.48)]

$$\omega \equiv \left(\frac{k}{m}\right)^{1/2} \tag{37.14}$$

is the frequency of the oscillator. In fact, this does provide the correct quantization prescription except for the zero-point energy.

As a second example, consider a particle moving in the xy plane under the influence of a central force (see Fig. 3.1), where the angular momentum is a constant of the motion

$$p_\phi = l_z = \text{const} \tag{37.15}$$

The action variable for one period of rotation can be evaluated immediately

$$J_\phi = \oint p_\phi \, d\phi = 2\pi l_z \tag{37.16}$$

and the Sommerfeld-Wilson quantization prescription then implies that the z component of the angular momentum for motion in a central force is quantized according to

$$l_z = n\frac{h}{2\pi} = nh \tag{37.17}$$

This is a correct quantum-mechanical result. We note, however, that l_z is also the *total* angular momentum in the classical theory, and Eq. (37.17) thus implies (*incorrectly†*) that the total angular momentum also should occur in integral multiples of $\hbar$.

The Sommerfeld-Wilson prescription provides a concise derivation of Bohr's *correspondence principle* for a separable periodic system of the sort considered in Sec. 36. Substitution of Eq. (37.11) into Eq. (36.6) shows that the allowed energy levels are specified by a set of integers $n_1\ n_2, \ldots, n_n$

$$E = H(n_1 h, \ldots, n_n h) \tag{37.18}$$

Correspondingly, the change in energy in the transition $n_\sigma \to n_\sigma + 1$ is just

$$E_{n_\sigma \to n_\sigma + 1} = E(\ldots, n_\sigma h + h, \ldots) - E(\ldots, n_\sigma h, \ldots) \tag{37.19}$$

For large n_σ, Taylor's theorem allows an approximate evaluation of the right-hand side, and we therefore obtain

$$E_{n_\sigma \to n_\sigma + 1} \approx h \frac{\partial E}{\partial J_\sigma} \tag{37.20a}$$

$$= h\nu_\sigma \tag{37.20b}$$

where the last form follows from Eq. (36.16). Thus the transition energy between adjacent quantum states is h times the classical oscillation frequency ν_σ. This result is just Bohr's correspondence principle.

To remedy the defects of the Sommerfeld-Wilson description, the canonical quantization procedure was developed subsequently by Schrödinger, Dirac, and others.‡ It applies to *any* classical mechanical system, not only a system of point particles of the sort discussed so far in this book but also a continuous medium described by a field theory, as discussed in the remainder of this text. Physicists now believe that the canonical quantization procedure provides the correct formulation of quantum mechanics, consistent with all known microscopic phenomena. For discrete systems, this procedure starts from a lagrangian

$$L \equiv L(q_1, \ldots, q_n, \dot{q}_1, \ldots, \dot{q}_n, t) \tag{37.21}$$

The canonical momenta are defined by

$$p_\sigma = \frac{\partial L}{\partial \dot{q}_\sigma} \qquad \sigma = 1, \ldots, n \tag{37.22}$$

and the hamiltonian is given by

$$H(p_1, \ldots, p_n, q_1, \ldots, q_n, t) \equiv \sum_\sigma p_\sigma \dot{q}_\sigma - L \tag{37.23}$$

To quantize the system, we first define the *commutator* of two quantities according to

$$[A, B] = AB - BA \tag{37.24}$$

---

† Schiff [2], sec. 14.
‡ Schiff [2], sec. 24.

The canonical quantization prescription is to *replace the corresponding classical Poisson bracket by the quantum-mechanical commutator* according to

$$[A, B]_{PB} \to \frac{1}{i\hbar}[A, B] \tag{37.25}$$

In particular, Eqs. (37.9) and (37.10) give the basic relations

$$[p_\alpha, q_\beta] = \frac{\hbar}{i}\,\delta_{\alpha\beta} \qquad [p_\alpha, p_\beta] = [q_\alpha, q_\beta] = 0 \tag{37.26}$$

known as the *canonical commutation relations*. It is clear that this quantization procedure can use any set of generalized coordinates and corresponding canonical momenta.

The introduction of commutators (37.24) takes us far beyond the realm of classical mechanics, for the quantities A and B must now be interpreted as non-commuting operators acting on a Hilbert space of state vectors. Observable quantities must be represented by hermitian operators, and one immediate general consequence of the commutation relations in Eqs. (37.26) is the *uncertainty principle* (stated for one mode)

$$(\Delta p)(\Delta q) \geq \tfrac{1}{2}\hbar \tag{37.27}$$

The left-hand side is the product of the root-mean-square deviations in the measurements of the quantities p and q; thus, in quantum mechanics, one can no longer simultaneously specify momenta and coordinates independently. This view represents a complete logical break with classical mechanics, which starts from the assumption that the equations of motion completely specify the time development of the system for given initial values of p and q. In this sense, the quantum-mechanical theory is no longer deterministic.

Invoking the canonical quantization prescription of Eq. (37.25), we infer the following quantum-mechanical equations of motion from Eq. (37.5)

$$\frac{d\hat{F}}{dt} = \frac{i}{\hbar}[\hat{H}, \hat{F}] + \frac{\partial \hat{F}}{\partial t} \tag{37.28}$$

where a caret indicates that the quantity is an operator in the abstract Hilbert space of state vectors. Equation (37.28) is the *Heisenberg operator equation of motion* in quantum mechanics. It guarantees, through Ehrenfest's theorem,[†] that the quantum-mechanical equations of motion reproduce the classical ones in the classical limit of well-localized wave packets. Equation (37.28) is also equivalent to the abstract Schrödinger equation

$$\hat{H}|\Psi\rangle = i\hbar\frac{\partial}{\partial t}|\Psi\rangle \tag{37.29}$$

In the Heisenberg picture of quantum mechanics,[‡] one puts all of the time depen-

[†] Schiff [2], sec. 7.
[‡] Schiff [2], sec. 24.

dence into the operators, as in Eq. (37.28), and the state vectors are time-independent. In the Schrödinger picture, however, all the time dependence is in the state vector, as in Eq. (37.29), and the operators are time-independent. These two pictures of quantum mechanics are fully equivalent and differ only by a unitary transformation.

We examine two examples of the canonical quantization procedure. Consider a mechanical system described by a wave function $\Psi(q_1, \ldots, q_n, t)$. To satisfy the quantization prescription (37.26), we can define the canonical momenta according to

$$p_\alpha = \frac{\hbar}{i} \frac{\partial}{\partial q_\alpha} \tag{37.30}$$

Physically observable quantities must be described by *hermitian* operators O, which means that they satisfy the relation

$$\int \Psi_1^* O \Psi_2 (d\tau) = \int (O\Psi_1)^* \Psi_2 (d\tau) \tag{37.31}$$

The operator of Eq. (37.30) will be hermitian if the appropriate partial integrations indicated in Eq. (37.31) can be carried out. This can always be done in cartesian coordinates (assuming appropriate boundary conditions so that surface terms can be discarded) since there the volume element is simply

$$(d\tau) = dq_1 \cdots dq_n \tag{37.32}$$

Confining ourselves to cartesian coordinates (although this quantization prescription is, in fact, more general), we therefore accept Eq. (37.30) as the correct quantization prescription in nonrelativistic quantum mechanics. The coordinate-space representation of the Schrödinger equation (37.29) then becomes [compare Eq. (35.28)]

$$\left[H\left(\frac{\hbar}{i} \frac{\partial}{\partial q_1}, \ldots, \frac{\hbar}{i} \frac{\partial}{\partial q_n}; q_1, \ldots, q_n; t \right) - i\hbar \frac{\partial}{\partial t} \right] \Psi(q_1, \ldots, q_n, t) = 0 \tag{37.33}$$

which is the most familiar representation of the abstract quantum-mechanical problem in Eq. (37.29).

As a second example, consider any classical mechanical system that can be described with *normal modes*. For such a system, the lagrangian is a sum of uncoupled harmonic oscillators

$$L = \frac{1}{2} \sum_\sigma (\dot{\zeta}_\sigma^2 - \omega_\sigma^2 \zeta_\sigma^2) \tag{37.34}$$

where $\{\omega_\sigma\}$ denotes the normal-mode frequencies. The canonical momenta

$$p_\sigma = \dot{\zeta}_\sigma \qquad \sigma = 1, \ldots, n \tag{37.35}$$

are defined by Eq. (37.22), the hamiltonian is given by

$$H = \frac{1}{2} \sum_\sigma (p_\sigma^2 + \omega_\sigma^2 \zeta_\sigma^2) \tag{37.36}$$

and the canonical commutation relations (37.26) require

$$[p_\sigma, \zeta_\lambda] = \frac{\hbar}{i} \delta_{\sigma\lambda} \tag{37.37}$$

Equations (37.36) and (37.37) completely define the quantum-mechanical problem,† which is just that of n uncoupled simple harmonic oscillators. In general, the normal coordinates ζ_σ with $\sigma = 1, \ldots, n$ are complicated combinations of the actual particle displacements (or the Fourier amplitudes in the continuum problem); nevertheless, Eqs. (37.36) and (37.37) and our knowledge of the quantum-mechanical one-dimensional simple harmonic oscillator‡ immediately imply that the energy spectrum of the many-body system must take the form

$$E = \sum_\sigma \hbar\omega_\sigma(n_\sigma + \tfrac{1}{2}) \qquad n_\sigma = 1, 2, \ldots, \infty \tag{37.38}$$

Thus the energy of this many-particle system is restricted to integral units of $\hbar\omega_\sigma$. These quantum units of the normal-mode energies are called *phonons* when describing elastic waves in solids, *photons* when describing the normal modes of the free electromagnetic field, *surfons* or *ripplons* when describing surface waves on continuous systems, and so on.

This brief discussion of the transition to quantum mechanics has emphasized the direct relation between classical mechanics and the formal description of quantum mechanics. In particular, the connection made through Eqs. (37.21) to (37.26) applies to *any* classical mechanical system. As seen in our examples, it provides a convenient basis for quantizing its behavior.

PROBLEMS

6.1 A particle moves in a central potential $V(r)$. Use three-dimensional spherical polar coordinates (r, θ, ϕ) to obtain the hamiltonian. Derive the equations of motion and relate the trajectory to that found previously in Chap. 1.

6.2 A spherical pendulum consists of a particle of mass m in a gravitational field constrained to move on the surface of a sphere of radius l. Use the polar angle θ (measured from the downward vertical) and the azimuthal angle ϕ to obtain the equations of motion in the hamiltonian formulation. Expand the hamiltonian to second order about uniform circular motion with $\theta = \theta_0$ and show that the resulting expression is just that for a simple harmonic oscillator with $\omega^2 = [g/l \cos\theta_0](1 + 3\cos^2\theta_0)$.

6.3 Construct the hamiltonian for a heavy symmetrical top with one point fixed in a gravitational field and derive Hamilton's equations of motion. Relate these equations to those derived in Sec. 31 and show how the solution can be reduced to definite integrals.

6.4 The relativistic motion of a particle in a static potential $V(\mathbf{r})$ can be obtained from the lagrangian $L = -mc^2(1 - v^2/c^2)^{1/2} - V(\mathbf{r})$.

    (*a*) Write out Lagrange's equations and verify the above assertion.

    (*b*) Find the canonical momentum $\mathbf{p}$ and show that the relativistic hamiltonian becomes $H = (m^2c^4 + p^2c^2)^{1/2} + V(\mathbf{r})$. Is H a constant of the motion?

    (*c*) Assume V is spherically symmetric. Show that $\mathbf{r} \times \mathbf{p}$ is a constant of motion. Hence reduce H to the form $H = c(m^2c^2 + p_r^2 + r^{-2}p_\phi^2)^{1/2} + V(r)$.

---

† We must aso assume that there is a vacuum or no-particle state in the theory.
‡ Schiff [2], sec. 13.

6.5 A particle with charge e and mass m moves in a specified electromagnetic field with $\Phi = 0$ and $\mathbf{A} = \hat{z}A_z(x, y, t)$.

(a) Use the Lorentz-force equation (33.8) to construct the explicit first integral $\dot{z} + eA_z/mc = C$ (a constant).

(b) Show that the x and y equations can then be written $\ddot{\mathbf{r}}_\perp = -\nabla_\perp \frac{1}{2}(C - eA_z/mc)^2$.

(c) Consider a uniform static magnetic field $\mathbf{B} = B_0 \hat{x}$. Integrate these equations to show that the particle executes a helix about the $\hat{x}$ direction with an angular frequency eB_0/mc.

6.6 Construct from first principles the hamiltonian for a one-dimensional harmonic oscillator of mass m and spring constant k. Determine the value of the constant C such that the following equations define a canonical transformation from the old variables (q, p) to the new variables (Q, P):

$$Q = C(p + im\omega q) \quad \text{and} \quad P = C(p - im\omega q)$$

where $\omega = (k/m)^{1/2}$. What is the generating function $S(q, P)$ for this transformation? Find Hamilton's equations of motion for the new variables and integrate them. Hence find the solution to the original problem. (This transformation defines the creation and annihilation operators of quantum theory.)

6.7 (a) Prove that the function $S_0(q, P) \equiv \sum_\sigma q_\sigma P_\sigma$ generates the identity transformation $Q_\sigma = q_\sigma, P_\sigma = p_\sigma$.

(b) Prove that the function $S_0 + H\,dt$ generates the dynamical transformation from t to $t + dt$, with $Q_\sigma = q_\sigma(t + dt), P_\sigma = p_\sigma(t + dt)$. Discuss the corollary that the time development of any mechanical system (no matter how complicated) is itself a canonical transformation.

6.8 For a system described by cartesian coordinates, use the results of Prob. 6.7 to prove that the functions $S_0 + \mathbf{P} \cdot d\mathbf{r}$ and $S_0 + \hat{n} \cdot \mathbf{L}\,d\phi$ generate infinitesimal translations $d\mathbf{r}$ and rotations $\hat{n}\,d\phi$, respectively, where $\mathbf{P}$ and $\mathbf{L}$ are the total linear and angular momenta.

6.9 Consider the physical system described by the following kinetic energy T and potential energy V

$$T = \tfrac{1}{2}(\dot{q}_1^2 + \dot{q}_2^2)(q_1^2 + q_2^2) \qquad V = (q_1^2 + q_2^2)^{-1}$$

where q_1 and q_2 are the generalized coordinates. What is the Hamilton-Jacobi equation for this system? Solve this equation to find Hamilton's principal function. Hence deduce the dynamical motion of the system (you need not evaluate any definite integrals).

6.10 Consider the bead on a rotating wire hoop, studied in Sec. 16.

(a) Find the generalized momentum and the hamiltonian expressed in terms of p and θ. Derive Hamilton's equations and show that they are equivalent to Eq. (16.12).

(b) Use the Hamilton-Jacobi technique to construct an explicit integral for the trajectory $\theta(t)$.

6.11 Consider the one-dimensional motion of a particle of mass m subject to a uniform force At that increases linearly with time. Find the hamiltonian of the system. What is the corresponding Hamilton-Jacobi equation? Show that Hamilton's principal function S can be written in the form

$$S = \tfrac{1}{2}At^2x + \alpha x - \phi(t)$$

where α is a constant and ϕ is a function of the time. Solve the resulting equation for ϕ. Hence find the position and momentum as functions of the time.

6.12 A point mass m with initial position $\mathbf{r}_0$ and velocity $\mathbf{v}_0$ moves in a uniform gravitational field. Use the Hamilton-Jacobi method to find both the time-dependent trajectory and the orbit.

6.13 A symmetric top with one fixed point is located in a uniform gravitational field (see Fig. 31.1). Use the Hamilton-Jacobi method to reduce the motion to definite integrals.

6.14 (a) Use the Hamilton-Jacobi technique to construct explicit integrals for the trajectory $r(t)$ and orbit $r(\phi)$ for the relativistic particle in Prob. 6.4.

(b) For the newtonian potential $V = -GMm/r$ with $p_\phi > GMm/c$, show that the bound orbits $(E < mc^2)$ are precessing ellipses. If $p_\phi \gg GMm/c$, prove that the precession per cycle is $\pi(GMm/p_\phi c)^2$ and evaluate this quantity for the orbit of mercury about the sun (Prob. 1.12).

(c) Discuss briefly what happens if $p_\phi < GMm/c$.

6.15 Use the action-angle variables to find the frequencies of a three-dimensional harmonic oscillator with unequal spring constants.

6.16 Consider a particle of mass m moving in three dimensions in the attractive potential $V(r) = -k/r$ with $k = $ const. Use spherical polar coordinates (r, θ, ϕ).

(a) Prove that the energy can be expressed in terms of the action variables as

$$ E = -\frac{2\pi^2 mk^2}{(J_r + J_\theta + J_\phi)^2} $$

(b) Discuss the frequencies of the system.

(c) Use the Sommerfeld-Wilson quantization conditions to derive the Bohr theory of the hydrogen atom.

6.17 Consider a canonical transformation generated by the function

$$ S(q_1, \ldots, q_n, P_1, \ldots, P_n, t) = \sum_\sigma q_\sigma P_\sigma + \varepsilon G(q_1, \ldots, q_n, P_1, \ldots, P_n, t) $$

where ε is an infinitesimal quantity.

(a) Show that the resulting canonical transformation differs from the identity by terms of order ε with

$$ P_\sigma = p_\sigma - \varepsilon \frac{\partial G(q, p, t)}{\partial q_\sigma} + O(\varepsilon^2) $$

$$ Q_\sigma = q_\sigma + \varepsilon \frac{\partial G(q, p, t)}{\partial p_\sigma} + O(\varepsilon^2) $$

(b) Under this canonical transformation, show that an arbitrary function $F(q_1, \ldots, q_n, p_1, \ldots, p_n)$ changes by an amount $dF = \varepsilon[F, G]_{PB}$.

(c) If G is a constant of the motion, prove that the hamiltonian H is invariant under such transformations. Discuss the necessary symmetry of H if G is the total linear or angular momentum of a system of particles interacting through two-body potentials.

6.18 Use the fundamental Poisson-bracket relations (37.9) and (37.10) to derive the following Poisson brackets involving the components of angular momentum

$$ [L_i, L_j]_{PB} = L_k \qquad i, j, k \text{ in cyclic order} $$

$$ [L^2, L_i]_{PB} = 0 $$

In the treatment of central forces, we took L_z as the canonical momentum p_ϕ for the variable ϕ. Can the set L_x, L_y, L_z serve as canonical momenta for some set of generalized coordinates?

6.19 (a) A particle of mass m moves in an axially symmetric potential $V[(x^2 + y^2)^{1/2}, z]$. Express the lagrangian $\tilde{L}$ in terms of the cartesian coordinates (x', y', z') with respect to a system of axes rotating with constant angular velocity Ω about the z axis. Hence find the new hamiltonian $\tilde{H}$. Is $\tilde{H}$ a constant of the motion? Is it the energy? Derive Hamilton's equations of motion in the rotating system and interpret the various terms.

(b) Express the components of position and momentum $(\mathbf{r}', \mathbf{p}')$ viewed in the rotating frame in terms of those $(\mathbf{r}, \mathbf{p})$ viewed in the inertial frame. Use Poisson brackets to verify explicitly that the transformation relating them is canonical. Construct the generating function $S(\mathbf{r}, \mathbf{p}', t)$ for this transformation and verify that $\tilde{H} = H + \partial S/\partial t$.

(c) Discuss *briefly* the necessary modifications for a system of identical particles of mass m interacting through two-body central potentials $v(|\mathbf{r}_i - \mathbf{r}_j|)$ and placed in the same external potential V.

6.20 The one-dimensional hamiltonian $H = c|p| + f|x|$ describes a relativistic particle in an attractive potential $f|x|$.

(a) Obtain the dynamical equations and integrate them explicitly for the initial conditions $x = x_0 > 0$, $p = 0$. Show from first principles that the motion is periodic with period $4E/fc$, where $E = fx_0$ is the initial energy.

(b) Verify that the trajectory in phase space is a rhombus and use the action-angle formalism to rederive the same result.

(c) Apply the Sommerfeld-Wilson quantization prescription to obtain the allowed energy levels for the corresponding quantum system.

SELECTED ADDITIONAL READINGS

1. J. D. Jackson, "Classical Electrodynamics," Wiley, New York, 1962.
2. L. I. Schiff, "Quantum Mechanics," 3d ed., McGraw-Hill, New York, 1968.

SEVEN

STRINGS

In this chapter, we begin the treatment of continuum mechanics with a detailed study of a stretched string. This familiar physical system provides a concrete example for introducing the concepts and techniques needed to describe the dynamics of a general continuous medium.

38 REVIEW OF FIELD THEORY

We start by reviewing the field description (Sec. 25) of the small-amplitude displacement $u(x, t)$ of a string with specified endpoint conditions (Figs. 25.1 and 25.2). The lagrangian for a string has the form

$$L = \int_0^l \mathscr{L} \, dx = T - V \tag{38.1}$$

where the lagrangian density is given by

$$\mathscr{L} = \tfrac{1}{2}\sigma(x)\left(\frac{\partial u}{\partial t}\right)^2 - \tfrac{1}{2}\tau(x)\left(\frac{\partial u}{\partial x}\right)^2 \tag{38.2}$$

The coefficients in this expression

$$\sigma(x) = \text{mass density} \tag{38.3}$$

$$\tau(x) = \text{tension} \tag{38.4}$$

may vary along the string but are assumed to be time-independent.

Hamilton's principle, the basic principle of mechanics, states that

$$\delta \int_{t_1}^{t_2} L \, dt = 0 \qquad \text{fixed endpoints in time} \qquad (38.5)$$

Substitution of Eq. (38.1) yields

$$\delta \int_{t_1}^{t_2} dt \int_0^l dx \; \mathscr{L}\left(u, \frac{\partial u}{\partial x}, \frac{\partial u}{\partial t}; x, t\right) = 0 \qquad \text{fixed endpoints in time} \quad (38.6)$$

We perform the indicated variation by varying the generalized coordinate $u(x, t)$ at each (x, t). The variation of the first term in Eq. (38.2) can be integrated by parts with respect to time, using the condition $\delta u(x, t_1) = \delta u(x, t_2) = 0$ in Hamilton's principle (38.5) to discard the contribution from the endpoints. The variation of the second term in (38.2) involves a similar partial integration with respect to the spatial coordinate, and the contribution from the spatial endpoints can be discarded if the following expression vanishes:

$$\left[\delta u \; \tau(x) \frac{\partial u}{\partial x}\right]_0^l = 0 \qquad (38.7)$$

We restrict our discussion to *boundary conditions* that automatically satisfy this condition. For example, a string with fixed endpoints (Fig. 25.2) ensures Eq. (38.7), for the variations δu must vanish at $x = 0$ and l, giving fixed endpoints in space as well as time. Other possible spatial boundary conditions satisfying Eq. (38.7) will be discussed shortly.

The Euler-Lagrange equation for the variational problem in Eq. (38.6) thus takes the general form [compare Eq. (25.59)]

$$\frac{\partial}{\partial t} \frac{\partial \mathscr{L}}{\partial(\partial u/\partial t)} + \frac{\partial}{\partial x} \frac{\partial \mathscr{L}}{\partial(\partial u/\partial x)} - \frac{\partial \mathscr{L}}{\partial u} = 0 \qquad (38.8)$$

This relation is Lagrange's equation for a one-dimensional continuous system. The general one-dimensional string equation follows from the specific form of the lagrangian density (38.2) [compare Eq. (25.13)]

$$\sigma(x) \frac{\partial^2 u}{\partial t^2} = \frac{\partial}{\partial x}\left[\tau(x) \frac{\partial u}{\partial x}\right] \qquad (38.9)$$

The special case

$$\sigma = \text{const} \qquad \tau = \text{const} \qquad (38.10)$$

and

$$c^2 \equiv \frac{\tau}{\sigma} \qquad (38.11)$$

yields the one-dimensional wave equation [compare Eq. (25.62)]

$$\frac{1}{c^2} \frac{\partial^2 u(x, t)}{\partial t^2} = \frac{\partial^2 u(x, t)}{\partial x^2} \qquad (38.12)$$

To solve Eq. (38.12), we seek normal modes of the form

$$u(x, t) = C\rho(x) \cos (\omega t + \phi) \qquad (38.13)$$

where each point of the string oscillates with the same given frequency. Substitution of Eq. (38.13) into (38.12) yields the ordinary differential equation

$$\frac{d^2\rho(x)}{dx^2} + k^2\rho(x) = 0 \qquad (38.14a)$$

with

$$k \equiv \frac{\omega}{c} \qquad (38.14b)$$

In general, only certain values of k^2 will permit solutions that satisfy the boundary conditions of the appropriate physical problem. Equation (38.14) is known as an *eigenfunction* equation, and the acceptable values of k^2 are the *eigenvalues*. For example, fixed endpoints (Fig. 25.2) require the boundary conditions

$$\rho(0) = \rho(l) = 0 \qquad (38.15)$$

The solutions to Eq. (38.14) that satisfy the conditions (38.15) are

$$\rho_n(x) = \left(\frac{2}{l\sigma}\right)^{1/2} \sin \frac{n\pi x}{l} \qquad (38.16a)$$

where

$$k_n = \frac{n\pi}{l} \qquad n = 1, 2, 3, \ldots, \infty \qquad (38.16b)$$

These eigenfunctions obey the orthonormality condition [see Eq. (25.34)]

$$\int_0^l \rho_n(x)\rho_m(x)\sigma \, dx = \delta_{nm} \qquad (38.17)$$

The general solution to the wave equation is found by superposing all the normal modes

$$u(x, t) = \sum_{n=1}^{\infty} \rho_n(x)C_n \cos (\omega_n t + \phi_n) \qquad (38.18)$$

Each term in this expansion satisfies the wave equation and the boundary conditions (38.15). It is convenient to rewrite Eq. (38.18) in the form

$$u(x, t) = \sum_{n=1}^{\infty} \left(\frac{2}{l\sigma}\right)^{1/2} \sin \frac{n\pi x}{l} \left[a_n\left(\cos \frac{n\pi ct}{l}\right) + b_n\left(\sin \frac{n\pi ct}{l}\right)\right] \qquad (38.19)$$

where the new coefficients are related to the old ones by

$$a_n \equiv C_n \cos \phi_n \qquad (38.20a)$$

$$b_n \equiv -C_n \sin \phi_n \qquad (38.20b)$$

To determine the solution for all subsequent times, it only remains to satisfy the initial conditions. Since we are still solving Newton's equation of motion, we must specify the initial position and initial velocity at all points along the string

$$u(x, 0) = f(x) \tag{38.21a}$$

$$\dot{u}(x, 0) = g(x) \tag{38.21b}$$

Application of these conditions to Eq. (38.19) leads to a Fourier-series representation of the functions $f(x)$ and $g(x)$, whose coefficients can be obtained by using the orthonormality condition of Eq. (38.17),

$$a_n = \left(\frac{2}{l\sigma}\right)^{1/2} \int_0^l \left(\sin \frac{n\pi x}{l}\right) f(x)\sigma \, dx \tag{38.22a}$$

$$\frac{n\pi c}{l} b_n = \left(\frac{2}{l\sigma}\right)^{1/2} \int_0^l \left(\sin \frac{n\pi x}{l}\right) g(x)\sigma \, dx \tag{38.22b}$$

The solutions (38.19) and (38.22) to the problem of a uniform string with fixed endpoints are called *Bernoulli's solution.* In Sec. 25 this result was derived by taking the limit $N \to \infty$ of the solution for N mass points on a stretched massless string. For finite N, the general solution is well defined at every step because the number of normal modes is finite and the manipulations on Eq. (38.19) always involve a finite series. We may question, however, whether the limit $N \to \infty$ still yields the general solution, because the series in Eq. (38.19) is now infinite. Does it converge? In addition, verifying that Eq. (38.19) satisfies the wave equation (38.12) involves two derivatives, each of which introduces an additional power of n that decreases the convergence of the series. Thus we must also consider whether the second derivative of Eq. (38.19) converges and whether the infinite sum in Eq. (38.19) in fact explicitly satisfies Eq. (38.12). Finally, the functions specifying the initial conditions in Eqs. (38.21) can be chosen arbitrarily, subject only to some very general restrictions such as continuity of the physical string. Is the Fourier series obtained from the superposition of normal modes (38.19) at $t = 0$ sufficiently general to match these arbitrary initial conditions? [That is, do the eigenfunctions of Eq. (38.14) provide a *complete* set of functions in the spatial coordinate?] These questions show that even the most elementary continuum problem immediately raises complex difficulties of mathematical analysis. Our subsequent discussion attempts to answer some of them.

Before proceeding, it is first useful to derive a general expression for the energy of the string. Recall the definition of the normal coordinates [compare Eq. (25.46)]

$$\zeta_n \equiv C_n \cos(\omega_n t + \phi_n) \tag{38.23}$$

which were shown in Sec. 25 to diagonalize the lagrangian for the uniform string [see Eq. (25.50)]

$$L = \frac{1}{2} \sum_n (\dot{\zeta}_n^2 - \omega_n^2 \zeta_n^2) \tag{38.24}$$

We now take this denumerably infinite set of coordinates (38.23) as the generalized coordinates that completely specify the behavior of the string in Eq. (38.18). The hamiltonian can be immediately constructed through the application of Eqs. (32.9) and (32.11)

$$H = \frac{1}{2} \sum_n (\dot{\zeta}_n^2 + \omega_n^2 \zeta_n^2) \tag{38.25}$$

Equations (38.24) and (38.25) represent a sum of uncoupled harmonic oscillators. Evidently, the hamiltonian (38.25) is simply the sum of the energies of the individual oscillators and is therefore the total energy of the system [compare Eqs. (25.39) and (25.42)]

$$H = T + V = \tfrac{1}{2}\sigma \int_0^l dx \left(\frac{\partial u}{\partial t}\right)^2 + \tfrac{1}{2}\tau \int_0^l dx \left(\frac{\partial u}{\partial x}\right)^2 \tag{38.26}$$

Insertion of the definition (38.23) into Eq. (38.25) yields an explicit expression for the energy in terms of the amplitudes of the normal coordinates

$$H = E = \frac{1}{2} \sum_{n=1}^{\infty} \omega_n^2 C_n^2 \tag{38.27}$$

Note that the energy in (38.26) is proportional to the *square* of the displacement u and that Eq. (38.27) is indeed a time-independent constant of the motion.

39 D'ALEMBERT'S SOLUTION TO THE WAVE EQUATION

In connection with the general solution to the one-dimensional wave equation (38.12), it will be instructive to consider another (apparently unrelated) way of solving the same equation.

Solution for an Infinite String

We start with an infinite string and later superpose solutions to match a given set of spatial boundary conditions, such as those in Fig. 25.2. Introduce the new variables

$$r \equiv x - ct \tag{39.1a}$$

$$s \equiv x + ct \tag{39.1b}$$

and write the solution to the wave equation in terms of those new variables as

$$u(x, t) \equiv U(r, s) \tag{39.2}$$

The chain rule for differentiation of an implicit function implies

$$\frac{\partial u}{\partial x} = \frac{\partial U}{\partial r}\frac{\partial r}{\partial x} + \frac{\partial U}{\partial s}\frac{\partial s}{\partial x} = \frac{\partial U}{\partial r} + \frac{\partial U}{\partial s} \tag{39.3}$$

A second application of this result yields

$$\frac{\partial^2 u}{\partial x^2} = \frac{\partial}{\partial r}\left(\frac{\partial U}{\partial r} + \frac{\partial U}{\partial s}\right) + \frac{\partial}{\partial s}\left(\frac{\partial U}{\partial r} + \frac{\partial U}{\partial s}\right) \tag{39.4}$$

The derivative with respect to time can be similarly evaluated

$$\frac{\partial u}{\partial t} = \frac{\partial U}{\partial r}\frac{\partial r}{\partial t} + \frac{\partial U}{\partial s}\frac{\partial s}{\partial t} = -c\frac{\partial U}{\partial r} + c\frac{\partial U}{\partial s} \tag{39.5}$$

and a second differentiation results in

$$\frac{1}{c^2}\frac{\partial^2 u}{\partial t^2} = -\frac{\partial}{\partial r}\left(-\frac{\partial U}{\partial r} + \frac{\partial U}{\partial s}\right) + \frac{\partial}{\partial s}\left(-\frac{\partial U}{\partial r} + \frac{\partial U}{\partial s}\right) \tag{39.6}$$

Substitution of Eqs. (39.4) and (39.6) in (38.12) transforms the wave equation into that for the new function U in terms of the new variables (r, s)

$$4\frac{\partial^2 U}{\partial r\,\partial s} = 0 \tag{39.7}$$

fully equivalent to the original form (38.12).

We shall construct the general solution to this equation. If the variable s is first kept fixed, Eq. (39.7) takes the form

$$\frac{\partial}{\partial r}\frac{\partial U}{\partial s} = 0 \tag{39.8}$$

Integration with respect to the variable r gives

$$\frac{\partial U}{\partial s} = \phi'(s) \tag{39.9}$$

where $\phi'(s)$ is an arbitrary function of s that is *independent of r*. Equation (39.9) can now be integrated with respect to the variable s to yield

$$U = \phi(s) + \psi \tag{39.10}$$

Here ψ must be independent of s, but it may be an arbitrary function of r and still satisfy Eq. (39.9). Thus we arrive at the general solution to the second-order partial differential equation (39.8)

$$U(r, s) = \psi(r) + \phi(s) \tag{39.11}$$

To verify that Eq. (39.11) solves Eq. (39.8), ψ and ϕ need merely be *any* once-differentiable functions of the indicated variables, since the result of performing one partial derivative on Eq. (39.11) is then independent of the other variable and immediately satisfies Eq. (39.8). Use of Eqs. (39.1) and (39.2) reexpresses Eq. (39.11) in terms of the original variables x and t

$$u(x, t) = \psi(x - ct) + \phi(x + ct) \tag{39.12}$$

This solution is easy to interpret: the first term $\psi(x - ct)$ is a function that retains

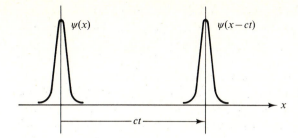

Figure 39.1 The function $\psi(x - ct)$ that satisfies the one-dimensional wave equation and the initial value $\psi(x)$.

its identical shape with respect to a moving coordinate system with origin at $x_0 = ct$, as illustrated in Fig. 39.1. Thus $\psi(x - ct)$ is a *disturbance moving with fixed shape to the right with velocity c*. Similarly, $\phi(x + ct)$ is a disturbance moving with fixed shape to the left.

To prove that Eq. (39.12) is the general solution to the one-dimensional wave equation (39.8) for an infinite string, it is still necessary to fit the arbitrary initial conditions (38.21), where the infinite length now requires that the functions $f(x)$ and $g(x)$ be specified *for all x*. The first condition (38.21a) can be matched with a solution of the form (39.12) by introducing the function

$$u_1(x, t) \equiv \tfrac{1}{2}[f(x - ct) + f(x + ct)] \tag{39.13}$$

which indeed satisfies the wave equation provided only that the function f is once differentiable; it has the initial behavior

$$u_1(x, 0) = f(x) \tag{39.14a}$$

$$\dot{u}_1(x, 0) = 0 \tag{39.14b}$$

Similarly, the function

$$u_2(x, t) \equiv \frac{1}{2c}\left[\int_A^{x+ct} g(\xi)\, d\xi - \int_A^{x-ct} g(\xi)\, d\xi \right] \tag{39.15}$$

satisfies the one-dimensional wave equation for arbitrary fixed A because it also has the correct form (39.12). In addition, it has the initial behavior

$$u_2(x, 0) = 0 \tag{39.16a}$$

$$\dot{u}_2(x, 0) = g(x) \tag{39.16b}$$

Since the wave equation is linear, these two solutions can be superposed

$$u(x, t) \equiv u_1(x, t) + u_2(x, t) \tag{39.17}$$

to yield *d'Alembert's solution*

$$u(x, t) = \tfrac{1}{2}[f(x - ct) + f(x + ct)] + \frac{1}{2c}\int_{x-ct}^{x+ct} g(\xi)\, d\xi \tag{39.18}$$

This expression solves the wave equation, and it is evident from Eqs. (39.14) and (39.16) that it also satisfies the initial conditions (38.21). Thus we have indeed constructed another form of the general solution to the one-dimensional wave equation (38.12). In contrast to Bernoulli's solution, (39.18) has a nice clear physical interpretation. Suppose that the string has a specified initial displacement $f(x)$ and is released from rest, so that the initial velocity $g(x)$ vanishes. Then Eq. (39.18) implies that at a later time we shall find one wave of the original shape moving to the right with velocity c and another wave of the same shape moving to the left with the same velocity, the amplitude of each wave being precisely half the original amplitude. A similar interpretation obtains for an initial velocity distribution (see Prob. 7.2).

Solution for a Finite String

Equation (39.18) describes an infinite string, and it is natural to consider next the more complicated situation of a finite string, e.g., one with fixed endpoints (Fig. 39.2). Such a solution to the wave equation is completely specified by the *boundary conditions*

$$u(0, t) = 0 \tag{39.19a}$$

$$u(l, t) = 0 \tag{39.19b}$$

and the *initial conditions* for the finite segment

$$u(x, 0) = f(x) \tag{39.20a}$$
$$\dot{u}(x, 0) = g(x) \qquad 0 \le x \le l \tag{39.20b}$$

where f and g must vanish at each end. Any solution to the wave equation that matches the boundary conditions (39.19) and initial conditions (39.20) necessarily solves the finite-string problem for all times.

To use d'Alembert's solution (39.18) for this finite string, it is necessary to know the functions $f(x)$ and $g(x)$ *for all* x because the arguments $x \pm ct$ tend toward $\pm \infty$ with increasing time. For a finite string, however, these functions are given only on the interval $0 \le x \le l$, and the relations (39.20) must therefore be extended to all x for d'Alembert's solution to apply. In particular, if we can *choose* $f(x)$ *and* $g(x)$ *outside of the interval* $0 \le x \le l$ *so that the boundary conditions*

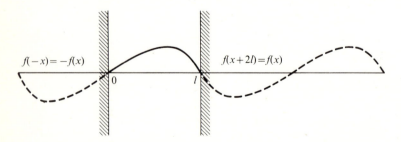

$$f(-x) = -f(x) \qquad \qquad f(x+2l) = f(x)$$

Figure 39.2 Appropriate extension of the initial displacement of a finite string with fixed endpoints to use in d'Alembert's solution.

(39.19) *hold for all times*, then the arguments just given ensure that we have found the complete solution to the wave equation.

How do we extend the definitions of $f(x)$ and $g(x)$ to yield the appropriate initial conditions for an equivalent infinite string? Evidently, if the functions f and g are odd about the origin, the solution (39.18) will satisfy the first boundary condition (39.19a) at all times. Thus we require

$$f(-x) = -f(x) \qquad g(-x) = -g(x) \tag{39.21}$$

Similarly, if the functions f and g are odd about the point $x = l$, the solution (39.18) will satisfy the boundary condition (39.19b). Thus we also require

$$f[l + (x - l)] = -f[l - (x - l)] \qquad g[l + (x - l)] = -g[l - (x - l)] \tag{39.22}$$

These relations can be combined to yield an equivalent set

$$f(x + 2l) = f(x) \qquad g(x + 2l) = g(x) \tag{39.23a}$$

$$f(-x) = -f(x) \qquad g(-x) = -g(x) \tag{39.23b}$$

which state that the extended functions f and g defined for all x must be odd about the origin and *periodic* with period $2l$ (see Fig. 39.2).

Equivalence of d'Alembert's and Bernoulli's Solutions

We can expand the function $f(x)$ defined in Eq. (39.20a) in a Fourier series of the eigenfunctions in Eq. (38.16)

$$f(x) = \sum_{n=1}^{\infty} \left(\frac{2}{l\sigma}\right)^{1/2} a_n \sin \frac{n\pi x}{l} \tag{39.24a}$$

The orthonormality (38.17) of the eigenfunctions directly provides the coefficients a_n, according to

$$a_n = \left(\frac{2}{l\sigma}\right)^{1/2} \int_0^l \left(\sin \frac{n\pi \xi}{l}\right) f(\xi)\sigma \, d\xi \tag{39.24b}$$

A similar Fourier expansion of the function $g(x)$ yields

$$g(x) = \sum_{n=1}^{\infty} \left(\frac{2}{l\sigma}\right)^{1/2} B_n \sin \frac{n\pi x}{l} \tag{39.25a}$$

$$B_n = \left(\frac{2}{l\sigma}\right)^{1/2} \int_0^l \left(\sin \frac{n\pi \xi}{l}\right) g(\xi)\sigma \, d\xi \tag{39.25b}$$

Comparison with Eq. (38.22b) implies the identification

$$B_n \equiv \frac{n\pi c}{l} b_n \tag{39.26}$$

We now make the crucial observation that *the Fourier series of the functions $f(x)$ and $g(x)$, originally defined over the finite interval $0 \le x \le l$ by Eqs. (39.24) and*

(39.25), *immediately give an extension to all x that is odd about the origin and periodic with period 2l. These series thus satisfy the required extension conditions* (39.23). *When inserted in Eq.* (39.18), *they provide a general solution to the one-dimensional wave equation satisfying the boundary conditions of Eqs.* (39.19) *and the initial conditions of Eqs.* (39.20). The resulting solution again has a simple physical interpretation similar to that for an infinite string. For definiteness, let the finite string be released from rest with an initial displacement $f(x)$ for $0 \leq x \leq l$. Equations (39.23) then define an equivalent infinite string with an initial displacement obtained from the odd, periodic extension of $f(x)$ to all x. D'Alembert's solution (39.18) with $g(x) = 0$ expresses the subsequent motion of the equivalent infinite string as the sum of two disturbances that move rigidly to the right and left, each with half the initial amplitude. Owing to the periodic extension of the function $f(x)$, successive repetitions of the initial waveform enter and leave the physical domain with a period $2l/c$ in time. These represent the multiple reflections from the fixed ends, which occur each time the wave has traveled twice the string's length $2l$. Note that Bernoulli's form of the solution (38.19) and (38.22) completely obscures this simple interpretation, although the two solutions are, in fact, identical.

Before proving this equivalence, we must raise the following questions:

1. What kinds of functions can be represented through Fourier series of the form in Eqs. (39.24) and (39.25)?
2. What are the properties of the resulting Fourier series?

It is here sufficient to consider the following class of functions:

> $f(x)$ is *continuous* and $f'(x)$ is *piecewise continuous and differentiable* on the required interval, with $f(x)$ obeying the relevant boundary conditions (39.27)

An example of such a function is illustrated in Fig. 39.3.

To answer the preceding questions, we state two theorems concerning Fourier series of such functions.

Theorem 1 The Fourier series in Eq. (39.24) is complete.

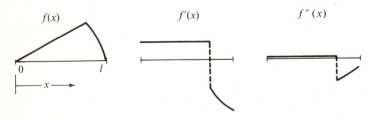

Figure 39.3 Example of a function for which $f(x)$ is *continuous* and $f'(x)$ is *piecewise continuous and differentiable* on the interval $0 \leq x \leq l$.

To define the notion of completeness, introduce the integrated mean-square deviation between the function $f(x)$ and the first N terms in the series (39.24)

$$\delta_N \equiv \int_0^l \left[f(x) - \sum_{n=1}^N a_n \left(\frac{2}{l\sigma} \right)^{1/2} \sin \frac{n\pi x}{l} \right]^2 \sigma \, dx \geq 0 \qquad (39.28)$$

The statement of *completeness* then asserts

$$\lim_{N \to \infty} \delta_N = 0 \qquad (39.29)$$

ensuring that the integrated mean-square deviation between the function $f(x)$ and its Fourier series vanishes in the limit that the series (39.24) has an infinite number of terms. If the function $f(x)$ is continuous, as has been assumed, then the theorem implies that $f(x)$ must be *identical* with its Fourier series, for the positive-definite integrand in (39.28) requires that the series cannot differ from $f(x)$ throughout any finite interval.

PROOF We shall not prove this theorem here because Sec. 41 demonstrates the completeness of the eigenfunctions for general equations of the Sturm-Liouville type.

Theorem 2 $n^2 a_n$ is bounded as $n \to \infty$ $\qquad (39.30)$

This theorem describes the large-n behavior of the Fourier coefficients defined in Eq. (39.24b) for the class of functions in Eq. (39.27).

PROOF To prove this relation, partially integrate Eq. (39.24b) twice. The first partial integration gives

$$\left(\frac{l}{2\sigma} \right)^{1/2} a_n = \left[-\frac{l}{n\pi} \left(\cos \frac{n\pi\xi}{l} \right) f(\xi) \right]_0^l + \frac{l}{n\pi} \int_0^l \left(\cos \frac{n\pi\xi}{l} \right) f'(\xi) \, d\xi \quad (39.31)$$

The endpoints make no contribution because the function $f(x)$ vanishes there by assumption. Assume that the derivative $f'(x)$ has one point of discontinuity, say at the point x_0 (the argument can easily be generalized to include any finite number of such points). Break the remaining integral into two parts $[0, x_0]$ and $[x_0, l]$. A second partial integration on each of these integrals then leads to the expression

$$\left(\frac{l}{2\sigma} \right)^{1/2} a_n = \left(\frac{l}{n\pi} \right)^2 \left\{ \left[\left(\sin \frac{n\pi\xi}{l} \right) f'(\xi) \right]_0^{x_0} \right.$$

$$+ \left[\left(\sin \frac{n\pi\xi}{l} \right) f'(\xi) \right]_{x_0}^l - \left(\int_0^{x_0} + \int_{x_0}^l \right)$$

$$\left. \times \left[\left(\sin \frac{n\pi\xi}{l} \right) f''(\xi) \, d\xi \right] \right\} \qquad (39.32)$$

Since $f'(x)$ is differentiable except at x_0, the function $f''(x)$ is continuous in each of the integrals in the last term in Eq. (39.32). As n becomes large, the Fourier eigenfunctions $\sin (n\pi\xi/l)$ oscillate rapidly as a function of ξ, forcing the integrals in the last term in Eq. (39.32) to zero for $n \to \infty$.† Moreover, the contributions from the endpoints 0 and l in the first pair of terms in Eq. (39.32) also vanish because the Fourier eigenfunctions $\sin (n\pi\xi/l)$ satisfy the boundary conditions. Thus the only remaining contribution to (39.32) for large n arises from the point x_0, where the derivative $f'(x)$ is discontinuous. Hence we find

$$n^2 a_n \xrightarrow[n \to \infty]{} \left(\frac{2\sigma}{l}\right)^{1/2} \frac{l^2}{\pi^2}\left(\sin \frac{n\pi x_0}{l}\right)[f'(x_0 - 0) - f'(x_0 + 0)] \quad (39.33)$$

and the quantity in brackets is the discontinuity in the derivative at the point x_0. Since $|\sin (n\pi x_0/l)|$ is always less than 1, the relation (39.30) has been established. It has the following corollary:

Corollary The Fourier series (39.24) is uniformly convergent $\qquad$ (39.34)

which has two consequences:

1. The sum of the series is continuous and hence identical with $f(x)$ everywhere.
2. The series can be integrated term by term.

PROOF Equation (39.30) shows that the absolute value of the nth term in the Fourier series (39.24) is less than const$/n^2$ for large n. Since the series $\sum_{n=1}^{\infty} (\text{const}/n^2)$ converges absolutely and is independent of x, the Weierstrass comparison test‡ immediately proves the result (39.34).

Consequences 1 and 2 of the corollary stated above are simply the basic properties of uniformly convergent series:§

1. A uniformly convergent series of continuous functions is continuous.
2. A uniformly convergent series of continuous functions can be integrated term by term.

We now apply these results to establish the equivalence of the d'Alembert and Bernoulli solutions to the one-dimensional wave equation. Substitute the Fourier-series representations of $f(x)$ and $g(x)$ [Eqs. (39.24) and (39.25)] into d'Alembert's solution (39.18) to the wave equation. The Fourier series in the last term in Eq. (39.18) can be integrated term by term using consequence 2 established under Eq. (39.34):

$$u(x, t) = \left(\frac{2}{l\sigma}\right)^{1/2} \sum_{n=1}^{\infty} \left\{\frac{a_n}{2}\left[\sin \frac{n\pi(x - ct)}{l} + \sin \frac{n\pi(x + ct)}{l}\right]\right.$$
$$\left. + \frac{B_n}{2c}\left(-\frac{l}{n\pi}\right)\left[\cos \frac{n\pi(x + ct)}{l} - \cos \frac{n\pi(x - ct)}{l}\right]\right\} \quad (39.35)$$

---

† Although this relation is sufficiently evident to state without proof at this point, it is in fact the celebrated Riemann-Lebesgue lemma, proved, for example, in Whittaker and Watson [1], sec. 9.41.

‡ Whittaker and Watson [1], sec. 3.34.

§ Whittaker and Watson [1], secs. 3.32 and 4.7.

Some elementary trigonometry and the definition in Eq. (39.26) lead to

$$u(x, t) = \sum_{n=1}^{\infty} \left(\frac{2}{l\sigma}\right)^{1/2} \left(\sin \frac{n\pi x}{l}\right) \left[a_n\left(\cos \frac{n\pi ct}{l}\right) + b_n\left(\sin \frac{n\pi ct}{l}\right)\right] \quad (39.36)$$

which is precisely Bernoulli's solution (38.19).

We may comment briefly on the class of functions $f(x)$ and $g(x)$ that occur for a physical string. Continuity of the string requires that $u(x, t)$ be continuous at every instant t, and we now show that $\partial u(x, t)/\partial x$ must also be continuous. For simplicity, assume a constant tension τ and mass density σ, and consider a small element ds of the string (Fig. 25.1). For small displacements, Newton's second law reads

$$\tau\left(\left.\frac{\partial u}{\partial x}\right|_+ - \left.\frac{\partial u}{\partial x}\right|_-\right) = (\sigma \, ds)\frac{\partial^2 u}{\partial t^2} \xrightarrow[ds \to 0]{} 0 \quad (39.37)$$

The final limit in Eq. (39.37) follows because the mass density σ is constant, and no physical system can experience infinite accelerations. As a result, the function $f(x)$ in any physical-string problem is both continuous and differentiable, implying the same behavior for the first term in d'Alembert's solution (39.18) [recall the discussion below Eq. (39.11)]. Note that this condition eliminates an initial form like that in Fig. 39.3, so that such functions are *more* restricted than the class in (39.27). The situation is different for an initial velocity distribution $g(x)$, where functions of the form (39.27) can indeed occur. The integral in the second term of d'Alembert's solution (39.18) accounts for this difference, because continuity of the integrand $g(x)$ itself suffices to ensure that this integral is once differentiable and hence acceptable as a solution of the one-dimensional wave equation. It is also clear on physical grounds that $g(x)$ must be continuous if the string itself is to remain continuous at all subsequent times.

40 EIGENFUNCTION EXPANSIONS

The normal-mode solutions (38.13) to the one-dimensional wave equation for the uniform string (38.12) led to the eigenfunction equation (38.14). Imposing the boundary condition of fixed endpoints $\rho(0) = \rho(l) = 0$ (see Fig. 25.2) then restricted the solutions to the discrete set given in (38.16).

More generally, Eq. (38.14) has different solutions for each type of spatial boundary conditions. Besides the fixed-endpoint condition, another very important example is *periodic boundary conditions* (illustrated in Fig. 40.1). Here we consider a string of length $2l$ and demand that a translation by a distance $L = 2l$ from any point in the original interval $-l \le x \le l$ return the solution of Eq. (38.14) to its original value

$$\rho(x + L) = \rho(x) \quad (40.1)$$

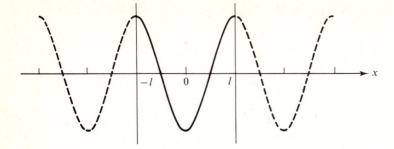

Figure 40.1 Periodic boundary conditions for a string of length $2l$.

Solutions to Eq. (38.14) obeying the periodic boundary condition (40.1) can be written

$$\rho_n(x) = (L\sigma)^{-1/2} e^{ik_n x} \qquad (40.2a)$$

$$k_n = \frac{2\pi n}{L} \qquad n = 0, \pm 1, \pm 2, \ldots, \pm\infty \qquad (40.2b)$$

The integer n in Eq. (40.2) must have both signs to allow for waves moving in both directions along the string.† Although the boundary condition in Eq. (40.1) may seem somewhat artificial, it is readily realized by joining together the two ends of the string, as illustrated in Fig. 24.4 for a string loaded with mass points. In this configuration, if one starts at any point on the string and translates by a distance L, the result is to return to the original position, satisfying the boundary condition (40.1) identically. Direct integration immediately verifies that the solutions in Eq. (40.2a) satisfy the generalized orthonormality condition

$$\int_{-l}^{l} \rho_n(x)^* \rho_m(x) \sigma \, dx = \delta_{nm} \qquad (40.3)$$

where the asterisk denotes complex conjugation required by the form of Eq. (40.2a). We shall show in Sec. 41 that the solutions to Eq. (38.14) satisfying periodic boundary conditions (40.1) are also *complete;* hence any function of the type in Eq. (39.27) can be expanded in terms of these eigenfunctions according to

$$f(x) = (L\sigma)^{-1/2} \sum_{n=-\infty}^{\infty} C_n \exp \frac{2\pi n i x}{L} \qquad (40.4a)$$

$$C_n \equiv (L\sigma)^{-1/2} \int_{-l}^{l} \exp\left(-\frac{2\pi n i \xi}{L}\right) f(\xi) \sigma \, d\xi \qquad (40.4b)$$

where the orthonormality condition (40.3) has been used to solve for the Fourier

---

† Alternatively, we could use the real solutions $(2/L\sigma)^{1/2} \cos k_n x$ and $(2/L\sigma)^{1/2} \sin k_n x$, with $n = 1$, $2, 3, \ldots, \infty$ and $(1/L\sigma)^{1/2}$ for $n = 0$. It is instructive to rewrite Eqs. (40.4) in terms of these functions.

coefficients. Equations (40.4) are the complex form of the *general Fourier series*. If the function $f(x)$ is real, it is apparent from Eq. (40.4b) that

$$C_n^* = C_{-n} \qquad (40.5)$$

Substitution of this relation in the right-hand side of Eq. (40.4a) leads to an *explicitly real* series on combining the terms with n and $-n$.

To motivate our discussion of the general Sturm-Liouville eigenfunction theory, we first derive a more *general* string equation by including an additional restoring force per unit length of the form $-v(x)u(x, t)$ along the string. This can be realized physically, for example, by attaching springs continuously along the string, as illustrated in Fig. 40.2. Here $v(x)$ is the force constant per unit length, and the Hooke's-law restoring force is proportional to the displacement from equilibrium and opposes it. The lagrangian density for this system is obtained by generalizing Eq. (38.2) to the form

$$\mathcal{L} = \mathcal{T} - \mathcal{V} \qquad (40.6a)$$

$$\mathcal{T} = \tfrac{1}{2}\sigma(x)\left(\frac{\partial u}{\partial t}\right)^2 \qquad (40.6b)$$

$$\mathcal{V} = \tfrac{1}{2}\tau(x)\left(\frac{\partial u}{\partial x}\right)^2 + \tfrac{1}{2}v(x)u^2 \qquad (40.6c)$$

where, as previously, the mass density $\sigma(x)$ and the tension $\tau(x)$ may vary along the string. In Eq. (40.6), $\mathcal{L}$ has been separated into a kinetic-energy density $\mathcal{T}$ and a potential-energy density $\mathcal{V}$ that includes the extra restoring force. The Euler-Lagrange equation (38.8) then leads to the equation of motion

$$\sigma(x)\frac{\partial^2 u}{\partial t^2} = \frac{\partial}{\partial x}\left[\tau(x)\frac{\partial u}{\partial x}\right] - v(x)u \qquad (40.7)$$

which will now be called the *general string equation*. We again seek normal-mode solutions

$$u(x, t) = \rho(x)\cos(\omega t + \phi) \qquad (40.8)$$

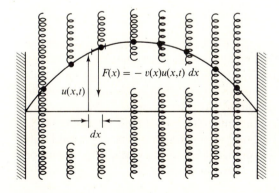

$$F(x) = -v(x)u(x,t)\,dx$$

$u(x,t)$

dx

Figure 40.2 Physical realization of additional restoring force $-v(x)u(x, t)$ along the string. Here $v(x)$ is the force constant per unit length.

and Eq. (40.7) reduces to the form

$$-\frac{d}{dx}\left[\tau(x)\frac{d\rho}{dx}\right] + v(x)\rho = \omega^2\sigma(x)\rho \qquad (40.9)$$

known as the *Sturm-Liouville equation.*

Sturm-Liouville theory is the general analysis of the differential equation (40.9) and its solutions. Motivated by our discussion of the general string problem, we consistently make the following assumptions about the coefficients appearing in Eq. (40.9)

$$\tau(x), v(x), \sigma(x) \text{ are } real \qquad \begin{array}{l}\text{in the appropriate}\\ \text{interval } a \le x \le b\end{array} \qquad (40.10a)$$

$$\tau(x) > 0 \qquad \text{in the open interval } a < x < b \qquad (40.10b)$$

$$\sigma(x) > 0 \qquad \text{in the open interval } a < x < b \qquad (40.10c)$$

Note that since (40.10b) and (40.10c) hold in the open interval, $\tau(x)$ and $\sigma(x)$ may vanish at the endpoints. Although the motivation of the string problem might leads us to insist on positive $v(x)$ ($\equiv$ force constant per unit length), we shall for generality admit both signs of $v(x)$ in discussing the general Sturm-Liouville equation (40.9). In the string problem a negative $v(x)$ will produce an equilibrium *distortion* of the string. Provided that the transverse force associated with $v(x)$ is much smaller than the tension $\tau(x)$ in the string, it is still permissible to talk about small transverse oscillations of an essentially straight string.

We shall consider solutions to the Sturm-Liouville equation corresponding to different types of spatial boundary conditions:

1. *Fixed endpoint:*

$$\rho = 0 \qquad (40.11)$$

We have already used this boundary condition several times, in particular in generating the eigenfunctions in Sec. 38 and in discarding the boundary term in Eq. (38.7) by imposing the condition that the variation δu vanish at the endpoint.

2. *Natural boundary condition:*

$$\tau\frac{d\rho}{dx} = 0 \qquad (40.12)$$

It will become clear in the following discussion how the natural boundary condition enters the theory and just what physical situation it refers to. Here we merely point out that it has already appeared in deriving the general Euler-Lagrange equation (38.8) from Hamilton's principle for a continuous system (38.6). If the natural boundary condition (40.12) is satisfied, the boundary term in Eq. (38.7) coming from the spatial integration in Hamilton's principle vanishes for *arbitrary variations δu* at the spatial endpoint. Thus a natural boundary condition ensures that the spatial partial integration in Hamilton's principle produces no contribution from that endpoint.

3. *General homogeneous boundary condition:*

$$\alpha \frac{d\rho}{dx} - \beta\rho = 0 \qquad (40.13)$$

where α and β are real constants. This relation can alternatively be written

$$\frac{\rho'}{\rho} = \frac{\beta}{\alpha} \qquad (40.14)$$

and thus the general homogeneous boundary condition corresponds to a fixed ratio of slope to value at an endpoint.

4. *Periodic boundary conditions.* We have already met these conditions in Eq. (40.1). Direct evaluation of that relation and its derivative at the endpoints of the physical interval yields an equivalent condition on the function and its derivative. Thus the condition of periodic boundary conditions requires

$$\rho(b) = \rho(a) \qquad (40.15a)$$

$$\rho'(b) = \rho'(a) \qquad (40.15b)$$

and we also must assume that all the coefficients in (40.10a) obey the same periodicity (40.1).

If any combination of the boundary conditions 1 to 3 or the boundary conditions 4 are imposed on the Sturm-Liouville equation (40.9), solutions to this problem will exist only for a certain set of *eigenvalues* ω_n^2, with $n = 1, 2, \ldots, \infty$. The corresponding *eigenfunctions* will be denoted by $\rho_n(x)$ and satisfy

$$-\frac{d}{dx}\left[\tau(x)\frac{d\rho_n(x)}{dx}\right] + v(x)\rho_n(x) = \omega_n^2\sigma(x)\rho_n(x)$$

$$n = 1, 2, \ldots, \infty \qquad (40.16)$$

We make the following assumptions about these eigenvalues, again motivated by the discussion of the string problem.

Assumption 1 The eigenvalues are assumed to increase without bound as $n \to \infty$

$$\omega_n^2 \to \infty \qquad \text{as } n \to \infty \qquad (40.17)$$

This is a general property of solutions to the Sturm-Liouville equation,† although we shall not give a proof here. We merely note that it can be verified in any given example by explicit construction of the solutions to the Sturm-Liouville equation subject to the appropriate set of boundary conditions. It certainly holds for the uniform string problem, as is evident from Eqs. (38.14b) and (38.16b). It also applies for a mass point m attached to the midpoint of a uniform string, as shown in Prob. 7.4.

---

† Morse and Feshbach, [2], pp. 739–743; Courant and Hilbert [3], pp. 336–339 and 412–415.

Assumption 2 The eigenvalues ω_n^2 are assumed nonnegative

$$\omega_n^2 \geq 0 \qquad (40.18)$$

For a string, Eq. (40.18) is the physical condition for *stability*, because negative ω_n^2 correspond to imaginary frequencies and runaway solutions.

We now establish some general properties of the eigenfunctions in Eq. (40.16). Consider two different solutions to this equation corresponding to two different eigenvalues

$$\frac{d}{dx}\tau\rho_p' - v\rho_p = -\omega_p^2\sigma\rho_p \qquad (40.19a)$$

$$\frac{d}{dx}\tau\rho_q' - v\rho_q = -\omega_q^2\sigma\rho_q \qquad (40.19b)$$

Multiply the first equation by ρ_q^* and integrate over the interval, multiply the complex conjugate of the second by ρ_p and integrate over the interval, and subtract. The potential term v cancels in this operation, leaving the result

$$\int_a^b \frac{d}{dx}(\rho_q^*\tau\rho_p' - \rho_p\tau\rho_q'^*)\,dx = [(\omega_q^2)^* - \omega_p^2]\int_a^b \rho_q^*\sigma\rho_p\,dx \qquad (40.20)$$

Note that the integrand on the left-hand side has been rewritten as a total spatial derivative. This transformation is possible because the terms $\rho_q'^*\tau\rho_p'$ cancel when taking the derivative, reducing the left-hand side of Eq. (40.20) to the required expression. The total differential on the left-hand side can now be immediately integrated to yield the result

$$\int_a^b \frac{d}{dx}(\rho_q^*\tau\rho_p' - \rho_p\tau\rho_q'^*)\,dx = [\rho_q^*\tau\rho_p' - \rho_p\tau\rho_q'^*]_a^b = 0 \qquad (40.21)$$

Here, we have made explicit that *this expression vanishes for any combination of the boundary conditions under consideration in Eqs.* (40.11) *to* (40.14) *and for the periodic boundary conditions in Eqs.* (40.15). The reader is urged to verify this in detail for each boundary condition. It is important to remember that a given set of boundary conditions defines and applies to all the eigenfunctions. Comparison of Eqs. (40.20) and (40.21) immediately yields

$$[(\omega_q^2)^* - \omega_p^2]\int_a^b \rho_q^*\sigma\rho_p\,dx = 0 \qquad (40.22)$$

Equation (40.22) allows us to demonstrate that the eigenvalues in Eq. (40.16) are *real*. We merely set $q = p$ to obtain

$$[(\omega_p^2)^* - \omega_p^2]\int_a^b \rho_p^*\rho_p\,dm = 0 \qquad (40.23)$$

where the *measure dm* has been defined by

$$dm \equiv \sigma(x)\,dx > 0 \qquad (40.24)$$

Since the integrand is positive definite, the remaining factor must vanish, which gives

$$(\omega_p^2)^* = \omega_p^2 \tag{40.25}$$

and establishes our result.

For $\omega_q^2 \neq \omega_p^2$, Eq. (40.22) shows that if the boundary conditions are satisfied, the eigenfunctions obey the *orthogonality* condition

$$\int_a^b \rho_q^* \rho_p \, dm = 0 \qquad \text{if } \omega_q^2 \neq \omega_p^2 \tag{40.26}$$

The expression in Eq. (40.26) is sometimes known as the *inner product* of the two eigenfunctions ρ_p and ρ_q and is written in the form

$$\int_a^b \rho_q^* \rho_p \, dm \equiv \langle \rho_q | \rho_p \rangle \tag{40.27}$$

Equation (40.16) and the boundary conditions (40.11) to (40.15) are homogeneous, which allows us to *scale* the solutions to Eq. (40.16) by any constant factor. Once the eigenfunctions have been constructed, they can be *normalized* in any convenient manner. In connection with (40.27) we impose the general *orthonormality* condition

$$\int_a^b \rho_p(x)^* \rho_q(x) \, dm = \langle \rho_p | \rho_q \rangle = \delta_{pq} \tag{40.28}$$

Although Eq. (40.22) contains no information about the orthogonality of the eigenfunctions if p and q correspond to the *same* eigenvalue $\omega_p^2 = \omega_q^2$, it is always possible to orthogonalize these solutions using the Gram-Schmidt orthogonalization procedure (see Prob. 4.10), and we can therefore assert that Eq. (40.28) holds for *all* the solutions to Eq. (40.16).

We illustrate the Sturm-Liouville theory with the example of a mass point m attached at the center of a string with uniform tension τ, uniform mass density σ, and fixed endpoints (see Fig. 40.3). The eigenfunctions and eigenvalues for this

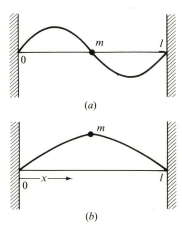

(a)

(b)

Figure 40.3 Mass point at center of string with uniform tension, uniform mass density, and fixed endpoints: (a) odd and (b) even modes.

system were obtained in Prob. 4.17. The eigenfrequencies for the *odd modes* illustrated in Fig. 40.3*a* are given by†

$$\frac{\omega_n}{c} = \frac{n\pi}{l} \qquad n = 2, 4, 6, \ldots, \infty \tag{40.29}$$

In these modes the mass m never moves from its nodal position, and the frequencies and eigenfunctions are just those of the odd modes for a uniform string (38.16). In contrast, the mass m is actually displaced in the *even* modes, and the corresponding eigenfrequencies are given as solutions to the following transcendental equation (see Prob. 4.17)

$$\frac{2c}{\omega l} \cot \frac{\omega l}{2c} = \frac{m}{\sigma l} = \frac{m}{M} \tag{40.30}$$

where

$$\sigma l = M_{\text{string}} \equiv M \tag{40.31}$$

is the mass of the string and $c^2 = \tau/\sigma$. The eigenfunctions corresponding to these eigenvalues are nontrivial linear combinations of the eigenfunctions for the free string. Nevertheless, it was shown in Prob. 4.17 that all the eigenfunctions of this problem are orthonormal with respect to the measure

$$dm = \sigma \left[1 + \frac{m}{\sigma} \delta(x - \tfrac{1}{2}l) \right] dx \tag{40.32}$$

The coefficient of dx is the mass density of the string, including the mass point at the center; thus this measure is exactly that of Eq. (40.24).

We now claim that the set of orthonormal eigenfunctions satisfying Eq. (40.16) subject to any combination of the boundary conditions in Eqs. (40.11) to (40.14) or the periodic boundary conditions in Eq. (40.15) are *complete* in the sense of Eqs. (39.28) and (39.29) for the class of functions defined in Eq. (39.27) as long as the eigenvalues in the Sturm-Liouville equation satisfy conditions (40.17) and (40.18). To establish this result, we first show that the Sturm-Liouville equation can be derived from a variational principle and then use it to prove the completeness of the Sturm-Liouville eigenfunctions.

41 VARIATIONAL PRINCIPLE

We now establish a variational principle for the Sturm-Liouville equation.‡ It will provide an extremely powerful calculational tool as well as the crucial step in the proof of completeness of the eigenfunction expansion.

---

† These modes are *odd* about the center of the string. They correspond to even values of the integer n. Modes that are *even* about the center of the string correspond to odd values of n.

‡ For a very general discussion of Sturm-Liouville theory, see Morse and Feshbach [2], sec. 6.3.

Basic Formulation

To be specific, we assert that Eq. (40.9) is the Euler-Lagrange equation of the following variational problem:

Find the ρ that makes the following functional stationary

$$\omega^2[\rho] \equiv \frac{\frac{1}{2}\int_a^b \left[\tau(x)\left(\frac{d\rho}{dx}\right)^2 + v(x)\rho^2 \right] dx}{\frac{1}{2}\int_a^b [\sigma(x)\rho^2]\, dx} \equiv \frac{I_1}{I_2} \qquad (41.1)$$

subject to arbitrary variations $\delta\rho(x)$ consistent with a given set of boundary conditions chosen from Eq. (40.11), (40.12), or (40.15).†

For simplicity, we here and henceforth assume that all the relevant functions are real.

PROOF We use the notation of the calculus of variations and make $\omega^2[\rho]$ stationary by setting its first variation equal to zero. Direct calculation gives

$$\delta\omega^2 = \frac{\delta I_1}{I_2} - \frac{I_1\, \delta I_2}{I_2^2} = \frac{1}{I_2}\left(\delta I_1 - \frac{I_1}{I_2}\, \delta I_2 \right)$$

$$= \frac{1}{I_2}(\delta I_1 - \omega^2\, \delta I_2) = 0 \qquad (41.2)$$

where Eq. (41.1) has been introduced in obtaining the third equality. Since I_2 is positive, the numerator in this expression must vanish. Written in terms of ρ, it becomes [recall that $\delta\rho' = d(\delta\rho)/dx$]

$$\int_a^b \left[\tau(x)\frac{d\rho}{dx}\frac{d}{dx}\, \delta\rho + v(x)\rho\, \delta\rho - \omega^2\sigma(x)\rho\, \delta\rho \right] dx = 0 \qquad (41.3)$$

A partial integration on the first term yields

$$\int_a^b dx\, \tau(x)\frac{d\rho}{dx}\frac{d}{dx}\, \delta\rho = \left[\delta\rho\, \tau(x)\frac{d\rho}{dx} \right]_a^b - \int_a^b \delta\rho\, \frac{d}{dx}\left[\tau(x)\frac{d\rho}{dx} \right] dx \qquad (41.4)$$

and the contributions from the endpoints vanish whenever the following relation is satisfied

$$[\delta\rho\, \tau\rho']_a^b = 0 \qquad (41.5)$$

We observe that *this relation holds for all the boundary conditions listed above*:

1. For fixed endpoints, $\delta\rho = 0$.
2. For the natural boundary conditions, $\tau\rho' = 0$ with arbitrary $\delta\rho$.

---

† The case of general homogeneous boundary conditions (40.13) requires a slight modification, to be considered subsequently.

3. For periodic boundary conditions, $\delta\rho(a) = \delta\rho(b)$, and hence the contributions from the two endpoints in Eq. (41.5) cancel.

Thus the condition that the variation of Eq. (41.1) vanish can be written in the form

$$I_2\ \delta\omega^2[\rho] = 0 = \int_a^b \left\{ -\frac{d}{dx}\left[\tau(x)\frac{d\rho}{dx}\right] + v(x)\rho - \omega^2\sigma(x)\rho \right\} \delta\rho(x)\ dx \quad (41.6)$$

The variational principle requires that this relation hold for arbitrary $\delta\rho(x)$, implying that the integrand must vanish identically

$$-\frac{d}{dx}\left[\tau(x)\frac{d\rho}{dx}\right] + v(x)\rho = \omega^2\sigma(x)\rho \quad (41.7)$$

This Euler-Lagrange equation is precisely the Sturm-Liouville equation.

It only remains to consider the Sturm-Liouville equation subject to the general homogeneous boundary condition (40.13), which may be imposed at one or both endpoints. For definiteness, suppose that it holds at both ends

$$\alpha\rho' = \beta\rho \qquad \text{at } x = a \text{ and } x = b \quad (41.8)$$

In this case,† it is merely necessary to replace I_1 in Eq. (41.1) by

$$\tilde{I}_1 \equiv I_1 - \frac{\beta}{2\alpha}\tau(b)[\rho(b)]^2 + \frac{\beta}{2\alpha}\tau(a)[\rho(a)]^2 \quad (41.9)$$

requiring that $\tilde{I}_1/I_2$ be stationary under arbitrary variations of ρ *including the endpoints*. The only change in the calculation is the appearance of two extra terms

$$-\frac{\beta}{\alpha}\tau(b)\rho(b)\ \delta\rho(b) + \frac{\beta}{\alpha}\tau(a)\rho(a)\ \delta\rho(a) \equiv -\frac{\beta}{\alpha}\left[\tau\rho\ \delta\rho\right]_a^b \quad (41.10)$$

on the left-hand side of (41.3). When these terms are combined with the endpoint contributions in (41.4), we find that Eq. (41.6) is replaced by

$$\delta\omega^2[\rho] \equiv \delta\frac{\tilde{I}_1}{I_2} = 0 = \int_a^b \left\{ -\frac{d}{dx}\left[\tau(x)\frac{d\rho}{dx}\right] + v(x)\rho \right.$$
$$\left. - \omega^2\sigma(x)\rho \right\} \delta\rho(x)\ dx + \left[\tau\left(\rho' - \frac{\beta}{\alpha}\rho\right)\delta\rho\right]_a^b \quad (41.11)$$

The boundary conditions ensure that the last term vanishes, independent of $\delta\rho$, and we again obtain the Sturm-Liouville equation (41.7), even for these more general boundary conditions.

To summarize, we have obtained an important variational principle. The Sturm-Liouville equation and an acceptable set of boundary conditions are wholly equivalent to the following variational problem: the quantity $\omega^2[\rho]$

---

† Courant and Hilbert [3], pp. 402–405.

defined in Eq. (41.1) [or (41.9)] must be stationary subject to arbitrary variations of ρ consistent with the boundary conditions. In this derivation, the Sturm-Liouville equation is just the corresponding Euler-Lagrange equation.

This equivalence suggests a change in our way of finding solutions to the differential equation (41.7). Suppose that an expression for ρ is inserted in the functional $\omega^2[\rho]$ and allowed to vary in an *arbitrary* fashion consistent with the boundary conditions of the problem. If the functional $\omega^2[\rho]$ actually becomes stationary under arbitrary variations $\delta\rho(x)$ at each point in space, the above argument implies that this ρ is indeed a solution to the Sturm-Liouville equation. Furthermore, the quantity $\omega^2[\rho]$ is the eigenvalue for those functions ρ which render $\omega^2[\rho]$ stationary, as is evident from Eqs. (41.2) and (41.3). This result is also obtained directly from the differential equation (41.7) by multiplying by ρ and integrating over the interval a to b. The first term on the left can be partially integrated, and the boundary contributions vanish with our choice of boundary conditions. If the resulting expression is then solved for ω^2, we immediately obtain Eq. (41.1) or its extension (41.9) for general homogeneous boundary conditions (40.13).

Since the quantity $\omega^2[\rho]$ is independent of the normalization of ρ, we are free to choose any normalization we wish. It is convenient for the present discussion to assume

$$\int_a^b [\sigma(x)\rho^2]\, dx \equiv \int_a^b \rho^2\, dm = 1 \tag{41.12}$$

which eliminates the denominator in Eq. (41.1). This choice in no way restricts the variational principle, for the normalization of ρ does not affect the quantity $\omega^2[\rho]$. Hence our original variational principle is fully equivalent to the requirement that $\delta\omega^2[\rho] = 0$ under arbitrary variations $\delta\rho(x)$ consistent with the boundary conditions *and* consistent with the normalization statement (41.12).

Minimum Character of the Functional

We now make the important observation that the *eigenfunction corresponding to the lowest eigenvalue actually minimizes the functional* $\omega^2[\rho]$. This assertion evidently assumes that the functional $\omega^2[\rho]$ indeed possesses a true minimum. To justify this claim, we examine two simple cases. First consider the original uniform string with fixed endpoints and no linear restoring force, so that $v = 0$. In this case, the functional in Eq. (41.1) reduces to an integral along the string of the quantity $(d\rho/dx)^2$, which is a measure of the string's *curvature*. The more the function $\rho(x)$ varies in space, the larger the integral of this positive-definite quantity $(d\rho/dx)^2$ will be. It is evident on physical grounds that there will be some minimum value to this integrated curvature consistent with the boundary conditions and the normalization condition (41.12). In addition, the function $\rho(x)$ which minimizes $\omega^2[\rho]$ must satisfy the Euler-Lagrange equation (41.7) to guarantee that the functional $\omega^2[\rho]$ is stationary. Thus the curve $\rho_1(x)$ that produces a *true minimum* of the functional in Eq. (41.1) not only must be a normalized solution to the

Sturm-Liouville equation (41.7) consistent with the proper boundary conditions but must also minimize the integrated curvature $\int_a^b (d\rho/dx)^2 \, dx$. This solution is just the lowest eigenfunction of the Sturm-Liouville equation, and the corresponding value of the functional $\omega^2[\rho_1]$ is the lowest eigenvalue. Other functions $\rho(x)$ will render the functional $\omega^2[\rho]$ in Eq. (41.1) *stationary*, but they do not produce an absolute *minimum*. For example, Fig. 41.1 shows a second stationary solution, which will have higher integrated curvature and hence higher $\omega^2[\rho]$. The general argument leading to Eq. (40.28) proves that the second solution must be orthogonal to the first; hence it must have at least one node, since the first eigenfunction has a definite sign.

As the second example, consider the uniform string with a distributed linear restoring force $v(x)$ indicated in Fig. 41.2. The functional in Eq. (41.1) now consists of two terms

$$\omega^2 = \bar{c} + \bar{v} \tag{41.13}$$

associated with the curvature energy and the potential energy, respectively. We still argue that adding more nodes for the function ρ increases the integrated mean-square curvature $\bar{c} \equiv \int_a^b \tau(d\rho/dx)^2 \, dx$. On the other hand, decreasing the amount of ρ^2 where the potential is high reduces the integrated potential energy $\bar{v} = \int_a^b v\rho^2 \, dx$. If $v(x)$ is negative, this argument is reversed, for we then want to increase the amount of ρ^2 in the region of low potential. The deformation of ρ^2 is not arbitrary, however, owing to the normalization condition (41.12). It is again evident that the functional $\omega^2[\rho]$ has a minimum, obtained by concentrating ρ^2 in the region of low potential, which decreases $\bar{v}$, without introducing too much additional curvature, which increases $\bar{c}$. Since the minimum in the functional $\omega^2[\rho]$ again represents a stationary point of this functional, the corresponding function ρ_1 must satisfy the Euler-Lagrange equation (41.7). Thus the true minimum of the functional (41.1) is just the lowest eigenvalue of Eq. (41.7), and the solution $\rho_1(x)$ is the corresponding eigenfunction.

The above physical observation can be expressed in the following analytical form

$$\omega^2[\rho] \geq \omega^2[\rho_1] \tag{41.14}$$

implying that the functional $\omega^2[\rho]$ possesses a true minimum. *For any function ρ consistent with the boundary conditions of the problem, Eq. (41.1) will always lead to*

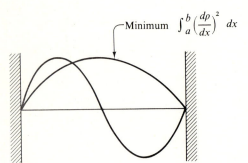

Minimum $\int_a^b \left(\dfrac{d\rho}{dx}\right)^2 \, dx$

Figure 41.1 Uniform string with fixed endpoints and fixed norm (41.12), showing two curves that make the Sturm-Liouville functional $\omega^2[\rho]$ stationary.

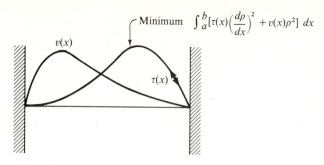

$$\text{Minimum} \quad \int_a^b [\tau(x)\left(\frac{d\rho}{dx}\right)^2 + v(x)\rho^2] \, dx$$

Figure 41.2 Uniform string with fixed endpoints and fixed norm, with an additional linear restoring force constant per unit length $v(x)$. For clarity, only the lowest eigenfunction is shown.

a value of the functional $\omega^2[\rho]$ that is greater than or equal to its absolute minimum. This absolute-minimum value of the functional is the lowest eigenvalue in the Sturm-Liouville equation (41.7).

The inequality (41.14), together with the identification of the Sturm-Liouville equation (41.7) as the Euler-Lagrange equation of the variational problem (41.1), suggests a systematic way of solving the Sturm-Liouville problem without ever integrating the differential equation. We can simply carry out the following series of steps:

1. First find the function ρ that *minimizes* the functional $\omega^2[\rho]$ under arbitrary variations. Call this function ρ_1. Equation (41.14) shows that

$$\omega^2[\rho_1] = \omega_1^2 \tag{41.15}$$

   is the lowest eigenvalue in the Sturm-Liouville equation, and ρ_1 is the corresponding eigenfunction.
2. To obtain the *next lowest eigenvalue* and corresponding eigenfunction, *repeat the above procedure with a function that is orthogonal to ρ_1*. For any $\rho \neq \rho_1$, such a function can be constructed according to the prescription

$$\rho \to \rho - \langle \rho_1 | \rho \rangle \rho_1 \tag{41.16}$$

   in the notation of Eq. (40.27).
3. Now search this restricted class for the ρ that minimizes the functional $\omega^2[\rho]$. The above arguments make clear that the functional $\omega^2[\rho]$ again possesses an absolute minimum in this restricted space of functions orthogonal to the lowest eigenfunction ρ_1. In Fig. 41.1, for example, all the functions orthogonal to ρ_1 must have at least one node and change sign in the interval $[a, b]$. Thus they will always have a higher integrated mean-square curvature than ρ_1. Once the functional $\omega^2[\rho]$ has been minimized, and therefore made stationary, we know that the corresponding function ρ must satisfy the Sturm-Liouville equation (41.7). Consequently, this $\rho(x)$ is again an eigenfunction. It cannot be ρ_1, as is evident from Eq. (41.16), so that it must be the eigenfunction ρ_2 corresponding to the next lowest eigenvalue with

$$\omega^2[\rho_2] = \omega_2^2 \tag{41.17}$$

4. This procedure can be repeated. At each step, search for a minimum of the functional (41.1) in the restricted space of functions *orthogonal to all the previously obtained solutions*. For example, after having obtained the first N eigenvalues ω_n^2 and eigenfunctions ρ_n, with $n = 1, \ldots, N$, we seek the next eigenvalue and eigenfunction by making the functional stationary under the restricted class of functions

$$\rho \to \rho - \sum_{n=1}^{N} \langle \rho_n | \rho \rangle \rho_n \tag{41.18}$$

In this way, all the solutions to the Sturm-Liouville equation (41.7) can be generated systematically without solving a differential equation. It is only necessary to perform the integrations in Eqs. (41.1) and (41.12), allowing enough freedom to make the functional $\omega^2[\rho]$ truly stationary under arbitrary variations of this function ρ.

Completeness of Eigenfunctions

The above procedure generates an orthonormal set of eigenfunctions satisfying

$$\int_a^b \rho_p(x)\rho_q(x) \, dm = \delta_{pq} \tag{41.19}$$

where, for simplicity, the eigenfunctions are here and henceforth assumed real.†
We now demonstrate that this set of eigenfunctions is also complete in the sense of Eqs. (39.28) and (39.29). Thus we shall prove that a function $f(x)$ from the class defined in Eq. (39.27) can be expanded in an infinite series of the eigenfunctions $\rho_n(x)$ according to

$$f(x) = \sum_{n=1}^{\infty} a_n \rho_n(x) \tag{41.20}$$

If this expansion is valid, the orthonormality condition Eq. (41.19) immediately yields the coefficients

$$a_n = \int_a^b \rho_n(\xi) f(\xi) \, \sigma(\xi) \, d\xi \tag{41.21}$$

To proceed with the proof, first approximate $f(x)$ with the *finite series* $\sum_{n=1}^{N} \alpha_n \rho_n(x)$. What is the best choice for the coefficients α_n? To answer this question, consider the integrated mean-square deviation between the function $f(x)$ and the finite approximating series

$$\delta_N \equiv \int_a^b \left[f(x) - \sum_{n=1}^{N} \alpha_n \rho_n(x) \right]^2 dm \geq 0 \tag{41.22}$$

---

† Our discussion of eigenvectors in Chap. 4 shows that, in fact, there is no loss of generality here (see also the footnote on page 220).

It is evident that this quantity is positive semidefinite, and we shall minimize its value with respect to the coefficients α_n with $n = 1, 2, \ldots, N$. Since (41.22) contains a finite series, the integral can be evaluated term by term to yield

$$\delta_N = \int_a^b [f(x)]^2 \, dm - 2 \sum_{n=1}^N \alpha_n a_n + \sum_{n=1}^N \alpha_n^2 \tag{41.23}$$

where the coefficients a_n have been identified through Eq. (41.21). To minimize this expression, set the partial derivative with respect to each of the α_n equal to zero

$$\frac{\partial \delta_N}{\partial \alpha_n} = 2(\alpha_n - a_n) = 0 \tag{41.24}$$

This set of equations has the solution

$$\alpha_n = a_n \tag{41.25}$$

Given our criterion, the best choice for the coefficients in the *finite series* is precisely the set of expansion coefficients (41.21). This choice obviously minimizes the integrated mean-square deviation in Eq. (41.22), because the set of conditions (41.24) has a unique solution.

Now consider the following function, which will be used to evaluate the functional (41.1),

$$g_N(x) \equiv \delta_N^{-1/2} \left[f(x) - \sum_{n=1}^N a_n \rho_n(x) \right] \tag{41.26}$$

This function has the following two properties:

1. It is orthogonal to the first N eigenfunctions, as is evident from Eqs. (41.19) and (41.21)

$$\int_a^b g_N(x) \rho_n(x) \, dm = 0 \qquad n = 1, 2, \ldots, N \tag{41.27}$$

2. It is normalized to unity [see Eq. (41.12)]

$$\int_a^b [g_N(x)]^2 \, dm = 1 \tag{41.28}$$

because of the factor $\delta_N^{-1/2}$, which is evaluated with Eqs. (41.22) and (41.25). If $g_N(x)$ is inserted into the functional (41.1), the conditions (41.27) and (41.28) and the arguments following Eq. (41.14) can be used to conclude that the functional $\omega^2[g_N]$ must equal or exceed the value ω_{N+1}^2:

$$\omega^2[g_N] \geq \omega_{N+1}^2 \tag{41.29}$$

This last result now allows us to demonstrate completeness merely by an explicit evaluation of $\omega^2[g_N]$. Substitute Eq. (41.26) into Eq. (41.1) and observe that the normalization condition (41.28) already provides the denominator of this expression. All operations involve finite series, and the result is

$$\omega^2[g_N] = \frac{1}{\delta_N} \int_a^b \left[\tau(f')^2 + vf^2\right] dx$$

$$- \frac{2}{\delta_N} \sum_{n=1}^{N} a_n \int_a^b (\tau\rho_n' f' + v\rho_n f) \, dx$$

$$+ \frac{1}{\delta_N} \sum_{n=1}^{N} \sum_{q=1}^{N} a_n a_q \int_a^b (\tau\rho_n'\rho_q' + v\rho_n\rho_q) \, dx \qquad (41.30)$$

The first term is proportional to the functional $\omega^2[f]$ and can be rewritten

$$\frac{1}{\delta_N} \int_a^b \left[\tau(f')^2 + vf^2\right] dx = \frac{1}{\delta_N} \omega^2[f] \int_a^b f^2 \, dm \qquad (41.31)$$

The integral in the third term in Eq. (41.30) can be partially integrated; the boundary conditions from Sec. 40 ensure that the endpoint contributions vanish†

$$\left[\rho_q \tau \rho_n'\right]_a^b = 0 \qquad (41.32)$$

Use of the eigenvalue equation (40.16) and the orthonormality (41.19) then reduces the integral to the form

$$\int_a^b \rho_q\left[-(\tau\rho_n')' + v\rho_n\right] dx = \omega_n^2 \int_a^b \rho_q\rho_n \, dm = \omega_n^2 \, \delta_{nq} \qquad (41.33)$$

In this way, the last term in Eq. (41.30) becomes

$$\frac{1}{\delta_N} \sum_{n=1}^{N} \sum_{q=1}^{N} a_n a_q \int_a^b (\tau\rho_n'\rho_q' + v\rho_n\rho_q) \, dx = \frac{1}{\delta_N} \sum_{n=1}^{N} \omega_n^2 a_n^2 \qquad (41.34)$$

The second term in Eq. (41.30) can also be partially integrated. Since the function $f(x)$ satisfies the same boundary conditions as the eigenfunction $\rho_n(x)$, the boundary terms again vanish

$$\left[f\tau\rho_n'\right]_a^b = 0 \qquad (41.35)$$

As a result, the integral in the second term in Eq. (41.30) simplifies to

$$\int_a^b f\left[-(\tau\rho_n')' + v\rho_n\right] dx = \omega_n^2 \int_a^b f\rho_n \, dm = \omega_n^2 a_n \qquad (41.36)$$

where the eigenvalue equation has again been used and the expansion coefficient a_n has been identified with Eq. (41.21). Thus the second term in Eq. (41.30) can be written as

$$- \frac{2}{\delta_N} \sum_{n=1}^{N} a_n \int_a^b (\tau f'\rho_n' + vf\rho_n) \, dx = - \frac{2}{\delta_N} \sum_{n=1}^{N} \omega_n^2 a_n^2 \qquad (41.37)$$

---

† The case of a general homogeneous boundary condition at one or both ends is treated as in Eq. (41.9), the extra terms in I_1 again eliminating the boundary contributions.

A combination of these results gives

$$\omega^2[g_N] = \frac{1}{\delta_N}\left\{\omega^2[f]\int_a^b f^2\, dm - \sum_{n=1}^{N}\omega_n^2 a_n^2\right\} \geq \omega_{N+1}^2 \geq 0 \tag{41.38}$$

where the last inequality follows from Eq. (40.18). These relations show that the quantity in braces in Eq. (41.38) must be positive. Its second term is manifestly positive definite and can therefore be discarded without affecting the remaining inequality in Eq. (41.38). Consequently, we obtain the simple result

$$\frac{1}{\delta_N}\omega^2[f]\int_a^b f^2\, dm \geq \omega_{N+1}^2 \tag{41.39}$$

It can be solved for the integrated mean-square deviation

$$\delta_N \leq \frac{\omega^2[f]\int_a^b f^2\, dm}{\omega_{N+1}^2} \tag{41.40}$$

which bounds the mean-square difference in (41.22) between the function $f(x)$ and the approximating N-term series, calculated with the coefficients a_n defined in (41.21). Now observe that the numerator in Eq. (41.40) is a *positive finite number, independent of N* [we use here the assumed differentiability (39.27) of f]. It has also been *assumed in Eq. (40.17) that the eigenvalues in the Sturm-Liouville problem increase without bound*

$$\omega_N^2 \to \infty \qquad \text{as } N \to \infty \tag{41.41}$$

Thus we conclude that

$$\lim_{N\to\infty}\delta_N = 0 \tag{41.42}$$

which is the desired statement (39.29) of completeness. These results can be summarized by saying that any Sturm-Liouville equation of the type (41.7) together with an appropriate choice of boundary conditions from the set described in Eqs. (40.11) to (40.15) generates a *complete orthonormal set of eigenfunctions*.

The uniform convergence of the series in Eq. (41.20), and hence the continuity of the sum, can be established from the explicit form of the eigenfunctions, like the Fourier series examined in Eq. (39.24). If Eq. (41.21) is substituted in Eq. (41.20), that expression can be formally manipulated to yield the following remarkable result:

$$f(x) = \int_a^b\left[\sum_{n=1}^{\infty}\rho_n(x)\rho_n(\xi)\sigma(\xi)\right]f(\xi)\, d\xi \tag{41.43}$$

For any function $f(\xi)$, it states that the integral with the function of x and ξ given in brackets in Eq. (41.43) *simply picks out the value of f at the point $\xi = x$, indepen-*

dent of $f(\xi)$ elsewhere in the integral. Such a function of x and ξ is known as a *Dirac delta function,* written†

$$\sigma(\xi) \sum_{n=1}^{\infty} \rho_n(x)\rho_n(\xi) = \delta(x - \xi) \tag{41.44}$$

This relation is merely an abbreviation for *completeness* as stated in Eq. (41.43), which itself is just the property summarized in Eqs. (41.20) and (41.21) and demonstrated in this section. Note that the factor in front of the sum on the left-hand side of Eq. (41.44) can be written $\sigma(\xi)$ or $\sigma(x)$ or $[\sigma(\xi)\sigma(x)]^{1/2}$ because the only contribution of Eq. (41.44) to Eq. (41.43) comes at the point $\xi = x$.

42 ESTIMATES OF LOWEST EIGENVALUES; THE RAYLEIGH-RITZ APPROXIMATION METHOD

We have shown that the functional‡

$$\omega^2[\rho] = \frac{\dfrac{1}{2} \displaystyle\int_a^b \left[\tau(x)\left(\dfrac{d\rho}{dx}\right)^2 + v(x)\rho^2 \right] dx}{\dfrac{1}{2} \displaystyle\int_a^b [\sigma(x)\rho^2] \, dx} \tag{42.1}$$

provides a variational basis for Sturm-Liouville theory. If this functional is stationary under arbitrary variations of ρ consistent with the appropriate boundary conditions, the Euler-Lagrange equation is precisely the Sturm-Liouville equation

$$-\frac{d}{dx}\left[\tau(x)\frac{d\rho}{dx}\right] + v(x)\rho = \omega^2\sigma(x)\rho \tag{42.2}$$

Conversely, if Eq. (42.2) is satisfied, the functional $\omega^2[\rho]$ is stationary.

---

† The delta function evidently vanishes for $x \neq \xi$, and becomes infinite at $x = \xi$ in such a way that its integral is 1. Thus it has the following properties:

(i) $\qquad\qquad\qquad\qquad \delta(x - \xi) = \delta(\xi - x) \qquad$ symmetric in its arguments

(ii) $\qquad\qquad\qquad\qquad \delta(x - \xi) = 0 \qquad\qquad$ if $x \neq \xi$

(iii) $\qquad\qquad \displaystyle\int_a^b \delta(x - \xi) \, d\xi = \qquad \begin{cases} 1 & a < x < b \\ 0 & x < a \text{ or } b < x \end{cases}$

(iv) $\qquad\qquad \displaystyle\int_a^b \delta(x - \xi)f(\xi) \, d\xi = \qquad \begin{cases} f(x) & a < x < b \\ 0 & x < a \text{ or } b < x \end{cases}$

Sometimes the delta function contains a more complicated argument of the form $\delta[f(x)]$. In that case, a simple change of variables leads to

(v) $\qquad\qquad\qquad \displaystyle\int_a^b \delta[f(x)] \, dx = \sum_i \frac{1}{|df/dx|_i}$

where the sum runs over all roots x_i of $f(x) = 0$ lying in the interval (a, b).

‡ If one or both boundary conditions are of the form (40.13), the functional must be modified according to Eq. (41.9).

We have also argued that *minimization* of the functional $\omega^2[\rho]$ is fully equivalent to finding the lowest eigenvalue and corresponding eigenfunction of the Sturm-Liouville problem. This observation served to prove *completeness*: if $\rho(x)$ is any continuous piecewise differentiable function satisfying the same boundary conditions as $\rho_n(x)$, it can be expanded according to

$$\rho(x) = \sum_{n=1}^{\infty} a_n \rho_n(x) \tag{42.3}$$

General Theory

It is useful to verify the *consistency* of these arguments as follows. The first term in the numerator of Eq. (42.1) can be partially integrated and the boundary terms discarded through the choice of the appropriate Sturm-Liouville boundary conditions. Now substitute the series expansion (42.3) into the functional $\omega^2[\rho]$ defined in Eq. (42.1). We assume that the operations required in Eq. (42.1) can be carried out term by term on the series (42.3). This procedure of course depends on the convergence properties of the series and can be checked in detail in any particular example. These steps lead to the expression

$$\omega^2[\rho] = \frac{\sum\limits_{n=1}^{\infty} \sum\limits_{q=1}^{\infty} a_n a_q \int_a^b \rho_q \left[-\frac{d}{dx}\left(\tau \frac{d\rho_n}{dx}\right) + v\rho_n \right] dx}{\sum\limits_{n=1}^{\infty} \sum\limits_{q=1}^{\infty} a_n a_q \int_a^b \rho_q \sigma \rho_n \, dx} \tag{42.4}$$

Since the ρ_n are *eigenfunctions* according to Eq. (40.16) and *orthonormal* according to Eq. (41.19), the double sums in (42.4) reduce to

$$\omega^2[\rho] = \frac{\sum\limits_{n=1}^{\infty} \omega_n^2 a_n^2}{\sum\limits_{n=1}^{\infty} a_n^2} \tag{42.5}$$

This important result expresses the quantity $\omega^2[\rho]$ as a *weighted mean* of the actual eigenvalues of the Sturm-Liouville problem ω_n^2, with the positive weight factors $a_n^2 / \sum_{n=1}^{\infty} a_n^2$.

Suppose that ω_1^2 is the lowest eigenvalue in the Sturm-Liouville problem. Then Eq. (42.5) immediately implies

$$\omega^2[\rho] = \frac{\sum\limits_{n=2}^{\infty} (\omega_n^2 - \omega_1^2) a_n^2}{\sum\limits_{n=1}^{\infty} a_n^2} + \omega_1^2 \geq \omega_1^2 \tag{42.6}$$

Furthermore, assume that the eigenvalues in the Sturm-Liouville problem are nondegenerate. Then the equality can hold only if all the coefficients except a_1 vanish, and the normalization requires that a_1 must equal 1:

$$a_1 = 1$$

$$a_n = 0 \qquad n = 2, 3, \ldots, \infty \tag{42.7}$$

This observation demonstrates the consistency of our argument that the functional in Eq. (42.1) possesses an absolute minimum, that the minimum is the lowest eigenvalue, and that ρ_1 is precisely the corresponding eigenfunction. Suppose, however, that $a_1 = 0$ in the expansion (42.3); this condition may be guaranteed by *orthogonalizing* the function ρ to the function ρ_1, as in Eq. (41.16). It immediately follows from Eq. (42.5) that $\omega^2[\rho]$ is then greater than or equal to the *next* lowest eigenvalue

$$\omega^2[\rho] \geq \omega_2^2 \tag{42.8}$$

This was the argument used below Eq. (41.14) to develop a systematic way of finding the eigenvalues and eigenfunctions of the Sturm-Liouville equation and was also the basis for our discussion of completeness.

Equation (42.5) serves to establish a result that is central to many calculations. Note first that the functional (42.1) is homogeneous in ρ and is therefore independent of the choice of normalization. We can simplify the present discussion by choosing *a different normalization* and requiring [compare Eq. (40.28)]

$$\langle \rho_1 | \rho \rangle = 1 \tag{42.9}$$

instead of the previous convention (41.12). In this way, Eq. (42.3) takes the form

$$\rho(x) = \rho_1(x) + \sum_{n=2}^{\infty} \epsilon_n \rho_n(x) \tag{42.10}$$

with

$$a_1 = 1$$

$$a_n = \epsilon_n \qquad n = 2, 3, \ldots, \infty \tag{42.11}$$

We next assume that the quantities ϵ_n are small, so that the function ρ differs only slightly from the exact eigenfunction ρ_1, by terms that are *first order in the small quantities* ϵ_n. Substitution of (42.11) into Eq. (42.5) yields

$$\omega^2[\rho] = \frac{\omega_1^2 + \sum_{n=2}^{\infty} \epsilon_n^2 \omega_n^2}{1 + \sum_{n=2}^{\infty} \epsilon_n^2} \approx \omega_1^2 + \sum_{n=2}^{\infty} (\omega_n^2 - \omega_1^2)\epsilon_n^2 \tag{42.12}$$

This relation states that *the functional $\omega^2[\rho]$ differs from the exact lowest eigenvalue ω_1^2 only by terms that are of second order in the small quantities ϵ_n.* Consequently, a crude approximation to the lowest eigenfunction can give an accurate value for the lowest eigenvalue. Moreover, the corrections in Eq. (42.12) are all positive so that *any approximation for the lowest eigenfunction gives a value for the* *functional $\omega^2[\rho]$ that lies above the true lowest eigenvalue. Thus any alteration in ρ that lowers the value of the functional $\omega^2[\rho]$ in Eq. (42.1) improves the approxima-tion both for the lowest eigenvalue and for the corresponding lowest eigenfunction.* By orthogonalizing with respect to the (true) eigenfunction corresponding to the lowest eigenvalue as in Eq. (41.16) one can repeat the procedure to obtain similar estimates for the next eigenvalue and eigenfunction according to Eq. (42.8), and so on.

This method for estimating the lowest eigenvalue in any Sturm-Liouville problem, no matter how complicated, is one of the most powerful techniques in theoretical physics. Physical intuition can serve to generate a trial function that resembles the true eigenfunction. Equation (42.12) shows that such a trial function can yield accurate values for the lowest eigenvalue quite readily, even for problems where an exact solution is unknown. This general procedure is often known as the *Rayleigh-Ritz approximation method*. Note that the trial function need not, in fact, be normalized according to Eq. (42.9) since the functional $\omega^2[\rho]$ is independent of the choice of normalization of ρ.

Example: Mass Point on a String

We give a simple example of this approximation method. Consider the even modes of oscillation for a mass point m at the middle of a stretched uniform string, solved in Prob. 4.17, summarized in Eqs. (40.29) to (40.32), and illustrated in Fig. 40.3. The exact eigenvalue equation (40.30) for the even modes can be rewrit-ten in the form

$$\xi \tan \xi = \frac{M}{m} \tag{42.13}$$

where the dimensionless normal-mode frequency ξ has been defined as

$$\xi \equiv \frac{\omega l}{2c} \tag{42.14}$$

and M is the mass of the string, defined in Eq. (40.31). In this problem, the quantities appearing in the Sturm-Liouville equation (42.2) and in the functional (42.1) take the form [see Eq. (40.32)]

$$\tau(x) = \tau \qquad \text{tension} \tag{42.15a}$$

$$\sigma(x) = \sigma\left[1 + \frac{m}{\sigma}\delta(x - \tfrac{1}{2}l)\right] \qquad \text{mass density} \tag{42.15b}$$

with no imposed restoring force ($v = 0$). The functional in Eq. (42.1) thus becomes

$$\frac{\omega^2[\rho]}{c^2} = \frac{\int_0^l \left(\frac{d\rho}{dx}\right)^2 dx}{\int_0^l \rho^2 \left[1 + \frac{m}{\sigma}\delta(x - \tfrac{1}{2}l)\right] dx} = \frac{\int_0^l \left(\frac{d\rho}{dx}\right)^2 dx}{\int_0^l \rho^2 \, dx + \frac{m}{\sigma}[\rho(\tfrac{1}{2}l)]^2} \tag{42.16}$$

Note that the only complication arises from the mass density, which contains a term $m\delta(x - \tfrac{1}{2}l)$ from the point mass at the center of the string. This extra term merely evaluates the function $\rho(x)$ at the center of the string, as is evident from the last term in the denominator in Eq. (42.16).

To investigate the accuracy of the variational approach, we have chosen a problem whose exact eigenvalue follows from numerical solution of Eq. (42.13). Suppose, however, that the true lowest eigenvalue had not been found. We know that the even modes of the string must have the shape indicated in Fig. 40.3b, and we attempt to represent this shape analytically with the following function:

$$\rho(x) = \begin{cases} Ax^\alpha & 0 \le x \le \tfrac{1}{2}l \\ A(l - x)^\alpha & \tfrac{1}{2}l \le x \le l \end{cases} \tag{42.17}$$

which is even about the midpoint $\tfrac{1}{2}l$. In this expression, A is an irrelevant normalization constant and α is a parameter that can be used to describe various forms of the function in Fig. 40.3b. Substitute Eq. (42.17) in Eq. (42.16). Since the function ρ is even about the point $\tfrac{1}{2}l$, each of the integrals is just twice the integral from zero to $\tfrac{1}{2}l$, and we find

$$\frac{\omega^2[\rho]}{c^2} = \frac{2\displaystyle\int_0^{l/2} \alpha^2 x^{2\alpha-2}\, dx}{2\displaystyle\int_0^{l/2} x^{2\alpha}\, dx + \frac{m}{\sigma}(\tfrac{1}{2}l)^{2\alpha}} \tag{42.18}$$

The integral in the numerator converges only if

$$\alpha > \tfrac{1}{2} \tag{42.19a}$$

which will always be assumed to hold. We also expect that

$$\alpha \le 1 \tag{42.19b}$$

since otherwise ρ''/ρ would be positive, in contradiction to Eqs. (42.2) and (42.15). The functional (42.18) can now be simplified to give

$$\frac{\omega^2[\rho]}{c^2}\frac{l^2}{4} = \frac{\alpha^2(2\alpha + 1)}{(2\alpha - 1)[1 + (m/M)(2\alpha + 1)]} \tag{42.20a}$$

or, equivalently,

$$(\xi^2)_{\text{var}} = f\!\left(\alpha, \frac{m}{M}\right) \equiv \frac{\alpha^2(2\alpha + 1)}{(2\alpha - 1)[1 + (m/M)(2\alpha + 1)]} \tag{42.20b}$$

in terms of the dimensionless variable defined in Eq. (42.14).

Figure 42.1 sketches this expression as a function of α for a fixed ratio of the point mass to the mass of the string m/M. Evidently, $f \to \infty$ as $\alpha \to \tfrac{1}{2}$ from the right. Furthermore, f also increases as $\alpha \to \infty$, implying that f has some minimum value $(\xi^2)_{\text{min}}$ in the intermediate region. Finally, our general discussion guarantees that the value of ξ^2 calculated from our functional in Eq. (42.16) *always lies above* the true minimum eigenvalue ξ_1^2, as illustrated in Fig. 42.1. *Thus the best estimate for*

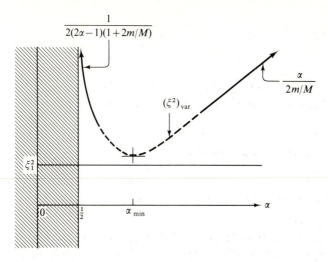

Figure 42.1 Sketch of variational functional $(\xi^2)_{\text{var}} = f(\alpha, m/M)$ in Eq. (42.20b) as a function of α for a fixed ratio m/M. The exact lowest eigenvalue ξ_1^2 is also indicated schematically. Note that both $\alpha_{\min}$ and ξ_1^2 depend on the value of m/M.

the lowest eigenvalue ξ_1^2 is obtained by minimizing f with respect to α, which is the only parameter remaining in our functional. If Eq. (42.20) is to be a minimum, its derivative with respect to α must vanish

$$\frac{df}{d\alpha} = 0 \qquad (42.21a)$$

Straightforward algebra yields

$$4\alpha^2 - 2\alpha - 1 + \frac{m}{M}(2\alpha + 1)^2(\alpha - 1) = 0 \qquad (42.21b)$$

The value of α that minimizes $f(\alpha, m/M)$ must satisfy this cubic equation.

Consider three special cases representing different values of m/M.

Case 1 Suppose that the mass at the center of the string is very light, so that

$$\frac{m}{M} \to 0 \qquad (42.22)$$

The additional mass will then have only a small effect on the motion of the string, which will move essentially as if the small mass were absent (see Fig. 42.2). In the limit (42.22), Eq. (42.13) shows that the exact eigenvalue is

$$\xi_1 = \tfrac{1}{2}\pi \qquad (42.23a)$$

or, using the definition (42.14),

$$\omega_1 = \frac{\pi c}{l} \qquad (42.23b)$$

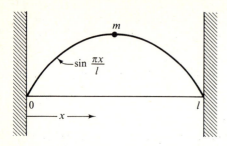

Figure 42.2 Lowest even mode in limit $m/M \to 0$.

As expected, this is the lowest frequency of the uniform string. The corresponding exact eigenfunction is given in Eq. (38.16)

$$\rho_1 = A \sin \frac{\pi x}{l} \qquad (42.24)$$

In effect, our approximation scheme seeks to reproduce this exact answer with the form $A x^\alpha$, given in Eq. (42.17). In the limit (42.22), the minimization condition (42.21) reduces to

$$4\alpha^2 - 2\alpha - 1 = 0 \qquad (42.25)$$

Of the two solutions to this equation

$$\alpha = \tfrac{1}{8}(2 \pm \sqrt{20}) = \tfrac{1}{4}(1 \pm \sqrt{5}) \qquad (42.26)$$

only the plus sign satisfies the condition (42.19a). This value indeed minimizes the function f (see Fig. 41.1), giving the value

$$f(\alpha_{\min}, 0) = \left[\frac{\alpha^2(2\alpha + 1)}{2\alpha - 1} \right]_{\min} = \frac{1}{8} \frac{(3 + \sqrt{5})^2}{\sqrt{5} - 1} \qquad (42.27)$$

Column 1 of Table 42.1 compares the numerical value obtained from this expression with the exact answer. Observe that the approximate value $\xi_{\text{var}} \approx 1.6651$ *lies above* the true value $\xi_1 = \tfrac{1}{2}\pi \approx 1.5708$, as it must; on the other hand, the simple

Table 42.1 Comparison of variational calculation with exact results for the lowest even mode of system illustrated in Fig. 40.3b

The dimensionless parameter ξ is defined in Eq. (42.14); note that $\alpha_{\min}$ always lies between $\tfrac{1}{2}$ and 1, in accordance with Eq. (42.19).

	(1)	(2)	(3)
m/M	$\to 0$	1	$\to \infty$
$\alpha_{\min}$	0.8090	0.9279	1
ξ_{var}	1.6651	0.8632	$(M/m)^{1/2}$
ξ_1	$\tfrac{1}{2}\pi = 1.5708$	0.8603	$(M/m)^{1/2}$

one-parameter variational trial function (42.17) yields a result within 6 percent of the exact answer.

Case 2 Suppose that the mass of the string is negligible with respect to the mass of the point particle

$$\frac{m}{M} \to \infty \qquad (42.28)$$

In this case, the point mass governs the dynamics of the coupled system, and the lowest-frequency mode of the system will be as illustrated in Fig. 42.3. In the limit (42.28) the tangent in Eq. (42.13) can be replaced by its argument, and the exact lowest eigenfrequency becomes

$$\xi_1 = \left(\frac{M}{m}\right)^{1/2} \qquad (42.29)$$

This answer can readily be understood by referring to Fig. 42.3. The string effectively has no intrinsic inertia, responding instantaneously by deforming into a straight line. Newton's second law for the displacement u of the mass m in the lowest mode can be written as

$$m\ddot{u} = -m\omega^2 u = -2\tau \sin\theta \approx -2\tau \frac{u}{\frac{1}{2}l} \qquad (42.30)$$

where τ is the tension. The first equality assumes harmonic motion, and the last assumes small displacements. The eigenvalue equation (42.30) thus yields the lowest normal-mode frequency

$$\frac{\omega^2 l^2}{4c^2} = \frac{M}{m} \qquad (42.31)$$

which is precisely the exact result (42.29) [see Eq. (42.14)].

In the limit (42.28), the minimization condition (42.21) becomes

$$(2\alpha + 1)^2(\alpha - 1) = 0 \qquad (42.32)$$

The only solution of this equation satisfying the condition (42.19a) is evidently

$$\alpha_{\min} = 1 \qquad (42.33)$$

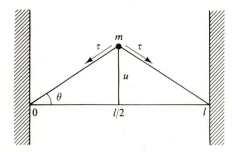

Figure 42.3 Lowest even mode in limit $m/M \to \infty$.

Correspondingly, the functional in Eq. (42.20) has the value

$$f(\alpha_{min}) = \frac{M}{m}\left(\frac{\alpha^2}{2\alpha - 1}\right)_{min} = \frac{M}{m} \tag{42.34}$$

so that the *variational method here gives the exact answer*, as shown in column 3 of Table 42.1. Figure 42.3 explains this result, for the exact lowest eigenfunction is even and linear in the limit (42.28), with $\rho_1 = Ax$ for $0 \le x \le \frac{1}{2}l$. The variational trial function (42.17) has sufficient flexibility to reproduce the exact solution at $\alpha = 1$ [see Eqs. (42.17) and (42.33)].

Case 3 If

$$\frac{m}{M} = 1 \tag{42.36}$$

Eq. (42.21b) reduces to the cubic equation

$$\alpha^3 + \alpha^2 - \tfrac{5}{4}\alpha - \tfrac{1}{2} = 0 \tag{42.37}$$

which must be solved numerically. There is only one root satisfying the condition (42.19a), and it is given in column 2 of Table 42.1. The corresponding value of ξ_{var} obtained from this value through Eq. (42.20) is also shown in column 2, as is the exact lowest eigenvalue obtained numerically from Eq. (42.13). Again the variational eigenvalue lies *above* the true lowest eigenvalue; in this case the accuracy is quite remarkable, for the variational solution is within 0.3 percent of the exact value.

Figure 42.4 exhibits both the exact lowest eigenfrequency ξ_1 and the quantity obtained from the variational method with the trial function (42.17) as a function of the quantity m/M. For large values of m/M, the variational calculation is exact. This figure shows that the largest discrepancy between the two results occurs in the limit $m/M \to 0$ [see Eq. (42.22)], when the variational calculation is still within 6 percent of the true eigenfrequency.

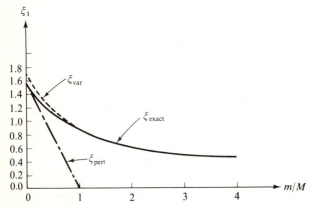

Figure 42.4 Comparison of exact, variational, and perturbative solutions for the lowest eigenfrequency of a point mass m added to the center of a uniform string of mass M. Here, $\xi_{pert} = \frac{1}{2}\pi(1 - m/M)$, from Eq. (44.36).

43 GREEN'S FUNCTION IN ONE DIMENSION

This chapter has concentrated on the eigenfunctions and eigenfrequencies of a general one-dimensional string, emphasizing their properties and their role in solving the initial-value problem. We now turn to a slightly different aspect of the same physical system, in which the string is subjected to a periodic force Re $\sigma(x)f(x)e^{-i\omega t}$ per unit length that may vary along the string. Equivalently, Re $f(x)e^{-i\omega t}$ is the force per unit mass. The dynamical equation of the string must now be modified to include the external force. A simple analysis shows that we merely add an extra term to the right-hand side of the general one-dimensional string equation (40.7), which becomes

$$\sigma(x)\frac{\partial^2 u}{\partial t^2} = \frac{\partial}{\partial x}\left[\tau(x)\frac{\partial u}{\partial x}\right] - v(x)u + \text{Re }\sigma(x)f(x)e^{-i\omega t} \tag{43.1}$$

In addition, $u(x, t)$ must satisfy appropriate boundary conditions of the sort discussed in Eqs. (40.11) to (40.15). The linearity of Eq. (43.1) permits us to write $u(x, t)$ in the form

$$u(x, t) = \text{Re } u(x)e^{-i\omega t} \tag{43.2}$$

and the function $u(x)$ then satisfies an *inhomogeneous Sturm-Liouville* equation

$$-\frac{d}{dx}\left[\tau(x)\frac{du(x)}{dx}\right] + v(x)u(x) - \omega^2\sigma(x)u(x) = \sigma(x)f(x) \tag{43.3}$$

Note that ω here is an arbitrary frequency, different from any of the eigenfrequencies ω_n given in Eq. (40.16). It will be convenient to abbreviate the first two terms in Eq. (43.3) as $L_0 u$; in this case,

$$L_0 = -\frac{d}{dx}\tau(x)\frac{d}{dx} + v(x) \tag{43.4a}$$

is a *differential operator*, and Eq. (43.3) becomes

$$L_0 u - \omega^2\sigma u = \sigma f \tag{43.4b}$$

This equation can be solved in at least two ways. We shall first use an eigenfunction expansion that applies in many different situations. We then consider a special one-dimensional technique that uses the solutions of the homogeneous equation (43.4b) for $f = 0$.

Eigenfunction Expansion

The resemblance of Eq. (43.4b) to the Sturm-Liouville equation [see Eq. (40.16)]

$$L_0\rho_n = \omega_n^2\sigma\rho_n \tag{43.5}$$

with the same boundary conditions as in Eqs. (43.1) and (43.3) suggests that we seek the solution $u(x)$ as an expansion in the complete set of eigenfunctions

$$u(x) = \sum_{n=1}^{\infty} c_n \rho_n(x) \tag{43.6}$$

Here $\{c_n\}$ denotes a set of coefficients that can be determined by substituting (43.6) into the differential equation (43.4b), and the set $\{\rho_n\}$ is again assumed real. Use of Eq. (43.5) immediately yields

$$\sum_{n=1}^{\infty} c_n(\omega_n^2 - \omega^2)\sigma\rho_n = \sigma f$$

Multiply this equation by $\rho_m(x)$ and integrate over the allowed interval $a \le x \le b$. The orthonormality of the eigenfunctions (40.28) leads to the explicit solution [see Eq. (40.27)]

$$c_m = \frac{1}{\omega_m^2 - \omega^2} \int_a^b \rho_m(y')\sigma(y')f(y')\,dy' = \frac{\langle \rho_m | f \rangle}{\omega_m^2 - \omega^2} \tag{43.7}$$

and Eq. (43.6) then becomes

$$u(x) = \sum_{n=1}^{\infty} \int_a^b \frac{\rho_n(x)\rho_n(y')\sigma(y')f(y')}{\omega_n^2 - \omega^2}\,dy'$$

$$= \sum_{n=1}^{\infty} \frac{\rho_n(x)\langle \rho_n | f \rangle}{\omega_n^2 - \omega^2} \tag{43.8}$$

Note that u generally diverges as the driving frequency ω approaches any of the normal-mode frequencies, which is just the resonant behavior familiar in simple mechanical systems. The only exception occurs if the driving force $f(x)$ happens to be orthogonal to a particular eigenfunction ρ_m, say, in which case c_m vanishes identically and u remains finite as $\omega \to \omega_m$.

For many purposes, it is useful to rewrite Eq. (43.8) in the form

$$u(x) = \int_a^b G_\omega(x, y')\sigma(y')f(y')\,dy' \tag{43.9}$$

where $G_\omega(x, y)$ is called the *Green's function* for the operator $L_0 - \omega^2\sigma$ and the particular boundary conditions in question; comparison with Eq. (43.8) yields the explicit expression

$$G_\omega(x, y) = \sum_{n=1}^{\infty} \frac{\rho_n(x)\rho_n(y)}{\omega_n^2 - \omega^2} \tag{43.10}$$

It has several important properties.

1. $G_\omega(x, y)$ is symmetric under interchange of the spatial variables x and y:

$$G_\omega(x, y) = G_\omega(y, x) \tag{43.11}$$

It also depends parametrically on the applied frequency ω. In addition, G automatically satisfies the appropriate boundary conditions at $x = a$ and $x = b$ because each eigenfunction $\rho_n(x)$ does so.

2. Suppose that the applied force is localized at y ($a < y < b$) with the form $\sigma(x)f(x) = \delta(x - y)$ [compare Eq. (41.44) and the associated discussion of Dirac delta functions]. In this case, Eq. (43.9) shows that $u(x) = G_\omega(x, y)$, and the corresponding differential equation (43.4b) becomes

$$(L_0 - \omega^2\sigma)G_\omega(x, y) = \delta(x - y) \tag{43.12}$$

Thus $G_\omega(x, y)$ satisfies the same differential equation as $u(x)$ but with a *particular inhomogeneous term* on the right-hand side, i.e., a delta function localized at $x = y$. In this way, the solution of the complete equation (43.3) separates into two distinct steps. First, solve Eq. (43.12) for $G_\omega(x, y)$, which represents the response to a unit force with frequency ω localized at y. Second, the linearity of the original differential equation allows us to construct the full solution $u(x)$ by superposing the response for forces at different points y weighted with the strength $\sigma(y)f(y)$. This step provides the physical interpretation of the definite integral (43.9).

3. Substitute (43.10) into (43.12) and use Eq. (43.5). The resulting relation is just the formal expression of completeness (41.44).

4. Equation (43.10) exhibits the analytic structure of $G_\omega(x, y)$ as a function of the complex variable ω^2. Evidently, it has simple poles at each of the eigenvalues ω_n^2, with a residue $-\rho_n(x)\rho_n(y)$ that is minus the product of the normalized eigenfunctions.

Construction from Solutions to Homogeneous Equations

The preceding method for constructing the Green's function as an eigenfunction expansion is very general, but it has the drawback of expressing G as an infinite series. In the special case of a one-dimensional equation, however, we can use an alternative procedure, based directly on the form of Eq. (43.12). If $x \neq y$, the right-hand side vanishes, and $G_\omega(x, y)$ therefore satisfies a *homogeneous equation* in the two intervals $a \leq x < y$ and $y < x \leq b$, which must be treated separately. In the first interval, we denote the solution $G_\omega^<(x, y)$; it satisfies the homogeneous equation

$$(L_0 - \omega^2\sigma)G_\omega^<(x, y) = 0 \qquad a \leq x < y \tag{43.13}$$

subject to an appropriate boundary condition at $x = a$. Apart from a multiplicative constant, Eq. (43.13) has a unique solution $u_1(x)$ that satisfies the boundary condition, so that $G^<$ must have the form

$$G_\omega^<(x, y) = Au_1(x) \tag{43.14}$$

with A depending parametrically on y. Similarly, if $G_\omega^>(x, y)$ denotes the solution for $y < x \leq b$, it also satisfies a homogeneous equation

$$(L_0 - \omega^2\sigma)G_\omega^>(x, y) = 0 \qquad y < x \leq b \tag{43.15}$$

subject to an appropriate boundary condition at $x = b$. Let $u_2(x)$ denote the unique solution of the homogeneous equation that satisfies this boundary condition; $G^>$ then takes the form

$$G_\omega^>(x, y) = Bu_2(x) \qquad (43.16)$$

where B also depends parametrically on y. Note that u_1 and u_2 also depend on the frequency ω through the differential equations (43.13) and (43.15).

It is now necessary to join the two parts of G at the point y. To achieve this, we write out Eq. (43.12) in more detail

$$-\frac{d}{dx}\left[\tau(x)\frac{dG_\omega(x, y)}{dx}\right] + v(x)G_\omega(x, y) - \omega^2\sigma(x)G_\omega(x, y) = \delta(x - y) \qquad (43.17)$$

Integrate this equation with respect to x over an infinitesimal region of length 2ϵ about the point y from $x = y - \epsilon$ to $y = y + \epsilon$

$$-\int_{y-\epsilon}^{y+\epsilon} dx \frac{d}{dx}\left[\tau(x)\frac{dG_\omega(x, y)}{dx}\right] + \int_{y-\epsilon}^{y+\epsilon} dx\, [v(x) - \omega^2\sigma(x)]G_\omega(x, y)$$

$$= \int_{y-\epsilon}^{y+\epsilon} dx\, \delta(x - y) \qquad (43.18)$$

Since the second derivative of $G_\omega(x, y)$ has a delta-function contribution at $x = y$, its first derivative will have a finite discontinuity there. As discussed in point 2 under Eq. (43.9), $G_\omega(x, y) = u(x)$ is the value of the physical disturbance, e.g., displacement of a string, for an applied force $\delta(x - y)$ at y. Thus G_ω will itself be continuous at $x = y$. The second term on the left-hand side of Eq. (43.18) is therefore of order ϵ and vanishes as $\epsilon \to 0$. The first integral can be evaluated immediately to give

$$-\tau(y)\left[\frac{dG_\omega(x, y)}{dx}\right]_{y-\epsilon}^{y+\epsilon} = -\tau(y)\left[\frac{dG_\omega^>(x, y)}{dx} - \frac{dG_\omega^<(x, y)}{dx}\right]_{x=y} \qquad (43.19)$$

where $G^>$ and $G^<$ have now been introduced because $y + \epsilon$ and $y - \epsilon$ definitely lie above and below y. By definition, the right-hand side of (43.18) is unity, and we therefore obtain the *jump condition*

$$\left[\frac{dG_\omega^>(x, y)}{dx} - \frac{dG_\omega^<(x, y)}{dx}\right]_{x=y} = -\frac{1}{\tau(y)} \qquad (43.20)$$

In addition, as noted above, G_ω is continuous at $x = y$

$$[G_\omega^>(x, y) - G_\omega^<(x, y)]_{x=y} = 0 \qquad (43.21)$$

If Eqs. (43.14) and (43.16) are substituted into Eqs. (43.20) and (43.21), we obtain two linear equations for the coefficients A and B. A straightforward solution yields

$$A = \frac{-u_2(y)}{\tau(y)W[u_1(y), u_2(y)]} \qquad B = \frac{-u_1(y)}{\tau(y)W[u_1(y), u_2(y)]} \qquad (43.22)$$

where W denotes the *wronskian* of the two solutions

$$W[u_1(y), u_2(y)] \equiv u_1(y)u_2'(y) - u_2(y)u_1'(y) \qquad (43.23)$$

To complete the solution, we shall prove that the wronskian has the following property:

$$\tau(y)W[u_1(y), u_2(y)] = \text{const independent of } y \qquad (43.24)$$

PROOF Both u_1 and u_2 satisfy the same homogeneous equation

$$(L_0 - \omega^2\sigma)u_1 = 0 \qquad (L_0 - \omega^2\sigma)u_2 = 0 \qquad (43.25)$$

Multiply the first equation by u_2 and the second by u_1 and take their difference; the term $\omega^2\sigma u_1 u_2$ cancels identically, as does the part of L_0 containing v. The remaining contribution becomes

$$-u_2(x)\frac{d}{dx}\left[\tau(x)\frac{du_1(x)}{dx}\right] + u_1(x)\frac{d}{dx}\left[\tau(x)\frac{du_2(x)}{dx}\right] = 0 \qquad (43.26a)$$

which can be rewritten as a total derivative

$$\frac{d}{dx}\left\{\tau(x)\left[u_1(x)\frac{du_2(x)}{dx} - u_2(x)\frac{du_1(x)}{dx}\right]\right\} = 0 \qquad (43.26b)$$

The quantity in brackets is just the wronskian, and we therefore infer that

$$\tau(x)W[u_1(x), u_2(x)] = C \qquad (43.27)$$

where C is a constant independent of the variable in question, as asserted in (43.24).

Substitution of Eq. (43.27) into Eqs. (43.22) and then into (43.14) and (43.16) yields the final simple expression

$$G_\omega^<(x, y) = \frac{-u_1(x)u_2(y)}{C} \qquad G_\omega^>(x, y) = \frac{-u_1(y)u_2(x)}{C} \qquad (43.28)$$

It is often written in the more compact form

$$G_\omega(x, y) = \frac{-u_1(x_<)u_2(x_>)}{C} \qquad (43.29)$$

where $x_<$ ($x_>$) denotes the smaller (larger) of the two variables x and y. Note that this solution for $G_\omega(x, y)$ automatically satisfies the correct boundary conditions at $x = a$ and $x = b$; it also satisfies the symmetry relation (43.11) because $G_\omega^<(x, y) = G_\omega^>(y, x)$.

Example: Uniform String with Fixed Endpoints

To illustrate these various solutions, we consider a uniform string with constant mass density σ and tension τ, clamped at the ends $x = 0$ and $x = l$. In this case, the differential operator is just

$$L_0 = -\tau\frac{d^2}{dx^2} \qquad (43.30)$$

and the Green's function $G_\omega(x, y)$ obeys the inhomogeneous differential equation

$$-\left(\tau \frac{d^2}{dx^2} + \omega^2 \sigma\right) G_\omega(x, y) = \delta(x - y) \tag{43.31}$$

subject to the boundary conditions

$$G_\omega(0, y) = G_\omega(l, y) = 0 \tag{43.32}$$

To construct the eigenfunction expansion, we notice that the solutions of Eq. (43.5) with $\rho_n(0) = \rho_n(l) = 0$ are just the eigenfunctions obtained in Eq. (38.16)

$$\rho_n(x) = \left(\frac{2}{\sigma l}\right)^{1/2} \sin \frac{n\pi x}{l} \qquad n = 1, 2, \ldots, \infty \tag{43.33a}$$

$$\omega_n = \frac{c n\pi}{l} \qquad n = 1, 2, \ldots, \infty \tag{43.33b}$$

where $c = (\tau/\sigma)^{1/2}$. We now normalize according to our standard convention (40.28); the Green's function follows immediately from Eq. (43.10)

$$G_\omega(x, y) = \frac{2}{\sigma l} \sum_{n=1}^{\infty} \frac{\sin (n\pi x/l) \sin (n\pi y/l)}{(cn\pi/l)^2 - \omega^2} \tag{43.34}$$

Substitution into the differential equation (43.31) leads to the relation

$$\frac{2}{l} \sum_{n=1}^{\infty} \sin \frac{n\pi x}{l} \sin \frac{n\pi y}{l} = \delta(x - y) \tag{43.35}$$

which is the appropriate expression of completeness (41.44).

We can also construct $G_\omega(x, y)$ from the solutions $u_1(x)$ and $u_2(x)$ of the homogeneous equations

$$L_0 u_1 = \omega^2 \sigma u_1 \qquad \text{with} \quad u_1(0) = 0 \tag{43.36a}$$

$$L_0 u_2 = \omega^2 \sigma u_2 \qquad \text{with} \quad u_2(l) = 0 \tag{43.36b}$$

Equation (43.30) shows that these solutions are just trigonometric functions, and we can evidently take the specific forms

$$u_1(x) = \sin \frac{\omega x}{c} \tag{43.37a}$$

$$u_2(x) = \sin \left[\frac{\omega}{c}(l - x)\right] \tag{43.37b}$$

which satisfy the correct boundary conditions. The corresponding wronskian becomes

$$W[u_1(x), u_2(x)] = -\frac{\omega}{c}\left\{\sin \frac{\omega x}{c} \cos \left[\frac{\omega}{c}(l - x)\right] + \cos \frac{\omega x}{c} \sin \left[\frac{\omega}{c}(l - x)\right]\right\}$$

$$= -\frac{\omega}{c} \sin \frac{\omega l}{c} \tag{43.38}$$

where the last line follows from the standard trigonometric identities. Since $\tau(x)$ is here constant, Eq. (43.38) explicitly verifies Eq. (43.27), with

$$C = -\frac{\tau\omega}{c}\sin\frac{\omega l}{c} \tag{43.39}$$

In this way, we obtain the simple closed-form solution

$$G_\omega(x, y) = \frac{\sin(\omega x_</c)\sin[(\omega/c)(l - x_>)]}{(\tau\omega/c)\sin(\omega l/c)} \tag{43.40}$$

where $x_<$ and $x_>$ are the smaller and larger of the two variables x and y.

Equations (43.34) and (43.40) provide two very different representations for G_ω, and it is instructive to reconcile them explicitly. We first note that the numerator and denominator of (43.40) are each entire functions of ω, so that the only singularities in the ω plane can arise from zeros of the denominator. In addition, $\omega = 0$ is not a singularity of G_ω because the numerator vanishes there too. Thus the only singularities of G_ω can occur at the remaining zeros of $\sin(\omega l/c)$, which *occur precisely at the values* $\pm\omega_n$ *given in* (43.33b). It is a simple exercise in complex analysis to verify that $G_\omega(x, y)$ in Eq. (43.40) in fact has *simple poles* at $\omega = \pm\omega_n = \pm cn\pi/l$, with just the residue given in Eq. (43.34). Thus the two representations are wholly equivalent, for they agree at an infinite sequence of points that has a limit point at infinity.

The demonstration of this equivalence depends on the observation that the two distinct solutions $u_1(x)$ and $u_2(x)$ become proportional to each other at the special values $\omega = \pm\omega_n$. This property is quite general, for the only singularities of G_ω in Eq. (43.28) can arise from the zeros of C, or equivalently, from the zeros of the wronskian (43.23). But this is exactly the condition for the two solutions to be linearly dependent and hence proportional to each other. This observation suggests yet another means of determining the normal-mode frequencies of a one-dimensional mechanical system. Apply a driving force f at arbitrary frequency and construct the appropriate solutions, namely, $u_1(x)$ that satisfies the boundary condition at $x = a$ and $u_2(x)$ that satisfies the boundary condition at $x = b$. For general ω, u_1 and u_2 are linearly independent, and their wronskian is nonzero, allowing the explicit construction of the Green's function with Eq. (43.29). For certain frequencies $\pm\omega_n$, however, their wronskian vanishes, implying that they are linearly dependent, with $u_1(x)$ and $u_2(x)$ proportional to a single function $\rho_n(x)$ that *satisfies both boundary conditions simultaneously*. Evidently $\rho_n(x)$ is just the desired eigenfunction corresponding to the eigenfrequency $\pm\omega_n$. The appropriate normalization constant for $\rho_n(x)$ can be determined either directly from Eq. (40.28) or by examining the residue of Eq. (43.40) at $\omega = \pm\omega_n$.

44 PERTURBATION THEORY

In many physical situations, the full dynamical equations are too complicated for a complete solution. It oftens happens, however, that the equations contain two contributions, one of which is exactly solvable and the other of which incorporates the added physical complexity of the true physical system. To illustrate this behavior, we consider two Sturm-Liouville problems subject to the same domain $a \leq x \leq b$ and the same boundary conditions

$$-\frac{d}{dx}\left[\tau(x)\frac{d\rho_n}{dx}\right] + v(x)\rho_n = \omega_n^2\sigma(x)\rho_n \qquad (44.1)$$

$$-\frac{d}{dx}\left[\tau(x)\frac{d\rho}{dx}\right] + v(x)\rho + \epsilon v(x)\sigma(x)\rho = \omega^2\sigma(x)\rho \qquad (44.2)$$

Assume that the first problem can be solved for the exact eigenfunctions and eigenvalues. The second problem differs only in the presence of an extra potential $\epsilon v(x)\sigma(x)$, where the form is chosen to exhibit the integration measure $\sigma(x)$ and a dimensionless coupling strength ϵ. We aim to express the solution to Eq. (44.2) in terms of the (assumed known) solutions to Eq. (44.1).

General Theory

In the preceding section, we introduced the abbreviation L_0 for the Sturm-Liouville operator (43.4a), in which case, Eq. (44.1) takes the form [see Eq. (43.5)]

$$L_0\rho_n = \omega_n^2\sigma(x)\rho_n \qquad (44.3)$$

Correspondingly, Eq. (44.2) can be written

$$(L_0 + L_1)\rho = \omega^2\sigma(x)\rho \qquad (44.4)$$

where the additional interaction term is defined

$$L_1 \equiv \epsilon v(x)\sigma(x) \qquad (44.5)$$

It is helpful to convert Eq. (44.4) into an integral equation in the following manner. First rewrite it as

$$[L_0 - \omega^2\sigma(x)]\rho(x) = -L_1\rho(x) \qquad (44.6)$$

This equation now has precisely the inhomogeneous structure of Eq. (43.4b), and it can be solved with the same Green's-function technique [compare Eq. (43.9)]

$$\rho(x) = -\int_a^b G_\omega(x, y)L_1\rho(y)\,dy \qquad (44.7)$$

It is evident that this expression satisfies the appropriate boundary conditions in x since these have been built into $G_\omega(x, y)$ according to Eq. (43.10). Furthermore, Eq. (44.7) satisfies Eq. (44.2), which can be proved by applying the operator $L_0 - \omega^2\sigma$ to (44.7) and using the basic property of the Green's function (43.12) on the right-hand side

$$[L_0 - \omega^2\sigma(x)]\rho(x) = -\int_a^b \delta(x - y)L_1\rho(y)\,dy = -L_1\rho(x) \qquad (44.8)$$

This relation is exactly that in Eq. (44.6). In contrast to Eq. (43.9), however, Eq. (44.7) is still an *integral equation* because the unknown quantity ρ appears under the integral on the right-hand side.

If the eigenfunction expansion (43.10) for the Green's function is substituted into the integral equation (44.7), the result is

$$\rho(x) = \sum_{m=1}^{\infty} \rho_m(x) \frac{1}{\omega^2 - \omega_m^2} \int_a^b \rho_m(y) L_1 \rho(y)\, dy \qquad (44.9)$$

which is an expression of the form

$$\rho(x) = \sum_{m=1}^{\infty} c_m \rho_m(x) \qquad (44.10)$$

It relates the function $\rho(x)$ to the eigenfunctions defined in Eqs. (44.1) and (44.3). In addition, Eqs. (44.2), (44.4), and (44.7) are all homogeneous in the function ρ, allowing us to *normalize* ρ in any convenient way. For the following discussion, we choose the normalization [compare Eq. (42.9)]

$$\langle \rho_n | \rho \rangle = 1 \qquad \text{one particular } n \qquad (44.11a)$$

where n denotes some specified eigenfunction of Eq. (44.1). Comparison with Eq. (44.10) yields the equivalent condition

$$c_n = 1 \qquad \text{one particular } n \qquad (44.11b)$$

Equations (44.9), (44.10), and (44.11b) require that

$$\omega^2 - \omega_n^2 = \int_a^b \rho_n(y) L_1 \rho(y)\, dy \qquad (44.12)$$

which merely expresses our particular choice of normalization. The series (44.9) then becomes

$$\rho(x) = \rho_n(x) + \sum_{q \neq n} \rho_q(x) \frac{1}{\omega^2 - \omega_q^2} \int_a^b \rho_q(y) L_1 \rho(y)\, dy \qquad (44.13)$$

where the sum runs over all q *except* for the value $q = n$.

We have now achieved our aim of expressing the solution of the more complicated Sturm-Liouville equation (44.2) in terms of the complete set of solutions of the (assumed solvable) Sturm-Liouville equation (44.1). These relations are exact, for they involve no approximations, apart from the assumption that the solution $\rho(x)$ contains *some nonzero component* of the particular eigenfunction $\rho_n(x)$ in the expansion (44.10); otherwise, it is not possible to impose the normalization condition (44.11a). Note that the resulting integral equation for $\rho(x)$ involves the additional potential L_1 [see Eq. (44.5)]. If L_1 vanishes, the solutions (44.12) and (44.13) explicitly reduce to the original eigenvalue ω_n^2 and eigenfunction ρ_n.

Expansion for Small Coupling Strength

Equations (44.11) to (44.13) reformulate the Sturm-Liouville equation (44.2) in terms of the solvable problem (44.1) subject to the same domain and same set of boundary conditions. They are extremely important, for they allow a systematic study of the effect of the additional potential (44.5) *as a power series in the strength parameter* ϵ. Assume that ϵ (and hence L_1) is small. The corrections to the eigenvalue ω_n^2 in Eq. (44.12) and to the eigenfunction ρ_n in Eq. (44.13) are both explicitly proportional to ϵ. Hence the first correction on the right-hand side of Eq. (44.13) is obtained by the substitution $\rho \to \rho_n$ in the integral and $\omega^2 \to \omega_n^2$ in the denominator, since the remaining corrections vanish faster than ϵ. In this way, Eq. (44.13) becomes

$$\rho(x) = \rho_n(x) + \epsilon \sum_{q \neq n}^{\infty} \rho_q(x) \frac{1}{\omega_n^2 - \omega_q^2} \langle q|v|n \rangle + O(\epsilon^2) \qquad (44.14)$$

where the shorthand notation

$$\langle q|v|n \rangle \equiv \int_a^b \rho_q(y) v(y) \rho_n(y) \, dm \equiv \langle n|v|q \rangle \qquad (44.15)$$

has been introduced. Note that the metric

$$dm \equiv \sigma(y) \, dy \qquad (44.16)$$

has absorbed the factor $\sigma(y)$ in the potential (44.5). Evidently, Eq. (44.14) constitutes the leading term in our expansion for the particular eigenfunction ρ of the Sturm-Liouville equation (44.2) that *reduces to ρ_n as $\epsilon \to 0$*.

The expansion (44.14) for the eigenfunction of Eq. (44.2) can be substituted into Eq. (44.12) to yield an expression for the eigenvalue ω^2 correct through order ϵ^2

$$\omega^2 - \omega_n^2 = \epsilon \langle n|v|n \rangle + \epsilon^2 \sum_{q \neq n}^{\infty} \langle n|v|q \rangle \frac{1}{\omega_n^2 - \omega_q^2} \langle q|v|n \rangle + O(\epsilon^3) \quad (44.17)$$

In writing relations (44.14) and (44.17), we have tacitly assumed *nondegenerate perturbation theory*, namely

$$\omega_q^2 \neq \omega_n^2 \qquad \text{if } q \neq n \qquad (44.18)$$

This condition guarantees that all the denominators in (44.14) and (44.17) are *finite* and contain *corrections* which themselves are of order ϵ, as indicated in Eq. (44.12).† In nondegenerate perturbation theory, a calculation of the eigenfunction ρ through any given order in ϵ always yields an expression for the eigenvalue (44.12) through one higher power in ϵ. In first order, this feature already appeared in our discussion of Eqs. (42.9) to (42.12).

---

† *Degenerate* perturbation theory is more complicated; see Prob. 7.15.

Example: Mass Point on a String Revisited

As an example of perturbation theory, consider our model "laboratory" problem of a point mass m at the center of a stretched uniform string (see Fig. 40.3). In this case, the one-dimensional wave equation becomes

$$\sigma(x)\frac{\partial^2 u}{\partial t^2} = \tau\frac{\partial^2 u}{\partial x^2} \tag{44.19}$$

where [see Eqs. (40.31) and (40.32)]

$$\sigma(x) = \sigma\left[1 + \frac{m}{\sigma}\,\delta(x - \tfrac{1}{2}l)\right] \tag{44.20a}$$

$$\sigma = \text{const} \qquad \tau = \text{const} \qquad c^2 = \frac{\tau}{\sigma} \tag{44.20b}$$

If we seek normal-mode solutions to the wave equation

$$u(x, t) = \rho(x)\cos(\omega t + \phi) \tag{44.21}$$

the resulting eigenvalue equation is

$$\tau\frac{d^2\rho}{dx^2} = -\omega^2\sigma(x)\rho \tag{44.22}$$

Equation (44.22) can thus be written in the standard form (44.2)

$$-\tau\frac{d^2\rho}{dx^2} - \left[\omega^2\frac{m}{\sigma}\,\delta(x - \tfrac{1}{2}l)\right]\sigma\rho = \omega^2\sigma\rho \tag{44.23}$$

Comparison with Eq. (44.2) yields the expression for the perturbation

$$\epsilon v(x) = -\omega^2\frac{m}{\sigma}\,\delta(x - \tfrac{1}{2}l) \tag{44.24}$$

where we have chosen the uniform string as the unperturbed problem

$$-\tau\frac{d^2\rho_n}{dx^2} = \omega_n^2\sigma\rho_n \tag{44.25}$$

The corresponding unperturbed eigenfunctions and eigenvalues are [see Eq. (43.33) and the subsequent discussion]

$$\rho_n(x) = \left(\frac{2}{\sigma l}\right)^{1/2}\sin\frac{n\pi x}{l} \tag{44.26a}$$

$$\omega_n^2 = \left(\frac{n\pi c}{l}\right)^2 \tag{44.26b}$$

where $\{\rho_n\}$ are orthonormal with respect to the unperturbed measure $dm = \sigma\,dx$, as in Eq. (40.28).

We shall now compute the change in the eigenvalues of Eq. (44.22) to first order in the small quantity

$$\frac{m}{M} = \frac{m}{\sigma l} \ll 1 \tag{44.27}$$

where $M = \sigma l = M_{\text{string}}$ is the mass of the string. The ratio m/M plays the role of the dimensionless expansion parameter ϵ in the perturbation series. The first-order shifts follow immediately from the first term on the right-hand side of Eq. (44.17)

$$\omega^2 - \omega_n^2 \approx -\omega^2 \frac{m}{\sigma} \int_0^l \rho_n(x)\, \delta(x - \tfrac{1}{2}l)\rho_n(x)\sigma \, dx \tag{44.28}$$

which is just the average value of the perturbation taken over the unperturbed eigenfunctions. The present example is, in fact, somewhat unusual because the correction on the right-hand side of Eq. (44.28) is itself proportional to the eigenvalue ω^2. Since the correction on the right-hand side of Eq. (44.28) is already small, however, we can use the relation

$$\omega = \omega_n + \delta\omega_n \tag{44.29}$$

to replace its explicit factor ω^2 by ω_n^2. This step neglects higher-order corrections in ϵ. The integral in Eq. (44.28) merely evaluates the eigenfunctions (44.26a) at the center of the string, and the shift in the normal-mode frequencies to the first order in $m/\sigma l = m/M$ becomes

$$\omega^2 - \omega_n^2 \approx -\omega_n^2 \frac{2m}{\sigma l} \sin^2 \frac{n\pi}{2} \tag{44.30}$$

The left-hand side can be rewritten $(\omega + \omega_n)(\omega - \omega_n) \approx 2\omega_n\, \delta\omega_n$ with Eq. (44.29). We therefore obtain the fractional shift

$$\frac{\delta\omega_n}{\omega_n} = \frac{\delta\xi_n}{\xi_n} = \begin{cases} -\dfrac{m}{M} & n \text{ odd} \qquad (44.31a) \\[2mm] 0 & n \text{ even} \qquad (44.31b) \end{cases}$$

where the dimensionless parameter (42.14)

$$\xi \equiv \frac{\omega l}{2c} \tag{44.32}$$

has again been introduced.

It is interesting to compare the results (44.31) obtained in first-order perturbation theory with the exact solutions given in Eqs. (40.29) to (40.32) and (42.13). For n even, the eigenfunctions are odd about the midpoint (see Fig. 40.3a), with the added mass located at a node and therefore stationary. These eigenfunctions and eigenfrequencies are wholly unaffected by the mass m [see Eq. (40.29)], in agreement with the first-order perturbation expression (44.31b). The situation is more complicated for n odd, when the eigenfunctions are even about the midpoint (see Fig. 40.3b) and the mass m moves with a nonzero amplitude. The exact eigenvalue equation (42.13) for these modes can be rewritten as

$$\frac{1}{\xi}\cot\xi = \frac{m}{M} \tag{44.33}$$

Since the exact eigenvalues differ from the unperturbed eigenvalues by terms of first order in the small expansion parameter (44.27), the desired solution for odd n can be expressed in a form analogous to (44.29)

$$\xi = \xi_n + \delta\xi_n = \tfrac{1}{2}n\pi + \delta\xi_n \tag{44.34}$$

where Eqs. (44.26b) and (44.32) have been used to identify ξ_n. The Taylor-series expansion of the cotangent in Eq. (44.33) then yields (recall that n is odd here)

$$\cot\left(\tfrac{1}{2}n\pi + \delta\xi_n\right) \approx (\cot\tfrac{1}{2}n\pi) - \delta\xi_n(\csc^2\tfrac{1}{2}n\pi) \approx -\delta\xi_n \tag{44.35}$$

Substitution of Eqs. (44.34) and (44.35) into the exact eigenvalue equation (44.33) yields the expression

$$-\frac{\delta\xi_n}{\xi_n} = \frac{m}{M} \qquad n \text{ odd} \tag{44.36}$$

through terms of first order in the small quantity (44.27). Comparison of Eqs. (44.36) and (44.31a) confirms that perturbation theory reproduces the exact eigenfrequencies through the first order in m/M for the even modes as well as the odd ones.†

Figure 42.4 compares the perturbation approximation for the lowest eigenvalue with the exact one obtained by numerical analysis of Eq. (44.33) and also with the corresponding variational calculation. Perturbation theory allows us to compute the change in the eigenvalue as the parameter m/M increases from zero. It is evident from Eq. (44.36) that our first-order calculation gives the exact slope at the origin $m/M = 0$, not only for the lowest even mode shown in Fig. 42.4 but also for all even modes. In fact, the perturbation analysis for this particular problem nicely complements the variational calculation carried out in Sec. 42 and illustrated in Fig. 42.4, for that particular variational calculation was least accurate as $m/M \to 0$. Other trial functions, $\sin(\pi x/l)$ for example, could yield improved variational estimates in this regime.

In summary, perturbation theory provides a powerful theoretical tool. It allows one to find approximate eigenfunctions and eigenvalues in a problem whose exact solution may be hard. The only difficulty is to construct a corresponding model problem with known eigenfunctions and eigenvalues that do not differ too greatly from those of the exact problem.

---

† The extension of these results to second order is discussed in Prob. 7.14.

45 ENERGY FLUX

We conclude this chapter with an analysis of the continuity equation for the hamiltonian density and the energy flux in continuum problems. This discussion is not only relevant to strings but also forms a basis for our subsequent work in continuum mechanics.

Continuity Equation for the Hamiltonian Density

Suppose we have a lagrangian density $\mathscr{L}(u, \nabla u, \partial u/\partial t; \mathbf{x})$ that is a function of a generalized coordinate $u(\mathbf{x}, t)$, the spatial derivative of the generalized coordinate at a fixed time, and the first time derivative of the generalized coordinate at a fixed point $\mathbf{x}$. As in Eq. (38.2) for an inhomogeneous string, the lagrangian density may depend on the spatial coordinate $\mathbf{x}$, but *we assume that the lagrangian density has no explicit time dependence*. For generality, the present discussion will allow an arbitrary number of spatial dimensions, and we use $d\mathbf{x}$ to denote the appropriate volume element. Hamilton's principle, which constitutes the general equation of mechanics, states

$$\delta \int_{t_1}^{t_2} dt \int d\mathbf{x} \; \mathscr{L}\left(u, \nabla u, \frac{\partial u}{\partial t}; \mathbf{x}\right) = 0$$

Fixed endpoints in time: $\qquad\qquad \delta u(t_1) = \delta u(t_2) = 0$ (45.1)

Lagrange's equation follows as the Euler-Lagrange equation

$$\frac{\partial}{\partial t}\left[\frac{\partial \mathscr{L}}{\partial(\partial u/\partial t)}\right] + \sum_i \frac{\partial}{\partial x_i}\left[\frac{\partial \mathscr{L}}{\partial(\partial u/\partial x_i)}\right] - \frac{\partial \mathscr{L}}{\partial u} = 0$$ (45.2)

Here the sum over i includes the appropriate number of spatial dimensions. Once again we emphasize that *all the other quantities* in the lagrangian density are held fixed when evaluating the partial derivatives in brackets in Eq. (45.2). On the other hand, the final space and time derivatives are carried out keeping fixed only the other independent variables among the set $\{\mathbf{x}, t\}$ [see the discussion following Eq. (25.59)]. In the derivation of Lagrange's equation (45.2) from Hamilton's principle (45.1), it has been assumed that the spatial boundary conditions allow a spatial integration by parts. The evident generalization of Eq. (38.7) is that the following quantity must vanish on the bounding surface of the region under consideration (see Fig. 45.1)

$$\delta u \left[\sum_i \hat{n}_i \frac{\partial \mathscr{L}}{\partial(\partial u/\partial x_i)}\right] = 0$$ (45.3)

Here $\hat{n}_i$ is the appropriate component of the outward unit normal to the bounding surface.

The *canonical momentum density* is defined by the following relation [compare Eq. (20.2) for a discrete system]

$$\mathscr{P} \equiv \frac{\partial \mathscr{L}}{\partial(\partial u/\partial t)}$$ (45.4)

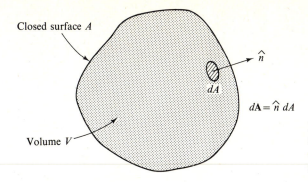

Closed surface A

$\hat{n}$

dA

$d\mathbf{A} = \hat{n}\, dA$

Volume V

Figure 45.1 Quantities appearing in the discussion of the continuity equation for the hamiltonian density and corresponding flux.

where, as usual, the partial derivative means to keep all the other variables in the lagrangian density $\mathscr{L}$ fixed. By analogy with Eq. (20.12), we also define *the hamiltonian density*

$$\mathscr{H} = \mathscr{P}\frac{\partial u}{\partial t} - \mathscr{L} \tag{45.5}$$

Consider the time rate of change of the hamiltonian density $\partial\mathscr{H}(\mathbf{x}, t)/\partial t$ at a fixed point in space $\mathbf{x}$. Since the lagrangian density depends on t implicitly through the quantities $(u, \nabla u, \partial u/\partial t)$, the partial derivative of Eq. (45.5) with respect to time yields the expression

$$\frac{\partial\mathscr{H}}{\partial t} = \frac{\partial\mathscr{P}}{\partial t}\frac{\partial u}{\partial t} + \mathscr{P}\frac{\partial^2 u}{\partial t^2} - \left[\frac{\partial\mathscr{L}}{\partial u}\frac{\partial u}{\partial t} + \frac{\partial\mathscr{L}}{\partial(\partial u/\partial t)}\frac{\partial}{\partial t}\frac{\partial u}{\partial t} + \sum_i \frac{\partial\mathscr{L}}{\partial(\partial u/\partial x_i)}\frac{\partial}{\partial t}\frac{\partial u}{\partial x_i}\right] \tag{45.6}$$

The quantity in brackets follows from the general chain rule for differentiating an implicit function. The definition of the canonical momentum density in Eq. (45.4) can be used to cancel the second and fourth terms in Eq. (45.6), leading to the simpler form

$$\frac{\partial\mathscr{H}}{\partial t} = \frac{\partial\mathscr{P}}{\partial t}\frac{\partial u}{\partial t} - \left[\frac{\partial\mathscr{L}}{\partial u}\frac{\partial u}{\partial t} + \sum_i \frac{\partial\mathscr{L}}{\partial(\partial u/\partial x_i)}\frac{\partial^2 u}{\partial t\,\partial x_i}\right] \tag{45.7}$$

Furthermore, Lagrange's equation (45.2) provides an expression for the time derivative of the canonical momentum density (45.4). Substitution into the first term in Eq. (45.7) gives

$$\frac{\partial\mathscr{H}}{\partial t} = \left[\frac{\partial\mathscr{L}}{\partial u} - \sum_i \frac{\partial}{\partial x_i}\frac{\partial\mathscr{L}}{\partial(\partial u/\partial x_i)}\right]\frac{\partial u}{\partial t} - \left[\frac{\partial\mathscr{L}}{\partial u}\frac{\partial u}{\partial t} + \sum_i \frac{\partial\mathscr{L}}{\partial(\partial u/\partial x_i)}\frac{\partial^2 u}{\partial t\,\partial x_i}\right] \tag{45.8}$$

The first and third quantities in this expression cancel, and the remaining terms combine to yield

$$\frac{\partial\mathscr{H}}{\partial t} = -\sum_i \frac{\partial}{\partial x_i}\left[\frac{\partial\mathscr{L}}{\partial(\partial u/\partial x_i)}\frac{\partial u}{\partial t}\right] \tag{45.9}$$

The right-hand side of this equation has the form of a *total divergence*, so that Eq. (45.9) can be written as a continuity equation

$$\frac{\partial \mathscr{H}}{\partial t} + \mathbf{V} \cdot \mathbf{S} = 0 \tag{45.10}$$

where the vector $\mathbf{S}$ is defined according to

$$S_i \equiv \frac{\partial \mathscr{L}}{\partial(\partial u/\partial x_i)} \frac{\partial u}{\partial t} \tag{45.11}$$

Equations (45.10) and (45.11) constitute a central theorem, although the derivation may initially seem somewhat obscure. It becomes less mysterious when the quantities $\mathscr{H}$ and $\mathbf{S}$ are recognized as just those needed to treat the continuum analogs of the discussion of the hamiltonian for discrete systems in Secs. 20 and 32. For example, consider the generalization of the theorem stated in Eq. (20.13). The integral of the hamiltonian density over the fixed volume is now taken to define the *total* hamiltonian of the system†

$$H \equiv \int_V \mathscr{H}(\mathbf{x}, t) \, d\mathbf{x} \tag{45.12}$$

It can only depend on the time because all the spatial variables have been integrated out. Its total time derivative becomes

$$\frac{dH}{dt} = \int_V \frac{\partial \mathscr{H}(\mathbf{x}, t)}{\partial t} \, d\mathbf{x} \tag{45.13}$$

where the time derivative under the integral in (45.13) is written as a partial derivative since the spatial coordinates are held fixed. Substitution of the continuity equation (45.10) in the right-hand side of Eq. (45.13) yields

$$\frac{dH}{dt} = -\int_V (\mathbf{V} \cdot \mathbf{S}) \, d\mathbf{x} \tag{45.14}$$

Use of Gauss' theorem to convert the integral on the right-hand side of Eq. (45.14) into a surface integral over the closed surface A bounding the volume V (see Fig. 45.1) leads to the final result

$$\frac{dH}{dt} = -\int_A \mathbf{S} \cdot d\mathbf{A} \tag{45.15}$$

This equation has numerous applications. We first consider two simple possibilities:

1. The disturbance under consideration occurs in a finite region, and the surrounding surface is sufficiently far away for $\hat{n} \cdot \mathbf{S}$ on the right-hand side of

---

† This is the appropriate continuum limit of the hamiltonian defined for point particles; see Prob. 7.16.

Eq. (45.15) to vanish at these distances. A physical example would be a localized disturbance on an infinite string.

2. The boundary conditions of the problem ensure that the quantity $\hat{n} \cdot \mathbf{S}$ vanishes on the boundary of the system. As we shall demonstrate, a string with fixed endpoints provides a physical example.

If either of these conditions is satisfied, the right-hand side of Eq. (45.15) vanishes, implying that

$$\frac{dH}{dt} = 0 \tag{45.16}$$

It follows from this relation that the total hamiltonian of the system is a constant of the motion

$$H = \text{const} \tag{45.17}$$

just as in Eq. (20.13) for point mechanics.

Equation (45.15) also applies to a volume V that is part of a larger isolated system. In this case, the quantity H in (45.12) need not be constant, owing to the interaction across the bounding surface A. Nevertheless, Eq. (45.15) still has the simple interpretation that the total amount of H within the volume V can increase only from inward flow through the surface. In this way, $\mathcal{H}$ is identified as the density of the conserved quantity H and $\mathbf{S}$ as the flux of the same quantity H. Since $\mathbf{S} \cdot d\mathbf{A}$ measures the amount of H crossing the oriented area $d\mathbf{A}$ per unit time, $\mathbf{S}$ has the dimension of H per unit area per unit time [compare Eq. (5.17) for the particle flux].

What is the continuum analog of Eq. (20.16) for point mechanics? Suppose that

$$\mathcal{L} = \mathcal{T} - \mathcal{V} \tag{45.18}$$

where the kinetic-energy density $\mathcal{T}$ is quadratic in the generalized velocity

$$\mathcal{T} = \tfrac{1}{2}\sigma(\mathbf{x})\left(\frac{\partial u}{\partial t}\right)^2 \tag{45.19}$$

and the potential-energy density is independent of the generalized velocity

$$\frac{\partial \mathcal{V}}{\partial(\partial u/\partial t)} = 0 \tag{45.20}$$

In this case, the canonical momentum density defined by Eq. (45.4) becomes

$$\mathcal{P} \equiv \frac{\partial \mathcal{L}}{\partial(\partial u/\partial t)} = \sigma(\mathbf{x})\frac{\partial u}{\partial t} \tag{45.21}$$

and the hamiltonian density (45.5) is given by

$$\mathcal{H} = \mathcal{P}\frac{\partial u}{\partial t} - \mathcal{L} = 2\mathcal{T} - (\mathcal{T} - \mathcal{V}) = \mathcal{T} + \mathcal{V} \tag{45.22}$$

Thus, just as in Eq. (20.16), the hamiltonian density is then the energy density

$$\mathcal{H} = \mathcal{T} + \mathcal{V} \tag{45.23}$$

Throughout this discussion, note the central role of the basic assumption that the lagrangian density in Eq. (45.1) *contains no explicit dependence on the time.*

Example: One-dimensional String

We conclude this analysis with a treatment of the one-dimensional string. Here there is only one spatial coordinate, with

$$x_i = x \qquad d\mathbf{x} = dx \tag{45.24}$$

Equation (38.2) gives the lagrangian density for the string

$$\mathcal{L} = \tfrac{1}{2}\sigma(x)\left(\frac{\partial u}{\partial t}\right)^2 - \tfrac{1}{2}\tau(x)\left(\frac{\partial u}{\partial x}\right)^2 \tag{45.25}$$

The corresponding canonical momentum density becomes

$$\mathcal{P} = \frac{\partial \mathcal{L}}{\partial(\partial u/\partial t)} = \sigma(x)\frac{\partial u}{\partial t} \tag{45.26}$$

Note that the quantity

$$\mathcal{P}\, dx = \sigma(x)\, dx\, \frac{\partial u}{\partial t} \tag{45.27}$$

here is the actual transverse momentum density in the string arising from the transverse motion of the mass element $\sigma\, dx$. The hamiltonian density is obtained from Eqs. (45.5), (45.25), and (45.26)

$$\mathcal{H} = \tfrac{1}{2}\sigma(x)\left(\frac{\partial u}{\partial t}\right)^2 + \tfrac{1}{2}\tau(x)\left(\frac{\partial u}{\partial x}\right)^2 \tag{45.28}$$

In this case, $\mathcal{H}$ in fact represents the energy density of the system [see Eq. (45.23)].

Equation (45.11) now provides the *energy-flux vector* **S** along the string. It is first necessary to compute

$$\frac{\partial \mathcal{L}}{\partial(\partial u/\partial x)} = -\tau(x)\frac{\partial u}{\partial x} \tag{45.29}$$

and the energy-flux vector (45.11) then takes the form

$$\mathbf{S} = -\tau(x)\frac{\partial u}{\partial x}\frac{\partial u}{\partial t}\,\hat{x} \tag{45.30}$$

As an application of Eq. (45.30), consider a wave

$$u(x, t) = A\cos\left[k(x - ct)\right] = A\cos\left(kx - \omega t\right) \tag{45.31}$$

moving to the right along a uniform string (see Fig. 45.2). Substitution of

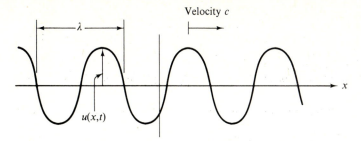

Figure 45.2 Wave $u(x, t) = A \cos[k(x - ct)] = A \cos[(2\pi/\lambda)(x - ct)]$ propagating along an infinite uniform string.

Eq. (45.31) in Eq. (45.30) yields an expression for the x component of the instantaneous energy flux at the point x and time t

$$S_x = A^2(\tau k^2 c) \sin^2 [k(x - ct)] = A^2(\tau k^2 c) \sin^2 (kx - \omega t) \qquad (45.32)$$

Although S_x varies with time at a given x, it always remains positive, implying that energy indeed propagates to the right for the waveform (45.31). For most purposes, it is sufficient to consider the time average of Eq. (45.32) at any fixed point x

$$\langle S_x \rangle = \tfrac{1}{2} A^2(\tau k^2 c) = \tfrac{1}{2} A^2 \sigma c \omega^2 \qquad (45.33)$$

Equations (45.32) and (45.33) also follow directly by calculating the energy passing a given point per unit time. In this case, we compute the instantaneous energy density in the wave (45.31) from Eq. (45.28) and then multiply by the distance the wave travels in unit time, which is just the velocity of propagation c.

Finally, note that the following relations are satisfied for a string with fixed endpoints

$$\left.\frac{\partial u}{\partial t}\right|_{x=0} = \left.\frac{\partial u}{\partial t}\right|_{x=l} = 0 \qquad (45.34)$$

Equation (45.30) then shows that these boundary conditions ensure that the energy flux vanishes at each end

$$\left. S_x \right|_{x=0} = \left. S_x \right|_{x=l} = 0 \qquad (45.35)$$

Thus we conclude from Eq. (45.15) that the total hamiltonian is a constant of the motion and from (45.28) that it is also the total energy of the string. These results agree with our previous normal-mode analysis in Eqs. (38.25) to (38.27). Note also that Eq. (45.35) implies that the energy is totally reflected at each end, in accordance with the qualitative discussion below Eq. (39.26).

Transmission and Reflection at a Discontinuity in Density

The preceding analysis of energy flux S_x on strings provides a natural introduction to scattering theory, which, in the present one-dimensional context, reduces to

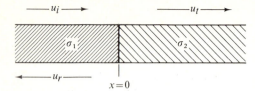

Figure **45.3** Two semi-infinite strings with mass densities σ_1 and σ_2 and a common tension τ joined at $x = 0$. The incident wave u_i from the left gives rise to a reflected wave u_r and a transmitted wave u_t.

transmission and reflection of an incident wave. For definiteness, we consider a specific example, in which two uniform semi-infinite strings with mass density σ_1 and σ_2 and common tension τ are joined at $x = 0$ (see Fig. 45.3). The same techniques will also apply in more general situations.

Suppose a traveling wave u_i with frequency ω approaches from the left. Since the mass density is different on the two sides, the propagation velocity and wave numbers also differ, and we therefore write

$$u_i(x, t) = A \cos (k_1 x - \omega t) = \text{Re } A e^{i(k_1 x - \omega t)} \qquad x \le 0 \qquad (45.36)$$

which depends on $x - c_1 t$ and evidently moves to the right. Although it is possible to work consistently with a real representation, the complex representation simplifies the analysis considerably, and we shall use this example to demonstrate its power. The incident wave gives rise to a reflected wave u_r moving to the left in medium 1 and a transmitted wave u_t moving to the right in medium 2. Since the ends of the strings remain joined, the motions at $x = 0$ all oscillate at the common frequency ω, implying

$$u_r(x, t) = \text{Re } B e^{-i(k_1 x + \omega t)} \qquad x \le 0 \qquad (45.37)$$

$$u_t(x, t) = \text{Re } C e^{i(k_2 x - \omega t)} \qquad x \ge 0 \qquad (45.38)$$

where B and C are constants that will be determined subsequently.

Since the string remains unbroken, the displacement at $x = 0$ must be continuous. Thus we require

$$[u_i(x, t) + u_r(x, t)]_{x=0} = [u_t(x, t)]_{x=0} \qquad (45.39a)$$

and Eqs. (45.36) to (45.38) yield the specific relation

$$A + B = C \qquad (45.39b)$$

Similarly, the instantaneous slope must be equal on each side of the origin because the acceleration would otherwise be infinite there [see Eq. (39.37)]

$$\frac{\partial}{\partial x} [u_i(x, t) + u_r(x, t)]_{x=0} = \frac{\partial}{\partial x} [u_t(x, t)]_{x=0} \qquad (45.40a)$$

Substitution of Eqs. (45.36) to (45.38) provides the second relation

$$ik_1(A - B) = ik_2 C \qquad (45.40b)$$

The linear equations (45.39b) and (45.40b) are readily solved to give

$$\frac{B}{A} = \frac{k_1 - k_2}{k_1 + k_2} \tag{45.41a}$$

$$\frac{C}{A} = \frac{2k_1}{k_1 + k_2} \tag{45.41b}$$

which completely characterize the motion.

We now show that this solution has the simple interpretation that a fraction R of the incident energy is reflected and the remainder T is transmitted. For definiteness, consider the incident wave (45.36), which we shall write as

$$u_i(x, t) = \text{Re } u_i(x)e^{-i\omega t} \tag{45.42a}$$

where

$$u_i(x) = Ae^{ik_1 x} \tag{45.42b}$$

The associated energy flux can be computed as in Eq. (45.32), but it is simpler to realize that its time average is usually the most relevant quantity. In that case, we can use the following trick to work directly with the complex representation. Take two oscillatory quantities with a common frequency

$$\text{Re } \alpha e^{-i\omega t} = \tfrac{1}{2}(\alpha e^{-i\omega t} + \alpha^* e^{i\omega t})$$
$$\text{Re } \beta e^{-i\omega t} = \tfrac{1}{2}(\beta e^{-i\omega t} + \beta^* e^{i\omega t}) \tag{45.43}$$

Their product oscillates, but its time average, denoted by angular brackets, is given by

$$\langle(\text{Re } \alpha e^{-i\omega t})(\text{Re } \beta e^{-i\omega t})\rangle = \tfrac{1}{4}\langle\alpha\beta e^{-2i\omega t} + \alpha\beta^* + \alpha^*\beta + \alpha^*\beta^* e^{2i\omega t}\rangle \tag{45.44}$$

Here, the first and last terms have zero time average, and we therefore obtain the simple result

$$\langle(\text{Re } \alpha e^{-i\omega t})(\text{Re } \beta e^{-i\omega t})\rangle = \tfrac{1}{4}(\alpha\beta^* + \alpha^*\beta) = \tfrac{1}{2} \text{ Re } (\alpha\beta^*) \tag{45.45}$$

expressed solely in terms of the complex coefficients of the oscillatory factor $e^{-i\omega t}$.

This analysis applies directly to the energy flux (45.30) in the incident wave; its time average becomes

$$\langle S_x^{\text{inc}}\rangle = -\tau\left\langle\frac{\partial u_i(x, t)}{\partial x}\frac{\partial u_i(x, t)}{\partial t}\right\rangle$$

$$= -\tau\left\langle\left|\text{Re }\frac{\partial}{\partial x}[u_i(x)e^{-i\omega t}]\right|\left|\text{Re }\frac{\partial}{\partial t}[u_i(x)e^{-i\omega t}]\right|\right\rangle \tag{45.46}$$

Use of Eq. (45.45) and the explicit form (45.42b) readily yields

$$\langle S_x^{\text{inc}}\rangle = -\tfrac{1}{2}\tau \text{ Re }\left|\frac{\partial u_i(x)}{\partial x}[-i\omega u_i(x)]^*\right| = \tfrac{1}{2}\tau k_1 \omega |A|^2 \tag{45.47}$$

in agreement with Eq. (45.33). A similar analysis for the reflected and transmitted waves gives

$$\langle S_x^{\text{refl}}\rangle = -\tfrac{1}{2}\tau k_1 \omega |B|^2 \tag{45.48}$$

$$\langle S_x^{\text{trans}}\rangle = \tfrac{1}{2}\tau k_2 \omega |C|^2 \tag{45.49}$$

The reflection coefficient R is defined as the ratio of the reflected energy flux to the incident flux

$$R = -\frac{\langle S_x^{\text{refl}}\rangle}{\langle S_x^{\text{inc}}\rangle} \tag{45.50}$$

where the minus sign arises from the reversed direction. Use of Eqs. (45.41a), (45.47), and (45.48) gives the simple expression

$$R = \frac{|B|^2}{|A|^2} = \left(\frac{k_1 - k_2}{k_1 + k_2}\right)^2 \tag{45.51}$$

Correspondingly, the transmission coefficient T is defined as

$$T = \frac{\langle S_x^{\text{trans}}\rangle}{\langle S_x^{\text{inc}}\rangle} \tag{45.52}$$

and a similar calculation yields

$$T = \frac{k_2}{k_1}\frac{|C|^2}{|A|^2} = \frac{4k_1 k_2}{(k_1 + k_2)^2} \tag{45.53}$$

These expressions have several interesting features.

1. Since $R + T = 1$, energy is explicitly conserved. Note the crucial role of the factor k_2/k_1 in Eq. (45.53), which arises from the difference in density on the two sides of the origin.
2. Since the tension τ is the same on both sides, the wave numbers $k_1 = \omega/c_1$ and $k_2 = \omega/c_2$ can be written in terms of $\sigma_1^{1/2}$ and $\sigma_2^{1/2}$ to give the ratio

$$\frac{B}{A} = \frac{\sigma_1^{1/2} - \sigma_2^{1/2}}{\sigma_1^{1/2} + \sigma_2^{1/2}} \tag{45.54}$$

It vanishes identically for $\sigma_1 = \sigma_2$, in which case the string is continuous. If $\sigma_1 > \sigma_2$, the reflected wave has the same phase as the incident wave, whereas it differs by 180° if $\sigma_1 < \sigma_2$. In the same way, the ratio C/A reduces to

$$\frac{C}{A} = \frac{2\sigma_1^{1/2}}{\sigma_1^{1/2} + \sigma_2^{1/2}} \tag{45.55}$$

It vanishes for $\sigma_1/\sigma_2 \to 0$ (a fixed end) and reduces to 1 for $\sigma_1 = \sigma_2$, in agreement with physical expectations.
3. The reflection and transmission coefficients become

$$R = \left(\frac{\sigma_1^{1/2} - \sigma_2^{1/2}}{\sigma_1^{1/2} + \sigma_2^{1/2}}\right)^2 \qquad T = \frac{4\sigma_1^{1/2}\sigma_2^{1/2}}{(\sigma_1^{1/2} + \sigma_2^{1/2})^2} \tag{45.56}$$

which again exhibit the expected behavior: $R \to 1$ and $T \to 0$ for $\sigma_1/\sigma_2 \ll 1$ and $\sigma_1/\sigma_2 \gg 1$; $R \to 0$ and $T \to 1$ as $\sigma_1/\sigma_2 \to 1$.

PROBLEMS

7.1 A uniform string under tension τ with fixed endpoints $\rho(0) = \rho(l) = 0$ is plucked in the middle, giving rise to an initial displacement which we approximate by (here, $h \ll l$)

$$u(x, 0) = \begin{cases} \dfrac{2hx}{l} & 0 \le x \le \tfrac{1}{2}l \\[2ex] \dfrac{2h(l-x)}{l} & \tfrac{1}{2}l \le x \le l \end{cases}$$

and released from rest.

(a) Determine the Fourier amplitudes and the energy in each mode.

(b) Construct the Bernoulli and d'Alembert solutions to the wave equation and show that they coincide.

(c) Discuss the uniform convergence of the Fourier series for $u(x, t)$.

7.2 Repeat Prob. 7.1 for an initially undeformed string that is struck at its midpoint, giving rise to an initial velocity distribution which can be approximated by

$$\dot{u}(x, 0) = v_0 \, l \delta(x - \tfrac{1}{2}l)$$

7.3 A uniform string of length $3l$ under tension τ with fixed endpoints is plucked a distance h ($\ll l$) at a point of trisection and released from rest. Find the energy in each normal mode. Verify that the total energy is just that in the initial stretching (you may quote the known result from Prob. A.12 that $\sum_{n=1}^{\infty} n^{-2} = \tfrac{1}{6}\pi^2$).

7.4 The normal modes of a uniform stretched string with fixed endpoints and a mass m at its center can be characterized by an integer n, where $n - 1$ is the number of nodes (see Fig. 40.3). The odd modes have frequencies $\omega_2, \omega_4, \ldots$, given in Eq. (40.29), and the even modes have frequencies $\omega_1, \omega_3, \ldots$, given by successive roots of Eq. (40.30).

(a) For the even modes, show that ω_n lies between $(n-1)\pi c/l$ and $n\pi c/l$. Hence conclude that the even and odd modes are interlaced.

(b) For the even modes and large n, prove that

$$\omega_n \approx \frac{(n-1)\pi c}{l}\left[1 + \frac{4M}{m\pi^2(n-1)^2} + \cdots\right]$$

(c) Hence verify Eq. (40.17) for this system.

7.5 A string of uniform mass density and length l hangs under its own weight in the earth's gravitational field. Consider small transverse displacements $u(x, t)$ in a plane.

(a) Compute the equilibrium tension in the string $\tau(x)$, where x is the distance from the point of suspension.

(b) Show that the normal modes satisfy Bessel's equation. *Hint:* See Appendix C and make the substitution $s^2 \equiv l - x$.

(c) What are the boundary conditions?

(d) What are the normal-mode frequencies? (Find the first three from tables.)

(e) What are the normal modes? (Sketch the first three.)

(f) Construct the general solution to the initial-value problem. Justify your result.

7.6 One end of a uniform rope of length l and mass density σ is attached to a vertical rod that rotates with a constant angular velocity Ω. Neglect the effect of gravity so that the rope sweeps out a horizontal circle.

(a) Transform to the co-rotating frame of reference and derive the equations of motion for small transverse displacements $u(x, t)$ in the horizontal plane and $v(x, t)$ in the vertical plane, where x is measured from the rotation axis. Why are they different?

(b) Show that the solutions can be expressed in terms of Legendre polynomials. Apply the boundary conditions to find the equation for the eigenfrequencies. Determine the three lowest eigenfrequencies and sketch the corresponding eigenfunctions.

7.7 Consider the linear second-order differential operator $L \equiv -a(x)(d^2/dx^2) - b(x)(d/dx) + c(x)$, with a and b positive. Show that L can be written as

$$L = -\frac{1}{\sigma(x)}\frac{d}{dx}\tau(x)\frac{d}{dx} + c(x)$$

where $\tau(x) = \exp\left[\int^x dy\, b(y)/a(y)\right]$ and $\sigma(x) = \tau(x)/a(x)$. Explain why the eigenfunctions of the equation $L\rho = \omega^2\rho$ will be orthogonal with respect to the weight function $\sigma(x)$.

7.8 Consider the eigenvalue functional (42.1) for the Sturm-Liouville equation.

(a) Expand the function $\rho(x)$ in a truncated basis of known (real) wave functions ϕ_n

$$\rho(x) = \sum_{n=1}^{N} a_n \phi_n(x)$$

and treat the coefficients $(a_1, \ldots, a_n)$ as variational parameters. Show that the Euler-Lagrange equations for this variational problem take the form

$$\sum_{p=1}^{N} [\langle \phi_n | C + V | \phi_p \rangle - \omega^2 \langle \phi_n | \phi_p \rangle] a_p = 0 \qquad n = 1, \ldots, N$$

where

$$\langle \phi_n | C | \phi_p \rangle \equiv \int_a^b \frac{\tau(x)}{\sigma(x)} \frac{d\phi_n}{dx}\frac{d\phi_p}{dx}\, dm$$

$$\langle \phi_n | V | \phi_p \rangle \equiv \int_a^b \frac{v(x)}{\sigma(x)} \phi_n \phi_p\, dm$$

$$\langle \phi_n | \phi_p \rangle \equiv \int_a^b \phi_n \phi_p\, dm$$

[The wave functions need not be orthonormal with respect to $dm \equiv \sigma(x)\,dx$.]

(b) Discuss the solution to these equations. *Hint:* Recall the normal-mode analysis in Sec. 22.

7.9 A string of length $2a$ is stretched to a constant tension τ with its ends fixed. The density of the string is given by $\sigma(x) = \sigma_0(1 - |x|/a)$ for $|x| < a$.

(a) Use a zero-parameter trial function to derive a variational estimate of the lowest resonant frequency ω_1. Compare with the numerical value $\omega_1^2 \approx 3.477\tau/a^2\sigma_0$.

(b) Devise a one-parameter trial function and show that it leads to a better (lower) estimate.

(c) Repeat part (a) for the next eigenfrequency ω_2, whose numerical value is $\omega_2^2 \approx 18.956\tau/a^2\sigma_0$.

7.10 In some cases, the closed form (43.29) of G_ω can be constructed only for $\omega = 0$. This problem illustrates that the resulting G_0 is still useful.

(a) Consider the Sturm-Liouville equation $L_0\rho = \omega^2\sigma\rho$, where L_0 is given in Eq. (43.4a) and ρ satisfies appropriate boundary conditions at the ends of the interval $x = a$ and $x = b$. Define a Green's function $G_0(x, x')$ by the equation [compare Eq. (43.12) for $\omega = 0$]

$$L_0 G_0(x, x') = \delta(x - x')$$

plus appropriate boundary conditions. Show that ρ satisfies a *homogeneous* integral equation

$$\rho(x) = \omega^2 \int_a^b G_0(x, x')\sigma(x')\rho(x')\, dx'$$

(b) Verify that the functional

$$\omega^2[\rho] = \frac{\int dx\, \sigma(x)\rho^2(x)}{\iint dx\, dx'\, \rho(x)\sigma(x)G_0(x, x')\sigma(x')\rho(x')}$$

is stationary for small variations in ρ about the exact solution.

(c) Apply this variational principle to the system in Prob. 7.9 and obtain an upper bound for the lowest frequency using your zero-parameter trial function. How does the estimate compare with those obtained in Prob. 7.9?

7.11 Consider the differential operator [compare Eq. (C.22)]

$$L_0 = -(1-x^2)\frac{d^2}{dx^2} + 2x\frac{d}{dx} = -\frac{d}{dx}\left[(1-x^2)\frac{d}{dx}\right]$$

and the Green's function $G_\omega(x, x')$ that satisfies Eq. (43.12) subject to the boundary conditions $G_\omega(0, x') = 0$, $G_\omega(1, x')$ finite.

(a) Construct $G_0(x, x')$ directly from the solutions to the homogeneous differential equation (see Appendix D2).

(b) Show that the eigenfunctions from Eq. (43.5) are Legendre polynomials. Hence construct the bilinear expansion (43.10) of $G_\omega(x, x')$. Comment briefly on its analytic structure as a function of ω.

(c) Take the limit $\omega \to 0$ and hence derive an identity involving a sum of products of Legendre polynomials. In particular, show that

$$\tfrac{1}{2}\ln\frac{1+x}{1-x} = \sum_{n\ \text{odd}} \frac{2n+1}{n(n+1)} P_n(x)$$

7.12 Prove that the new normal-mode amplitudes given to order ϵ in nondegenerate perturbation theory by Eq. (44.14) and here denoted by $\rho_n^{(1)}$ are *orthonormal* to this order, that is, $\langle \rho_n^{(1)} | \rho_q^{(1)} \rangle = \delta_{nq} + O(\epsilon^2)$.

7.13 A uniform string with fixed endpoints is subjected to a weak additional restoring force $v(x) = v_0(x - \tfrac{1}{2}l)/l$, where $v_0 = \text{const}$ [see Eq. (40.7) and Fig. 40.2]. Use first-order nondegenerate perturbation theory to evaluate the shifts in the normal-mode frequencies and the normal-mode amplitudes.

7.14 A stretched uniform string of length l has a point mass m at its center. Use second-order nondegenerate perturbation theory [Eq. (44.17)] to find the shift in the allowed frequencies to second order in the small parameter m/M. Verify that your result reproduces the corresponding expansion of the exact eigenvalue equation (44.33), $\delta\xi_n = \tfrac{1}{2}n\pi[-m/M + (m/M)^2 + \cdots]$, where $n = 1, 3, \ldots$ (recall that $\xi \equiv \omega l/2c$).

7.15 Consider a system with *degeneracy*, which means that modes with different values of q and n have $\omega_n^2 = \omega_q^2$.

(a) Assume that the perturbation in Eq. (44.15) is diagonal in the degenerate subspaces

$$\langle n|v|q \rangle = \delta_{nq}\langle n|v|q \rangle \qquad \text{for all } n, q \text{ such that } \omega_n^2 = \omega_q^2$$

Iterate the exact integral equation (44.13) in two steps and justify your procedure: (1) substitute the lowest-order expression for ρ in the numerator, retaining the exact ω^2 in the denominator; (2) substitute the appropriate expression for ω^2. Thus show that with a degeneracy of this type, the results of Eqs. (44.14) and (44.17) hold to order ϵ with the sum in Eq. (44.14) only receiving contributions from modes with $\omega_n^2 \neq \omega_q^2$.

(b) Explain how you would proceed if the perturbation is not originally diagonal in the degenerate subspaces.

7.16 Consider the lagrangian for N identical mass points connected by springs [Eq. (24.1)] or stretched strings [Eq. (24.11)].

(a) Find the associated canonical momenta and the hamiltonian.

(b) Carry out the transition from a discrete to a continuous system and hence rederive Eqs. (45.26) and (45.28) for a uniform string.

7.17 A wave travels along an infinite string stretched to a tension τ. The string has a segment of length $2a$ for $|x| < a$ and density σ_1 differing from the density σ_0 of the remaining parts.

(a) Solve the differential equations for $|x| < a$ and $|x| > a$ to find exact expressions for the transmission and reflection *amplitudes*.

(b) Show that the energy transmission coefficient is given by

$$T = 1 - R = \left[1 + \frac{1}{4}\left(\frac{k_1^2 - k_0^2}{k_1 k_0}\right)^2 \sin^2(2k_1 a)\right]^{-1}$$

where $k_i = \omega(\sigma_i/\tau)^{1/2}$. Discuss the frequency dependence of T, including the position and widths of the transmission resonances (where $T = 1$).

7.18 (*a*) An infinite string of mass density σ, stretched to a tension τ, has a point mass m located at $x = 0$. Find *separately* the coefficients T and R for transmission and reflection of energy for an incident wave with wave number k. Verify that $T + R = 1$. Sketch T as a function of k for various values of m. Discuss this result.

(*b*) A second identical mass is added to the string at the position $x = d$. Show that

$$T = 1 - R = [1 + \gamma^2(\cos kd - \tfrac{1}{2}\gamma \sin kd)^2]^{-1}$$

where $\gamma = mk/\sigma$. Sketch T as a function of k, including the transmission resonances, for various values of m. Compare this behavior with that in part (*a*). Discuss the applicability of the large-k limit to a real physical system.

7.19 Start from the general form (40.4) of the complex Fourier series for a periodic function of period L and take the limit as $L \to \infty$ (compare the discussion in Sec. 25). Hence derive the *Fourier-transform* relations, which state that if $f(x)$ has the representation

$$f(x) = (2\pi)^{-1/2} \int_{-\infty}^{\infty} e^{ikx}\phi(k)\,dk$$

its Fourier transform is given by

$$\phi(k) = (2\pi)^{-1/2} \int_{-\infty}^{\infty} e^{-ik\xi}f(\xi)\,d\xi$$

and vice versa. Using your derivation as a basis, discuss the class of functions $f(x)$ for which these relations can be expected to hold.

SELECTED ADDITIONAL READINGS

1. E. T. Whittaker and G. N. Watson, "A Course of Modern Analysis," 4th ed., Cambridge University Press, London, 1969.
2. P. M. Morse and H. Feshbach, "Methods of Mathematical Physics," McGraw-Hill, New York, 1953.
3. R. Courant and D. Hilbert, "Methods of Mathematical Physics," Interscience, New York, 1953, vol. 1.

EIGHT

MEMBRANES

The analysis developed in Chap. 7 has general applicability in theoretical physics. Here we use it to discuss the small-amplitude behavior of two-dimensional elastic membranes.

46 GENERAL FORMULATION

Consider a stretched elastic membrane, as illustrated in Fig. 46.1. Let $\sigma(x, y)$ denote the areal mass density (mass per unit area) of the membrane and $u(x, y, t)$ the vertical displacement of the membrane from its equilibrium configuration. For simplicity, the equilibrium configuration is assumed flat and can be taken to define the xy plane. The kinetic energy of an element of the membrane with equilibrium area dA arises from its motion in the z direction

$$\mathcal{T} \, dA = \tfrac{1}{2}\sigma \, dA \left(\frac{\partial u}{\partial t}\right)^2 \tag{46.1}$$

Since the displacement of the membrane u is assumed small, the actual stretched area dS (see Fig. 46.1) in Eq. (46.1) has been replaced by the equilibrium area dA to leading order in u. In contrast, the potential energy is due entirely to the actual increase in area. If $\tau(x, y)$ is the surface tension or surface energy density (dimensions of force per unit length or energy per unit area), the potential energy is just the work done in stretching the membrane. By the definition of surface tension, this work is given by

$$\mathcal{V} \, dA = \tau(dS - dA) \tag{46.2}$$

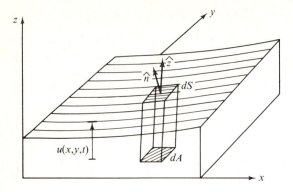

Figure 46.1 Instantaneous configuration of elastic membrane.

where the assumption of small deformations allows us to use the equilibrium value for τ.

To evaluate this expression, we need to compute the actual area dS for an arbitrary instantaneous configuration of the membrane. If the small element of actual area is considered to be a vector $d\mathbf{S} = \hat{n}\, dS$ with magnitude dS and direction along the normal $\hat{n}$ to the plane tangent to the surface at the point (x, y), the equilibrium area dA is the projection of this vector in the z direction (see Fig. 46.1)

$$dA = \hat{z} \cdot (\hat{n}\, dS) \tag{46.3}$$

This relation is exact in the limit of infinitesimal elements when the area dS lies entirely in the tangent plane. Now the general deformed surface is described by the relation

$$F(x, y, z, t) \equiv z - u(x, y, t) = 0 \tag{46.4}$$

If we move from one point on the instantaneously fixed surface to another, it is evident from Eq. (46.4) that F does not change. This observation has the differential statement

$$dF = \nabla F \cdot d\mathbf{r} = 0 \tag{46.5}$$

where the total differential has been expressed in terms of the gradient of F. Since Eq. (46.5) holds for arbitrary displacements $d\mathbf{r}$ in the surface, we conclude that ∇F must lie along the *normal* $\hat{n}$ to the surface, which is therefore given by

$$\hat{n} = \frac{\nabla F}{[(\nabla F)^2]^{1/2}} = \frac{\hat{z} - \nabla u}{[1 + (\nabla u)^2]^{1/2}} \tag{46.6}$$

It follows that the z projection of the normal is

$$\hat{z} \cdot \hat{n} = \frac{1}{[1 + (\nabla u)^2]^{1/2}} \tag{46.7}$$

Thus the potential energy in Eq. (46.2) can be rewritten with Eq. (46.3) as

$$\mathscr{V}\, dA = \tau\, dA\{[1 + (\nabla u)^2]^{1/2} - 1\} \tag{46.8a}$$

$$\approx \tfrac{1}{2}\tau\, dA(\nabla u)^2 \tag{46.8b}$$

where the last relation holds for small displacements.

The lagrangian density of the membrane is obtained by combining Eqs. (46.1) and (46.8)

$$\mathscr{L} = \mathscr{T} - \mathscr{V} \tag{46.9a}$$

In the limit of small displacements of the membrane, it takes the form [compare Eq. (25.60) for a string]

$$\mathscr{L} = \tfrac{1}{2}\sigma\left(\frac{\partial u}{\partial t}\right)^2 - \tfrac{1}{2}\tau(\nabla u)^2 \tag{46.9b}$$

The total lagrangian for the system is the integral of the lagrangian density

$$L = \int \mathscr{L}\, dA \tag{46.10}$$

The equations of motion of the membrane follow as an application of Hamilton's principle to the lagrangian density (46.9b). It requires that the action be stationary for variations of the generalized coordinate u about the actual motion, subject to the condition of vanishing variations at the initial and final times

$$\delta \int_{t_1}^{t_2} L\, dt = \delta \int_{t_1}^{t_2} dt \int dA\; \mathscr{L}\left(u, \frac{\partial u}{\partial x}, \frac{\partial u}{\partial y}, \frac{\partial u}{\partial t}; x, y, t\right) = 0 \tag{46.11}$$

The steps leading to the Euler-Lagrange equation for this variational problem should now be familiar. We assume that the boundary conditions on the membrane allow us to discard the boundary contribution from the partial integrations of the spatial derivatives. This condition requires

$$(\delta u\; \tau\; \nabla u) \cdot \hat{n}_{\text{bndry}} = 0 \tag{46.12}$$

where $\hat{n}_{\text{bndry}}$ is the outward normal to the boundary curve of the membrane. The Euler-Lagrange equation for this problem then takes the form [see Eq. (45.2)]

$$\frac{\partial}{\partial t}\frac{\partial \mathscr{L}}{\partial(\partial u/\partial t)} + \frac{\partial}{\partial x}\frac{\partial \mathscr{L}}{\partial(\partial u/\partial x)} + \frac{\partial}{\partial y}\frac{\partial \mathscr{L}}{\partial(\partial u/\partial y)} - \frac{\partial \mathscr{L}}{\partial u} = 0 \tag{46.13}$$

Inserting the lagrangian density Eq. (46.9b), we arrive at the general two-dimensional wave equation

$$\sigma(\mathbf{x})\frac{\partial^2 u}{\partial t^2} = \nabla \cdot [\tau(\mathbf{x})\, \nabla u] \tag{46.14}$$

which governs the behavior of the membrane. In the usual case of constant mass density σ and constant surface tension τ it reduces to the two-dimensional wave equation

$$\frac{1}{c^2}\frac{\partial^2 u}{\partial t^2} = \nabla^2 u \qquad \begin{array}{l} \sigma = \text{const} \\[4pt] \tau = \text{const} \end{array} \tag{46.15}$$

where the velocity of propagation is defined by

$$c^2 \equiv \frac{\tau}{\sigma} \tag{46.16}$$

The two-dimensional wave equation can be analyzed with the techniques developed in the previous chapter. We first seek normal-mode solutions in which all elements of the membrane oscillate with the same frequency

$$u(\mathbf{x}, t) = \rho(\mathbf{x}) \cos{(\omega t + \phi)} \qquad (46.17)$$

Substitution into the wave equation leads to the eigenvalue equation

$$(\nabla^2 + k^2)\rho(\mathbf{x}) = 0 \qquad (46.18)$$

where the wave number is defined by

$$k \equiv \frac{\omega}{c} \qquad (46.19)$$

Equation (46.18) is known as the scalar *Helmholtz equation*; it describes the scalar function $\rho(\mathbf{x})$. As in the case of a string, its solutions depend on the spatial boundary conditions of the problem. Once the eigenfunctions have been obtained, the general time-dependent solution $u(\mathbf{x}, t)$ can be constructed from Eq. (46.17) by superposition.

47 SPECIFIC GEOMETRIES

We shall illustrate these results with a few specific applications.

Rectangular Membrane

Consider the simplest case of a rectangular membrane with clamped boundaries ($u = 0$), illustrated in Fig. 47.1a. In cartesian coordinates, the Helmholtz equation (46.18) takes the form

$$\frac{\partial^2 \rho}{\partial x^2} + \frac{\partial^2 \rho}{\partial y^2} + k^2\rho = 0 \qquad (47.1)$$

Furthermore, the boundary conditions here separate in the x and y directions,

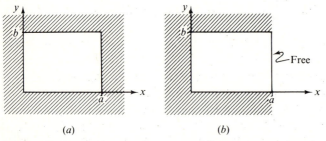

(a)   (b)

Figure 47.1 Rectangular membrane with (a) four clamped boundaries and (b) three clamped boundaries.

which suggests a separation of variables in cartesian coordinates. We thus look for factored solutions

$$\rho(x, y) = X(x)Y(y) \tag{47.2}$$

Substitution into Eq. (47.1) and division by ρ yields

$$k^2 + \frac{1}{X}\frac{d^2X}{dx^2} = -\frac{1}{Y}\frac{d^2Y}{dy^2} \tag{47.3}$$

The left-hand side is a function only of x, and the right-hand side is a function only of y. Hence this relation can only hold if both sides are in fact a constant, which we denote k_y^2. Thus we must have

$$\frac{d^2X}{dx^2} = -k_x^2 X \tag{47.4a}$$

$$\frac{d^2Y}{dy^2} = -k_y^2 Y \tag{47.4b}$$

where the separation constants are related by

$$k_x^2 + k_y^2 = k^2 \tag{47.5}$$

The two-dimensional problem has now been reduced to two one-dimensional problems; the solutions to Eqs. (47.4) that vanish at the endpoints are precisely those obtained in the previous chapter for a string with fixed endpoints

$$X(x) = \left(\frac{2}{a}\right)^{1/2} \sin k_x x \tag{47.6a}$$

$$Y(y) = \left(\frac{2}{b}\right)^{1/2} \sin k_y y \tag{47.6b}$$

where the eigenvalues are given by

$$k_x = \frac{m\pi}{a} \qquad m = 1, 2, 3, \ldots, \infty \tag{47.7a}$$

$$k_y = \frac{n\pi}{b} \qquad n = 1, 2, 3, \ldots, \infty \tag{47.7b}$$

and it is convenient here to omit the constant factor σ in the normalization integral. A combination with Eqs. (47.5) and (46.19) yields the normal-mode eigenfunctions and frequencies of a uniform stretched rectangular membrane with clamped boundaries

$$\rho_{mn}(x, y) = \left(\frac{4}{ab}\right)^{1/2} \sin \frac{m\pi x}{a} \sin \frac{n\pi y}{b} \tag{47.8a}$$

$$\frac{\omega_{mn}^2}{c^2} = \left(\frac{m\pi}{a}\right)^2 + \left(\frac{n\pi}{b}\right)^2 \tag{47.8b}$$

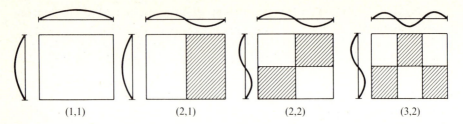

(1,1)	(2,1)	(2,2)	(3,2)

Figure 47.2 Normal modes (m,n) for a rectangular membrane with clamped sides. The shaded areas represent negative displacements, and the x and y profiles are also indicated.

Figure 47.2 sketches the first few eigenfunctions ρ_{mn}. It indicates the regions where the eigenfunctions are positive and negative, as well as the *nodal* lines, where the normal-mode eigenfunctions vanish. Note that the normal-mode frequencies, or overtones, are not simple multiples of the fundamental frequency

$$\frac{\omega_{11}}{c} = \left(\frac{\pi^2}{a^2} + \frac{\pi^2}{b^2}\right)^{1/2} \tag{47.9}$$

as they are for a one-dimensional string [compare Eq. (38.16b)]. This explains why a string is more pleasing musically than a plate or a rectangular drum.

The general solution to the two-dimensional wave equation can be obtained by superposition of the normal modes; in this case, it becomes

$$u(x, y, t) = \sum_{m=1}^{\infty} \sum_{n=1}^{\infty} \rho_{mn}(x, y)(a_{mn} \cos \omega_{mn} t + b_{mn} \sin \omega_{mn} t) \tag{47.10}$$

As in our previous analysis, it is also useful to define the normal coordinates [see Eq. (38.23), but note the different normalization]

$$\zeta_{mn}(t) \equiv \sigma^{1/2}(a_{mn} \cos \omega_{mn} t + b_{mn} \sin \omega_{mn} t)$$
$$\equiv \sigma^{1/2} c_{mn} \cos (\omega_{mn} t + \phi_{mn}) \tag{47.11}$$

To determine the behavior of the membrane for all subsequent times, the initial displacement and velocity of the membrane must be specified

$$u(x, y, 0) \equiv f(x, y) \tag{47.12a}$$
$$\dot{u}(x, y, 0) \equiv g(x, y) \tag{47.12b}$$

Substitution of the general solution (47.10) into the initial conditions leads to two-dimensional Fourier series, which can be inverted through the familiar orthogonality relation for the solutions (47.6) to give the coefficients

$$a_{mn} = \int_0^a dx \int_0^b dy \, \rho_{mn}(x, y) f(x, y) \tag{47.13a}$$

$$\omega_{mn} b_{mn} = \int_0^a dx \int_0^b dy \, \rho_{mn}(x, y) g(x, y) \tag{47.13b}$$

Since (47.10) is a solution to Eq. (46.15) with enough freedom to match an arbitrary set of initial conditions, it is indeed the general solution to the two-dimensional wave equation.

The total kinetic energy in the membrane is obtained from Eqs. (47.10), (47.11), and (46.1) as

$$T = \int_0^a dx \int_0^b dy \, \tfrac{1}{2}\sigma \left(\frac{\partial u}{\partial t}\right)^2 = \frac{1}{2} \sum_{mn} \dot{\zeta}_{mn}^2 \tag{47.14}$$

Similarly, Eq. (46.8) gives the total potential energy

$$V = \int_0^a dx \int_0^b dy \, \tfrac{1}{2}\tau(\nabla u)^2 = -\tfrac{1}{2}\tau \int_0^a dx \int_0^b dy \, u \, \nabla^2 u \tag{47.15}$$

where the last form follows from a partial integration and the boundary conditions eliminate the surface term. Use of the eigenvalue relation (46.18), the orthonormality of the eigenfunctions (47.8a), and Eq. (46.16) yields

$$V = \frac{1}{2} \sum_{mn} \omega_{mn}^2 \, \zeta_{mn}^2 \tag{47.16}$$

As anticipated, the normal coordinates defined in Eq. (47.11) diagonalize the lagrangian

$$L = T - V = \frac{1}{2} \sum_{mn} (\dot{\zeta}_{mn}^2 - \omega_{mn}^2 \, \zeta_{mn}^2) \tag{47.17}$$

Since this expression describes a set of uncoupled simple harmonic oscillators, it is also evident that the hamiltonian is both the energy and a constant of the motion

$$H = E = T + V = \text{const} \tag{47.18}$$

What happens if we alter the boundary conditions? To indicate the various possibilities, we consider a general boundary condition, shown in Fig. 47.3. Although this figure illustrates a one-dimensional string, the extension to a two-dimensional membrane merely requires adding a transverse dimension. In this example, a massless spring with force constant κ constrains the left-hand end of the string to move vertically. The string acts on the spring with a vertical component of force

$$F_z = \tau \sin \phi \approx \tau \frac{\partial u(x, t)}{\partial x}\bigg|_{x=0} \tag{47.19}$$

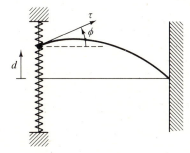

Figure 47.3 Illustration of homogeneous boundary condition. A string with tension τ is attached to a spring with force constant κ constrained to move in vertical direction.

where the last form assumes small displacements of the string. By Hooke's law, this force is proportional to the displacement d of the spring, and we have

$$F_z = \kappa d = \kappa u(0, t) \tag{47.20}$$

since $d = u(0, t)$ is the displacement of the end of the string. These relations can be combined to give the boundary condition

$$\tau \frac{\partial u(x, t)}{\partial x}\bigg|_{x=0} = \kappa u(0, t) \tag{47.21}$$

It provides a physical example of the *general homogeneous boundary condition* (40.13) discussed in the previous chapter. For normal-mode solutions

$$u(x, t) = \rho(x) \cos (\omega t + \phi) \tag{47.22}$$

the homogeneous boundary condition becomes

$$\tau \rho'(0) = \kappa \rho(0) \tag{47.23}$$

In particular, a free end is simply the limiting case as the force constant of the spring goes to zero

Free end: $$\kappa \to 0 \tag{47.24}$$

Thus *a free end requires the natural boundary condition* (40.12)

$$\tau \rho'(0) = 0 \tag{47.25}$$

As a specific example, consider a rectangular membrane with three edges clamped and one free (Fig. 47.1b). The only modification of the previous analysis of Eq. (47.1) is that the separated solutions to Eqs. (47.4) must now satisfy Eq. (47.25) to guarantee zero slope at the free end

$$X'(a) = 0 \tag{47.26}$$

The corresponding normal-mode eigenfunctions take the form

$$\rho_{mn}(x, y) = \left(\frac{4}{ab}\right)^{1/2} \sin\left(\frac{2m + 1}{2}\frac{\pi x}{a}\right) \sin\frac{n\pi y}{b} \quad \begin{array}{l} m = 0, 1, 2, \ldots, \infty \\ n = 1, 2, \ldots, \infty \end{array} \tag{47.27}$$

and the normal-mode frequencies are given by

$$\frac{\omega_{mn}^2}{c^2} = \left(\frac{2m + 1}{2}\frac{\pi}{a}\right)^2 + \left(\frac{n\pi}{b}\right)^2 \tag{47.28}$$

Note that the fundamental frequency $\omega_{01} = c\pi[(4a^2)^{-1} + b^{-2}]^{1/2}$ now lies lower than Eq. (47.9) for a fully clamped membrane. This comparison confirms the intuition that removing constraints on a mechanical system lowers the resonant frequencies. We leave the sketch of the normal modes and nodal surfaces as an exercise for the reader.

Circular Membrane

As a second example, consider a circular membrane of radius a with clamped boundaries, illustrated in Fig. 47.4. The boundary condition obviously separates in plane polar coordinates (r, ϕ) since it involves only the radial variable r and is independent of the angle ϕ. Hence it is appropriate to seek solutions to Eq. (46.18) that also separate in plane polar coordinates. The only difficulty arises in transforming the laplacian. We recall from Eq. (B.26) that it takes the form

$$\nabla^2 = \frac{1}{r}\frac{\partial}{\partial r}r\frac{\partial}{\partial r} + \frac{1}{r^2}\frac{\partial^2}{\partial \phi^2} \tag{47.29a}$$

in plane polar coordinates, where we now use r to avoid confusion with the eigenfunction ρ. The scalar Helmholtz equation (46.18) then becomes

$$\left(\frac{1}{r}\frac{\partial}{\partial r}r\frac{\partial}{\partial r} + \frac{1}{r^2}\frac{\partial^2}{\partial \phi^2} + k^2\right)\rho(r, \phi) = 0 \tag{47.29b}$$

To solve this equation by separation of variables, we assume

$$\rho(r, \phi) = R(r)\Phi(\phi) \tag{47.30}$$

Substitution into Eq. (47.29b) (see Appendix C) leads to a polar-angle equation of the Sturm-Liouville form

$$-\frac{d^2\Phi}{d\phi^2} = m^2\Phi \tag{47.31}$$

with the separation constant m^2 acting as the eigenvalue. Physically acceptable solutions of Eq. (47.31) must be *single-valued*, which means that the disturbance must return to its original value if ϕ increases by 2π:

$$\Phi(\phi + 2\pi) = \Phi(\phi) \tag{47.32}$$

This requirement can be recognized as specifying periodic boundary conditions (40.15) in the variable ϕ. The complex Fourier series [see Eq. (40.2)] provides just such a complete set of periodic solutions to (47.31) on the interval $0 \leq \phi \leq 2\pi$

$$\Phi_m(\phi) = (2\pi)^{-1/2}e^{im\phi} \qquad m = 0, \pm 1, \pm 2, \ldots, \pm \infty \tag{47.33}$$

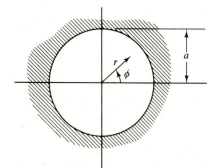

Figure 47.4 Circular membrane with clamped boundary.

and the periodicity (47.32) selects the integral values of m in Eq. (47.33). These functions satisfy the orthonormality condition

$$\langle \Phi_m | \Phi_{m'} \rangle = (2\pi)^{-1} \int_0^{2\pi} e^{i(m'-m)\phi} \, d\phi = \delta_{mm'} \tag{47.34}$$

Separation of variables also gives the radial equation [see Eq. (C.7)]

$$r^2 \frac{d^2 R}{dr^2} + r \frac{dR}{dr} + (k^2 r^2 - m^2)R = 0 \tag{47.35}$$

which is Bessel's equation (see Appendix D3). It can be recast in Sturm-Liouville form (40.9)

$$-\frac{d}{dr}\left(r \frac{dR}{dr}\right) + \frac{m^2}{r} R = k^2 r R \tag{47.36a}$$

leading to the identification of the Sturm-Liouville parameters

$$x = r \qquad \tau(x) = \sigma(x) = r \qquad v(x) = m^2 r^{-1} \tag{47.36b}$$

and k^2 is the eigenvalue of the equivalent radial Sturm-Liouville problem. A fundamental system of solutions to Bessel's equation (47.35) for integral m is given by the Bessel and Neumann functions $J_m(z)$ and $N_m(z)$ with $z = kr$. We note, however, that the function $N_m(z)$ is singular as $z \to 0$ [see Eqs. (D3.22) and (D3.23)]. Since the membrane cannot undergo an infinite displacement at the origin, we must discard the functions $N_m(z)$ and retain only $J_m(z)$.

The condition of a clamped boundary at $r = a$ (Fig. 47.4) requires that the acceptable radial solution vanish at this radius [see Eq. (C.9)]

$$J_m(ka) = 0 \tag{47.37}$$

We let $\alpha_{m,n}$ denote the nth zero of the mth Bessel function

$$J_m(\alpha_{m,n}) = 0 \tag{47.38}$$

A comparison of Eqs. (47.37) and (47.38) yields the eigenvalues

$$k_{mn}^2 = \frac{\alpha_{m,n}^2}{a^2} = \frac{\omega_{mn}^2}{c^2} \tag{47.39}$$

For given m, the zeros of the mth Bessel function determine the eigenvalues. Since $J_m(z)$ oscillates for $z \to \infty$ [see Eq. (D3.28)], the eigenvalues increase without bound. Moreover, the general Sturm-Liouville theory demonstrates that the solutions to Eq. (47.35) corresponding to different eigenvalues n are orthonormal with respect to the metric [see Eq. (47.36)]

$$dm = \sigma(x) \, dx = r \, dr \tag{47.40}$$

Thus we have

$$\int_0^a r \, dr \, J_m\left(\alpha_{m,n} \frac{r}{a}\right) J_m\left(\alpha_{m,n'} \frac{r}{a}\right) = \delta_{nn'} \tfrac{1}{2} a^2 [J_{m\pm1}(\alpha_{m,n})]^2 \tag{47.41}$$

where the normalization constant for $n = n'$ on the right-hand side of Eq. (47.41) requires an explicit evaluation of the indicated integral and the choice between $\pm$ is arbitrary because of the standard recurrence relation

$$2mz^{-1}J_m(z) = J_{m+1}(z) + J_{m-1}(z) \tag{47.42}$$

Finally, the general Sturm-Liouville theory also proves that these eigenfunctions of Eq. (47.36) form a complete set *for each value of the separation constant m.*

It is instructive to verify explicitly that the products of our separated solutions

$$\rho_{mn}(r, \phi) = \frac{1}{\pi^{1/2}a|J_{m\pm 1}(\alpha_{m, n})|} J_m\left(\alpha_{m, n}\frac{r}{a}\right)e^{im\phi} \qquad \begin{array}{l} m = 0, \pm 1, \ldots, \pm\infty \\ n = 1, 2, \ldots, \infty \end{array} \tag{47.43}$$

provide a complete basis for describing an arbitrary function of (r, ϕ) that vanishes for $r = a$. First suppose that the coordinate r is fixed. The completeness of the Fourier series (47.33) guarantees that any function $f(r, \phi)$ that is periodic in ϕ can be expanded according to

$$f(r, \phi) = \sum_{m=-\infty}^{\infty} c_m(r)e^{im\phi} \tag{47.44a}$$

Furthermore, the completeness of the solutions $J_m(\alpha_{m, n}r/a)$ to the radial Sturm-Liouville equation for each m ensures that any function of r, in particular the coefficients $c_m(r)$, can be expanded in terms of the eigenfunctions of the radial Sturm-Liouville equation according to

$$c_m(r) = \sum_{n=1}^{\infty} c_{mn}J_m\left(\alpha_{m, n}\frac{r}{a}\right) \tag{47.44b}$$

Substitution of Eq. (47.44b) into Eq. (47.44a) gives the desired expansion

$$f(r, \phi) = \sum_{m=-\infty}^{\infty}\sum_{n=1}^{\infty} c_{mn}J_m\left(\alpha_{m, n}\frac{r}{a}\right)e^{im\phi} \tag{47.45}$$

for continuous piecewise differentiable functions.

The general solution to the two-dimensional wave equation again represents a superposition of the separated normal-mode solutions

$$u(r, \phi, t) = \sum_{m=-\infty}^{\infty}\sum_{n=1}^{\infty} \rho_{mn}(r, \phi)(a_{mn}\cos\omega_{mn}t + b_{mn}\sin\omega_{mn}t) \tag{47.46}$$

As in previous examples, the coefficients allow just enough flexibility to match the initial conditions

$$u(r, \phi, 0) = f(r, \phi) \tag{47.47a}$$

$$\dot{u}(r, \phi, 0) = g(r, \phi) \tag{47.47b}$$

The orthonormality of the separated solutions determines these expansion coefficients a_{mn} and b_{mn} explicitly. First project out a given value of m by integrating with an angular eigenfunction from Eq. (47.33). Then project out a given value of n by using the orthonormality (47.41) of the radial eigenfunctions. (Note the importance of carrying out the operations in this particular order.) In this way, the coefficients in Eq. (47.46) follow by inverting the initial conditions (47.47)

$$a_{mn} = \int_0^a r\, dr \int_0^{2\pi} d\phi\, \rho_{mn}(r,\,\phi)^* f(r,\,\phi) \qquad (47.48a)$$

$$\omega_{mn} b_{mn} = \int_0^a r\, dr \int_0^{2\pi} d\phi\, \rho_{mn}(r,\,\phi)^* g(r,\,\phi) \qquad (47.48b)$$

We now use Eq. (47.43) and the following property of Bessel functions

$$J_{-m}(z) = (-1)^m J_m(z) \qquad m \text{ an integer} \qquad (47.49)$$

to note that

$$\rho_{mn}(r,\,\phi)^* = (-1)^m \rho_{-m,\,n}(r,\,\phi) \qquad (47.50)$$

Hence if the initial values in Eq. (47.47) are *real*, the complex conjugate of Eqs. (47.48) shows that the expansion coefficients in Eq. (47.46) must satisfy the conditions

$$a_{mn}^* = (-1)^m a_{-m,\,n} \qquad b_{mn}^* = (-1)^m b_{-m,\,n} \qquad (47.51)$$

Conversely, if the expansion coefficients in Eq. (47.46) satisfy these conditions, the displacement u is explicitly real.

Figure 47.5 sketches the first four normal-mode eigenfunctions Re $\rho_{mn}(r,\,\phi)$, again indicating regions of positive and negative displacement and the nodal lines. It also gives the corresponding radial functions normalized according to Eq. (47.41) and the dimensionless frequency $\alpha_{m,\,n} \equiv \omega_{mn}\, a/c$. As in Eq. (47.9), these frequencies are not simple multiples of the fundamental $\omega_{01} \approx 2.405 c/a$. Note that the nodal lines may in fact have any orientation, for a real superposition

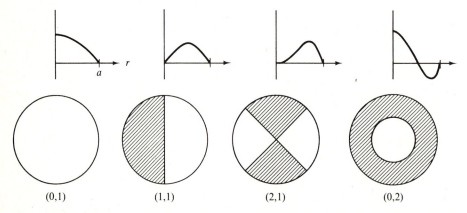

Figure 47.5 Lowest four normal modes $J_m(\alpha_{m,n} r/a) \cos m\phi$ for circular drumhead with clamped boundaries. The shaded areas indicate negative displacements for the mode (m,n). The associated frequency is $\omega_{mn} a/c \equiv \alpha_{m,n}$ with $\alpha_{0,1} = 2.405$, $\alpha_{1,1} = 3.832$, $\alpha_{2,1} = 5.136$, and $\alpha_{0,2} = 5.520, \ldots$. The radial eigenfunctions are also shown.

of the degenerate separated solutions $(e^{im\phi}, e^{-im\phi})$ is equivalent to a normal-mode solution of the form $\cos(m\phi + \epsilon_{mn})$, according to the relations

$$a_{mn}e^{im\phi} + a_{mn}^{*}e^{-im\phi} = c_{mn}\cos(m\phi + \epsilon_{mn}) \qquad a_{mn} \equiv \tfrac{1}{2}c_{mn}e^{i\epsilon_{mn}} \qquad (47.52)$$

Variational Estimate of Lowest Drumhead Mode

As a simple example of the Rayleigh-Ritz approximation method developed in Sec. 42, we use the variational principle to estimate the lowest eigenvalue for a circular drumhead. The Sturm-Liouville parameters are identified in Eq. (47.36), and the lowest mode is assumed independent of angle $(m = 0)$ so that $v = 0$ in this case. The variational estimate (42.1) for the eigenvalue k^2 is thus given by

$$k^2[\rho] = \frac{\int_0^a \left(\dfrac{d\rho}{dr}\right)^2 r\,dr}{\int_0^a \rho^2 \, r\,dr} \qquad (47.53)$$

Take the following form [compare Eq. (42.17)]

$$\rho(r) = 1 - \left(\frac{r}{a}\right)^{\beta} \qquad (47.54)$$

for the variational trial function. We require

$$\beta > 1 \qquad (47.55)$$

to ensure that the function has vanishing slope at the origin; otherwise, the membrane would have a cusp at $r = 0$, with a finite radial slope for all values of ϕ. Furthermore, the trial function (47.54) vanishes on the boundary $r = a$. Use of this variational form and the substitution $r/a = x$ reduces Eq. (47.53) to

$$a^2k^2(\beta) = \frac{\beta^2 \int_0^1 x^{2\beta-1}\,dx}{\int_0^1 x(1 - 2x^{\beta} + x^{2\beta})\,dx} \qquad (47.56)$$

Note that the integral in the numerator exists for all positive β. The integrals are elementary, and straightforward algebra gives the desired upper bound for the lowest eigenvalue of the circular drumhead

$$a^2k^2(\beta) = \beta + 3 + 2\beta^{-1} \qquad (47.57)$$

The best estimate for the lowest eigenvalue follows by minimizing this expression with respect to the parameter β. Setting the derivative with respect to β equal to zero, we obtain

$$1 - \frac{2}{\beta^2} = 0 \qquad (47.58)$$

The solution to this equation that satisfies the condition (47.55) is

$$\beta_{min} = \sqrt{2} \qquad (47.59)$$

It is evident by inspection that this value indeed minimizes the expression (47.57), and our estimate for the lowest eigenvalue therefore becomes

$$a^2 k^2(\beta_{\min}) = 3 + 2\sqrt{2} = 5.828 \cdots \qquad (47.60)$$

It should be compared with the exact answer

$$a^2 k_{01}^2 = (\alpha_{0,1})^2 = (2.4048 \cdots)^2 = 5.783 \cdots \qquad (47.61)$$

where $\alpha_{0,1}$ is the first zero of $J_0(z)$. The variational estimate is high, as it must be, but it differs from the exact value by less than 1 percent. The exact lowest-mode eigenfunction

$$\rho_{01}(r) \propto J_0\left(\alpha_{0,1} \frac{r}{a}\right) \qquad (47.62)$$

can be written explicitly with the power series (see Appendix D3)

$$J_0(z) = \sum_{p=0}^{\infty} \frac{(-1)^p}{(p!)^2} (\tfrac{1}{2}z)^{2p} \approx 1 - \tfrac{1}{4}z^2 + \tfrac{1}{64}z^4 + \cdots \qquad (47.63)$$

We have evidently tried to approximate this infinite series with a two-term expression of the form (47.54).

Perturbation Theory for Nearly Circular Boundary

As an illustration of perturbation theory, we consider the normal-mode frequencies and eigenfunctions for a uniform membrane whose boundary differs in a small, but otherwise arbitrary, amount from a circle. In polar coordinates an arbitrary single-valued curve has the equation (see Fig. 47.6)

$$R(\phi) = a\left[1 + \sum_{p=1}^{\infty} (\epsilon_p \cos p\phi + \bar{\epsilon}_p \sin p\phi)\right] \qquad (47.64)$$

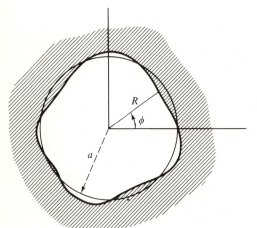

Figure 47.6 Membrane with clamped boundary that differs slightly from a circle.

which is just an expansion in a (real) Fourier series. Here the angle-independent term has been absorbed into the definition of the mean radius of the figure

$$\langle R \rangle = a \tag{47.65}$$

where angular brackets here denote an angular average. Furthermore, the quantities $\{\epsilon_p, \bar{\epsilon}_p\}$, denoted generically by ϵ throughout this discussion, are small and will be treated in first order. In this way, we analyze small distortions of the bounding curve from a circle of radius a, as indicated in Fig. 47.6.

We now show that the terms with $p = 1$ in the sum (47.64) can be discarded, for they correspond to a uniform translation of the original circle that does not affect the properties of the normal modes. Let $\mathbf{a} = a\hat{r}$ be the vector from the origin to a general point on the original circle. Suppose that this vector is displaced by a constant small vector $\boldsymbol{\eta}$. The distance from the original origin to the new displaced point is given by

$$R = |\mathbf{a} + \boldsymbol{\eta}| = (a^2 + 2\mathbf{a} \cdot \boldsymbol{\eta} + \eta^2)^{1/2} \tag{47.66}$$

and an expansion through first order in the small displacement $\boldsymbol{\eta}$ yields

$$R \approx a + \hat{a} \cdot \boldsymbol{\eta} = a + \eta \cos \psi \tag{47.67}$$

where ψ is the angle between the unit vectors $\hat{a}$ ($= \hat{r}$) and $\hat{\eta}$. It is evident that the correction term in Eq. (47.67) is precisely the $p = 1$ term in (47.64) [compare Eq. (47.52)].

The *area* of the deformed figure will play a central role in our discussion. It can be computed in the usual fashion

$$A = \int_0^{2\pi} d\phi \int_0^{R(\phi)} r\, dr = \int_0^{2\pi} d\phi\, \tfrac{1}{2} R^2(\phi) \tag{47.68}$$

where the radial integration has been carried out *first*, at a fixed angle ϕ. Substitution of Eq. (47.64) and expansion to first order in ϵ gives the expression

$$A = \tfrac{1}{2} a^2 \int_0^{2\pi} d\phi \left[1 + 2 \sum_{p=2}^{\infty} (\epsilon_p \cos p\phi + \bar{\epsilon}_p \sin p\phi) \right] + O(\epsilon^2) \tag{47.69}$$

Since each term in the sum over p has zero angular average, we conclude that the new figure described by Eq. (47.64) *has the same area* as the original circle to first order in ϵ

$$A = \pi a^2 + O(\epsilon^2) \tag{47.70}$$

The normal-mode eigenfunctions of the membrane with deformed boundary in Fig. 47.6 still satisfy the scalar Helmholtz equation (46.18). Although the boundary conditions are no longer separable in polar coordinates, we can still construct a solution as a superposition of those solutions that are periodic in ϕ and nonsingular at the origin

$$\rho(r, \phi) = \sum_{m=0}^{\infty} J_m(kr)(A_m \cos m\phi + B_m \sin m\phi) \tag{47.71}$$

Here, B_0 in fact never appears because $\sin m\phi = 0$ for $m = 0$. The remaining coefficients must be chosen to satisfy the boundary condition

$$\rho[R(\phi), \phi] = 0 \tag{47.72}$$

which, for a general deformation of the bounding contour, leads to a complicated set of coupled equations. It is, however, relatively simple to solve Eq. (47.72) to first order in ϵ.

If $\epsilon = 0$, the modes are just those of a circular membrane, and we therefore must consider how they change for small but nonzero ϵ. For simplicity, we shall study the behavior of a mode that was originally circularly symmetric, assuming that the coefficients in Eq. (47.71) have the following orders of magnitude

$$A_0 = 1 \tag{47.73a}$$

$$A_m, B_m = O(\epsilon) \qquad m = 1, 2, 3, \ldots, \infty \tag{47.73b}$$

The choice in Eq. (47.73a) merely defines a convenient normalization. If Eqs. (47.71) and (47.64) are substituted in Eq. (47.72), the result can be rewritten as

$$J_0 \left\{ ka \left[1 + \sum_{p=2}^{\infty} (\epsilon_p \cos p\phi + \bar{\epsilon}_p \sin p\phi) \right] \right\}$$

$$+ \sum_{m=1}^{\infty} J_m(kR)(A_m \cos m\phi + B_m \sin m\phi) = 0 \tag{47.74}$$

Now observe that we can make the following simplifications to first order in ϵ:

1. Expand the first term of (47.74) in a Taylor series and keep only the first correction of order ϵ
2. Replace R by a in the second term of (47.74) because the coefficients A_m and B_m are already of order ϵ

As a result, the boundary condition (47.74) has the simpler form

$$J_0(ka) + kaJ'_0(ka) \sum_{p=2}^{\infty} (\epsilon_p \cos p\phi + \bar{\epsilon}_p \sin p\phi)$$

$$+ \sum_{m=1}^{\infty} J_m(ka)(A_m \cos m\phi + B_m \sin m\phi) = 0 \tag{47.75}$$

Since this Fourier series vanishes identically for all ϕ, each Fourier coefficient must vanish separately

$$J_0(ka) = 0 \tag{47.76a}$$

$$A_1 J_1(ka) = B_1 J_1(ka) = 0 \tag{47.76b}$$

$$kaJ'_0(ka)\epsilon_m = -J_m(ka)A_m \tag{47.76c}$$

$$kaJ'_0(ka)\bar{\epsilon}_m = -J_m(ka)B_m \tag{47.76d}$$

$$m = 2, 3, \ldots, \infty$$

Equation (47.76a) determines the actual eigenfrequencies for the modes that were circularly symmetric in the original unperturbed problem. In the notation of Eqs. (47.38) and (47.39), we have

$$k_{0n}a = \alpha_{0,n} \qquad n = 1, 2, 3, \ldots, \infty \tag{47.77}$$

In addition, $J_1(x)$ does not vanish at those values $\alpha_{0,n}$ where $J_0(x) = 0$, and Eq. (47.76b) then requires that

$$A_1 = B_1 = 0 \tag{47.78}$$

Finally, the Bessel functions satisfy

$$J_0'(x) = -J_1(x) \tag{47.79}$$

and Eqs. (47.76c) and (47.76d) can be rewritten as

$$A_m = \epsilon_m \frac{kaJ_1(ka)}{J_m(ka)} \tag{47.80a}$$

$$m = 2, 3, \ldots, \infty$$

$$B_m = \bar{\epsilon}_m \frac{kaJ_1(ka)}{J_m(ka)} \tag{47.80b}$$

When inserted in Eq. (47.71), these results provide solutions that satisfy both the scalar Helmholtz equation for a uniform membrane and the condition (47.72) on the boundary (47.64) correct to first order in ϵ. Furthermore, it is evident from Eq. (47.73) that these solutions reduce to the following expression as the bounding contour approaches the circle ($\epsilon \to 0$)

$$\rho(r, \phi) \xrightarrow[\epsilon \to 0]{} J_0\left(\alpha_{0,n}\frac{r}{a}\right) \tag{47.81}$$

The normal-mode frequencies of the membrane with deformed boundary are now readily obtained from Eq. (47.77)

$$k_{0n}^2 = \frac{\omega_{0n}^2}{c^2} = \left(\frac{\alpha_{0,n}}{a}\right)^2 = \frac{\pi\alpha_{0,n}^2}{A} \tag{47.82}$$

where the area of the actual membrane has been identified through Eq. (47.70). This remarkable result demonstrates that these normal-mode frequencies depend only on the *total area* A of the membrane. Hence, to first order in ϵ, they are identical with those of a circular membrane of the same area, even though the bounding contour (47.64) contains all Fourier components and is therefore essentially arbitrary.† The lowest normal-mode frequency for a nearly circular membrane of area A is given by [see Eq. (47.61)]

$$\frac{\omega_{01}}{c} \approx \left(\frac{\pi}{A}\right)^{1/2}(2.4048 \cdots) \approx \frac{4.2624 \cdots}{A^{1/2}} \tag{47.83}$$

---

† This calculation and conclusion are due to Rayleigh (1945) secs. 209–211.

Although a square cannot be considered a small deformation of a circle, it is amusing to express its exact lowest normal-mode frequency (47.9) in the same form

$$\frac{\omega_{11}}{c} = \frac{(2\pi^2)^{1/2}}{A^{1/2}} \approx \frac{4.4429\cdots}{A^{1/2}} \tag{47.84}$$

This value exceeds the approximate perturbative prediction (47.83) by only about 4 percent.

The preceding expansion can be extended to order ϵ^2. If the deformation of the bounding contour preserves the area of the membrane, a straightforward analysis eventually shows that the lowest normal-mode frequency is a positive-definite quadratic form in the parameters ϵ (Prob. 8.12). Consequently, any alteration in the shape (but not area) of an originally circular membrane necessarily raises the fundamental frequency; conversely, of all membranes with a given area, the circular form has the lowest fundamental frequency. Equation (47.84) evidently exemplifies this last relation.

PROBLEMS

8.1 Derive the general two-dimensional wave equation (46.14) directly from Newton's second law applied to a small arbitrary element of the membrane.

8.2 A rectangular membrane has two clamped sides and two free sides. Find the normal modes and frequencies if the free sides are opposite and if the free sides are adjacent. Compare the low-lying modes with those of a membrane with three or four sides clamped.

8.3 A rectangular membrane with sides a and b is stretched so that the tension in the x direction is τ_1 and in the y direction is τ_2.

(a) Show that the equation of motion is

$$\tau_1 \frac{\partial^2 u}{\partial x^2} + \tau_2 \frac{\partial^2 u}{\partial y^2} = \sigma \frac{\partial^2 u}{\partial t^2}$$

(b) Change variables to $x/\sqrt{\tau_1}$, $y/\sqrt{\tau_2}$ and find the eigenfrequencies and eigenfunctions.

8.4 A rectangular membrane of sides a and b has clamped boundaries. Show that the number of normal modes with frequency less than $v_{max} \equiv \omega_{max}/2\pi$ is approximately equal to the area of the first quadrant of the ellipse

$$\frac{x^2}{a^2} + \frac{y^2}{b^2} = \frac{4\sigma}{\tau} v_{max}^2$$

Thus show that this number is approximately $\pi\sigma ab v_{max}^2/\tau$.

8.5 A square membrane with area $A = a^2$ is clamped at its edges.

(a) Express the first four *distinct* eigenfrequencies both as multiples of $c/A^{1/2}$ and of the fundamental; determine the corresponding eigenfunctions $\rho_{mn}(x, y)$. What is the degeneracy of each of these modes [see Eq. (47.8)]?

(b) Sketch the displacement and nodal lines for each of these pure modes $\rho_{mn}(x, y)$. In addition, find the nodal lines for the sums and differences $\rho_{mn} \pm \rho_{nm}$ of the degenerate eigenfunctions.

(c) Compare with the first four frequencies and normal modes of a clamped circular membrane of equal area.

8.6 The modes of a clamped membrane whose boundary is an isosceles right triangle are identical with those of a clamped square membrane with a diagonal nodal line. Use the results of Prob. 8.5 to find the two lowest eigenfrequencies and sketch the corresponding displacements. Compare the fundamental frequency with Eq. (47.83) for a circle of the same area.

8.7 An annular membrane is stretched between two fixed concentric circles of radii a and b $(a < b)$.

(a) Find an exact condition for the eigenfrequencies.

(b) Consider the limit $a \to 0$ and compare with the circular membrane discussed in Sec. 47. Consider the limit $b - a \ll b$ and compare with the rectangular membrane clamped on two opposite sides and free on the other two sides (see Prob. 8.2).

8.8 Consider the wave equation $\nabla^2 u = c^{-2}\, \partial^2 u/\partial t^2$ for a membrane in the shape of a *sector of a circle* with opening angle γ and radius a, subject to the condition $u = 0$ on the boundary.

(a) Show that the problem is separable in plane polar coordinates and that the eigenfunctions are

$$\rho_{mn}(r, \phi) \propto J_{m\pi/\gamma}\left(\frac{X_{mn} r}{a}\right) \sin \frac{m\pi\phi}{\gamma} \qquad m = 1, 2, 3, \ldots, \infty$$

with eigenvalues determined by

$$J_{m\pi/\gamma}(X_{mn}) = 0 \qquad n = 1, 2, \ldots, \infty$$

(b) Sketch the normal modes and nodal lines for a few typical low-lying modes.

(c) Use tables to find the three lowest eigenfrequencies for each of the special cases $\gamma = \tfrac{1}{2}\pi, \pi, 2\pi$.

(d) Construct the general solution to the initial-value problem

$$u(r, \phi, 0) = f(r, \phi) \qquad \dot{u}(r, \phi, 0) = g(r, \phi)$$

8.9 (a) Generalize Eq. (42.1) to obtain a variational functional for the normal-mode function $\rho(\mathbf{x})$ of the two-dimensional wave equation (46.14). Verify that it has the correct Euler-Lagrange equation.

(b) Use the variational trial function $\rho(\mathbf{r}) = r(a - r) \sin(\pi\phi/\gamma)$ to estimate the lowest eigenvalue for general γ in Prob. 8.8.

(c) Compare with the exact result for the special cases $\gamma = \tfrac{1}{2}\pi, \pi, 2\pi$. Discuss the relative accuracy for the three values of γ.

8.10 A uniform circular membrane of radius a, areal mass density σ, and tension τ has the normalized eigenfunctions and eigenvalues of Eqs. (47.43) and (47.39).

(a) A point mass m is attached at the center of the membrane. Show that the total density is now $\sigma + (m/\pi r)\, \delta(r)$, where $\delta(r)$ is a one-dimensional delta function. [Note $\int_0^\infty \delta(r)\, dr = \tfrac{1}{2}$.]

(b) Use first-order perturbation theory to show that only the circularly symmetric modes are affected, in which case

$$k^2 a^2 = \alpha_{0,n}^2 \left[1 - \frac{m}{\pi\sigma a^2 J_1^2(\alpha_{0,n})}\right]$$

where $J_0(\alpha_{0,n}) = 0$. Use tables to evaluate the numerical coefficients for $n = 1, 2, 3$. Discuss the behavior for large n; compare with the corresponding case of a point mass on a string.

8.11 (a) Use the variational principle (Prob. 8.9a) with the separate trial functions $\rho = 1 - (r/a)^2$ and $\rho = J_0(\alpha_{0,1} r/a)$ to estimate $k^2 a^2$ for the lowest symmetric mode in Prob. 8.10. Which trial function gives a lower value for $m/\pi\sigma a^2 \to 0$? For $m/\pi\sigma a^2 \to \infty$?

(b) Expand each approximate eigenvalue to order $m/\pi\sigma a^2$ and verify explicitly that each provides an upper bound to the true first-order solution.

8.12 Consider a nearly circular membrane of specified area A, bounded by the closed curve (47.64). Prove that the frequencies of the modes that are nearly independent of angle are shifted from the unperturbed values $\omega_{0n} = c\alpha_{0,n}(\pi/A)^{1/2}$ by an amount

$$\frac{\delta\omega_{0n}}{\omega_{0n}} = \frac{1}{2} \sum_{p=2}^\infty \left[1 + \frac{\alpha_{0,n} J_p'(\alpha_{0,n})}{J_p(\alpha_{0,n})}\right](\epsilon_p^2 + \bar{\epsilon}_p^2)$$

correct to second order in ϵ. Here $\alpha_{0,n}$ is the nth zero of J_0. Use the form of the functions $J_p(x)$ to verify that $\delta\omega_{01}$ is a positive-definite function of the deformation parameters.

NINE

SOUND WAVES IN FLUIDS

The preceding two chapters introduced the basic concepts of continuum mechanics in the familiar context of strings and membranes. These systems are special in at least two distinct ways, however, for the lower dimensionality reduces the number of independent variables, and the elastic properties ensure that two initially adjacent elements remain so forever. In contrast, an element in a bulk fluid requires a three-dimensional vector to specify its position, and convective flow eventually can separate neighboring elements by arbitrary amounts. This last feature significantly affects the conservation of momentum.

48 GENERAL EQUATIONS OF HYDRODYNAMICS

As for a string or membrane, we shall treat a fluid as a continuous medium, ignoring its detailed microscopic structure. Consider a small element of the fluid. The concept of *pressure* allows us to incorporate the influence of the surrounding medium. If one introduces a gauge that measures force per unit area in the interior of a fluid at rest (see Fig. 48.1), the resulting pressure is given by

$$p \equiv \frac{\text{force}}{\text{area}} \tag{48.1}$$

Experiments verify the following characteristics:

1. The pressure is normal to the area.
2. The pressure is independent of orientation at any point x in the fluid.

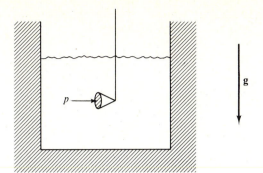

Figure 48.1 Measurement of pressure in a fluid at rest in a terrestrial laboratory.

Formulation of Newton's Second Law

In general, the pressure varies throughout the fluid. For example, if the fluid experiences a force that acts on each element, such as the gravitational force in a terrestrial laboratory, then a stationary fluid must establish a balancing pressure *gradient*. A mathematical formulation of this idea follows by considering the pressure force on a small element of the fluid, as indicated in Fig. 48.2. Since this force will turn out to be proportional to the volume of the fluid element, we define the pressure force as

$$\text{Pressure force on infinitesimal element of fluid} \equiv \mathbf{F}_{pr}\, dV \qquad (48.2)$$

where $\mathbf{F}_{pr}$ is a *force density* with dimension $f l^{-3}$. The observations following Eq. (48.1) above permit us to write the x component of the force acting on this small element of fluid in the following manner

$$F_x = p(x)\, dy\, dz - p(x + dx)\, dy\, dz \qquad (48.3)$$

Here the right-hand side is already proportional to $dy\, dz$, and we therefore keep y and z fixed when evaluating this expression. A Taylor-series expansion then yields

$$F_x \approx -\frac{\partial p}{\partial x}\, dV \qquad (48.4)$$

where we have invoked the usual definition of a partial derivative and have identified the infinitesimal volume element $dV = dx\, dy\, dz$. The same arguments

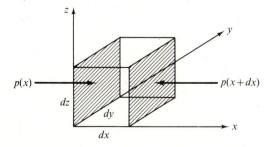

Figure 48.2 Calculation of pressure force on a small element of fluid.

apply in the other two orthogonal directions, and a comparison of Eqs. (48.2) and (48.4) allows us to identify the force exerted by the pressure on the small volume element indicated in Fig. 48.2 as

$$\mathbf{F}_{pr} = -\nabla p \tag{48.5}$$

Assume that an external force acts throughout the volume of the fluid. It will be written

$$\text{Applied volume force} \equiv \mathbf{f}_{app}\rho \, dV \tag{48.6}$$

where ρ is the *mass density* of the fluid. Thus $\mathbf{f}_{app}$ is explicitly the applied force per unit mass of fluid. In the case of *hydrostatic equilibrium*, the total force on the volume element in Fig. 48.2 must vanish. From Eqs. (48.2), (48.5), and (48.6), this condition takes the form

$$(\rho\mathbf{f}_{app} - \nabla p) \, dV = 0 \tag{48.7}$$

We conclude that

$$\mathbf{f}_{app} = \frac{1}{\rho} \nabla p \tag{48.8}$$

which is the fundamental equation of hydrostatics.

If the fluid is in motion, we assign a vector field $\mathbf{v}(\mathbf{x}, t)$ that specifies the velocity of a small element of fluid at each point in space (see Fig. 48.3). Newton's second law for a small mass element $\rho \, dV$ in the fluid states that

$$(\rho \, dV)\frac{d\mathbf{v}}{dt} = -\nabla p \, dV + \rho\mathbf{f}_{app} \, dV \tag{48.9}$$

where the forces acting on the fluid have been identified through Eqs. (48.2), (48.5), and (48.6). The derivative on the left-hand side is the *time rate of change of velocity for the given small element of fluid as observed in the laboratory (inertial) frame.* Now the total differential of the ith component of the velocity of the fluid has the familiar form

$$dv_i(\mathbf{x}, t) = \sum_{j=1}^{3} \frac{\partial v_i}{\partial x_j} dx_j + \frac{\partial v_i}{\partial t} dt \tag{48.10}$$

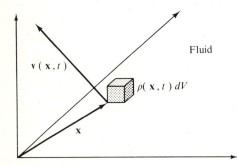

Fluid

$\mathbf{v}(\mathbf{x}, t)$

$\rho(\mathbf{x}, t) \, dV$

$\mathbf{x}$

Figure 48.3 Motion of a small mass element $\rho \, dV$ of fluid.

Dividing by dt, we have

$$\frac{dv_i}{dt} = \sum_{j=1}^{3} \frac{dx_j}{dt}\frac{\partial v_i}{\partial x_j} + \frac{\partial v_i}{\partial t} = \sum_{j=1}^{3} v_j \frac{\partial v_i}{\partial x_j} + \frac{\partial v_i}{\partial t} \qquad (48.11)$$

which can be rewritten as a vector relation

$$\frac{d\mathbf{v}}{dt} = (\mathbf{v} \cdot \mathbf{\nabla})\mathbf{v} + \frac{\partial \mathbf{v}}{\partial t} \qquad (48.12)$$

It expresses the rate of change of the velocity of a *given element* of a fluid; the velocity changes both because of the time dependence at a given point in space and because of the displacement to a new point in space. This quantity is known variously as the *total derivative*, the *material derivative*, or the *hydrodynamic derivative* and is sometimes denoted $D\mathbf{v}/Dt$. Substituting Eq. (48.12) into Eq. (48.9) and canceling common factors, we arrive at Newton's second law for an ideal fluid

$$\frac{d\mathbf{v}}{dt} = \frac{\partial \mathbf{v}}{\partial t} + (\mathbf{v} \cdot \mathbf{\nabla})\mathbf{v} = \mathbf{f}_{app} - \frac{1}{\rho}\mathbf{\nabla}p \qquad (48.13)$$

Here the fluid is called ideal because neither tangential shear forces nor viscosity appears in Eq. (48.9). The inclusion of such effects will be discussed in Chap. 12. We emphasize that Eq. (48.13) involves no new physics. It merely makes use of the elementary concept of the pressure in the fluid to rewrite Newton's second law for a continuous medium.

Equation (48.13) can be transformed further with the vector identity

$$(\mathbf{v} \cdot \mathbf{\nabla})\mathbf{v} \equiv \mathbf{\nabla}(\tfrac{1}{2}v^2) - [\mathbf{v} \times (\mathbf{\nabla} \times \mathbf{v})] \qquad (48.14)$$

which is established in the following fashion. The ith component of this relation can be rewritten as

$$(\mathbf{v} \cdot \mathbf{\nabla})v_i = \sum_{j=1}^{3} v_j \frac{\partial v_i}{\partial x_j} = \sum_{j=1}^{3} v_j \left[\frac{\partial v_j}{\partial x_i} - \left(\frac{\partial v_j}{\partial x_i} - \frac{\partial v_i}{\partial x_j}\right)\right] \qquad (48.15)$$

A little manipulation reduces this expression to

$$\text{rhs} \equiv \sum_{j=1}^{3} \frac{\partial}{\partial x_i}(\tfrac{1}{2}v_j v_j) - \sum_{j=1}^{3}\sum_{k=1}^{3} \epsilon_{ijk} v_j (\mathbf{\nabla} \times \mathbf{v})_k \qquad (48.16)$$

where we have introduced the curl and the totally antisymmetric tensor ϵ_{ijk} in three dimensions

$$\epsilon_{ijk} = \begin{cases} 1 & ijk = \text{any even permutation of 123,} \\ & \text{i.e., 123, 231, 312} \\ -1 & ijk = \text{any odd permutation of 123,} \\ & \text{i.e., 321, 213, 132} \\ 0 & \text{if any two indices the same} \end{cases} \qquad (48.17)$$

Equation (48.16) evidently has the equivalent form

$$\text{rhs} = \nabla_i(\tfrac{1}{2}v^2) - [\mathbf{v} \times (\nabla \times \mathbf{v})]_i \tag{48.18}$$

which establishes the vector identity. A combination of Eqs. (48.13) and (48.14) leads to

$$\frac{\partial \mathbf{v}}{\partial t} + \nabla(\tfrac{1}{2}v^2) - \mathbf{v} \times (\nabla \times \mathbf{v}) = \mathbf{f}_{\text{app}} - \frac{1}{\rho}\nabla p \tag{48.19}$$

which is another form of Newton's second law for an ideal fluid. Note that the fundamental differential equation of hydrodynamics is intrinsically *nonlinear* in the velocity field, even though the dynamics is simply Newton's second law. This complication can lead both to interesting physical behavior and to great mathematical difficulty. Except in very special circumstances, direct solution of these equations is impossible.

Conservation of Matter: The Continuity Equation

Throughout our study of mechanics, *conservation laws* have provided a natural framework for examining the dynamical behavior of mechanical systems. Unlike the situation for discrete point masses and rigid bodies, the conservation of matter in a fluid is no longer trivial. The formulation of this principle yields one of the fundamental dynamical laws and also serves as a model for other more complicated conservation laws associated with momentum and energy.

We treat the fluid as a continuous medium specified by a mass density $\rho(\mathbf{x}, t)$, a velocity $\mathbf{v}(\mathbf{x}, t)$, and a pressure $p(\mathbf{x}, t)$. An alternative to the preceding *lagrangian description* of the dynamical behavior, which focuses on a *given infinitesimal element of fluid*, is the *eulerian description*, which focuses on the transport *through a small fixed volume*. This viewpoint will prove extremely useful in hydrodynamics. Let $d\mathbf{A} = \hat{n}\, dA$ be an element of surface area with the normal $\hat{n}$ oriented outward (Fig. 48.4) and let $\mathbf{v}$ be the velocity of the fluid at that point. In a time interval Δt, all the fluid in a cylinder of length $v\,\Delta t$ and cross-sectional area $|\hat{v} \cdot \hat{n}|\, dA = -\hat{v} \cdot d\mathbf{A}$ will flow *into* the enclosed volume. Correspondingly, the mass flowing inward through $d\mathbf{A}$ in time Δt is $-\rho \mathbf{v} \cdot d\mathbf{A}\, \Delta t$.

Consider now the whole closed volume V (Fig. 45.1). The mass enclosed is $\int_V d^3x\, \rho(\mathbf{x}, t)$, and its increase in a time Δt is

$$\Delta t\, \frac{d}{dt} \int_V d^3x\, \rho(\mathbf{x}, t) = \Delta t \int_V d^3x\, \frac{\partial}{\partial t} \rho(\mathbf{x}, t) \tag{48.20}$$

Here the fixed location of the volume element and its bounding surface permits us to differentiate under the integral sign, and the partial derivative is taken at fixed $\mathbf{x}$. Evidently, any increase in mass must reflect a net inflow. Integrating over the surface, we readily find the integral form for the conservation law of matter

$$\frac{d}{dt} \int_V d^3x\, \rho = \int_V d^3x\, \frac{\partial \rho}{\partial t} = -\int_A d\mathbf{A} \cdot \mathbf{v}\rho \tag{48.21}$$

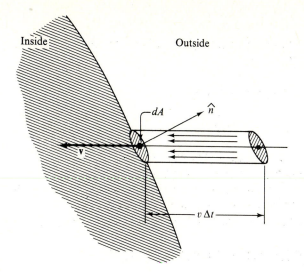

Inside

Outside

dA

$\hat{n}$

v

$v\,\Delta t$

Figure 48.4 Behavior at surface of a small volume element.

This form of the conservation law will prove useful when discussing the behavior of fluid surfaces (Chap. 10). The right-hand side of Eq.(48.21) can be rewritten with the divergence theorem

$$-\int_A d\mathbf{A} \cdot \mathbf{v}\rho = -\int_V d^3x\, \mathbf{\nabla} \cdot (\rho\mathbf{v}) \tag{48.22}$$

and substitution into (48.21) yields

$$\int_V d^3x \left[\frac{\partial \rho}{\partial t} + \mathbf{\nabla} \cdot (\rho\mathbf{v})\right] = 0 \tag{48.23}$$

This equation holds for an arbitrary volume V, and the integrand must therefore vanish identically, providing the (differential) *continuity equation*

$$\frac{\partial \rho}{\partial t} + \mathbf{\nabla} \cdot (\rho\mathbf{v}) = 0 \tag{48.24}$$

It has the typical form [see Eq. (45.10)] encountered earlier in studying energy flow in strings, relating the time derivative of some quantity (here, the mass density) to the divergence of the corresponding current (here, the mass current density)

$$\mathbf{j}_m = \rho\mathbf{v} \tag{48.25}$$

For many purposes, it is preferable to expand the divergence of $\rho\mathbf{v}$, rewriting the continuity equation as

$$\frac{\partial \rho}{\partial t} + \mathbf{v} \cdot \mathbf{\nabla}\rho + \rho\mathbf{\nabla} \cdot \mathbf{v} = 0 \tag{48.26}$$

It is evident from Eqs. (48.10) to (48.12) that the first two terms in this expression are just the *total* time derivative of ρ (that is, the change in ρ of a given moving fluid element)

$$\frac{d\rho}{dt} = \frac{\partial\rho}{\partial t} + \mathbf{v} \cdot \nabla\rho \tag{48.27}$$

Thus the continuity equation has the equivalent form

$$\frac{d\rho}{dt} + \rho\nabla \cdot \mathbf{v} = 0 \tag{48.28}$$

In the particular case of a uniform incompressible fluid, both $\partial\rho/\partial t$ and $d\rho/dt$ vanish identically, and (48.28) reduces to the simpler expression

$$\nabla \cdot \mathbf{v} = 0 \qquad \text{for incompressible flow} \tag{48.29}$$

We see that incompressible flow is characterized by a solenoidal (divergenceless) velocity field.

Conservation of Momentum: Stress Tensor and Euler's Equation

The next question in studying the dynamics of fluids is the transport of momentum, whose vector nature makes it convenient to consider a particular component (the ith, say). We again follow the eulerian description and concentrate on a *fixed* volume element V in the fluid. At a given instant, this volume element contains the total momentum $\int_V d^3x\, \rho v_i$, and its increase in an infinitesimal time interval Δt is

$$\Delta t \frac{d}{dt} \int_V d^3x\, \rho v_i = \Delta t \int_V d^3x\, \frac{\partial}{\partial t} \rho v_i \tag{48.30}$$

We defer considering viscous fluids until Chap. 12, so that the total momentum can change in only three ways:

1. There may be a volume force density ρf_i, where f_i is the force per unit mass defined in Eq. (48.6) (we now suppress the subscript "app"), and the factor ρ merely recognizes that bulk forces like gravity act on each individual atom to yield a total force proportional to ρ. In time Δt, such forces increase the total momentum by $\Delta t \int_V d^3x\, \rho f_i$.
2. The pressure p represents the effect of the surrounding medium; it exerts a net inward force $-p\, d\mathbf{A}$ on each element of the surface enclosing V. In an interval Δt, its total contribution (impulse) to the change in the ith component of momentum in V is $-\Delta t \int_A dA_i\, p$. This result follows from Newton's second law and the definition of pressure.
3. The two preceding contributions simply enumerate the impulses imparted to the fluid in the volume V. Because these impulses are proportional to Δt, all factors can be evaluated in the initial configuration of the fluid. We now use

Newton's second law to relate the impulse to the change in momentum of the fluid. If we concentrate on a *fixed volume* V in the fluid, however, the momentum contained in that volume also changes because of the net intrinsic momentum $\rho\mathbf{v}$ of the fluid passing in and out of V. We refer to this as the *convective transport* of momentum. Since the convective flow of the ith component of the momentum density ρv_i through $d\mathbf{A}$ in a time Δt is $-\rho v_i \mathbf{v} \cdot d\mathbf{A}\, \Delta t$, the corresponding integrated increase in this momentum component in V in the time Δt is

$$-\Delta t \int_A d\mathbf{A} \cdot \mathbf{v}\rho v_i = -\Delta t \sum_{j=1}^{3} \int dA_j\, v_i v_j \rho$$

Any *change* in the quantities appearing in the integrand of this expression during the time interval Δt makes a contribution of order $(\Delta t)^2$ and is therefore negligible.

A combination of the above terms gives the basic equation for the time rate of change of the momentum in a given volume V of a nonviscous fluid

$$\frac{d}{dt}\int_V d^3x\, \rho v_i = -\int_A dA_i\, p - \sum_{j=1}^{3} \int_A dA_j\, v_i v_j \rho + \int_V d^3x\, \rho f_i \qquad (48.31)$$

We shall refer to this result as the *statement of momentum conservation*. It can be rewritten in several distinct forms that emphasize special aspects. First, we define the (symmetric) stress tensor for nonviscous fluids

$$T_{ij} \equiv p\delta_{ij} + \rho v_i v_j \qquad (48.32)$$

where δ_{ij} is the Kronecker delta. Substitution into Eq. (48.31) provides an integral form of momentum conservation

$$\frac{d}{dt}\int_V d^3x\, \rho v_i = -\sum_{j=1}^{3} \int_A dA_j\, T_{ij} + \int_V d^3x\, \rho f_i \qquad (48.33)$$

showing that T_{ij} is the flux of the ith component of momentum density across a surface oriented along the unit vector $\hat{j}$; this interpretation is analogous to that of $\rho\mathbf{v}$ as the flux of mass [compare Eq. (48.21)]. The stress tensor is much more than a formal construct, for the right-hand side of (48.33) provides a convenient basis for computing the net force on a given volume. In particular, if $\mathbf{f} = 0$, the surface integral of T fixes the total force, independent of the interior configuration.

It is sometimes more useful to transform the surface term in Eq. (48.33) with the divergence theorem; an argument identical with that leading to (48.24) yields the differential form of the momentum conservation law

$$\frac{\partial(\rho v_i)}{\partial t} + \sum_{j=1}^{3} \frac{\partial T_{ij}}{\partial x_j} = \rho f_i \qquad (48.34)$$

which is just another rewriting of Newton's second law. Note that Eqs. (48.33) and (48.34) are very general; with suitable interpretation of the stress tensor, they hold even for viscous fluids or elastic media (see Chaps. 12 and 13).

An alternative differential formulation of the conservation of momentum follows by combining Eqs. (48.32) and (48.34) and then expanding the derivatives of products. Comparison with Eq. (48.24) shows that several terms cancel, leaving the celebrated *Euler's equation* of nonviscous hydrodynamics; it can be expressed either in components

$$\frac{\partial v_i}{\partial t} + \sum_{j=1}^{3} v_j \frac{\partial v_i}{\partial x_j} = -\frac{1}{\rho} \frac{\partial p}{\partial x_i} + f_i \qquad (48.35a)$$

or in vector form

$$\frac{d\mathbf{v}}{dt} \equiv \frac{\partial \mathbf{v}}{\partial t} + (\mathbf{v} \cdot \nabla)\mathbf{v} = -\frac{1}{\rho} \nabla p + \mathbf{f} \qquad (48.35b)$$

where the total derivative is that introduced in Eq. (48.12). This equation is precisely our previous result (48.13). Note again the intrinsic nonlinearity, which accounts for both the difficulty and the fascination of hydrodynamics.

The preceding exact equations for nonviscous fluids must be supplemented by an equation of state for the fluid relating the density, pressure, and a third state variable such as temperature or entropy. If this third state variable can be considered constant during the motion, and we shall discuss the appropriate situations below, then Euler's equation (48.35) (namely Newton's second law), the continuity equation (48.24), and the equation of state provide five equations for the unknown functions ρ, p, and $\mathbf{v}$.

Conservation of Energy

Nonviscous flow is typically nondissipative, either because there is no mechanism for heat generation or because the motion is too rapid for heat conduction. In both cases, the motion is "reversible" in the thermodynamic sense, and the absence of local heat flow ensures that the local *entropy* remains constant. Equivalently, nonviscous flow is *isentropic*. In this case, conservation of energy takes a particularly simple form, with the energy density consisting of two separate contributions, the *kinetic energy density* $\frac{1}{2}\rho v^2$ associated with the center-of-mass motion and the *internal energy density* $\rho\epsilon$, where ϵ is the internal energy per unit mass. Here, the "internal energy" $\rho\epsilon \, dV$ of the fluid in a volume element dV is used in the thermodynamic sense of the first law. It is the energy stored in the various degrees of freedom of the fluid's elementary constituents and is therefore the energy measured by a comoving observer at rest with respect to the center of mass of the volume element. The energy in a fixed volume then increases at a rate

$$\frac{d}{dt} \int_V d^3x \, (\tfrac{1}{2}\rho v^2 + \rho\epsilon) = \int_V d^3x \left(\frac{\partial}{\partial t} \tfrac{1}{2}\rho v^2 + \frac{\partial}{\partial t} \rho\epsilon \right) \qquad (48.36)$$

The first term on the right-hand side is easily rewritten by differentiating the product and using Eqs. (48.24) and (48.35a); a little manipulation gives

$$\frac{\partial}{\partial t} \tfrac{1}{2}\rho v^2 = -\nabla \cdot (\tfrac{1}{2}\rho v^2 \mathbf{v}) - \mathbf{v} \cdot \nabla p + \rho \mathbf{v} \cdot \mathbf{f} \qquad (48.37)$$

where the last two terms have a simple interpretation as the power supplied by the force density $\rho\mathbf{f} - \nabla p$ [see Eq. (48.9)].

To rewrite the remaining term in Eq. (48.36), consider a given element of fluid with fixed total mass M and internal energy $M\epsilon$ occupying a volume V. For any reversible isentropic process, the *first law of thermodynamics* relates the increase in the internal energy to the work done on the element, as seen by a comoving observer. In the present case of an ideal fluid, the only contribution arises from a change in the volume

$$M\,d\epsilon = -p\,dV = \frac{Mp}{\rho^2}\,d\rho \tag{48.38}$$

where the last relation follows from the definition $V = M/\rho$. Thus the internal energy per unit mass becomes an indefinite integral

$$\epsilon(s, \rho) = \int^{\rho} d\rho' \frac{p(s, \rho')}{\rho'^2} \tag{48.39}$$

where s is the entropy per unit mass. The equation of state $p = p(s, \rho)$ is here defined in the comoving frame, as is the resulting internal energy $\epsilon(s, \rho)$. For an isentropic process, Eq. (48.39) shows that

$$\frac{\partial\epsilon}{\partial t} = \left(\frac{\partial\epsilon}{\partial\rho}\right)_s \frac{\partial\rho}{\partial t} = \frac{p}{\rho^2}\frac{\partial\rho}{\partial t} \tag{48.40}$$

A straightforward calculation with Eq. (48.24) then yields

$$\frac{\partial}{\partial t}\rho\epsilon = -\left(\epsilon + \frac{p}{\rho}\right)\nabla\cdot(\rho\mathbf{v})$$

$$= -\nabla\cdot[(\rho\epsilon + p)\mathbf{v}] + \rho\mathbf{v}\cdot\nabla\left(\epsilon + \frac{p}{\rho}\right) \tag{48.41}$$

Note that $\epsilon + p/\rho$ is the *enthalpy*† per unit mass; at fixed s, Eq. (48.39) shows that its gradient is given by

$$\nabla\left(\epsilon + \frac{p}{\rho}\right) = \frac{1}{\rho}\nabla p \tag{48.42}$$

and a combination with Eq. (48.41) yields

$$\frac{\partial}{\partial t}\rho\epsilon = -\nabla\cdot[(\rho\epsilon + p)\mathbf{v}] + \mathbf{v}\cdot\nabla p \tag{48.43}$$

---

† Given the internal energy E of a system, the enthalpy H is defined by the Legendre transformation $H \equiv E + pV$. The differential relation $dE = T\,dS - p\,dV$ shows that $dH = T\,dS + V\,dp$. Thus H has the natural variables S and p; it is especially useful for isentropic processes at specified p (compare Sec. 57). See Reif [1], chap. 5.

This equation, the divergence theorem, and Eqs. (48.36) and (48.37) lead to the integral form of energy conservation

$$\frac{d}{dt}\int_V d^3x \left(\tfrac{1}{2}\rho v^2 + \rho\epsilon\right) = -\int_A dA \cdot \mathbf{v}\left(\tfrac{1}{2}\rho v^2 + \rho\epsilon\right) - \int_A dA \cdot p\mathbf{v} + \int_V d^3x \,\rho\mathbf{f}\cdot\mathbf{v}$$

(48.44)

Thus three sources contribute to the increase of the total energy in V: convection of kinetic and internal energy, work done by the pressure force on the surface, and work done by the volume force $\mathbf{f}$. The conservation law has the equivalent differential form

$$\frac{\partial}{\partial t}\left(\tfrac{1}{2}\rho v^2 + \rho\epsilon\right) + \nabla \cdot \left[\left(\tfrac{1}{2}\rho v^2 + \rho\epsilon + p\right)\mathbf{v}\right] = \rho\mathbf{f}\cdot\mathbf{v}$$

(48.45)

which identifies the energy-flux vector

$$\mathbf{j}_e = \left(\tfrac{1}{2}\rho v^2 + \rho\epsilon + p\right)\mathbf{v}$$

(48.46)

Note that $\mathbf{j}_e$ contains an additional pressure contribution, as did the stress tensor (48.32).

Bernoulli's Theorem

Despite the nonlinearity of Euler's equation (48.35), there exists an exact first integral (Bernoulli's theorem) in the important case of irrotational flow $(\nabla \times \mathbf{v} = 0)$. To derive this theorem, we recall the mathematical identity (48.14) that holds for any vector field $\mathbf{v}$ expressed in cartesian components. When applied to the velocity field $\mathbf{v}$, it leads to the equivalent exact form of Euler's equation (48.19)

$$\frac{\partial \mathbf{v}}{\partial t} + \nabla(\tfrac{1}{2}v^2) - \mathbf{v} \times (\nabla \times \mathbf{v}) = -\frac{1}{\rho}\nabla p + \mathbf{f}$$

(48.47)

We now impose the following restrictions:

1. The motion is irrotational

$$\nabla \times \mathbf{v} = 0 \qquad \text{irrotational motion}$$

(48.48)

so that $\mathbf{v}$ can be derived from a scalar velocity potential $\bar{\Phi}$ according to the relation

$$\mathbf{v} = -\nabla\bar{\Phi}$$

(48.49)

2. The external force is *conservative* in the sense that

$$\mathbf{f} = -\nabla U(\mathbf{x}, t) \qquad \text{conservative applied force}$$

(48.50)

where U is the potential energy per unit mass. Note that the external potential is still allowed to have an *arbitrary time dependence*.

3. The fluid is incompressible with fixed constant density

$$\rho = \text{const} \qquad \text{incompressible fluid} \qquad (48.51)$$

These three restrictions permit us to rewrite Eq. (48.47) as

$$\nabla\left(\frac{p}{\rho} + U + \tfrac{1}{2}v^2 - \frac{\partial\bar\Phi}{\partial t}\right) = 0 \qquad (48.52)$$

and direct integration yields

$$\frac{p}{\rho} + U + \tfrac{1}{2}v^2 - \frac{\partial\bar\Phi}{dt} = C(t) \qquad (48.53)$$

where $C(t)$ is a function of time only. In fact, $C(t)$ may be taken to vanish, for we can always make the following *gauge transformation*† from the given (old) velocity potential $\bar\Phi$ to a new one

$$\Phi(\mathbf{x}, t) \equiv \bar\Phi(\mathbf{x}, t) + \int^t dt'\, C(t') \qquad (48.54)$$

Evidently, Φ and $\bar\Phi$ both yield the same velocity by Eq. (48.49) and are therefore equally acceptable. Reexpressing (48.53) in terms of Φ, we obtain the usual form of *Bernoulli's theorem*

$$\frac{p}{\rho} + U + \tfrac{1}{2}v^2 - \frac{\partial\Phi}{\partial t} = 0 \qquad \text{incompressible irrotational flow} \qquad (48.55)$$

This exact first integral correlates changes in p, U, $\mathbf{v}$, and $\partial\Phi/\partial t$ to maintain the constant value in (48.55). In the special case of steady flow $[U(\mathbf{x}, t) = U(\mathbf{x})$ and $\partial\Phi/\partial t = 0]$, Bernoulli's theorem merely expresses the conservation of energy (recall that the zero of the potential U is arbitrary), but the general form is more widely applicable.

This discussion depends crucially on the irrotational nature of the flow, but the restriction to an incompressible fluid is readily relaxed to include compressible isentropic flow of the sort considered previously. In this case, Eq. (48.42) expresses $\rho^{-1}\nabla p$ in Eq. (48.47) as a gradient, and Bernoulli's theorem for isentropic irrotational flow therefore has the more general form

$$\epsilon + \frac{p}{\rho} + U + \tfrac{1}{2}v^2 - \frac{\partial\Phi}{\partial t} = 0 \qquad \text{isentropic irrotational flow} \qquad (48.56)$$

Here $\epsilon + p/\rho$ depends on the isentropic equation of state; for an incompressible fluid, the constancy of ρ and ϵ immediately reproduces (48.55) once the constant ϵ has been absorbed into the velocity potential according to Eq. (48.54).

† Similar gauge transformations are familiar in electrodynamics; see, for example, Jackson [2], chap. 6.

Thomson's (Lord Kelvin's) Theorem on Circulation

Consider an isentropic fluid in a conservative force field $\mathbf{f} = -\nabla U(\mathbf{x}, t)$; substitution into Eq. (48.35b) and use of (48.42) produces yet another form of Euler's equation

$$\frac{d\mathbf{v}}{dt} = \frac{\partial \mathbf{v}}{\partial t} + (\mathbf{v} \cdot \nabla)\mathbf{v} = -\nabla\left(\epsilon + \frac{p}{\rho} + U\right) \tag{48.57}$$

where the left-hand side is again the rate of change of $\mathbf{v}$ for a given element of fluid (the total derivative). The *circulation* Γ around some closed contour C lying wholly in the fluid is defined by the instantaneous line integral (Fig. 48.5a)

$$\Gamma = \oint_C d\mathbf{s} \cdot \mathbf{v} \tag{48.58}$$

Suppose that the contour C is permanently attached to the same set of fluid elements and follows their time-dependent motion. How does Γ change with time?

To answer this question, we study Γ at times t and $t + \Delta t$. In the small interval Δt, both the velocity field $\mathbf{v}$ and the contour C change, with the change in Γ given by

$$\Gamma(t + \Delta t) - \Gamma(t) = \oint_{C(t + \Delta t)} d\mathbf{s} \cdot \mathbf{v} \bigg|_{t + \Delta t} - \oint_{C(t)} d\mathbf{s} \cdot \mathbf{v} \bigg|_t \tag{48.59}$$

Let $\mathbf{x}_1$ and $\mathbf{x}_2$ be two nearby points on the contour, with $d\mathbf{s} = \mathbf{x}_2 - \mathbf{x}_1$ at time t (Fig. 48.5b). In the interval Δt, the fluid element at $\mathbf{x}_1$ has moved to $\mathbf{x}_1 + \mathbf{v}(\mathbf{x}_1)\,\Delta t$ and similarly for $\mathbf{x}_2$. As a result, the line element $d\mathbf{s}$ at $t + \Delta t$ becomes

$$
\begin{aligned}
d\mathbf{s}(t + \Delta t) &= \mathbf{x}_2(t + \Delta t) - \mathbf{x}_1(t + \Delta t) \\
&= \mathbf{x}_2(t) - \mathbf{x}_1(t) + [\mathbf{v}(\mathbf{x}_2) - \mathbf{v}(\mathbf{x}_1)]\,\Delta t \\
&= d\mathbf{s}(t) + [\mathbf{v}(\mathbf{x}_1 + d\mathbf{s}) - \mathbf{v}(\mathbf{x}_1)]\,\Delta t \\
&\approx d\mathbf{s}(t) + \Delta t\,(d\mathbf{s} \cdot \nabla)\mathbf{v}
\end{aligned}
\tag{48.60a}
$$

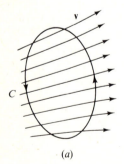

(a)

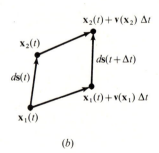

(b)

Figure 48.5 (a) Closed contour used to evaluate the circulation Γ and (b) time evolution for an element $d\mathbf{s}$ of the contour C.

where we have used Taylor's theorem in obtaining the last line. Furthermore, the velocity $\mathbf{v}(t + \Delta t)$ of a given element of fluid in the first term of (48.59) differs from $\mathbf{v}(t)$ by $(d\mathbf{v}/dt)\, \Delta t$, and the change in circulation becomes

$$\Gamma(t + \Delta t) - \Gamma(t) = \oint \left\{ \left(\mathbf{v} + \frac{d\mathbf{v}}{dt}\, \Delta t \right) \cdot [d\mathbf{s} + \Delta t\, (d\mathbf{s} \cdot \nabla)\mathbf{v}] - \mathbf{v} \cdot d\mathbf{s} \right\}$$

$$\approx \Delta t \oint \left\{ \frac{d\mathbf{v}}{dt} \cdot d\mathbf{s} + \mathbf{v} \cdot [(d\mathbf{s} \cdot \nabla)\mathbf{v}] \right\} \tag{48.60b}$$

Use of Euler's equation (48.57) for isentropic flow reduces the first term to $-\Delta t \oint d\mathbf{s} \cdot \nabla(\epsilon + \rho^{-1}p + U)$, which vanishes identically *because* $\epsilon + \rho^{-1}p + U$ *is single-valued in the fluid.* Moreover, the second term is easily rewritten as $\Delta t \oint d\mathbf{s} \cdot \nabla(\tfrac{1}{2}v^2)$, which vanishes for the same reason. Thus we are left with Thomson's theorem: *the circulation in a nonviscous fluid is constant about any closed contour that moves with the fluid,* assuming only that the external forces are conservative

$$\frac{d\Gamma}{dt} = 0 \qquad \text{Thomson's theorem} \tag{48.61}$$

This celebrated relation has the following important corollary. Suppose that the fluid is initially stationary and then subjected to conservative forces [Eq. (48.50)]. Clearly Γ vanishes for any initial path, and Thomson's theorem implies that Γ remains zero forever. On the other hand, Stokes' theorem permits us to rewrite Eq. (48.58) as

$$\Gamma = \oint_C d\mathbf{s} \cdot \mathbf{v} = \int_A d\mathbf{A} \cdot (\nabla \times \mathbf{v}) \tag{48.62}$$

where the surface integral is over any surface bounded by the contour C and lying wholly within the fluid. Since Γ vanishes for any closed path C, we infer that $\nabla \times \mathbf{v}$ is permanently zero; hence the *flow in a simply connected region necessarily remains irrotational under the action of arbitrary conservative forces.*

Lagrangian for Isentropic Irrotational Flow

The existence of a first integral (48.56) for irrotational isentropic flow suggests a close relation with the mechanics of point masses studied in Chaps. 1 and 3. To confirm this notion, we recast the theory in lagrangian form, exactly as in Sec. 25 for a string. The restriction to irrotational isentropic flow simplifies the equations considerably, for it leaves only two independent field variables, which we shall take as the density ρ and the velocity potential Φ. In this context, the equation of continuity (48.24) and Bernoulli's theorem (48.56) become

$$\frac{\partial \rho}{\partial t} - \nabla \cdot (\rho\, \nabla\Phi) = 0 \tag{48.63}$$

$$\epsilon + \frac{p}{\rho} + U + \tfrac{1}{2}(\nabla\Phi)^2 - \frac{\partial \Phi}{\partial t} = 0 \tag{48.64}$$

where we again require conservative external forces as in Eq. (48.50). It is remarkable that these two equations are precisely the Euler-Lagrange equations of the lagrangian density

$$\mathscr{L} = \rho \frac{\partial \Phi}{\partial t} - \tfrac{1}{2}\rho(\nabla\Phi)^2 - \rho U - \rho\epsilon(\rho) \tag{48.65}$$

where $\mathscr{L}$ is considered a function of the generalized fields Φ and ρ. To verify this assertion, note that the canonical momentum density

$$\mathscr{P}_\Phi \equiv \frac{\partial \mathscr{L}}{\partial(\partial\Phi/\partial t)} = \rho \tag{48.66}$$

obeys the equation of motion

$$\frac{\partial}{\partial t}\mathscr{P}_\Phi + \nabla \cdot \frac{\partial \mathscr{L}}{\partial(\nabla\Phi)} = \frac{\partial \mathscr{L}}{\partial\Phi} \tag{48.67}$$

which is just the continuity equation (48.63). The other canonical momentum density $\mathscr{P}_\rho$ vanishes identically because $\mathscr{L}$ does not contain $\partial\rho/\partial t$. Furthermore, the corresponding dynamical equation

$$\frac{\partial}{\partial t}\frac{\partial \mathscr{L}}{\partial(\partial\rho/\partial t)} + \nabla \cdot \frac{\partial \mathscr{L}}{\partial(\nabla\rho)} = \frac{\partial \mathscr{L}}{\partial\rho} \tag{48.68}$$

is just Bernoulli's equation (48.64), because Eq. (48.38) shows that $\partial(\rho\epsilon)/\partial\rho = \epsilon + \rho^{-1}p$. If the external force is time-independent, so that $U(\mathbf{x}, t) = U(\mathbf{x})$, the absence of explicit time dependence in $\mathscr{L}$ implies that the volume integral of the hamiltonian density

$$\mathscr{H} = \mathscr{P}_\Phi \frac{\partial \Phi}{\partial t} + \mathscr{P}_\rho \frac{\partial \rho}{\partial t} - \mathscr{L} = \tfrac{1}{2}\rho(\nabla\Phi)^2 + \rho U + \rho\epsilon(\rho) \tag{48.69}$$

is a constant of the motion. For point masses, the presence of linear velocity terms in the lagrangian generally implied that the hamiltonian was not the total energy. In the present case of isentropic flow, however, $\mathscr{H}$ is clearly the energy density, for it contains the kinetic, potential, and internal energy density.

As in Sec. 45, the lagrangian density (48.65) provides an alternative derivation of the energy flux vector $\mathbf{j}_e$ defined in Eq. (48.46). The general formalism of Sec. 45 immediately yields (note that there are now two generalized coordinates ρ and Φ)

$$\mathbf{j}_e = \frac{\partial \mathscr{L}}{\partial(\nabla\Phi)}\frac{\partial \Phi}{\partial t} + \frac{\partial \mathscr{L}}{\partial(\nabla\rho)}\frac{\partial \rho}{\partial t} = -\rho\,\nabla\Phi\frac{\partial \Phi}{\partial t}$$

$$= -\rho\,\nabla\Phi\,[\epsilon + \rho^{-1}p + U + \tfrac{1}{2}(\nabla\Phi)^2] \tag{48.70}$$

where the last line has made use of Bernoulli's theorem (48.64). The identification $\mathbf{v} = -\nabla\Phi$ shows that (48.69) and (48.70) are just the energy density and energy flux from (48.45), modified to incorporate the static conservative force $\mathbf{f} = -\nabla U(\mathbf{x})$ on the left-hand side. This last rearrangement follows because $\rho\mathbf{v} \cdot \mathbf{f} = -\rho\mathbf{v} \cdot \nabla U$ is easily rewritten $-\nabla \cdot (\rho\mathbf{v}U) - \partial(\rho U)/\partial t$.

49 SOUND WAVES

In principle, the equations of mass conservation (48.24), momentum conservation (48.35), and the equation of state $p = p(s, \rho)$ describe most of nonviscous isentropic hydrodynamics. Apart from the general theorems obtained in Sec. 48, however, further restrictions are necessary, and we now treat two specific phenomena (sound waves and surface waves) that illustrate many of the fundamental aspects of fluid dynamics. The remainder of the present chapter deals with compressional waves, which will be seen to represent a natural extension of the ideas developed in Chaps. 7 and 8. In contrast, the behavior of incompressible fluids with a free surface involves several new concepts that merit a separate discussion (Chap. 10).

Fundamental Equations

Consider a uniform stationary fluid with constant density ρ_0 and pressure p_0, neglecting gravity which is irrelevant in most aspects of sound propagation. If the fluid experiences a small perturbation, the density, velocity, and pressure acquire first-order (primed) contributions

$$\rho = \rho_0 + \rho' \qquad \mathbf{v} = \mathbf{v}' \qquad p = p_0 + p' \tag{49.1}$$

and we shall linearize in the primed quantities. The equation of continuity reduces to

$$\frac{1}{\rho_0} \frac{\partial \rho'}{\partial t} + \nabla \cdot \mathbf{v}' = 0 \tag{49.2}$$

Similarly, the linearized Euler's equation becomes

$$\frac{\partial \mathbf{v}'}{\partial t} = -\frac{1}{\rho_0} \nabla p' \tag{49.3}$$

If the thermal conductivity of the fluid is sufficiently low, the heat conduction during a cycle of the acoustic disturbance becomes negligible. Section 62 discusses the criterion for this condition (see also Prob. 12.13), which we here assume to hold. In this case, the motion is reversible in the thermodynamic sense, and the local entropy remains constant. The isentropic equation of state can then be expanded with Taylor's theorem

$$p_0 + p' = p(s, \rho_0 + \rho') = p(s, \rho_0) + \rho' \left(\frac{\partial p}{\partial \rho} \right)_s + \cdots \tag{49.4}$$

and it is convenient to introduce the abbreviation

$$c^2 \equiv \left(\frac{\partial p}{\partial \rho} \right)_s \tag{49.5}$$

where the partial derivative is evaluated at constant entropy and at the equilibrium density ρ_0. In this way, small deviations in pressure and density are related by

$$p' = c^2 \rho' \tag{49.6}$$

The curl of Eq. (49.3) shows that $\mathbf{V} \times \mathbf{v}'$ is time-independent, and we shall assume that the flow is irrotational with

$$\mathbf{v}' = -\nabla \Phi \tag{49.7}$$

This choice is consistent with Thomson's theorem if the initial disturbance is generated by conservative forces of the form (48.50). Note that Φ here is a first-order quantity although the prime will be omitted for simplicity. Bernoulli's theorem for irrotational isentropic flow (48.56) can also be expanded to first order. The constant term $\epsilon(\rho_0) + p_0/\rho_0$ can be eliminated with a gauge transformation of the form (48.54), and use of the isentropic relation $(\partial\epsilon/\partial\rho)_s = \rho^{-2}p$ from Eq. (48.38) gives

$$p' = \rho_0 \frac{\partial \Phi}{\partial t} \tag{49.8a}$$

or, with the aid of Eq. (49.6),

$$c^2 \rho' = \rho_0 \frac{\partial \Phi}{\partial t} \tag{49.8b}$$

The basic dynamical equations for the isentropic propagation of small irrotational [Eq. (49.7)] disturbances through an ideal fluid are then Eq. (49.2), which describes the conservation of mass; Eq. (49.8a), which is the dynamical statement of Newton's second law; and the equation of state (49.6). These last two relations are combined in Eq. (49.8b). The time derivative of (49.8b) can now be combined with the continuity equation (49.2), and use of (49.7) then yields the familiar wave equation for the velocity potential

$$\nabla^2 \Phi = \frac{1}{c^2} \frac{\partial^2 \Phi}{\partial t^2} \tag{49.9}$$

analogous to that for the displacement of a string or membrane; this equation shows that the constant c is indeed the propagation speed. It is evident that ρ' and p' satisfy the same wave equations, but it is usually easier to solve Eq. (49.9) for Φ and then obtain ρ' and p' from Eqs. (49.8).

The wave equation has numerous solutions, and boundary conditions determine the relevant one in a particular situation. The simplest case is a fluid bounded by an impenetrable surface that moves with a prescribed velocity $\mathbf{V}$ (Fig. 49.1a). To have the fluid follow the boundary smoothly, the normal component of the fluid's velocity must match that of the boundary, and Φ therefore obeys the boundary condition

$$-\hat{n} \cdot \nabla \Phi = \mathbf{v}' \cdot \hat{n} = \mathbf{V} \cdot \hat{n} \tag{49.10a}$$

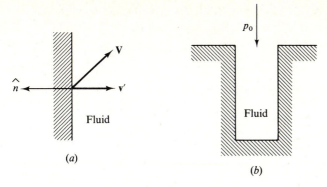

Figure 49.1 Boundary conditions for sound waves: (*a*) impenetrable surface and (*b*) free surface.

where $\hat{n}$ is the unit vector normal to the surface, here taken to be directed *away* from the fluid. In the special case of a stationary boundary, this condition has the simpler form

$$\hat{n} \cdot \nabla\Phi = 0 \qquad \text{fixed rigid boundary} \tag{49.10b}$$

The behavior at other surfaces is more complicated, and we shall treat only the limiting case of a free surface. For definiteness, consider an acoustic cavity with a small aperture in the boundary, connecting the interior with a large reservoir at constant pressure p_0 (Fig. 49.1*b*). This open region in the wall may be considered a free surface, and continuity of pressure dictates that $p' = 0$ there. Equation (49.8*a*) then provides the equivalent boundary condition for the velocity potential

$$\frac{\partial\Phi}{\partial t} = 0 \qquad \text{on free surface} \tag{49.11}$$

This condition is obviously an idealization, for it wholly confines the acoustic disturbance to the original cavity, with no coupling to the large reservoir. To clarify this point, we recall the energy-flux vector for isentropic motion (48.70). In the present small-amplitude limit, the leading term is

$$\mathbf{j}_e = -\rho_0 \, \nabla\Phi \frac{\partial\Phi}{\partial t} = \mathbf{v}'p' \tag{49.12}$$

which is manifestly *second* order in the small amplitudes. This expression is the power transmitted per unit area by the acoustic pressure acting on the moving particles. If $p' = 0$ at a free surface, no energy passes through the aperture. Practical systems naturally violate this strict condition; for example, an audible organ pipe with an open end certainly radiates some acoustic energy. Nevertheless, the fractional loss per cycle is usually small, and Eq. (49.11) remains a useful first approximation for a free surface. In other cases, more complicated conditions may be required, as shown in Prob. 9.22, in connection with the concept of acoustic impedance.

These expressions apply directly to plane waves in an unbounded fluid, where we assume $\Phi(\mathbf{r}, t) = \text{Re }\Phi_0 e^{i(\mathbf{k}\cdot\mathbf{r}-\omega t)}$ with $\omega = c|\mathbf{k}|$ to satisfy (49.9). The corresponding pressure and velocity become

$$p' = -\text{Re }i\omega\rho_0\Phi_0 e^{i(\mathbf{k}\cdot\mathbf{r}-\omega t)} \tag{49.13a}$$

$$\mathbf{v}' = -\text{Re }i\mathbf{k}\Phi_0 e^{i(\mathbf{k}\cdot\mathbf{r}-\omega t)} \tag{49.13b}$$

confirming that the sound waves are indeed longitudinal ($\mathbf{v}'\|\mathbf{k}$). Moreover, the time-average energy-flux vector is obtained in terms of the complex quantities $\mathbf{v}'$ and p' as [see Eqs. (45.45) and (49.12)]

$$\langle \mathbf{j}_e \rangle = \tfrac{1}{2}\text{ Re }\mathbf{v}'p'^* = \tfrac{1}{2}ck^2\rho_0|\Phi_0|^2\hat{k} \tag{49.14}$$

It lies along the propagation vector $\hat{k}$ with magnitude proportional to $|\Phi_0|^2$.

Standing Waves in Cavities

As a first detailed application of this formalism, consider a closed cavity filled with a compressible fluid at equilibrium density ρ_0 and pressure p_0. The velocity potential $\Phi(\mathbf{r}, t)$ for small-amplitude motion obeys the wave equation (49.9) subject to appropriate boundary conditions on the surrounding surface. We seek oscillatory solutions of the form† $\Phi(\mathbf{r}, t) = \text{Re }\Phi(\mathbf{r})e^{-i\omega t}$, and the wave equation then becomes a three-dimensional Helmholtz equation

$$\nabla^2\Phi(\mathbf{r}) + k^2\Phi(\mathbf{r}) = 0 \tag{49.15}$$

where $k \equiv \omega/c$. It has the form of our familiar eigenvalue problem. In the typical case of a rectangular cavity bounded by rigid walls, the solution proceeds exactly as in Chap. 8. The boundary conditions are quite different, however, for the acoustic velocity potential has zero normal derivative whereas the displacement itself vanishes for a clamped membrane. The altered boundary conditions are readily accommodated, merely by replacing the sines in Eqs. (47.6) with cosines. It is then straightforward to determine the frequency of the low-lying normal modes and the location of the pressure, density, and velocity nodes and antinodes (Prob. 9.1).

The analysis becomes more complicated for a cylindrical cavity of length L enclosed by rigid walls of radius a. This configuration serves as a model for an organ pipe, and we consider it in some detail. Suppose first that both ends are closed (Fig. 49.2a), so that Φ satisfies the boundary conditions

$$\frac{\partial\Phi}{\partial r} = 0 \qquad r = a \tag{49.16a}$$

$$\frac{\partial\Phi}{\partial z} = 0 \qquad z = 0 \text{ and } z = L \tag{49.16b}$$

---

† This notation for the normal-mode amplitudes differs from that in Chaps. 7 and 8 to avoid confusion with the density ρ.

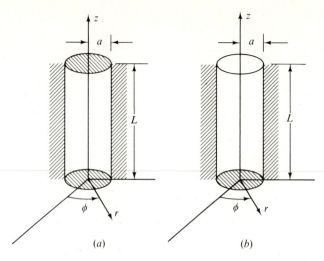

Figure 49.2 Cylindrical acoustic cavity with (a) two closed ends and (b) one closed end and one open end.

In cylindrical polar coordinates (r, ϕ, z), Eq. (49.15) becomes [see Appendix B and Eq. (47.29)]

$$\frac{1}{r}\frac{\partial}{\partial r}r\frac{\partial \Phi}{\partial r} + \frac{1}{r^2}\frac{\partial^2 \Phi}{\partial \phi^2} + \frac{\partial^2 \Phi}{\partial z^2} + k^2\Phi = 0 \qquad (49.17)$$

allowing a factored solution of the form

$$\Phi(r, \phi, z) = R(r)F(\phi)Z(z) \qquad (49.18)$$

It is easy to verify that F and Z must be sinusoidal:

$$F(\phi) = e^{\pm im\phi} \qquad m = 0, \pm 1, \pm 2, \ldots \qquad (49.19a)$$

$$Z(z) = e^{\pm i\alpha z} \qquad (49.19b)$$

where m is an integer to ensure a single-valued solution and α must be real to satisfy the boundary condition (49.16b) at both ends. The appropriate linear combination with $Z'(0) = 0$ is clearly $\cos \alpha z$, and the condition $Z'(L) = 0$ requires that

$$\alpha = \frac{p\pi}{L} \qquad p = 0, 1, 2, \ldots, \infty \qquad (49.20)$$

with p an integer including zero.

The remaining function satisfies Bessel's equation [Appendix C and Eq. (47.35)]

$$\frac{1}{r}\frac{d}{dr}r\frac{dR}{dr} - \frac{m^2}{r^2}R + \kappa^2 R = 0 \qquad (49.21)$$

where
$$\kappa^2 = k^2 - \left(\frac{p\pi}{L}\right)^2 \tag{49.22}$$

The solution must be bounded at the origin and have zero slope at $r = a$. These conditions cannot be satisfied for $\kappa^2 < 0$, because $J_m(i|\kappa|r)$ increases monotonically with increasing r. Hence $\kappa^2 > 0$ and $R(r) = J_m(\kappa r)$. The boundary condition $J'_m(\kappa a) = 0$ then fixes κ as one of the discrete set

$$\kappa_{mn} = a^{-1}\alpha'_{m,n} \tag{49.23}$$

where $\alpha'_{m,n}$ is the nth root of the equation

$$J'_m(\alpha'_{m,n}) = 0 \tag{49.24}$$

In this way, the complete solution can be constructed from the eigenfunctions

$$\Phi(\mathbf{r}, t) = \text{Re } \Phi_0 J_m\left(\alpha'_{m,n}\frac{r}{a}\right)\cos\frac{p\pi z}{L}\exp\left(im\phi - i\omega_{mnp}t\right) \tag{49.25}$$

with the triply infinite discrete set of frequencies

$$\omega_{mnp}^2 = c^2 k_{mnp}^2 = c^2\left[\left(\frac{\alpha'_{m,n}}{a}\right)^2 + \left(\frac{p\pi}{L}\right)^2\right] \tag{49.26}$$

These three-dimensional relations have a clear analogy with those in Chap. 8 for a two-dimensional circular membrane.

As a specific example, suppose the cavity is long ($L \gg a$). The lowest modes are those with the smallest value of $\alpha'_{m,n}$, and inspection of the various Bessel functions shows that this situation occurs for $m = 0$, $n = 1$, when $\alpha'_{0,1} = 0$. Since $J_0(0) = 1$, the corresponding amplitudes have no radial dependence, reducing to simple axial standing waves (we assume here that Φ_0 is real)

$$\Phi(z, t) = \Phi_0 \cos\frac{p\pi z}{L}\cos\omega_p t \tag{49.27a}$$

with equally spaced harmonics

$$\omega_p \left(\equiv \omega_{01p}\right) = c\frac{p\pi}{L} \qquad p = 1, 2, 3, \ldots, \infty \tag{49.27b}$$

Note that $p = 0$ is now excluded for it corresponds to a constant Φ. The associated acoustic pressure and velocity are

$$p'(z, t) = -\rho_0\omega_p\Phi_0\cos\frac{p\pi z}{L}\sin\omega_p t \tag{49.28a}$$

$$v'_z(z, t) = \frac{\omega_p}{c}\Phi_0\sin\frac{p\pi z}{L}\cos\omega_p t \tag{49.28b}$$

showing that the motion is purely axial with p' and $\mathbf{v}'$ out of phase, transferring energy between kinetic and compressional forms. Moreover, the ends are velocity nodes and pressure antinodes, with the cavity containing p half wavelengths (see

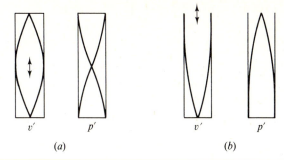

v' $\qquad$ p' $\qquad\qquad$ v' $\qquad$ p'

(a) $\qquad\qquad\qquad\qquad$ (b)

Figure 49.3 Nodes and antinodes for the fundamental mode in an organ pipe: (a) two closed ends and (b) one open end.

Fig. 49.3a for the fundamental). The simplicity of the long tube explains its frequent appearance in elementary texts; even in this case, however, the higher modes become quite complicated.

It is interesting to compare these results with those for a cylinder with one open end (at $z = L$, see Fig. 49.2b). The requirement $Z'(0) = 0$ still demands the cosine solution $Z(z) = \cos \alpha z$, but the free-end condition $Z(L) = 0$ [see Eq. (49.11)] now replaces (49.20) by

$$\alpha = \frac{2p + 1}{2} \frac{\pi}{L} \qquad p = 0, 1, 2, \ldots, \infty \qquad (49.29)$$

The remaining analysis proceeds exactly as before, and the allowed normal-mode frequencies are given by

$$\omega_{mnp}^2 = c^2 \left[\left(\frac{\alpha'_{m,n}}{a} \right)^2 + \left(\frac{2p + 1}{2} \frac{\pi}{L} \right)^2 \right] \qquad (49.30)$$

In the case of a long pipe $(L \gg a)$, the low-lying modes $(m = 0, n = 1)$ again become equally spaced harmonics, but with Eq. (49.27b) replaced by

$$\omega_p = c \left(\frac{2p + 1}{2} \frac{\pi}{L} \right) \qquad (49.31)$$

Since $p = 0$ is now allowed, the fundamental frequency of an organ pipe with one open end is one-half that of a pipe closed at both ends. More generally, the open end is a pressure node and a velocity antinode, with the cylinder containing $2p + 1$ quarter wavelengths (see Fig. 49.3b for the fundamental).

50 FOURIER TRANSFORMS AND GREEN'S FUNCTIONS IN THREE DIMENSIONS

Our treatment of sound waves in Sec. 49 considered the normal modes in cavities, which involves only the homogeneous Helmholtz equation. The inhomogeneous form will also be important, as already seen in the discussion of strings subject to external forces (Sec. 43). Once again, a Green's function provides a simple and flexible method for solving such equations, and we now develop the necessary techniques.

Screened Poisson Equation

To introduce this subject, we consider the purely mathematical problem of solving the inhomogeneous partial differential equation

$$\nabla^2 u(\mathbf{r}) - \lambda^2 u(\mathbf{r}) = -f(\mathbf{r}) \tag{50.1}$$

in unbounded space, subject to the condition that $u(\mathbf{r})$ vanishes sufficiently rapidly as $r \to \infty$ in any direction. Here we assume that $f(\mathbf{r}) = 0$ outside some large sphere of radius R_0; it represents a localized source for the field $u(\mathbf{r})$. Note that Eq. (50.1) differs from the Helmholtz equation in that $k^2 \to -\lambda^2$; this equation occurs rather frequently in physics, e.g., in the Yukawa theory of mesons or the Debye-Hückel screening in ionic solutions.

The unbounded domain $-\infty < x, y, z < \infty$ greatly simplifies the treatment of (50.1) because it permits a solution with three-dimensional Fourier transforms. To be precise, the spatial Fourier transform $\tilde{f}(\mathbf{p})$ of a function $f(\mathbf{r})$ is defined as

$$\tilde{f}(\mathbf{p}) = \int d^3 r \, e^{-i\mathbf{p}\cdot\mathbf{r}} f(\mathbf{r}) \tag{50.2}$$

and the inversion formula follows from Fourier's theorem (see Prob. 7.19)

$$f(\mathbf{r}) = (2\pi)^{-3} \int d^3 p \, e^{i\mathbf{p}\cdot\mathbf{r}} \tilde{f}(\mathbf{p}) \tag{50.3}$$

Alternatively, these relations follow from the representations of the one- and three-dimensional delta functions

$$\delta(x) = (2\pi)^{-1} \int_{-\infty}^{\infty} dp \, e^{ipx} \tag{50.4a}$$

$$\delta(\mathbf{r}) = (2\pi)^{-3} \int d^3 p \, e^{i\mathbf{p}\cdot\mathbf{r}} \tag{50.4b}$$

If $\tilde{f}(\mathbf{p})$ is known, Eq. (50.2) can be considered an integral equation for $f(\mathbf{r})$, and (50.3) provides the explicit solution.

To solve the original equation with a Fourier transform, we multiply by $e^{-i\mathbf{p}\cdot\mathbf{r}}$ and integrate over all space. The first term $\int d^3 r \, e^{-i\mathbf{p}\cdot\mathbf{r}} \nabla^2 u(\mathbf{r})$ can be rewritten

$$\int d^3 r \, \nabla \cdot (e^{-i\mathbf{p}\cdot\mathbf{r}} \nabla u) - \int d^3 r \, \nabla u \cdot \nabla e^{-i\mathbf{p}\cdot\mathbf{r}}$$

and the divergence theorem transforms the first term into a surface integral at infinity that vanishes owing to the boundary condition on u. The second term is just $i\mathbf{p} \cdot \int d^3 r \, e^{-i\mathbf{p}\cdot\mathbf{r}} \nabla u$, and a repetition of this procedure yields the relation

$$\int d^3 r \, e^{-i\mathbf{p}\cdot\mathbf{r}} \nabla^2 u = -p^2 \tilde{u}(\mathbf{p}) \tag{50.5}$$

Evaluation of the remaining Fourier transforms in (50.1) is elementary, and we obtain an *algebraic* equation

$$(p^2 + \lambda^2)\tilde{u}(\mathbf{p}) = \tilde{f}(\mathbf{p}) \tag{50.6}$$

for the transformed function $\tilde{u}(\mathbf{p})$. This simplification constitutes the great power of transform methods, for Eq. (50.6) has the trivial solution

$$\tilde{u}(\mathbf{p}) = \frac{\tilde{f}(\mathbf{p})}{p^2 + \lambda^2} \tag{50.7}$$

Substitution into the inversion formula (50.3) for $u(\mathbf{r})$ provides an explicit integral representation for the desired solution

$$u(\mathbf{r}) = \int \frac{d^3 p}{(2\pi)^3} \frac{\tilde{f}(\mathbf{p})}{p^2 + \lambda^2} e^{i\mathbf{p}\cdot\mathbf{r}} \tag{50.8}$$

with a known integrand.

Given any particular $f(\mathbf{r})$, it is straightforward in principle to compute $\tilde{f}(\mathbf{p})$ and then evaluate $u(\mathbf{r})$ directly from (50.8). A more illuminating approach is to express $\tilde{f}(\mathbf{p})$ explicitly as a Fourier transform, combining Eqs. (50.2) and (50.8) to give

$$u(\mathbf{r}) = \int \frac{d^3 p\, d^3 r'}{(2\pi)^3} \frac{e^{i\mathbf{p}\cdot(\mathbf{r}-\mathbf{r}')}}{p^2 + \lambda^2} f(\mathbf{r}')$$

This expression has the form familiar from Chap. 7

$$u(\mathbf{r}) = \int d^3 r'\, G(\mathbf{r}, \mathbf{r}') f(\mathbf{r}') \tag{50.9}$$

and it therefore provides an integral representation for the Green's function for Eq. (50.1) in unbounded space,

$$G(\mathbf{r}, \mathbf{r}') = G(\mathbf{r} - \mathbf{r}') = \int \frac{d^3 p}{(2\pi)^3} \frac{e^{i\mathbf{p}\cdot(\mathbf{r}-\mathbf{r}')}}{p^2 + \lambda^2} \tag{50.10}$$

If the inhomogeneous function is localized at $\mathbf{r}_0$ with $f(\mathbf{r}) = \delta(\mathbf{r} - \mathbf{r}_0)$, Eq. (50.9) verifies the expected relation

$$(\nabla^2 - \lambda^2)G(\mathbf{r} - \mathbf{r}_0) = -\delta(\mathbf{r} - \mathbf{r}_0) \tag{50.11}$$

The appearance of the difference variables $\mathbf{r} - \mathbf{r}_0$ reflects the translational invariance of the unbounded domain, with the disturbance depending only on the relative separation from the source.

It is interesting to construct the Green's function explicitly from Eq. (50.10). Once the integrations have been evaluated, the resulting function can depend only on the magnitude (and not direction) of the vector $\mathbf{R} \equiv \mathbf{r} - \mathbf{r}'$ because of the isotropy of Eq. (50.1) and the associated boundary conditions; in addition, the appearance of p^2 in the integrand suggests an evaluation in spherical polar coordinates with the polar axis along $\hat{\mathbf{R}}$

$$G(R) = \frac{1}{(2\pi)^3} \int_0^\infty p^2\, dp \int_0^\pi \sin\theta\, d\theta \int_0^{2\pi} d\phi\, \frac{e^{ipR\cos\theta}}{p^2 + \lambda^2}$$

The angular integrals are readily performed to give

$$G(R) = \int_0^\infty \frac{p^2 \, dp}{2\pi^2} \frac{\sin pR}{pR(p^2 + \lambda^2)} = \frac{1}{4\pi^2 R} \int_{-\infty}^\infty \frac{p \, dp \sin pR}{p^2 + \lambda^2} \qquad (50.12a)$$

where the second form follows because the integrand is an even function of p. This integral is easily evaluated with contour techniques, writing $\sin pR = (2i)^{-1}(e^{ipR} - e^{-ipR})$ and noting the presence of simple poles at $p = \pm i\lambda$. Closing the contour respectively above and below for the terms containing e^{ipR} and e^{-ipR}, we use Jordan's lemma (Prob. A.5) and the residue theorem to obtain

$$G(R) = G(|\mathbf{r} - \mathbf{r}'|) = \frac{e^{-\lambda|\mathbf{r}-\mathbf{r}'|}}{4\pi|\mathbf{r} - \mathbf{r}'|} \qquad (50.12b)$$

This solution is the celebrated Yukawa potential of meson physics, where λ^{-1} is the pion Compton wavelength, $\hbar/m_\pi c \approx 1.4 \times 10^{-13}$ cm, which characterizes the finite range of nuclear forces. If $\lambda \to 0$, Eq. (50.12b) reproduces the familiar Coulomb potential that is the Green's function for Laplace's and Poisson's equation of electrostatics.

Helmholtz Equation: Causality and Analyticity

We now return to the study of sound waves, where the preceding methods will permit us to solve the inhomogeneous wave equation in unbounded space. To be precise, consider

$$\nabla^2 \Phi(\mathbf{r}, t) - \frac{1}{c^2} \frac{\partial^2 \Phi(\mathbf{r}, t)}{\partial t^2} = -f(\mathbf{r}, t) \qquad (50.13)$$

where $f(\mathbf{r}, t)$ is a time-dependent source function and $-\infty < x, y, z < \infty$. It is convenient first to apply a Fourier transform in the time variable, using the relations

$$\Phi(\mathbf{r}, t) = (2\pi)^{-1} \int_{-\infty}^\infty d\omega \, e^{-i\omega t} \tilde{\Phi}(\mathbf{r}, \omega)$$

$$f(\mathbf{r}, t) = (2\pi)^{-1} \int_{-\infty}^\infty d\omega \, e^{-i\omega t} \tilde{f}(\mathbf{r}, \omega) \qquad (50.14)$$

and their inverses. Multiply Eq. (50.13) by $e^{i\omega t}$ and integrate over all t. Two integrations by parts yield the inhomogeneous Helmholtz equation

$$\nabla^2 \tilde{\Phi}(\mathbf{r}, \omega) + \left(\frac{\omega}{c}\right)^2 \tilde{\Phi}(\mathbf{r}, \omega) = -\tilde{f}(\mathbf{r}, \omega) \qquad (50.15)$$

where we assume that $f(\mathbf{r}, t)$ and hence $\Phi(\mathbf{r}, t)$ vanish sufficiently rapidly for large $|t|$. In direct analogy with the solution (50.9) for the inhomogeneous screened Poisson equation, Eq. (50.15) has a solution of the form

$$\tilde{\Phi}(\mathbf{r}, \omega) = \int d^3r' \, \tilde{G}(\mathbf{r} - \mathbf{r}', \omega)\tilde{f}(\mathbf{r}', \omega) \qquad (50.16)$$

where the Green's function satisfies the simpler equation

$$\left[\nabla^2 + \left(\frac{\omega}{c}\right)^2\right]\tilde{G}(\mathbf{r} - \mathbf{r}', \omega) = -\delta(\mathbf{r} - \mathbf{r}') \tag{50.17}$$

As in Eq. (50.10), $\tilde{G}$ here can depend only on $\mathbf{r} - \mathbf{r}'$, because the inhomogeneous term $-\delta(\mathbf{r} - \mathbf{r}')$ does so and there are no boundaries to destroy the translational invariance.

Before constructing the explicit form of $\tilde{G}(\mathbf{r} - \mathbf{r}', \omega)$, it is helpful to invert the temporal Fourier transform to obtain the full solution

$$\Phi(\mathbf{r}, t) = (2\pi)^{-1} \int_{-\infty}^{\infty} d\omega \, e^{-i\omega t} \int d^3r' \, \tilde{G}(\mathbf{r} - \mathbf{r}', \omega)\tilde{f}(\mathbf{r}', \omega)$$

Since $\tilde{f}(\mathbf{r}', \omega)$ is itself a Fourier integral, simple manipulations give the equivalent expression

$$\Phi(\mathbf{r}, t) = \int d^3r' \int_{-\infty}^{\infty} dt' \, G(\mathbf{r} - \mathbf{r}', t - t')f(\mathbf{r}', t') \tag{50.18}$$

where

$$G(\mathbf{r}, t) = (2\pi)^{-1} \int_{-\infty}^{\infty} d\omega \, e^{-i\omega t}\tilde{G}(\mathbf{r}, \omega) \tag{50.19}$$

is the time-dependent Green's function for the inhomogeneous wave equation in unbounded space. Choosing $f(\mathbf{r}, t) = \delta(\mathbf{r} - \mathbf{r}_0) \, \delta(t - t_0)$ in (50.13) and (50.18) shows that the full space- and time-dependent $G(\mathbf{r}, t)$ satisfies the inhomogeneous equation

$$\left(\nabla^2 - \frac{1}{c^2}\frac{\partial^2}{\partial t^2}\right)G(\mathbf{r} - \mathbf{r}_0, t - t_0) = -\delta(\mathbf{r} - \mathbf{r}_0) \, \delta(t - t_0) \tag{50.20}$$

with delta functions on the right-hand side. Given any source function $f(\mathbf{r}, t)$, these relations solve the inhomogeneous wave equation through the principle of superposition, which holds for linear equations.

The Green's function $G(\mathbf{r}, t)$ has a direct physical interpretation as the response to an instantaneous disturbance at $t = 0$ localized at the origin. Now wave signals obey the condition of *causality*; i.e., no signal appears until after the disturbance has taken place. Causality therefore requires that G *vanish* for all $t < 0$, and we now show that this apparently innocuous condition has profound consequences for the corresponding temporal transform

$$\tilde{G}(\mathbf{r}, \omega) = \int_{-\infty}^{\infty} dt \, e^{i\omega t}G(\mathbf{r}, t) = \int_{0}^{\infty} dt \, e^{i\omega t}G(\mathbf{r}, t) \tag{50.21}$$

Assume that $G(\mathbf{r}, t)$ is absolutely integrable, so that $\tilde{G}(\mathbf{r}, \omega = 0)$ exists and is bounded. For general complex values of $\omega \equiv \omega_1 + i\omega_2$, we have

$$\tilde{G}(\mathbf{r}, \omega_1 + i\omega_2) = \int_{0}^{\infty} dt \, e^{i\omega_1 t}e^{-\omega_2 t}G(\mathbf{r}, t) \tag{50.22}$$

In particular, $\tilde{G}(\mathbf{r}, \omega)$ is bounded throughout the upper half ω plane ($\omega_2 \geq 0$) because $|\tilde{G}(\mathbf{r}, \omega)|$ then satisfies the condition

$$|\tilde{G}(\mathbf{r}, \omega)| \leq \int_0^\infty dt \, e^{-\omega_2 t} |G(\mathbf{r}, t)| \leq \int_0^\infty dt \, |G(\mathbf{r}, t)| < \infty \qquad (50.23)$$

Furthermore, all derivatives $d^n\tilde{G}(\mathbf{r}, \omega)/d\omega^n$ exist for positive ω_2; we can therefore construct a Taylor series at every point in the upper half plane, and it converges inside any circle that does not cut the real axis. This assertion proves that $\tilde{G}(\mathbf{r}, \omega)$ is *analytic* in the upper half plane, which evidently depends only on the causal nature of $G(\mathbf{r}, t)$. We infer the following general principle: *the Fourier transform $\tilde{F}(\omega)$ of any causal response function $F(t)$ is an analytic function of ω throughout the upper half plane.*

The inverse statement follows immediately from Eq. (50.19), assuming that $G(\mathbf{r}, t)$ exists at $t = 0$. Suppose t is negative; the exponential factor becomes small in the upper half ω plane, permitting us to close the contour above with a large semicircle. Since $\tilde{G}(\mathbf{r}, \omega)$ is assumed analytic throughout that region, Cauchy's theorem ensures that $G(\mathbf{r}, t < 0)$ indeed vanishes, proving its causal character. In any specific example, we should verify the convergence of $G(\mathbf{r}, t = 0)$ and $\tilde{G}(\mathbf{r}, \omega = 0)$, but this process is usually straightforward.

This discussion applies directly to the Green's function for the Helmholtz equation (50.17). To appreciate the consequences of analyticity for Im $\omega > 0$, it is convenient to use a three-dimensional spatial Fourier transform, defining

$$\tilde{\tilde{G}}(\mathbf{p}, \omega) = \int d^3r \, e^{-i\mathbf{p} \cdot \mathbf{r}} \tilde{G}(\mathbf{r}, \omega) \qquad (50.24)$$

The same analysis used for Eq. (50.1) converts the second-order partial differential equation into an algebraic equation

$$\left[-p^2 + \left(\frac{\omega}{c} \right)^2 \right] \tilde{\tilde{G}}(\mathbf{p}, \omega) = -1$$

with the immediate solution

$$\tilde{\tilde{G}}(\mathbf{p}, \omega) = \frac{1}{p^2 - (\omega/c)^2} \qquad (50.25)$$

As written, this function has simple poles on the real ω axis at $\omega = \pm cp$; to ensure the proper analyticity, we must displace them an infinitesimal amount $-i\eta$ into the *lower* half ω plane, placing them at $\omega = \pm cp - i\eta$ (Fig. 50.1a). In this way, Eq. (50.25) is replaced by

$$\tilde{\tilde{G}}(\mathbf{p}, \omega) = \frac{1}{p^2 - (\omega + i\eta)^2/c^2} \qquad (50.26)$$

where the limit $\eta \to 0$ is to be taken at the end of all manipulations. We see that (50.26) follows from (50.25) with the substitution $\omega \to \omega + i\eta$, and this simple prescription suffices in most cases of practical interest.

The remaining inversion of Eq. (50.24) for $\tilde{G}(\mathbf{r}, \omega)$ merely involves careful

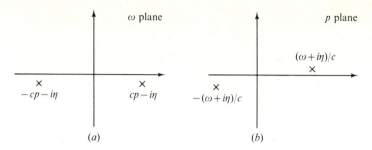

Figure 50.1 Analytic structure of $\tilde{\tilde{G}}(p, \omega)$ dictated by causality: (a) in complex ω plane and (b) in complex p plane for fixed real ω.

attention to the infinitesimal $i\eta$. As in Eq. (50.10), the three-dimensional $\mathbf{p}$ integration can be performed in spherical polar coordinates with the polar axis along $\hat{r}$. Evaluating the angular integrals, we obtain a form similar to (50.12a)

$$\tilde{G}(r, \omega) = \int \frac{d^3p}{(2\pi)^3} e^{i\mathbf{p} \cdot \mathbf{r}} \tilde{\tilde{G}}(\mathbf{p}, \omega) \qquad (50.27a)$$

$$= \frac{1}{8\pi^2 ir} \int_{-\infty}^{\infty} \frac{p \, dp \, (e^{ipr} - e^{-ipr})}{p^2 - (\omega + i\eta)^2/c^2} \qquad (50.27b)$$

where the even integrand again allows the "radial" integral to run from $-\infty$ to ∞. Although the term $i\eta$ was originally introduced to ensure analyticity in the upper half ω plane, it now serves to render the p integration well defined. The integrand in (50.27) has two terms, each with simple poles at $p = \pm(\omega + i\eta)/c$, as shown in Fig. 50.1b. Since the quantity r is nonnegative, the contour in the first term (containing e^{ipr}) must be closed in the upper half p plane, enclosing the pole at $(\omega + i\eta)/c$ in the positive sense, whereas that in the second term (containing e^{-ipr}) must be closed in the lower half p plane, enclosing the pole at $-(\omega + i\eta)/c$ in the negative sense. Use of Jordan's lemma and the residue theorem yields the fundamental result

$$\tilde{G}(r, \omega) = \frac{e^{i\omega r/c}}{4\pi r} \qquad (50.28)$$

as the causal Green's function for the three-dimensional Helmholtz equation; as expected, it is bounded for Im $\omega > 0$. This function characterizes the response of an acoustic medium to a harmonic perturbation with frequency ω. This restriction to a single frequency implies no loss of generality, of course, because the full solution to the wave equation merely requires an inverse temporal Fourier transform (50.14).

Boundaries and the Method of Images

The previous analysis involved the special restriction to unbounded space, which permitted signals from a localized source to propagate unimpeded to infinity.

Unfortunately, practical acoustic phenomena involve boundaries at finite distances, and we now turn to this more realistic but difficult situation. In particular, consider a harmonic source function $f(\mathbf{r}, t) = \text{Re } f(\mathbf{r})e^{-i\omega t}$; the induced velocity potential will have the same form $\Phi(\mathbf{r}, t) = \text{Re } \Phi(\mathbf{r})e^{-i\omega t}$, where the complex function $\Phi(\mathbf{r})$ satisfies the inhomogeneous Helmholtz equation

$$\nabla^2 \Phi(\mathbf{r}) + \left(\frac{\omega}{c}\right)^2 \Phi(\mathbf{r}) = -f(\mathbf{r}) \tag{50.29}$$

In addition, we assume the presence of bounding surfaces A on which Φ must satisfy linear homogeneous conditions of the form [compare Eq. (40.13)]

$$\alpha \hat{n} \cdot \nabla \Phi(\mathbf{r}) - \beta \Phi(\mathbf{r}) = 0 \qquad \text{for } \mathbf{r} \text{ on } A \tag{50.30}$$

where $\hat{n}$ is the unit normal vector directed out of the fluid and α and β are constants. This condition encompasses both previous limiting cases of a fixed rigid surface ($\beta = 0$) and a free surface ($\alpha = 0$).

The formal solution to this problem is easily obtained by defining a Green's function $\tilde{G}(\mathbf{r}, \mathbf{r}')$ that satisfies the partial differential equation [compare (50.17)]

$$\left(\nabla^2 + \frac{\omega^2}{c^2}\right)\tilde{G}(\mathbf{r}, \mathbf{r}') = -\delta(\mathbf{r} - \mathbf{r}') \tag{50.31}$$

and the boundary conditions

$$\alpha \hat{n} \cdot \nabla \tilde{G}(\mathbf{r}, \mathbf{r}') - \beta \tilde{G}(\mathbf{r}, \mathbf{r}') = 0 \qquad \text{for } \mathbf{r} \text{ on } A \tag{50.32}$$

For simplicity, we here suppress the additional dependence on ω, although $\tilde{G}$ must still be analytic in the upper half ω plane. By inspection, the integral

$$\Phi(\mathbf{r}) = \int d^3r' \; \tilde{G}(\mathbf{r}, \mathbf{r}')f(\mathbf{r}') \tag{50.33}$$

solves both the original differential equation and the boundary conditions. Unfortunately, the actual construction of $\tilde{G}$ is extremely difficult except in certain very special geometries.

To illustrate the available techniques, we consider the simplest case of a half space $z > 0$ bounded by the xy plane as illustrated in Fig. 50.2. The desired Green's function satisfies Eq. (50.31) throughout the domain $-\infty < x, y < \infty$, and $z > 0$, subject to the boundary condition

$$\left[\alpha \frac{\partial \tilde{G}(\mathbf{r}, \mathbf{r}')}{\partial z} + \beta \tilde{G}(\mathbf{r}, \mathbf{r}')\right]_{z=0} = 0 \tag{50.34}$$

since $\hat{n} = -\hat{z}$. This problem has an explicit solution because the boundary involves only the single variable z, leaving the other two variables unaffected. If we assume that $\tilde{G}$ has the representation

$$\tilde{G}(\mathbf{x}, z; \mathbf{x}', z') = (2\pi)^{-2} \int d^2p \; e^{i\mathbf{p}\cdot(\mathbf{x}-\mathbf{x}')}\tilde{G}(\mathbf{p}, z, z') \tag{50.35}$$

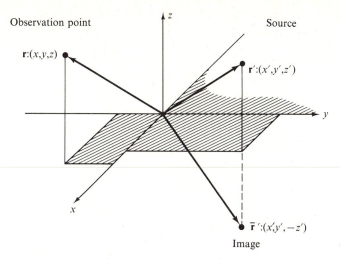

Figure 50.2 Construction of the Green's function $\tilde{G}(\mathbf{r},\mathbf{r}')$ for the half space $z > 0$.

with $\mathbf{x}$ and $\mathbf{x}'$ denoting two-dimensional vectors in the xy plane, then a two-dimensional spatial Fourier transform in x and y reduces Eq. (50.31) to a one-dimensional ordinary differential equation, and $\tilde{\tilde{G}}(\mathbf{p}, z, z')$ can be constructed with the one-dimensional techniques introduced in Sec. 43. The inverse transform then yields an integral representation of $\tilde{G}(\mathbf{x}, z; \mathbf{x}', z')$, but the detailed integrations become very intricate.

As a result, we shall further limit our scope and consider only the particular cases $\alpha = 0$ and $\beta = 0$, where special tricks are available. For definiteness, consider first a rigid boundary $(\beta = 0)$, where G satisfies (50.31) and the boundary condition

$$\left. \frac{\partial \tilde{G}(\mathbf{r}, \mathbf{r}')}{\partial z} \right|_{z=0} = 0 \qquad \text{rigid boundary} \qquad (50.36)$$

Let the source point $\mathbf{r}'$ have the cartesian components $(x', y'\ z')$ and define another vector $\bar{\mathbf{r}}' = (x', y', -z')$ that is the unique mirror image of $\mathbf{r}'$ in the xy plane (Fig. 50.2). We now assert that the desired solution is just

$$\tilde{G}(\mathbf{r}, \mathbf{r}') = \frac{e^{ik|\mathbf{r}-\mathbf{r}'|}}{4\pi|\mathbf{r}-\mathbf{r}'|} + \frac{e^{ik|\mathbf{r}-\bar{\mathbf{r}}'|}}{4\pi|\mathbf{r}-\bar{\mathbf{r}}'|} \qquad \text{rigid boundary} \qquad (50.37)$$

where $k \equiv \omega/c$. To prove this result, note first that

$$(\nabla^2 + k^2)\tilde{G}(\mathbf{r}, \mathbf{r}') = -\delta(\mathbf{r} - \mathbf{r}') - \delta(\mathbf{r} - \bar{\mathbf{r}}') \qquad (50.38)$$

Since z and z' are necessarily positive for any physical source and observation point, the second delta function never contributes. Thus (50.37) satisfies the correct equation in the domain $z > 0$. Second, $\tilde{G}$ has the form $F(|\mathbf{r} - \mathbf{r}'|) + F(|\mathbf{r} - \bar{\mathbf{r}}'|)$, and the z derivative becomes

$$\frac{\partial \tilde{G}}{\partial z} = F'(|\mathbf{r} - \mathbf{r}'|)\frac{z - z'}{|\mathbf{r} - \mathbf{r}'|} + F'(|\mathbf{r} - \bar{\mathbf{r}}'|)\frac{z - \bar{z}'}{|\mathbf{r} - \bar{\mathbf{r}}'|}$$

where F' denotes the derivative of F. By construction, we see that $\bar{z}' = -z'$ and that $|\mathbf{r} - \mathbf{r}'| = |\mathbf{r} - \bar{\mathbf{r}}'|$ at $z = 0$, ensuring that $\tilde{G}$ also satisfies the rigid boundary condition (50.36). The first term of (50.37) is just the disturbance seen at $\mathbf{r}$ arising from a unit positive source located at $\mathbf{r}'$ in the absence of the boundary, and the second term can be interpreted as an induced unit positive source located at the image point $\bar{\mathbf{r}}'$. In this way, the whole effect of the infinite wall is replaced by an equivalent problem involving a source and a single image in unbounded space but with the solution restricted to $z > 0$.

A similar solution applies for a free planar surface ($\alpha = 0$), where the corresponding Green's function is

$$\tilde{G}(\mathbf{r}, \mathbf{r}') = \frac{e^{ik|\mathbf{r} - \mathbf{r}'|}}{4\pi|\mathbf{r} - \mathbf{r}'|} - \frac{e^{ik|\mathbf{r} - \bar{\mathbf{r}}'|}}{4\pi|\mathbf{r} - \bar{\mathbf{r}}'|} \qquad \text{free boundary} \qquad (50.39)$$

and the image now has the opposite sign from the original source. For $k = 0$, this latter situation is familiar from electrostatics. Unfortunately, these two special boundary conditions exhaust the applicability of the method of images for the Helmholtz equation in a half space. Although the exact Green's function (50.35) for the more general boundary condition (50.34) still consists of two terms, one being the disturbance generated by the unit source in unbounded space and the second including the effect of the wall, the remaining term cannot be interpreted as a single fictitious image source.

51 RADIATION, DIFFRACTION, AND SCATTERING

The preceding section concentrated on the mathematical question of solving the Helmholtz equation, and we now consider a few of the numerous physical applications of these techniques. One interesting problem is the detailed form of the acoustic disturbance generated by a specific source. A second class of questions concerns the subsequent propagation of sound waves through a uniform medium and the associated diffraction or scattering by boundaries. Since the sound generator is typically less interesting in these latter applications, the incident wave is generally approximated by a plane wave with the source at infinity.

Radiation from a Piston in a Wall

As a first example, we study a half space ($z > 0$) bounded by a rigid wall containing a flush circular piston of radius a that oscillates vertically with small-amplitude motion Re $\epsilon a e^{-i\omega t}$, where $|\epsilon| \ll 1$ (see Fig. 51.1). The velocity of the

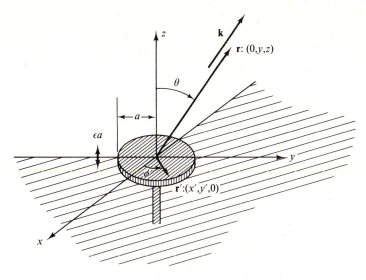

Figure 51.1 Radiation from a circular piston in a wall. The axial symmetry of $\Phi(\mathbf{r})$ has been used to place $\mathbf{r}$ in the yz plane.

piston is evidently Re $(-i\omega\epsilon a e^{-i\omega t})$ in the $\hat{z}$ direction. The resulting acoustic disturbance $\Phi(\mathbf{r}, t) = \text{Re } \Phi(\mathbf{r})e^{-i\omega t}$ satisfies

$$\nabla^2\Phi(\mathbf{r}) + k^2\Phi(\mathbf{r}) = 0 \tag{51.1}$$

subject to the boundary condition (49.10a) obtained by matching the velocity to that of the boundary at $z = 0$:

$$-\frac{\partial\Phi}{\partial z}\bigg|_{z=0} = \begin{cases} 0 & x^2 + y^2 > a \\ -i\omega\epsilon a & x^2 + y^2 < a \end{cases} \tag{51.2}$$

Strictly, the boundary condition on the piston holds at the displaced surface, but the difference makes only a second-order correction in ϵ and is negligible in linearized theory.

The present problem differs significantly from that in Sec. 50, where we constructed a Green's function for an *inhomogeneous* differential equation subject to *homogeneous* boundary conditions. Here, in contrast, Φ satisfies a *homogeneous* equation with *inhomogeneous* boundary conditions. The two problems are closely related, however, and we now prove the remarkable result that the same Green's function solves them both. Consider two arbitrary functions $\Phi(\mathbf{r})$ and $\Psi(\mathbf{r})$ defined in a volume V enclosed by a surface A with outward normal $\hat{n}$. Green's identity asserts that

$$\int_V d^3r \, (\Phi \, \nabla^2\Psi - \Psi \, \nabla^2\Phi) = \int_A d\mathbf{A} \cdot (\Phi \, \nabla\Psi - \Psi \, \nabla\Phi) \tag{51.3}$$

which follows immediately from the divergence theorem and elementary manipulations. Let Φ be a solution to the homogeneous Helmholtz equation (51.1) subject to the inhomogeneous boundary condition

$$\hat{n} \cdot \nabla\Phi = F(\mathbf{r}) \qquad \text{for } \mathbf{r} \text{ on } A \tag{51.4}$$

and let $\Psi(\mathbf{r})$ be the Green's function $\tilde{G}(\mathbf{r}, \mathbf{r}')$ defined by Eq. (50.31) with the homogeneous boundary condition

$$\hat{n} \cdot \nabla\tilde{G}(\mathbf{r}, \mathbf{r}') = 0 \qquad \text{for } \mathbf{r} \text{ on } A \tag{51.5}$$

By inspection, the left-hand side of (51.3) is $-\Phi(\mathbf{r}')$, and the boundary conditions on Φ and $\tilde{G}$ simplify the right-hand side to yield the desired solution

$$\Phi(\mathbf{r}') = \int_A dA \; \tilde{G}(\mathbf{r}, \mathbf{r}')F(\mathbf{r}) \tag{51.6}$$

This relation expresses Φ throughout the region V in terms of the known boundary function F and the previously derived Green's function for the inhomogeneous equation with homogeneous boundary conditions. With slight alterations, the same method applies for the more general boundary conditions (50.32), including the limit of a free surface ($\alpha = 0$).

In the present planar configuration, the relevant Green's function is just (50.37). It is convenient to interchange primed and unprimed variables in (51.6) to obtain the explicit solution to (51.1) and (51.2) [note that $\hat{n} = -\hat{z}$ and that $\tilde{G}(\mathbf{r}, \mathbf{r}') = \tilde{G}(\mathbf{r}' \; \mathbf{r})$]

$$\Phi(\mathbf{r}) = -i\omega\epsilon a \int dx' \, dy' \; \tilde{G}(\mathbf{r}, \mathbf{r}') = -\frac{i\omega\epsilon a}{2\pi} \int dx' \, dy' \frac{e^{ik|\mathbf{r}-\mathbf{r}'|}}{|\mathbf{r}-\mathbf{r}'|} \tag{51.7}$$

where $z' = 0$ and the integral is over the region $x'^2 + y'^2 \leq a^2$. This expression is easily interpreted as a superposition of outgoing spherical waves generated by each element of the piston, with strength proportional to the vertical velocity $-i\omega\epsilon a$ [compare Eq. (50.28)].

The general behavior of (51.7) can be rather complicated, and it is simplest to consider only the far-field behavior when the observation point $\mathbf{r}$ (see Fig. 51.1) is many wavelengths from the piston ($kr \gg 1$) and its distance far exceeds the radius a, still allowing ka to be arbitrary. The approximate relation $|\mathbf{r} - \mathbf{r}'| \approx r - \hat{r} \cdot \mathbf{r}' + \cdots$ permits us to expand (51.7) to leading order

$$\Phi(\mathbf{r}) \sim -\frac{i\omega\epsilon a}{2\pi} \frac{e^{ikr}}{r} \int_{(r' < a)} dx' \, dy' \; e^{-ik\hat{r}\cdot\mathbf{r}'} \tag{51.8}$$

where the terms omitted are of higher order in r^{-1}. This two-dimensional integration can be performed in plane polar coordinates (Fig. 51.1)

$$\Phi(\mathbf{r}) \sim -\frac{i\omega\epsilon a}{2\pi} \frac{e^{ikr}}{r} \int_0^a r' \, dr' \int_0^{2\pi} d\phi' \; e^{-ikr' \sin\theta \sin\phi'} \tag{51.9}$$

and we have used the axial symmetry to choose $\mathbf{r}$ in the yz plane. Reference to

Eq. (D3.26) shows that the Bessel function $J_n(z)$ for integral n has the simple representation

$$J_n(z) = \frac{1}{2\pi} \int_{-\pi}^{\pi} d\phi \; e^{i(z \sin \phi - n\phi)} \tag{51.10}$$

In addition, the power series (D3.8) shows that J_n satisfies the obvious relation

$$J_n(-z) = (-1)^n J_n(z) \tag{51.11}$$

A combination of these properties gives

$$J_0(z) = J_0(-z) = (2\pi)^{-1} \int_0^{2\pi} d\phi' \; e^{\pm iz \sin \phi'} \tag{51.12}$$

We therefore identify the angular integral in Eq. (51.9) as a Bessel function of order zero, obtaining the asymptotic form

$$\Phi(\mathbf{r}) \sim -\frac{i\omega\epsilon}{2\pi} \frac{e^{ikr}}{r} 2\pi \int_0^a r' \, dr' \, J_0(kr' \sin \theta) \tag{51.13}$$

The remaining radial integral then follows from the standard recursion relation

$$\frac{d}{dz} z^n J_n(z) = z^n J_{n-1}(z) \tag{51.14}$$

which for $n = 1$ yields

$$\int_0^z J_0(\rho)\rho \, d\rho = z J_1(z) \tag{51.15}$$

Thus we find

$$\Phi(\mathbf{r}) \sim -\frac{i\omega\epsilon}{2\pi} \frac{e^{ikr}}{r} \pi a^2 \frac{2 J_1(ka \sin \theta)}{ka \sin \theta} \tag{51.16}$$

Note that this far-field dependence is essentially that of a point source ($\propto r^{-1} e^{ikr}$), with a strength proportional to the piston's area and velocity but modulated by an angular factor $2 J_1(ka \sin \theta)/ka \sin \theta$ with absolute value ≤ 1. This factor has the limiting behavior (Appendix D)

$$2x^{-1} J_1(x) \approx \begin{cases} 1 - \frac{1}{8}x^2 & x \ll 1 \\ \left(\frac{8}{\pi x^3}\right)^{1/2} \cos\left(x - \frac{3}{4}\pi\right) & x \gg 1 \end{cases} \tag{51.17}$$

The most interesting physical quantity is the time-average energy flux, which follows directly from Eqs. (45.45) and (49.12) [recall Eqs. (49.7) and (49.8)]

$$\langle \mathbf{j}_e \rangle = \tfrac{1}{2} \operatorname{Re} \mathbf{v}' p'^* \tag{51.18a}$$

$$= \tfrac{1}{2}\rho_0 \operatorname{Re} \left[-\nabla\Phi(-i\omega\Phi)^* \right] \tag{51.18b}$$

In the asymptotic limit, it reduces to

$$\langle \mathbf{j}_e \rangle \approx \frac{\rho_0 \epsilon^2 c^3 k^4 a^6}{8r^2} \left[\frac{2J_1(ka \sin \theta)}{ka \sin \theta} \right]^2 \hat{r} \qquad (51.18c)$$

which varies inversely with r^2. Thus it is more convenient to consider the mean acoustic power $\langle dP \rangle = \langle \mathbf{j}_e \rangle \cdot \hat{r} \, dA = \langle \mathbf{j}_e \rangle \cdot \hat{r} \, r^2 \, d\Omega$ crossing a surface element that subtends a solid angle $d\Omega$. The corresponding differential power radiated

$$\left\langle \frac{dP}{d\Omega} \right\rangle = \tfrac{1}{8}\rho_0 \epsilon^2 c^3 k^4 a^6 \left[\frac{2J_1(ka \sin \theta)}{ka \sin \theta} \right]^2 \qquad (51.19)$$

has the angular dependence of the last factor, which is shown in Fig. 51.2. Since $|\sin \theta| \le 1$, the radiation is isotropic for long wavelengths ($ka \ll 1$), whereas it exhibits a characteristic angular distribution for $ka \gg 1$. The intensity has a maximum in the forward direction $\theta = 0$, where $2x^{-1}J_1(x) \to 1$, and the first minimum, or zero, in the radiated power occurs at an angle θ_1 determined by $ka \sin \theta_1 = \alpha_{1,1} = 3.83 \cdots$ or $a \sin \theta_1 \approx 0.61\lambda$, where $\lambda = 2\pi/k$. The total average power radiated is obtained by integrating $\langle dP/d\Omega \rangle$ over the hemisphere

$$\langle P \rangle = \tfrac{1}{8}\rho_0 \epsilon^2 c^3 k^4 a^6 2\pi \int_0^{\pi/2} d\theta \, \sin \theta \left[\frac{2J_1(ka \sin \theta)}{ka \sin \theta} \right]^2 \qquad (51.20)$$

which must be evaluated numerically for various values of ka. Note that a larger piston radiates more power and that a given piston radiates more efficiently at higher frequency and for larger amplitudes.

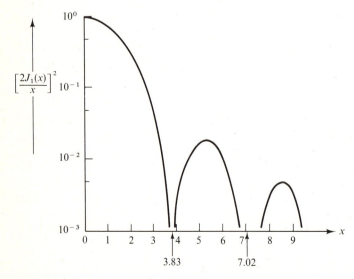

Figure 51.2 Intensity pattern $[2J_1(x)/x]^2$ for radiation from a circular piston of radius a in a wall. Here $x = ka \sin \theta$ [see Eq. (51.19) and Fig. 51.1]. This graph also gives the diffraction pattern for a circular aperture where $x = q_\perp a = ka \sin \theta'$ for normal incidence [see Eqs. (51.41) and (51.42)].

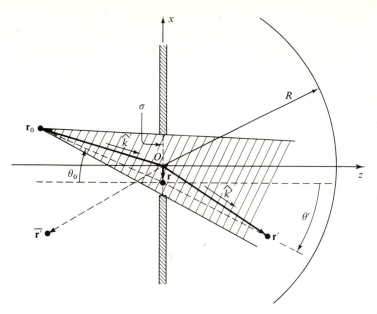

Figure 51.3 Point source at $\mathbf{r}_0$, rigid screen with aperture σ, and observer at $\mathbf{r}'$.

Diffraction in Kirchhoff's Approximation

Consider a rigid infinite plane surface $z = 0$ containing an aperture σ and place a unit source of acoustic waves at the point $\mathbf{r}_0$ in the region $z < 0$. What is the disturbance throughout the half space $z > 0$? This problem illustrates the characteristic features of *diffraction*. In the limit of purely geometric propagation, the wave for $z > 0$ would be confined to the shaded region in Fig. 51.3. The inclusion of wave properties alters this simple picture, for the acoustic disturbance is diffracted into the shadow zone with finite amplitude throughout the whole region $z > 0$.

A precise formulation of this question involves the inhomogeneous Helmholtz equation

$$(\nabla^2 + k^2)\Phi(\mathbf{r}) = -\delta(\mathbf{r} - \mathbf{r}_0) \tag{51.21}$$

subject to the homogeneous boundary condition $\hat{n} \cdot \nabla\Phi = 0$ on the rigid plane, *omitting* the aperture σ. The solution $\Phi(\mathbf{r})$ characterizes the radiation from a unit source at $\mathbf{r}_0$ in the presence of the boundary; it is the exact Green's function. Unfortunately, constructing this quantity is not feasible except in a few special cases.† Consequently, we shall introduce an approximate description originally suggested by Kirchhoff; it employs the simpler (method-of-images) Green's

---

† Sommerfeld obtained a rigorous solution for a half plane, in which σ is the other half plane. See, for example, Sommerfeld [3], sec. 38, or Landau and Lifshitz [4], secs. 74 and 75.

function (50.37) that satisfies the condition $\hat{n} \cdot \nabla \tilde{G} = 0$ on the whole plane *including* the aperture σ. This quantity has the great advantage of being known explicitly.

We start with Green's identity (51.3) applied to half space $z > 0$; the closed surface required in this relation is taken as the plane $z = 0$ and a large hemisphere of radius R, as illustrated in Fig. 51.3. We identify Φ with the desired solution and Ψ with the known Green's function (50.37). Use of Eqs. (50.31), (50.36), and (51.21) and the observation that $\mathbf{r}_0$ is outside the region in question then leads to

$$\Phi(\mathbf{r}') = \int_\sigma dA \; \tilde{G}(\mathbf{r}, \mathbf{r}')\hat{n} \cdot \nabla\Phi(\mathbf{r})$$

$$+ \int_R dA \; [\tilde{G}(\mathbf{r}, \mathbf{r}')\hat{n} \cdot \nabla\Phi(\mathbf{r}) - \Phi(\mathbf{r})\hat{n} \cdot \nabla\tilde{G}(\mathbf{r}, \mathbf{r}')] \qquad (51.22a)$$

or $$\Phi(\mathbf{r}') \equiv \Phi_\sigma(\mathbf{r}') + \Phi_R(\mathbf{r}') \qquad (51.22b)$$

where the two terms denote the integrals over σ and R. Now $\mathbf{r}'$ denotes the observation point, and $\mathbf{r}$ lies on the bounding surface (Fig. 51.3). The first surface integral is restricted to the *aperture* σ because $\hat{n} \cdot \nabla\Phi = 0$ on the rest of the plane $z = 0$. Note that we have not *solved* the problem because the unknown quantity Φ and its normal derivatives appear on the right-hand side of these relations. Nevertheless, physical intuition indicates that $\Phi(\mathbf{r}')$ in the region $z > 0$ represents the diffraction pattern determined by the disturbance coming from the aperture $\Phi_\sigma(\mathbf{r}')$, the contribution $\Phi_R(\mathbf{r}')$ from the large hemisphere being negligible in the limit $R \to \infty$. To support this picture, we now demonstrate that it is consistent to ignore $\Phi_R(\mathbf{r}')$ in this limit. Assume that $\Phi(\mathbf{r}') = \Phi_\sigma(\mathbf{r}')$ in the interior of the hemisphere. The evaluation of $\Phi_\sigma(\mathbf{r}')$ requires the asymptotic form of the Green's function (50.37) for $\mathbf{r}$ confined to the aperture σ and $r' \to \infty$. Since $|\mathbf{r}' - \mathbf{r}| \to r' - \mathbf{r} \cdot \hat{r}'$ and $|\bar{\mathbf{r}}' - \mathbf{r}| \to r' - \mathbf{r} \cdot \hat{\bar{r}}'$ in this limit, we conclude that

$$\Phi_\sigma(\mathbf{r}') \xrightarrow[|\mathbf{r}'| \to \infty]{} \frac{e^{ikr'}}{4\pi r'} \int_\sigma dA \; (e^{-i\mathbf{k}' \cdot \mathbf{r}} + e^{-i\bar{\mathbf{k}}' \cdot \mathbf{r}})\hat{n} \cdot \nabla\Phi_\sigma(\mathbf{r}) \qquad (51.23)$$

where $\mathbf{k}' \equiv k\hat{r}'$ and $\bar{\mathbf{k}}' \equiv k\hat{\bar{r}}'$; thus the asymptotic behavior of $\Phi_\sigma(\mathbf{r}')$ for large r' is given by const $\times r'^{-1}e^{ikr'}$.

We next consider $\Phi_R(\mathbf{r}')$; it involves the asymptotic form of the Green's function for $\mathbf{r}'$ at a fixed point in the *interior* of the hemisphere and $|\mathbf{r}| \to \infty$ (on the *surface* of the hemisphere). As above, we conclude that

$$\tilde{G}(\mathbf{r}, \mathbf{r}') \xrightarrow[|\mathbf{r}| \to \infty]{} \frac{e^{ikr}}{4\pi r} (e^{-i\mathbf{k} \cdot \mathbf{r}'} + e^{-i\mathbf{k} \cdot \bar{\mathbf{r}}'}) \qquad (51.24)$$

where $\mathbf{k} \equiv k\hat{r}$. It is now possible to evaluate $\Phi_R(\mathbf{r}')$ for large R. We note that $\hat{n} \cdot \nabla = \partial/\partial r$ on the hemisphere and write

$$\Phi_R(\mathbf{r}') \approx \int_R dA \left[\frac{e^{ikr}}{4\pi r}(e^{-i\mathbf{k} \cdot \mathbf{r}'} + e^{-i\mathbf{k} \cdot \bar{\mathbf{r}}'}) \frac{\partial}{\partial r} \Phi_\sigma(\mathbf{r}) \right.$$

$$\left. - \Phi_\sigma(\mathbf{r}) \frac{\partial}{\partial r} \frac{e^{ikr}}{4\pi r} (e^{-i\mathbf{k} \cdot \mathbf{r}'} + e^{-i\mathbf{k} \cdot \bar{\mathbf{r}}'}) \right] \qquad (51.25)$$

Retaining the leading terms as $R \to \infty$, we find

$$\Phi_R(\mathbf{r}') \approx \int_R dA(e^{-i\mathbf{k}\cdot\mathbf{r}'} + e^{-i\mathbf{k}\cdot\bar{\mathbf{r}}})\frac{e^{ikr}}{4\pi r}\left[\frac{\partial\Phi_\sigma(\mathbf{r})}{\partial r} - ik\Phi_\sigma(\mathbf{r})\right] \tag{51.26}$$

Equation (51.23) shows that this expression vanishes asymptotically because

$$\left(\frac{\partial}{\partial r} - ik\right)\frac{e^{ikr}}{r} \approx \frac{1}{r}\left(\frac{\partial}{\partial r} - ik\right)e^{ikr} \approx 0 \qquad \text{as } r \to \infty$$

Since these arguments are exact in the limit $R \to \infty$, we conclude that

$$\lim_{R\to\infty}\Phi_R(\mathbf{r}') = 0 \tag{51.27}$$

Hence we can consistently retain just $\Phi_\sigma(\mathbf{r}')$ and write (now dropping the extraneous subscript)

$$\Phi(\mathbf{r}') = \int_\sigma dA\,\tilde{G}(\mathbf{r},\mathbf{r}')\hat{n}\cdot\nabla\Phi(\mathbf{r}) \tag{51.28}$$

Substitution of Eq. (50.37) then yields the *exact* representation (cf. Fig. 51.3)

$$\Phi(\mathbf{r}') = -\int_\sigma dA\left[\frac{e^{ik|\mathbf{r}-\mathbf{r}'|}}{2\pi|\mathbf{r}-\mathbf{r}'|}\frac{\partial\Phi(\mathbf{r})}{\partial z}\right]_{z=0} \tag{51.29}$$

expressing Φ throughout the right-hand half space in terms of its (still unknown) normal derivative in the aperture. Although this relation does not solve the problem, it does provide a useful basis for short-wavelength approximations, and we now consider a modified form of Kirchhoff's analysis.

The fundamental assumption is that the aperture has large dimensions compared with the wavelength $\lambda = 2\pi/k$ of the disturbance. Since λ acts as the characteristic scale of length, the propagating wave in the aperture is largely unaffected by the screen except within a few wavelengths of the boundary. This idea motivates Kirchhoff's approximation of replacing the exact $\partial\Phi/\partial z$ in the aperture by the value [see Eq. (50.28)] that it would have if there were no screen at all. Moreover, the assumption of short wavelengths implies that this derivative can be written

$$\frac{\partial\Phi}{\partial z}\bigg|_{z=0} \approx \left(\frac{ike^{ik|\mathbf{r}-\mathbf{r}_0|}}{4\pi|\mathbf{r}-\mathbf{r}_0|}\frac{z-z_0}{|\mathbf{r}-\mathbf{r}_0|}\right)_{z=0}$$

apart from terms of order $\lambda|\mathbf{r}-\mathbf{r}_0|^{-1}$. A combination with (51.29) gives Kirchhoff's approximate formula

$$\Phi(\mathbf{r}') \approx \frac{ik}{8\pi^2}\int_\sigma dx\,dy\,\frac{e^{ik|\mathbf{r}-\mathbf{r}_0|}}{|\mathbf{r}-\mathbf{r}_0|}\frac{e^{ik|\mathbf{r}-\mathbf{r}'|}}{|\mathbf{r}-\mathbf{r}'|}\frac{z_0}{|\mathbf{r}-\mathbf{r}_0|} \tag{51.30}$$

where $\mathbf{r}$ lies in the xy plane. This expression has the following simple physical interpretation, embodying the essential features of *Huygens' principle*. The source at $\mathbf{r}_0$ emits spherical waves with an amplitude proportional to $|\mathbf{r}-\mathbf{r}_0|^{-1}\exp(ik|\mathbf{r}-\mathbf{r}_0|)$ in the aperture. Each element on that wavefront acts

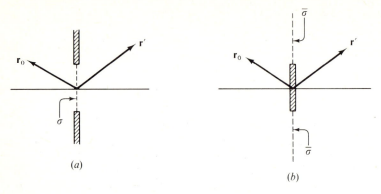

(a)

(b)

Figure 51.4 Relation between transmission and diffraction by complementary apertures (Babinet's principle).

as a new source of secondary spherical waves with amplitude at $\mathbf{r}'$ proportional to $|\mathbf{r} - \mathbf{r}'|^{-1} \exp(ik|\mathbf{r} - \mathbf{r}'|)$. The total disturbance at $\mathbf{r}'$ is obtained by a coherent superposition of all these elementary wavelets. The integrand of (51.30) also contains an *obliquity factor* $-z_0|\mathbf{r} - \mathbf{r}_0|^{-1} = \cos\theta_0$, where θ_0 is the angle of incidence between the normal $\hat{n}$ and the incident wave (see Fig. 51.3); this factor vanishes if the source is located near the screen ($\theta_0 \approx \frac{1}{2}\pi$). Various other formulations of Kirchhoff's formula replace $\cos\theta_0$ by $\cos\theta'$ (Fig. 51.3) or their average,† but the difference is not very significant because the description is most applicable near the forward directions, where θ_0 and θ' are both small. Equation (51.30) has an additional factor $-i$, indicating that the diffracted wave differs in phase by $\frac{1}{2}\pi$ from the incident wave in the absence of the screen.

Kirchhoff's formula implies an interesting relation between the waves Φ *transmitted* through the aperture σ and the waves $\overline{\Phi}$ *diffracted* by a planar obstacle of the same shape when the original screen becomes the complementary aperture $\bar{\sigma}$ (see Fig. 51.4). In each case, use of Kirchhoff's approximation in Eq. (51.28) gives the formulas

$$\Phi(\mathbf{r}') = -\int_\sigma dA \; \tilde{G}(\mathbf{r}, \mathbf{r}') \frac{\partial}{\partial z} \frac{e^{ik|\mathbf{r} - \mathbf{r}_0|}}{4\pi|\mathbf{r} - \mathbf{r}_0|} \tag{51.31a}$$

$$\overline{\Phi}(\mathbf{r}') = -\int_{\bar{\sigma}} dA \; \tilde{G}(\mathbf{r}, \mathbf{r}') \frac{\partial}{\partial z} \frac{e^{ik|\mathbf{r} - \mathbf{r}_0|}}{4\pi|\mathbf{r} - \mathbf{r}_0|} \tag{51.31b}$$

and the sum of these two terms is just an integral over the whole plane

$$\Phi(\mathbf{r}') + \overline{\Phi}(\mathbf{r}') = -\int_{\sigma+\bar{\sigma}} dA \; \tilde{G}(\mathbf{r}, \mathbf{r}') \frac{\partial}{\partial z} \frac{e^{ik|\mathbf{r} - \mathbf{r}_0|}}{4\pi|\mathbf{r} - \mathbf{r}_0|} \tag{51.32}$$

Application of Green's identity (51.3) with

$$\Phi(\mathbf{r}) = \Phi_{\text{inc}}(\mathbf{r}) \equiv (4\pi|\mathbf{r} - \mathbf{r}_0|)^{-1} \exp(ik|\mathbf{r} - \mathbf{r}_0|)$$

---

† Coulson (1958), secs. 86 and 87; Sommerfeld [3], sec. 34.

and $\Psi(\mathbf{r}) = \tilde{G}(\mathbf{r}, \mathbf{r}')$ from (50.37) shows that this integral is just the incident wave in the *absence* of the screen

$$\Phi(\mathbf{r}') + \bar{\Phi}(\mathbf{r}') = \Phi_{inc}(\mathbf{r}') \equiv \frac{e^{ik|\mathbf{r}' - \mathbf{r}_0|}}{4\pi|\mathbf{r}' - \mathbf{r}_0|} \tag{51.33}$$

For the configuration of $\mathbf{r}_0$ and $\mathbf{r}'$ shown in Fig. 51.4, the screen blocks out the incident wave (Fig. 51.4a), and Φ contains only the diffracted wave Φ_{diff}; on the other hand (Fig. 51.4b), $\bar{\Phi}$ contains both the direct wave Φ_{inc} and the diffracted wave $\bar{\Phi}_{diff}$. Comparison with (51.33) yields *Babinet's principle*

$$\Phi_{diff} = -\bar{\Phi}_{diff} \tag{51.34}$$

stating that the amplitude diffracted by a planar obstacle is the negative of that diffracted by the complementary aperture.

Among the numerous applications of Kirchhoff's formula, we shall consider only the simplest case of a small aperture of typical dimension a, with distant source and observation points $r_0 \gg a$, $r' \gg a$. Since the integration variable $\mathbf{r}$ necessarily lies in the aperture, the quantity $|\mathbf{r} - \mathbf{r}_0|$ can be expanded in inverse powers of r_0:

$$|\mathbf{r} - \mathbf{r}_0| \approx r_0 - \hat{r}_0 \cdot \mathbf{r} + (2r_0)^{-1}[r^2 - (\hat{r}_0 \cdot \mathbf{r})^2] + \cdots$$

with a similar form for $|\mathbf{r} - \mathbf{r}'|$. The leading terms suffice in the denominator of (51.30), but the condition of short wavelengths $ka \gg 1$ requires the additional terms in the exponential. It is convenient to introduce the unit propagation vectors $\hat{k} = -\hat{r}_0$ and $\hat{k}' = \hat{r}'$ for the incident waves from $\mathbf{r}_0$ and the diffracted waves from the aperture to $\mathbf{r}'$ (see Fig. 51.3). In this way, Eq. (51.30) becomes

$$\Phi(\mathbf{r}') = -\frac{ik}{8\pi^2} \frac{e^{ik(r_0+r')}}{r_0 r'} \int_\sigma dx\, dy\, e^{ik\mathbf{r} \cdot (\hat{k}-\hat{k}')} \cos\theta_0$$

$$\times \exp\left\{\frac{ik}{2r_0}[r^2 - (\hat{k} \cdot \mathbf{r})^2] + \frac{ik}{2r'}[r^2 - (\hat{k}' \cdot \mathbf{r})^2]\right\} \tag{51.35}$$

where $\mathbf{r}$ lies in the xy plane, so that only the transverse components of $\hat{k}$ and $\hat{k}'$ contribute. In practice, the angles θ_0 and θ' are both small, so that $r^2 - (\hat{k} \cdot \mathbf{r})^2 \approx r^2 - (\hat{k}' \cdot \mathbf{r})^2 \approx x^2 + y^2$, and the last exponential contains terms of order ka^2/r_0 and ka^2/r'. If either is large, the resulting *Fresnel diffraction* requires careful attention to the interference between these rapidly varying exponentials. On the other hand, for $ka^2/r_0 \ll 1$ and $ka^2/r' \ll 1$, the resulting *Fraunhofer diffraction* is effectively that for infinitely distant source and observer, with the spherical waves replaced by plane waves.

For simplicity, we henceforth consider only the latter case of Fraunhofer diffraction with near normal incidence and transmission, when Eq. (51.35) reduces to the form

$$\Phi(\mathbf{r}') = -\frac{ik}{8\pi^2} \frac{e^{ik(r_0+r')}}{r_0 r'} \int_\sigma dA\, e^{-i\mathbf{q} \cdot \mathbf{r}} \tag{51.36}$$

where

$$\mathbf{q} \equiv \mathbf{k}' - \mathbf{k} \equiv k(\hat{k}' - \hat{k}) \tag{51.37}$$

characterizes the change in direction of the plane wave on transmission through the aperture. Note that the transmitted amplitude is proportional to the two-dimensional Fourier transform of the aperture shape, which thus becomes directly measurable. In particular, a simple integration for a rectangular opening $|x| < a$, $|y| < b$ immediately gives

$$\Phi(\mathbf{r}') = -\frac{ik}{8\pi^2}\frac{e^{ik(r_0+r')}}{r_0 r'}4ab\frac{\sin q_x a}{q_x a}\frac{\sin q_y b}{q_y b} \tag{51.38}$$

and the corresponding intensity [see Eq. (51.18)] becomes

$$I(\mathbf{r}') = I_0\left(\frac{\sin q_x a}{q_x a}\right)^2\left(\frac{\sin q_y b}{q_y b}\right)^2 \propto |\Phi(\mathbf{r}')|^2 \tag{51.39}$$

where I_0 is the intensity of the sound on a direct line through the aperture ($\mathbf{q} = 0$). Nonzero values of q_x and q_y imply that the vector $\mathbf{r}' - \mathbf{r}_0$ deviates from this straight path in the xz plane and yz plane, respectively. The function $x^{-2}\sin^2 x = [j_0(x)]^2$ is plotted in Fig. 51.5; it has a central peak at $x = 0$, small subsidiary maxima near $(n + \frac{1}{2})\pi$, and zeros at $n\pi$ for $n = 1, 2, \ldots$. Consequently, the intensity has a principal maximum along the direct path and falls to zero along directions specified by $q_x a = m\pi$ and $q_y b = n\pi$, with m and n nonzero integers. If $\mathbf{r}_0$ and $\mathbf{r}'$ lie in the xz plane so that q_y vanishes, then the pattern is that of a single slit of width $2a$, with the associated zeros of intensity occurring at the values $k'_x - k_x \equiv k(\hat{k}'_x - \hat{k}_x) = m\pi/a$. Figure 51.6 illustrates the equivalent relation $a(\sin\theta' - \sin\theta_0) = \frac{1}{2}m\lambda$, verifying the elementary condition that extinction occurs when the path difference between the rays passing through the edge and center differs by an integral number of half wavelengths.

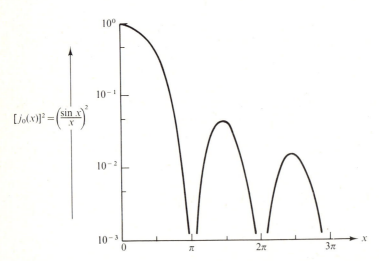

Figure 51.5 Intensity pattern $[(\sin x)/x]^2$ for diffraction through a slit of width $2a$. Here $x = qa = ka\sin\theta'$ for normal incidence [see Eq. (51.39) and following discussion].

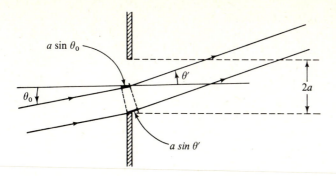

Figure 51.6 Computation of difference in optical-path length for single-slit diffraction. As drawn, θ_0 and θ' are both positive.

A very similar treatment describes a circular aperture of radius a when the relevant integral becomes [compare Eqs. (51.9) to (51.16)]

$$\int_\sigma dA\, e^{-i\mathbf{q}\cdot\mathbf{r}} = \int_0^a r\, dr \int_0^{2\pi} d\phi\, e^{-iq_\perp r\,\cos\phi}$$

$$= \int_0^a r\, dr \int_0^{2\pi} d\phi\, e^{-iq_\perp r\,\sin\phi}$$

$$= 2\pi a^2 \frac{J_1(q_\perp a)}{q_\perp a} \tag{51.40}$$

where $q_\perp^2 = q_x^2 + q_y^2$. In this way, the associated intensity becomes

$$I = I_0 \left[\frac{2 J_1(q_\perp a)}{q_\perp a} \right]^2 \tag{51.41}$$

Figure 51.2 shows that the intensity is largely confined to a cone about the forward direction, with a series of subsidiary rings of nearly equal width. If we simplify to the case of normal incidence ($\theta_0 = 0$), then

$$q_\perp a = ka \sin \theta' \qquad \text{normal incidence} \tag{51.42a}$$

where θ' is now the polar angle of the diffracted wave, measured from the normal $\hat{z}$. The angular width of the central *Airy disk* is thus determined by the condition

$$\sin \theta' \approx 0.61 \frac{\lambda}{a} \tag{51.42b}$$

There is a striking resemblance between the acoustic power (51.19) radiated by a circular piston set flush in a wall and the diffracted intensity (51.41) by a circular opening for normal incidence. This correspondence holds more generally, as can be seen from Eq. (51.30). Suppose that the source is located far away on the negative z axis, so that

$$-\frac{e^{ik|\mathbf{r}-\mathbf{r}_0|}}{|\mathbf{r}-\mathbf{r}_0|} \frac{z_0}{|\mathbf{r}-\mathbf{r}_0|} \approx \frac{e^{ikr_0}}{r_0}$$

Kirchhoff's formula then has the simpler form

$$\Phi(\mathbf{r}') \approx -\frac{ik}{8\pi^2} \frac{e^{ikr_0}}{r_0} \int_\sigma dx\, dy\, \frac{e^{ik|\mathbf{r}-\mathbf{r}'|}}{|\mathbf{r}-\mathbf{r}'|} \qquad (51.43)$$

which differs from the acoustic amplitude radiated by a piston of shape σ [compare Eq. (51.7)] only in overall factors. Since each element of the piston's surface serves as a source of new spherical wavefronts, this correspondence justifies Huygens' principle. It must be remembered, however, that Kirchhoff's formula and the approximate form (51.43) require short wavelengths $(ka \gg 1)$, whereas (51.7) describes the acoustic radiation for *arbitrary values* of ka.

Radiation from an Oscillating Sphere

Our previous treatment of acoustic phenomena involving planar boundaries can be generalized to include other simple shapes like spheres and cylinders. These geometries illustrate the use of important mathematical functions, and we shall study one example of each. For definiteness, we first consider the acoustic radiation from a rigid sphere of radius a executing small transverse oscillations, in which the moving surface acts as the source. Let the center of sphere lie at the point $\epsilon a \cos \omega t\, \hat{z} = \text{Re}\ (\epsilon a e^{-i\omega t})\hat{z}$, where $0 < \epsilon \ll 1$ (see Fig. 51.7). Each point on the surface of the sphere evidently oscillates with velocity $\mathbf{V}_{\text{sphere}} = \text{Re}\ (-i\omega\epsilon a e^{-i\omega t})\hat{z}$. The induced velocity potential $\Phi(\mathbf{r}, t) = \text{Re}\ \Phi(\mathbf{r})e^{-i\omega t}$ oscillates at the same frequency ω, and the amplitude $\Phi(\mathbf{r})$ satisfies the rigid boundary condition (49.10a) on the instantaneous surface

$$-\hat{n} \cdot \nabla\Phi(\mathbf{r}) = -i\omega\epsilon a\hat{n} \cdot \hat{z} \qquad (51.44)$$

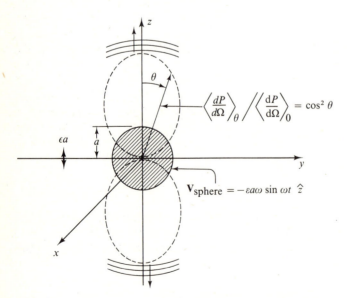

Figure 51.7 Radiation from sphere oscillating along $\hat{z}$ axis with displacement $\epsilon a \cos \omega t$.

where $\hat{n}$ is the *inward* normal to the moving sphere. Since Φ itself is already of order ϵ, the boundary condition can be evaluated on the unperturbed surface $r = a$, apart from corrections of order ϵ^2. We therefore take $\hat{n} = -\hat{r}$ and obtain the final boundary condition

$$\left.\frac{\partial \Phi(\mathbf{r})}{\partial r}\right|_{r=a} = i\omega\epsilon a \cos \theta \tag{51.45}$$

where θ is the polar angle measured from $\hat{z}$. In addition, $\Phi(\mathbf{r})$ satisfies the homogeneous Helmholtz equation (49.15) for $r > a$ with outgoing waves at infinity.

One approach to this problem is to construct the Green's function that satisfies the inhomogeneous differential equation (50.17) for $r > a$ and the homogeneous boundary conditions $\partial \tilde{G}(\mathbf{r}, \mathbf{r}')/\partial r = 0$ at $r = a$ (Prob. 9.21). As for a planar piston, Green's identity then provides the complete solution. Such a procedure is unnecessarily cumbersome here, and instead we use the particularly simple boundary condition (51.45) to construct the solution directly.

In spherical polar coordinates (r, θ, ϕ), the Helmholtz equation has the familiar form (Appendix B)

$$\left(\frac{1}{r^2}\frac{\partial}{\partial r}r^2\frac{\partial}{\partial r} - \frac{L^2}{r^2} + k^2\right)\Phi(r, \theta, \phi) = 0 \tag{51.46}$$

where L^2 denotes the angular differential operator

$$L^2 \equiv -\frac{1}{\sin \theta}\frac{\partial}{\partial \theta}\sin \theta \frac{\partial}{\partial \theta} - \frac{1}{\sin^2 \theta}\frac{\partial^2}{\partial \phi^2} \tag{51.47}$$

The corresponding solutions separate into radial functions and angular functions

$$\Phi(r, \theta, \phi) = R(r)Y_{lm}(\theta, \phi) \tag{51.48}$$

with $Y_{lm}(\theta, \phi)$ the normalized spherical harmonics (see Appendix D2). Substitution into (51.46) yields the radial equation (see Appendix C)

$$\left[\frac{1}{r^2}\frac{d}{dr}r^2\frac{d}{dr} - \frac{l(l+1)}{r^2} + k^2\right]R(r) = 0 \tag{51.49}$$

which is independent of the azimuthal index m. This equation is a modified form of Bessel's equation, with solutions $j_l(kr)$ and $n_l(kr)$, the spherical Bessel functions. They are products of polynomials and trigonometric functions (see Appendix D3) with $j_l(x)$ bounded and $n_l(x)$ divergent at $x = 0$. Alternatively, we often prefer the linear combination $h_l^{(1)}(kr) \equiv j_l(kr) + in_l(kr)$ that contains only the outgoing wave e^{ikr}, ensuring causal behavior.

The boundary condition (51.45) involves only the single angular function $\cos \theta$, which is just $P_1(\cos \theta) \propto Y_{10}$. Since $\Phi(\mathbf{r})$ satisfies a homogeneous equation, the oscillating sphere excites only that particular component, and we can therefore assume a solution with the simple form

$$\Phi(\mathbf{r}) = CP_1(\cos \theta)f(kr) = C \cos \theta f(kr) \tag{51.50}$$

where C is a constant and $f(kr)$ is a linear combination of $j_1(kr)$ and $n_1(kr)$, *both of which are now admissible because the origin is excluded.* We therefore use the causal requirement of outgoing waves to choose the particular linear combination

$$f(kr) = h_1^{(1)}(kr) \equiv -(kr + i)(kr)^{-2} e^{ikr} \tag{51.51}$$

Application of the boundary condition (51.45) fixes the remaining coefficient

$$C = \frac{i\omega\epsilon a}{kf'(ka)} = \frac{ic\epsilon a}{f'(ka)} \tag{51.52}$$

where the prime denotes differentiation with respect to the argument. In this way we obtain the full solution

$$\Phi(r, \theta, t) = \mathrm{Re} \; \frac{ic\epsilon a \cos \theta \; f(kr)e^{-i\omega t}}{f'(ka)} \tag{51.53}$$

where $f(x) \equiv h_1^{(1)}(x)$.

This mathematical expression has several interesting physical consequences, and we first examine the asymptotic energy flux radiated by the oscillating sphere. At large distances $(kr \gg 1)$, the solution has the approximate form

$$\Phi(r, \theta, t) \sim -\mathrm{Re} \left[C \cos \theta \frac{e^{i(kr - \omega t)}}{kr} \right] \tag{51.54}$$

confirming the outgoing nature of the waves. The corresponding asymptotic pressure [Eq. (49.8)] and radial velocity field [Eq. (49.7)] become

$$p' = \rho_0 \frac{\partial \Phi}{\partial t} \sim ic\rho_0 C \cos \theta \frac{e^{i(kr - \omega t)}}{r} \tag{51.55a}$$

$$v_r = -\frac{\partial \Phi}{\partial r} \sim iC \cos \theta \frac{e^{i(kr - \omega t)}}{r} \tag{51.55b}$$

where the physical quantities are, of course, the real parts of these expressions. Both vanish in the equatorial plane $(\theta = \frac{1}{2}\pi)$ owing to the longitudinal character of the sound waves. The power radiated through a radial surface element $\hat{r} \, dA = \hat{r} \, r^2 \, d\Omega$ is just $\mathbf{j}_e \cdot \hat{r} \, dA$, with the time-average differential power given in terms of the complex pressure and radial velocity by [compare Eqs. (51.18) and (51.19)]

$$\left\langle \frac{dP}{d\Omega} \right\rangle = \tfrac{1}{2} r^2 \, \mathrm{Re} \; v_r p'^* = \tfrac{1}{2} c |C|^2 \rho_0 \cos^2 \theta \tag{51.56}$$

It exhibits a two-lobed pattern, with maxima along the direction of motion and a node in the equatorial plane (Fig. 51.7). A simple calculation yields the final result

$$\left\langle \frac{dP}{d\Omega} \right\rangle = \frac{1}{2} \frac{c\rho_0 (ka)^6 (ca\epsilon)^2 \cos^2 \theta}{4 + k^4 a^4} \tag{51.57}$$

for the radiation seen by a distant observer $(kr \gg 1)$ but with no restriction on the parameter $ka = \omega a/c$. For long and short wavelengths, respectively, we have the approximate form

$$\left\langle \frac{dP}{d\Omega} \right\rangle \approx \begin{cases} \frac{1}{8}c\rho_0 \cos^2 \theta \, (ca\epsilon)^2(ka)^6 & ka \ll 1 \\ \frac{1}{2}c\rho_0 \cos^2 \theta \, (ca\epsilon)^2(ka)^2 & ka \gg 1 \end{cases} \tag{51.58}$$

and it is easy to verify that $\langle dP/d\Omega \rangle$ increases monotonically with frequency and wave number. In addition, the differential intensity is proportional to ϵ^2, just as in the case of the piston (51.19). Similar remarks describe the total power radiated

$$\langle P \rangle = \int d\Omega \left\langle \frac{dP}{d\Omega} \right\rangle = \frac{2\pi}{3} \frac{c\rho_0(ca\epsilon)^2(ka)^6}{4 + k^4a^4} \tag{51.59}$$

When a rigid sphere moves through a fluid, it experiences a reaction force that affects its own dynamical properties. As a second application of Eq. (51.53), we evaluate this reaction force exactly for small-amplitude motion. The complex amplitude of the pressure throughout all space is given by

$$p'(\mathbf{r}) = -i\omega\rho_0 \, \Phi(\mathbf{r}) = \frac{\rho_0 \, \omega ca\epsilon \, \cos \theta \, f(kr)}{f'(ka)} \tag{51.60}$$

with the surface value $p'(r = a, \theta)$. The corresponding force on the sphere is just $\mathbf{F}^{(r)} = \int d\mathbf{A} \, p'(r = a, \theta)$, where the integral is over the (to this approximation, undisplaced) surface of the sphere and here $d\mathbf{A} = -\hat{r} \, dA$. In particular, the $\hat{z}$ component is

$$F_z^{(r)} = -a^2 \int d\Omega \, p'(r = a, \theta)\hat{z} \cdot \hat{r}$$

$$= -\frac{f(ka)}{kaf'(ka)} \omega^2 a\epsilon\rho_0 V$$

$$= \omega^2 a\epsilon\rho_0 V \frac{2 + k^2a^2 + ik^3a^3}{4 + k^4a^4} \tag{51.61}$$

where $V = 4\pi a^3/3$ is the sphere's volume and $\rho_0 V$ is the mass of fluid displaced. Note that the reaction force oscillates harmonically at frequency ω, with both real (reactive) and imaginary (dissipative) parts. Suppose we apply an external force $\hat{z}F^{(e)}e^{-i\omega t}$ to initiate the motion. The dynamical equation (Newton's second law) for the sphere has the z component

$$F^{(e)} + F_z^{(r)} = -M\omega^2 a\epsilon \tag{51.62a}$$

where M is the sphere's mass. This relation can be rewritten

$$F^{(e)} = -M^*\omega^2 a\epsilon \tag{51.62b}$$

where the *effective mass* M^* is defined by

$$M^* = M + \rho_0 V \frac{2 + k^2a^2 + ik^3a^3}{4 + k^4a^4} \tag{51.63}$$

For slow motion ($ka \ll 1$) in a nonviscous fluid, the sphere acts as if its mass were increased by half the mass of fluid displaced. With increasing frequency, however, M^* becomes complex because the acoustic radiation carries energy away to infinity. A similar calculation for a long cylinder (Prob. 9.20) shows that the corresponding effective mass for $ka \ll 1$ exceeds the intrinsic value by the full mass of fluid displaced.

Scattering by a Rigid Cylinder

As a final topic in small-amplitude acoustic phenomena, we turn to *scattering* of sound waves by fixed obstacles. Such an investigation differs from the previous radiation problems in that the source of the waves is now irrelevant; instead, scattering is more closely related to diffraction, but the latter concept is generally restricted to short wavelengths, as seen in the derivation of Kirchhoff's formula.

Given the form of the obstacle, the problem consists in solving the Helmholtz equation subject to the fixed surface condition (49.10b). If the source is at a finite distance from the scattering object, it can be incorporated as an inhomogeneity in Helmholtz's equation for $\Phi(\mathbf{r})$. More commonly, however, the source is assumed infinitely distant and replaced by the physical condition that the solution $\Phi(\mathbf{r})$ contain two terms

$$\Phi(\mathbf{r}) = \Phi_{\text{inc}}(\mathbf{r}) + \Phi_{\text{sc}}(\mathbf{r}) \tag{51.64}$$

where

$$\Phi_{\text{inc}}(\mathbf{r}) = e^{i\mathbf{k}\cdot\mathbf{r}} \tag{51.65}$$

represents an incident plane wave and Φ_{sc} is an outgoing scattered wave.

For definiteness, consider an infinite rigid cylinder of radius a with its axis along $\hat{z}$ and the incident wave directed along $\hat{x}$, that is, $\mathbf{k} = k\hat{x}$, as in Fig. 51.8. In cylindrical polar coordinates (r, ϕ, z), the velocity potential satisfies the Helmholtz equation (Appendix B)

$$(\nabla^2 + k^2)\Phi = \left(\frac{1}{r}\frac{\partial}{\partial r}r\frac{\partial}{\partial r} + \frac{1}{r^2}\frac{\partial^2}{\partial \phi^2} + \frac{\partial^2}{\partial z^2} + k^2\right)\Phi = 0 \tag{51.66}$$

for $r > a$, subject to the boundary condition

$$\frac{\partial\Phi}{\partial r} = 0 \qquad \text{at } r = a \tag{51.67}$$

Since Φ_{inc} and the scattering surface are both independent of z, the scattered wave will be so too, and we therefore seek solutions that depend only on r and ϕ. Equation (51.66) has separated solutions (Appendix C) of the form $J_m(kr)e^{\pm im\phi}$, $N_m(kr)e^{\pm im\phi}$, considered previously in Eqs. (47.30) and (49.18).

If there were no cylinder, Φ_{inc} would be the full solution to Helmholtz's equation. Since $\Phi_{\text{inc}} \equiv e^{i\mathbf{k}\cdot\mathbf{r}}$ is bounded everywhere, its expansion in the separated solutions cannot contain N_m. Thus it must have the form

$$\Phi_{\text{inc}} \equiv e^{ikr\cos\phi} = \sum_{m=-\infty}^{\infty} e^{im\phi}c_m J_m(kr) \tag{51.68}$$

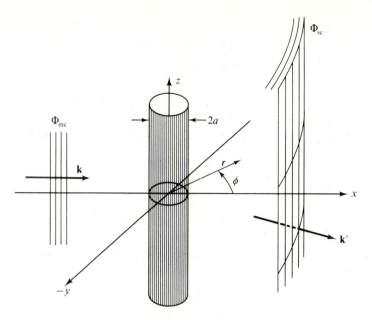

Figure 51.8 Scattering from a rigid cylinder.

which can be considered the Fourier expansion of a function of ϕ with period 2π [compare Eq. (40.4)]. The corresponding Fourier coefficients can be obtained from Eq. (51.10), and we therefore derive the full Fourier series

$$\Phi_{\text{inc}} \equiv e^{ikr\cos\phi} = \sum_{m=-\infty}^{\infty} e^{im\phi} i^m J_m(kr) \qquad (51.69)$$

The presence of the cylinder alters this simple picture by producing outgoing scattered waves proportional to e^{ikr}, giving rise to an additional contribution of the form $H_m^{(1)} \equiv J_m + iN_m$. As a result, the relevant radial function $R_m(r)$ for the mth angular term must be a linear combination of the incident wave $i^m J_m$ and the scattered wave proportional to $H_m^{(1)}$. The coefficient of this term must be chosen to satisfy the boundary condition (51.67). Thus we find

$$R_m(r) = i^m \left[J_m(kr) - H_m^{(1)}(kr) \frac{J_m'(ka)}{H_m^{(1)\prime}(ka)} \right]$$

The full solution follows by combining all angular contributions

$$\Phi(r, \phi) = \sum_{m=-\infty}^{\infty} e^{im\phi} i^m \left[J_m(kr) - H_m^{(1)}(kr) \frac{J_m'(ka)}{H_m^{(1)\prime}(ka)} \right] \qquad (51.70)$$

Comparison of Eqs. (51.64), (51.69), and (51.70) allows us to identify the scattered wave

$$\Phi_{\text{sc}}(r, \phi) = - \sum_{m=-\infty}^{\infty} e^{im\phi} i^m \frac{J_m'(ka)}{H_m^{(1)\prime}(ka)} H_m^{(1)}(kr) \qquad (51.71)$$

which is now known explicitly.

The most interesting phenomena occur at large distances from the cylinder $(kr \gg 1)$. We therefore use the asymptotic approximation (obtained as with the results in Appendix D)

$$i^m H_m^{(1)}(x) \sim \left(\frac{2}{\pi x}\right)^{1/2} \exp\left[i(x - \tfrac{1}{4}\pi)\right] \qquad x \to \infty \tag{51.72}$$

to express $\Phi_{\text{sc}}(\mathbf{r})$ in the general asymptotic form

$$\Phi_{\text{sc}}(r, \phi) \sim f(\phi) r^{-1/2} e^{ikr} \qquad kr \to \infty \tag{51.73}$$

where $f(\phi)$ is known as the scattering amplitude. In the present problem, it has the explicit Fourier representation

$$f(\phi) = -\sum_{m=-\infty}^{\infty} \exp[i(m\phi - \tfrac{1}{4}\pi)]\left(\frac{2}{\pi k}\right)^{1/2} \frac{J'_m(ka)}{H_m^{(1)'}(ka)} \tag{51.74}$$

As is evident from Eqs. (51.18), the incident plane wave carries a time-average energy flux $\mathbf{F} = F_0 \hat{x}$, where $F_0 = \tfrac{1}{2}\omega\rho_0 k$. Correspondingly, the scattered wave transports energy outward, and the time-average power crossing an area $d\mathbf{A}$ is $\tfrac{1}{2}\operatorname{Re}(\mathbf{v}_{\text{sc}} p'^*_{\text{sc}}) \cdot d\mathbf{A}$. Let $d\mathbf{A} = \hat{r} \, r \, d\phi$ be an element of a cylindrical surface of radius r with unit length along the axis and subtending an angular interval $d\phi$; a simple calculation gives the associated time-average power $\tfrac{1}{2}\omega\rho_0 k |f(\phi)|^2 \, d\phi = F_0 |f(\phi)|^2 \, d\phi$. To interpret this expression, it is convenient to imagine a strip perpendicular to $\hat{x}$, of width $d\sigma$ and unit length along $\hat{z}$, such that all the power incident on the strip is scattered into the interval between ϕ and $\phi + d\phi$. Just as in Sec. 5, this power is $F_0 \, d\sigma$, and comparison with the alternative expression $F_0 |f(\phi)|^2 \, d\phi$ identifies the *differential cross section*

$$\frac{d\sigma}{d\phi} = |f(\phi)|^2 \tag{51.75}$$

Note that $|f|^2$ here has the dimensions of a length, and $d\sigma/d\phi$ might more properly be called the scattering width. These relations hold for any two-dimensional scattering configuration.

To illustrate these very general expressions, we consider the limit of a small cylinder $ka \ll 1$, when the exact scattering amplitude (51.74) simplifies considerably to contain only three terms $(m = 0, \pm 1)$. A straightforward calculation with the collected relations in Appendix E yields the approximate long-wavelength limit

$$f(\phi) \approx \left(\frac{2}{\pi ki}\right)^{1/2} \frac{\pi k^2 a^2}{4i}(1 - 2\cos\phi) \qquad \text{for } ka \ll 1 \tag{51.76}$$

where the remaining terms for $|m| \geq 2$ contribute only terms of order $(ka)^{|2m|}$. The corresponding differential cross section becomes

$$\frac{d\sigma}{d\phi} = \tfrac{1}{8}\pi k^3 a^4 (1 - 2\cos\phi)^2 \qquad \text{for } ka \ll 1 \tag{51.77}$$

It is extremely anisotropic, with the backward intensity 9 times the forward inten-

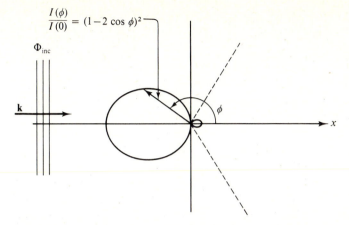

Figure 51.9 Polar plot of relative scattered acoustic intensity from a small rigid cylinder with $ka \ll 1$. The nodal lines occur at $\phi = \pm 60°$, and the forward (backward) lobe lies in the region $|\phi| < 60°$ ($|\phi| > 60°$).

sity and a node at $\phi = \pm 60°$ (see Fig. 51.9). An angular integration yields the total cross section

$$\sigma = \int_0^{2\pi} \frac{d\sigma}{d\phi}\, d\phi = 2a(\tfrac{3}{8}\pi^2 k^3 a^3) \qquad \text{for } ka \ll 1 \tag{51.78}$$

of which the forward lobe contributes only about 6 percent. This expression has a simple form; it is the geometric width $2a$, multiplied by a small factor proportional to $(ka)^3 \propto (a/\lambda)^3 \ll 1$. Since audible sound wavelengths are typically measured in meters, this small factor accounts for the ease with which sound passes through screens and grilles.

52 NONLINEAR PHENOMENA AND SHOCK WAVES

This chapter has concentrated primarily on the linearized equations for small-amplitude motion of a fluid. Since linear equations appear in many branches of physics, techniques developed in one such field frequently prove useful in attacking problems that arise in another. For example, the scattering of electromagnetic radiation by fixed conductors is formally similar to the scattering of sound treated in the previous section.

Mathematical physics involves more than just linear equations, however, and the convective terms in Euler's equation provide a fascinating example of intrinsic nonlinearity. Unfortunately, a thorough study of such nonlinear phenomena would be prohibitive. We here restrict ourselves to the one-dimensional motion of compressible fluids,† which turns out to allow an essentially complete solution.

† We have found the following references particularly helpful: Landau and Lifshitz (1959), chaps. 9 and 10; Sommerfeld (1950), sec. 37; Zel'dovich and Raizer [5], chap. 1.

Traveling Waves

For an isentropic small-amplitude wave, we have seen that the squared speed of sound c^2 is just the adiabatic compressibility $(\partial p/\partial \rho)_s$. Furthermore, the quantity $(\partial^2 p/\partial \rho^2)_s$ is generally positive, so that c^2 increases with increasing density. This physical fact suggests a nonlinearity for finite-amplitude disturbances; sound should propagate faster in a compressed region than in a rarefied one, with the high-density crests of a wave tending to overtake the low-density troughs. This distortion can ultimately produce a *shock front* when the flow ceases to be isentropic; a fundamentally new description is then required.

We now verify these qualitative arguments in the special case of a one-dimensional traveling wave in a compressible fluid. As in the previous sections, the basic equations describe the conservation of mass

$$\frac{\partial \rho}{\partial t} + \nabla \cdot (\rho \mathbf{v}) = 0 \tag{52.1}$$

and momentum

$$\frac{d\mathbf{v}}{dt} + \frac{1}{\rho} \nabla p = 0 \tag{52.2}$$

which reduce in one dimension to the respective equations

$$\frac{\partial \rho}{\partial t} + v\frac{\partial \rho}{\partial x} + \rho\frac{\partial v}{\partial x} = 0 \tag{52.3}$$

and

$$\frac{\partial v}{\partial t} + v\frac{\partial v}{\partial x} + \frac{1}{\rho}\frac{\partial p}{\partial x} = 0 \tag{52.4}$$

Here all nonlinear terms are retained. Assume an adiabatic equation of state $p = p(\rho)$ at fixed entropy s_0 and define

$$c^2(\rho) = \frac{dp}{d\rho} \tag{52.5}$$

which is now a function of density. Substitution into (52.4) yields

$$\frac{\partial v}{\partial t} + v\frac{\partial v}{\partial x} + \frac{c^2}{\rho}\frac{\partial \rho}{\partial x} = 0 \tag{52.6}$$

which must be solved simultaneously with Eq. (52.3) for the unknown functions $\rho(x, t)$ and $v(x, t)$ subject to suitable boundary conditions.

In analogy with d'Alembert's solution to the wave equation, we here seek a traveling wave of the form

$$\rho(x, t) = f(x - ut) + \rho_0 \tag{52.7}$$

where

$$u = u(\rho) \tag{52.8}$$

is a self-consistent propagation velocity that depends on the density itself and ρ_0 is a constant. The fluid velocity $v(x, t)$ must have a similar form, for otherwise the coupled equations could never have solutions valid for finite intervals of x and t. As a result, the same combination $x - ut$ determines both v and ρ, or, equivalently, knowledge of ρ at any given (x, t) suffices to determine the corresponding v. We therefore assume

$$v = v(\rho) \tag{52.9}$$

and the wave character [Eq. (52.7)] of $\rho(x, t)$ ensures a similar form for v. It is now straightforward to evaluate the various partial derivatives appearing in Eqs. (52.3) and (52.6). For example, we have

$$\frac{\partial \rho}{\partial t} = \left(-u - t \frac{du}{d\rho} \frac{\partial \rho}{\partial t} \right) f'$$

where f' denotes the derivative with respect to the argument $x - ut$; a little manipulation yields

$$\frac{\partial \rho}{\partial t} = \frac{-u f'}{1 + t(du/d\rho)f'}$$

A similar calculation gives

$$\frac{\partial \rho}{\partial x} = \frac{f'}{1 + t(du/d\rho)f'}$$

and since $\partial v/\partial x = (dv/d\rho)(\partial \rho/\partial x)$, Eq. (52.3) reduces to

$$\left(-u + v + \rho \frac{dv}{d\rho} \right) \frac{f'}{1 + t(du/d\rho)f'} = 0 \tag{52.10}$$

The quantity f' does not vanish identically, so that this equation implies the simpler condition

$$-u + v + \rho \frac{dv}{d\rho} = 0 \tag{52.11}$$

An analogous treatment of (52.6) leads to the second condition

$$(-u + v) \frac{dv}{d\rho} + \frac{c^2(\rho)}{\rho} = 0 \tag{52.12}$$

and the simultaneous solution of these relations determines both $v(\rho)$ and $u(\rho)$ in terms of the function $c(\rho)$, which is assumed given. Furthermore, if Eqs. (52.11) and (52.12) are satisfied, Eqs. (52.7) to (52.9) provide a traveling-wave solution to the full nonlinear wave Eqs. (52.3) and (52.4) *for any function f in Eq. (52.7)*.

To solve these equations, multiply Eq. (52.11) by $dv/d\rho$. Comparison with (52.12) implies that

$$\left(\frac{dv}{d\rho} \right)^2 = \frac{c^2(\rho)}{\rho^2}$$

or, equivalently,

$$\frac{dv}{d\rho} = \pm \frac{c(\rho)}{\rho} \tag{52.13}$$

where the right-hand side is a known function of ρ. Thus $v(\rho)$ is given by an indefinite integral

$$v(\rho) = \pm \int_{\rho_0}^{\rho} d\rho' \frac{c(\rho')}{\rho'} \tag{52.14}$$

where the density ρ_0 is fixed by the condition

$$v(\rho_0) = 0 \tag{52.15}$$

Substitution into (52.11) then provides the remaining quantity

$$u(\rho) = \pm \left[c(\rho) + \int_{\rho_0}^{\rho} d\rho' \frac{c(\rho')}{\rho'} \right] \tag{52.16}$$

with the upper (lower) sign describing a wave propagating to the right (left). In this way, we obtain an *exact one-dimensional traveling-wave solution* to the full nonlinear equations of compressible hydrodynamics

$$\rho(x, t) = \rho_0 + f\{x \mp [c(\rho) + v(\rho)]t\} \tag{52.17a}$$

$$v(x, t) = \pm v[\rho(x, t)] \tag{52.17b}$$

where $u(\rho)$ and $v(\rho)$ henceforth denote the positive functions in Eqs. (52.14) and (52.16), and f is an arbitrary function. Note that the local propagation speed is not $c(\rho)$ but $c(\rho) + v(\rho)$.

To analyze this situation in detail, we first recover our previous results by treating a small-amplitude disturbance $\rho = \rho_0 + \delta\rho$, with $|\delta\rho| \ll \rho_0$. To lowest order in $\delta\rho$, we have $c(\rho) \approx c_0 = [(\partial p/\partial \rho)_0]^{1/2}$ and $v(\rho) = c_0 \, \delta\rho/\rho_0$, so that $\delta\rho(x, t) \approx f(x \mp c_0 t)$, in agreement with the linearized solution obtained in Sec. 49. Note that the linearized wave travels without change of shape, whereas the general situation is more complicated owing to the nonlinearity implicit in $u(\rho)$.

Given the function f at some instant of time, the instantaneous density profile (52.17a) determines the local propagation speed $u(\rho)$ at each point x through Eqs. (52.16) and (52.17b). As a result, each element of the profile travels a distance $u(\rho) \, dt$ in an interval dt. As shown below, $u(\rho)$ typically increases with increasing ρ, so that regions of high density move faster than those of low density. Thus the wavefront steepens and ultimately becomes singular (Fig. 52.1). The demonstration that $du/d\rho > 0$ is straightforward, for Eqs. (52.5) and (52.13) simplify this condition to

$$\frac{du}{d\rho} = \frac{dc}{d\rho} + \frac{dv}{d\rho} = \frac{1}{\rho} \frac{d}{d\rho} \rho c = \frac{1}{\rho} \frac{d}{d\rho} \left(\rho^2 \frac{dp}{d\rho} \right)^{1/2} \tag{52.18}$$

Moreover, ρ^{-1} is just the specific volume V per unit mass, with $\rho^2 \, dp/d\rho = -dp/dV$. Substitution into Eq. (52.18) readily yields the final isentropic relation

$$\frac{du(\rho)}{d\rho} = \frac{1}{2c\rho^4} \left(\frac{\partial^2 p}{\partial V^2} \right)_s \tag{52.19}$$

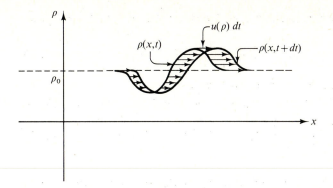

Figure 52.1 Propagation of finite-amplitude one-dimensional disturbance in a compressible fluid.

Although a negative value of this thermodynamic derivative is compatible with general principles, it turns out to be positive in all cases of practical interest.

Example: Ideal Gas

The adiabatic equation of state $p = p(\rho)$ plays a central role in the nonlinear distortion of a one-dimensional traveling wave, and it is therefore helpful to study a specific example that permits a complete integration. In particular, consider an ideal classical gas with specific heats c_p and c_V per unit mass. The familiar equation of state is $p\mathscr{V} = nRT$, where n is the number of moles of gas, $R \approx 8.3 \times 10^7$ ergs K^{-1} is the gas constant, and $\mathscr{V}$ is the total volume occupied by the gas. This equation can be rewritten in terms of the specific volume V (the volume per gram) as

$$pV = \frac{RT}{M_0}$$

$$p = \frac{RT\rho}{M_0} \tag{52.20}$$

where M_0 is now the molecular weight in grams. In addition, the molecules in an ideal classical gas are assumed noninteracting, so that the only energy arises from thermal motion. For an ideal nonrelativistic classical gas with f internal degrees of freedom including translations of the molecule, the equipartition theorem yields†

$$E = \tfrac{1}{2}fnRT = \tfrac{1}{2}fp\mathscr{V}$$

for the total internal energy, or

$$\epsilon = \tfrac{1}{2}f\frac{RT}{M_0} = \tfrac{1}{2}fpV \tag{52.21}$$

---

† We assume that the reader is familiar with these elementary properties of the ideal gas (see, for example Reif [1], chaps. 5 and 7).

for the internal energy per unit mass. In the particular case of a monatomic gas, we have only translations and $f = 3$; therefore

$$\epsilon = \frac{3}{2}\frac{RT}{M_0} = \frac{3}{2}pV \qquad \text{ideal monatomic gas} \qquad (52.22)$$

These two equations (52.20) and (52.21) provide a complete description of the system

Application of the *first and second laws of thermodynamics* to a unit mass of an ideal gas yields the result

$$T\,ds = d\epsilon + p\,dV \qquad (52.23)$$

The specific heats at constant volume and pressure are given by

$$c_V = T\left(\frac{\partial s}{\partial T}\right)_V = \left(\frac{\partial \epsilon}{\partial T}\right)_V \qquad (52.24a)$$

$$c_p = T\left(\frac{\partial s}{\partial T}\right)_p = \left(\frac{\partial \epsilon}{\partial T}\right)_p + p\left(\frac{\partial V}{\partial T}\right)_p \qquad (52.24b)$$

Use of Eqs. (52.21) and (52.20) immediately leads to the relations

$$c_V = \frac{f}{2}\frac{R}{M_0} \qquad (52.25a)$$

$$\text{ideal gas}$$

$$c_p = \frac{f+2}{2}\frac{R}{M_0} \qquad (52.25b)$$

We let γ denote the ratio c_p/c_V

$$\gamma \equiv \frac{c_p}{c_V} = \frac{f+2}{f} = 1 + \frac{2}{f} \qquad (52.26)$$

and note that $\gamma > 1$. For an ideal monatomic gas, it follows from Eq. (52.22) that

$$c_V = \left(\frac{d\epsilon}{dT}\right)_V = \frac{3}{2}\frac{R}{M_0} \qquad \text{ideal monatomic gas} \qquad (52.27)$$

and from Eq. (52.26) that

$$\gamma = \frac{5}{3} \qquad \text{ideal monatomic gas} \qquad (52.28)$$

We observe from Eq. (52.25) that the specific heats c_V and c_p for an ideal gas are constants, independent of the temperature.

A combination of Eqs. (52.24a), (52.25a), (52.26), and (52.20) then yields the integrated relation

$$\epsilon(T) = c_V\,T = \frac{RT}{(\gamma - 1)M_0} = \frac{p}{(\gamma - 1)\rho} \qquad (52.29)$$

Furthermore, Eq. (52.23) can be rewritten with these relations

$$ds = \frac{1}{T}\left[d\epsilon + pd\left(\frac{1}{\rho}\right)\right] = \frac{p}{T\rho(\gamma-1)}\left[\frac{dp}{p} - \gamma\frac{d\rho}{\rho}\right]$$

$$= c_V\, d(\ln p\rho^{-\gamma}) \tag{52.30}$$

and an elementary integration gives the entropy

$$s = c_V \ln p\rho^{-\gamma} + \text{const} \tag{52.31}$$

In particular, the isentropic equation of state acquires a *polytropic form*

$$\frac{p}{p_0} = \left(\frac{\rho}{\rho_0}\right)^{\gamma} \qquad \text{constant } s = s_0 \tag{52.32a}$$

where p_0 and ρ_0 denote some standard set of conditions.

The dynamical equations for a one-dimensional traveling wave can be integrated exactly for a polytrope, since the quantity $c^2(\rho)$ is just†

$$c^2(\rho) = \left(\frac{dp}{d\rho}\right)_s = \frac{\gamma p}{\rho} \tag{52.32b}$$

The integration in (52.14) and (52.16) then gives

$$v(\rho) = \frac{2}{\gamma-1}[c(\rho) - c_0] \tag{52.33a}$$

$$u(\rho) = \frac{\gamma+1}{\gamma-1}c(\rho) - \frac{2c_0}{\gamma-1} \tag{52.33b}$$

where $c_0^2 = \gamma p_0/\rho_0$. In practice $c(\rho)$ is not directly observable, and it is preferable to combine the last relations to obtain

$$u(\rho) = c_0 + \tfrac{1}{2}(\gamma+1)v(\rho) \tag{52.34}$$

where
$$v(\rho) = \frac{2}{\gamma-1}c_0\left[\left(\frac{\rho}{\rho_0}\right)^{(\gamma-1)/2} - 1\right] \tag{52.35}$$

As a result of Eq. (52.34), the nonlinear traveling wave in Eq. (52.17a) has the form

$$\rho(x,\,t) = \rho_0 + f\{x - [c_0 + \tfrac{1}{2}(\gamma+1)v(\rho)]t\} \tag{52.36}$$

For some purposes, it is more convenient to consider v the fundamental quantity. It has the form [compare Eq. (52.17b)]

$$v(x,\,t) = g\{x - [c_0 + \tfrac{1}{2}(\gamma+1)v]t\} \tag{52.37}$$

---

† Note the corollary that the ordinary sound velocity of an ideal gas at an equilibrium pressure p_0, density ρ_0, and temperature T is given by $c_0 = (\gamma p_0/\rho_0)^{1/2} = (\gamma RT/M_0)^{1/2}$, where the equation of state (52.20) has been used.

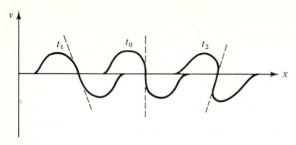

Figure 52.2 Fluid velocity of finite-amplitude one-dimensional wave in an ideal gas, with $t_1 < t_0 < t_2$.

with g another *arbitrary* function and ρ given by Eq. (52.35)

$$\rho(v) = \rho_0 \left[1 + \frac{(\gamma - 1)v}{2c_0} \right]^{2/(\gamma - 1)} \tag{52.38}$$

Alternatively, Eq. (52.37) can be inverted to yield

$$x - [c_0 + \tfrac{1}{2}(\gamma + 1)v]t = g^{-1}(v) \tag{52.39}$$

This relation shows that a point on the wavefront associated with a given fluid velocity v [recall Eqs. (52.1) and (52.2)] moves to the right with speed $c_0 + \tfrac{1}{2}(\gamma + 1)v$. Thus regions with higher fluid velocity propagate faster, and eventually v ceases to be a single-valued function of x (see Fig. 52.2). This catastrophic situation occurs when the slope $(dv/dx)_t$ becomes infinite or, equivalently, when $(dx/dv)_t = 0$. Furthermore this transition from single-valued behavior to triple-valued behavior must occur at an inflection point (Fig. 52.2), where $(d^2x/dv^2)_t = 0$. Application of these conditions to Eq. (52.39) yields the time t_0 for the onset of shock formation

$$t_0 = -\frac{2}{\gamma + 1} \frac{d}{dv} g^{-1}(v) \tag{52.40}$$

where the appropriate value of v is determined by

$$\frac{d^2}{dv^2} g^{-1}(v) = 0 \tag{52.41}$$

A cosine wave

$$v = v_0 \cos k(x - ut) \tag{52.42a}$$

provides an interesting explicit example in Eq. (52.37). In this case, Eq. (52.39) becomes

$$x - [c_0 + \tfrac{1}{2}(\gamma + 1)v]t = \frac{1}{k} \arccos \frac{v}{v_0} \tag{52.42b}$$

An elementary calculation then shows that the singularity first appears for $v = 0$ (the nodes of the fluid velocity) at a time

$$t_0 = \frac{2}{(\gamma + 1)kv_0} = \frac{\lambda}{\pi v_0(\gamma + 1)} \tag{52.43}$$

As expected from dimensional considerations, t_0 is of order λ/v_0 and becomes large for small-amplitude motion. We have thus also arrived at the fascinating result that *any* traveling sound wave in an ideal gas, no matter how small the initial amplitude, will *eventually* form a shock front if we wait long enough.†

Shock Waves

The preceding analysis has shown how the nonlinearity in the hydrodynamic equations changes the form of a finite-amplitude sound wave. Eventually, the wavefront becomes discontinuous, and the previous description fails because the flow *ceases to be isentropic*. This transition signals the formation of a shock wave, which then propagates in a characteristic way, independent of its detailed history. We now demonstrate that these general properties of shock waves are readily understood in terms of the conservation of mass, momentum, and energy.

For definiteness, consider a one-dimensional shock wave propagating to the right with speed u in a fluid initially at rest, as indicated in Fig. 52.3. If the wave preserves its form, the density, pressure, and velocity must propagate as

$$p(x, t) = p(x - ut) \qquad \rho(x, t) = \rho(x - ut) \qquad v(x, t) = v(x - ut) \quad (52.44)$$

and we require that the conditions become steady far ahead ($x \to \infty$, denoted by subscript 1) and behind ($x \to -\infty$, denoted by subscript 2). We assume that once the shock wave is formed, it will propagate through the uniform medium with a given velocity, and thus we treat u as a *constant*. In addition, the shock front is assumed thin, with the transition between regions 1 and 2 occurring in a surface layer that depends on more complicated thermodynamic and hydrodynamic behavior than we have yet discussed, precluding an analysis with the present

---

† Certain techniques (originated by Riemann) have been developed to construct more general solutions to Eqs. (52.3) and (52.6). See, for example, Landau and Lifshitz (1959), secs. 96–98; Sommerfeld (1950), sec. 37; Zel'dovich and Raizer [5], chap. 1, secs. 5–12.

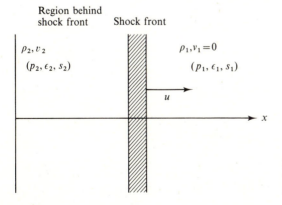

Figure **52.3** A one-dimensional shock wave moving through a fluid initially at rest.

techniques.† We can, however, still sensibly talk about the fluid velocity v, pressure p, and energy density ϵ everywhere in the fluid and derive some properties of the shock wave from general principles. For functions of the form (52.44), the time derivatives $\partial/\partial t$ in the conservation equations are equivalent to spatial derivatives $-u\partial/\partial x$. The conservation of mass (52.3), which must hold everywhere, thus becomes

$$-u\frac{\partial \rho}{\partial x} + \frac{\partial}{\partial x} v\rho = \frac{\partial}{\partial x}(v\rho - u\rho) = 0 \tag{52.45}$$

We integrate this equation across the shock front, evaluating the resulting quantities in the steady regions to the left (2) and right (1); in this way we obtain the jump condition

$$-u\rho_2 + v_2\rho_2 = -u\rho_1 + v_1\rho_1 \tag{52.46}$$

where we have left $v_1 \neq 0$ for generality. Similarly, the conservation of momentum (48.34) and energy (48.45) are also general results; for $\mathbf{f} = 0$, they imply

$$-u\rho_2 v_2 + p_2 + \rho_2 v_2^2 = -u\rho_1 v_1 + p_1 + \rho_1 v_1^2 \tag{52.47}$$

$$-u\rho_2(\tfrac{1}{2}v_2^2 + \epsilon_2) + \rho_2 v_2 \left(\tfrac{1}{2}v_2^2 + \epsilon_2 + \frac{p_2}{\rho_2}\right)$$

$$= -u\rho_1(\tfrac{1}{2}v_1^2 + \epsilon_1) + \rho_1 v_1\left(\tfrac{1}{2}v_1^2 + \epsilon_1 + \frac{p_1}{\rho_1}\right) \tag{52.48}$$

where the stress tensor (48.32) has been used explicitly. Three quantities (p_1, ρ_1, and v_1, say) suffice to specify the initial conditions, and another (p_2, say) characterizes the strength of the shock front. The three equations (52.46) to (52.48) then determine the three remaining unknowns u, ρ_2, and v_2 where we assume the energies are given by an equation of state $\epsilon = \epsilon(p, \rho)$. As seen below, the entropy increases on crossing the front from region 1 to region 2, so that $s_2 > s_1$; for this reason, the previous assumption of a universal isentropic relation $p = p(\rho)$ becomes invalid.

These general relations become simpler when expressed in relative variables $\bar{v}_i = v_i - u$ measured with respect to the moving shock front. If $v_1 = 0$, an observer moving to the right with the velocity u of the shock wave sees the fluid approach from the right with speed u and recede to the left with speed $\bar{v}_2$. Simple manipulations with Eqs. (52.46) to (52.48) yield the desired relations

$$\rho_2 \bar{v}_2 = \rho_1 \bar{v}_1 \tag{52.49a}$$

$$p_2 + \rho_2 \bar{v}_2^2 = p_1 + \rho_1 \bar{v}_1^2 \tag{52.49b}$$

$$\epsilon_2 + \tfrac{1}{2}\bar{v}_2^2 + \frac{p_2}{\rho_2} = \epsilon_1 + \tfrac{1}{2}\bar{v}_1^2 + \frac{p_1}{\rho_1} \tag{52.49c}$$

---

† It involves irreversible processes of heat conduction and viscosity. See, for example, Landau and Lifshitz (1959), sec. 87; Zel'dovich and Raizer [5], chap. 1, sec. 23.

which also follow directly from the steady form of the conservation laws applied in the comoving frame. Suppose the initial pressure and density (p_1, ρ_1) and the pressure behind the shock front p_2 are given. The first two equations are readily solved for $\bar{v}_1^2$ and $\bar{v}_2^2$ in terms of the known quantities and the density behind the shock front ρ_2, which will be determined subsequently,

$$\bar{v}_1^2 = \frac{\rho_2}{\rho_1} \frac{p_2 - p_1}{\rho_2 - \rho_1} \tag{52.50a}$$

$$\bar{v}_2^2 = \frac{\rho_1}{\rho_2} \frac{p_2 - p_1}{\rho_2 - \rho_1} \tag{52.50b}$$

If the pressure and density both increase behind the shock front, as we might expect, we see that $|\bar{v}_1|$ exceeds $|\bar{v}_2|$ by the factor ρ_2/ρ_1. Substitution into Eq. (52.49c) gives the final relation

$$2\left(\epsilon_2 + \frac{p_2}{\rho_2} - \epsilon_1 - \frac{p_1}{\rho_1}\right) = (p_2 - p_1)\left(\frac{1}{\rho_2} + \frac{1}{\rho_1}\right) \tag{52.51}$$

where $\epsilon + p/\rho$ can again be identified as the specific enthalpy in the fluid (see footnote on page 299). Since ϵ can be expressed in terms of p and ρ through the equation of state and thermodynamic measurements, Eq. (52.51) indeed fixes ρ_2 in terms of the remaining variables. For given initial conditions p_1 and ρ_1, this relation determines a curve in the $p_2 \rho_2$ plane known as the *shock adiabatic*, first determined by Hugoniot in the late nineteenth century. Substitution back into Eq. (52.50a) then gives the speed $|\bar{v}_1|$ of the shock front for a laboratory observer at rest with respect to the initial fluid $(v_1 = 0)$. This speed u can vary and depends on the initial conditions leading to the formation of the shock wave, in particular, on the amplitude of the initial disturbance.

To clarify these general expressions, it is helpful to evaluate them for the ideal gas considered previously in Eqs. (52.20) and (52.29), when the specific enthalpy becomes

$$\epsilon + \frac{p}{\rho} = \frac{\gamma}{\gamma - 1} \frac{p}{\rho} \tag{52.52}$$

Substitution into (52.51) yields the explicit solution

$$\frac{\rho_2}{\rho_1} = \left(\frac{\gamma + 1}{\gamma - 1} \frac{p_2}{p_1} + 1\right)\left(\frac{\gamma + 1}{\gamma - 1} + \frac{p_2}{p_1}\right)^{-1} \tag{52.53}$$

for ρ_2 in terms of the initial parameters p_1 and ρ_1 and the "strength of the shock" p_2/p_1. Note that p_2/p_1 can be arbitrarily large, but the ratio ρ_2/ρ_1 cannot exceed $(\gamma + 1)(\gamma - 1)^{-1}$. Use of the equation of state (52.20) determines the ratio of the *temperatures* on each side

$$\frac{T_2}{T_1} = \frac{p_2 \rho_1}{p_1 \rho_2} = \frac{p_2}{p_1}\left(\frac{\gamma + 1}{\gamma - 1} + \frac{p_2}{p_1}\right)\left(\frac{\gamma + 1}{\gamma - 1} \frac{p_2}{p_1} + 1\right)^{-1} \tag{52.54}$$

which approaches $(\gamma - 1)(\gamma + 1)^{-1}(p_2/p_1)$ for a strong shock. Another interesting quantity is the fluid velocity, and simple manipulations with Eqs. (52.32), (52.50a), and (52.53) give

$$\frac{\bar{v}_1^2}{c_1^2} = \frac{1}{2\gamma}\left[\gamma - 1 + (\gamma + 1)\frac{p_2}{p_1}\right] \tag{52.55}$$

where $c_1 = (\gamma p_1/\rho_1)^{1/2}$ is the adiabatic speed of sound on the right. This ratio always exceeds unity and approaches $(2\gamma)^{-1}(\gamma + 1)p_2/p_1$, which may again be arbitrarily large† for a strong shock. Thus the comoving observer sees the fluid approach from the right at supersonic speed; equivalently, the laboratory observer $(v_1 = 0)$ finds that the shock front itself propagates through the fluid at the supersonic speed $u = |\bar{v}_1| > c_1$, which depends on the strength of the shock. Similarly, Eq. (52.50b) implies

$$\frac{\bar{v}_2^2}{c_2^2} = \frac{1}{2\gamma}\left[\gamma - 1 + (\gamma + 1)\frac{p_1}{p_2}\right] \tag{52.56}$$

where $c_2^2 = \gamma p_2/\rho_2$. This ratio is always less than unity and approaches the constant value $(2\gamma)^{-1}(\gamma - 1)$ for a strong shock. Consequently, the comoving observer sees the fluid recede to the left at subsonic speed. A laboratory observer finds that the fluid behind the front moves to the right with a speed given by Eqs. (52.32b), (52.46), (52.53), and (52.55) as (recall that $\bar{v}_1 = -u$)

$$\frac{v_2^2}{c_2^2} = \frac{2}{\gamma}\frac{(1 - p_1/p_2)^2}{(\gamma + 1)(p_1/p_2) + \gamma - 1} \tag{52.57}$$

In the strong-shock limit, this ratio approaches the value $2/\gamma(\gamma - 1)$. It is thus clear that the passage of the shock wave strongly affects the fluid.

Finally, we consider the entropy difference on the two sides of the shock front. Entropy is a state variable and for an ideal gas is given in terms of the density and pressure by Eq. (52.31). Thus the required entropy difference is

$$s_2 - s_1 = c_V \ln\left[\frac{p_2}{p_1}\left(\frac{\rho_1}{\rho_2}\right)^{\gamma}\right] \tag{52.58}$$

with ρ_2/ρ_1 given by (52.53). This quantity is positive definite and approaches the limiting value $c_V\{\ln(p_2/p_1) - \gamma \ln[(\gamma + 1)/(\gamma - 1)]\}$ for strong shocks. We thus conclude that *these singular phenomena are always irreversible, as shown by the increase in entropy.*

It is also interesting to study the behavior of these quantities in the limit of a weak shock front, where $p_2/p_1 = 1 + \delta$ with $|\delta| \ll 1$. A straightforward expansion shows that $\rho_2/\rho_1 \approx 1 + \delta/\gamma$ and that $(c_2/c_1)^2 \approx 1 + (\gamma - 1)\delta/\gamma$, so that the disturbance produces first-order changes in these physical parameters. Furthermore, $(\bar{v}_1/c_1)^2 \approx 1 + \frac{1}{2}(\gamma + 1)\delta/\gamma$ and $(\bar{v}_2/c_2)^2 \approx 1 - \frac{1}{2}(\gamma + 1)\delta/\gamma$. Thus, a weak shock front propagates in the laboratory to the right with a speed $u = |\bar{v}_1|$ nearly

---

† Note that we always consider only nonrelativistic motion.

that of a sound wave c_1 and only weakly supersonic; such small-amplitude motion is little different from an acoustic wave. Behind the front, the laboratory motion is subsonic, with $v_2 \approx \delta c_2/\gamma$ from Eq. (52.57). If δ is defined to be $p_2/p_1 - 1$, an expansion of Eqs. (52.53) and (52.58) to *third* order in δ shows that $s_2 - s_1 \approx (c_V/12\gamma^2)(\gamma^2 - 1)\,\delta^3$; this small deviation from isentropic flow confirms that a weak shock is essentially acoustic in nature.

PROBLEMS

9.1 A rectangular cavity with dimensions $a < b < c$ is bounded by rigid walls.

(a) Determine the eigenfrequencies and eigenfunctions for the acoustic normal modes.

(b) Find the position of the pressure and velocity nodes and antinodes for the first few low-lying modes.

(c) If $c \gg b$, compare your results with those for a long cylindrical cavity with closed ends.

9.2 Consider sound waves in a gas confined between rigid concentric spheres of radii a and b ($a < b$).

(a) What are the normal modes of the velocity potential? (Express your answer in terms of spherical Bessel functions.)

(b) What are the boundary conditions?

(c) Derive an eigenvalue equation for the normal-mode frequencies.

(d) Sketch the first few normal modes.

9.3 A gas is contained in a rigid sphere of radius b.

(a) Show that the fundamental acoustic mode has $l = 1$ with frequency $\omega = x_0 c/b$, where $x_0 \approx 2.08$ is the lowest root of the equation $\cot x = x^{-1} - \tfrac{1}{2}x$.

(b) Discuss qualitatively the angular and radial dependence of the three distinct lowest-frequency modes.

9.4 A compressible fluid moves with uniform velocity **u**. Prove that the velocity potential for small-amplitude acoustic disturbances satisfies the modified wave equation

$$\left[\nabla^2 - \frac{1}{c^2}\left(\frac{\partial}{\partial t} + \mathbf{u}\cdot\nabla\right)^2\right]\Phi(\mathbf{r}, t) = 0$$

9.5 The vorticity $\boldsymbol{\zeta}$ is defined by the equation $\boldsymbol{\zeta} \equiv \nabla \times \mathbf{v}$.

(a) Show that the field lines of $\boldsymbol{\zeta}$ are continuous and can only end on the boundaries.

(b) For nonviscous isentropic flow in conservative fields, derive the dynamical equation

$$\frac{\partial \boldsymbol{\zeta}}{\partial t} = \nabla \times (\mathbf{v} \times \boldsymbol{\zeta})$$

(c) For incompressible flow, reduce this equation to

$$\frac{d\boldsymbol{\zeta}}{dt} \equiv \frac{\partial \boldsymbol{\zeta}}{\partial t} + (\mathbf{v}\cdot\nabla)\boldsymbol{\zeta} = (\boldsymbol{\zeta}\cdot\nabla)\mathbf{v}$$

Hence show that the field lines of $\boldsymbol{\zeta}$ move as if rigidly attached to the fluid particles. Relate this picture to Thomson's theorem.

(d) For compressible isentropic flow, show that the preceding equation must be altered to

$$\frac{d}{dt}\frac{\boldsymbol{\zeta}}{\rho} = \left(\frac{\boldsymbol{\zeta}}{\rho}\cdot\nabla\right)\mathbf{v}$$

Interpret and discuss.

9.6 A steady axisymmetric flow $\mathbf{v}(\mathbf{r}) = V(r)\hat{\phi}$ in cylindrical coordinates is established in an incompressible nonviscous fluid with density ρ in a gravitational field $-g\hat{z}$.

(a) Verify that any flow of this form satisfies the equations of hydrodynamics for all time. Find the pressure throughout the fluid.

(b) If $V(r) = \Omega r$ (solid-body rotation), determine the vorticity $\zeta = \text{curl } \mathbf{v}$ throughout the medium, and the circulation $\Gamma \equiv \oint \mathbf{v} \cdot d\mathbf{s}$ around a concentric circular contour of radius a. Find the shape of a free surface passing through the center at $z = 0$. Sketch the resulting profile.

(c) Repeat part (b) for $V(r) = \kappa/2\pi r$ (a vortex), assuming that the free surface approaches $z = 0$ as $r \to \infty$. What happens for $r \to 0$?

9.7 Consider the wave equation (49.9) in n dimensions ($n = 1, 2, 3$). Use a spatial Fourier transform to determine the solution for all $t > 0$ and all $\mathbf{r}$ subject to the initial conditions $\Phi(\mathbf{r}, 0) = f(\mathbf{r})$, $\dot{\Phi}(\mathbf{r}, 0) = g(\mathbf{r})$. Consider the special case $f(\mathbf{r}) = \delta(\mathbf{r})$ and $g(\mathbf{r}) = 0$. [For $n = 2$, see part (b) Prob. D.5; discuss the presence of a long-time tail in this case.]

9.8 (a) Use a Fourier transform to solve the one-dimensional equation $u''(x) - \lambda^2 u(x) = -f(x)$ subject to the boundary condition that $u(x)$ vanishes as $|x| \to \infty$. Express your answer in the form $\int_{-\infty}^{\infty} G(x, x') f(x') \, dx'$; hence construct the appropriate Green's function. Discuss the limit $\lambda \to 0$.

(b) Relate your answer for G to that obtained by solving the homogeneous equation for $x \neq x'$.

9.9 Use a Fourier transform to solve the Helmholtz equation $(\nabla^2 + k^2)u(\mathbf{x}) = -f(\mathbf{x})$ in one and two dimensions. Show that the outgoing-wave Green's functions are

$$\tilde{G}(x, x') = \frac{i}{2k} e^{ik|x - x'|} \qquad \text{one dimension}$$

$$\tilde{G}(\mathbf{x}, \mathbf{x}') = \tfrac{1}{4} i H_0^{(1)}(k|\mathbf{x} - \mathbf{x}'|) \qquad \text{two dimensions}$$

Hint: For two dimensions, use Eq. (51.12) and show that

$$\int_0^{\infty} p \, dp \, J_0(pr) \tilde{f}(p) = \frac{1}{2} \int_C z \, dz \, H_0^{(1)}(zr) \tilde{f}(z)$$

where C runs just above the real axis and $\tilde{f}(p)$ is a suitable even function.

9.10 The Green's function $\tilde{G}(\mathbf{r}, \mathbf{r}')$ satisfies Poisson's equation $\nabla^2 \tilde{G}(\mathbf{r}, \mathbf{r}') = -\delta(\mathbf{r} - \mathbf{r}')$ in the half space $z \geq 0$ subject to the homogeneous boundary condition

$$\alpha \hat{n} \cdot \nabla \tilde{G}(\mathbf{r}, \mathbf{r}') - \beta \tilde{G}(\mathbf{r}, \mathbf{r}') = 0 \qquad \text{for } z = 0$$

(a) Use a Fourier transform (50.35) in x and y to obtain a one-dimensional differential equation for $\tilde{G}(\mathbf{p}, z, z')$, and construct $\tilde{G}$ with the methods from Sec. 43.

(b) Use the Fourier inversion theorem to obtain an integral representation for $\tilde{G}(\mathbf{r}, \mathbf{r}')$. Show that $\tilde{G}(\mathbf{r}, \mathbf{r}')$ can be written $(4\pi|\mathbf{r} - \mathbf{r}'|)^{-1}$ plus a correction that reflects the presence of the boundary at $z = 0$. [See part (a) Prob. D.5.]

(c) Prove that this correction is just that obtained with the method of images if $\alpha = 0$ or if $\beta = 0$ but that the general case ($\alpha \neq 0$, $\beta \neq 0$) is more complicated.

9.11 The addition theorem for spherical harmonics states that

$$P_l(\hat{v}_1 \cdot \hat{v}_2) = \frac{4\pi}{2l + 1} \sum_{m=-l}^{l} Y_{lm}(\hat{v}_1) Y_{lm}^*(\hat{v}_2)$$

where the polar and azimuthal angles of the unit vectors $\hat{v}_1$ and $\hat{v}_2$ are measured with respect to a common reference frame. Starting from Eq. (50.27a) and the plane-wave expansion of Prob. D.7, show through contour integration that the Green's function for the three-dimensional scalar Helmholtz equation in free space has the following representation

$$\tilde{G}(\mathbf{r} - \mathbf{r}', k) = \frac{e^{ik|\mathbf{r} - \mathbf{r}'|}}{4\pi|\mathbf{r} - \mathbf{r}'|} = ik \sum_{l=0}^{\infty} \frac{2l + 1}{4\pi} j_l(kr_<) h_l^{(1)}(kr_>) P_l(\hat{r} \cdot \hat{r}')$$

where $r_<$ ($r_>$) is the smaller (larger) of $|\mathbf{r}|$ and $|\mathbf{r}'|$ and $k \equiv \omega/c$.

9.12 Consider the Green's function $\tilde{G}(\mathbf{r}, \mathbf{r}')$ for Helmholtz's equation (50.31) in the interior of a sphere of radius a subject to the general homogeneous boundary condition (50.32).

(a) Use an expansion in spherical harmonics to reduce the problem to a one-dimensional equation and solve that with the techniques of Sec. 43. Hence construct an explicit solution for $\tilde{G}(\mathbf{r}, \mathbf{r}')$ (compare Prob. 9.11).

(b) If $\alpha = 0$ *and* $\omega = 0$, show that the solution is just that obtained with the method of images (as is familiar from electrostatics). In all other cases, however, show that the method of images fails (even for $\beta = 0$ and $\omega = 0$).

9.13 Consider the system in Prob. 9.4.

(a) A point source with frequency ω is placed at the origin. Show that the Green's function has the spatial Fourier transform

$$\tilde{G}_\omega(\mathbf{k}) = [k^2 - c^{-2}(\omega + i\eta - \mathbf{k} \cdot \mathbf{u})^2]^{-1}$$

(b) If $\mathbf{u} = u\hat{z}$ and $u < c$, use a change of variables to derive

$$\tilde{G}_\omega(\mathbf{r}) = \frac{c}{(c^2 - u^2)^{1/2}} \frac{1}{4\pi R} \exp\left[\frac{i\omega R}{(c^2 - u^2)^{1/2}} - \frac{i\omega uz}{c^2 - u^2}\right]$$

where $R^2 \equiv x^2 + y^2 + c^2 z^2 (c^2 - u^2)^{-1}$. Verify that the surfaces of constant phase emitted at $r = 0$ and $t = 0$ are specified by $x^2 + y^2 + (z - ut)^2 = c^2 t^2$. Sketch these surfaces and interpret them.

(c) If $u > c$, prove that $\tilde{G}_\omega(\mathbf{r}) = 0$ for $cz \le [(x^2 + y^2)(u^2 - c^2)]^{1/2}$ and conclude that the signal is confined to a downstream cone with apex angle $\arcsin(c/u)$. In this region, prove that $\tilde{G}_\omega(\mathbf{r})$ has the form

$$\tilde{G}_\omega(\mathbf{r}) = \frac{c}{(u^2 - c^2)^{1/2}} \frac{1}{2\pi|R|} \exp\frac{i\omega uz}{u^2 - c^2} \cos\frac{\omega|R|}{(u^2 - c^2)^{1/2}}$$

9.14 (a) An acoustic point source with frequency ω is placed in an unbounded medium. Find the mean differential and total power radiated $\langle dP/d\Omega \rangle_0$ and $\langle P \rangle_0$.

(b) Two such sources are placed at the points $\pm d\hat{z}$ and oscillate with phases $\pm \alpha$. For a distant observer at $\mathbf{r}$, show that $\langle dP/d\Omega \rangle = 2\langle dP/d\Omega \rangle_0[1 + \cos(2kd \cos\theta - 2\alpha)]$, where $\cos\theta = \hat{z} \cdot \hat{r}$ and that $\langle P \rangle = 2\langle P \rangle_0[1 + (2kd)^{-1} \sin 2kd \cos 2\alpha]$. Discuss.

9.15 (a) An acoustic source is located a distance d above an infinite rigid plane and an observer is a distance l away and a height h above the plane. If $l \gg h$ and $l \gg d$, how does the intensity vary with h?

(b) Repeat for a source and observer located beneath the (free) surface of the sea.

9.16 Consider a gas contained in a half space $y \ge 0$.

(a) A portion of the wall consisting of the infinite strip $|x| < a$ oscillates along $\hat{y}$ with amplitude $\epsilon e^{-i\omega t}$, radiating sound into the medium. Integrate the analog of (51.7) over z to obtain

$$\Phi(x, y) = \tfrac{1}{2}\omega\epsilon \int_{-a}^{a} dx' \, H_0^{(1)}(k[(x - x')^2 + y^2]^{1/2})$$

Hint: It is simplest to use the Fourier-integral representation obtained from (50.26) for the three-dimensional Green's function.

(b) Expand the integrand for large $r = (x^2 + y^2)^{1/2}$ and hence obtain the mean power $\langle dP/d\phi \rangle$ radiated into a unit angular interval. Sketch the angular dependence for $ka \ll 1$, for $ka \gg 1$. Compare briefly with the Fraunhofer diffraction by a slit of width $2a$.

9.17 The half space $y \ge 0$ is filled with compressible gas flowing in the positive x direction with uniform speed u. A section of the boundary $(y = 0)$ between $0 < x < a$ executes small vibrations in the y direction with amplitude $\epsilon e^{-i\omega t}$. Use the result of Prob. 9.4 to conclude that the velocity potential $\Phi e^{-i\omega t}$ satisfies the equation $L(\mathbf{x}, u)\Phi(\mathbf{x}) = 0$, where

$$L(\mathbf{x}, u) \equiv \frac{\partial^2}{\partial x^2} + \frac{\partial^2}{\partial y^2} + \frac{1}{c^2}\left(\omega + iu\frac{\partial}{\partial x}\right)^2$$

Generalize the left-hand side of Green's identity (51.3) to the form

$$\int d^2x \; [\Phi(\mathbf{x})L(\mathbf{x}, -u)\Psi(\mathbf{x}) - \Psi(\mathbf{x})L(\mathbf{x}, u)\Phi(\mathbf{x})]$$

and obtain a corresponding generalization for the right-hand side. Hence derive the integral representation for the solution

$$\Phi(x, y) = -i\omega\epsilon \int_0^a \tilde{G}(x - x', y) \; dx'$$

where
$$\tilde{G}(x - x', y) = \int \frac{d^2k}{(2\pi)^2} \frac{e^{ik_x(x - x')}}{k^2 - c^{-2}(\omega + i\eta - k_x u)^2} \left(e^{ik_y(y - y')} + e^{ik_y(y + y')}\right)_{y' = 0}$$

9.18 (a) Evaluate $\tilde{G}(x, y)$ in Prob. 9.17 explicitly. The cases $u < c$ and $u > c$ require separate treatments. Use your answer to compute $\Phi(x, y)$ in the limit $x^2 + y^2 \gg a^2$ (the far field). Discuss the physical and mathematical differences between the two cases $u < c$ and $u > c$.

(b) Rewrite the integral representation from Prob. 9.17 by substituting $x = r \cos \theta$; $y = r \sin \theta$. Evaluate the asymptotic form of $\Phi(r, \theta)$ in the limit $r \to \infty$ with the method of stationary phase (see Sec. 55). The cases $u < c$ and $u > c$ must be treated separately. For simplicity, the angle may be restricted to the domains $\theta \ll 1$ and $\pi - \theta \ll 1$. Discuss the relation of your answer to that obtained in part (a).

9.19 An infinite cylinder (lying along the z axis) with fixed center and oscillating radius $\text{Re} \, a(1 + \epsilon e^{-i\omega t})$ emits sound waves into a surrounding gas, where $\epsilon \ll 1$. Find an expression for the velocity potential throughout the gas. Compute the power radiated per unit length of cylinder into an angular interval $d\phi$, where (r, ϕ) are polar coordinates in the xy plane. Discuss the angular distribution and find the total power radiated per unit length. Consider both limiting cases $ka \ll 1$ and $ka \gg 1$, where $k = \omega/c$ and c is the speed of sound.

9.20 (a) Repeat Prob. 9.19 for an infinite oscillating cylinder of fixed radius a whose center is located at $\text{Re} \, a\hat{\mathbf{x}}e^{-i\omega t}$. Compare the two systems. Which is the more efficient radiator for long wavelengths $(ka \ll 1)$?

(b) Use the results of part (a) to show that the effective mass of a long cylinder executing slow transverse oscillations in an incompressible fluid exceeds its bare mass by the mass of fluid displaced.

9.21 A point source of acoustic disturbance with frequency ω is located at $\mathbf{r}'$ outside a sphere of radius a.

(a) Separate variables in spherical polar coordinates to find the outgoing-wave Green's function $\tilde{G}(\mathbf{r}, \mathbf{r}')$ (compare Prob. 9.11).

(b) Use Green's identity (51.3) to solve the corresponding radiation problem with $(\nabla^2 + c^{-2}\omega^2)\Phi(\mathbf{r}) = 0$ and $\partial\Phi/\partial r$ given on the spherical surface $r = a$.

(c) Specialize to a rigid sphere executing transverse oscillations and rederive the results of Sec. 51.

9.22 For many purposes, a surface can be characterized by its acoustic impedance Z, defined as the ratio of the complex magnitudes $p'/\mathbf{v}' \cdot \hat{\mathbf{n}}$ omitting the common factors $e^{-i\omega t}$.

(a) Consider a plane sound wave in a gas with mean mass density ρ_0 and propagation speed c incident on such a plane surface at an angle θ relative to the perpendicular. Explain clearly the boundary condition on the sound wave. Verify that the fraction of energy reflected is $R = |Z \cos \theta - c\rho_0|^2/|Z \cos \theta + c\rho_0|^2$. Discuss the relative phase and amplitude of the reflected and incident wave for large and small values of $|Z \cos \theta/c\rho_0|$.

(b) Evaluate explicitly the energy flux into the wall and verify that energy is conserved.

(c) As a simple model, assume that the wall consists of simple harmonic oscillators with areal mass density σ, subject to areal force densities $-\kappa\xi - \dot{\xi}\sigma/T$, where ξ is the normal displacement of the surface and T is a damping time. Find $Z(\omega)$ in this example. For normal incidence $(\theta = 0)$ determine $R(\omega)$ and sketch its frequency dependence. Discuss the various special cases of large and small $\sigma/c\rho_0 T$ and large and small $\omega^2\sigma/\kappa$.

9.23 A sphere of radius a and acoustic impedance Z (see Prob. 9.22) scatters an incident sound wave with frequency ω propagating in a medium of mean density ρ_0 and wave speed c.

(a) Use the plane-wave expansion of Prob. D.7 to find the scattered wave $\Phi_{sc}(\mathbf{r})^{-i\omega t}$ throughout all space.

(b) Verify that $\Phi_{sc}(\mathbf{r})$ behaves like $r^{-1}e^{ikr}f(\theta)$ for large kr, where $\cos\theta = \hat{k}\cdot\hat{r}$. Hence show that the differential cross section $d\sigma/d\Omega$ is given by $|f(\theta)|^2$. Evaluate $f(\theta)$ for $ka \ll 1$ in terms of Z, c, and ρ_0.

(c) For a rigid sphere $(Z \to \infty)$, deduce the long-wavelength expression $d\sigma/d\Omega \approx \frac{1}{9}k^4 a^6 (1 - \frac{3}{2}\cos\theta)^2$. Sketch the angular dependence and evaluate the total cross section.

9.24 An infinite string stretched to a tension τ has a mass density $\sigma_0 + \sigma(x)$, where $\sigma(x)$ is a localized inhomogeneity.

(a) If a wave Re $Ae^{i(kx-\omega t)}$ is incident from the left (here $k = \omega/c_0$ and $c_0^2 = \tau/\sigma_0$), use the Green's function from Prob. 9.9 to rewrite Helmholtz's equation as an *integral* equation [compare Eq. (51.64)]

$$u(x) = Ae^{ikx} + k^2\sigma_0^{-1}\int_{-\infty}^{\infty} dx' \, \tilde{G}(x, x')\sigma(x')u(x')$$

Interpret the various terms.

(b) For $x \to -\infty$, show that $u(x)$ has the form (45.36) and (45.37) for an incident and reflected wave, with

$$B = \tfrac{1}{2}ik\sigma_0^{-1}\int_{-\infty}^{\infty} dx' \, e^{ikx'}\sigma(x')u(x')$$

(c) Similarly, show that the solution for $x \to \infty$ is a transmitted wave (45.38) with

$$C = A + \tfrac{1}{2}ik\sigma_0^{-1}\int_{-\infty}^{\infty} dx' \, e^{-ikx'}\sigma(x')u(x')$$

(d) Find the corresponding exact transmission and reflection coefficients T and R.

9.25 The formulas in Prob. 9.24 for the transmission and reflection amplitudes require the *exact* solution in the region where $\sigma(x) \neq 0$.

(a) In the *Born approximation*, $u(x)$ under the integral is replaced by the incident wave Ae^{ikx}. Discuss the validity of this approximation and find the corresponding transmission and reflection coefficients T and R.

(b) If $\sigma(x) = \sigma$ $(\ll \sigma_0)$ for $|x| < a$ and zero elsewhere, relate the resulting expressions to those found in Prob. 7.17.

9.26 A point source of acoustic vibrations is located a distance d_0 behind a disk of radius a, on its axis of symmetry. The disturbance is observed on the symmetry axis a distance d' in front of its plane. In the Kirchhoff approximation, prove that:

(a) If $d' = d_0$, the intensity at d' is equal to $I(d')/I_0 = \frac{1}{4}d'^2(a^2 + d'^2)^{-1}$, where I_0 is the incident intensity at the edge of the disk.

(b) If the disk is replaced by the complementary aperture in an infinite screen, the intensity at $d' = d_0$ (for $d' \gg a$) is given by $I(d')/I_0 \approx \sin^2(ka^2/2d')$. Explain carefully how Babinet's principle holds.

(c) Repeat (a) and (b) for $d_0 \to \infty$ with d' fixed. For a disk, verify that its effect disappears in the limit $d' \gg a$ (this phenomenon is known as *Poisson's spot*).

9.27 An incompressible nonviscous fluid is viewed from a frame of reference rotating with constant velocity $\mathbf{\Omega}$ relative to an inertial frame.

(a) Show that the Euler equation in the rotating frame becomes

$$\frac{\partial\mathbf{v}}{\partial t} + (\mathbf{v}\cdot\nabla)\mathbf{v} = -\nabla(\rho^{-1}p - \tfrac{1}{2}|\mathbf{\Omega}\times\mathbf{r}|^2) - 2\mathbf{\Omega}\times\mathbf{v} + \mathbf{f}$$

where $\mathbf{r}$ and $\mathbf{v}$ are the position and velocity seen in the rotating frame.

(b) If $\mathbf{f} = -\nabla U$, prove that the "vorticity" $\boldsymbol{\zeta} \equiv \mathrm{curl}\, \mathbf{v}$ obeys the equation $\partial \boldsymbol{\zeta}/\partial t = \mathrm{curl}\, [\mathbf{v} \times (\boldsymbol{\zeta} + 2\boldsymbol{\Omega})]$. For steady slow ($|\boldsymbol{\zeta}| \ll \Omega$) motion relative to the rotating frame, deduce that $(\boldsymbol{\Omega} \cdot \nabla)\mathbf{v} = 0$. Hence explain why small steady disturbances of a uniformly rotating nonviscous fluid lie in the plane perpendicular to $\boldsymbol{\Omega}$ and thus are two-dimensional (Taylor-Proudman theorem).

(c) Prove that a uniformly rotating incompressible nonviscous fluid with $\mathbf{f} = 0$ can support transverse circularly polarized waves of the form $\mathbf{v} = \mathbf{v}_0\, e^{i(\mathbf{k} \cdot \mathbf{r} - \omega t)}$, with $\omega = \pm 2\boldsymbol{\Omega} \cdot \hat{k}$ and *arbitrary amplitude*. What is the sense of polarization? Discuss the phase velocity and group velocity (see Sec. 54) for different orientations of $\hat{k}$. (These *inertial waves* are direct analogs of Alfvén waves in hydromagnetics.)

SELECTED ADDITIONAL READINGS

1. F. Reif, "Fundamentals of Statistical and Thermal Physics," McGraw-Hill, New York, 1965.
2. J. D. Jackson, "Classical Electrodynamics," Wiley, New York, 1962.
3. A. Sommerfeld, "Optics," Academic, New York, 1954.
4. L. D. Landau and E. M. Lifshitz, "Electrodynamics of Continuous Media," Addison-Wesley, Reading, Mass., 1960.
5. Ya. B. Zel'dovich and Yu. P. Raizer, "Physics of Shock Waves and High-Temperature Hydrodynamic Phenomena," Academic, New York, 1966, vol. 1.

TEN

SURFACE WAVES ON FLUIDS

We now turn to phenomena involving fluid surfaces. These have direct physical interest, exemplifying some of the most familiar classical behavior, like waves on the sea.

53 TIDAL WAVES†

Consider an incompressible fluid subject to a uniform gravitational field. It is convenient to deal first with *tidal waves*, or *canal waves*, namely, waves in a channel or basin where the wavelength is large relative to the depth. Thus we assume that the fluid is shallow in a sense to be made more precise later in this section. In this limit, the principal particle motion is *horizontal*, so that the fluid moves back and forth. When fluid accumulates at any point, the surface rises and produces a hydrostatic pressure. The fluid then responds to the unbalanced force by flowing horizontally to relieve this pressure head.

Equations of Motion

Consider the situation illustrated in Fig. 53.1. Let $\zeta(x, y, t)$ be the displacement of the surface from the free *equilibrium* surface, which is taken to define the xy plane. Assume that the fluid moves slowly and that its displacement from equilibrium is

---

† This choice of title follows Lamb (1945), p. 250, who uses it to describe "gravitational oscillations" similar to those "produced by the action of the sun and moon." A better but more cumbersome designation might be "long waves on shallow water."

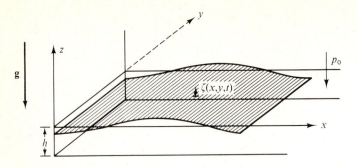

Figure 53.1 Tidal wave in a tank or channel.

small. We shall subsequently formulate this condition in detail, but for the present we merely linearize the equations in the small displacements and velocities. The basic equation (48.13) of fluid motion is just Newton's second law. It states that

$$\frac{d\mathbf{v}}{dt} = \mathbf{f}_{app} - \frac{1}{\rho}\nabla p \tag{53.1}$$

In the present case, the applied force is the earth's gravity, which takes the form

$$\mathbf{f}_{app}\rho \, dV = -g\hat{z}\rho \, dV \tag{53.2}$$

or, equivalently,
$$\mathbf{f}_{app} = -g\hat{z} \equiv \mathbf{g} \tag{53.3}$$

Our basic assumption (to be justified below) is that the velocity in the $\hat{z}$ direction is small, which implies that the corresponding vertical acceleration is also small. As a result, the term dv_z/dt is negligible in the z component of Eq. (53.1), which then reads

$$-g - \frac{1}{\rho}\frac{\partial p}{\partial z} \approx 0 \tag{53.4}$$

Since the free surface of the fluid is always at atmospheric pressure p_0 (see Fig. 53.1), direct integration of Eq. (53.4) in the z direction at any fixed (x, y) immediately gives the instantaneous pressure throughout the fluid

$$p(x, y, z, t) = p_0 + g\rho[\zeta(x, y, t) - z] \tag{53.5}$$

This expression satisfies the linear differential equation (53.4) and reduces to atmospheric pressure at the instantaneous free surface of the fluid $z = \zeta(x, y, t)$, as illustrated in Fig. 53.1. Equation (53.5) demonstrates that, in the present approximation, the pressure at any point comes simply from the piled-up fluid.

Recall from Sec. 48 that the total derivative of the velocity entering into the basic law of hydrodynamics (53.1) can be written in the form

$$\frac{d\mathbf{v}}{dt} = \frac{\partial\mathbf{v}}{\partial t} + O(\mathbf{v}^2) \tag{53.6}$$

Under the condition of small velocities, we retain only the linear term. Hence the equations of horizontal motion for the fluid in the xy plane follow directly from Eq. (53.1) as

$$\frac{dv_x}{dt} \approx \frac{\partial v_x}{\partial t} = -\frac{1}{\rho}\frac{\partial p}{\partial x} \tag{53.7a}$$

$$\frac{dv_y}{dt} \approx \frac{\partial v_y}{\partial t} = -\frac{1}{\rho}\frac{\partial p}{\partial y} \tag{53.7b}$$

Use of Eq. (53.5) allows us to eliminate the gradients of the pressure, and we therefore obtain the basic equations for *horizontal motion*

$$\frac{\partial v_x}{\partial t} = -g\frac{\partial \zeta}{\partial x} \tag{53.8a}$$

$$\frac{\partial v_y}{\partial t} = -g\frac{\partial \zeta}{\partial y} \tag{53.8b}$$

Note that the displacement of the surface ζ depends only on the variables (x, y, t), so that the left-hand sides of Eqs. (53.8) are *independent of z*. This leads to the important observation that the *horizontal accelerations are the same at all depths in the fluid*, in the present approximation of negligible vertical acceleration. If the initial fluid is at rest or in uniform motion, the horizontal *velocities also remain the same at all depths*. We shall assume this to be the case.

One-dimensional Waves

To illustrate the behavior, we start by considering a *one-dimensional* wave $\zeta(x, t)$ in a channel bounded by rigid walls (the generalization to two dimensions will be studied subsequently). In this case, the displacement of the surface depends only on the distance x along the channel (see Fig. 53.1). Equation (53.8a) constitutes the dynamical equation for the fluid, and we must next analyze the continuity equation. Consider a fixed transverse section of the channel of height $h(x)$, breadth $b(x)$, and thickness dx, as shown in Fig. 53.2. The fluid flows in and out of this

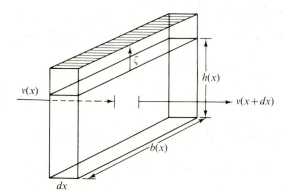

Figure 53.2 Fixed transverse section of the channel used in derivation of the continuity equation. The equilibrium cross-sectional area of the channel is $S(x) \equiv b(x)h(x)$.

region due to its motion in the horizontal direction, and the height of the fluid (characterized by ζ) correspondingly moves up and down. The integral form (48.21) of the continuity equation equates the net inward flow of fluid with the total rate of increase of the fluid's mass in the enclosed volume

$$-\int_A \rho\mathbf{v} \cdot d\mathbf{A} = \frac{d}{dt}\int_V \rho \, dV \tag{53.9}$$

Since the velocity has been shown to be the same at all depths in the fluid, the surface integral appearing in Eq. (53.9) follows immediately from Fig. 53.2 as

$$-\int_A \rho\mathbf{v} \cdot d\mathbf{A} = -[\rho(S + b\zeta)v_x]_{x+dx} + [\rho(S + b\zeta)v_x]_x \tag{53.10}$$

where the equilibrium cross-sectional surface area of the channel is $S(x) \equiv b(x)h(x)$ (see Fig. 53.2). A Taylor-series expansion of Eq. (53.10) to first order in dx gives

$$-\int \rho\mathbf{v} \cdot d\mathbf{A} = -\rho \, dx \frac{\partial}{\partial x}[(S + b\zeta)v_x] \approx -\rho \, dx \frac{\partial}{\partial x} Sv_x \tag{53.11}$$

where we have discarded the term containing ζv_x as being of explicitly higher order. Similarly, the right-hand side of Eq. (53.9) can be written

$$\frac{d}{dt}\int_V \rho \, dV = \frac{\partial}{\partial t}[\rho(h + \zeta)b(x) \, dx] \tag{53.12}$$

where the volume has been determined from Fig. 53.2 and the partial derivative indicates that this expression is evaluated at a given point x. Only the height of the surface ζ varies with time, and we conclude that

$$\frac{d}{dt}\int_V \rho \, dV = \rho b(x) \, dx \frac{\partial \zeta}{\partial t} \tag{53.13}$$

Comparison of Eqs. (53.9), (53.11), and (53.13) allows us to write the *continuity equation* for the one-dimensional flow in the following form (recall that ρ is constant)

$$-\frac{\partial}{\partial x} Sv_x = b(x)\frac{\partial \zeta}{\partial t} \tag{53.14}$$

This equation relates the rate of change of the height of the free surface to the flow in and out of a given transverse slice of the channel, as illustrated in Fig. 53.2. It is important to emphasize that the basic fluid motion is horizontal, while the free surface simply bobs up and down on top of this region.

The dynamical equation of motion (53.8a) can now be combined with the continuity equation (53.14) to yield a one-dimensional wave equation. Multiply Eq. (53.8a) by $S(x) \equiv b(x)h(x)$ and differentiate partially with respect to x; comparison with the partial time derivative of Eq. (53.14) yields the relation

$$b(x)\frac{\partial^2 \zeta(x, t)}{\partial t^2} = g\frac{\partial}{\partial x}\left[S(x)\frac{\partial}{\partial x}\zeta(x, t)\right] \tag{53.15}$$

We recognize this expression as a *generalized one-dimensional wave equation*. In fact, the following identification provides an exact analogy with the general wave equation (38.9) for a one-dimensional string studied in Chap. 7:

$$\sigma(x) \rightarrow b(x) \qquad \text{"mass density"}$$

$$\tau(x) \rightarrow gS(x) \qquad \text{"tension"} \tag{53.16}$$

If the channel under consideration has both constant height and constant breadth

$$b = \text{const} \qquad h = \text{const} \qquad S = bh = \text{const} \tag{53.17}$$

then (53.15) reduces to the more familiar one-dimensional wave equation

$$\frac{1}{c^2}\frac{\partial^2 \zeta}{\partial t^2} = \frac{\partial^2 \zeta}{\partial x^2} \tag{53.18}$$

where the square of the velocity appearing in this equation is obtained from (53.15) and (53.17) as $c^2 = gS/b$, or

$$c^2 = gh \tag{53.19}$$

Recall that g is the gravitational constant and h is the height of the channel. Thus we have obtained the fascinating result that the deeper the channel the greater the wave velocity.† Throughout this section on small-amplitude shallow-water waves, it is important to remember that the fluid actually flows in the horizontal direction with the same velocity at all depths.

Consider the solution to the wave equation (53.18) in a closed rectangular tank, obtained by inserting rigid vertical ends in Fig. 53.1. The boundary conditions at the ends of the tank require that the velocity of fluid should vanish there at all times

$$v_x = 0 \qquad \text{at fixed end} \tag{53.20}$$

It is then evident from Eq. (53.8a) that the displacement of the free surface must correspondingly satisfy

$$\frac{\partial \zeta}{\partial x} = 0 \qquad \text{at fixed end} \tag{53.21}$$

at the closed ends of the tank. We seek normal-mode solutions to Eq. (53.18) of the form

$$\zeta(x, t) = u(x) \cos (\omega t + \phi) \tag{53.22}$$

Substitution of this expression yields

$$\frac{d^2 u}{dx^2} + k^2 u = 0 \tag{53.23a}$$

$$k^2 \equiv \frac{\omega^2}{c^2} \tag{53.23b}$$

---

† This observation provides a *qualitative* explanation of the breaking surf on a beach, for the crests travel slightly faster than the troughs. The full nonlinear theory confirms this picture and even predicts the formation of a surface discontinuity, similar to the shock fronts studied in Sec. 52 (see Prob. 10.7).

To satisfy the boundary condition (53.21) at $x = 0$ and l, we must evidently take the normalized cosine solutions to Eq. (53.23)

$$u(x) = \left(\frac{2}{l}\right)^{1/2} \cos \frac{n\pi x}{l} \tag{53.24a}$$

$$k_n = \frac{n\pi}{l} \qquad n = 1, 2, 3, \ldots, \infty \tag{53.24b}$$

where the value $n = 0$ is omitted because it represents a uniform displacement and violates conservation of matter. Thus the nth normal mode takes the form

$$\zeta^{(n)}(x, t) = \zeta_0^{(n)} \cos k_n x \cos (k_n ct + \phi_n) \tag{53.25}$$

where $\zeta_0^{(n)}$ is a constant. Note that the rigid walls perpendicular to the y axis in Fig. 53.1 require the boundary condition $v_y = 0$ at all times. This restriction is here satisfied identically since $\zeta^{(n)}$ is independent of y. Thus $\partial \zeta / \partial y = 0$, and Eq. (53.8b) then implies that v_y indeed vanishes in these normal modes.

The velocity of the fluid corresponding to (53.25) is obtained from Eq. (53.14) and the boundary condition (53.20) as follows:

$$v_x^{(n)}(x, t) = \frac{c}{h} \zeta_0^{(n)} \sin k_n x \sin (k_n ct + \phi_n) \tag{53.26}$$

The displacement of the fluid is assumed small compared to the height of the channel, so that $|\zeta_0| \ll h$. Equation (53.26) immediately implies that $|v_x| \ll c$, where v_x is the actual velocity of fluid flow and c is the propagation speed of the wave that appears on the surface of the fluid.

We can now state more precisely the conditions for our approximations. At the free surface of the fluid, the vertical motion of the fluid itself must be the same as the motion of the surface elevation.† Thus we require

$$v_z = \dot{\zeta} \qquad \text{at free surface} \tag{53.27}$$

to first order in small quantities. Equations (53.27) and (53.25) then show that

$$\frac{dv_z^{(n)}}{dt} \approx \frac{\partial v_z^{(n)}}{\partial t} \approx -k_n^2 c^2 \zeta_0^{(n)} \cos k_n x \cos (k_n ct + \phi_n) \tag{53.28}$$

We first compare the order of magnitude of the accelerations of the fluid in the z and x directions [obtained from Eq. (53.26)]

$$\frac{|dv_z^{(n)}/dt|}{|dv_x^{(n)}/dt|} = O\left(\frac{k_n^2 c^2}{c^2 k_n / h}\right) = O(hk_n) = O\left(\frac{2\pi h}{\lambda_n}\right) \tag{53.29}$$

This expression is small if

$$\frac{h}{\lambda_n} = \frac{hn}{2l} \ll 1 \tag{53.30}$$

---

† Note this is actually the maximum value of $|v_z|$, for the boundary condition at the bottom of the tank evidently requires $v_z = 0$.

which holds when the wavelength of the disturbance is much larger than the depth of the channel. Thus these are "long waves in shallow water." If condition (53.30) is satisfied, the motion of the fluid is primarily in the horizontal plane, as asserted below Eq. (53.8).

We can also compare the vertical acceleration of the fluid, which was neglected in writing Eq. (53.4), with the acceleration of gravity g, which was retained in that expression. Equation (53.28) gives the order of magnitude of this ratio [recall $|\zeta_0^{(n)}| \ll h$ and Eq. (53.19)]

$$\frac{|dv_z^{(n)}/dt|}{g} = O\left(\frac{k_n^2 c^2 |\zeta_0^{(n)}|}{g}\right) = O(hk_n^2|\zeta_0^{(n)}|) \ll \left(\frac{2\pi h}{\lambda_n}\right)^2 \qquad (53.31)$$

This quantity also is small if Eq. (53.30) is satisfied. Hence we conclude that

$$\frac{dv_z}{dt} \text{ has negligible effect on the pressure if } \left(\frac{h}{\lambda}\right)^2 \ll 1 \qquad (53.32)$$

and that Eq. (53.5) then provides a valid expression for the pressure.

The conditions $|\zeta| \ll h \ll \lambda$ for the present theory of small-amplitude waves in shallow channels appear quite restrictive. Indeed, they fail whenever the depth of the fluid is comparable with the largest horizontal linear dimension. On the other hand, the theory applies directly to many cases of physical interest, such as long-wavelength disturbances in a canal or river. It also explains the extremely rapid propagation of long-wavelength tsunami (also loosely known as "tidal waves") on the open ocean, where Eq. (53.19) yields $c \approx 0.2$ km s^{-1} $\approx$ 800 km h^{-1} for the typical depth $h \approx 5$ km.

The general solution to the wave equation (53.18) can be obtained by superposing the normal modes (53.25)

$$\zeta(x, t) = \sum_{n=1}^{\infty} C_n \left(\frac{2}{l}\right)^{1/2} \cos \frac{n\pi x}{l} \cos \left(\frac{n\pi ct}{l} + \phi_n\right) \qquad (53.33)$$

Although this expression is mathematically correct, it is physically sensible only if the small wavelengths (those with large n) play a negligible role. Otherwise the physical assumptions in the derivation of the wave equation, i.e., the conditions (53.30) and (53.32), necessarily fail as $n \to \infty$.

The preceding explicit solution (53.25) describes the one-dimensional surface waves in a channel with constant height and breadth. More generally, Eq. (53.15) applies to the similar propagation of long-wavelength disturbances in channels of variable height and breadth, again assuming $|\zeta| \ll h \ll \lambda$ (see Probs. 10.3 to 10.6). In addition, the derivation of Eq. (53.14) requires that $h(x)$ and $b(x)$ vary slowly $[|h'(x)| \ll 1$ and $|b'(x)| \ll 1]$, for otherwise the unit normals on the bottom or the side walls would acquire an appreciable x component. In that case, the velocities v_y and v_z and the corresponding accelerations would no longer be negligible.

Two-dimensional Waves

The previous discussion is readily generalized to the case of tidal waves in two dimensions. We still assume the conditions (53.30) and (53.32), so that the motion

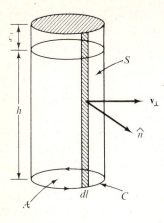

Figure 53.3 Cylinder for discussion of continuity equation in two dimensions. [Also used for computation of the gravitational potential energy of a fluid with surface waves (Sec. 54) where the area $\mathscr{A} \equiv dA$ is differentially small.]

of the fluid is again horizontal and the pressure is just that given in Eq. (53.5). As a result, the dynamical equations (53.8) continue to describe the horizontal motion. If we use the subscript $\perp$ to indicate a quantity confined to the horizontal direction and write

$$\mathbf{v}_\perp \equiv \mathbf{v}_\perp(x, y, t) = v_x \hat{x} + v_y \hat{y} \tag{53.34}$$

Eqs. (53.8) can be summarized in this obvious shorthand as

$$\frac{\partial \mathbf{v}_\perp}{\partial t} = -g\, \mathbf{\nabla}_\perp \zeta \tag{53.35}$$

The remaining problem is to generalize the *continuity equation* to two-dimensional flow. Consider the cylindrical region illustrated in Fig. 53.3. The integral form of mass conservation again states that

$$-\int_A \rho \mathbf{v} \cdot d\mathbf{A} = \frac{d}{dt} \int_V \rho \, dV \tag{53.36}$$

where the first integral is over the entire surface in Fig. 53.3. The rate of increase of the mass of fluid contained in this given region is

$$\frac{d}{dt} \int_V \rho \, dV = \frac{d}{dt} \int_{\mathscr{A}} d\mathscr{A} \, \rho(h + \zeta) = \int_{\mathscr{A}} d\mathscr{A} \, \rho \frac{\partial \zeta}{\partial t} \tag{53.37}$$

where the surface integral is over the area $\mathscr{A}$ at the base of this cylindrical region. Since the flow is horizontal and independent of the depth, as pointed out in the discussion following Eqs. (53.8), the net inward flux of fluid is evidently given by

$$-\int_A \rho \mathbf{v}_\perp \cdot d\mathbf{A} = -\rho \oint_C (h + \zeta) \, dl \, (\hat{n} \cdot \mathbf{v}_\perp) \tag{53.38}$$

Here the line integral is taken around the closed curve C encircling the base $\mathscr{A}$ in

Fig. 53.3, and $\hat{n}$ is the outward normal to C. Gauss' theorem in two dimensions allows us to rewrite the right-hand side of Eq. (53.38) as

$$\oint_C (h + \zeta)\mathbf{v}_\perp \cdot (\hat{n} \, dl) = \int_{\mathscr{A}} d\mathscr{A} \, \mathbf{V}_\perp \cdot [(h + \zeta)\mathbf{v}_\perp] \approx \int_{\mathscr{A}} d\mathscr{A} \, \mathbf{V}_\perp \cdot (h\mathbf{v}_\perp) \quad (53.39)$$

where this surface integral is also taken over the area $\mathscr{A}$ at the base of the cylindrical region in Fig. 53.3. A comparison of Eqs. (53.36) to (53.39) gives the integral form of the linearized continuity equation. If we observe that these equalities must hold for an *arbitrary* region $\mathscr{A}$, we obtain the corresponding differential form

$$\frac{\partial \zeta}{\partial t} = -\mathbf{V}_\perp \cdot (h\mathbf{v}_\perp) \quad (53.40)$$

Multiply Eq. (53.35) by h and take the two-dimensional divergence; the result can be combined with the time derivative of (53.40) to yield the generalized two-dimensional wave equation for tidal motion

$$\frac{\partial^2 \zeta}{\partial t^2} = g \, \mathbf{V}_\perp \cdot (h\mathbf{V}_\perp \, \zeta) \quad (53.41)$$

In cartesian coordinates, it has the explicit form

$$\frac{\partial}{\partial x}\left(h\frac{\partial \zeta}{\partial x}\right) + \frac{\partial}{\partial y}\left(h\frac{\partial \zeta}{\partial y}\right) = \frac{1}{g}\frac{\partial^2 \zeta}{\partial t^2} \quad (53.42)$$

If the depth of the basin is constant, this expression reduces to the more familiar two-dimensional wave equation

$$\frac{\partial^2 \zeta}{\partial x^2} + \frac{\partial^2 \zeta}{\partial y^2} = \frac{1}{c^2}\frac{\partial^2 \zeta}{\partial t^2} \quad (53.43a)$$

$$c^2 = gh \quad (53.43b)$$

In the special case that the free surface varies in only one spatial dimension $\zeta = \zeta(x, t)$, the previous results follow immediately [see Eq. (53.18)].

The boundary condition at the fixed walls clearly requires that the normal component of the velocity of the fluid vanish there

$$\hat{n} \cdot \mathbf{v} = 0 \qquad \text{at fixed wall} \quad (53.44)$$

Since this equation holds for all times, Eq. (53.35) can be used to convert this condition into a more convenient one involving the free surface

$$\hat{n} \cdot \mathbf{V}\zeta = 0 \qquad \text{at fixed wall} \quad (53.45)$$

In addition, the depth $h(x, y)$ must vary slowly ($|\mathbf{V}_\perp h| \ll 1$); otherwise, the approximation of neglecting the vertical accelerations necessarily breaks down.

Using the analysis developed in Chap. 8, we can immediately write down the solutions to the wave equation (53.43) satisfying the boundary conditions (53.45) in some special geometries. Consider a shallow rectangular tank with uniform depth h, length a, and width b. The normal-mode solutions then take the form

$$\zeta_{mn}(x, y, t) = c_{mn} \cos \frac{m\pi x}{a} \cos \frac{n\pi y}{b} \cos (\omega_{mn} t + \phi_{mn}) \qquad (53.46)$$

where c_{mn} is a normalization constant and the normal-mode frequencies are given by

$$k_{mn}^2 = \frac{\omega_{mn}^2}{c^2} = \left(\frac{m\pi}{a}\right)^2 + \left(\frac{n\pi}{b}\right)^2 \qquad \begin{matrix} m, n = 0, 1, 2, \ldots, \infty \\ (\text{not } m = n = 0) \end{matrix} \qquad (53.47)$$

Here the value $m = 0$, $n = 0$ is excluded because an incompressible fluid cannot undergo a steady uniform shift of the free surface. Note that these solutions are similar to those for the vibrating membrane (see Prob. 10.1) and that the behavior of the surface resembles a choppy sea.

A second example is a shallow circular tank with uniform depth h and radius a. In this case, the solution of Eq. (53.43) contains only Bessel functions, since the Neumann functions are singular at the origin and are therefore excluded. Thus the normal-mode solutions take the form

$$\zeta_{mn}(r, \phi, t) = c_{mn} e^{im\phi} J_m(k_{mn} r) \cos (\omega_{mn} t + \phi_{mn}) \qquad (53.48)$$

and the boundary condition (53.45) at the fixed circular wall requires

$$J_m'(k_{mn} a) = 0 \qquad (53.49)$$

In this way, the normal-mode frequencies are given by

$$k_{mn} a = \alpha_{m, n}' = n\text{th extremum of } m\text{th Bessel function} \qquad (53.50)$$

$$\frac{\omega_{mn}}{c} = k_{mn} = \frac{\alpha_{m, n}'}{a} \qquad \begin{matrix} m = 0, 1, 2, \ldots, \infty \\ n = 1, 2, \ldots, \infty \end{matrix} \qquad (53.51)$$

Although these solutions are again similar to those [Eqs. (47.43) and (47.46)] for a clamped circular membrane, the altered boundary condition (vanishing normal derivative instead of vanishing value) changes the detailed ordering and values of the wave numbers k_{mn}. In particular, the fundamental mode here has $m = n = 1$, representing a sloshing mode, because the symmetric mode $m = 0$, $n = 1$ found for a circular drumhead is here excluded (see Prob. 10.2).

54 SURFACE WAVES

The preceding section presented an approximate theory of long-wavelength small-amplitude surface waves under gravity. We now consider a more general description of two-dimensional surface waves on a three-dimensional fluid, as illustrated in Fig. 54.1. Here we place no restriction on the relative magnitudes of the height of the fluid h and the wavelength of the disturbance λ, in contrast to the tidal waves of the previous section, where it was assumed that $h \ll \lambda$. Discarding this

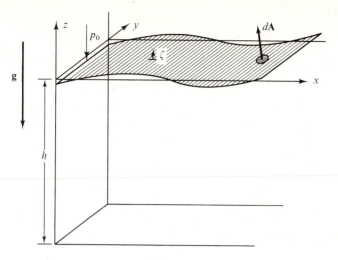

Figure 54.1 Surface waves on a fluid.

assumption complicates the mathematics but also leads to some interesting new physical phenomena. We start by formulating the more general problem.

Formulation for Arbitrary Depths

Consider an incompressible fluid, so that the exact continuity equation becomes [compare Eq. (48.29)]

$$\mathbf{V} \cdot \mathbf{v} = 0 \tag{54.1}$$

Consistent with Thomson's theorem (Sec. 48), we assume that the motion of the fluid is irrotational

$$\mathbf{V} \times \mathbf{v} = 0 \tag{54.2}$$

This relation implies that the velocity can be determined from a potential [compare Eq. (48.49)]

$$\mathbf{v} = -\mathbf{V}\bar{\Phi} \tag{54.3}$$

In terms of this potential, the continuity equation then states

$$\mathbf{V}^2\bar{\Phi} = 0 \tag{54.4}$$

Thus the velocity potential for an *incompressible* fluid undergoing irrotational motion necessarily satisfies *Laplace's equation in three dimensions* throughout the interior of the fluid at every instant.

The dynamics of this fluid follows from the basic law of hydrodynamics, which is just Newton's second law [compare Eq. (48.13)]

$$\frac{d\mathbf{v}}{dt} = \mathbf{f}_{\mathrm{app}} - \frac{1}{\rho}\mathbf{V}p \tag{54.5}$$

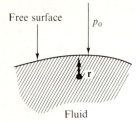

Free surface

p_0

$\mathbf{r}$

Fluid

Figure 54.2 Evaluation of Bernoulli's equation as one moves to the free surface of the fluid.

In the present case of an incompressible fluid undergoing irrotational motion subject to an applied force derivable from a potential U, Eq. (54.5) implies Bernoulli's equation (48.55)

$$-\frac{\partial \bar{\Phi}}{\partial t} + \tfrac{1}{2}v^2 + U + \frac{p}{\rho} = 0 \tag{54.6}$$

as was shown in Sec. 48. This relation must hold everywhere inside the fluid. In particular, it must be satisfied as we move from a point in the interior of the fluid to the free surface, as illustrated in Fig. 54.2. To evaluate Eq. (54.6) at this free surface, we make the following observations:

1. At the free surface, the pressure reduces to

$$p = p_0 = \text{const} \tag{54.7}$$

where p_0 is the external (atmospheric) pressure (see Fig. 54.1).

2. For small amplitudes and velocities of motion, we have

$$\mathbf{v}^2 \text{ is of second order} \tag{54.8}$$

allowing us to neglect the term $\tfrac{1}{2}\mathbf{v}^2$.

3. We assume that the applied force arises solely from a uniform gravitational field and choose to measure the potential U from the equilibrium free surface. Thus the potential at the actual instantaneous free surface of the fluid is given by

$$U = g\zeta \tag{54.9}$$

where ζ is the displacement of the surface from its equilibrium value.

4. Bernoulli's equation at the free surface now reads

$$-\frac{\partial \bar{\Phi}}{\partial t} + g\zeta + \frac{p_0}{\rho} = 0 \tag{54.10}$$

We shall make a change of gauge [compare Eq. (48.54)] and introduce a new velocity potential Φ by the relation

$$\Phi \equiv \bar{\Phi} - \frac{p_0 t}{\rho} \tag{54.11}$$

The velocity defined in Eq. (54.3) is clearly invariant under this transformation. In terms of this new velocity potential Φ, Bernoulli's equation on the free surface now takes the form

$$-\frac{\partial \Phi}{\partial t} + g\zeta = 0 \tag{54.12}$$

This equation has the desirable feature that $\partial \Phi/\partial t$ vanishes when there is no displacement of the equilibrium free surface ($\zeta = 0$). Equivalently, Eq. (54.12) can be rewritten

$$\zeta = \frac{1}{g}\frac{\partial \Phi}{\partial t} \qquad \text{at free surface} \tag{54.13a}$$

or, explicitly,
$$\zeta(x, y, t) = \frac{1}{g}\left[\frac{\partial}{\partial t}\Phi(x, y, z, t)\right]_{z=\zeta} \tag{54.13b}$$

Since Φ and ζ are assumed small, this last relation can be written

$$\zeta(x, y, t) \approx \frac{1}{g}\frac{\partial}{\partial t}\Phi(x, y, 0, t) \tag{54.14}$$

correct to first order. In this way, Bernoulli's equation provides a fundamental boundary condition at the free surface.

Equation (54.13) specifies one relation between the displacement of the surface ζ and the velocity potential Φ at the surface. To obtain an additional boundary condition between these quantities we consider the equation [compare Eq. (46.4)]

$$F(\mathbf{r}, t) = F(x, y, z, t) \equiv z - \zeta(x, y, t) = 0 \tag{54.15}$$

that specifies the instantaneous shape of the free surface. In an interval dt, the surface moves because of the fluid's *normal* component of velocity $\mathbf{v}_n = \hat{n}(\hat{n} \cdot \mathbf{v})$, where $\hat{n} = \nabla F|\nabla F|^{-1}$ is the unit vector normal to the free surface (see the discussion in Sec. 46). At a later time $t + dt$, the element of the surface at $\mathbf{r}$ has moved to $\mathbf{r} + \mathbf{v}_n\, dt$, and the displaced surface is given by

$$F(\mathbf{r} + \mathbf{v}_n\, dt, t + dt) = 0 \tag{54.16}$$

A Taylor expansion to first order in dt yields the equivalent condition

$$\left[\sum_{i=1}^{3}(v_n)_i\frac{\partial F}{\partial x_i} + \frac{\partial F}{\partial t}\right]dt = 0 \tag{54.17a}$$

or
$$\mathbf{v}_n \cdot \nabla F + \frac{\partial F}{\partial t} = 0 \tag{54.17b}$$

In addition, we notice that $\mathbf{v}_n \cdot \nabla F = \mathbf{v} \cdot \nabla F$. Equation (54.17b) can therefore be rewritten

$$\mathbf{v} \cdot \nabla F + \frac{\partial F}{\partial t} = 0 \tag{54.18}$$

or, with the aid of Eq. (54.15),

$$v_z - v_x \frac{\partial \zeta}{\partial x} - v_y \frac{\partial \zeta}{\partial y} - \frac{\partial \zeta}{\partial t} = 0 \qquad \text{at free surface} \qquad (54.19)$$

The combination on the left-hand side of (54.18) can be recognized as the hydrodynamic derivative dF/dt [see Eq. (48.12)] that gives the rate of change of F associated with a particular element of fluid at the surface. Thus Eq. (54.18) indicates that F is a constant of the motion for such a fluid element. Now the location of the free surface is characterized by $F = 0$ *at all* times, which implies that an element of fluid initially on the free surface remains there forever.

The general nonlinear condition (54.19) can be simplified considerably for small-amplitude motion, when the quadratic terms become negligible. A combination of Eqs. (54.3), (54.11), and (54.19) yields the approximate linearized free-surface boundary condition

$$v_z = -\frac{\partial \Phi}{\partial z} = \frac{\partial \zeta}{\partial t} \qquad \text{at free surface} \qquad (54.20)$$

It merely states that, for small displacements, the velocity of the fluid in the z direction must be identical with the actual velocity of the surface in that direction.

At the fixed surface, the normal component of the fluid velocity vanishes

$$\hat{n} \cdot \nabla \Phi = 0 \qquad \text{at fixed surface} \qquad (54.21)$$

just as in Eq. (49.10b). Thus the basic problem of small-amplitude surface waves on an incompressible irrotational fluid reduces to solving Laplace's equation

$$\nabla^2 \Phi = 0 \qquad (54.22)$$

subject to the boundary conditions (54.13), (54.20), and (54.21). Note that the time appears only through the free-surface conditions, in contrast to the intrinsic time dependence of the two-dimensional wave equation (46.15) for a membrane or the three-dimensional wave equation (49.9) for sound waves. A second new feature is the appearance of two qualitatively distinct functions, for the velocity potential Φ is defined throughout the three-dimensional fluid whereas the elevation ζ is defined on a two-dimensional manifold. Furthermore, Eq. (54.20) relates $\partial \zeta / \partial t$ to the spatial derivatives of Φ evaluated on the free surface, thus requiring a solution to the full three-dimensional form of Laplace's equation (54.22). Consequently, surface phenomena on fluids necessarily involve the dynamics of the underlying medium, making them inherently more complex than a surface disturbance on a membrane or a volume disturbance in a compressible fluid.

Dispersion Relation

As a specific example, we seek a solution to these equations describing a one-dimensional wave traveling with speed c on the surface of a fluid in an infinite channel with constant breadth b and depth h (Fig. 54.1). In this case, the solution is independent of the transverse variable y and depends only on the combination $x - ct$ and the depth z. Guided by our previous experience, we attempt to separate variables in the form of a harmonic wave

$$\Phi(x, z, t) = Z(z) \cos [k(x - ct)] \tag{54.23}$$

where $k \equiv 2\pi/\lambda$ is the wave number, $ck \equiv \omega$ is the angular frequency, and $Z(z)$ characterizes the velocity potential as a function of depth. Can a solution of this form satisfy the equations of motion? Substitution into Laplace's equation (54.22) gives

$$\frac{d^2Z}{dz^2} \cos [k(x - ct)] - k^2 Z(z) \cos [k(x - ct)] = 0 \tag{54.24}$$

or, upon canceling the common factor,

$$\frac{d^2Z}{dz^2} - k^2 Z = 0 \tag{54.25}$$

Thus the expression (54.23) is a solution of Laplace's equation throughout the interior of the fluid only if the function Z satisfies this ordinary differential equation. Note the sign in this relation, which differs from the one-dimensional Helmholtz equation. Correspondingly, the general solution to Eq. (54.25) is a linear combination of *real* exponentials

$$Z(z) = A e^{kz} + B e^{-kz} \tag{54.26}$$

The boundary condition (54.21) at the bottom of the channel

$$Z'(-h) = 0 \tag{54.27}$$

implies that the appropriate combination of exponentials in Eq. (54.26) must be taken as

$$Z(z) = C \cosh k(z + h) \tag{54.28}$$

where C is now a single overall constant. As in the one-dimensional solution from Sec. 53, the boundary condition (54.21) holds automatically on the walls perpendicular to the y axis (see Fig. 54.1) because the solution (54.23) is independent of y.

The remaining boundary conditions (54.13) and (54.20) can be combined to yield a single boundary condition for the velocity potential on the free surface of the fluid

$$-\frac{\partial \Phi}{\partial z} = \frac{1}{g} \frac{\partial^2 \Phi}{\partial t^2} \qquad \text{on free surface} \tag{54.29}$$

Substitution of Eqs. (54.28) and (54.23) into this relation gives

$$-kC \sinh k(\zeta + h) = -\frac{1}{g} (kc)^2 C \cosh k(\zeta + h) \tag{54.30}$$

We shall assume that

$$|k\zeta| \ll 1 \tag{54.31}$$

which merely states that the *amplitude* of the surface disturbance is small compared with the characteristic wavelength $\lambda \equiv 2\pi/k$. In this limit, the boundary condition (54.29) can be evaluated on the unperturbed surface $z = 0$ [compare the discussion of Eqs. (51.2) and (54.14)], giving the simple result

$$\frac{g}{k} \tanh kh = \frac{g\lambda}{2\pi} \tanh \frac{2\pi h}{\lambda} = c^2 \qquad (54.32)$$

This relation between the speed of propagation c and the wave number k or wavelength λ of the disturbance is known as a *dispersion relation*. It must hold if the velocity potential in Eq. (54.23) is to satisfy Laplace's equation and the appropriate dynamical boundary conditions at the free surface.

If the depth of the fluid is much less than the wavelength of the disturbance, we recover our previous result (53.19) for the propagation velocity of tidal waves

$$gh = c^2 \qquad \text{if } h \ll \lambda \qquad (kh \ll 1) \qquad (54.33)$$

In the opposite case of a deep channel, where the depth of the fluid is much greater than the wavelength of the surface wave, the dispersion relation (54.32) takes the approximate form

$$\frac{g}{k} = \frac{g\lambda}{2\pi} = c^2 \qquad \text{if } h \gg \lambda \qquad (kh \gg 1) \qquad (54.34)$$

The velocity of wave propagation now depends on the wavelength, and there is *strong dispersion*. In particular, the long wavelengths travel fastest on the surface of the fluid and the short wavelengths travel slowest. It is again evident that the velocity potential in Eq. (54.23) does not satisfy the simple wave equation, where all disturbances travel with a common speed.

The overall constant C in Eq. (54.28) can be eliminated in favor of the maximum displacement ζ_0 of the surface of the fluid through the definition [compare Eq. (54.13)]

$$\zeta_0 \equiv C \frac{kc}{g} \cosh kh \qquad (54.35)$$

A combination with Eqs. (54.28) and (54.23) gives

$$\Phi(x, z, t) = \frac{g\zeta_0}{kc} \frac{\cosh [k(z + h)]}{\cosh kh} \cos [k(x - ct)] \qquad (54.36)$$

Equation (54.13) and the approximation (54.31) then allows us to conclude that the displacement of the surface satisfies the relation

$$\zeta(x, t) = \zeta_0 \sin [k(x - ct)] \qquad (54.37)$$

This equation describes a simple harmonic wave of small amplitude ζ_0 and wavelength $\lambda = 2\pi/k$ moving to the right with speed c. Equation (54.36) demonstrates that the actual velocity of the fluid decreases exponentially toward the interior of the fluid, with the motion confined to a characteristic surface layer of thickness $\approx \lambda$. It is now evident that Eqs. (54.36) and (54.37) satisfy Eqs. (54.22), (54.21), (54.20), and (54.13) provided that the velocity and wavelength are related through the dispersion relation (54.32) and condition (54.31) holds. Thus we have obtained a particular solution that describes a wave running to the right in Fig. 54.1; for standing waves, we can superpose the running waves because we are still solving a set of linear equations.

The physical nature of this wave motion is clarified by considering the actual displacement of a small element of the fluid. This displacement must take the form

$$x - x_0 = \frac{g\zeta_0}{kc^2} \frac{\cosh [k(z + h)]}{\cosh kh} \cos [k(x - ct)] \qquad (54.38a)$$

$$z - z_0 = \frac{g\zeta_0}{kc^2} \frac{\sinh [k(z + h)]}{\cosh kh} \sin [k(x - ct)] \qquad (54.38b)$$

to ensure that the velocity components of the fluid $\mathbf{v} = d\mathbf{r}/dt$ obtained from these relations match those obtained from the velocity potential (54.36). Since the amplitude of the disturbance $|\zeta_0|$ is small, the right-hand side of Eqs. (54.38) can be evaluated at the equilibrium point (x_0, z_0) in the fluid, and these equations reduce to

$$x - x_0 = \frac{g\zeta_0}{kc^2} \frac{\cosh [k(z_0 + h)]}{\cosh kh} \cos [k(x_0 - ct)] \qquad (54.39a)$$

$$z - z_0 = \frac{g\zeta_0}{kc^2} \frac{\sinh [k(z_0 + h)]}{\cosh kh} \sin [k(x_0 - ct)] \qquad (54.39b)$$

These time-dependent relations evidently imply the time-independent orbit

$$\left(\frac{x - x_0}{a}\right)^2 + \left(\frac{z - z_0}{b}\right)^2 = 1 \qquad (54.40)$$

which describes an ellipse with semimajor and semiminor axes given by

$$a = \frac{g\zeta_0}{kc^2} \frac{\cosh [k(z_0 + h)]}{\cosh kh} \qquad (54.41a)$$

$$b = \frac{g\zeta_0}{kc^2} \frac{\sinh [k(z_0 + h)]}{\cosh kh} \qquad (54.41b)$$

Figure 54.3 illustrates this elliptical motion, in which the ratio of semiminor to semimajor axes has the simple expression

$$\frac{b}{a} = \tanh [k(z_0 + h)] \qquad (54.42)$$

For deep water $(kh \gg 1)$ away from the bottom $(|z_0| \ll h)$, this ratio reduces to $b/a \approx 1$. Thus we obtain the interesting result that in deep water the actual motion of the fluid is circular except for a thin layer near the bottom of thickness approximately equal to λ. Moreover, the radius of the circular orbit decreases exponen-

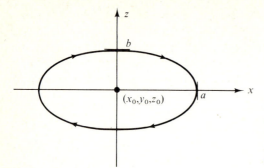

Figure 54.3 Elliptical motion of particles in interior of incompressible, irrotational fluid with a harmonic surface wave traveling in $\hat{x}$ direction (Fig. 54.1). The semimajor and semiminor axes are given in Eqs. (54.41).

tially with depth like $\exp(-2\pi|z_0|/\lambda)$. In contrast, the motion in shallow water $(kh \ll 1)$ is effectively horizontal at all depths because $b/a \ll 1$. This result confirms our discussion of tidal waves in Sec. 53.

Energy

It is instructive to consider the energy in the surface waves. The kinetic energy of irrotational motion in an incompressible fluid is most easily written in terms of the velocity potential. The relations

$$\mathbf{v} = -\nabla\Phi \tag{54.43a}$$

and

$$\nabla^2\Phi = 0 \tag{54.43b}$$

allow us to express the kinetic energy in the equivalent forms

$$T = \frac{1}{2}\int_V \rho v^2 \, dV = \frac{1}{2}\int_V \rho \, \nabla\Phi \cdot \nabla\Phi \, dV = \frac{1}{2}\int_V \rho \, \mathbf{V} \cdot (\Phi \, \nabla\Phi) \, dV \tag{54.44}$$

Gauss' theorem then transforms this last relation into a surface integral

$$T = \frac{1}{2}\int_V \rho \mathbf{V} \cdot (\Phi \, \nabla\Phi) \, dV = \frac{1}{2}\int_A \rho\Phi \, \nabla\Phi \cdot d\mathbf{A} \tag{54.45}$$

since ρ is constant. Here the surface integral is taken over the entire bounding surface of the fluid, as illustrated in Fig. 54.1, but the boundary condition (54.21) ensures that a fixed boundary makes no contribution to this expression. Thus the expression (54.45) in fact runs only over the free surface of the fluid. In this form, the kinetic energy is usually very simple, being now explicitly second order in the small quantity Φ. To leading order, the integral can therefore be taken over the equilibrium free surface.

The potential energy can be computed in the following fashion. Since the fluid is incompressible, no pressure-volume work is ever done in the interior of the fluid. Hence the potential energy is all gravitational. We shall measure the gravitational potential from some fixed horizontal reference plane that always remains in the interior of the fluid, a depth h below the equilibrium surface. The fluid lying below the plane clearly experiences no change in gravitational potential energy. To compute the gravitational potential energy of the fluid lying above the plane, consider a small column of fluid with base area dA located on the reference plane, as illustrated in Fig. 53.3. The gravitational potential energy dU of this column of fluid is given by

$$dU = \left(\int_0^{h+\zeta} g\rho z \, dz \right) dA = \tfrac{1}{2}\rho g (h + \zeta)^2 \, dA \qquad (54.46)$$

where again ζ is the height above the equilibrium free surface of the fluid. At equilibrium, the fluid would have $\zeta = 0$, and the gravitational potential energy would be

$$dU_{eq} = \tfrac{1}{2}\rho g h^2 \, dA \qquad (54.47)$$

The difference between Eqs. (54.46) and (54.47) gives the change in potential energy when the surface is displaced

$$\delta(dU) \equiv dU - dU_{eq} = (\rho g h \zeta + \tfrac{1}{2}\rho g \zeta^2) \, dA \qquad (54.48)$$

As it stands, this expression is linear in small quantities, whereas it should be quadratic for a true displacement from equilibrium. Note, however, that Eq. (54.48) can be integrated over the entire area, and the conservation of volume of the fluid guarantees that

$$\int h \, dA = \int (h + \zeta) \, dA \qquad (54.49)$$

As a result, we conclude that the *integral* of the linear term in Eq. (54.48) does indeed vanish

$$\int \zeta \, dA = 0 \qquad (54.50)$$

Thus the total change in gravitational potential energy of the fluid is given by

$$\delta U = \tfrac{1}{2} \int \rho g \zeta^2 \, dA \qquad (54.51)$$

which is now explicitly of second order in small quantities (here ζ). For this reason the integral in Eq. (54.51) can again be taken over the free equilibrium surface of the fluid. Equations (54.45) and (54.51) provide simple expressions for the total energy in the surface wave. They do not yield a lagrangian or hamiltonian formulation of the problem, however, because they are not expressed consistently in terms of a single set of generalized coordinates (but see Prob. 10.14). Nevertheless, they are readily evaluated in any specific case. For the traveling wave specified by Eqs. (54.36) and (54.37), for example, it is easily shown that the kinetic and potential energies are equal when integrated over one wavelength (see Prob. 10.11).

Group Velocity

Dispersive waves with a dispersion relation such as that in Eq. (54.32) have some novel features. Consider first a monochromatic surface wave of the type (54.37), which can be written in complex form as [compare Eqs. (45.36) and (49.13)]

$$Z(x, t) = \zeta_0 e^{ik(x - ct)} = \zeta_0 e^{i(kx - \omega t)} \tag{54.52}$$

where

$$\omega(k) \equiv kc(k) \tag{54.53}$$

for the particular wave number k. Here the physical disturbance is given by

$$\zeta = \text{Re } Z \tag{54.54}$$

and we have just adjusted the phase by 90° relative to that in (54.37). Since the expression (54.52) depends only on the variable $(x - ct)$, this wave maintains a constant form with respect to a coordinate system whose origin moves to the right with velocity c, as seen previously in Sec. 39 and illustrated in Fig. 45.2. A given crest or trough of this monochromatic wave evidently travels with a velocity c, which is known as the phase velocity v_p of the wave:

$$v_p \equiv \frac{\omega}{k} \qquad \text{phase velocity of wave} \tag{54.55}$$

In addition, Fig. 45.2 shows that a monochromatic wave has an amplitude that extends throughout all space.

This last feature is clearly unphysical, for any real wave necessarily has a finite extent. To describe such a spatially localized disturbance, we must make a *wave packet* by superposing various monochromatic waves. Thus we consider

$$Z(x, t) \equiv (2\pi)^{-1} \int_{-\infty}^{\infty} \tilde{f}(k) e^{i(kx - \omega t)} \, dk \tag{54.56}$$

Since this expression merely superposes solutions with different values of k, it still satisfies the equations of motion, provided they are linear. By changing the amplitude $\tilde{f}(k)$, we can describe various forms of spatially localized disturbance. In particular, the disturbance $\zeta(x, 0)$ at the initial time $t = 0$ is simply the real part of the Fourier transform of the amplitude $\tilde{f}(k)$

$$Z(x, 0) = (2\pi)^{-1} \int_{-\infty}^{\infty} \tilde{f}(k) e^{ikx} \, dk \tag{54.57}$$

As an elementary example of this situation, consider the Fourier amplitude

$$\tilde{f}(k) = e^{-\alpha^2(k - k_0)^2} \tag{54.58}$$

which is localized about the wave number k_0 with a width $\Delta k \approx \alpha^{-1}$, as illustrated in Fig. 54.4a. The corresponding initial spatial disturbance is obtained from the Fourier transform (54.57), and the change of variable $l = k - k_0$ transforms it to

$$Z(x, 0) = (2\pi)^{-1} e^{ik_0 x} \int_{-\infty}^{\infty} e^{ilx} e^{-\alpha^2 l^2} \, dl \tag{54.59}$$

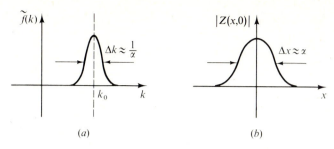

Figure 54.4 (a) Typical Fourier amplitude $\tilde{f}(k)$ [Eq. (54.58)] of width $\Delta k \approx \alpha^{-1}$ and (b) corresponding spatial modulation envelope $|Z(x, 0)|$ [Eq. (54.63)] of width $\Delta x \approx \alpha$.

This integral can be evaluated by completing the square in the exponent

$$Z(x, 0) = (2\pi)^{-1} e^{ik_0 x} \int_{-\infty}^{\infty} e^{-\alpha^2(l - ix/2\alpha^2)^2} e^{-x^2/4\alpha^2} \, dl \tag{54.60}$$

Introducing the complex variable $z \equiv l - ix/2\alpha^2$ reduces the integral to the form

$$Z(x, 0) = (2\pi)^{-1} e^{ik_0 x} e^{-x^2/4\alpha^2} \int_C e^{-\alpha^2 z^2} \, dz \tag{54.61}$$

where the contour C runs parallel to the real axis a distance $x/2\alpha^2$ below it. Since the integrand is analytic, this contour can be moved up to the real axis,† and the resulting integral immediately gives

$$\int_{-\infty}^{\infty} e^{-\alpha^2 x^2} \, dx = \frac{\sqrt{\pi}}{\alpha} \tag{54.62}$$

Thus the initial complex spatial disturbance corresponding to the simple gaussian amplitude in Eq. (54.58) is

$$Z(x, 0) = \frac{1}{2\alpha\sqrt{\pi}} e^{ik_0 x} e^{-x^2/4\alpha^2} \tag{54.63a}$$

and the associated real part becomes

$$\zeta(x, 0) = \text{Re } Z(x, 0) = \frac{1}{2\alpha\sqrt{\pi}} e^{-x^2/4\alpha^2} \cos k_0 x \tag{54.63b}$$

The plane wave $\cos k_0 x$ is now modulated with a spatial envelope proportional to $e^{-x^2/4\alpha^2}$, localized at $x \approx 0$ with a width of order $\Delta x \approx \alpha$. This situation is illustrated in Fig. 54.4b, where we have sketched the quantity $|Z(x, 0)|$. It is evident from Fig. 54.4 that a sharply localized disturbance in coordinate space requires a broad distribution in wave-number space and vice versa. More precisely, the product of the widths $\Delta k \, \Delta x$ is a constant of order unity, and this

---

† This contour deformation involves two segments that are perpendicular to the real axis; their contributions are easily shown to be exponentially small.

relation holds for quite general choices of the Fourier amplitude function $\tilde{f}(k)$ (see Prob. 10.15).

We now return to the general time-dependent wave packet (54.56). Ignoring the details of the amplitude $\tilde{f}(k)$, we merely assume that it is sufficiently well localized in wave-number space (see Fig. 54.4a) to permit a Taylor-series expansion about $k \approx k_0$ of the quantities appearing in the exponential

$$k = k_0 + (k - k_0) \tag{54.64}$$

$$\omega(k) = \omega(k_0) + (k - k_0)\left[\frac{\partial\omega(k)}{\partial k}\right]_{k_0} + \cdots \tag{54.65}$$

A change of variable to $l = k - k_0$ then reduces Eq. (54.56) to the form

$$Z(x, t) = (2\pi)^{-1}e^{i[k_0x - \omega(k_0)t]} \int_{-\infty}^{\infty} \tilde{f}(k_0 + l)e^{il(x - v_g t)}\, dl \tag{54.66}$$

where the quantity v_g is defined as

$$v_g \equiv \left[\frac{\partial\omega(k)}{\partial k}\right]_{k_0} \tag{54.67}$$

We can rewrite this expression as

$$Z(x, t) = a(x - v_g t)e^{i[k_0x - \omega(k_0)t]} \tag{54.68}$$

where $a(x)$ is the initial envelope function

$$a(x) \equiv (2\pi)^{-1} \int_{-\infty}^{\infty} \tilde{f}(k_0 + l)e^{ilx}\, dl \tag{54.69}$$

and $|a(x)| = |Z(x, 0)|$. It is clear that Eq. (54.68) is an amplitude-modulated plane wave, and the modulus of the disturbance has the character illustrated in Fig. 54.5. The envelope a depends on x and t only through the variable $x - v_g t$; just as in our previous discussion of Fig. 39.1, it therefore represents a disturbance that moves to the right with a velocity v_g, known as the *group velocity* of the wave. Evidently, v_g is the velocity at which the amplitude-modulation envelope of the wave propagates. It also represents the velocity of any true physical disturbance, for the transmission of a signal necessarily involves the construction of a spatially

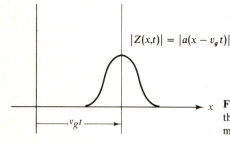

$$|Z(x,t)| = |a(x - v_g t)|$$

Figure 54.5 Spatial motion of a wave packet having the characteristics illustrated in Fig. 54.4. The displacement defines the group velocity [see Eq. (54.67)].

localized disturbance at one point and its subsequent reception at another. In the present linear approximation (54.65), the envelope propagates without change of shape (compare the discussion in Sec. 39). Higher-order corrections tend to degrade the signal, however, because they act to broaden the peak and decrease its height (see Prob. 10.16).

Numerous examples from Chaps. 7 to 9 show that solutions to the ordinary wave equation have the dispersion relation

$$\omega = kc \qquad c = \text{const} \tag{54.70}$$

where c is a *constant* that depends on the particular system. We conclude from Eqs. (54.55) and (54.67) that

$$v_p = v_g = c \qquad \text{solutions to the wave equation} \tag{54.71}$$

so that the group velocity and phase velocity are both identical to the characteristic speed c. If the dispersion relation is more complicated than Eq. (54.70), however, the group velocity and phase velocity differ. For example, the speed of a monochromatic surface wave in a deep channel satisfies Eq. (54.34), so that

$$c(k) = \left(\frac{g}{k}\right)^{1/2} = \left(\frac{g\lambda}{2\pi}\right)^{1/2} \tag{54.72a}$$

This quantity is just the phase velocity v_p. Equation (54.72a) can be rewritten in the equivalent form

$$\omega \equiv kc(k) = (gk)^{1/2} \tag{54.72b}$$

The group velocity (54.67) is then given by

$$v_g = \frac{\partial \omega}{\partial k} = \frac{1}{2}\left(\frac{g}{k}\right)^{1/2} = \frac{1}{2}\left(\frac{g\lambda}{2\pi}\right)^{1/2} \tag{54.73}$$

which is evidently half the phase velocity

$$v_g = \tfrac{1}{2}v_p = \tfrac{1}{2}c \tag{54.74}$$

We conclude from Eq. (54.73) that the velocity of propagation of a disturbance on the surface of a deep incompressible fluid increases with the square root of its wavelength. Anyone who has thrown a stone into a lake will have observed that the long-wavelength part of the disturbance propagates fastest away from the point of impact.

Inclusion of Surface Tension

The preceding analysis has omitted the important physical phenomenon of surface tension, and we now investigate its effect on surface waves. We first recall from Chap. 8 the equation of small-amplitude motion (46.14) of a two-dimensional membrane with constant surface tension τ and mass density σ

$$\tau \nabla^2 u = \sigma \frac{\partial^2 u}{\partial t^2} \tag{54.75}$$

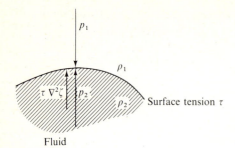

p_1

p_1

$\tau \nabla^2 \zeta$ | p_2

p_2

Surface tension τ

Fluid

Figure 54.6 Additional pressure exerted across a curved surface in the presence of surface tension.

This relation merely expresses Newton's second law for the membrane. It shows that the effect of a surface tension τ is to exert a force per unit area of magnitude $\tau \nabla^2 u$ in the direction of positive displacement

$$\frac{F_{\text{surface tension}}}{\text{Area}} = \tau \nabla^2 u \tag{54.76}$$

where the displacement of the surface is assumed small. By the definition (48.1) of pressure, Eq. (54.76) implies that the surface tension exerts an additional pressure as one crosses a curved surface of fluid. Figure 54.6 illustrates the situation, and the difference in pressure across the curved fluid surface has the analytical expression

$$p_2 + \tau \nabla^2 \zeta = p_1 \tag{54.77}$$

where ζ is the displacement of the fluid surface from its equilibrium value. Note that if the surface of the fluid is concave downward, as illustrated in Fig. 54.6, $\nabla^2 \zeta$ is negative and the pressure p_2 inside the fluid must be *larger* than the pressure p_1 outside.

Bernoulli's equation (54.6) must still hold inside the fluid

$$-\frac{\partial \bar{\Phi}}{\partial t} + \frac{p}{\rho} + gz + \tfrac{1}{2}v^2 = 0 \tag{54.78}$$

Evaluating this expression as one approaches the free surface from inside the fluid and neglecting the second-order term v^2, we find

$$-\frac{\partial \bar{\Phi}}{\partial t} + \frac{p_2}{\rho} + g\zeta = 0 \qquad \text{at free surface} \tag{54.79}$$

Equivalently, this relation expresses the pressure just inside the free surface in terms of the time derivative of the velocity potential and the displacement of the surface

$$p_2 = \rho \frac{\partial \bar{\Phi}}{\partial t} - \rho g\zeta \qquad \text{at free surface} \tag{54.80}$$

Equation (54.77) then serves to relate the pressure inside the fluid to the pressure outside the fluid, and a combination with Eq. (54.80) leads to the expression

$$p_1 - \tau \, \nabla^2 \zeta = \rho \frac{\partial \bar{\Phi}}{\partial t} - \rho g \zeta \qquad \text{at free surface} \qquad (54.81)$$

If we now assume that the pressure p_1 outside the fluid is simply *atmospheric pressure* p_0, the constant p_0 can be included in the velocity potential through the change of gauge in Eq. (54.11) and Eq. (54.81) reduces to

$$\rho \frac{\partial \Phi}{\partial t} - \rho g \zeta = -\tau \, \nabla^2 \zeta \qquad \text{at free surface} \qquad (54.82)$$

This equation can be rewritten as

$$\frac{1}{g} \frac{\partial \Phi}{\partial t} = \zeta - \frac{\tau \, \nabla^2 \zeta}{\rho g} \qquad \text{at free surface} \qquad (54.83)$$

which is the desired modification of the boundary condition (54.13) arising from the inclusion of surface tension. The problem then is to solve Laplace's equation (54.22) subject to the previous boundary conditions (54.20) and (54.21) and this new boundary condition (54.83).

For simplicity, we consider traveling waves on deep water, where $h \to \infty$ (see Prob. 10.9 for the general case of finite h). We thus seek solutions to these equations of the form

$$\Phi = \text{Re} \; \phi_0 \, e^{i(kx - \omega t)} e^{kz} \qquad (54.84a)$$

$$\zeta = \text{Re} \; \zeta_0 \, e^{i(kx - \omega t)} \qquad (54.84b)$$

in which the velocity potential decreases exponentially for negative z. The boundary condition (54.20) that the z component of the velocity of the surface be the z component of the velocity of the fluid requires that

$$\frac{\partial \zeta}{\partial t} = - \left[\frac{\partial \Phi}{\partial z} \right]_{z=0} \qquad (54.85a)$$

or, equivalently,

$$\phi_0 = \frac{i \omega \zeta_0}{k} \qquad (54.85b)$$

Equation (54.83) and assumption (54.31) then yield the condition

$$-\frac{i \omega}{g} \phi_0 = \zeta_0 + \frac{\tau k^2 \zeta_0}{\rho g} \qquad (54.86)$$

Substitution of Eq. (54.85b) into this expression and cancellation of the overall amplitude ζ_0 provide the dispersion relation that the wave must satisfy in order to solve the equations of motion

$$\omega^2 = gk + \frac{\tau k^3}{\rho} \qquad (54.87)$$

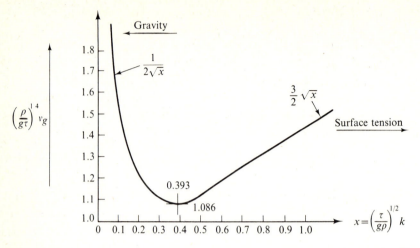

Figure 54.7 Group velocity for surface waves on fluid in presence of surface tension [Eq. (54.92b)].

This equation evidently generalizes Eq. (54.72) for deepwater waves to include the effect of surface tension. In the long-wavelength limit $(k \to 0)$, surface tension becomes unimportant, and we recover Eq. (54.72b)

$$\omega \approx (gk)^{1/2} \qquad \text{long wavelengths } (k \to 0) \tag{54.88a}$$

In the short-wavelength limit $(k \to \infty)$, however, surface tension dominates the problem, and the dispersion relation becomes

$$\omega \approx \left(\frac{\tau k^3}{\rho}\right)^{1/2} \qquad \text{short wavelengths } (k \to \infty) \tag{54.88b}$$

The group velocity is given by the general relation (54.67). In the two specific limits (54.88), v_g takes the form

$$v_g = \frac{\partial \omega}{\partial k} = \begin{cases} \dfrac{1}{2}\left(\dfrac{g}{k}\right)^{1/2} & k \to 0 \qquad\qquad (54.89a)\\[3mm] \dfrac{3}{2}\left(\dfrac{\tau k}{\rho}\right)^{1/2} & k \to \infty \qquad\qquad (54.89b) \end{cases}$$

These limiting cases are sketched in Fig. 54.7. It is clear that the group velocity becomes large in the long-wavelength limit because of gravity and in the short-wavelength limit because of surface tension. It is also evident from Fig. 54.7 that the group velocity must have a minimum. To determine this minimum value, we introduce the dimensionless variable x through the definition

$$k \equiv \left(\frac{g\rho}{\tau}\right)^{1/2} x \tag{54.90}$$

in which case Eq. (54.87) takes the form

$$\omega(x) = \left(\frac{g^3\rho}{\tau}\right)^{1/4}(x + x^3)^{1/2} \tag{54.91}$$

The general expression for the group velocity (sketched in Fig. 54.7) follows from

$$v_g = \frac{\partial\omega}{\partial k} = \left(\frac{g\rho}{\tau}\right)^{-1/2}\frac{\partial\omega}{\partial x} \tag{54.92a}$$

or

$$v_g = \left(\frac{g\tau}{\rho}\right)^{1/4}\frac{1 + 3x^2}{2(x + x^3)^{1/2}} \tag{54.92b}$$

The position of the minimum in the group velocity is obtained by minimizing this expression with respect to x. Elementary manipulations yield

$$x^2_{min} = \frac{2}{\sqrt{3}} - 1 \approx 0.1547 \tag{54.93}$$

which gives the corresponding wave number from Eq. (54.90)

$$k_{min} \approx 0.393\left(\frac{g\rho}{\tau}\right)^{1/2} \tag{54.94}$$

The minimum group velocity is obtained by substituting (54.93) into (54.92b)

$$(v_g)_{min} \approx 1.086\left(\frac{g\tau}{\rho}\right)^{1/4} \tag{54.95}$$

Recall from our discussion of group velocity that (54.95) is the *minimum velocity* with which a signal can propagate on the surface of the fluid.

As an example, take the case of deep *water* in a terrestrial laboratory where the parameters appearing in (54.87) have the following values

$$\rho = 1 \text{ g cm}^{-3} \qquad g = 980 \text{ cm s}^{-2} \qquad \tau = 74 \text{ ergs cm}^{-2} \tag{54.96}$$

This yields the numerical quantities

$$k_{min} = 1.43 \text{ cm}^{-1} \qquad \lambda_{min} = 4.39 \text{ cm} \qquad (v_g)_{min} = 17.8 \text{ cm s}^{-1} \tag{54.97}$$

Thus no signal on such a water surface can propagate *more slowly* than 17.8 cm s^{-1}, and the wavelength of the signal corresponding to this minimum velocity is 4.39 cm.

55 INITIAL-VALUE PROBLEM

In Sec. 39 and Prob. 9.7, we studied d'Alembert's solution to the initial-value problem for the wave equation in one, two, and three dimensions. Despite the long-time tail in two dimensions, the absence of dispersion in Eq. (54.70) implies that an initial pulse gives rise to a signal that propagates outward at speed c. In this section, we now consider the corresponding behavior of an initial surface configuration on an incompressible, irrotational fluid in a uniform gravitational field. This problem illustrates the complications introduced by dispersion and serves to develop an extremely powerful mathematical technique, the method of stationary phase. To simplify the discussion, we treat only the case of one-dimensional surface waves on deep water, where the dispersion relation is given by Eq. (54.72b).

Surface Waves on Deep Water

We first review the basic equations of motion:

1. The fluid is incompressible and irrotational, so that Laplace's equation must hold throughout the fluid

$$\nabla^2 \Phi = 0 \qquad (55.1)$$

2. On any fixed boundaries of the fluid, the normal component of the velocity must vanish

$$\hat{n} \cdot \nabla \Phi = 0 \qquad \text{on fixed boundaries} \qquad (55.2)$$

3. For small displacements of the surface, the velocity of the surface must equal the z component of the velocity of the fluid

$$-\frac{\partial \Phi}{\partial z} = \frac{\partial \zeta}{\partial t} \qquad \text{on free surface} \qquad (55.3)$$

4. Newton's second law, or Bernoulli's equation, becomes the free-surface boundary condition

$$\zeta = \frac{1}{g}\frac{\partial \Phi}{\partial t} \qquad \text{on free surface} \qquad (55.4)$$

Each of these coupled, linear differential equations is at most first order in the time. As a result, the values of Φ and ζ at a given instant suffice to determine the solution at all subsequent times. Note, however, that the initial value of Φ must be specified *throughout* the fluid and is not completely arbitrary, for it must satisfy Laplace's equation (55.1) and the boundary condition (55.2) on the fixed surfaces.

In the present case of surface waves on deep water ($h \gg \lambda$ or $kh \to \infty$) the dispersion relation (54.72b) takes the form

$$\omega^2 = g|k| \qquad (55.5)$$

where the absolute-value symbol now allows for waves moving to the left as well as to the right. For deepwater waves ($|k|h \to \infty$), the monochromatic traveling-wave solutions (54.36) and (54.37) can be written (in complex form) as

$$\zeta = \text{Re } Z = \text{Re } (-i\zeta_0 e^{i(kx - \omega t)}) \qquad (55.6)$$

$$\Phi = \text{Re } \phi_0 e^{|k|z} e^{i(kx - \omega t)} \qquad (55.7)$$

Since Eq. (55.7) satisfies Eqs. (55.1) and (55.2) throughout the region $z < 0$, any linear superposition of Eqs. (55.6) and (55.7) for different k again satisfies the same equations of motion. Thus let us try to construct a general solution to the problem of the form

$$\zeta(x, t) = \text{Re } Z(x, t) = \text{Re } (2\pi)^{-1} \int_{-\infty}^{\infty} \tilde{f}(k) e^{i(kx - \omega t)} \, dk \qquad (55.8)$$

$$\Phi(x, z, t) = \text{Re } (2\pi)^{-1} \int_{-\infty}^{\infty} \tilde{g}(k) e^{|k|z} e^{i(kx - \omega t)} \, dk \qquad (55.9)$$

where the functions $\tilde{f}(k)$ and $\tilde{g}(k)$ remain to be determined. Evidently, (55.9) satisfies Laplace's equation (55.1) throughout the fluid and the boundary condition (55.2) on the bottom of the tank because (55.9) vanishes exponentially as $z \to -\infty$. Equations (55.8) and (55.9) also satisfy the boundary condition (55.3) provided that

$$-|k| \tilde{g}(k) = -i\omega \tilde{f}(k) \qquad (55.10)$$

where the displacement of the fluid's surface is assumed small compared with the wavelength of the disturbance

$$|k\zeta| \ll 1 \qquad (55.11)$$

In this same limit, the remaining equation of motion (55.4) imposes the following condition on Eqs. (55.8) and (55.9):

$$\tilde{f}(k) = -\frac{i\omega}{g} \tilde{g}(k) \qquad (55.12)$$

Equations (55.10) and (55.12) are compatible only if the dispersion relation (55.5) is satisfied, in which case they reduce to the same result

$$\tilde{g}(k) = i\left(\frac{g}{|k|}\right)^{1/2} \tilde{f}(k) \qquad (55.13)$$

Thus Eqs. (55.8) and (55.9) take the form

$$\zeta(x, t) = \text{Re } Z(x, t) = \text{Re } (2\pi)^{-1} \int_{-\infty}^{\infty} \tilde{f}(k) e^{i(kx - \omega t)} \, dk \qquad (55.14)$$

$$\Phi(x, z, t) = \text{Re } (2\pi)^{-1} \int_{-\infty}^{\infty} i\left(\frac{g}{|k|}\right)^{1/2} \tilde{f}(k) e^{|k|z} e^{i(kx - \omega t)} \, dk \qquad (55.15)$$

For *any* reasonable Fourier amplitude $\tilde{f}(k)$, these equations provide a solution to the problem posed in Eqs. (55.1) to (55.4) for deepwater waves. The reader can indeed directly verify that Eqs. (55.14) and (55.15) satisfy Eqs. (55.1) to (55.4) if the dispersion relation (55.5) is satisfied as well as condition (55.11).

The integrands of Eqs. (55.14) and (55.15) have similar structures, and we shall concentrate on the simpler and more readily observable quantity $\zeta(x, t)$. It is convenient to make explicit the real part as follows:

$$\zeta(x, t) = \tfrac{1}{2}(Z + Z^*)$$
$$= \tfrac{1}{2}(2\pi)^{-1} \int_{-\infty}^{\infty} [\tilde{f}(k) e^{i(kx - \omega t)} + \tilde{f}^*(k) e^{-i(kx - \omega t)}] \, dk \qquad (55.16)$$

If we change variables $k \rightarrow -k$ in the second integral and note that the dispersion relation (55.5) depends only on the magnitude of k, Eq. (55.16) can be rewritten in the explicitly real form

$$\zeta(x, t) = \tfrac{1}{2}(2\pi)^{-1} \int_{-\infty}^{\infty} e^{ikx}[\tilde{f}(k)e^{-i\omega t} + \tilde{f}*(-k)e^{i\omega t}] \, dk \qquad (55.17)$$

Consider now the *initial value* of the surface displacement at $t = 0$

$$\zeta(x, 0) = \tfrac{1}{2}(2\pi)^{-1} \int_{-\infty}^{\infty} e^{ikx}[\tilde{f}(k) + \tilde{f}*(-k)] \, dk \qquad (55.18)$$

This simple one-dimensional Fourier transform can be inverted to give

$$\tfrac{1}{2}[\tilde{f}(k) + \tilde{f}*(-k)] = \int_{-\infty}^{\infty} d\xi \, e^{-ik\xi}\zeta(\xi, 0) \qquad (55.19)$$

Similarly, the initial value of the time derivative of the surface displacement follows from Eq. (55.17) according to

$$\dot{\zeta}(x, 0) = \tfrac{1}{2}(2\pi)^{-1} \int_{-\infty}^{\infty} e^{ikx}[-i\omega\tilde{f}(k) + i\omega\tilde{f}*(-k)] \, dk \qquad (55.20)$$

A Fourier inversion gives

$$-\tfrac{1}{2}i\omega[\tilde{f}(k) - \tilde{f}*(-k)] = \int_{-\infty}^{\infty} d\xi \, e^{-ik\xi}\dot{\zeta}(\xi, 0) \qquad (55.21)$$

Equations (55.19) and (55.21) can be solved to yield

$$\tilde{f}(k) = \int_{-\infty}^{\infty} d\xi \, e^{-ik\xi}\left[\zeta(\xi, 0) + \frac{i}{\omega} \dot{\zeta}(\xi, 0)\right] \qquad (55.22a)$$

$$\tilde{f}*(-k) = \int_{-\infty}^{\infty} d\xi \, e^{-ik\xi}\left[\zeta(\xi, 0) - \frac{i}{\omega} \dot{\zeta}(\xi, 0)\right] \qquad (55.22b)$$

Note that (55.22b) in fact follows from (55.22a) because $\zeta(x, 0)$ and $\dot{\zeta}(x, 0)$ must be real. The essential point here is that a complete determination of $\tilde{f}(k)$ requires both $\zeta(x, 0)$ and $\dot{\zeta}(x, 0)$. Conversely, once we have specified the initial conditions $\zeta(x, 0)$ and $\dot{\zeta}(x, 0)$, Eq. (55.22a) uniquely fixes the function $\tilde{f}(k)$.

At first sight, it seems to require a formidable sequence of steps to solve the initial-value problem for a one-dimensional small-amplitude surface disturbance. To clarify the logic, we summarize as follows. By construction, Eq. (55.14) with $\tilde{f}(k)$ given by (55.22) satisfies the basic equations (55.3) and (55.4) for all time $t \geq 0$. At the initial time $t = 0$, it reproduces both the initial surface displacement $\zeta(x, 0)$ and the initial surface velocity $\dot{\zeta}(x, 0)$. In addition, Eq. (55.15) satisfies Laplace's equation (55.1) and the boundary condition (55.2) at all times including $t = 0$. By the discussion given below Eq. (55.4), we have therefore constructed the general solution to the equations of small-amplitude motion arising from an arbitrary initial configuration (displacement and velocity) of the surface. Conversely, the initial configuration of the surface suffices to determine the solution for all subsequent time. Note, however, that the initial conditions cannot contain arbitrarily high Fourier components, for otherwise Eq. (55.11) fails to hold for any fixed small amplitude.

Substitution of Eq. (55.22) into (55.17) yields the expression

$$\zeta(x, t) = \int_{-\infty}^{\infty} d\xi \left[\int_{-\infty}^{\infty} \frac{dk}{2\pi} e^{ik(x-\xi)} \tfrac{1}{2}(e^{i\omega t} + e^{-i\omega t}) \right] \zeta(\xi, 0)$$

$$+ \int_{-\infty}^{\infty} d\xi \left[\int_{-\infty}^{\infty} \frac{dk}{2\pi} e^{ik(x-\xi)} \frac{1}{2i\omega} (e^{i\omega t} - e^{-i\omega t}) \right] \dot{\zeta}(\xi, 0) \qquad (55.23)$$

It is evidently appropriate to introduce the following Green's function†

$$\mathscr{G}(x, t) \equiv \int_{-\infty}^{\infty} \frac{dk}{2\pi} e^{ikx} \frac{\sin \omega t}{\omega} \qquad (55.24a)$$

and its time derivative

$$\frac{\partial}{\partial t} \mathscr{G}(x, t) \equiv \dot{\mathscr{G}}(x, t) = \int_{-\infty}^{\infty} \frac{dk}{2\pi} e^{ikx} \cos \omega t \qquad (55.24b)$$

where ω is the known function of k from Eq. (55.5). Equation (55.23) can then be rewritten compactly as

$$\zeta(x, t) = \int_{-\infty}^{\infty} d\xi \, \dot{\mathscr{G}}(x - \xi, t)\zeta(\xi, 0) + \int_{-\infty}^{\infty} d\xi \, \mathscr{G}(x - \xi, t)\dot{\zeta}(\xi, 0) \qquad (55.25)$$

This extremely valuable relation determines the displacement of the surface at all subsequent time arising from an initial surface displacement $\zeta(x, 0)$ and surface velocity $\dot{\zeta}(x, 0)$ at the initial time $t = 0$.

An explicit evaluation of the Green's function (55.24) is tedious, even for the relatively simple dispersion relation (55.5). It is, however, possible to extract the asymptotic form for very large times $t \to \infty$, using the mathematical approximation method of *stationary phase*. We proceed to discuss this technique.

Method of Stationary Phase

Consider an integral of the form

$$I(t) = (2\pi)^{-1} \int_a^b dk \, g(k) e^{it\, h(k)} \qquad (55.26)$$

where the functions $g(k)$ and $h(k)$ are assumed real and well behaved in the closed interval $a \le k \le b$. We seek an asymptotic evaluation of this integral in the limit $t \to \infty$, in which case the argument of the exponential varies rapidly and the

---

† Note that $\mathscr{G}$ and $\dot{\mathscr{G}}$ differ from our previous Green's functions in that they do not satisfy inhomogeneous differential equations. Instead, they become singular as $t \to 0$, reflecting their interpretation as the response to a localized initial disturbance.

oscillations of the exponential damp the integral to zero. There are two distinct cases of interest, depending on the derivative of $h(k)$. We first assume that the derivative of $h(k)$ does not vanish in the region of integration, i.e.,

$$h'(k) \neq 0 \quad \text{in the interval } a \leq k \leq b \tag{55.27}$$

In this case, Eq. (55.26) can be integrated by parts as follows

$$du \equiv e^{it\,h(k)}h'(k)\,dk \qquad u = \frac{e^{it\,h(k)}}{it} \tag{55.28a}$$

$$v \equiv \frac{g(k)}{h'(k)} \qquad dv = \left| \frac{g'(k)}{h'(k)} - \frac{g(k)h''(k)}{[h'(k)]^2} \right| dk \tag{55.28b}$$

Note that condition (55.27) is essential in writing (55.28). Partial integration of Eq. (55.26) thus yields

$$I(t) = \frac{1}{t} \left[\frac{g(k)}{2\pi i h'(k)} e^{it\,h(k)} \right]_a^b - \frac{1}{t} \int_a^b \left| \frac{g'(k)}{h'(k)} - \frac{g(k)h''(k)}{[h'(k)]^2} \right| e^{it\,h(k)} \frac{dk}{2\pi i} \tag{55.29}$$

Since all the contributions in this expression are now *explicitly of order* t^{-1}, we conclude that

If $h'(k) \neq 0$ in the interval $[a, b]$, then

$$I(t) \sim O(t^{-1}) \qquad \text{as } t \to \infty \tag{55.30}$$

As shown below, the dominant contributions are typically of order $t^{-1/2}$, so that Eq. (55.30) will generally permit us to neglect contributions to the integral I from any region of integration where $h'(k) \neq 0$.

Let us now assume that the derivative of h vanishes somewhere in the region of integration

$$h'(k_0) = 0 \qquad \text{in interval } a < k_0 < b \tag{55.31}$$

We refer to the point k_0 as a point of *stationary phase* since the argument of the exponential does not change for k near k_0. A Taylor-series expansion of the function $h(k)$ about this point of stationary phase yields

$$h(k) = h(k_0) + (k - k_0)h'(k_0) + \tfrac{1}{2}(k - k_0)^2 h''(k_0) + \cdots$$
$$= h(k_0) + \tfrac{1}{2}(k - k_0)^2 h''(k_0) + \cdots \tag{55.32}$$

The change of variable $l \equiv k - k_0$ now reduces the integral I to the form

$$I(t) = (2\pi)^{-1} \int_{-\epsilon}^{\epsilon} g(k_0 + l) \, \exp\{it[h(k_0) + \tfrac{1}{2}l^2 h''(k_0)]\} \, dl + O(t^{-1}) \tag{55.33}$$

Here we have isolated a small region of *fixed* width 2ϵ about the point of stationary phase k_0 because the above arguments show that the remaining regions where $h'(k) \neq 0$ can only produce terms of order t^{-1} as $t \to \infty$. In contrast, the contribution of the stationary-phase point will turn out to be of order $t^{-1/2}$, which dominates for large t.

It is convenient to introduce the new variable $p = t^{1/2}l$, in which case Eq. (55.33) becomes

$$I(t) = \frac{e^{it\,h(k_0)}}{t^{1/2}} \int_{-t^{1/2}\epsilon}^{t^{1/2}\epsilon} g(k_0 + t^{-1/2}p) e^{(1/2)ip^2\,h''(k_0)}\,\frac{dp}{2\pi} + O(t^{-1})$$

$$\sim \frac{e^{it\,h(k_0)}g(k_0)}{t^{1/2}} \int_{-\infty}^{\infty} e^{(1/2)ip^2\,h''(k_0)}\,\frac{dp}{2\pi} + O(t^{-1}) \qquad t \to \infty \qquad (55.34)$$

This argument evidently assumes that

$$h''(k_0) \neq 0 \qquad (55.35)$$

Otherwise it is necessary to include higher terms in the Taylor series (55.32). A simple change of variable casts the remaining integral in (55.34) in the form of a Fresnel integral, which can be evaluated with contour methods (see Prob. A.11)

$$\int_{-\infty}^{\infty} dz\, e^{\pm iz^2} = \sqrt{\pi}\, e^{\pm i\pi/4} \qquad (55.36)$$

A combination of Eqs. (55.34) and (55.36) yields

$$I(t) \sim \frac{\sqrt{\pi}}{2\pi} \frac{g(k_0)e^{it\,h(k_0)}}{[\frac{1}{2}t\,|h''(k_0)|]^{1/2}}\, e^{\pm i\pi/4} + O(t^{-1}) \qquad h''(k_0) \gtrless 0 \qquad (55.37a)$$

or, upon simplification,

$$I(t) \sim \frac{1}{t^{1/2}} \frac{g(k_0)e^{it\,h(k_0)}}{[2\pi\,|h''(k_0)|]^{1/2}}\, e^{\pm i\pi/4} + O(t^{-1}) \qquad (55.37b)$$

where $\pm$ is the sign of $h''(k_0)$. This asymptotic result for $t \to \infty$ has the crucial feature that its leading term (of order $t^{-1/2}$) comes from the immediate vicinity of the stationary-phase point and that all other contributions are of order t^{-1}. Equation (55.37) considers only a single stationary-phase point; if there are several stationary-phase points, each one makes an additive contribution [see, for example, Eq. (D3.28)].

Application to Surface Waves

This approximation method readily determines the asymptotic form of the Green's function (55.24a) in the large-t limit. We must evaluate

$$\mathcal{G}(x, t) = \int_{-\infty}^{\infty} \frac{dk}{2\pi} e^{ikx}\, \frac{e^{i\omega t} - e^{-i\omega t}}{2i\omega} \qquad (55.38)$$

where ω satisfies the dispersion relation

$$\omega = (|k|g)^{1/2} \qquad (55.39)$$

First split the integral into two parts

$$\mathcal{G}(x, t) = \frac{1}{4\pi i}\left[\int_0^\infty dk\ e^{ikx}\frac{e^{i(kg)^{1/2}t} - e^{-i(kg)^{1/2}t}}{(kg)^{1/2}}\right.$$
$$\left. + \int_{-\infty}^0 dk\ e^{ikx}\frac{e^{i(-kg)^{1/2}t} - e^{-i(-kg)^{1/2}t}}{(-kg)^{1/2}}\right] \qquad (55.40a)$$

and change variables $k \to -k$ in the second integral to obtain

$$\mathcal{G}(x, t) = \frac{1}{4\pi i}\int_0^\infty dk\left[\frac{e^{ikx}}{(kg)^{1/2}}\left(e^{i(kg)^{1/2}t} - e^{-i(kg)^{1/2}t}\right)\right.$$
$$\left. + \frac{e^{-ikx}}{(kg)^{1/2}}\left(e^{i(kg)^{1/2}t} - e^{-i(kg)^{1/2}t}\right)\right] \qquad (55.40b)$$

This expression can be rewritten as the sum of four integrals

$$\mathcal{G}(x, t) = \frac{1}{4\pi i}\int_0^\infty \frac{dk}{(kg)^{1/2}}\left(e^{it[(kg)^{1/2} + k\xi]} - e^{-it[(kg)^{1/2} - k\xi]}\right.$$
$$\left. + e^{it[(kg)^{1/2} - k\xi]} - e^{-it[(kg)^{1/2} + k\xi]}\right)$$
$$\equiv I_1 + I_2 + I_3 + I_4 \qquad (55.40c)$$

where we have introduced the definition

$$\xi \equiv \frac{x}{t} \qquad (55.40d)$$

We now seek the asymptotic form of (55.40c) as $t \to \infty$, keeping the parameter ξ defined in (55.40d) fixed. Our stationary-phase methods then apply directly. Note that holding ξ fixed implies that the variable x tends to infinity in a fixed ratio to the variable t. It is sufficient to evaluate (55.40) for $x > 0$, or equivalently $\xi > 0$, because the contribution for negative x can be obtained from the relation

$$\mathcal{G}(-x, t) = \mathcal{G}^*(x, t) \qquad (55.41)$$

which follows from Eq. (55.24a). For $\xi > 0$, only the second and third integrals in (55.40c) have stationary-phase points. If we define

$$h_3(k) \equiv (kg)^{1/2} - k\xi \qquad (55.42)$$

the condition of stationary phase requires

$$h_3'(k_0) = \frac{1}{2}\left(\frac{g}{k_0}\right)^{1/2} - \xi = 0 \qquad (55.43)$$

whose solution gives the stationary phase point k_0 according to

$$k_0 = \frac{g}{4\xi^2} \qquad (55.44)$$

The second derivative of $h_3(k)$ is

$$h_3''(k) = -\frac{1}{4}\left(\frac{g}{k^3}\right)^{1/2} \qquad (55.45)$$

and the values of h and h'' at the stationary-phase point (55.44) yield

$$h_3(k_0) = \frac{g}{4\xi} \qquad (55.46)$$

$$h_3''(k_0) = -\frac{2\xi^3}{g} \qquad (55.47)$$

The asymptotic formula (55.37b) now provides an approximate evaluation of the third integral in Eq. (55.40c)

$$I_3 \sim \frac{1}{t^{1/2}}\frac{1}{2i(k_0 g)^{1/2}}\frac{e^{it(g/4\xi)}}{[2\pi(2\xi^3/g)]^{1/2}}e^{-i\pi/4} \qquad (55.48)$$

To evaluate the second integral in (55.40c), the appropriate phase function is just the negative of the above

$$h_2(k) = -h_3(k) \qquad (55.49)$$

Thus the second integral in (55.40c) is immediately evaluated to yield

$$I_2 \sim -\frac{1}{t^{1/2}}\frac{1}{2i(k_0 g)^{1/2}}\frac{e^{-it(g/4\xi)}}{[2\pi(2\xi^3/g)]^{1/2}}e^{i\pi/4} \qquad (55.50)$$

Since the first and fourth integrals in (55.40c) contain no stationary-phase points, they must be of order t^{-1} by our general arguments

$$I_1, I_4 = O(t^{-1}) \qquad (55.51)$$

These results can be combined with the aid of Eq. (55.44) to yield the asymptotic form of the Green's function (55.40c)

$$\mathcal{G}(x, t) \sim \frac{1}{2i(tg\pi\xi)^{1/2}}e^{i[tg/4\xi - \pi/4]} - \frac{1}{2i(tg\pi\xi)^{1/2}}e^{-i(tg/4\xi - \pi/4)} \qquad \begin{array}{l} t \to \infty \\ \xi = x/t \text{ fixed} \end{array} \qquad (55.52)$$

or, equivalently,

$$\mathcal{G}(x, t) \sim \frac{1}{(gt\pi\xi)^{1/2}}\sin\left(\frac{tg}{4\xi} - \tfrac{1}{4}\pi\right) \qquad \begin{array}{l} t \to \infty \\ \xi = x/t \text{ fixed} \end{array} \qquad (55.53)$$

This is the desired relation. It can be rewritten in terms of the more familiar variables x and t by invoking (55.40d)

$$\mathcal{G}(x, t) = \frac{1}{(g\pi x)^{1/2}}\sin\left(\frac{gt^2}{4x} - \tfrac{1}{4}\pi\right) \qquad (55.54a)$$

$$\begin{array}{l} t \to \infty \\ x/t \text{ fixed} \end{array}$$

$$\dot{\mathcal{G}}(x, t) = \left(\frac{gt^2}{4\pi x^3}\right)^{1/2}\cos\left(\frac{gt^2}{4x} - \tfrac{1}{4}\pi\right) \qquad (55.54b)$$

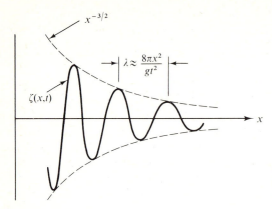

Figure 55.1 Behavior for large (fixed) t and large x of a surface disturbance that is initially $\zeta(x, 0) \approx \delta(x)$ with $\dot{\zeta}(x, 0) = 0$.

The second line, which is simply the Green's function of Eq. (55.24b), is obtained by taking the partial time derivative of the first relation. A combination with Eq. (55.25) determines how an initial surface disturbance propagates for large times.

Let us discuss these results. Assume for simplicity that there is only an initial localized displacement of the surface $\zeta(x, 0) = \delta(x)$ and that the initial velocity of the surface vanishes $\dot{\zeta}(x, 0) = 0$. In this case, only the second Green's function (55.54b) is relevant to Eq. (55.25). Imagine taking a snapshot at fixed large t. How does the disturbance behave as a function of position x far from the initial disturbance? It is evident from Eq. (55.54b) that the amplitude of the disturbance decreases with an envelope proportional to $x^{-3/2}$, as sketched in Fig. 55.1. In the vicinity of a point x, the wavelength λ of one oscillation of the cosine in (55.54b) is obtained from the relation

$$-\frac{gt^2}{4(x + \lambda)} + \frac{gt^2}{4x} \approx 2\pi \tag{55.55a}$$

For large x, this approximate relation becomes

$$\frac{gt^2}{4x^2}\lambda \approx 2\pi \tag{55.55b}$$

giving

$$\lambda \approx \frac{8\pi x^2}{gt^2} \qquad \begin{array}{l} t \text{ fixed} \\ x \to \infty \end{array} \tag{55.55c}$$

Thus the wavelength of the disturbance at fixed t increases like x^2. At a given time, the greater the distance from the initial impact the longer the wavelength of the local disturbance. This result is consistent with our previous observation that the long-wavelength surface waves travel the fastest. More precisely, the group velocity of the deepwater waves was shown in Eq. (54.73) to be

$$v_g = \frac{1}{2}\left(\frac{g}{k}\right)^{1/2} = \frac{1}{2}\left(\frac{g\lambda}{2\pi}\right)^{1/2} \tag{55.56}$$

The distance x that a long-wavelength disturbance has traveled in the time t is given by

$$x \approx v_g t = \tfrac{1}{2}t\left(\frac{g\lambda}{2\pi}\right)^{1/2} \tag{55.57}$$

Correspondingly, the approximate wavelength of the disturbance observed after a large elapsed time t and at a point x far from the initial source is given by

$$\lambda \approx \frac{8\pi x^2}{gt^2} \tag{55.58}$$

in agreement with Eq. (55.55c).

It is also interesting to consider the time-dependent signal at a fixed point x far from the initial disturbance. At time t, the approximate period τ of the disturbance can be obtained from (55.54b) through the relation

$$\frac{g(t+\tau)^2}{4x} - \frac{gt^2}{4x} \approx 2\pi \tag{55.59a}$$

For large t, this equation gives the approximate expression

$$\tau \approx \frac{4\pi x}{gt} \qquad \begin{matrix} x \text{ fixed} \\ t \to \infty \end{matrix} \tag{55.59b}$$

The local period thus *decreases* inversely with increasing t, again consistent with the picture that the long-wavelength, low-frequency components travel fastest. The local oscillations far from the initial disturbance start slowly with small amplitude, but their frequency and amplitude [see Eq. (55.54b)] both increase linearly with time.

The preceding analysis has considered only a one-dimensional disturbance, but the same basic approach also applies to two-dimensional surface waves. In particular, a straightforward generalization describes the outgoing rings caused by a stone dropped in a deep pool (see Prob. 10.18). More interesting, but also more intricate, is the surface pattern on deep water generated by a moving source like a boat. A lengthy analysis (due to Kelvin) shows the decisive role of the dispersion relation (55.5). In particular, the wave pattern is confined within an angle arccos $(\tfrac{2}{3}\sqrt{2}) \approx 19.5°$ of the direction of the boat's motion, *independent of its forward velocity*.† This last conclusion again emphasizes the difference between the dispersive surface waves and the nondispersive solutions to the wave equation; in the latter situation, an object moving with a velocity $u > c$ generates a shock wave with an opening angle given by the Mach relation arcsin (c/u) (see Prob. 9.13).

56 SOLITARY WAVES

In Sec. 53 we discussed tidal waves, or "long waves in shallow water" where the condition $kh \ll 1$ held. In that case, any surface displacement (as illustrated in

† See, for example, Sommerfeld (1950), sec. 28.

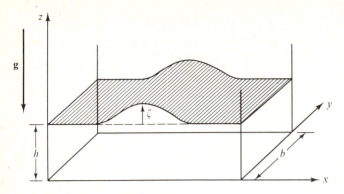

Figure 56.1 Solitary wave in a channel.

Fig. 53.1, for example) satisfies a simple linear wave equation and travels without distortion at a velocity $c = (gh)^{1/2}$. General surface waves on an incompressible irrotational fluid (Sec. 54) are more complicated because their dynamics cannot be expressed as a simple wave equation for the surface displacement. In the linearized approximation, however, simple monochromatic waves can still propagate along the surface, although the strong dispersion rapidly distorts a traveling wave packet. In this section, we first extend the discussion of tidal waves by keeping the first (dispersive) correction terms in the small parameter kh. This procedure will include some of the effects obtained for surface waves on a deeper channel, but it still leads to *a linear effective wave equation*. We then further generalize the analysis to include the leading *nonlinear* contribution to this effective wave equation. Finally, the *extended, nonlinear wave equation* is shown to have a remarkable solution consisting of a single solitary wave of a special shape that propagates along a channel *without distortion*, as illustrated in Fig. 56.1.†

We first recall the general equations that describe an incompressible fluid undergoing irrotational motion.

1. The velocity potential satisfies Laplace's equation throughout the fluid

$$\nabla^2\Phi = 0 \qquad (56.1)$$

2. Bernoulli's equation evaluated at the free surface becomes

$$-\frac{\partial\Phi}{\partial t} + \tfrac{1}{2}(\nabla\Phi)^2 + g\zeta = 0 \qquad \text{on free surface} \qquad (56.2)$$

where we have now explicitly written $\mathbf{v} = -\nabla\Phi$.

3. On any fixed boundary, the condition

$$\hat{n} \cdot \nabla\Phi = 0 \qquad \text{on fixed boundary} \qquad (56.3)$$

must hold.

---

† Such solitary waves on narrow channels were first reported by Russell [1]. Modern discussions can be found in Whitham [2], secs. 13.11–13.12, and in the review article by Scott, Chu, and McLaughlin [3].

4. The free surface of the fluid is determined by the equation

$$z - \zeta(x, y, t) = 0 \tag{56.4}$$

As shown in Sec. 54, this relation implies that a given element of fluid remains at the surface forever. Differentiation with respect to time then yields [see Eq. (54.19)]

$$v_z - \left(\frac{\partial \zeta}{\partial x} v_x + \frac{\partial \zeta}{\partial y} v_y + \frac{\partial \zeta}{\partial t} \right) = 0 \tag{56.5}$$

or, in terms of the velocity potential,

$$-\frac{\partial \Phi}{\partial z} + \nabla\Phi \cdot \nabla\zeta = \frac{\partial \zeta}{\partial t} \qquad \text{on free surface} \tag{56.6}$$

Equations (56.1) to (56.6) are the general nonlinear equations for an incompressible, irrotational fluid.

Extended Equation for Tidal Waves

Consider a fluid in a channel of constant depth h confined by rigid parallel walls a distance b apart and placed in a uniform gravitational field $\mathbf{g}$ (Fig. 56.1). A one-dimensional wave propagates in the x direction with the velocity potential $\Phi(x, z, t)$. This velocity potential automatically satisfies the boundary condition (56.3) on the parallel walls at $y = 0$ and $y = b$, since Φ is independent of y. The results from Eqs. (54.36) and (55.9) provide a complex velocity potential that satisfies both Laplace's equation (56.1) and the boundary condition (56.3) at the bottom of the tank

$$\Phi(x, z, t) = (2\pi)^{-1} \int_{-\infty}^{\infty} \cosh\left[k(z + h)\right] e^{ikx} \tilde{f}(k, t) \, dk \tag{56.7}$$

where we here denote the remaining Fourier amplitude simply by $\tilde{f}(k, t)$. This solution represents an arbitrary localized disturbance on the surface of the fluid. We assume that the factor $\tilde{f}(k, t)$ falls off fast enough for large $|k|$ for us to be able to differentiate and expand freely under the integral sign.

The linearized theory of tidal waves emerges directly from the following approximations:

1. Drop all nonlinear terms
2. Set $k\zeta \to 0$ in the expression for Φ on the free surface
3. Take the limit $kh \to 0$ in Φ

We shall show immediately that these three approximations indeed reproduce the simple linear wave equation for tidal waves. It is instructive, however, to improve on approximation 3 by expanding the integrand in Eq. (56.7) in powers of kh, retaining the first correction. In this way, we shall derive an *effective wave equation* that incorporates the leading dispersive correction for the propagation of surface waves. On the other hand, we shall still *linearize* the equations with approximations 1 and 2 above. On the free surface, the velocity potential thus takes the form

$$\Phi \approx (2\pi)^{-1} \int_{-\infty}^{\infty} \cosh{(kh)} \, e^{ikx} \tilde{f}(k, t) \, dk \tag{56.8a}$$

An expansion of the first factor in the integrand gives

$$\Phi \approx (2\pi)^{-1} \int_{-\infty}^{\infty} \left[1 + \frac{1}{2!} (kh)^2 + \cdots \right] e^{ikx} \tilde{f}(k, t) \, dk \tag{56.8b}$$

The term k^2 in the expansion can be converted into a derivative with respect to x and removed from the integral. The remaining factor is just the Fourier transform of $\tilde{f}(k, t)$, which we shall call $f(x, t)$:

$$(2\pi)^{-1} \int_{-\infty}^{\infty} e^{ikx} \tilde{f}(k, t) \, dk \equiv f(x, t) \tag{56.9}$$

To this approximation, we therefore have

$$\Phi \approx \left(1 - \frac{h^2}{2!} \frac{\partial^2}{\partial x^2} \right) f(x, t) \qquad \text{on free surface} \tag{56.10}$$

Similarly, it follows from Eq. (56.7) that $\partial\Phi/\partial z$ on the free surface becomes

$$\frac{\partial\Phi}{\partial z} \approx (2\pi)^{-1} \int_{-\infty}^{\infty} k \sinh{(kh)} \, e^{ikx} \tilde{f}(k, t) \, dk \tag{56.11a}$$

$$\approx (2\pi)^{-1} \int_{-\infty}^{\infty} k \left[kh + \frac{1}{3!} (kh)^3 + \cdots \right] e^{ikx} \tilde{f}(k, t) \, dk \tag{56.11b}$$

A treatment analogous to that above yields

$$\frac{\partial\Phi}{\partial z} \approx \left(-h \frac{\partial^2}{\partial x^2} + \frac{h^3}{3!} \frac{\partial^4}{\partial x^4} \right) f(x, t) \qquad \text{on free surface} \tag{56.11c}$$

where we again retain the first correction in the small parameter $(kh)^2$.

These relations (56.10) and (56.11) can be substituted into the basic equations (56.6) and (56.2). Neglecting nonlinear terms, we obtain the *extended tidal-wave equations*

$$\frac{\partial\zeta}{\partial t} = h \frac{\partial^2 f}{\partial x^2} - \frac{h^3}{6} \frac{\partial^4 f}{\partial x^4} \tag{56.12a}$$

$$g\zeta = \frac{\partial f}{\partial t} - \frac{h^2}{2} \frac{\partial^3 f}{\partial x^2 \, \partial t} \tag{56.12b}$$

If we had kept only the leading term in the expansion of Eq. (56.7) in powers of $(kh)^2$, Eqs. (56.12) would reduce to

$$\frac{\partial\zeta}{\partial t} = h \frac{\partial^2 f}{\partial x^2} \tag{56.13a}$$

$$g\zeta = \frac{\partial f}{\partial t} \tag{56.13b}$$

The derivative of the first of these with respect to time and two derivatives of the second with respect to x immediately yield the simple wave equation (53.18) for the surface displacement of tidal waves

$$\frac{1}{c^2}\frac{\partial^2 \zeta}{\partial t^2} = \frac{\partial^2 \zeta}{\partial x^2} \qquad (56.14a)$$

with the appropriate velocity

$$c \equiv (gh)^{1/2} \qquad (56.14b)$$

We now consider the extended tidal wave equations (56.12), seeking plane-wave solutions of the form

$$\zeta = \zeta_0 e^{i(kx - \omega t)} \qquad (56.15a)$$

$$f = f_0 e^{i(kx - \omega t)} \qquad (56.15b)$$

Substitution into Eqs. (56.12) yields

$$-i\omega\zeta_0 + hk^2 f_0 + \tfrac{1}{6}h^3 k^4 f_0 = 0 \qquad (56.16a)$$

$$i\omega f_0 + g\zeta_0 + \tfrac{1}{2}ih^2 k^2 \omega f_0 = 0 \qquad (56.16b)$$

which can be rewritten

$$\zeta_0 = \frac{f_0}{i\omega} hk^2 [1 + \tfrac{1}{6}(hk)^2] \qquad (56.17a)$$

$$g\zeta_0 = -i\omega f_0 [1 + \tfrac{1}{2}(hk)^2] \qquad (56.17b)$$

The ratio of these equations leads to a new dispersion relation

$$\omega^2 = (gh)k^2 \frac{1 + \tfrac{1}{6}(hk)^2}{1 + \tfrac{1}{2}(hk)^2} \qquad (56.18a)$$

Retaining only the leading correction in the small parameter $(kh)^2$, we find

$$\omega^2 = (gh)k^2 [1 - \tfrac{1}{3}(kh)^2 + \cdots] \qquad (56.18b)$$

where the first term is just that for tidal waves and reproduces the common phase and group velocity (56.14b). For comparison, recall the exact dispersion relation for small-amplitude surface waves [see Eq. (54.32) and recall $\omega \equiv kc(k)$]

$$\omega^2 = gk \tanh kh \qquad (56.19)$$

An expansion of this result in powers of kh gives the same expression

$$\omega^2 = (gh)k^2 [1 - \tfrac{1}{3}(kh)^2 + \cdots] \qquad (56.20)$$

The extended wave equations (56.12) are coupled linear differential equations for the surface displacement $\zeta(x, t)$ and the function $f(x, t)$ appearing in the velocity potential in Eqs. (56.7) and (56.9); these equations correctly reproduce the dispersion relation (56.20) for surface waves through the first correction in the small parameter $(kh)^2$. In this approximation, the phase velocity and group velocity of the waves will differ, and any initial waveform changes its shape with increasing time, as shown in Sec. 55 for deepwater waves.

Effective Nonlinear Wave Equation

We shall now further generalize the effective wave equations (56.12) by including the first *nonlinear* terms in the exact hydrodynamic equations (56.1) to (56.6). This procedure requires the leading nonlinear correction both in Bernoulli's equation at the free surface (56.2) and in the exact boundary condition defining the free surface (56.6). To be consistent, we also must keep the leading nonlinear term in the expansion of the velocity potential [Eq. (56.7)] when evaluated at the free surface. Since the displacements of the surface are less than (or the order of) the height of the channel in Fig. 56.1, it is natural to expand the integrand in Eq. (56.7) in powers of a new small parameter $k(h + \zeta)$ and retaining the leading correction term in this quantity. A simple generalization of (56.10) and (56.11c) immediately yields the free-surface relations

$$\Phi \approx \left[1 - \frac{(h + \zeta)^2}{2!} \frac{\partial^2}{\partial x^2} \right] f \qquad (56.21a)$$

on free surface

$$\frac{\partial \Phi}{\partial z} \approx \left[-(h + \zeta) \frac{\partial^2}{\partial x^2} + \frac{(h + \zeta)^3}{3!} \frac{\partial^4}{\partial x^4} \right] f \qquad (56.21b)$$

Just as before, substitute these results into Eqs. (56.2) and Eqs. (56.6) and now make the following approximations:

1. Keep the leading *nonlinear* correction terms, of order $\{f^2, f\zeta, \zeta^2\}$
2. Neglect the nonlinear contributions in small terms that are already of order $(kh)^2$ in the extended tidal-wave equations

Under these approximations, Eq. (56.6) reads

$$\frac{\partial \zeta}{\partial t} = (h + \zeta) \frac{\partial^2 f}{\partial x^2} - \frac{h^3}{6} \frac{\partial^4 f}{\partial x^4} + \frac{\partial \zeta}{\partial x} \frac{\partial f}{\partial x} \qquad (56.22a)$$

which can be rewritten as

$$\frac{\partial \zeta}{\partial t} - \frac{\partial}{\partial x} \left[(h + \zeta) \frac{\partial f}{\partial x} \right] + \frac{h^3}{6} \frac{\partial^4 f}{\partial x^4} = 0 \qquad (56.22b)$$

Similarly, Bernoulli's equation at the free surface [Eq. (56.2)] becomes

$$-\frac{\partial f}{\partial t} + g\zeta + \frac{h^2}{2} \frac{\partial^3 f}{\partial x^2 \partial t} + \frac{1}{2} \left(\frac{\partial f}{\partial x} \right)^2 = 0 \qquad (56.23)$$

Equations (56.22) and (56.23) generalize Eqs. (56.12) to form the *extended nonlinear wave equations* for tidal waves. By construction, they include both dispersion and nonlinearity.

Solitary Waves

Let us seek a traveling-wave solution to the extended nonlinear wave equations (56.22) and (56.23)

$$\zeta(x, t) = \zeta(\xi) \qquad f(x, t) = f(\xi) \tag{56.24a}$$

where

$$\xi \equiv x - ct \tag{56.24b}$$

and the velocity c is a constant to be determined. Substitution of the assumed form (56.24) into Eqs. (56.22) and (56.23) yields the expressions

$$-c\zeta' - [(h + \zeta)f']' + \tfrac{1}{6}h^3 f'''' = 0 \tag{56.25a}$$

$$cf' + g\zeta - \tfrac{1}{2}ch^2 f''' + \tfrac{1}{2}(f')^2 = 0 \tag{56.25b}$$

where a prime now denotes differentiation with respect to the single argument ξ. The first crucial observation in solving these coupled nonlinear differential equations is that (56.25a) is already a *perfect differential*. It can be integrated immediately to give

$$-c\zeta - (h + \zeta)f' + \tfrac{1}{6}h^3 f''' = \text{const} \tag{56.26a}$$

This equation must be solved along with (56.25b), which can be rewritten

$$cf' = -g\zeta + \tfrac{1}{2}ch^2 f''' - \tfrac{1}{2}(f')^2 \tag{56.26b}$$

We now combine these equations to obtain a single nonlinear differential equation for the surface displacement ζ by systematically eliminating the function f through substitution of Eq. (56.26b) into Eq. (56.26a). The zero-order term $-hf'$ in Eq. (56.26a) requires the full expression (56.26b), but the resulting equation then contains f only in the higher-order correction terms that reflect either the dispersion or nonlinearity. In these terms, it is permissible to make two approximations consistent with those made throughout this section:

1. Keep only the leading contribution to Eq. (56.26b), which is

$$cf' = -g\zeta \qquad \text{leading order} \tag{56.27a}$$

2. Replace the exact velocity of propagation by the zero-order approximation [compare Eq. (56.14b)]

$$c^2 = gh \qquad \text{leading order} \tag{56.27b}$$

In this way, Eqs. (56.26) can be reduced to

$$\zeta\left(1 - \frac{gh}{c^2}\right) - \frac{3}{2h}\zeta^2 - \frac{h^2}{3}\zeta'' = \text{const} \tag{56.28}$$

which is the first integral of the effective nonlinear wave equations (56.22) and (56.23).

Among the various solutions to (56.28), we shall exhibit only *solitary waves* ("solitons"), which are isolated disturbances traveling down the channel, as illustrated in Fig. 56.1. Such solutions clearly satisfy

$$\zeta \to 0 \text{ as } x \to \pm\infty \qquad \text{solitary waves} \qquad (56.29)$$

and these boundary conditions ensure that the constant appearing on the right-hand side of Eq. (56.28) must vanish

$$\zeta\left(1 - \frac{gh}{c^2}\right) - \frac{3}{2h}\zeta^2 - \frac{h^2}{3}\zeta'' = 0 \qquad (56.30)$$

Equation (56.30) is still a nonlinear second-order differential equation, but we already know how to construct a first integral of an equation of this form. The procedure is formally identical with finding the energy integral of Newton's second law for one-dimensional motion in a time-independent potential. Multiply Eq. (56.30) by ζ', converting it into a perfect differential

$$\frac{d}{d\xi}\left[\tfrac{1}{2}\zeta^2\left(1 - \frac{gh}{c^2}\right) - \frac{1}{2h}\zeta^3 - \frac{h^2}{6}(\zeta')^2\right] = 0 \qquad (56.31)$$

Direct integration gives

$$\tfrac{1}{2}\zeta^2\left(1 - \frac{gh}{c^2}\right) - \frac{1}{2h}\zeta^3 - \frac{h^2}{6}(\zeta')^2 = 0 \qquad (56.32)$$

where the integration constant again vanishes by the condition (56.29). Consider a solution that is symmetric in x at $t = 0$, as illustrated in Fig. 56.2. In this case we have

$$\zeta(x = 0, t = 0) = \zeta(\xi = 0) = \zeta_0 \qquad (56.33a)$$

$$\text{initial condition at } t = 0$$

$$\left.\frac{\partial \zeta(x, t = 0)}{\partial x}\right|_{x=0} = \zeta'(\xi = 0) = 0 \qquad (56.33b)$$

where ζ_0 is the maximum amplitude of the wave at the time $t = 0$. Evaluation of Eq. (56.32) at $\xi = 0$ leads to

$$1 - \frac{gh}{c^2} = \frac{\zeta_0}{h} \qquad (56.34a)$$

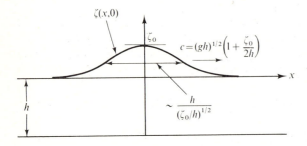

Figure 56.2 Sketch of initial configuration at $t = 0$ used in solving Eq. (56.32) for a solitary wave.

This relation can be rewritten to first order in the parameter ζ_0/h as

$$c = (gh)^{1/2}\left(1 + \frac{\zeta_0}{2h}\right) \tag{56.34b}$$

Evidently, the finite amplitude of wave affects the velocity of propagation.

Substitution of Eq. (56.34) into Eq. (56.32) provides another first integral of the effective wave equation

$$\frac{\zeta_0}{2h}\zeta^2 - \frac{1}{2h}\zeta^3 - \frac{h^2}{6}(\zeta')^2 = 0 \tag{56.35}$$

It can be solved for the derivative ζ' to yield

$$\zeta' = \pm\left(\frac{3}{h^3}\right)^{1/2}(\zeta_0 - \zeta)^{1/2}\zeta \tag{56.36}$$

Since ζ is expected to decrease with increasing positive $\xi \equiv x$ at $t = 0$ (see Fig. 56.2), we take the minus sign in Eq. (56.36). A simple rearrangement gives

$$\frac{d\zeta}{\zeta(\zeta_0 - \zeta)^{1/2}} = -\left(\frac{3}{h^3}\right)^{1/2} d\xi \tag{56.37}$$

This equation can be integrated to yield

$$\left(\frac{3}{h^3}\right)^{1/2}\xi = -\int_{\zeta_0}^{\zeta} \frac{d\zeta}{\zeta(\zeta_0 - \zeta)^{1/2}} \tag{56.38a}$$

where we have noted that $\zeta = \zeta_0$ at $\xi = 0$. The new variable $v^2 = \zeta_0/\zeta$ reduces Eq. (56.38a) to the form

$$\left(\frac{3}{h^3}\right)^{1/2}\xi = \frac{2}{\zeta_0^{1/2}}\int_1^{(\zeta_0/\zeta)^{1/2}} \frac{dv}{(v^2 - 1)^{1/2}} \tag{56.38b}$$

which is readily evaluated explicitly

$$\left(\frac{3}{h^3}\right)^{1/2}\xi = \frac{2}{\zeta_0^{1/2}}\operatorname{arccosh}\left(\frac{\zeta_0}{\zeta}\right)^{1/2} \tag{56.38c}$$

Straightforward manipulations yield the final result for the desired function

$$\zeta(\xi) = \zeta_0 \operatorname{sech}^2\left[\frac{\xi}{2h}\left(\frac{3\zeta_0}{h}\right)^{1/2}\right] \tag{56.39}$$

which is the surface displacement at $t = 0$ (see Fig. 56.2). The solution to the extended nonlinear equations (56.22) and (56.23) *for all times* follows immediately from Eq. (56.24) as

$$\zeta(x, t) = \zeta_0 \operatorname{sech}^2\left[\left(\frac{3\zeta_0}{h}\right)^{1/2}\frac{x - ct}{2h}\right] \tag{56.40}$$

with the velocity of propagation now determined by Eq. (56.34*b*)

$$c = (gh)^{1/2}\left(1 + \frac{\zeta_0}{2h}\right) \tag{56.41}$$

These remarkable relations show that a finite-amplitude solitary wave of the specific form (56.40) can propagate without distortion at a velocity c that depends on the amplitude ζ_0 through Eq. (56.41). It is clear from Eqs. (56.35) and (56.36) that ζ_0 must be positive, so that the solitary wave is an *elevation* (not a depression) traveling at a speed exceeding $(gh)^{1/2}$. This situation is very different from the tidal waves studied in Sec. 53, where *any* initial infinitesimal-amplitude disturbance propagates unchanged at the speed $(gh)^{1/2}$. Here, in contrast, the only allowed disturbance has a width of order $(h/\zeta_0)^{1/2}h$ and falls off exponentially in both the positive and negative directions (see Fig. 56.2). As the amplitude ζ_0 tends to zero, the width of the solitary waves becomes infinite and the whole effect in fact disappears. Thus this solitary wave is an intrinsically nonlinear phenomenon that also requires the presence of dispersion in Eq. (56.18*b*). Although either of these effects separately would distort the traveling wave (compare Secs. 52 and 55), it is remarkable that the combination of dispersion [Eq. (56.18*b*)] and nonlinearity [Eqs. (56.22) and (56.23)] together yields a solitary wave that moves without change of shape.

PROBLEMS

10.1 (*a*) Evaluate the normalization constants c_{mn} for the normal modes (53.46) in a shallow rectangular tank.

   (*b*) Determine the first few low-lying eigenfrequencies and sketch the corresponding displacements. Compare with those for a rectangular membrane with clamped edges (Fig. 47.2). Discuss.

10.2 Repeat Prob. 10.1 for a shallow circular tank of radius a [see Eq. (53.48) and Fig. 47.5].

10.3 A semi-infinite channel ($x \geq 0$) of large constant width has a depth that increases linearly with x according to $h = \kappa x$ (compare Fig. 53.1). Prove that the corresponding tidal waves have a surface displacement

$$\zeta(x, t) = CJ_0\left[2\omega\left(\frac{x}{\kappa g}\right)^{1/2}\right]\cos \omega t$$

where J_0 is the Bessel function of order zero. Show that the wavelength increases with increasing values of x. Sketch and discuss, including the behavior for small x. Discuss practical applications.

10.4 An inlet of the sea of length L has a depth $h_0(x/L)^m$ and breadth $b_0(x/L)^n$. The shore lies at $x = 0$, and the sea lies at $x = L$, where the boundary condition on the shallow-water amplitude is $\zeta(L, t) = \zeta_0 \cos \omega t$. Show that the corresponding small-amplitude tidal waves are expressible in terms of Bessel functions. Find the amplitude $\zeta(x, t)$, sketch the shape of the free surface, and discuss the changes in wavelength and amplitude as the wave approaches the shore for the following cases:

   (*a*) Constant depth and breadth ($m = n = 0$), terminated by a vertical cliff at $x = 0$
   (*b*) Constant depth, linearly varying breadth ($m = 0, n = 1$)
   (*c*) Linearly varying depth and breadth ($m = 1, n = 1$)
The case of linearly varying depth and constant breadth ($m = 1, n = 0$) is discussed in Prob. 10.3.

10.5 A shallow channel has a V-shaped cross section with breadth $2a$ and depth h_0 ($\ll 2a$) at the center. Show that the transverse tidal modes that flow back and forth across the channel and are uniform along the channel are either even or odd about the center of the channel. Determine the equation for the corresponding eigenvalues and discuss the boundary condition at the edges. Use tables to find the two lowest frequencies of each class and sketch the associated free surface. Discuss the ordering of the low-lying modes.

10.6 Repeat Prob. 10.5 for a channel with parabolic cross section and depth given by $h(y) = h_0(1 - y^2/a^2)$.

10.7 Consider tidal motion in a channel of fixed breadth b and depth h.

(a) If vertical accelerations are neglected, obtain the following nonlinear equations for one-dimensional flow

$$\frac{\partial \bar{\zeta}}{\partial t} + \frac{\partial}{\partial x} v_x \bar{\zeta} = 0 \qquad \frac{\partial v_x}{\partial t} + v_x \frac{\partial v_x}{\partial x} + g \frac{\partial \bar{\zeta}}{\partial x} = 0$$

where $\bar{\zeta} = h + \zeta$. Compare with Eqs. (52.3) and (52.6).

(b) A *bore* is a propagating discontinuous surface profile with total depth h_1 for $x \to -\infty$ and h_2 ($< h_1$) for $x \to \infty$. If the fluid far to the right is stationary, prove that the discontinuity propagates to the right with a speed $[(gh_2/2h_1)(h_1 + h_2)]^{1/2}$ (neglect atmospheric pressure). Show that $E_1 < E_2$, and hence conclude that this phenomenon dissipates energy. Discuss.

10.8 Consider the interface between two fluids (1 for $z > 0$ and 2 for $z < 0$) of thickness h_1 and h_2, placed in a gravitational field $-g\hat{z}$, and bounded externally by fixed horizontal planes perpendicular to the z axis. Show that the frequency of a small-amplitude surface disturbance with wave number k is given by

$$\omega^2 = \frac{gk(\rho_2 - \rho_1)}{\rho_2 \coth kh_2 + \rho_1 \coth kh_1}$$

Discuss various limits.

10.9 A fluid with surface tension τ at its upper (free) surface is bounded below by a rigid plane at a depth h. Show that the frequency of small-amplitude disturbances with wave number k is given by

$$\omega^2 = \left(gk + \frac{\tau k^3}{\rho}\right) \tanh kh$$

Discuss various limits.

10.10 Two incompressible fluids of density ρ_1 and ρ_2 are superposed in an external gravitational field $-g\hat{z}$.

(a) In steady equilibrium, they occupy the semi-infinite regions $z > 0$ and $z < 0$ and move with uniform velocities $u_1 \hat{x}$ and $u_2 \hat{x}$, respectively. Find the equilibrium velocity potential and equilibrium pressure throughout all space.

(b) If the interface has a surface tension τ, derive the frequency of small surface oscillations of the form $\zeta(x, t) = \zeta_0 e^{i(kx - \omega t)}$ about the steady state in part (a).

(c) Show that the motion is stable for all k if and only if

$$(u_1 - u_2)^4 < 4g\tau(\rho_2 - \rho_1)(\rho_2 + \rho_1)^2/\rho_1^2\rho_2^2$$

Discuss various limiting cases with practical applications wherever possible.

10.11 Evaluate the kinetic and potential energy per wavelength for the one-dimensional surface wave in Eqs. (54.36) and (54.37). Hence verify the discussion below Eq. (54.51).

10.12 A cylindrical tank of radius a and depth h is filled with an ideal incompressible fluid.

(a) Show that the velocity potential has the form

$$\Phi(\mathbf{r}, t) = \text{Re } CJ_m(kr)e^{im\phi} \cosh[k(z + h)] \cos \omega t$$

with $\omega^2 = gk \tanh kh$. What are the allowed values of k? Find the associated free-surface profile [compare Eq. (53.48)].

(b) For a given normal mode, evaluate the total kinetic and potential energies; discuss the conservation of energy.

10.13 A traveling surface wave on an unbounded infinitely deep fluid impinges on a fixed vertical post of radius a. Show that the differential cross section for the scattered surface wave is precisely that given in Eqs. (51.74) and (51.75) for acoustic scattering by a cylinder.

10.14 For surface waves on a fluid of uniform depth h, the following steps verify that the quantity

$$\mathscr{L} = -\int_{-h}^{\zeta} dz \left[\tfrac{1}{2}(\nabla\Phi)^2 + gz - \frac{\partial\Phi}{\partial t}\right]$$

serves as an appropriate lagrangian density.

(a) For variations $\Phi \to \Phi + \delta\Phi$, show that Hamilton's principle requires that $\nabla^2\Phi = 0$ in the fluid, that $\hat{n} \cdot \nabla\Phi = 0$ at a fixed surface, and that

$$\frac{\partial\Phi}{\partial z} - \mathbf{V}_\perp \zeta \cdot \mathbf{V}_\perp \Phi + \frac{\partial\zeta}{\partial t} = 0 \qquad \text{at free surface}$$

(b) For variations $\zeta \to \zeta + \delta\zeta$, show that

$$\tfrac{1}{2}(\nabla\Phi)^2 + g\zeta - \frac{\partial\Phi}{\partial t} = 0 \qquad \text{at free surface}$$

10.15 For each of the following envelope functions $\tilde{f}(k)$ compute the Fourier transform $f(x)$. Sketch $\tilde{f}(k)$ and $f(x)$ and devise suitable quantitative measures of the widths Δk and Δx. Hence verify the discussion below Eq. (54.63).

(a) $\tilde{f}(k) = \begin{cases} 0 & |k|\alpha > 1 \\ 1 & |k|\alpha < 1 \end{cases}$

(b) $\tilde{f}(k) = e^{-|k|\alpha}$

10.16 For the gaussian waveform (54.58), retain the quadratic correction to Eq. (54.65) and study the time dependence of the propagating wave packet in (54.56).

10.17 A semi-infinite $(z < 0)$ incompressible fluid in a gravitational field $-g\hat{z}$ flows with uniform velocity $u\hat{x}$.

(a) Assume that the fluid experiences a small localized periodic surface pressure $p(x) \cos \omega t = \mathrm{Re}\ p(x)e^{-i\omega t}$. Find an integral representation for the velocity potential throughout the fluid and show that the resulting free surface is given by the equation

$$\zeta(x, t) = \mathrm{Re}\ \zeta(x)e^{-i\omega t} \qquad \text{where } \zeta(x) = -\frac{1}{\rho}\lim_{\epsilon\to 0}\int_{-\infty}^{\infty}\frac{dk}{2\pi}\frac{e^{ikx}|k|\tilde{p}(k)}{g|k| - (\omega + i\epsilon - ku)^2}$$

and $\tilde{p}(k)$ is the Fourier transform of $p(x)$.

(b) If $p(x) = p_0\ \delta(x)$ and $\omega \to 0$, use an appropriate contour deformation to obtain

$$\zeta(x) = \frac{p_0}{\rho u^2}\left[-2\theta(x) \sin \frac{gx}{u^2} + \frac{1}{\pi}\int_0^{\infty} d\kappa \frac{\kappa e^{-\kappa|x|}}{\kappa^2 + (g/u^2)^2}\right]$$

where $\theta(x)$ is the unit step function. Evaluate the integral approximately for large and small values of $g|x|/u^2$; discuss the resulting shape of the free surface and its physical interpretation.

10.18 Write down the equations for a two-dimensional traveling surface disturbance on deep water. Use a two-dimensional Fourier transform to find an integral representation for the velocity potential $\Phi(\mathbf{r}, z, t)$ throughout the fluid if the free surface satisfies the initial conditions

$$\zeta(\mathbf{r}, 0) = f(\mathbf{r}) \qquad \dot{\zeta}(\mathbf{r}, 0) = 0$$

In the special case that $f(|\mathbf{r}|)$ is axisymmetric with $\tilde{f}(|\mathbf{k}|)$ its two-dimensional Fourier transform, derive an asymptotic expression for $\Phi(\mathbf{r}, z, t)$; hence show that

$$\zeta(r, t) \sim \frac{gt^2}{4\pi\sqrt{2}\, r^3}\tilde{f}\left(\frac{gt^2}{4r^2}\right) \cos \frac{gt^2}{4r}$$

where J_0 under the integral sign has been approximated by its leading asymptotic term [see Eq. (D3.28)] (why?). Discuss the special case $f(\mathbf{r}) = (\pi\lambda^2)^{-1}e^{-r^2/\lambda^2}$ with sketches of appropriate functions.

10.19 Write down the equations for one-dimensional traveling waves on the surface of water of depth h. Use a Fourier transform to find the free surface $\zeta(x, t)$ for $t > 0$, subject to the initial conditions

$$\zeta(x, 0) = f(x) \qquad \dot\zeta(x, 0) = 0$$

In the approximation of very shallow water, show that the phase velocity and group velocity are equal, so that the initial waveform propagates without distortion. Consider what happens when you include the lowest-order correction to the dispersion relation for shallow water; in what circumstances will this represent a valid approximation? If $f(x) = \delta(x)$, use the method of stationary phase to find the approximate asymptotic form of $\zeta(x, t)$. What happens near the point $x = (gh)^{1/2}t$? Discuss this behavior of the free surface with appropriate sketches.

10.20 An infinite membrane with mass σ per unit area and tension τ is in contact on both sides with an infinite medium of density ρ and speed of sound c_0.

(a) The membrane is subjected to a transverse driving force $f(x, y)e^{-i\omega t}$ per unit area. Show that the equation of motion for simple harmonic transverse motion of the membrane is

$$\nabla^2\zeta(x, y) + K_m^2\zeta(x, y) + \frac{K_m^2\rho}{\pi\sigma}\int_0^{2\pi} d\phi \int_0^\infty \zeta(x_0, y_0)e^{iK_0R}\, dR = \frac{f(x, y)}{\tau}$$

where $K_0^2 = (\omega/c_0)^2$, $K_m^2 = (\omega/c_m)^2 = \sigma\omega^2/\tau$, $\zeta(x, y)$ is the transverse displacement, and $R^2 = (x - x_0)^2 + (y - y_0)^2$.

(b) Show that free plane waves of the form $e^{i\mathbf{k}\cdot\mathbf{r}-i\omega t}$ can travel along the membrane whenever the magnitude $|\mathbf{k}|$ satisfies the equation

$$k^2 = K_m^2\left[1 + \frac{2\rho/\sigma}{(k^2 - K_0^2)^{1/2}}\right]$$

(c) Show that there are two types of waves, one for k near K_0 and the other for k near K_m (if ρ/σ is small). Discuss their physical significance. Show that if $c_0 < c_m$, the latter type is attenuated but that if $c_0 > c_m$, neither type is attenuated. What is the physical reason for this?

10.21 Generalize the treatment of Sec. 56 to study nonlinear spatially periodic propagating waves with $\zeta(\xi + L) = \zeta(\xi)$.

(a) Let $\zeta' = 0$ at the crest ($\xi = 0$, $\zeta = \zeta_0$) and the trough ($\xi = \frac{1}{2}L$, $\zeta = 0$). Show that the period L can be written as

$$\tfrac{1}{2}L\left(\frac{3}{h^3}\right)^{1/2} = \int_0^{\zeta_0} d\zeta\, [(\zeta_0 - \zeta)\zeta(\zeta^* + \zeta)]^{-1/2}$$

where ζ^* determines the speed of propagation according to $c^2 = gh^2(h + \zeta^* - \zeta_0)^{-1}$.

(b) Show that the limit $\zeta^* \to 0$ reproduces the solitary wave found previously in Eq. (56.39).

(c) Consider the opposite limit $\zeta^* \gg \zeta_0$ and obtain an explicit solution for the surface profile.

SELECTED ADDITIONAL READINGS

1. J. S. Russell, Report on Waves, *Br. Assoc. Rep.*, 1844.
2. G. B. Whitham, "Linear and Nonlinear Waves," Wiley, New York, 1974.
3. A. C. Scott, F. Y. F. Chu, and D. W. McLaughlin, *Proc. IEEE*, **61**: 1443–1483 (1973).

CHAPTER
ELEVEN

HEAT CONDUCTION

The concept of temperature has played virtually no role in our preceding discussions of mechanics, where the energy consisted either of stored potential energy or organized kinetic energy (see, however, the propagating shock front, treated in Sec. 52). In practice, friction ultimately degrades the mechanical energy into random thermal motion and heat. Such behavior is most simply analyzed from a macroscopic viewpoint. For simplicity, the present chapter is restricted to solids, where there is no net mass transport. The more complicated case of fluids is discussed in Chap. 12 in connection with viscosity. Since the study of heat conduction necessarily involves elementary thermodynamics, we assume some acquaintance with these matters.†

57 BASIC EQUATIONS

Conservation of energy is the fundamental principle of heat conduction, and we first review its analytical formulation. Consider a macroscopic sample and imagine it divided into a large number of small volume elements, as illustrated in Fig. 57.1. We assume that each of these regions has reached local thermodynamic equilibrium, with some characteristic temperature T, entropy S, and pressure p. These quantities will vary smoothly with position, so that their values in adjacent regions will differ only by small amounts. As a result, any transfer of energy between them is reversible in the thermodynamic sense. Consider now one of these small regions, with fixed mass m and volume v as illustrated in Fig. 57.1. For this

---

† See, for example Reif [1], chaps. 4 and 5.

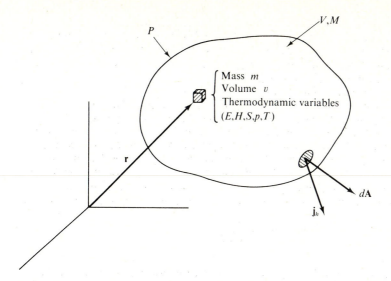

Figure 57.1 Spatial geometry for discussion of basic equations of heat conduction.

elementary system, the first and second laws of thermodynamics have the differential expression

$$dE = dQ + dW \qquad (57.1a)$$

$$dE = T \, dS - p \, dv \qquad (57.1b)$$

where dE = increase in internal energy for some elementary reversible process
$dQ = T \, dS$ = heat added
dS = increase in entropy
$dW = -p \, dv$ = external work done on system

Here, the symbol d indicates that dQ and dW are merely infinitesimal quantities, not generally expressible as differentials of some particular function. On the other hand, the principal content of the first and second laws is the assertion that dE and dS are indeed perfect differentials of functions E and S depending only on the state of the system.

As written, Eq. (57.1b) is convenient for a system at fixed *volume* because E then changes only because of heat transfer. Unfortunately, thermal expansion necessarily affects v, and practical experiments are generally performed at fixed pressure. Although this distinction is unimportant in practice for most solids, it is essential for gases and liquids (treated in Chap. 12). To allow for this more general view, we first perform a Legendre transformation from the internal energy E to the *enthalpy*†

$$H = E + pv \qquad (57.2)$$

---

† Despite the slight risk of confusion with the hamiltonian, we use the conventional notation H for the enthalpy; see also the footnotes below Eqs. (32.13) and (48.41).

which also depends on the state variables of the system. Substitution of Eqs. (57.1) into the differential of this relation gives

$$dH = d(E + pv) = dQ + v\, dp \tag{57.3a}$$

$$dH = T\, dS + v\, dp \tag{57.3b}$$

At fixed *pressure*, this equation shows that the enthalpy of a given amount of material changes only because of heat transfer dQ. Correspondingly the constant-pressure heat capacity of the given element of material has the equivalent forms [compare Eq. (52.24)]

$$C_p \equiv \left(\frac{dQ}{dT}\right)_p \tag{57.4a}$$

$$= \left(\frac{\partial H}{\partial T}\right)_p \tag{57.4b}$$

$$= T\left(\frac{\partial S}{\partial T}\right)_p \tag{57.4c}$$

Equation (57.4b) can be integrated at constant pressure to find the enthalpy of the given mass m

$$H(T, p) = H(T_0, p) + \int_{T_0}^{T} dT'\; C_p(T', p) \tag{57.5a}$$

In most cases of interest, C_p depends only weakly on T, in which case Eq. (57.5a) takes the approximate form

$$H(T, p) \approx H(T_0, p) + C_{p0}(T - T_0) \tag{57.5b}$$

with C_{p0} evaluated at T_0.

The thermodynamic functions for the entire macroscopic sample in Fig. 57.1 can be obtained by adding the contribution of each small subunit. It is appropriate to work with infinitesimal volume elements and to write the small element of mass m as

$$dm \equiv \rho\, d^3r \tag{57.6a}$$

where ρ is the mass density. Similarly, we introduce the constant-pressure heat capacity per unit mass c_p according to

$$C_{p0} = c_p \rho\, d^3r \tag{57.6b}$$

If there is a temperature *distribution* $T(\mathbf{r}, t)$ throughout the macroscopic sample, which we now assume is under constant pressure p (see Fig. 57.1), the total enthalpy obtained from Eqs. (57.5b) and (57.6) is

$$H_{\text{tot}} = \int_V \rho c_p [T(\mathbf{r}, t) - T_0]\, d^3r + H_{\text{tot}}(T_0, p) \tag{57.7a}$$

where $H_{\text{tot}}(T_0, p)$ is the total enthalpy of the macroscopic sample at a uniform temperature T_0 and pressure p. Equation (57.6a) allows us to rewrite this result as an integral over mass elements rather than volume elements

$$H_{\text{tot}} = \int_M c_p[T(\mathbf{r}, t) - T_0]\, dm + H_{\text{tot}}(T_0, p) \tag{57.7b}$$

Although the volume may change during isobaric processes, we assume that the mass in each small element in Fig. 57.1 does not; i.e., we assume no convective mass transport during the heating and cooling [see Eq. (60.47) for the more general case]. The equation of *heat conduction* at constant pressure now follows by considering the time derivative of this total enthalpy

$$\frac{dH_{\text{tot}}}{dt} = \frac{d}{dt} \int_M c_p T\, dm = \int_M c_p \frac{\partial T}{\partial t}\, dm \tag{57.8}$$

where the partial derivative $\partial T/\partial t$ indicates that the position is fixed.† Since we assume that no material passes through the surface A bounding the mass M at a given instant t (see Fig. 57.1), the enthalpy can change only by heat transfer dQ, either *conduction* of heat through the surface with flux vector $\mathbf{j}_h$ or *generation* of heat $\rho \dot{q}$ per unit volume in the interior. A familiar analysis (see Sec. 48) gives both the integral conservation law‡

$$\int_M c_p \frac{\partial T}{\partial t}\, dm = \int_V d^3r\, \rho c_p \frac{\partial T}{\partial t} \tag{57.9a}$$

$$= -\int dA \cdot \mathbf{j}_h + \int_V d^3r\, \rho \dot{q} \tag{57.9b}$$

and the equivalent differential one

$$\rho c_p \frac{\partial T}{\partial t} = -\nabla \cdot \mathbf{j}_h + \rho \dot{q} \tag{57.10}$$

In most cases, experiments show that the heat current $\mathbf{j}_h$ is proportional to ∇T, and we express this empirical fact as

$$\mathbf{j}_h = -k_{\text{th}} \nabla T \tag{57.11}$$

where k_{th} is the *thermal conductivity*. The minus sign reflects the physical observation that heat flows from a hot region to a cold one. A combination of these last equations yields the basic *equation of heat conduction in solids*

$$\frac{\partial T}{\partial t} = \kappa \nabla^2 T + \frac{\dot{q}}{c_p} \tag{57.12}$$

---

† The last equality in Eq. (57.8) omits the additional term $\mathbf{v} \cdot \nabla T$ in the integrand which could arise, for example, from thermal expansion; it is included in Eq. (60.47).

‡ Multiplication of both sides of Eq. (57.9) by dt gives the change in enthalpy of the mass M in the time interval dt. Cancellation of this factor then leads to the instantaneous conservation law (57.9). Since (57.9) holds for an arbitrary mass element M in the system, the differential statement (57.10) follows. Note that at fixed pressure, the heat flux $\mathbf{j}_h$ is identical with the enthalpy flux [recall Eq. (57.3b)].

where $\dot{q}$ is the rate of heat generated per unit mass, k_{th} has been assumed independent of position, and

$$\kappa \equiv \frac{k_{th}}{\rho c_p} \qquad (57.13)$$

is known as the *thermal diffusivity*. It has the dimensions $l^2 t^{-1}$, or cm^2 s^{-1} in cgs units. In addition, we shall neglect any weak temperature dependence of the diffusivity for a *solid* and henceforth treat κ as a *constant* for any given material. Equation (57.12) is also called the *diffusion equation* (see Probs. 11.3 and 11.4); it differs from the simple wave equation in the presence of the first rather than the second time derivative. This apparently minor difference completely alters the character of the phenomena. While the *wave equation* describes waves that propagate without attenuation, the diffusion equation describes disturbances that *decay* with time and propagate through space in the manner of a random walk.†
In the absence of a source term ($\dot{q} = 0$), Eq. (57.12) exhibits the following simple scaling behavior: if the thermal disturbance has a characteristic dimension L, then the term $\kappa \nabla^2 T$ is of order $\kappa T/L^2$, implying a characteristic time L^2/κ for the disturbance to decay. As expected, this time becomes large for a poor conductor ($\kappa \to 0$).

The boundary conditions on the heat equation are very similar to those encountered earlier. One common situation is that the temperature T takes a specified value T_{ex} on the boundary

$$T = T_{ex} \qquad \text{on boundary at temperature } T_{ex} \qquad (57.14)$$

A second possibility is an insulated boundary, where $\hat{n} \cdot \mathbf{j}_h$ vanishes. Use of Eq. (57.11) implies the equivalent form

$$\hat{n} \cdot \nabla T = 0 \qquad \text{on insulated boundary} \qquad (57.15)$$

A third, more intricate, case is a radiating surface, when the heat flow is taken as proportional to the difference between the actual surface temperature T and the temperature T_{ex} of the external reservoir

$$\hat{n} \cdot \mathbf{j}_h = \alpha(T - T_{ex}) \qquad (57.16a)$$

or, alternatively,

$$k_{th}\hat{n} \cdot \nabla T = -\alpha(T - T_{ex}) \qquad \text{on radiating boundary} \qquad (57.16b)$$

This represents an inhomogeneous form of the homogeneous boundary condition discussed in Eq. (40.13).

58 EXAMPLES

Most of the previous analytical techniques can be used in solving the equation of heat conduction. To illustrate the various possibilities, we consider a few specific examples.

† See Reif [1], chap. 1.

Separation of Variables

The simplest approach is to separate variables in the form

$$T(\mathbf{r}, t) = T(\mathbf{r})e^{-\lambda t} \tag{58.1}$$

Substitution into Eq. (57.12) gives an inhomogeneous Helmholtz equation

$$(\nabla^2 + k^2)T(\mathbf{r}) = -\frac{\dot{q}}{\kappa c_p} \tag{58.2}$$

where

$$k^2 \equiv \frac{\lambda}{\kappa} \tag{58.3}$$

determines the separation constant. This equation describes many situations. For definiteness, we study the homogeneous equation ($\dot{q} = 0$) with no heat generated in the medium. In this case, our previous analysis shows that the homogeneous Helmholtz equation becomes an eigenvalue equation with solutions only for certain values of k^2 that depend on the specific boundary conditions.

For example, consider a rectangular parallelepiped a by b by c, with the ends at $x = 0$ and $x = a$ maintained at a constant temperature T_0, the remaining faces being insulated (see Fig. 58.1). If the temperature has an initial distribution $f(\mathbf{r})$, how does it relax to its final value (which here follows by inspection) $T = T_0$ everywhere? It is first convenient to define the deviation from the final temperature

$$\delta T = T - T_0 \tag{58.4}$$

so that $\delta T \to 0$ as $t \to \infty$. Evidently, the allowed solutions for δT satisfy the homogeneous Helmholtz equation

$$(\nabla^2 + k^2)\,\delta T = 0 \tag{58.5}$$

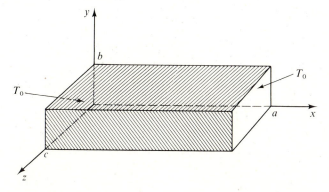

Figure 58.1 Heat conduction in a rectangular parallelepiped.

subject to the homogeneous boundary conditions [compare Eqs. (57.14) and (57.15)]

$$\delta T = 0 \qquad \text{at } x = 0 \text{ and } x = a$$

$$\frac{\partial}{\partial y} \delta T = 0 \qquad \text{at } y = 0 \text{ and } y = b \qquad (58.6)$$

$$\frac{\partial}{\partial z} \delta T = 0 \qquad \text{at } z = 0 \text{ and } z = c$$

This problem has the separated solution

$$\delta T(\mathbf{r}) = X(x)Y(y)Z(z) \qquad (58.7)$$

where [compare Eqs. (47.6) and (47.27) for a membrane and Eq. (53.46) for tidal waves] the (unnormalized) functions

$$X(x) = \sin \frac{m\pi x}{a} \qquad m = 1, 2, 3, \ldots$$

$$Y(y) = \cos \frac{n\pi y}{b} \qquad n = 0, 1, 2, \ldots \qquad (58.8)$$

$$Z(z) = \cos \frac{p\pi z}{c} \qquad p = 0, 1, 2, \ldots$$

satisfy the boundary conditions. Substitution into Eq. (58.5) gives the corresponding eigenvalue

$$k_{mnp}^2 \equiv \frac{\lambda_{mnp}}{\kappa} = \left(\frac{m\pi}{a}\right)^2 + \left(\frac{n\pi}{b}\right)^2 + \left(\frac{p\pi}{c}\right)^2 \qquad (58.9)$$

A linear combination then provides the general solution

$$\delta T(\mathbf{r}, t) = \sum_{mnp} C_{mnp} \sin \frac{m\pi x}{a} \cos \frac{n\pi y}{b} \cos \frac{p\pi z}{c} e^{-\lambda_{mnp}t} \qquad (58.10)$$

At $t = 0$, this solution is just a three-dimensional Fourier series

$$\delta T(\mathbf{r}, 0) = f(\mathbf{r}) - T_0$$

$$= \sum_{mnp} C_{mnp} \sin \frac{m\pi x}{a} \cos \frac{n\pi y}{b} \cos \frac{p\pi z}{c} \qquad (58.11)$$

and the coefficients C_{mnp} can be found with the orthogonality relations.

It is interesting to consider the long-time behavior of the solution (58.10), which is determined by the lowest eigenvalue ($m = 1, n = p = 0$)

$$T(\mathbf{r}, t) \sim T_0 + C_{100} \sin \frac{\pi x}{a} e^{-\lambda_{100}t} \qquad t \to \infty \qquad (58.12)$$

Any initial temperature distribution ultimately relaxes to this form, with the asymptotic characteristic time $\lambda_{100}^{-1} = a^2/\kappa\pi^2$, confirming the simple estimate below Eq. (57.13). If C_{100} happens to vanish, the next-lowest eigenvalue predominates, but the detailed behavior depends on the shape of the object, just as for a membrane or an acoustic cavity.

In general, the method of separation of variables encounters two difficulties. First, it is not always easy to find the final steady-state solution, which here followed by inspection. Second, the exact solution (58.10) is not very useful for short times because all the terms with $\lambda_{mnp} t \lesssim 1$ contribute to the sum. Moreover, it is not permissible to expand the exponentials in powers of t because the resulting series typically diverges, correctly suggesting that the short-time behavior is nonanalytic in t. Both these difficulties are avoided by using the more powerful Laplace-transform techniques, discussed in Sec. 59.

Thermal Waves in a Half Space

Consider a half space $z > 0$ subject to an oscillatory surface temperature

$$T(z = 0, t) = T_0 \cos \omega t = \text{Re } T_0 e^{-i\omega t} \tag{58.13}$$

with specified frequency ω (see Fig. 58.2). The temperature distribution satisfies the homogeneous equation

$$\frac{\partial^2 T}{\partial z^2} = \frac{1}{\kappa} \frac{\partial T}{\partial t} \tag{58.14}$$

and we again assume a separated solution

$$T(z, t) = \text{Re } T(z) e^{-i\omega t} \tag{58.15}$$

that oscillates at the applied frequency. Substitution into (58.14) yields an ordinary differential equation with constant coefficients for the complex amplitude $T(z)$

$$\frac{d^2 T}{dz^2} = -\frac{i\omega}{\kappa} T \tag{58.16}$$

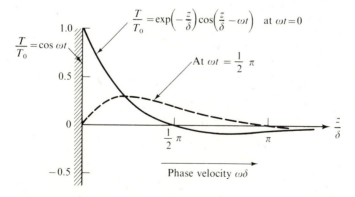

Figure 58.2 Thermal waves in a half space, where $\delta \equiv (2\kappa/\omega)^{1/2}$ is the penetration depth. In one quarter cycle, the crests and nodes move a distance $z = \frac{1}{2}\pi\delta$ at the velocity $\omega\delta$.

The corresponding solution has an exponential form $e^{\alpha z}$, with

$$\alpha^2 = -\frac{i\omega}{\kappa} = \frac{\omega}{\kappa} e^{-i\pi/2} \tag{58.17}$$

This equation has two roots

$$\alpha_{\pm} = \pm \left(\frac{\omega}{\kappa}\right)^{1/2} e^{-i\pi/4} = \pm \left(\frac{\omega}{2\kappa}\right)^{1/2} (1 - i) \tag{58.18}$$

and the general solution is therefore a linear combination of the two functions $e^{\alpha_+ z}$ and $e^{\alpha_- z}$. In the present case, the upper sign must be rejected because the corresponding solution diverges as $z \to \infty$. We therefore have

$$T(z) = A \exp\left[-\left(\frac{\omega}{2\kappa}\right)^{1/2} (1 - i)z\right] \equiv A \exp\left(\frac{iz}{\delta} - \frac{z}{\delta}\right) \tag{58.19}$$

where A is a constant and

$$\delta \equiv \left(\frac{2\kappa}{\omega}\right)^{1/2} \equiv \left(\frac{\kappa\tau}{\pi}\right)^{1/2} \tag{58.20}$$

is known as the *penetration depth* for the frequency ω and period $\tau = 2\pi/\omega$. Comparison with (58.13) shows that $A = T_0$, giving the final answer

$$T(z, t) = \text{Re } T(z)e^{-i\omega t} = T_0 e^{-z/\delta} \cos\left(\frac{z}{\delta} - \omega t\right) \tag{58.21}$$

This thermal disturbance is readily interpreted as an exponentially damped temperature wave, with a frequency-dependent decay length δ for an e-fold reduction in amplitude and a frequency-dependent phase velocity $\omega\delta = (2\kappa\omega)^{1/2}$ for propagation of maxima and minima.† Note that Eq. (58.20) again confirms the characteristic scaling relation that δ^2 is of order $\kappa\tau$. The cosine factor in (58.21) has a wavelength $2\pi\delta$, and the amplitude $T_0 e^{-z/\delta}$ attenuates by the factor $e^{-2\pi} \approx 0.0019$ in this distance (see Fig. 58.2).

Equation (58.20) characterizes a variety of periodic thermal phenomena. Here, we merely apply it to the propagation of annual and daily temperature variations into the ground, whose thermal diffusivity is of order 10^{-2} cm^2 s^{-1} in cgs units.‡ For $\omega_{ann} = 2\pi$ year^{-1}, the length $\delta_{ann} \approx 3.2$ m characterizes the depth at which the annual temperature variation is reduced to $e^{-1} \approx 0.368$ of its surface value. Equation (58.21) also exhibits a phase delay of magnitude z/δ. At the depth δ, for

---

† In a harmonic sound wave with wavelength λ, the local density variation gives rise to an oscillatory deviation from the ambient temperature. If the phase velocity c of the sound wave greatly exceeds that $[=(2\kappa\omega)^{1/2}]$ of the induced thermal wave, the motion is isentropic, as assumed in Sec. 49. For a simple example, the thermal diffusivity κ of a dilute gas is of order lc, where l is the mean free path (see Reif [1], sec. 12.4, but note his different notation); thus sound propagates isentropically when $l \ll \lambda$ (or, equivalently, when a particle undergoes many collisions per cycle of the sound wave). See Sec. 62 and also Prob. 12.13.

‡ See, for example, Carslaw and Jaeger [2], secs. 2.6 and 2.12–2.14.

example, the summer and winter extrema arrive about 2 months late. In a similar way, the penetration length for daily variations is smaller by a factor $(365)^{-1/2} \approx 0.052$. We may note that electromagnetic waves in a good conductor display a similar behavior; in particular, the components of the electromagnetic field obey a diffusion equation, and the characteristic skin depth also varies with frequency proportional to $\omega^{-1/2}$.†

Infinite Domain: Fourier Transform

For a final example, consider an effectively infinite uniform medium, with distant boundaries maintained at some fixed value ($T = 0$, say). Given an initial temperature distribution $f(\mathbf{r})$, what is the time-dependent solution? Since the temperature in the absence of heat sources satisfies the equation

$$\frac{\partial T}{\partial t} = \kappa \nabla^2 T \tag{58.22}$$

subject to the boundary condition $T(\mathbf{r}, t) \to 0$ for $|\mathbf{r}| \to \infty$, it is natural to introduce a three-dimensional Fourier transform

$$\tilde{T}(\mathbf{k}, t) = \int d^3r \, e^{-i\mathbf{k}\cdot\mathbf{r}} T(\mathbf{r}, t) \tag{58.23}$$

As in Sec. 50, we readily find that $\tilde{T}(\mathbf{k}, t)$ satisfies an ordinary differential equation

$$\frac{\partial}{\partial t} \tilde{T}(\mathbf{k}, t) = -\kappa k^2 \tilde{T}(\mathbf{k}, t) \tag{58.24}$$

with the initial condition

$$\tilde{T}(\mathbf{k}, t = 0) \equiv \tilde{f}(\mathbf{k}) = \int d^3r \, e^{-i\mathbf{k}\cdot\mathbf{r}} f(\mathbf{r}) \tag{58.25}$$

The solution is obviously

$$\tilde{T}(\mathbf{k}, t) = \tilde{T}(\mathbf{k}, t = 0)e^{-\kappa k^2 t} = \tilde{f}(\mathbf{k})e^{-\kappa k^2 t} \tag{58.26}$$

For any $f(\mathbf{r})$, and hence $\tilde{f}(\mathbf{k})$, Eq. (58.23) can now be inverted to find $T(\mathbf{r}, t)$, but it is often simpler to combine Eqs. (58.25) and (58.26) directly to give [compare Eqs. (50.8) and (50.9)]

$$T(\mathbf{r}, t) = \int \frac{d^3k}{(2\pi)^3} e^{i\mathbf{k}\cdot\mathbf{r}} e^{-\kappa k^2 t} \tilde{f}(\mathbf{k})$$

$$= \int \frac{d^3k}{(2\pi)^3} \int d^3r' \, e^{i\mathbf{k}\cdot(\mathbf{r}-\mathbf{r}')} e^{-\kappa k^2 t} f(\mathbf{r}') \tag{58.27}$$

† See, for example, Jackson [3], chap. 7.

This expression has the form of a Green's function

$$T(\mathbf{r}, t) = \int d^3r' \mathcal{G}(\mathbf{r} - \mathbf{r}', t) T(\mathbf{r}', 0) \tag{58.28}$$

that determines $T(\mathbf{r}, t)$ in terms of its initial value $f(\mathbf{r}) = T(\mathbf{r}, 0)$.† Here, the Green's function has the integral representation

$$\mathcal{G}(\mathbf{r}, t) = \int \frac{d^3k}{(2\pi)^3} e^{i\mathbf{k} \cdot \mathbf{r}} e^{-\kappa k^2 t} \tag{58.29}$$

It has a simple physical interpretation as the time-dependent temperature distribution arising from an initial temperature distribution $\delta(\mathbf{r}) = \delta(x)\,\delta(y)\,\delta(z)$ at the origin.

To construct $\mathcal{G}(\mathbf{r}, t)$ explicitly, we note that the integral is the product of three identical factors (recall $k^2 \equiv \mathbf{k}^2 = k_x^2 + k_y^2 + k_z^2$) and therefore

$$\mathcal{G}(\mathbf{r}, t) = I(x, t)I(y, t)I(z, t) \tag{58.30}$$

where

$$I(x, t) \equiv \int_{-\infty}^{\infty} \frac{dk}{2\pi} e^{ikx} e^{-\kappa k^2 t} \tag{58.31}$$

This integral can be evaluated by adding and subtracting the term $x^2/4\kappa t$ in the exponential to complete the square

$$I(x, t) = \exp\left(-\frac{x^2}{4\kappa t}\right) \int_{-\infty}^{\infty} \frac{dk}{2\pi} \exp\left(-\kappa k^2 t + ikx + \frac{x^2}{4\kappa t}\right) \tag{58.32}$$

A simple change of variables reduces the integral to a gaussian, and we find, as in Eq. (54.61),

$$I(x, t) = (4\pi\kappa t)^{-1/2} \exp\left(-\frac{x^2}{4\kappa t}\right) \tag{58.33}$$

To interpret this function, we note that Eq. (58.31) implies

$$\int_{-\infty}^{\infty} dx\, I(x, t) = 1 \tag{58.34}$$

for all t. Furthermore, Eq. (58.33) shows that $I(x, t)$ is peaked about the origin, with a width approximately equal to $(\kappa t)^{1/2}$ and a height approximately equal to $(\kappa t)^{-1/2}$; it approaches a one-dimensional delta function as $t \to 0$ and spreads out with increasing time.

The three-dimensional Green's function now follows immediately from (58.30) and (58.33)

$$\mathcal{G}(\mathbf{r}, t) = (4\pi\kappa t)^{-3/2} \exp\left(-\frac{r^2}{4\kappa t}\right) \tag{58.35}$$

---

† This solution is analogous to d'Alembert's solution (39.18) of the one-dimensional wave equation (see also Prob. 9.7). Note, however, that here the specification of the initial value $T(\mathbf{r}, 0)$ suffices to determine the solution $T(\mathbf{r}, t)$ at all subsequent times. This result follows because the basic differential equation (58.22) contains only the *first*, rather than the *second*, time derivative.

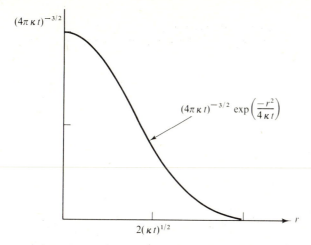

$(4\pi\kappa t)^{-3/2}$

$(4\pi\kappa t)^{-3/2} \exp\left(\dfrac{-r^2}{4\kappa t}\right)$

r

$2(\kappa t)^{1/2}$

Figure 58.3 Green's function $\mathcal{G}(\mathbf{r}, t) = (4\pi\kappa t)^{-3/2} e^{-r^2/4\kappa t}$ [Eq. (58.35)] for propagation of a thermal distribution in the infinite three-dimensional domain [Eq. (58.28)].

where $r^2 = x^2 + y^2 + z^2$. As $t \to 0$, it properly approaches the limit $\delta(\mathbf{r})$. For finite t, however, it spreads out (see Fig. 58.3) and ultimately tends to zero for all r as $t \to \infty$. In addition, the spatial variable appears only in the combination $r^2/4\kappa t$, again permitting a simple scaling of solutions for different t.

Although this solution describes an infinite domain, it also applies to a half space with specified temperature or heat current on the boundary, for the method of images can then provide the appropriate solutions just as in Sec. 50 (see Prob. 11.6).

59 LAPLACE TRANSFORM

We have seen in several examples that the Fourier transform typically works for a variable (either in space or time) defined on an infinite domain. If the domain is finite (or semi-infinite) owing to the presence of spatial boundaries or initial conditions, the problem is more complicated. In the latter case, however, when the time runs from 0 to ∞, the problem often can be solved with a Laplace transform in the time, *independent of the presence of spatial boundaries*. The theory of heat conduction provides many examples of such initial-value problems, which therefore serves as a convenient introduction to this technique.

Inversion Theorem

Consider a function $f(t)$ defined for $t \geq 0$. The Laplace transform $\bar{f}(s)$ is defined as

$$\bar{f}(s) = \int_0^\infty dt\, e^{-st} f(t) \tag{59.1}$$

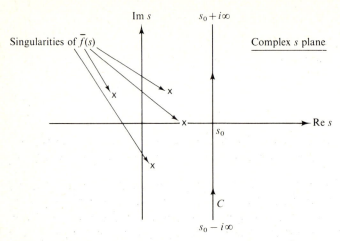

Figure 59.1 Contour C used for inverting of the Laplace transform.

for Re $s > 0$ and by analytic continuation elsewhere in the complex s plane. If $f(t)$ is integrable, so that $\bar{f}(0)$ is finite, $\bar{f}(s)$ and all its derivatives exist for Re $s > 0$. More generally, even if $f(t)$ grows like t^n as $t \to \infty$, the Laplace transform $\bar{f}(s)$ is analytic for Re $s > 0$.

As seen subsequently, we can often construct the Laplace transform $\bar{f}(s)$ of the solution to a given problem. The question then arises of the inversion to recover the original time-dependent solution $f(t)$. To analyze this problem, we consider the following contour integral (see Fig. 59.1)

$$I(t) = \int_C \frac{ds}{2\pi i} e^{st} \bar{f}(s) \tag{59.2}$$

where C runs from $s_0 - i\infty$ to $s_0 + i\infty$ *to the right* of all singularities of $\bar{f}(s)$. Substitute Eq. (59.1) into Eq. (59.2) and interchange the order of integrations

$$I(t) = \int_0^\infty dt'\, f(t') \int_C \frac{ds}{2\pi i} e^{s(t-t')} \tag{59.3}$$

The inner contour integral can be rewritten with the change of variable $s = s_0 + ix$ to give an integral representation of the delta function [see Eq. (50.4a)]

$$\int_C \frac{ds}{2\pi i} e^{s(t-t')} = e^{s_0(t-t')} \int_{-\infty}^\infty \frac{dx}{2\pi} e^{ix(t-t')}$$

$$= e^{s_0(t-t')} \delta(t-t') = \delta(t-t') \tag{59.4}$$

A combination with Eq. (59.3) yields the inversion formula [see footnote below Eq. (41.43)]

$$\int_C \frac{ds}{2\pi i} e^{st} \bar{f}(s) = \int_0^\infty dt'\, f(t')\delta(t-t') = \begin{cases} 0 & t < 0 \\ f(t) & t > 0 \end{cases} \tag{59.5}$$

Note that the contour C automatically guarantees that $f(t)$ vanishes for $t < 0$.

Example: Half Space at Fixed Surface Temperature

To exhibit the usefulness of the Laplace transform,† we consider a half space $z \geq 0$, initially at a uniform temperature $T = 0$. At $t = 0$, the surface $z = 0$ is brought in contact with a reservoir that maintains the constant surface temperature T_0 (see Fig. 59.2). What is the temperature throughout the half space for all subsequent time?

Evidently, we have a one-dimensional problem with the heat-conduction equation

$$\frac{\partial T(z, t)}{\partial t} = \kappa \frac{\partial^2 T(z, t)}{\partial z^2} \tag{59.6}$$

in the domain $0 \leq z < \infty$, $0 \leq t < \infty$, subject to the initial condition

$$T(z, 0) = 0 \tag{59.7a}$$

and boundary condition

$$T(0, t) = T_0 \tag{59.7b}$$

Introduce a Laplace transform in the time variable

$$\bar{T}(z, s) = \int_0^\infty dt \, e^{-st} T(z, t) \tag{59.8}$$

To determine the equation for $\bar{T}(z, s)$, we multiply Eq. (59.6) by e^{-st} and integrate over t from 0 to ∞. The right-hand side obviously gives $\kappa \, d^2\bar{T}/dz^2$; in contrast, the left-hand side must be integrated by parts

$$\int_0^\infty dt \, e^{-st} \frac{\partial}{\partial t} T(z, t) = e^{-st} T(z, t) \Big|_0^\infty + s \int_0^\infty dt \, e^{-st} T(z, t)$$

$$= - T(z, t = 0) + s\bar{T}(z, s) \tag{59.9}$$

Here we have observed that $e^{-st} T(z, t)$ vanishes as $t \to \infty$ for all $z \geq 0$, and that the integral remaining on the right-hand side of this relation is again $\bar{T}(z, s)$. In

† Other transform techniques are discussed in Tranter [4].

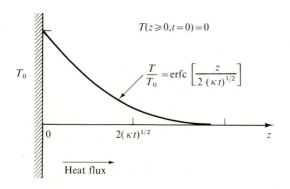

$T(z \geq 0, t = 0) = 0$

$\dfrac{T}{T_0} = \text{erfc}\left[\dfrac{z}{2(\kappa t)^{1/2}}\right]$

T_0

0 $2(\kappa t)^{1/2}$ z

Heat flux

Figure 59.2 Half space $z \geq 0$, initially at zero temperature, is brought into contact with a reservoir that maintains the constant surface temperature T_0 for $t > 0$.

this way, we obtain the following *ordinary, inhomogeneous differential equation* for $\bar{T}(z, s)$

$$\kappa \frac{d^2 \bar{T}(z, s)}{dz^2} = s\bar{T}(z, s) - T(z, t = 0) \tag{59.10}$$

that *automatically incorporates the initial conditions.* Note that the Laplace-transform variable s now simply appears as a *parameter* in this spatial differential equation. In the present case, Eq. (59.7a) simplifies the right-hand side to give the homogeneous equation

$$\left(\frac{d^2}{dz^2} - \frac{s}{\kappa} \right) \bar{T}(z, s) = 0 \tag{59.11}$$

whose solution is a linear combination of the two elementary solutions $\exp\left[\pm z(s/\kappa)^{1/2}\right]$. The physical condition that $\bar{T}(z, s)$ remain bounded for $z \to \infty$ eliminates the upper sign, and we therefore infer

$$\bar{T}(z, s) = A(s) \exp\left[-z\left(\frac{s}{\kappa}\right)^{1/2} \right] \tag{59.12}$$

where $A(s)$ is an arbitrary function of s that must be chosen to satisfy the *boundary condition at* $z = 0$. This is obtained from the Laplace transform of Eq. (59.7b)

$$\bar{T}(0, s) = \int_0^\infty dt\, e^{-st} T(0, t) = T_0 \int_0^\infty dt\, e^{-st} = \frac{T_0}{s} \tag{59.13}$$

since $T(0, t) = T_0$ for all times. It is convenient to introduce the abbreviation

$$\xi = z\kappa^{-1/2} \tag{59.14}$$

and Eqs. (59.12) to (59.14) give the result

$$\bar{T}(z, s) = \frac{T_0}{s} \exp\left(-s^{1/2}\xi \right) \tag{59.15}$$

It now remains to determine the full solution $T(z, t)$ with the inversion formula (59.5)

$$T(z, t) = \frac{T_0}{2\pi i} \int_C \frac{ds}{s} e^{st} \exp\left(-s^{1/2}\xi \right) \tag{59.16}$$

where, as previously, C runs to the right of all singularities in the complex s plane (see Fig. 59.3). In the present case, the integrand has a *branch point* at the origin $s = 0$, since an expansion near $s = 0$ gives $s^{-1} - \xi s^{-1/2} + \cdots$. To utilize the original contour C, the branch cut must lie in the left half s plane, and it is simplest to put it on the negative real axis. If $t < 0$, the contour C can be deformed far to the right without encountering any singularities, and the factor e^{st} ensures that $T(z, t)$ then vanishes.

For positive t, however, we deform the contour C into the *left half plane*

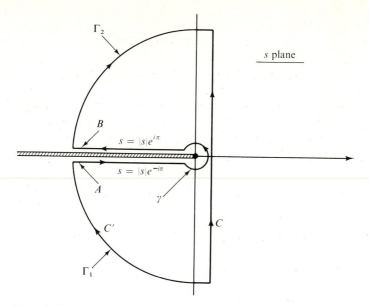

Figure 59.3 Contours C and C' used for evaluating the Laplace transforms (59.16) and (59.17).

(shown as C' in Fig. 59.3). Since the integrand is analytic in the region between C and C', Eq. (59.16) becomes

$$T(z, t) = \frac{T_0}{2\pi i} \int_{C'} \frac{ds}{s} e^{st} e^{-s^{1/2}\xi} \tag{59.17}$$

where C' consists of the two large arcs Γ_1 and Γ_2, the small circle γ about the origin, and the lines A and B below and above the branch cut. We consider these contributions separately.

1. Along Γ_1 and Γ_2, the integration variable is $s = Re^{i\phi}$, where the branch cut requires $|\phi| \le \pi$. The factor $s^{1/2}$ then has the form

$$s^{1/2} = R^{1/2} \exp i\tfrac{1}{2}\phi = R^{1/2} \cos \tfrac{1}{2}\phi + iR^{1/2} \sin \tfrac{1}{2}\phi \tag{59.18a}$$

and its exponential becomes

$$\exp\left(-s^{1/2}\xi\right) = \exp\left(-R^{1/2}\xi \cos \tfrac{1}{2}\phi\right) \exp\left(-iR^{1/2}\xi \sin \tfrac{1}{2}\phi\right) \tag{59.18b}$$

As a result, the quantity $s^{-1} \exp\left(-s^{1/2}\xi\right)$ satisfies the inequality on Γ_1 and Γ_2

$$\left| \frac{\exp\left(-s^{1/2}\xi\right)}{s} \right| \le \frac{\exp\left(-R^{1/2}\xi \cos \tfrac{1}{2}\phi\right)}{R} \tag{59.19}$$

and the right-hand side tends to zero as $R \to \infty$ because $\cos \tfrac{1}{2}\phi \ge 0$. This condition permits use of Jordan's lemma (see Prob. A.5) to conclude that the integrals along Γ_1 and Γ_2 indeed vanish as $R \to \infty$.

2. The integral along the small circle γ is easily evaluated with the substitution $s = \epsilon e^{i\phi}$, where $\epsilon \to 0$ and $|\phi| \leq \pi$. Since $s^{-1} ds = i \, d\phi$, we immediately obtain

$$\frac{T_0}{2\pi i} \int_\gamma \frac{ds}{s} e^{st} \exp\left(-s^{1/2}\xi\right) = \frac{T_0}{2\pi} \int_{-\pi}^{\pi} d\phi \left[1 + O(\epsilon^{1/2})\right] \to T_0 \qquad \text{as } \epsilon \to 0 \quad (59.20)$$

3. On the line B, we write $s = \sigma e^{i\pi}$, so that $ds = -d\sigma$ and $s^{1/2} = \sigma^{1/2} \exp i\tfrac{1}{2}\pi = i\sigma^{1/2}$. Similarly, $s = \sigma e^{-i\pi}$ on A, with $ds = -d\sigma$ and $s^{1/2} = \sigma^{1/2} \exp\left(-i\tfrac{1}{2}\pi\right) = -i\sigma^{1/2}$. Taking account of the opposite directions, we find

$$\frac{T_0}{2\pi i} \int_{A+B} \frac{ds}{s} e^{st} e^{-s^{1/2}\xi} = \frac{T_0}{2\pi i} \int_\epsilon^R \frac{d\sigma}{\sigma} e^{-\sigma t} (e^{-i\sigma^{1/2}\xi} - e^{i\sigma^{1/2}\xi}) = -T_0 J(\xi, t) \quad (59.21)$$

where $J(\xi, t)$ is a definite integral

$$J(\xi, t) \equiv \frac{1}{\pi} \int_0^\infty \frac{d\sigma}{\sigma} e^{-\sigma t} \sin \sigma^{1/2}\xi \qquad (59.22)$$

and we have now taken the limits $R \to \infty$, $\epsilon \to 0$. The substitution $\sigma = \lambda^2$ brings it to the form

$$J(\xi, t) = \frac{2}{\pi} \int_0^\infty \frac{d\lambda}{\lambda} e^{-\lambda^2 t} \sin \lambda\xi \qquad (59.23)$$

which could be integrated apart from the factor λ^{-1}. Thus we consider the derivative

$$\frac{\partial}{\partial \xi} J(\xi, t) = \frac{2}{\pi} \int_0^\infty d\lambda \, e^{-\lambda^2 t} \cos \lambda\xi = \frac{1}{\pi} \int_{-\infty}^\infty d\lambda \, e^{-\lambda^2 t} e^{i\lambda\xi} \qquad (59.24)$$

which is twice that evaluated in Eqs. (58.31) to (58.33)

$$\frac{\partial}{\partial \xi} J(\xi, t) = (\pi t)^{-1/2} e^{-\xi^2/4t} \qquad (59.25)$$

Integrate this relation from 0 to ξ

$$\int_0^\xi d\xi' \frac{\partial}{\partial \xi'} J(\xi', t) = J(\xi, t) - J(0, t)$$

$$= (\pi t)^{-1/2} \int_0^\xi d\xi' \, e^{-\xi'^2/4t} \qquad (59.26)$$

Equation (59.23) shows that $J(0, t)$ vanishes, and a simple change of variable yields the final form

$$J(\xi, t) = \frac{2}{\pi^{1/2}} \int_0^{\xi/2t^{1/2}} d\zeta \, e^{-\zeta^2} = \text{erf}\left(\frac{\xi}{2t^{1/2}}\right) \qquad (59.27)$$

Here, the error function erf (x) and its complement erfc (x) are defined by indefinite integrals†

$$\text{erf}\,(x) \equiv \frac{2}{\pi^{1/2}} \int_0^x d\zeta \, e^{-\zeta^2}$$

$$\text{erfc}\,(x) \equiv 1 - \text{erf}\,(x) = \frac{2}{\pi^{1/2}} \int_x^\infty d\zeta \, e^{-\zeta^2} \qquad (59.28)$$

They have the limiting forms

$$\text{erf}\,(x) \approx \frac{2}{\pi^{1/2}} (x - \tfrac{1}{3}x^3 + \cdots) \qquad x \to 0$$

$$\text{erfc}\,(x) \sim \frac{e^{-x^2}}{\pi^{1/2}x} \qquad x \to \infty \qquad (59.29)$$

A combination of Eqs. (59.20), (59.21), (59.27), and (59.28) with Eq. (59.14) gives the exact solution to the problem posed in this section and illustrated in Fig. 59.2

$$T(z, t) = T_0 \,\text{erfc}\left[\frac{z}{2(\kappa t)^{1/2}}\right] \qquad (59.30)$$

This expression has several notable features.

1. It satisfies the heat-conduction equation (59.6) in the domain $0 \le z < \infty$ and $0 \le t < \infty$, as the reader can readily verify. It depends only on the combination $z^2/\kappa t$ and again exhibits the characteristic scaling of heat conduction.
2. This result obviously satisfies the initial condition (59.7a) for $z \ne 0$ because erfc (x) vanishes exponentially as $x \to \infty$ [see Eq. (59.29)].
3. The result (59.30) also satisfies the boundary condition (59.7b) for all $t > 0$ because erfc $(0) = 1$. Evidently, $T(z, t)$ is nonanalytic for small z and t since it depends on the order of taking the limits; indeed, the $t^{-1/2}$ dependence in the argument of Eq. (59.30) exhibits an essential singularity as $t \to 0$.
4. Finally, note that T is everywhere positive for any finite time, implying that the disturbance spreads instantaneously throughout the whole sample. If $z^2/\kappa t$ is not too large, the classical continuum description indeed remains valid. Far out in the tail $(z^2/\kappa t \to \infty)$, however, the temperature gradient is so small that fluctuations become important. This latter situation represents a breakdown of the macroscopic picture, necessitating a wholly different microscopic analysis.‡

Given the analytic solution (59.30), it is instructive to consider its physical implications; as an example, we here evaluate the heat flux through the bounding plane

$$j_{hz} = -k_{\text{th}} \frac{\partial T}{\partial z}\bigg|_{z=0} \qquad (59.31)$$

† See, for example Abramowitz and Stegun [5], sec. 7.1.
‡ See, for example, Reif [1], secs. 15.5 and 15.6.

The integral representation (59.28) immediately yields

$$j_{hz}(t) = \frac{T_0 k_{th}}{(\pi \kappa t)^{1/2}} \tag{59.32}$$

which is manifestly nonanalytic in t. As might be inferred from the instantaneous propagation, j_{hz} diverges at $t \to 0$. At large times, in contrast, j_{hz} becomes small as the sample acquires the uniform temperature T_0 out to distances $z \approx 2(\kappa t)^{1/2}$ (see Fig. 59.2).

Although this problem also involves a semi-infinite *spatial* domain $0 \le z < \infty$, we may note that a Laplace transform in z fails to produce a solution. This difference arises because the ordinary differential equation for such a Laplace transform would involve both $T(0, t)$ and $[\partial T(z, t)/\partial z]_{z=0}$, but the boundary condition provides only $T(0, t)$. Nevertheless, this same problem can still be solved with other spatial transform techniques such as the method of images (see Prob. 11.6) or (essentially equivalent) a spatial sine transform (see Prob. 11.14).

Example: Sphere Heated Internally

In the preceding example, inversion of the Laplace transform produced an explicit solution for a boundary-value problem of heat conduction. More typically, however, the inversion integral cannot be evaluated in closed form. Nevertheless, the contour integral (59.5) provides an *exact integral representation* of the solution that is often more useful than the infinite series obtained with separation of variables. As an example, we treat a sphere of radius a, with its surface maintained at zero temperature. At $t = 0$, we insert a source of heat, e.g., a radioactive pellet, that emits $\dot{Q}_0$ ergs s^{-1} (see Fig. 59.4). The equation of heat conduction obtained from (57.9) to (57.11) in this case becomes

$$\frac{\partial T}{\partial t} = \kappa \nabla^2 T + \frac{\dot{Q}_0}{\rho c_p} \delta(\mathbf{r}) \tag{59.33}$$

where $\delta(\mathbf{r})$ is a three-dimensional delta function and $T(\mathbf{r}, t)$ satisfies the initial condition

$$T(\mathbf{r}, 0) = 0 \tag{59.34a}$$

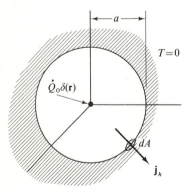

Fig. 59.4 Sphere with surface maintained at $T = 0$ and heat source $\dot{Q}_0 \delta(\mathbf{r})$ inserted at $t = 0$.

and the boundary condition

$$T(r = a, t) = 0 \tag{59.34b}$$

This problem is readily solved with a Laplace transform

$$\bar{T}(\mathbf{r}, s) = \int_0^\infty dt\, e^{-st} T(\mathbf{r}, t) \tag{59.35}$$

Multiplication of Eq. (59.33) by e^{-st} and integration over all positive time shows that the Laplace transform $\bar{T}(\mathbf{r}, s)$ satisfies the inhomogeneous differential equation

$$s\bar{T}(\mathbf{r}, s) = \kappa\, \nabla^2 \bar{T}(\mathbf{r}, s) + \frac{\dot{Q}_0}{\rho c_p s}\, \delta(\mathbf{r}) \tag{59.36}$$

where the first term is obtained through a partial integration with respect to time and incorporates the initial condition (59.34a). The factor of s in the denominator of the last term follows from $\int_0^\infty e^{-st}\, dt = s^{-1}$. In addition to the equation (59.36), $\bar{T}(\mathbf{r}, s)$ also satisfies the boundary condition

$$\bar{T}(r = a, s) = 0 \tag{59.37}$$

obtained as the Laplace transform of Eq. (59.34b).

Evidently, the spherical symmetry of the problem as formulated ensures that $\bar{T}(\mathbf{r}, s)$ depends only on the radial coordinate r. For any finite r, $\bar{T}$ satisfies the *homogeneous* differential equation [see Eq. (B.21)]

$$\frac{1}{r^2}\frac{d}{dr} r^2 \frac{d}{dr} \bar{T}(r, s) - \frac{s}{\kappa} \bar{T}(r, s) = 0 \qquad r \neq 0 \tag{59.38}$$

and it is convenient to replace the source term in Eq. (59.36) with an equivalent boundary condition at the origin, just as in the derivation of the one-dimensional Green's function in Sec. 43. Integrate Eq. (59.36) over the small sphere $r < \epsilon$

$$s\int_{r<\epsilon} d^3r\, \bar{T}(r, s) = \kappa \int_{r<\epsilon} d^3r\, \nabla^2 \bar{T}(r, s) + \frac{\dot{Q}_0}{\rho c_p s}\int_{r<\epsilon} d^3r\, \delta(\mathbf{r}) \tag{59.39}$$

The source emits heat at a constant rate, so that $\dot{Q}_0 \equiv \int d\mathbf{A} \cdot \mathbf{j}_h|_{r=\epsilon} = -k_{th}\int d\mathbf{A} \cdot \nabla T|_{r=\epsilon}$ must be finite on this small sphere. Thus we anticipate that $\bar{T}(r, s)$ diverges like r^{-1} as $r \to 0$, which renders the left-hand side of Eq. (59.39) of order ϵ^2 as $\epsilon \to 0$. On the right-hand side, the second integral is exactly 1, and the first term can be rewritten with the divergence theorem to yield the boundary condition for $\bar{T}(r, s)$ at the origin [see Eq. (57.13)]

$$\lim_{\epsilon \to 0}\left[\int_{r=\epsilon} d\mathbf{A} \cdot \nabla \bar{T}(r, s) + \frac{\dot{Q}_0}{\kappa\rho c_p s}\right] = 0$$

or
$$\lim_{\epsilon \to 0} 4\pi\epsilon^2 \left[\frac{d\bar{T}(r, s)}{dr}\right]_{r=\epsilon} + \frac{\dot{Q}_0}{k_{th} s} = 0 \tag{59.40}$$

For notational convenience, we again define new variables [compare Eq. (59.14)]

$$\xi = \frac{r}{\kappa^{1/2}} \qquad \xi_0 = \frac{a}{\kappa^{1/2}} \tag{59.41}$$

and $\bar{T}$ then satisfies the spherically symmetric equation [compare Eq. (59.38)]

$$\frac{1}{\xi^2} \frac{d}{d\xi} \xi^2 \frac{d\bar{T}}{d\xi} - s\bar{T} = 0 \qquad 0 < \xi \le \xi_0 \tag{59.42}$$

subject to the boundary conditions

$$\bar{T}(\xi_0, s) = 0 \tag{59.43a}$$

$$\left(\xi^2 \frac{d\bar{T}}{d\xi} \right)_{\xi=0} = -\frac{\dot{Q}_0}{4\pi k_{th} \kappa^{1/2} s} \tag{59.43b}$$

The independent solutions of Eq. (59.42) are spherical Bessel functions of order zero with imaginary argument $j_0(is^{1/2}\xi)$ and $n_0(is^{1/2}\xi)$ or, equivalently, $(s^{1/2}\xi)^{-1} \sinh s^{1/2}\xi$ and $(s^{1/2}\xi)^{-1} \cosh s^{1/2}\xi$. The boundary condition (59.43a) selects the particular linear combination of these latter solutions

$$\bar{T}(\xi, s) = A(s) \frac{\sinh [s^{1/2}(\xi_0 - \xi)]}{s^{1/2}\xi} \tag{59.44}$$

where $A(s)$ can still depend on s. An expansion for small ξ and comparison with Eq. (59.43b) fixes this coefficient, and we find

$$\bar{T}(\xi, s) = \frac{\dot{Q}_0}{4\pi k_{th} \kappa^{1/2}\xi} \frac{\sinh [(\xi_0 - \xi)s^{1/2}]}{s \sinh \xi_0 s^{1/2}} \tag{59.45}$$

Substitution into the inversion formula (59.5) gives an exact integral representation of the full solution

$$T(r, t) = \frac{\dot{Q}_0}{4\pi k_{th} r} \int_C \frac{ds}{2\pi i} \frac{e^{st}}{s} \frac{\sinh [(\xi_0 - \xi)s^{1/2}]}{\sinh \xi_0 s^{1/2}} \tag{59.46}$$

with ξ and ξ_0 given by Eqs. (59.41). It is easy to verify that the integrand is meromorphic in the s plane, for it has a simple pole at the origin with residue $\xi_0^{-1}(\xi_0 - \xi) = 1 - r/a$ and simple poles on the negative real axis at the values $s_n = -(n\pi/\xi_0)^2$ for $n = 1, 2, \ldots, \infty$. A straightforward deformation of the contour C into the left-hand plane gives the exact series solution

$$T(r, t) = \frac{\dot{Q}_0}{4\pi k_{th}} \left[\left(\frac{1}{r} - \frac{1}{a} \right) - \sum_{n=1}^{\infty} \frac{2}{n\pi r} \sin \frac{n\pi r}{a} \exp \left(-\frac{n^2\pi^2 \kappa t}{a^2} \right) \right] \tag{59.47}$$

which converges uniformly for all $t > 0$. Here, the first term in parentheses is the eventual steady-state distribution; it arises from the pole at $s = 0$. The remaining series contains the time-dependent transients that come from the remaining poles in the left-hand s plane. Note that the initial condition (59.34a) implies the nontrivial Fourier series

$$1 - \frac{r}{a} = \sum_{n=1}^{\infty} \frac{2}{n\pi} \sin \frac{n\pi r}{a} \qquad \text{for } 0 < r < a \qquad (59.48)$$

The same time-dependent solution (59.47) can also be obtained with separation of variables (see Prob. 11.2).

Approximation Methods for Long and Short Times

In the preceding example, evaluation of the integral (59.46) provided an infinite-series representation (59.47) of $T(r, t)$ that is valid for $0 < r \leq a$ and for all t. As with the rectangular parallelepiped [see Eq. (58.12)], the long-time behavior contains a constant component and a term that decays with a characteristic time $a^2/\kappa\pi^2$

$$T(r, t) \sim \frac{\dot{Q}_0}{4\pi k_{\text{th}}} \left[\left(\frac{1}{r} - \frac{1}{a} \right) - \frac{2}{\pi r} \sin \frac{\pi r}{a} e^{-\pi^2 \kappa t/a^2} + \cdots \right] \qquad t \to \infty \quad (59.49)$$

These terms arise from the poles at $s = 0$ and $s = -\kappa\pi^2/a^2$, respectively, with the remaining poles contributing more rapid transients. Such behavior is in fact quite general, and we therefore infer that

> The farthest right singularities of the Laplace transform $\bar{f}(s)$ (typically those nearest the imaginary s axis) control the asymptotic form of the solution $f(t)$ as $t \to \infty$
> (59.50)

In the present case, the existence of well-separated simple poles makes the inversion elementary, but the assertion also holds for more complicated singularities. Since $\bar{f}(s)$ is usually analytic in the right half s plane, it must contain one or more singularities in the left half s plane at $-a_n + ib_n$ with $a_n \geq 0$ and $n = 1, 2, \ldots$. Apart from algebraic factors in t, each of these contributes a term of the form $e^{-a_n t + ib_n t}$ to the function $f(t)$, and the one with the smallest a_n dominates as $t \to \infty$.

In this context, it is interesting to consider the slightly modified problem of a sphere with insulated surface. The corresponding Laplace transform turns out to have a second-order pole at $s = 0$, and the expansion of e^{st} to construct the residue yields a linearly increasing temperature, as expected on physical grounds (Prob. 11.13). The earlier example of a half space also exhibits analogous behavior (see Fig. 59.3), since the branch point at $s = 0$ produces the constant asymptotic term (59.20), with the time-dependent transient (59.22) coming from the rest of the branch cut.

We noted in connection with Eq. (58.10) that the separation-of-variables solution is awkward for small t when many terms contribute to the sum. In this limit, the inversion integral for the Laplace transform is invaluable because it provides an integral representation that is valid for all t. Moreover, this representation is easily manipulated by changing variables. In particular, we assert that

> The short-time behavior of $f(t)$ can be isolated by the substitution $\zeta = st$ in the inversion formula (59.5)   (59.51)

As a specific example, make this substitution in Eq. (59.46)

$$T(r, t) = \frac{\dot{Q}_0}{4\pi k_{th} r} \frac{1}{2\pi i} \int_C \frac{d\zeta}{\zeta} e^{\zeta} \frac{\sinh \left[(\xi_0 - \xi)(\zeta/t)^{1/2} \right]}{\sinh \left[\xi_0 (\zeta/t)^{1/2} \right]} \quad (59.52)$$

and observe that the dimensionless parameter $\xi_0 t^{-1/2}$ becomes large as $t \to 0$. Using the large-x expansion

$$\frac{1}{\sinh x} = \frac{2}{e^x - e^{-x}} = 2e^{-x} \sum_{p=0}^{\infty} e^{-2px} \quad (59.53)$$

we obtain the equivalent exact formula

$$T(r, t) = \frac{\dot{Q}_0}{4\pi k_{th} r} \frac{1}{2\pi i} \int_C \frac{d\zeta}{\zeta} e^{\zeta}$$
$$\times \sum_{p=0}^{\infty} \left\{ \exp \left[-(\xi + 2p\xi_0) \left(\frac{\zeta}{t} \right)^{1/2} \right] - \exp \left[-(2\xi_0 - \xi + 2p\xi_0) \left(\frac{\zeta}{t} \right)^{1/2} \right] \right\} \quad (59.54)$$

in which the integrand now has a completely different analytic structure from that in (59.52). Fortunately, each of the terms has just the form considered in Eqs. (59.16) and (59.14), allowing us to construct the alternative solution with Eqs. (59.30) and (59.41)

$$T(r, t) = \frac{\dot{Q}_0}{4\pi k_{th} r} \sum_{p=0}^{\infty} \left\{ \operatorname{erfc} \left[\frac{r + 2pa}{2(\kappa t)^{1/2}} \right] - \operatorname{erfc} \left[\frac{2a - r + 2pa}{2(\kappa t)^{1/2}} \right] \right\} \quad (59.55)$$

This expression is wholly equivalent to Eq. (59.47). It obviously satisfies the initial condition and boundary condition [Eqs. (59.34)], and it has the feature that it converges extremely rapidly for small t because of the exponential character of erfc for large arguments [see Eq. (59.29)]

$$T(r, t) \approx \frac{\dot{Q}_0}{4\pi k_{th} r} \left\{ \operatorname{erfc} \left[\frac{r}{2(\kappa t)^{1/2}} \right] - \operatorname{erfc} \left[\frac{2a - r}{2(\kappa t)^{1/2}} \right] \right.$$
$$\left. + \operatorname{erfc} \left[\frac{r + 2a}{2(\kappa t)^{1/2}} \right] + \cdots \right\} \quad t \to 0 \quad (59.56)$$

Such behavior confirms the general nonanalyticity for small time.

The technique of isolating the initial time dependence is particularly valuable for numerical studies. To illustrate its power, we consider the total heat transferred through the surface of the sphere

$$\dot{Q} \equiv \int_{r=a} d\mathbf{A} \cdot \mathbf{j}_h = -4\pi k_{th} a^2 \frac{\partial T}{\partial r} \bigg|_{r=a} \quad (59.57)$$

This quantity can be evaluated with either form of the solution. Substitution of (59.47) yields

$$\frac{\dot{Q}}{\dot{Q}_0} = 1 + 2 \sum_{n=1}^{\infty} (-1)^n e^{-n^2 \pi^2 \kappa t / a^2} \tag{59.58a}$$

which converges rapidly for $t \gtrsim a^2 / \kappa \pi^2$. Its limiting form

$$\frac{\dot{Q}}{\dot{Q}_0} \sim 1 - 2e^{-\pi^2 \kappa t / a^2} + 2e^{-4\pi^2 \kappa t / a^2} + \cdots \quad t \to \infty \tag{59.58b}$$

exhibits the steady state, when all the heat must pass outward through the surface, plus transient terms. In contrast, use of the series (59.55) yields the alternative formula

$$\frac{\dot{Q}}{\dot{Q}_0} = \frac{2a}{(\pi \kappa t)^{1/2}} \sum_{p=0}^{\infty} e^{-(2p+1)^2 a^2 / 4\kappa t} \tag{59.59a}$$

with the approximate short-time form

$$\frac{\dot{Q}}{\dot{Q}_0} \approx \frac{2a}{(\pi \kappa t)^{1/2}} e^{-a^2 / 4\kappa t} (1 + e^{-2a^2 / \kappa t} + \cdots) \quad t \to 0 \tag{59.59b}$$

This series converges rapidly for $t \lesssim a^2 / 4\kappa$, complementing the other form (59.58). To demonstrate this convergence, we may note that the simple approximations (59.58b) and (59.59b) give $\dot{Q}/\dot{Q}_0 \approx 0.587976$ and $\dot{Q}/\dot{Q}_0 \approx 0.587974$, respectively, for $t = a^2 / 2\pi\kappa$. Although these series can be obtained by other techniques (Prob. 11.15), the present approach leads naturally and directly to the answer.

In closing this section, we remark that these approximation methods also apply to Fourier integrals of the form

$$f(x) = \int_{-\infty}^{\infty} \frac{dk}{2\pi} e^{ikx} \tilde{f}(k) \tag{59.60}$$

If x is negative, the contour can be deformed into the lower half plane, and the corresponding singularities of $\tilde{f}(k)$ nearest the real axis determine the asymptotic behavior of $f(x)$ as $x \to -\infty$. Similarly, the singularities of $\tilde{f}(k)$ in the upper half plane nearest the real axis determine the asymptotic behavior as $x \to \infty$.† Finally, the substitution $kx = q$ frequently produces a useful form that can then be expanded to obtain an approximation for small $|x|$.

PROBLEMS

11.1 A long bar of thermal diffusivity κ has a square cross-sectional area with side a. The bar has an initial temperature T_0 and at time $t = 0$ is immersed in a thermal reservoir at zero temperature. The long dimension lies along the z axis.

(a) Show that the temperature T at a later time is given by $T = T_0 \, \phi(x, t)\phi(y, t)$, where

$$\phi(x, t) = \frac{4}{\pi} \sum_{n=0}^{\infty} \frac{1}{2n+1} \sin \left[(2n+1) \frac{\pi x}{a} \right] \exp \left[-\frac{\kappa \pi^2}{a^2} (2n+1)^2 t \right]$$

† For a detailed discussion, see Carrier, Krook, and Pearson [6], pp. 255–257.

(b) Discuss the uniform convergence of this series in x at $t = 0$ and for finite t.

(c) What is the large-t limit of this result?

11.2 Obtain the solution (59.47) for a sphere heated internally using the method of separation of variables.

11.3 (a) In a diffusive process, the flux of particles $\mathbf{j}$ is proportional to the concentration gradient. Hence $\mathbf{j} = -D\,\nabla n$, where D is the diffusion constant and n is the number density. Apply the conservation of particles to a fixed volume element and obtain the *diffusion equation* $\partial n/\partial t = D\,\nabla^2 n$.

(b) A cubical box with sides of length L is initially divided in half by a partition parallel to one pair of faces, with N atoms of isotope 1 on one side and N atoms of isotope 2 on the other. At $t = 0$, the partition is gently removed, and the otherwise identical species diffuse through each other. Derive an expression for the concentrations of the two species at subsequent times.

(c) How long does it take for the maximum relative concentration imbalance of isotope 1 to fall to the small value 0.01? Assuming the typical value $D = 0.1$ cm^2 s^{-1}, discuss the numerical result for various L.

11.4 (a) In certain reactor materials, neutrons propagate diffusively, so that the neutron flux $\mathbf{j}$ (number crossing unit area per unit time) is given by $\mathbf{j} = -D\,\nabla n$, where D is a diffusion constant and n is the neutron density. If the medium absorbs αn neutrons per unit volume per unit time, show that the conservation of neutrons is expressed by the equation $\partial n/\partial t = D\,\nabla^2 n - \alpha n$. How is this relation modified if the medium also emits p neutrons per unit volume per unit time?

(b) A rod of length L of such absorbing and emitting material has its sides and one end clad with an impenetrable barrier. The other end opens into a neutron bath with constant density n_0. Find the equilibrium (steady-state) neutron density at the closed end of the rod. Discuss the limiting cases of large and small $\alpha L^2/D$.

(c) Find the equilibrium value of the neutron influx through the open end, and verify by explicit calculation that it equals the total rate of neutron absorption and emission in the rod. Discuss the limiting cases as in part (b).

11.5 Consider one-dimensional heat conduction in a long rod with insulated sides. If the initial temperature $T(x, 0)$ is a given function $f(x)$ and $\dot{q}/c_p$ is a given time-independent function $g(x)$, show that the general solution for all $t \geq 0$ is

$$T(x, t) = \int_{-\infty}^{\infty} dx'\, I(x - x', t) f(x') + \int_{-\infty}^{\infty} dx' \int_0^t dt'\, I(x - x', t')\, g(x')$$

where $I(x, t)$ is the one-dimensional Green's function given in Eqs. (58.31) and (58.33).

11.6 Consider one-dimensional heat conduction in a semi-infinite rod $(x \geq 0)$ whose end is maintained at T_0 and whose initial temperature distribution is $f(x)$. Use the method of images to obtain the following solution for $t \geq 0$ (see Prob. 11.5)

$$T(x, t) = T_0 + \int_0^{\infty} dx'\, [I(x - x', t) - I(x + x', t)]\, [f(x') - T_0]$$

If $f(x) = 0$, rederive the solution obtained in (59.30).

11.7 A slab of thickness l perpendicular to the x axis has its two surfaces at $x = 0$ and $x = l$ maintained at zero temperature. Given the initial temperature $T(x, 0) = f(x)$, show that the general solution has the form [compare Eq. (58.28)]

$$T(x, t) = \int_0^l dx'\, \mathcal{G}(x, x', t) f(x')$$

and prove that $\mathcal{G}(x, x', t)$ has either of the alternative representations

$$\mathcal{G}(x, x', t) = \frac{2}{l} \sum_{n=1}^{\infty} \sin\frac{n\pi x}{l} \sin\frac{n\pi x'}{l}\, e^{-\kappa t(n\pi/l)^2}$$

or

$$\mathcal{G}(x, x', t) = U(x - x', t) - U(x + x', t)$$

where
$$U(x, t) = \frac{1}{(4\pi\kappa t)^{1/2}} \sum_{p=0}^{\infty} \left(e^{-(|x| + 2pl)^2/4\kappa t} + e^{-(2l - |x| + 2pl)^2/4\kappa t} \right)$$

Interpret the second form in terms of the method of images. Discuss the long- and short-time behavior. *Hint:* Use a Laplace transform and Eqs. (43.34) and (43.40).

11.8 A semi-infinite string ($x \geq 0$) of linear mass density σ is stretched to a tension τ. The string is placed in a frictional medium that resists transverse motion with a retarding force $-(\sigma/T)(\partial u/\partial t)$ per unit length, where T is a damping time. For $t < 0$, the string is undeformed, and at $t = 0$, the end at $x = 0$ is moved a transverse distance a and held there. Prove and discuss the following assertions:

(a) If $T \to \infty$, the deformed shape is $u(x, t) = a\theta(ct - x)$, where θ is the unit step function and $c = (\tau/\sigma)^{1/2}$.

(b) For any positive T, there is no motion at a point x if $t < x/c$.

(c) The slope at the displaced end is given asymptotically for large t by

$$-a(c^2\pi t T)^{-1/2}(1 - T/4t - 3T^2/32t^2 + \cdots)$$

11.9 Consider one-dimensional heat conduction in a rod of length l subject to the boundary conditions $T(0, t) = 0$ and $(\partial T/\partial x + \alpha T)_{x=l} = 0$ [see Eqs. (57.14) and (57.16b)] and the initial condition $T(x, 0) = f(x)$. Obtain an infinite-series solution analogous to that in Eq. (59.47). Discuss the orthogonality and completeness of the associated spatial eigenfunctions.

11.10 Consider heat conduction in an unbounded domain subject to an axisymmetric initial temperature distribution $T(r, 0) = f(r)$, where r is the radial distance in plane polar coordinates.

(a) Use a Laplace transform in t to derive an inhomogeneous ordinary differential equation for $\bar{T}(r, s)$.

(b) Solve this equation with an appropriate one-dimensional Green's function (see Sec. 43).

(c) Hence find an integral representation for $T(r, t)$. Take the limit $t \to 0$ and obtain the fundamental theorem for *Hankel transforms*

$$f(r) = \int_0^{\infty} p \, dp \, J_0(pr) \int_0^{\infty} r' \, dr' \, J_0(pr')f(r')$$

Explain why the integral over p expresses both orthogonality and completeness.

11.11 An infinite cylinder of radius a is initially at zero temperature. At $t = 0$, its surface temperature is raised to T_0 and maintained at that value.

(a) Use a Laplace transform to derive the integral representation

$$T(r, t) = \frac{T_0}{2\pi i} \int_c \frac{ds}{s} e^{st} \frac{I_0[r(s/\kappa)^{1/2}]}{I_0[a(s/\kappa)^{1/2}]}$$

where $I_0(x) = J_0(ix)$.

(b) For long times, show that

$$T(r, t) \sim T_0 - 2T_0 e^{-\alpha_{0,1}^2 t\kappa/a^2} \frac{J_0(\alpha_{0,1} r/a)}{\alpha_{0,1} J_1(\alpha_{0,1})} + \cdots$$

where $\alpha_{0,n}$ is the nth zero of $J_0(x)$. Discuss.

(c) For short times and $a - r \ll a$, show that

$$T(r, t) \approx T_0 \left(\frac{a}{r}\right)^{1/2} \operatorname{erfc}\left[\frac{a - r}{2(\kappa t)^{1/2}}\right]$$

Discuss.

(d) Find the heat flux through the surface in forms valid for long and short times.

11.12 A sphere of radius a initially at zero temperature is immersed at $t = 0$ in a bath with constant temperature T_0.

(a) Derive two equivalent solutions for $T(r, t)$ analogous to Eqs. (59.47) and (59.55). Compare with Eq. (59.30) for a planar geometry.

(b) Show that the central temperature can be expressed in series like (59.58a) and (59.59a) and sketch its time dependence.

(c) Obtain the analogous expressions for the total heat flow through the surface and compare with Eq. (59.32).

11.13 A sphere of radius a with an *insulated* surface is initially at zero temperature. At $t = 0$, a radioactive pellet that emits heat isotropically at a rate $\dot{Q}_0$ is inserted into the center of the sphere.

(a) Show that the Laplace transform of the temperature is given by

$$\bar{T}(r, s) = \frac{\dot{Q}_0}{4\pi k_{\text{th}}(\kappa s)^{1/2}} \left[u(\xi s^{1/2}) - \frac{v(\xi s^{1/2})u'(\xi_0 s^{1/2})}{v'(\xi_0 s^{1/2})} \right]$$

where $u(x) = x^{-1} \cosh x$, $v(x) = x^{-1} \sinh x$, $\xi = r\kappa^{-1/2}$ and $\xi_0 = a\kappa^{-1/2}$.

(b) For long times, show that the volume-averaged temperature is $\dot{Q}_0 t/C_p$, where $C_p = \rho c_p V$ is the sphere's heat capacity at constant pressure, and that the surface temperature lags behind by an amount $(\dot{Q}_0/C_p)(a^2/10\kappa)$.

11.14 Many other types of transform relations are useful in physics. For example, if the function $f(x)$ is defined only on the interval $0 \leq x \leq \infty$ we can write *sine-transform* relations

$$f(x) = \left(\frac{2}{\pi}\right)^{1/2} \int_0^\infty (\sin kx)\phi(k) \, dk$$

$$\phi(k) = \left(\frac{2}{\pi}\right)^{1/2} \int_0^\infty (\sin k\xi)f(\xi) \, d\xi$$

or, with obvious modifications, cosine transforms.

(a) Start with the results of Prob. 7.19 and derive these relations.

(b) Consider the problem of a half space at fixed surface temperature discussed in Sec. 59 (see Fig. 59.2). Introduce the sine transform $\phi(k, t)$ with respect to the spatial coordinate and derive the associated ordinary differential equation in the time, which incorporates the spatial *boundary conditions*. Why is the sine transform appropriate here?

(c) Solve this differential equation subject to the *initial condition* and invert the sine transform to recapture the result in Eq. (59.30).

11.15 The formulas in Eqs. (59.58a) and (59.59a) represent a special case of the Poisson sum formula. A direct derivation can proceed as follows.

(a) Given $f(x)$ and its Fourier transform $\tilde{f}(k) = \int_{-\infty}^{\infty} dx \, e^{-ikx}f(x)$, explain why the function $F(x) = \sum_{n=-\infty}^{\infty} f(x - na)$ has a Fourier-series expansion of the form

$$F(x) = \sum_{p=-\infty}^{\infty} e^{i2\pi px/a} g_p$$

and verify that $g_p = a^{-1}\tilde{f}(2\pi p/a)$. The limit of this relation for $x = 0$ constitutes the *Poisson sum formula*.

(b) Apply this expansion to the function $f(x) = e^{-\alpha^2 x^2 + iqx}$ and hence derive the remarkable Poisson sum formula for gaussians

$$\sum_{n=-\infty}^{\infty} e^{-n^2\alpha^2} = \frac{1}{\alpha}\pi^{1/2} \sum_{p=-\infty}^{\infty} e^{-\pi^2 p^2/\alpha^2}$$

which holds for $\alpha > 0$. Discuss the limiting forms for large and small α. Obtain an analogous relation for $\sum_{n=-\infty}^{\infty} (-1)^n e^{-n^2\alpha^2}$.

SELECTED ADDITIONAL READINGS

1. F. Reif, "Fundamentals of Statistical and Thermal Physics," McGraw-Hill, New York, 1965.
2. H. S. Carslaw and J. C. Jaeger, "Conduction of Heat in Solids," 2d ed., Oxford University Press, London, 1959.
3. J. D. Jackson, "Classical Electrodynamics," Wiley, New York, 1962.
4. C. J. Tranter, "Integral Transforms in Mathematical Physics," 3d ed., Methuen, London, 1966.
5. M. Abramowitz and I. A. Stegun (eds.), "Handbook of Mathematical Functions, with Formulas, Graphs, and Mathematical Tables," NBS Appl. Math. Ser. 55, Washington, 1964.
6. G. F. Carrier, M. Krook, and C. E. Pearson, "Functions of a Complex Variable," McGraw-Hill, New York, 1966

TWELVE

VISCOUS FLUIDS

Real fluids differ significantly from the "ideal" fluids considered in Chap. 9, most notably in their ability to resist dynamic tangential stress and in the existence of dissipation. These effects reflect the universal property of *viscosity*, and we now analyze its role in the general hydrodynamic equations.†

60 VISCOUS STRESS TENSOR

Consider a fluid in steady shear flow (Fig. 60.1), where

$$\mathbf{v} = v_x(y)\hat{x} \tag{60.1}$$

and v_x increases linearly with y. Clearly, this flow has $\mathbf{\nabla} \cdot \mathbf{v} = 0$ and is therefore permissible for an incompressible fluid; it is not irrotational, however, because $(\mathbf{\nabla} \times \mathbf{v})_z = -\partial v_x/\partial y$ is nonzero. Imagine a plane $y = y_0$ of constant flow velocity. Owing to the shear, the fluid above the surface exerts a force to the right on the fluid below. For small values of the shear, experiments show that the force per unit area is proportional to the velocity gradient $\partial v_x/\partial y$. This conclusion allows us to express the *drag force F_x acting on the fluid lying below the element of area dA* as‡

$$F_x = \eta \frac{\partial v_x}{\partial y} dA \tag{60.2}$$

---

† We have found the following books particularly useful: Batchelor [1]; Chandrasekhar [2]; Landau and Lifshitz (1959) chaps. 2, 5, and 8.

‡ The fluid below, of course, exerts an equal and opposite force on the fluid above, by Newton's third law.

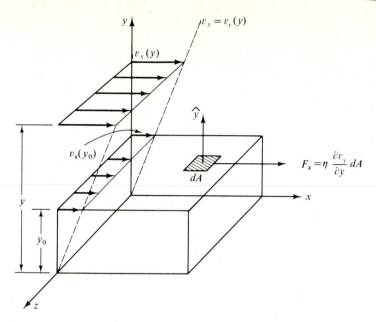

Figure 60.1 Viscous drag force exerted by fluid above surface $y = y_0$ on fluid below for linear shear flow.

where η is called the *viscosity*. It has the dimensions $ml^{-1}t^{-1}$ and is conventionally measured in the cgs unit grams per centimeter-second, or *poise*. Virtually all fluids exhibit this behavior, with η a characteristic constant that usually varies with temperature.

Basic Formulation

The first problem in the theory of viscous fluids is to incorporate the preceding elementary observation into the dynamical equations. Evidently, viscosity contributes to the surface stress, and it therefore fits naturally into the stress tensor T_{ij}. We first recall Eq. (48.33), which reexpresses Newton's second law for a small fixed volume V, subject to an external force $\mathbf{f}$ per unit mass

$$\int_V d^3x \frac{\partial}{\partial t} \rho v_k = -\sum_{l=1}^{3} \int_A dA_l\, T_{kl} + \int_V d^3x\, \rho f_k \tag{60.3}$$

Here the stress tensor T_{kl} includes the convective transport of momentum into and out of the volume V. In this form, Eq. (60.3) is very general, merely asserting that the rate of increase of the kth component of momentum is equal to the corresponding component of the integrated surface and volume forces, where

$$-\sum_{l=1}^{3} dA_l\, T_{kl} \equiv k\text{th component of force acting on oriented}$$

$$\text{surface } \hat{n}\, dA \text{ with outward normal } \hat{n} \tag{60.4}$$

Note the explicit minus signs in Eqs. (60.3) and (60.4), which ensure the usual form for a conservation law. If T_{kl} has the form (48.32)

$$T_{kl}^{\text{ideal}} = \rho v_k v_l + p\, \delta_{kl} \tag{60.5}$$

we recover the Euler equation (48.31) of nonviscous hydrodynamics. For a viscous fluid, we need only augment (60.5) with contributions that reflect the presence of viscous forces.

To formulate these concepts precisely, it is first convenient to prove that the stress tensor is symmetric. This property follows from the dynamics of the angular momentum $\mathbf{L}$ in a small fixed volume V, which we write either as a vector

$$\mathbf{L} = \int_V d^3x\, \rho \mathbf{r} \times \mathbf{v} \tag{60.6}$$

or in components

$$L_i = \sum_{jk=1}^{3} \int_V d^3x\, \epsilon_{ijk}\rho x_j v_k \tag{60.7}$$

Here, ϵ_{ijk} denotes the totally antisymmetric tensor defined in Eq. (48.17). From Newton's second law and the definition of angular momentum, L_i can increase only because of surface or volume *torques* exerted on V. It follows that the time rate of change of the angular momentum in V is given by†

$$
\begin{aligned}
\frac{dL_i}{dt} &= \sum_{jk=1}^{3} \int_V d^3x\, \epsilon_{ijk}\frac{\partial}{\partial t}\rho x_j v_k \\
&= -\sum_{jkl=1}^{3} \int_A dA_l\, \epsilon_{ijk} x_j T_{kl} + \sum_{jk=1}^{3} \int_V d^3x\, \epsilon_{ijk}\rho x_j f_k
\end{aligned} \tag{60.8}
$$

where the second line enumerates the torques. To simplify this expression, we use the differential form of Eq. (60.3)

$$\frac{\partial}{\partial t}\rho v_k = -\sum_{l=1}^{3} \frac{\partial T_{kl}}{\partial x_l} + \rho f_k \tag{60.9}$$

and take its cross product with $\mathbf{r}$

$$\sum_{jk=1}^{3} \epsilon_{ijk} x_j \frac{\partial}{\partial t}\rho v_k = -\sum_{jkl=1}^{3} \epsilon_{ijk} x_j \frac{\partial T_{kl}}{\partial x_l} + \sum_{jk=1}^{3} \epsilon_{ijk}\rho x_j f_k \tag{60.10}$$

On the left-hand side, the factor x_j can be brought inside the partial time derivative at fixed $\mathbf{r}$. On the right-hand side, the spatial derivative T_{kl} requires a little more work. We first rewrite the term as

$$-\sum_{jkl=1}^{3} \epsilon_{ijk} x_j \frac{\partial T_{kl}}{\partial x_l} = -\sum_{jkl=1}^{3} \epsilon_{ijk}\left(\frac{\partial}{\partial x_l} x_j T_{kl} - T_{kl}\frac{\partial x_j}{\partial x_l}\right)$$

---

† For a direct derivation from Newton's second law for the fluid including convective transport, see Prob. 12.1.

and then note that the partial derivative $\partial x_j/\partial x_l$ is just the Kronecker delta δ_{jl}

$$-\sum_{jkl=1}^{3}\epsilon_{ijk}x_j\frac{\partial T_{kl}}{\partial x_l}=-\sum_{jkl=1}^{3}\epsilon_{ijk}\frac{\partial}{\partial x_l}x_j T_{kl}+\sum_{jk=1}^{3}\epsilon_{ijk}T_{kj}\qquad(60.11)$$

Substitution of (60.11) into (60.10) and integration over V yield the relation

$$\sum_{jk=1}^{3}\int_V d^3x\,\epsilon_{ijk}\frac{\partial}{\partial t}\rho x_j v_k=-\sum_{jkl=1}^{3}\int_A dA_l\,\epsilon_{ijk}x_j T_{kl}$$

$$+\sum_{jk=1}^{3}\int_V d^3x\,\epsilon_{ijk}(T_{kj}+\rho x_j f_k)\quad(60.12)$$

Comparison with the angular-momentum equation (60.8) yields the general result

$$\sum_{jk=1}^{3}\int_V d^3x\,\epsilon_{ijk}T_{kj}=0\qquad(60.13a)$$

Since ϵ_{ijk} is totally antisymmetric and the volume V is arbitrary, we infer that

$$\sum_{jk=1}^{3}\epsilon_{ijk}\tfrac{1}{2}(T_{kj}-T_{jk})=0\quad i=1,2,3\qquad(60.13b)$$

Thus the stress tensor T_{kj} is symmetric under the interchange of its indices, for the linear equations (60.13b) reduce to

$$T_{kj}=T_{jk}\qquad(60.14)$$

To make use of Eq. (60.14), we note that the stress tensor (60.5) for an ideal nonviscous fluid is already symmetric. The discussion preceding Eq. (60.2) suggests that the viscous contribution T_{kl}^v should be linear in the velocity gradients $\partial v_k/\partial x_l$, and Eq. (60.14) requires that it also be symmetric. Since the stationary fluid is isotropic with no preferred spatial direction, the only two such tensors are

$$\frac{\partial v_k}{\partial x_l}+\frac{\partial v_l}{\partial x_k}\quad\text{and}\quad\delta_{kl}\sum_{j=1}^{3}\frac{\partial v_j}{\partial x_j}\equiv\delta_{kl}(\mathbf{\nabla}\cdot\mathbf{v})\qquad(60.15)$$

The first tensor has six elements, and it is convenient to separate it into a traceless part

$$\frac{\partial v_k}{\partial x_l}+\frac{\partial v_l}{\partial x_k}-\tfrac{2}{3}\delta_{kl}(\mathbf{\nabla}\cdot\mathbf{v})\qquad(60.16)$$

and a remainder $\tfrac{2}{3}\delta_{kl}(\mathbf{\nabla}\cdot\mathbf{v})$ that can be absorbed into the second tensor in (60.15). We therefore assume that the viscous stress tensor is a linear combination of these two symmetric tensors

$$T_{kl}^v=-\eta\left(\frac{\partial v_k}{\partial x_l}+\frac{\partial v_l}{\partial x_k}-\tfrac{2}{3}\delta_{kl}\mathbf{\nabla}\cdot\mathbf{v}\right)-\zeta\delta_{kl}(\mathbf{\nabla}\cdot\mathbf{v})\qquad(60.17)$$

where η and ζ are phenomenological constants known as the *viscosity* and the *bulk* (or *second*) *viscosity*. For incompressible flow ($\nabla \cdot \mathbf{v} = 0$), the bulk viscosity plays no role, and T_{kl}^v then becomes

$$T_{kl}^v = -\eta \left(\frac{\partial v_k}{\partial x_l} + \frac{\partial v_l}{\partial x_k} \right) \qquad \text{incompressible fluid} \qquad (60.18)$$

The minus sign in these equations agrees with our earlier definition (60.4), as seen by the following simple example. Consider the shear flow in Eq. (60.1) and Fig. 60.1, where $\nabla \cdot \mathbf{v} = 0$ and the only nonzero elements of the viscous stress tensor are

$$T_{xy}^v = T_{yx}^v = -\eta \frac{\partial v_x}{\partial y} \qquad (60.19)$$

The force on the fluid below an element of area $\hat{y}\,dA$ is

$$F_k = - \sum_{l=1}^{3} (dA\,\hat{y})_l T_{kl}^v = -dA\,T_{ky}^v \qquad (60.20)$$

Substitution of Eq. (60.19) confirms the form and sign of Eq. (60.2).

The symmetry of T_{ij}^v has the following important consequence. Consider a uniform rotation $\mathbf{v} = \mathbf{\Omega} \times \mathbf{r}$ or

$$v_i = \sum_{kl=1}^{3} \epsilon_{ikl} \Omega_k x_l \qquad \text{uniform rotation} \qquad (60.21)$$

The derivative $\partial v_i / \partial x_j$ becomes

$$\frac{\partial v_i}{\partial x_j} = \sum_{kl=1}^{3} \epsilon_{ikl} \Omega_k \frac{\partial x_l}{\partial x_j} = \sum_{k=1}^{3} \epsilon_{ikj} \Omega_k \qquad (60.22)$$

which is antisymmetric under interchange of i and j. As a result, T^v in Eq. (60.17) vanishes identically for a uniform rotation, in accordance with the intuitive notion that uniform (solid-body) rotation is an equilibrium nondissipative configuration for a viscous fluid. We may remark that T^v also vanishes for uniform flow, since in that case $\partial v_i / \partial x_j = 0$.

Navier-Stokes Equation

The total stress tensor for a viscous fluid is the sum of Eqs. (60.5) and (60.17)

$$T_{ij} = T_{ij}^{\text{ideal}} + T_{ij}^v \qquad (60.23a)$$

or, explicitly,

$$T_{ij} = \rho v_i v_j + p\delta_{ij} - \eta \left(\frac{\partial v_i}{\partial x_j} + \frac{\partial v_j}{\partial x_i} - \tfrac{2}{3}\delta_{ij} \nabla \cdot \mathbf{v} \right) - \zeta \delta_{ij}(\nabla \cdot \mathbf{v}) \qquad (60.23b)$$

Since Eq. (60.9) expresses Newton's second law for any particular T_{ij}, we readily find the dynamical equation

$$\frac{\partial}{\partial t} \rho v_i + \sum_{j=1}^{3} \frac{\partial}{\partial x_j} \rho v_i v_j = \rho f_i - \frac{\partial p}{\partial x_i} + \eta \sum_{j=1}^{3} \frac{\partial^2 v_i}{\partial x_j^2} + (\zeta + \tfrac{1}{3}\eta) \sum_{j=1}^{3} \frac{\partial^2 v_j}{\partial x_i\,\partial x_j} \qquad (60.24)$$

The left-hand side can be simplified with the equation of continuity [compare the similar treatment leading to Eq. (48.35)]

$$\frac{\partial \rho}{\partial t} + \sum_{j=1}^{3} \frac{\partial}{\partial x_j} \rho v_j = 0 \qquad (60.25)$$

and we therefore obtain the final equation

$$\frac{\partial v_i}{\partial t} + \sum_{j=1}^{3} v_j \frac{\partial v_i}{\partial x_j} = f_i - \frac{1}{\rho} \frac{\partial p}{\partial x_i} + \frac{\eta}{\rho} \sum_{j=1}^{3} \frac{\partial^2 v_i}{\partial x_j^2} + \frac{1}{\rho} (\zeta + \tfrac{1}{3}\eta) \sum_{j=1}^{3} \frac{\partial^2 v_j}{\partial x_i \, \partial x_j} \qquad (60.26a)$$

which also has the equivalent vector form

$$\frac{\partial \mathbf{v}}{\partial t} + (\mathbf{v} \cdot \nabla)\mathbf{v} = \mathbf{f} - \frac{1}{\rho} \nabla p + \frac{\eta}{\rho} \nabla^2 \mathbf{v} + \frac{1}{\rho} (\zeta + \tfrac{1}{3}\eta) \, \nabla(\nabla \cdot \mathbf{v}) \qquad (60.26b)$$

This equation is the desired generalization of Euler's equation (48.35) for a viscous fluid; it requires two additional phenomenological constants η and ζ, which here are assumed to be intrinsic properties of the fluid independent of $\mathbf{v}$ and ρ. In the special case of *incompressible flow*, the last term of (60.26) vanishes, giving the celebrated *Navier-Stokes equation*

$$\frac{\partial \mathbf{v}}{\partial t} + (\mathbf{v} \cdot \nabla)\mathbf{v} = \mathbf{f} - \frac{1}{\rho} \nabla p + v \, \nabla^2 \mathbf{v} \qquad (60.27)$$

where

$$v \equiv \frac{\eta}{\rho} \qquad (60.28)$$

is called the *kinematic viscosity*. Evidently, v has dimensions $l^2 t^{-1}$, just like the thermal diffusivity κ [see Eq. (57.13)]. Table 60.1 gives some typical values of η and v.

In a nonviscous incompressible fluid with $\mathbf{f} = 0$, dimensional considerations in Eq. (60.27) show that the hydrodynamic contribution to the pressure in steady flow is of order ρv_0^2, where v_0 is a characteristic velocity. For a viscous fluid, the

Table 60.1 Viscosity of typical fluids at 20°C and 1 atm*

Fluid	v (cm^2 s^{-1})	η (g cm^{-1} s^{-1})
Mercury	0.0011	0.0155
Water	0.0100	0.0100
Ethyl alcohol	0.0152	0.0120
Air	0.149	0.000180
Helium gas	1.166	0.000194
Glycerine	11.83	14.90

* Data from R. C. Weast (ed.), "CRC Handbook of Chemistry and Physics," CRC Press, West Palm Beach, Fla., 1978, sec. F.

Navier-Stokes equation (60.27) differs only in the final term $v\,\nabla^2\mathbf{v}$, which depends on the viscosity through the parameter v. The contribution of this term to the pressure will be of order $\rho v v_0 / l$, where v_0 and l are a typical velocity and length for the particular flow of interest. To characterize the relative effect of inertial forces and viscous forces, it is thus conventional to introduce a dimensionless parameter known as the *Reynolds number*†

$$\mathscr{R} \equiv \frac{\rho v_0^2}{\rho v v_0 / l} \equiv \frac{v_0 l}{v} \equiv \frac{v_0 l \rho}{\eta} \tag{60.29a}$$

For example, a sphere of radius a placed in an initially uniform flow $\mathbf{v}_0$ would have a Reynolds number

$$\mathscr{R} = \frac{v_0 a}{v} \tag{60.29b}$$

Different systems with the same geometry and the same $\mathscr{R}$ exhibit identical viscous flow. Experiments show that the motion is smooth for low values of $\mathscr{R}$ and v_0, but as $\mathscr{R}$ increases past a critical value, the flow suddenly undergoes a sequence of transitions, eventually leading to a chaotic state known as *full turbulence*. This topic remains only partially understood and will not be considered here.‡

The derivation of the equations of motion (60.26) and (60.27) assumed that the velocity field $\mathbf{v}(\mathbf{x}, t)$ and its derivatives $\partial v_i/\partial x_j$ are continuous and differentiable throughout the fluid. This picture should extend to the adjoining surfaces, and we now consider the conditions that must be imposed at various physical boundaries.

1. For sufficiently slow motion, a viscous fluid in contact with a rigid surface moves with the velocity $\mathbf{V}$ of the surface [compare Eq. (49.10a)]

$$\mathbf{v} = \mathbf{V} \qquad \text{on moving surface} \tag{60.30a}$$

   In particular, $\mathbf{v}$ vanishes at a fixed surface ($\mathbf{V} = 0$)

$$\mathbf{v} = 0 \qquad \text{on fixed surface} \tag{60.30b}$$

   Although this condition cannot be correct for arbitrarily high flow speeds, experiments show that it holds in most cases of interest [see the discussion below Eq. (61.21)].

2. A surface separating two fluids (1 and 2, say) is more complicated. The continuity of $\mathbf{v}$ and its derivatives requires

$$\mathbf{v}_1 = \mathbf{v}_2 \qquad \text{on surface separating two fluids} \tag{60.31a}$$

   which differs from that for nonviscous fluids in the continuity of tangential components.

---

† Batchelor [1] secs. 4.7, 4.12, and 5.11; Chandrasekhar [2], chaps. 7 and 8; Feynman, Leighton, and Sands [3], chap. 41; Landau and Lifshitz (1959) secs. 19, 20, and 26–29.

‡ Landau and Lifshitz (1959), chap. 3; Monin and Yaglom [4]; Orszag [5], p. 235.

3. In an ideal nonviscous fluid, the pressure was defined as the normal force per unit area measured by a comoving gauge. The resulting function $p(\mathbf{r}, t)$ was independent of the orientation of the surface and continuous and differentiable throughout the fluid. For a viscous fluid, however, such a comoving gauge measures the normal component of $p\delta_{ij} + T_{ij}^v$. In this way, the force per unit area on a surface oriented with normal $\hat{n}$ generally becomes anisotropic and differs from the pressure by terms proportional to the velocity gradients. As a result, the pressure could be measured only by extrapolating a series of observations with decreasing $\partial v_i / \partial x_j$. Alternatively, p is still defined by the equation of state $p(\epsilon, \rho)$ or $p(s, \rho)$ seen by a comoving observer. The physical condition of continuity of force [compare Eq. (39.37)] now requires in the comoving frame that

$$\sum_{j=1}^{3} (p\delta_{ij} + T_{ij}^v)\hat{n}_j \text{ must be continuous across any surface with normal } \hat{n},$$

including one separating two fluids $\qquad\qquad$ (60.31b)

Note that for a given $\hat{n}$, this boundary condition is a *vector* relation with three components.

4. We can combine these observations to observe that at a *free surface* adjoining a vacuum, the tangential component of velocity is unrestricted. In addition, the nonviscous condition of zero pressure must be extended in the comoving frame to

$$\sum_{j=1}^{3} (p\delta_{ij} + T_{ij}^v)\hat{n}_j = 0 \qquad \text{on free surface} \qquad (60.32)$$

To close this section, we remark that Eq. (60.26) has been derived in cartesian coordinates. In curvilinear systems, the vector operations $(\mathbf{v} \cdot \nabla)\mathbf{v}$ and $\nabla^2\mathbf{v}$ must be defined through the relations

$$(\mathbf{v} \cdot \nabla)\mathbf{v} = \nabla\tfrac{1}{2}v^2 - \mathbf{v} \times (\nabla \times \mathbf{v}) \qquad (60.33a)$$

$$\nabla^2\mathbf{v} = \nabla(\nabla \cdot \mathbf{v}) - \nabla \times (\nabla \times \mathbf{v}) \qquad (60.33b)$$

which are straightforward to evaluate (see Appendix B) in any particular coordinate system.

Energy Balance

We now consider the important role of viscosity in the energy balance. It is convenient to start with the kinetic-energy density $\tfrac{1}{2}\rho v^2$. For a nonviscous fluid, its time derivative has been calculated in Eq. (48.37) with Euler's equation, and the only new feature here is an additional viscous term on the right-hand side [compare Eqs. (60.9) and (60.23)]

$$\frac{\partial}{\partial t} \tfrac{1}{2}\rho v^2 = -\nabla \cdot (\tfrac{1}{2}\rho v^2 \mathbf{v}) - \mathbf{v} \cdot \nabla p + \rho \mathbf{v} \cdot \mathbf{f} - \sum_{ij=1}^{3} v_j \frac{\partial}{\partial x_i} T_{ij}^v \qquad (60.34)$$

Simple manipulations reexpress this equation as

$$\frac{\partial}{\partial t} \tfrac{1}{2}\rho v^2 = -\mathbf{V} \cdot [(\tfrac{1}{2}\rho v^2 + p)\mathbf{v}] - \sum_{ij=1}^{3} \frac{\partial}{\partial x_i} v_j T_{ij}^v$$

$$+ \rho \mathbf{v} \cdot \mathbf{f} + p(\mathbf{V} \cdot \mathbf{v}) + \sum_{ij=1}^{3} T_{ij}^v \frac{\partial v_j}{\partial x_i} \tag{60.35}$$

The last term on the right-hand side can be rewritten with the explicit equation (60.17) for T_{ij}^v

$$\sum_{ij=1}^{3} T_{ij}^v \frac{\partial v_j}{\partial x_i} = -\eta \sum_{ij=1}^{3} \left(\frac{\partial v_i}{\partial x_j} + \frac{\partial v_j}{\partial x_i} - \tfrac{2}{3}\delta_{ij}\mathbf{V}\cdot\mathbf{v} \right) \frac{\partial v_j}{\partial x_i} - \zeta \sum_{ij=1}^{3} \delta_{ij}\mathbf{V}\cdot\mathbf{v} \frac{\partial v_j}{\partial x_i}$$

$$= -\tfrac{1}{2}\eta \sum_{ij=1}^{3} \left(\frac{\partial v_i}{\partial x_j} + \frac{\partial v_j}{\partial x_i} - \tfrac{2}{3}\delta_{ij}\mathbf{V}\cdot\mathbf{v} \right)^2 - \zeta(\mathbf{V}\cdot\mathbf{v})^2 \tag{60.36}$$

where the last form is easily verified by expanding the right-hand side. To interpret this result, we integrate Eq. (60.35) over a fixed volume V. Use of the divergence theorem and Eq. (60.36) yields

$$\frac{d}{dt}\int_V d^3x \, \tfrac{1}{2}\rho v^2 = -\int_A d\mathbf{A} \cdot (\tfrac{1}{2}\rho v^2 + p)\mathbf{v} - \sum_{ij=1}^{3} \int_A dA_i \, v_j T_{ij}^v$$

$$+ \int_V d^3x \, (\rho\mathbf{v}\cdot\mathbf{f} + p\mathbf{V}\cdot\mathbf{v})$$

$$- \int_V d^3x \left[\tfrac{1}{2}\eta \sum_{ij=1}^{3} \left(\frac{\partial v_i}{\partial x_j} + \frac{\partial v_j}{\partial x_i} - \tfrac{2}{3}\delta_{ij}\mathbf{V}\cdot\mathbf{v} \right)^2 + \zeta(\mathbf{V}\cdot\mathbf{v})^2 \right] \tag{60.37}$$

This equation distinguishes the various contributions to the change in the kinetic energy in V. It becomes particularly simple for a localized disturbance in an *infinite incompressible fluid*, when $\mathbf{V}\cdot\mathbf{v}=0$, the volume V may be extended to infinity, and the surface terms vanish to give

$$\frac{d}{dt}\int_V d^3x \, \tfrac{1}{2}\rho v^2 = \int_V d^3x \left[\rho\mathbf{v}\cdot\mathbf{f} - \tfrac{1}{2}\eta \sum_{ij=1}^{3} \left(\frac{\partial v_i}{\partial x_j} + \frac{\partial v_j}{\partial x_i} \right)^2 \right] \tag{60.38}$$

In this special case, the kinetic energy changes only because of work done by the external force, which can have either sign, or because of viscous dissipation, which is intrinsically negative. Thus viscosity acts to degrade the organized kinetic energy, and we shall now show quite generally that it reappears as *heat*. Note that both terms of Eq. (60.36) are negative semidefinite, so that a similar dissipation also occurs for compressible fluids.

A complete treatment of this question necessarily involves other forms of energy, and we therefore consider the *total energy balance in a fixed volume V* [compare Eq. (48.44)]

$$\frac{d}{dt}\int_V d^3x \left(\tfrac{1}{2}\rho v^2 + \rho\epsilon\right) = -\int_A dA \cdot \mathbf{v}\left(\tfrac{1}{2}\rho v^2 + \rho\epsilon + p\right) + \int_V d^3x\, \rho\mathbf{v}\cdot\mathbf{f}$$

$$- \sum_{ij=1}^{3}\int_A dA_i\, T_{ij}^v v_j + \int_V d^3x\, \rho\dot{q}_{ex} - \int_A dA \cdot \mathbf{j}_h \quad (60.39)$$

This equation asserts that the total energy in V (the sum of the kinetic and internal energy) increases because of

1. Inward transport of energy with a flux [see Eq. (48.46)]

$$\mathbf{j}_e = \left(\tfrac{1}{2}\rho v^2 + \rho\epsilon + p\right)\mathbf{v} \quad (60.40)$$

2. Work done by the external force $\mathbf{f}$
3. Work done by the viscous stress on the surface
4. External volume sources of heat that release energy at a rate $\dot{q}_{ex}$ per unit mass
5. Inward transport of heat with a flux $\mathbf{j}_h$ [see Eq. (57.11)]

Although Eq. (60.39) in principle provides a complete description, it is clearer to compare it with the similar equation (60.37) that describes only the kinetic energy. The difference between the two equations gives the *integral relation for the internal energy*

$$\frac{d}{dt}\int_V d^3x\, \rho\epsilon = \int_V d^3x\, \frac{\partial}{\partial t}\rho\epsilon = -\int_A dA \cdot (\rho\mathbf{v}\epsilon + \mathbf{j}_h)$$

$$+ \int_V d^3x \left[\rho\dot{q}_{ex} - p\nabla\cdot\mathbf{v} + \tfrac{1}{2}\eta \sum_{ij=1}^{3}\left(\frac{\partial v_i}{\partial x_j} + \frac{\partial v_j}{\partial x_i} - \tfrac{2}{3}\delta_{ij}\nabla\cdot\mathbf{v}\right)^2 + \zeta(\nabla\cdot\mathbf{v})^2\right] \quad (60.41a)$$

and the equivalent differential form

$$\frac{\partial}{\partial t}\rho\epsilon + \nabla\cdot(\rho\epsilon\mathbf{v}) + p\nabla\cdot\mathbf{v} = -\nabla\cdot\mathbf{j}_h + \rho\dot{q}_{ex}$$

$$+ \tfrac{1}{2}\eta \sum_{ij=1}^{3}\left(\frac{\partial v_i}{\partial x_j} + \frac{\partial v_j}{\partial x_i} - \tfrac{2}{3}\delta_{ij}\nabla\cdot\mathbf{v}\right)^2 + \zeta(\nabla\cdot\mathbf{v})^2 \quad (60.41b)$$

To interpret this last equation, we recall the general thermodynamic relation (57.1). When applied to a fixed number of particles with unit mass, it becomes [compare Eqs. (48.38) and (52.23)]

$$d\epsilon = T\, ds + \frac{p}{\rho^2}\, d\rho \quad (60.42)$$

where ϵ and s are the internal energy and entropy per unit mass. As a result, the change in the internal-energy density becomes

$$d(\rho\epsilon) = \rho \, d\epsilon + \epsilon \, d\rho = T\rho \, ds + \left(\epsilon + \frac{p}{\rho}\right) d\rho \tag{60.43}$$

If we concentrate on a given location in the fluid, the partial derivative of the entropy per unit mass with respect to time can be determined from this total differential according to

$$\frac{\partial(\rho\epsilon)}{\partial t} = T\rho \frac{\partial s}{\partial t} + \left(\epsilon + \frac{p}{\rho}\right)\frac{\partial \rho}{\partial t} \tag{60.44a}$$

Similarly, use of Eq. (60.43) to obtain the spatial gradient of the entropy per unit mass results in the relation

$$\mathbf{\nabla} \cdot (\rho \mathbf{v}\epsilon) = \mathbf{v} \cdot \mathbf{\nabla}(\rho\epsilon) + \rho\epsilon \mathbf{\nabla} \cdot \mathbf{v}$$

$$= \rho T \mathbf{v} \cdot \mathbf{\nabla}s + \left(\epsilon + \frac{p}{\rho}\right)\mathbf{v} \cdot \mathbf{\nabla}\rho + \rho\epsilon \mathbf{\nabla} \cdot \mathbf{v}$$

$$= \rho T \mathbf{v} \cdot \mathbf{\nabla}s + \left(\epsilon + \frac{p}{\rho}\right)\mathbf{\nabla} \cdot (\rho \mathbf{v}) - p\mathbf{\nabla} \cdot \mathbf{v} \tag{60.44b}$$

It is important to emphasize that all the *processes under consideration are assumed to be locally reversible in the thermodynamic sense.* Substitution of Eqs. (60.44) into Eq. (60.41b) and use of the continuity equation (60.25) then provides the equation of local *entropy balance*

$$T\rho\left(\frac{\partial s}{\partial t} + \mathbf{v} \cdot \mathbf{\nabla}s\right) = -\mathbf{\nabla} \cdot \mathbf{j}_h + \rho \dot{q}_{\text{ex}}$$

$$+ \tfrac{1}{2}\eta \sum_{ij=1}^{3}\left(\frac{\partial v_i}{\partial x_j} + \frac{\partial v_j}{\partial x_i} - \tfrac{2}{3}\delta_{ij}\mathbf{\nabla} \cdot \mathbf{v}\right)^2 + \zeta(\mathbf{\nabla} \cdot \mathbf{v})^2 \tag{60.45}$$

In a nonviscous fluid without external heat sources or thermal conduction, the right-hand side vanishes, and each small element of fluid evidently moves with constant entropy $ds/dt = 0$. More generally, *the entropy changes by generation and conduction of heat (which can have either sign) and by viscous dissipation (which is intrinsically positive in this equation and thus increases the entropy).* As a result, viscous dissipation acts as a *positive* (internal) heat source, confirming the previous picture of production of random thermal motion.

Equation (60.45) can be further simplified if the pressure p is fixed and specified throughout the fluid. (This was the case, for example, in Chap. 11 on heat conduction.) Since the specific entropy is a state function, its independent variables can be taken as T and p. We then have the relations [see Eq. (57.4)]

$$\frac{\partial s}{\partial t} = \left(\frac{\partial s}{\partial T}\right)_p \frac{\partial T}{\partial t} = \frac{c_p}{T}\frac{\partial T}{\partial t} \qquad \text{fixed pressure}$$

$$\mathbf{\nabla}s = \frac{c_p}{T}\mathbf{\nabla}T \tag{60.46}$$

and Eq. (60.45) can thus be converted to an equation for the temperature $T(\mathbf{r}, t)$

$$\frac{dT}{dt} \equiv \frac{\partial T}{\partial t} + \mathbf{v} \cdot \nabla T = \kappa \nabla^2 T + \frac{\dot{q}_{ex}}{c_p}$$

$$+ \frac{\eta}{2\rho c_p} \sum_{ij=1}^{3} \left(\frac{\partial v_i}{\partial x_j} + \frac{\partial v_j}{\partial x_i} - \tfrac{2}{3}\delta_{ij} \nabla \cdot \mathbf{v} \right)^2 + \frac{\zeta}{\rho c_p} (\nabla \cdot \mathbf{v})^2 \qquad \text{fixed pressure} \quad (60.47)$$

where (57.11) and (57.13) have been used to reexpress the heat flux. This equation differs from our previous one (57.12) for solids only in the convective term $\mathbf{v} \cdot \nabla T$ and in the viscous source terms. It completely characterizes the conduction of heat in an isobaric viscous fluid and is therefore still quite complicated. For incompressible flow, it takes the somewhat simpler form

$$\frac{\partial T}{\partial t} + \mathbf{v} \cdot \nabla T = \kappa \nabla^2 T + \frac{\dot{q}_{ex}}{c_p} + \frac{\nu}{2c_p} \sum_{ij=1}^{3} \left(\frac{\partial v_i}{\partial x_j} + \frac{\partial v_j}{\partial x_i} \right)^2 \qquad (60.48)$$

61 EXAMPLES OF INCOMPRESSIBLE FLOW

To gain physical insight into viscous hydrodynamics, we present a few simple exact solutions of the Navier-Stokes equation (60.27) for incompressible flow.

Steady Flow in a Channel or Pipe

Consider a fluid executing steady flow in the z direction, with

$$\mathbf{v}(\mathbf{r}, t) = v_z(\mathbf{r})\hat{z}$$

$$\frac{\partial v_z}{\partial t} = 0 \qquad (61.1)$$

Assume, in addition, that the flow is invariant in the z direction, so that $dv_z/dt = 0$ or, equivalently,

$$\frac{\partial v_z}{\partial z} = 0 \qquad (61.2)$$

(see Fig. 61.1). The Navier-Stokes equation (60.27) then becomes a *linear* partial differential equation for the scalar function $v_z(x, y)$

$$\nabla^2 v_z = \frac{1}{\eta} \frac{\partial p}{\partial z} \qquad (61.3)$$

and assumption (61.2) requires that $\partial p/\partial z$ be independent of z. As a simple example, we take a uniform pressure gradient

$$\frac{\partial p}{\partial z} = -\frac{\Delta p}{L} \qquad (61.4)$$

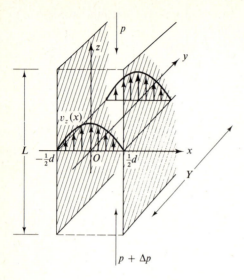

Figure 61.1 Steady viscous flow in a channel of width d.

where Δp is the total change in pressure over the length L and the minus sign reflects the decreased pressure along the direction of flow. In this way, $v_z(x, y)$ satisfies a two-dimensional Poisson equation

$$\nabla^2 v_z(x, y) = -\frac{\Delta p}{\eta L} \qquad (61.5)$$

subject to the viscous boundary condition that v_z vanish on the surface of the container [Eq. (60.30b)].

The simplest example is a channel of width d, bounded by the planes $x = \pm \tfrac{1}{2}d$ (see Fig. 61.1). In this geometry, v_z depends only on x, with $v_z(x)$ symmetric about the midpoint $x = 0$. The appropriate solution to Eq. (61.5) is

$$v_z(x) = -\frac{\Delta p}{2\eta L} x^2 + c \qquad (61.6)$$

where the constant c must be chosen to fit the boundary condition

$$v_z(\pm \tfrac{1}{2}d) = 0 \qquad (61.7)$$

As a result, we find

$$v_z(x) = \frac{\Delta p}{2\eta L} \left(\tfrac{1}{4}d^2 - x^2\right) \qquad (61.8)$$

The velocity is maximum at the center and decreases with a parabolic profile to zero at the walls (see Fig. 61.1).

It is instructive to evaluate the drag force F^d exerted by the walls on the fluid [see Eq. (60.4)]

$$F_i^d = -\sum_{j=1}^{3} \int_A dA_j\, T_{ij} = -\int dy\, dz\, (T_{ix}|_{x=d/2} - T_{ix}|_{x=-d/2}) \qquad (61.9)$$

where the surface integral is along the two bounding planes at $x = \pm d/2$ with outward normals $\pm \hat{x}$. The pressure contribution in Eq. (60.23b) (along $\hat{x}$) cancels identically, and only the viscous stress (60.18) remains. Use of Eq. (61.6) gives

$$T_{zx} = -\eta \frac{\partial v_z}{\partial x} = \frac{\Delta p}{L} x \tag{61.10}$$

and the total drag force becomes

$$F_z^d = -\frac{\Delta p}{L} A \left[\frac{d}{2} - \left(-\frac{d}{2} \right) \right] = -\Delta p \; Yd \tag{61.11}$$

where the area of the channel wall is $A = YL$ with Y the (large) dimension of the channel in the y direction. This force precisely balances the net pressure difference between the two ends (note that Yd is the cross-sectional area of the channel), accounting for the absence of accelerated motion.

Equation (61.8) also permits us to evaluate the total mass flow through the channel per unit length in the y direction. A simple calculation gives

$$\dot{M} = \rho \int_{-d/2}^{d/2} v_z(x)\, dx = \frac{\Delta p\; d^3}{12 v L} \tag{61.12}$$

where $v = \eta/\rho$ is the kinematic viscosity. In addition to the expected proportionality to Δp, this quantity varies as d^3 and involves the detailed properties of the fluid only through its inverse dependence on the parameter v. In principle, this expression provides a means of measuring v (and hence η), but the presence of boundaries in the y direction introduces inevitable edge corrections.

To avoid this last complication, it is common to use a long pipe with circular cross section of radius R, when the velocity $v_z(r)$ is axisymmetric, and Eq. (61.5) becomes [see Eq. (B.26)]

$$\nabla^2 v_z(r) = \frac{1}{r} \frac{d}{dr} r \frac{dv_z}{dr} = -\frac{\Delta p}{\eta L} \tag{61.13}$$

This equation is readily integrated to give

$$v_z(r) = -\frac{\Delta p r^2}{4\eta L} + c_1 \ln r + c_2 \tag{61.14}$$

where c_1 and c_2 are constants. The flow must be bounded everywhere, which requires $c_1 = 0$, and the boundary condition

$$v_z(R) = 0 \tag{61.15}$$

gives the final expression

$$v_z(r) = \frac{\Delta p}{4\eta L} (R^2 - r^2) \tag{61.16}$$

It again displays a parabolic profile.

The drag force exerted by the walls on the fluid can be evaluated as follows:

$$F_z^d = -\sum_{i=1}^{3} \int dA_i \, T_{zi}^v = -R \int dz \, d\phi \, (\cos\phi \, T_{zx}^v + \sin\phi \, T_{zy}^v)|_{r=R} \quad (61.17)$$

where the integral is over a cylindrical surface with outward normal $\hat{r} = \hat{x}\cos\phi + \hat{y}\sin\phi$. Equations (60.18) and (61.16) and the chain rule for partial derivatives lead to the result

$$(\cos\phi)T_{zx}^v + (\sin\phi)T_{zy}^v = -\eta\left(\cos\phi\frac{\partial v_z}{\partial x} + \sin\phi\frac{\partial v_z}{\partial y}\right)$$

$$= -\eta\left(\cos\phi\frac{x}{r} + \sin\phi\frac{y}{r}\right)\frac{dv_z}{dr} = -\eta\frac{dv_z}{dr} \quad (61.18)$$

A combination of Eqs. (61.16), (61.17), and (61.18) yields

$$F_z^d = \eta R \int dz \, d\phi \, \frac{dv_z}{dr}\bigg|_{r=R} \quad (61.19a)$$

$$= -\Delta p \, \pi R^2 \quad (61.19b)$$

which again balances the net pressure difference on the two ends. The same answer follows more directly from the results of Prob. B.8 for T_{zr}^v.

The total mass flow through the pipe is obtained by integrating over the cross-sectional area

$$\dot{M} = \rho \int r \, dr \, d\phi \, v_z(r) \quad (61.20)$$

Substitution of (61.16) gives the *Poiseuille formula*

$$\dot{M} = \frac{\pi R^4 \, \Delta p}{8\nu L} \quad (61.21)$$

It allows a simple and direct measurement of the viscosity of fluids by studying the steady efflux from a vertical tube under hydrostatic pressure. Experiments verify this formula in detail, which can be interpreted as confirming the no-slip boundary condition (60.30) and (61.15).

Tangential Flow in a Half Space

Consider the half space $z > 0$ bounded by a plane at $z = 0$ that can move rigidly in the x direction (see Fig. 61.2). Translational invariance in the xy plane implies that the motion has the form

$$p(\mathbf{r}, t) = p(z, t) \qquad \mathbf{v}(\mathbf{r}, t) = v_x(z, t)\hat{x} \quad (61.22)$$

with $\nabla \cdot \mathbf{v} = 0$. Once again, the nonlinear convective terms in the Navier-Stokes

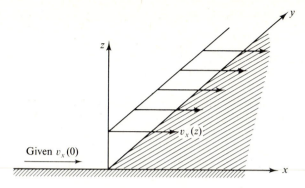

Figure 61.2 Shear flow $v_x(z)$ in the half space $z > 0$ subject to a moving wall with prescribed tangential velocity.

equation (60.27) vanish identically. The resulting z and x components yield the simple equations

$$0 = -\frac{1}{\rho}\frac{\partial p}{\partial z} \tag{61.23a}$$

$$\frac{\partial v_x}{\partial t} = v\frac{\partial^2 v_x}{\partial z^2} \tag{61.23b}$$

The first shows that p is in fact a spatial constant, whose time dependence has no effect on the flow. In addition, the velocity $v_x(z, t)$ obeys precisely the homogeneous one-dimensional heat-conduction equation (58.14) with the thermal diffusivity κ replaced by the kinematic viscosity v. As a result, we can apply all our previous expressions for heat conduction in a half space to the dynamic response of a viscous fluid.

As a simple example, suppose that the bounding plane is stationary for $t < 0$ and then suddenly moves with constant velocity v_0 for $t > 0$. The fluid velocity satisfies the conditions

$$v_x(z, 0) = 0 \qquad z > 0$$
$$v_x(0, t) = v_0 \qquad t > 0 \tag{61.24}$$

and the solution is identical with Eq. (59.30) for a half space at zero temperature suddenly subjected to a surface temperature T_0

$$v_x(z, t) = v_0 \operatorname{erfc}\left[\frac{z}{2(vt)^{1/2}}\right] \tag{61.25}$$

Note that the velocity is everywhere positive *for any* $t > 0$, which reflects the unphysical assumption of infinite acceleration of the bounding plane. Nevertheless the fluid at a depth z_0 remains essentially at rest until a time $z_0^2/4v$ that depends only on the kinematic viscosity.

The moving wall exerts a drag force on the fluid

$$F_x^d = -\sum_{i=1}^{3}\int_A dA_i\, T_{xi}^v = \int_A dA\, T_{xz}^v \tag{61.26}$$

where the normal $-\hat{z}$ points away from the fluid. Use of Eqs. (59.28), (60.18), and (61.25) gives the force per unit area

$$\frac{F_x^d}{A} = -\eta \left. \frac{\partial v_x}{\partial z} \right|_{z=0} = \frac{\eta v_0}{(\pi v t)^{1/2}} \qquad (61.27)$$

It diverges as $t \to 0$ and ultimately falls to zero as the motion penetrates deeper into the fluid. This expression is directly analogous to Eq. (59.32) for the heat transferred across the boundary. In other geometries, however, the relation between the two physical systems is less direct (see Prob. 12.9).

As a second example, suppose that the bounding plane oscillates tangentially with amplitude $\epsilon \hat{x} \cos \omega t = \text{Re } \epsilon \hat{x} e^{-i\omega t}$. We now must solve Eq. (61.23b) subject to the boundary condition

$$v_x(0, t) = -\omega \epsilon \sin \omega t = \text{Re} \left(-i\omega \epsilon e^{-i\omega t} \right) \qquad (61.28)$$

The solution is similar to Eq. (58.21) for penetration of temperature waves into the half space, and we obtain

$$v_x(z, t) = \text{Re} \left\{ -i\omega \epsilon \exp \left[(i-1)\frac{z}{\delta} - i\omega t \right] \right\} = \omega \epsilon e^{-z/\delta} \sin \left(\frac{z}{\delta} - \omega t \right) \qquad (61.29)$$

where [compare Eq. (58.20)]

$$\delta \equiv \left(\frac{2v}{\omega} \right)^{1/2} \qquad (61.30)$$

is the *viscous penetration depth*. Note that δ is proportional to $(v/\omega)^{1/2}$ and diverges as $\omega \to 0$. For water at 20°C ($v \approx 0.010$), a period of 1 s gives $\delta \approx 0.56$ mm.

As in the previous cases, the oscillating wall exerts a viscous drag force per unit area, which we here express in complex form

$$\frac{F_x^d}{A} = -\eta \left. \frac{\partial v_x}{\partial z} \right|_{z=0} = -\frac{\eta \omega \epsilon}{\delta} (1+i)e^{-i\omega t} \qquad (61.31)$$

The complex factor $1 + i$ indicates that the force is 45° out of phase with the impressed amplitude, and the real and imaginary parts act, respectively, as reactive and resistive components. The time-average power transferred to the fluid per unit area is

$$\left\langle \frac{F_x^d v_x}{A} \right\rangle \bigg|_{z=0} = \frac{1}{2A} \text{Re} \left. (F_x^d v_x^*) \right|_{z=0} = \frac{\omega^2 \epsilon^2 \eta}{2\delta} \qquad (61.32)$$

where Eqs. (45.45) and (61.28) have been used. This power increases with frequency like $\omega^{5/2}$ and with viscosity like $\eta^{1/2}$ [see Eq. (61.30)].

This example allows a direct verification of the conservation of energy. The viscous dissipation provides a heat source per unit volume [see Eq. (60.45)]

$$\frac{1}{2}\eta \sum_{ij=1}^{3} \left(\frac{\partial v_i}{\partial x_j} + \frac{\partial v_j}{\partial x_i} \right)^2 = \eta \left(\frac{\partial v_x}{\partial z} \right)^2 \qquad (61.33)$$

where a factor 2 arises from the two terms in the sum ($i = x, j = z$ and $i = z, j = x$). The time-average rate of heat generation becomes

$$\left\langle \eta \left(\frac{\partial v_x}{\partial z} \right)^2 \right\rangle = \tfrac{1}{2} \eta \ \text{Re} \left[\frac{\partial v_x}{\partial z} \left(\frac{\partial v_x}{\partial z} \right)^* \right] \tag{61.34}$$

and a simple calculation with Eq. (61.29) gives

$$\left\langle \eta \left(\frac{\partial v_x}{\partial z} \right)^2 \right\rangle = \frac{\eta \omega^2 \epsilon^2}{\delta^2} e^{-2z/\delta} \tag{61.35}$$

This heat generation is confined to a surface layer of approximate thickness $\tfrac{1}{2}\delta$, beyond which the induced motion becomes negligible. The total heat generated per unit area is the integral of (61.35) over all positive z; it precisely matches the mean power (61.32) exerted by the moving wall.

62 SOUND WAVES IN VISCOUS FLUIDS

The preceding section presented a few exact solutions of the Navier-Stokes equation; in each example, the finite viscosity was crucial to the physical behavior, since the solutions become singular as $\eta \to 0$. Such situations differ qualitatively from those considered earlier in Chaps. 9 and 10, where the *nonviscous* hydrodynamics yielded perfectly definite motion. Even these latter systems, however, are affected in some degree by viscosity, which typically acts as a *damping force*. As a specific example, we now study small-amplitude motion in an unbounded homogeneous compressible fluid, generalizing the theory of sound waves to include the effect of viscosity and thermal conduction.

The fundamental equations are the conservation of mass (60.25), momentum (60.26), and energy, which we here take in the form (60.45) describing the entropy balance. When the fluid experiences a small-amplitude disturbance, its ambient physical variables (denoted by a subscript zero) are modified to

$$\rho = \rho_0 + \rho' \qquad s = s_0 + s' \qquad \mathbf{v} = 0 + \mathbf{v}'$$

$$p = p_0 + p' \qquad T = T_0 + T' \tag{62.1}$$

Linearizing the dynamical equations in the small primed quantities, we find

$$\frac{\partial \rho'}{\partial t} + \rho_0 \nabla \cdot \mathbf{v}' = 0 \tag{62.2}$$

$$\frac{\partial \mathbf{v}'}{\partial t} = -\frac{1}{\rho_0} \nabla p' + \frac{\eta}{\rho_0} \nabla^2 \mathbf{v}' + \frac{1}{\rho_0} (\zeta + \tfrac{1}{3}\eta) \nabla (\nabla \cdot \mathbf{v}') \tag{62.3}$$

$$T_0 \rho_0 \frac{\partial s'}{\partial t} = k_{\text{th}} \nabla^2 T' \tag{62.4}$$

where k_{th} is the thermal conductivity and Eq. (57.11) has been used for $\mathbf{j}_h$. Note that the nonzero acoustic pressure p' necessitates using the full entropy equation (60.45) instead of the simpler equation (60.47) for isobaric heat conduction.

In addition to these dynamical equations, the four infinitesimal quantities T', s', ρ', and p' are connected by two thermodynamic relations, because the state of a one-component thermodynamic system with given mass is completely determined by only two independent state variables. As with sound waves in nonviscous fluids, it is convenient to treat s and ρ as independent variables. The primed quantities in Eq. (62.1) represent first-order deviations from the equilibrium values; the total differentials of the quantities $p(\rho, s)$ and $T(\rho, s)$ thus yield

$$p' = \left(\frac{\partial p}{\partial s}\right)_\rho s' + \left(\frac{\partial p}{\partial \rho}\right)_s \rho' \tag{62.5a}$$

$$T' = \left(\frac{\partial T}{\partial s}\right)_\rho s' + \left(\frac{\partial T}{\partial \rho}\right)_s \rho' \tag{62.5b}$$

Furthermore, since Eq. (60.42) is also a total differential, we have a relation between the thermodynamic coefficients appearing in Eq. (62.5). This linearized Maxwell relation is

$$\left(\frac{\partial T}{\partial \rho}\right)_s = \frac{1}{\rho_0^2}\left(\frac{\partial p}{\partial s}\right)_\rho \tag{62.6}$$

The remaining coefficients in Eq. (62.5) have the simple physical interpretations [see Eqs. (49.5) and (52.24a)]

$$\left(\frac{\partial p}{\partial \rho}\right)_s = c^2 \tag{62.7a}$$

$$\left(\frac{\partial T}{\partial s}\right)_\rho = \frac{T_0}{c_V} \tag{62.7b}$$

where c is the speed of sound in a nonviscous isentropic fluid and c_V is the specific heat per unit mass at constant volume. A combination of these relations gives the *following full set of coupled linear equations for the three variables s', ρ', and $\mathbf{v}'$ (the deviation from the equilibrium values) in a compressible fluid, including viscosity and heat conduction:*

$$\frac{\partial \rho'}{\partial t} + \rho_0 \mathbf{V} \cdot \mathbf{v}' = 0 \tag{62.8}$$

$$\frac{\partial \mathbf{v}'}{\partial t} + \frac{c^2}{\rho_0}\mathbf{V}\rho' = -\rho_0\left(\frac{\partial T}{\partial \rho}\right)_s \mathbf{V}s' + \frac{\eta}{\rho_0}\mathbf{V}^2\mathbf{v}' + \frac{1}{\rho_0}(\zeta + \tfrac{1}{3}\eta)\,\mathbf{V}(\mathbf{V} \cdot \mathbf{v}') \tag{62.9}$$

$$\frac{\partial s'}{\partial t} - \gamma\kappa\,\mathbf{V}^2 s' = \frac{c_p\kappa}{T_0}\left(\frac{\partial T}{\partial \rho}\right)_s \mathbf{V}^2\rho' \tag{62.10}$$

where κ is the thermal diffusivity (57.13) and [see Eq. (52.26)]

$$\gamma \equiv \frac{c_p}{c_V} \geq 1 \tag{62.11}$$

To interpret these equations, we note that $(\partial T/\partial \rho)_s = c^2(\partial T/\partial p)_s$ [see Eq. (62.7a)] and that $(\partial T/\partial p)_s = [\partial(1/\rho)/\partial s]_p$ by a Maxwell relation [see Eq. (57.3b)]. Simple manipulations yield $(\partial T/\partial \rho)_s = Tc^2\beta/\rho c_p$, where $\beta \equiv V^{-1}(\partial V/\partial T)_p$ is the coefficient of volume thermal expansion. In the absence of thermal expansion, the quantity $(\partial T/\partial \rho)_s$ vanishes identically, and we see that the equations then separate into a pair (62.8) and (62.9) describing the mechanical variations in ρ and $\mathbf{v}$, and the single equation (62.10) describing entropy propagation. Since β is frequently quite small in a liquid, this simpler picture often provides a useful first approximation for such systems.

To analyze the full set of Eqs. (62.8) to (62.10) in detail, we seek *normal modes* in the form of plane waves, with all primed quantities proportional to $e^{i(\mathbf{k} \cdot \mathbf{r} - \omega t)}$ and ω real. Retaining the same symbols for the complex amplitudes, we find

$$\omega \rho' = \rho_0 \mathbf{k} \cdot \mathbf{v}' \tag{62.12}$$

$$\omega \mathbf{v}' - \frac{c^2 \mathbf{k}}{\rho_0} \rho' = \rho_0 \left(\frac{\partial T}{\partial \rho}\right)_s \mathbf{k} s' - \frac{i\eta k^2}{\rho_0} \mathbf{v}' - \frac{i}{\rho_0}(\zeta + \tfrac{1}{3}\eta)\mathbf{k}(\mathbf{k} \cdot \mathbf{v}') \tag{62.13}$$

$$(\omega + i\gamma\kappa k^2)s' = -\frac{ic_p\kappa}{T_0}\left(\frac{\partial T}{\partial \rho}\right)_s k^2 \rho' \tag{62.14}$$

This set of coupled linear algebraic equations has two types of solutions. We can obtain solutions for *transverse* modes with $\mathbf{v}' \perp \mathbf{k}$ and, independently, for *longitudinal* modes with $\mathbf{v}' \| \mathbf{k}$. We consider first the transverse modes ($\mathbf{k} \cdot \mathbf{v}' = 0$) when the component of Eq. (62.13) perpendicular to $\mathbf{k}$ yields the dispersion relation

$$\omega = -\frac{i\eta k^2}{\rho_0} = -ivk^2 \qquad \text{transverse modes} \tag{62.15}$$

These transverse modes are exactly the viscous waves studied in the preceding section, since Eq. (62.15) has the equivalent form [see Eq. (61.30)]

$$k = \pm\left(\frac{i\omega}{v}\right)^{1/2} = \pm\frac{1 + i}{\delta} \tag{62.16}$$

Furthermore, Eqs. (62.12) and (62.14) show that ρ' and s' (and therefore p' and T') vanish for linearized transverse viscous waves, in accordance with the second-order character of the heat generated in (61.35).

The longitudinal modes are more interesting because they introduce several new physical phenomena. Taking the dot product of Eq. (62.13) with $\mathbf{k}$ and using (62.12) to eliminate $\mathbf{k} \cdot \mathbf{v}'$, we obtain the pair of coupled equations

$$\left[\omega^2 - c^2 k^2 + i\frac{\omega k^2}{\rho_0}(\tfrac{4}{3}\eta + \zeta)\right]\rho' - \rho_0^2 k^2\left(\frac{\partial T}{\partial \rho}\right)_s s' = 0 \tag{62.17}$$

$$\frac{i\kappa k^2 c_p}{T_0}\left(\frac{\partial T}{\partial \rho}\right)_s \rho' + (\omega + i\gamma\kappa k^2)s' = 0 \tag{62.18}$$

As mentioned previously, these modes would separate into density and entropy waves if $(\partial T/\partial \rho)_s = 0$. In general, however, they are coupled, the exact dispersion relation being given by equating the determinant of coefficients to zero. Among the various solutions, the most interesting is the sound wave, and we shall consider it in detail. Recall first that an isentropic sound wave in a nonviscous fluid has the exact dispersion relation $\omega = ck$. In the present case, we therefore assume the approximate form for the wave number in these normal modes†

$$k = \frac{\omega}{c} + i\alpha \tag{62.19}$$

and expand the determinant to first order in α. Keeping only the leading low-frequency contribution, we find

$$\alpha = \frac{\omega^2}{2c^3\rho_0}\left(\tfrac{4}{3}\eta + \zeta\right) + \frac{\kappa c_p \rho_0^2 \omega^2}{2T_0 c^5}\left(\frac{\partial T}{\partial \rho}\right)_s^2 \tag{62.20}$$

which is intrinsically positive. For a given real frequency ω, a wave along the z axis (say) becomes

$$\rho'(z, t) = \rho_0 e^{-\alpha z} \exp\left[i\frac{\omega}{c}(z - ct)\right] \tag{62.21}$$

Thus α is the spatial attenuation constant for the amplitude (the energy attenuates like $e^{-2\alpha z}$). Note that attenuation α increases like ω^2, so that the assumption of small α is a low-frequency approximation. The two terms in (62.20) arise from the different physical effects of viscosity and thermal conduction of the small induced temperature wave (62.5b), although both act to degrade the sound wave. Measurements of α can determine the bulk viscosity ζ, which is not readily accessible by other techniques.

For some purposes, it is helpful to rewrite the last term in (62.20) in an equivalent form. First note that

$$\frac{1}{c_p} = \frac{1}{T_0}\left(\frac{\partial T}{\partial s}\right)_p \tag{62.22}$$

and then change independent variables from (s, p) to the preferred set (s, ρ). A standard manipulation allows us to write

$$\frac{T_0}{c_p} = \frac{(\partial T/\partial s)_\rho(\partial p/\partial \rho)_s - (\partial T/\partial \rho)_s(\partial p/\partial s)_\rho}{(\partial p/\partial \rho)_s} \tag{62.23}$$

and use of Eqs. (62.6) and (62.7) gives

$$\frac{1}{c_p} = \frac{1}{c_V} - \frac{\rho_0^2}{T_0 c^2}\left(\frac{\partial T}{\partial \rho}\right)_s^2 \tag{62.24}$$

† Note that our assumption of real ω means that these linearized solutions are true dissipation-free normal modes, whose amplitude attenuates exponentially in space. An alternative approach is to assume real $\mathbf{k}$, in which case the frequency ω becomes complex with a negative imaginary part [compare Eq. (62.15)]. This latter approach is more appropriate for describing a specified initial configuration that subsequently decays in time.

Substitution into Eq. (62.20) yields the more readily interpreted expression

$$\alpha = \frac{\omega^2}{2\rho_0 c^3}\left[\tfrac{4}{3}\eta + \zeta + k_{\text{th}}\left(\frac{1}{c_V} - \frac{1}{c_p}\right)\right] \tag{62.25}$$

Equation (62.25) becomes particularly simple for a dilute gas when the quantity in square brackets is of order $c\rho_0 l$, where l is the mean free path [see the footnote below Eq. (58.21)]. The condition of small damping $(c\alpha \ll \omega)$ then becomes equivalent to the previous qualitative criterion $(l \ll \lambda)$ for isentropic sound propagation. In the opposite (high-frequency) limit, a medium with large thermal conductivity $(\kappa \gg v$ and $\kappa \gg \zeta/\rho_0)$ can still support weakly damped longitudinal density waves, but the motion is then isothermal with a speed $c\gamma^{-1/2} = [(\partial p/\partial \rho)_T]^{1/2}$ (see Prob. 12.13).

If the thermal expansion were negligible, Eq. (62.18) would describe heat conduction with a dispersion relation $\omega = -i\kappa k^2$ [cf. Eq. (62.15) and recall that $\gamma = 1$ if $(\partial T/\partial \rho)_s = 0$]. Simple manipulations with Eqs. (62.17), (62.18), and (62.24) show that this dispersion relation remains correct at long wavelengths for arbitrary γ, in accordance with Eq. (58.19).

PROBLEMS

12.1 (a) Assume an incompressible fluid and write Newton's second law for a given element of fluid including the viscous drag force of Eq. (60.2) and Fig. 60.1. Hence directly derive the Navier-Stokes equation (60.27).

(b) Assume a compressible fluid and generalize Eq. (48.31) to incorporate the effect of the viscous surface stress T_{kl}^v in Eq. (60.17) on the conservation of linear momentum in a fixed volume V. What is the corresponding differential form of this relation?

(c) Similarly construct the integral form of the conservation of angular momentum L_i in the fixed volume V [see Eq. (60.7)] by considering torques due to the volume force, the pressure, the viscous stress, and the convective transport of angular-momentum density $\sum_{jk} \epsilon_{ijk}\rho x_j v_k$. What is the corresponding differential form?

12.2 (a) In a viscous incompressible fluid, show that the vorticity $\boldsymbol{\zeta} \equiv \nabla \times \mathbf{v}$ obeys the dynamical equation (see Prob. 9.5)

$$\frac{d\boldsymbol{\zeta}}{dt} \equiv \frac{\partial \boldsymbol{\zeta}}{\partial t} + (\mathbf{v}\cdot\nabla)\boldsymbol{\zeta} = (\boldsymbol{\zeta}\cdot\nabla)\mathbf{v} + v\,\nabla^2\boldsymbol{\zeta}$$

(b) Illustrate the diffusive effect of viscosity by considering an axisymmetric flow $\mathbf{v}(\mathbf{r}, t) = \hat{\boldsymbol{\phi}}V(r, t)$. Show that $\boldsymbol{\zeta} = \hat{\mathbf{z}}\zeta(r, t)$, where $\zeta(r, t) = r^{-1}\,\partial(rV)/\partial r$, and that

$$\frac{\partial \zeta}{\partial t} = v\left(\frac{\partial^2 \zeta}{\partial r^2} + \frac{1}{r}\frac{\partial \zeta}{\partial r}\right)$$

(c) Consider the special initial configuration $\zeta(\mathbf{r}, 0) = \zeta_0\delta(\mathbf{r}) = \zeta_0\delta(x)\delta(y)$ and derive the time-dependent solution

$$\zeta(r, t) = \frac{\zeta_0}{4\pi vt}\,e^{-r^2/4vt}$$

where $r^2 = x^2 + y^2$. Discuss this result.

(d) Explicitly exhibit the violation of Thomson's theorem (Sec. 48) by proving that the circulation around a circle of radius a is given by $\Gamma_a(t) = \zeta_0(1 - e^{-a^2/4vt})$.

12.3 An incompressible viscous fluid in a long tube with rectangular cross section $a \times b$ flows under a uniform pressure gradient.

(a) Solve Eq. (61.5) by expanding in the eigenfunctions $\rho_{mn}(x, y)$ from Eq. (47.8a).

(b) Find an explicit expression for the total mass flow rate $\dot{M}$. Show that it reproduces Eq. (61.12) for $b \to \infty$.

(c) Discuss the mass flow rate for a square channel and compare with that for a circular one of equal area.

12.4 An incompressible viscous fluid flows under a uniform pressure gradient in the annular region between two concentric cylinders of radii R_1 and R_2, with $R_1 < R_2$. Use Eq. (61.13) to find the steady velocity field. Find the rate of mass flow and verify the limiting case (61.21) for $R_1 \to 0$.

12.5 An incompressible viscous fluid initially at rest occupies the region between two parallel planes $(z = 0$ and $z = d)$. At $t = 0$, the lower wall starts to move uniformly with velocity $v_0 \hat{x}$ while the upper wall remains at rest.

(a) Show that the velocity is given for all times by

$$v_x(z, t) = v_0 \frac{d - z}{d} - 2v_0 \sum_{n=1}^{\infty} \frac{1}{n\pi} e^{-n^2 \pi^2 vt/d^2} \sin \frac{n\pi z}{d}$$

(b) Evaluate the reaction forces on the upper and lower walls and discuss their limits for large times $(\gg d^2/v)$.

(c) Use Prob. 11.15 to find approximate expressions for these reaction forces at short times $(\ll d^2/v)$. Discuss and interpret.

12.6 An incompressible viscous fluid occupies the region between two parallel plates at $z = 0$ and $z = d$. The lower plate oscillates harmonically with amplitude $\epsilon \hat{x} \cos \omega t$, and the upper plate is stationary.

(a) Show that $v_x(z, t)$ is given by the expression

$$v_x(z, t) = \text{Re} \left\{ -i\omega\epsilon \frac{\sinh [(1 - i)(d - z)/\delta]}{\sinh [(1 - i)d/\delta]} e^{-i\omega t} \right\}$$

where δ is the viscous penetration length (61.30).

(b) Show that the drag force per unit area on the upper surface is

$$\text{Re} \left\{ -\frac{i\omega\epsilon\eta}{d} \frac{(1 - i)d/\delta}{\sinh [(1 - i)d/\delta]} e^{-i\omega t} \right\}$$

Discuss the limits $d \gg \delta$ and $d \ll \delta$.

(c) Find the corresponding reaction force on the bottom surface and show that the time-average work done by the moving wall per unit time is

$$\frac{\omega^2 \epsilon^2 \eta}{2d} \text{Re} \left[\frac{d}{\delta} (1 - i) \coth \frac{d}{\delta} (1 - i) \right]$$

Consider the limits $d \gg \delta$ and $d \ll \delta$. Verify that this quantity also equals the energy dissipated throughout the fluid.

12.7 A long tube of radius R and length L contains an incompressible viscous fluid with kinematic viscosity v. At $t = 0$, the initially stationary fluid experiences a uniform axial pressure gradient $-\Delta p/L$.

(a) Determine the fluid velocity $v_z(r, t)$ for all subsequent times.

(b) Show that the walls exert a drag force on the fluid given by [compare Eq. (61.19a)]

$$F_z^d = iR \, \Delta p \int_C \frac{ds}{s^{3/2}} e^{vst} \frac{I_1(Rs^{1/2})}{I_0(Rs^{1/2})}$$

where $I_n(x) = i^{-n} J_n(ix)$ is the Bessel function of imaginary argument and C is the Laplace-transform inversion contour.

(c) Evaluate the preceding expression for short and long times to obtain

$$F_z^d \approx -4R \, \Delta p (\pi v t)^{1/2} \qquad\qquad vt \ll R^2$$

$$F_z^d \approx -\pi R^2 \, \Delta p \left[1 - 4\alpha_{0,1}^{-2} \exp\left(-\alpha_{0,1}^2 \frac{vt}{R^2} \right) + \cdots \right] \qquad vt \gg R^2$$

where $\alpha_{0,1} \approx 2.4048$ is the first zero of $J_0(x)$.

12.8 A long cylinder of radius R is surrounded by initially stationary incompressible fluid with kinematic viscosity v. At $t = 0$, the cylinder starts to move uniformly parallel to its axis with speed v_0. Use a Laplace transform to show that the drag force per unit length on the cylinder has the magnitude [compare Eq. (61.19a)]

$$F_z^d = \frac{8\eta v_0}{\pi} \int_0^\infty \frac{d\zeta \, \exp\left(-\zeta^2 vt/R^2\right)}{\zeta \; J_0^2(\zeta) + N_0^2(\zeta)}$$

where J_0 and N_0 are the Bessel and Neumann functions of order zero. Obtain and discuss the following approximate expressions

$$F_z^d \approx \begin{cases} 2\eta v_0 \, R \left(\dfrac{\pi}{vt}\right)^{1/2} & t \ll \dfrac{R^2}{v} \\[2.5ex] 4\pi\eta v_0 \left(\ln \dfrac{vt}{R^2}\right)^{-1} & t \gg \dfrac{R^2}{v} \end{cases}$$

12.9 A long cylinder of radius a contains an incompressible fluid with viscosity η and density ρ. The fluid is initially at rest, and at $t = 0$ the walls suddenly rotate with constant angular velocity Ω.

(a) Using standard vector identities, show that the Navier-Stokes equations reduce to

$$\frac{\partial p}{\partial r} = \frac{\rho V^2}{r}$$

$$\frac{\partial V}{\partial t} = v\left(\frac{\partial^2 V}{\partial r^2} + \frac{1}{r}\frac{\partial V}{\partial r} - \frac{V}{r^2} \right)$$

where $v = \eta/\rho$ and $\mathbf{v}(\mathbf{r}, t) = \hat{\boldsymbol{\phi}} V(r, t)$ in cylindrical polar coordinates.

(b) Use a Laplace transform to derive the integral representation

$$V(r, t) = \frac{\Omega a}{2\pi i} \int_C \frac{ds}{s} \, e^{st} \, \frac{I_1[r(s/v)^{1/2}]}{I_1[a(s/v)^{1/2}]}$$

where C runs to the right of all singularities and $I_1(x) = -iJ_1(ix)$.

(c) For long and short times, derive results analogous to those in Prob. 11.11. Briefly compare these two physical systems.

12.10 Use the velocity field in Prob. 12.9 to study the time-dependent vorticity $\zeta \equiv \nabla \times \mathbf{v}$ and the circulation Γ_R about a concentric circular contour of radius R. Discuss the various limits.

12.11 An incompressible viscous fluid executes axisymmetric flow with $\mathbf{v} = \hat{\boldsymbol{\phi}} V(r, t)$, so that the particles move in concentric circles.

(a) Generalize the method used in deriving Eq. (61.19a) to show that the azimuthal component of the force exerted on a surface element $dA \, \hat{\mathbf{r}}$ is given by

$$dF_\phi = \eta \, dA \; r \frac{d(V/r)}{dr} \bigg|_{r=a}$$

Compare with the general result obtained from Prob. B.8.

(b) Show that the rotating cylinder in Prob. 12.9 exerts a torque

$$-ia^3\eta\Omega \int_C \frac{d\zeta}{\zeta^{1/2}} e^{\zeta vt} \frac{I_2(a\sqrt{\zeta})}{I_1(a\sqrt{\zeta})}$$

per unit length on the fluid. Evaluate the long- and short-time limits. Discuss.

12.12 Viscous fluid occupies the region between two long concentric cylinders of radius R_1 and R_2 $(R_1 < R_2)$.

(a) If the cylinders rotate with different angular velocities Ω_1 and Ω_2, find the steady-state velocity field, which is known as *Couette flow*.

(b) Use the result of part (a) of Prob. 12.11 to find the torque exerted by each cylinder on the fluid.

12.13 Consider the propagation of small-amplitude longitudinal waves in a fluid with high thermal conductivity, neglecting only the viscous contributions to the stress tensor [see Eqs. (62.17) and (62.18)].

(a) Show that the dispersion relation is given by the following quadratic equation in k^2:

$$k^4 - k^2\left(\frac{\gamma\omega^2}{c^2} + \frac{i\omega}{\kappa}\right) + \frac{i\omega^3}{\kappa c^2} = 0$$

[Make use of Eq. (62.24).]

(b) For low frequencies $(0 < \omega \ll c^2/\kappa$, or $\delta \ll \lambda)$, show that the two roots become

$$k_+^2 \approx \frac{i\omega}{\kappa} + \frac{(\gamma - 1)\omega^2}{c^2} \quad \text{and} \quad k_-^2 \approx \frac{\omega^2}{c^2} + \frac{i(\gamma - 1)\omega^3\kappa}{c^4}$$

Compare with the discussion below Eq. (62.25).

(c) For high frequencies $(c^2/\kappa \ll \omega$, or $\lambda \ll \delta)$, show that the two roots become

$$k_+^2 \approx \frac{\gamma\omega^2}{c^2} + \frac{i\omega}{\kappa}\frac{\gamma - 1}{\gamma} \quad \text{and} \quad k_-^2 \approx \frac{i\omega}{\gamma\kappa} + \frac{(\gamma - 1)c^2}{\kappa^2\gamma^3}$$

and that $c^2\gamma^{-1} = (\partial p/\partial\rho)_T$. Discuss, and compare with part (b).

SELECTED ADDITIONAL READINGS

1. G. K. Batchelor, "An Introduction to Fluid Dynamics," Cambridge University Press, London, 1967.
2. S. Chandrasekhar, "Hydrodynamic and Hydromagnetic Stability," Oxford University Press, London, 1961.
3. R. P. Feynman, R. B. Leighton, and M. Sands, "The Feynman Lectures on Physics," Addison-Wesley, Reading, Mass., 1964, vol. 2.
4. A. S. Monin and A. M. Yaglom, "Statistical Fluid Mechanics," MIT Press, Cambridge, Mass., 1971, vols. 1 and 2.
5. S. A. Orszag, in R. Balian and J.-L. Peube (eds.), "Fluid Dynamics," Gordon and Breach, New York, 1977.

THIRTEEN

ELASTIC CONTINUA

Our previous treatment of continuous media has considered either systems of reduced dimensionality (strings or membranes) or three-dimensional fluids. In the former case, the restoring forces arise solely from the tension τ, which acts along the string (or in the surface of the membrane). Since no additional elastic stress opposes the bending, these strings and membranes are, in effect, infinitely flexible. Fluids also entail some simplification, for they cannot sustain shear stress in static equilibrium. Hence if $\mathbf{v} = 0$, Eq. (60.23) shows that the stress tensor becomes isotropic with elements $p\delta_{ij}$. The situation is quite different for an elastic medium, where small applied shear forces produce a true static configuration that returns to the original undeformed state when the forces are removed. To analyze this behavior, it is first necessary to construct a mathematical description of deformations in a continuous medium.†

63 BASIC FORMULATION

Consider a stationary undeformed continuous medium, with each point labeled by its coordinate $\mathbf{x}$ relative to some fixed set of cartesian axes. Now deform the medium, with each point $\mathbf{x}$ moving to a new position $\mathbf{x}'$ (see Fig. 63.1). The displacement $\mathbf{x}' - \mathbf{x}$ will be denoted

$$\mathbf{u}(\mathbf{x}) \equiv \mathbf{x}' - \mathbf{x} \tag{63.1}$$

This vector $\mathbf{u}$ evidently depends on the original vector position $\mathbf{x}$.

---

† We have found the following books particularly useful: Joos [1], chap. 8; Landau and Lifshitz [2], chaps. 1 and 3; Sommerfeld (1950) chaps. 2, 3, and 8.

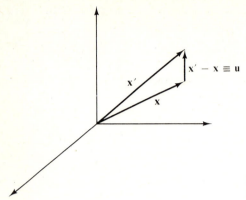

Figure 63.1 Displacement of a continuous medium.

Small Deformations

An elastic solid differs from a fluid in that two points in the solid that are initially close together remain so forever. Thus if **x** and **y** are neighboring in the initial configuration, the deformation moves them to similar nearby positions (see Fig. 63.2)

$$\mathbf{x}' = \mathbf{x} + \mathbf{u}(\mathbf{x}) \qquad \mathbf{y}' = \mathbf{y} + \mathbf{u}(\mathbf{y}) \tag{63.2}$$

These expressions are perfectly general and describe, for example, a uniform translation (**u** = const). In that case, however, the system moves rigidly without internal deformation. Since it is this latter contribution that involves the elastic properties of the medium, we instead consider the *vector separation* of the two nearby points

$$\mathbf{y}' - \mathbf{x}' = \mathbf{y} - \mathbf{x} + \mathbf{u}(\mathbf{y}) - \mathbf{u}(\mathbf{x}) \tag{63.3}$$

To first order in the small separation **y** − **x**, an expansion of the right-hand side gives the linear relation

$$(\mathbf{y}' - \mathbf{x}')_i - (\mathbf{y} - \mathbf{x})_i = \sum_{j=1}^{3} (\mathbf{y} - \mathbf{x})_j \frac{\partial u_i}{\partial x_j} \tag{63.4}$$

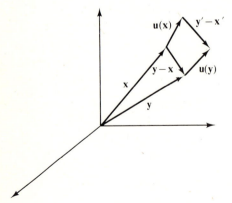

Figure 63.2 Deformation of a continuous medium.

where the partial derivative is evaluated at **x**. In addition, the displacements **u(x)** are themselves considered small, and we shall work to first order in these quantities, as in the discussion of strings and membranes.

Since $\partial u_i/\partial x_j$ vanishes for a uniform translation, it might initially seem suitable to characterize an elastic deformation. Such a procedure is incorrect, however, as seen by the following argument. The coefficients appearing in Eq. (63.4) can be separated uniquely into parts that are even and odd under the interchange of the indices i and j

$$\frac{\partial u_i}{\partial x_j} \equiv \frac{1}{2}\left(\frac{\partial u_i}{\partial x_j} + \frac{\partial u_j}{\partial x_i}\right) + \frac{1}{2}\left(\frac{\partial u_i}{\partial x_j} - \frac{\partial u_j}{\partial x_i}\right) \tag{63.5a}$$

$$\equiv \epsilon_{ij} + O_{ij} \tag{63.5b}$$

where the tensors ϵ_{ij} and O_{ij} are sets of real numbers that depend on the position **x**. We now demonstrate that the odd tensor O_{ij} corresponds to a *local rigid rotation* of the medium and hence cannot produce a true *internal deformation*. This is shown as follows: an antisymmetric tensor like O_{ij} has only three independent components; that is, $(i, j) = (1, 2)$, $(1, 3)$, and $(2, 3)$. They can be written quite generally in terms of three other real quantities $\delta\Omega_k$ with $k = 1, 2, 3$ according to [see Eq. (48.17)]

$$O_{ij} = -\sum_{k=1}^{3} \epsilon_{ijk}\, \delta\Omega_k \tag{63.6}$$

Furthermore, the three quantities $\delta\Omega_k$ can always be reexpressed as

$$\delta\Omega_k \equiv \delta\Omega\, n_k \qquad k = 1, 2, 3 \tag{63.7a}$$

with

$$\sum_{k=1}^{3} n_k^2 = 1 \tag{63.7b}$$

Here the coefficient $\delta\Omega$ can be treated as a small quantity because we are retaining only first-order terms in the displacements **u**. Consider now the transformation on the coordinate differences in the vicinity of the point **x** in Eq. (63.4) that is induced by the O_{ij} in Eqs. (63.5) and (63.6). It evidently takes the form

$$[(\mathbf{y}' - \mathbf{x}')_i - (\mathbf{y} - \mathbf{x})_i]_{\text{odd}} = -\sum_{jk=1}^{3} (\mathbf{y} - \mathbf{x})_j \epsilon_{ijk}\, \delta\Omega_k \tag{63.8a}$$

or, in vector notation,

$$[(\mathbf{y}' - \mathbf{x}') - (\mathbf{y} - \mathbf{x})]_{\text{odd}} = \delta\mathbf{\Omega} \times (\mathbf{y} - \mathbf{x}) \tag{63.8b}$$

The discussion in Sec. 7 shows that this transformation represents an infinitesimal *rigid rotation* of the medium, with magnitude $\delta\Omega$, about the $\hat{n}$ axis; hence the odd tensor O_{ij} does not represent an elastic distortion. The *true internal deformation* of the elastic continuum thus arises only from the even tensor ϵ_{ij} in Eq. (63.5), and we therefore define the *elastic strain tensor* as

$$\epsilon_{ij} \equiv \frac{1}{2}\left(\frac{\partial u_i}{\partial x_j} + \frac{\partial u_j}{\partial x_i}\right) \tag{63.9}$$

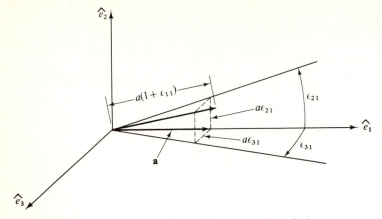

Figure 63.3 Elastic deformation of a vector $\mathbf{a} = a\hat{e}_1$. Note that $|\epsilon_{ij}| \ll 1$.

It vanishes identically for a rigid rotation as well as for a rigid translation. As a result, the *elastic deformation* of the vector $\mathbf{y} - \mathbf{x}$ joining two nearby points in the medium becomes

$$[(\mathbf{y}' - \mathbf{x}')_i - (\mathbf{y} - \mathbf{x})_i]_{el} \equiv \sum_{j=1}^{3} \epsilon_{ij}(\mathbf{y} - \mathbf{x})_j \qquad (63.10)$$

where we henceforth assume small strain ($|\epsilon_{ij}| \ll 1$) and work to lowest order in the dimensionless parameters ϵ_{ij}.

To clarify these rather formal arguments, we study a few simple cases of deformation under a *specified strain tensor* ϵ_{ij}.

1. Suppose that the undisplaced vector $\mathbf{y} - \mathbf{x}$ has a length a and lies along one of the coordinate axes $\hat{e}_1$ (say), with $\mathbf{y} - \mathbf{x} = a\hat{e}_1 \equiv \mathbf{a}$. Under the elastic deformation (63.10) specified by the given strain tensor ϵ_{ij}, this vector becomes (see Fig. 63.3)

$$\mathbf{a}' = \mathbf{a} + a(\epsilon_{11}\hat{e}_1 + \epsilon_{21}\hat{e}_2 + \epsilon_{31}\hat{e}_3)$$
$$= a[(1 + \epsilon_{11})\hat{e}_1 + \epsilon_{21}\hat{e}_2 + \epsilon_{31}\hat{e}_3] \qquad (63.11)$$

To first order in ϵ, it experiences two distinct alterations. Its length changes from a to

$$a' \equiv |\mathbf{a}' \cdot \mathbf{a}'|^{1/2} \approx a(1 + \epsilon_{11}) \qquad (63.12a)$$

which identifies the element

$$\epsilon_{11} = \frac{a' - a}{a} \qquad (63.12b)$$

of the strain tensor as the fractional change in length. In addition, $\mathbf{a}'$ has a different direction because it has nonzero projections $a\epsilon_{21}$ and $a\epsilon_{31}$ on the original axes $\hat{e}_2$ and $\hat{e}_3$.

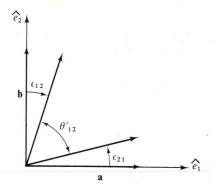

Figure 63.4 Change in angles under elastic deformation.

2. This latter observation has two important corollaries. Consider two vectors $\mathbf{a} = a\hat{e}_1$ and $\mathbf{b} = b\hat{e}_2$ that are originally orthogonal. Under the deformation, $\mathbf{a}$ changes as in Eq. (63.11), and $\mathbf{b}$ becomes

$$\mathbf{b}' = b[\epsilon_{12}\hat{e}_1 + (1 + \epsilon_{22})\hat{e}_2 + \epsilon_{32}\hat{e}_3] \tag{63.13}$$

The angle θ'_{12} between them is characterized by their dot product $\mathbf{a}' \cdot \mathbf{b}' = a'b' \cos \theta'_{12}$, which is given approximately by

$$\mathbf{a}' \cdot \mathbf{b}' \approx ab(\epsilon_{21} + \epsilon_{12}) \tag{63.14}$$

Thus the elastic deformation alters the original right angle to (see Fig. 63.4)

$$\theta'_{12} \approx \tfrac{1}{2}\pi - \epsilon_{21} - \epsilon_{12} = \tfrac{1}{2}\pi - 2\epsilon_{12} \tag{63.15}$$

Note that the symmetry $\epsilon_{12} = \epsilon_{21}$ indeed eliminates any net rotation of the two vectors. In a similar way, under the elastic deformation, the volume $V = abc$ bounded by the three orthogonal vectors $(a\hat{e}_1, b\hat{e}_2, c\hat{e}_3)$ changes to

$$V' = \mathbf{a}' \cdot \mathbf{b}' \times \mathbf{c}' = \det \begin{vmatrix} a'_1 & a'_2 & a'_3 \\ b'_1 & b'_2 & b'_3 \\ c'_1 & c'_2 & c'_3 \end{vmatrix}$$

$$\approx abc(1 + \epsilon_{11} + \epsilon_{22} + \epsilon_{33}) = V(1 + \text{Tr } \underline{\epsilon}) \tag{63.16}$$

Thus the trace of the strain tensor determines the fractional change in the volume. Use of Eq. (63.9) provides the equivalent form

$$\text{Tr } \underline{\epsilon} = \mathbf{\nabla} \cdot \mathbf{u} = \frac{V' - V}{V} = \frac{dV}{V} \tag{63.17}$$

expressed in terms of the displacement vector $\mathbf{u}$. Alternatively, this relation also determines the change in density.

$$\text{Tr } \underline{\epsilon} = \mathbf{\nabla} \cdot \mathbf{u} = -\frac{d\rho}{\rho} \tag{63.18}$$

Stress Tensor

As in our previous analysis of fluid dynamics, we define an elastic stress tensor T_{ij} [compare Eq. (60.4)]

$$-\sum_{j=1}^{3} T_{ij}\, dA_j = \text{ith component of force acting on a surface } d\mathbf{A} = dA\,\hat{n} \text{ with}$$

$$\text{outward normal } \hat{n} \tag{63.19}$$

The dynamics of the angular momentum again requires that T_{ij} be symmetric [see Eq. (60.14)]. Hence the equations of motion of an elastic continuum necessarily take the form (60.9), and the only new feature is the specific relation between the elastic stress tensor and the elastic strain. We recall that in an ideal fluid at rest, the stress tensor depends only on the *pressure*, reflecting the presence of the term $-\rho^{-1}\nabla p$ in the equations of motion. Viscous fluids have additional shear forces arising from the velocity gradients in the flow. In an elastic solid, there are still further forces because of the local *strains* characterized by the tensor ϵ_{ij}. The form of T_{ij} in an elastic solid will turn out to reflect the obvious physical distinction between a viscous fluid and an elastic solid.

Hooke's law governs the small deformation of an elastic medium. It asserts that the local stress is proportional to the elastic deformation (local strain) and acts to oppose further deformation. The only difference from the elementary law $F = -kx$ for a spring is the tensor character, which means that the coefficients in the linear relation between T_{ij} and ϵ_{ij} must, in general, be elements of a fourth-rank tensor. Symmetry considerations reduce the maximum number of independent elements to 21, and various crystalline symmetries introduce further restrictions.[†] For simplicity, we here consider only an *isotropic continuum*, where Hooke's law and the absence of a preferred direction in the undeformed state imply that T_{ij} must be a linear combination of the two symmetric tensors [see the discussion leading to Eq. (60.17)]

$$\epsilon_{ij} = \frac{1}{2}\left(\frac{\partial u_i}{\partial x_j} + \frac{\partial u_j}{\partial x_i}\right) \qquad \delta_{ij}\,\text{Tr}\,\underline{\epsilon} = \delta_{ij}\nabla\cdot\mathbf{u} \tag{63.20}$$

For formal analysis, Hooke's law is most frequently written

$$T_{ij} = -\lambda\delta_{ij}\,\text{Tr}\,\underline{\epsilon} - 2\mu\epsilon_{ij} \tag{63.21a}$$

$$T_{ij} = -\lambda\delta_{ij}\nabla\cdot\mathbf{u} - \mu\left(\frac{\partial u_i}{\partial x_j} + \frac{\partial u_j}{\partial x_i}\right) \tag{63.21b}$$

where λ and μ are known as the *Lamé coefficients* or *Lamé moduli*, and the minus sign reflects the restoring character of the forces. In practical calculations, it is often preferable to deal with the traceless tensor [compare Eq. (60.16)]

$$\epsilon_{ij} - \tfrac{1}{3}\delta_{ij}\,\text{Tr}\,\underline{\epsilon} = \frac{1}{2}\left(\frac{\partial u_i}{\partial x_j} + \frac{\partial u_j}{\partial x_i} - \tfrac{2}{3}\delta_{ij}\nabla\cdot\mathbf{u}\right) \tag{63.22}$$

† Landau and Lifshitz [2], sec. 10; Sommerfeld (1950), sec. 40.

In this form, Hooke's law becomes

$$T_{ij} = -K\delta_{ij}\,\mathrm{Tr}\,\underline{\epsilon} - 2\mu(\epsilon_{ij} - \tfrac{1}{3}\delta_{ij}\,\mathrm{Tr}\,\underline{\epsilon}) \tag{63.23a}$$

$$T_{ij} = -K\delta_{ij}\mathbf{V}\cdot\mathbf{u} - \mu\left(\frac{\partial u_i}{\partial x_j} + \frac{\partial u_j}{\partial x_i} - \tfrac{2}{3}\delta_{ij}\mathbf{V}\cdot\mathbf{u}\right) \tag{63.23b}$$

where
$$K = \lambda + \tfrac{2}{3}\mu \tag{63.24}$$

Evidently, the elastic moduli K, λ, and μ all have the dimensions of force per unit area.

Equations (63.21) and (63.23) characterize the stress resulting from a given strain, but we shall also want to invert these relations to find the strain induced by a given stress. Taking the trace of (63.23a), for example, we find

$$\mathrm{Tr}\,\underline{T} = -3K\,\mathrm{Tr}\,\underline{\epsilon} \tag{63.25}$$

and it is now simple to eliminate $\mathrm{Tr}\,\underline{\epsilon}$ from Eq. (63.23a) to obtain the desired form

$$\epsilon_{ij} = -\frac{1}{9K}\delta_{ij}\,\mathrm{Tr}\,\underline{T} - \frac{1}{2\mu}(T_{ij} - \tfrac{1}{3}\delta_{ij}\,\mathrm{Tr}\,\underline{T}) \tag{63.26}$$

Since our study of hydrodynamics dealt principally with the stress tensor, this latter relation will often prove easier to interpret. We now consider three specific examples.

1. Apply a small hydrostatic pressure dp to a uniform medium at rest. The stress tensor must reduce to that for an ideal stationary fluid [see Eq. (60.5)] with no convective terms

$$T_{ij} = dp\,\delta_{ij} \tag{63.27}$$

Equation (63.26) then shows that the corresponding induced strain is also isotropic

$$\epsilon_{ij} = -\frac{dp}{3K}\delta_{ij} \tag{63.28}$$

The absence of off-diagonal elements means that a small rectangular parallelepiped retains its shape under a hydrostatic pressure; on the other hand, its volume changes according to (63.17)

$$\frac{dV}{V} = \mathrm{Tr}\,\underline{\epsilon} = -\frac{dp}{K} \tag{63.29}$$

and we immediately identify K as the *bulk modulus*

$$K = -V\frac{\partial p}{\partial V} \tag{63.30}$$

Depending on the specific situation, K can be either the adiabatic or isothermal modulus.

2. Consider a system under a uniform axial stress, with

$$T_{zz} = p \qquad \text{all other elements of } T_{ij} = 0 \qquad (63.31)$$

This configuration is readily achieved by applying a pressure p to each end of a rod whose lateral sides remain unstressed. Equation (63.26) then shows that the medium undergoes a uniform axial compression

$$\epsilon_{zz} = -\left(\frac{1}{9K} + \frac{1}{3\mu}\right)p = -\left(\frac{1}{9K} + \frac{1}{3\mu}\right)T_{zz} \qquad (63.32)$$

with ϵ_{zz} negative for $p > 0$. This linear relation is usually written

$$T_{zz} \equiv -E\epsilon_{zz} \qquad (63.33a)$$

where $$E = \left(\frac{1}{9K} + \frac{1}{3\mu}\right)^{-1} = \frac{9K\mu}{3K + \mu} \qquad (63.33b)$$

is known as *Young's modulus*. In addition, the same axial stress induces a transverse *expansion*

$$\epsilon_{xx} = \epsilon_{yy} = \left(\frac{1}{6\mu} - \frac{1}{9K}\right)p = \left(\frac{1}{6\mu} - \frac{1}{9K}\right)T_{zz} \qquad (63.34)$$

with ϵ_{xx} and ϵ_{yy} positive for $p > 0$. The ratio of the transverse expansion to longitudinal compression defines *Poisson's ratio* σ through the relation

$$\epsilon_{xx} = \epsilon_{yy} \equiv -\sigma\epsilon_{zz} \qquad (63.35a)$$

where $$\sigma = \frac{1}{2}\frac{3K - 2\mu}{3K + \mu} \qquad (63.35b)$$

In practice, σ lies between 0 and $\frac{1}{2}$. Table 63.1 contains some typical values of these elastic constants.

Table 63.1 Elastic constants of selected isotropic materials†

| | Density ρ (g cm^{-3}) | Modulus (10^{11} dyn cm^{-2}) | | | Poisson's ratio σ | Propagation speed (10^5 cm s^{-1}) | |
		E	K	μ		c_l	c_t
Silver	10.4	7.5	10.3	2.7	0.38	3.65	1.61
Beryllium	1.87	30.8	11.4	14.7	0.05	12.89	8.88
Stainless steel	7.91	19.6	16.3	7.57	0.30	5.79	3.10
Tungsten carbide	13.8	53.4	31.7	21.95	0.22	6.66	3.98
Pyrex glass	2.32	6.2	4.0	2.5	0.24	5.64	3.28
Lucite	1.18	0.40	0.66	0.143	0.4	2.68	1.10

† Data from D. E. Gray (ed.), "American Institute of Physics Handbook," 3d ed., McGraw-Hill, New York, 1972, pp. 3-101 to 3-104.

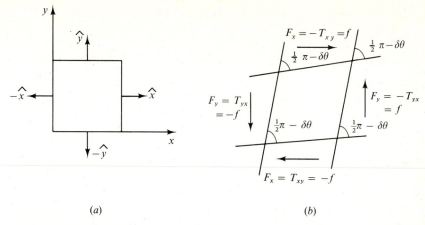

(a)   *(b)*

Figure 63.5 Effect of shear stress, showing induced force F_i per unit area on each face: *(a)* un-deformed and *(b)* deformed.

3. As a final example, consider a long rod with a square cross section whose sides experience a uniform shear stress (see Fig. 63.5)

$$T_{xy} = T_{yx} = -f \qquad \text{all other } T_{ij} = 0 \qquad (63.36)$$

The general relation (63.26) implies that

$$\epsilon_{xy} = \epsilon_{yx} = \frac{f}{2\mu} \qquad \text{all other } \epsilon_{ij} = 0 \qquad (63.37)$$

Comparison with Eq. (63.15) shows that the original right angles are deformed by an amount

$$\delta\theta = 2\epsilon_{xy} = \frac{f}{\mu} \qquad (63.38)$$

and the corresponding coefficient μ is known as the *shear modulus* or *modulus of rigidity*.

Evidently, only two of the four quantities E, K, μ, and σ are independent. It is often conventional to eliminate K and μ through the relations

$$K = \frac{1}{3} \frac{E}{1 - 2\sigma} \qquad (63.39a)$$

$$\mu = \frac{1}{2} \frac{E}{1 + \sigma} \qquad (63.39b)$$

and substitution into (63.26) yields the simpler form

$$\epsilon_{ij} = -\frac{1}{E} [(1 + \sigma)T_{ij} - \sigma\delta_{ij} \text{ Tr } \underline{T}] \qquad (63.40)$$

Elastic Energy

Consider an elastic medium subject to an external force field that exerts a force $\mathbf{F}^{(e)} \, d^3x$ on each small volume element d^3x throughout the medium. In static equilibrium, it assumes a deformed configuration with an induced stress T_{ij} that just balances the applied force through the relation [see Eq. (60.9)]

$$F_i^{(e)} = \sum_{j=1}^{3} \frac{\partial T_{ij}}{\partial x_j} \qquad (63.41)$$

Now suppose that the external force is altered by a small amount $\delta \mathbf{F}^{(e)}$, producing corresponding changes in the displacement $\delta \mathbf{u}(\mathbf{x})$ of each element of the medium. In this process, the external force does an amount of work on a volume V

$$\delta W = \int_V d^3x \, \mathbf{F}^{(e)}(\mathbf{x}) \cdot \delta \mathbf{u}(\mathbf{x}) \qquad (63.42)$$

Use of Eq. (63.41) and the divergence theorem yields

$$
\begin{aligned}
\delta W &= \int_V d^3x \sum_{ij=1}^{3} \frac{\partial T_{ij}}{\partial x_j} \delta u_i \\
&= \sum_{ij=1}^{3} \int_A dA_j \, T_{ij} \, \delta u_i - \sum_{ij=1}^{3} \int_V d^3x \, T_{ij} \frac{\partial}{\partial x_j} \delta u_i \\
&= \sum_{ij=1}^{3} \int_A dA_j \, T_{ij} \, \delta u_i - \sum_{ij=1}^{3} \int_V d^3x \, T_{ij} \frac{1}{2}\left(\frac{\partial}{\partial x_j} \delta u_i + \frac{\partial}{\partial x_i} \delta u_j \right) \qquad (63.43)
\end{aligned}
$$

where the last form follows from the symmetry of the stress tensor.

To interpret this result, consider a deformed medium, with displacement $\mathbf{u}(\mathbf{x})$, strain $\epsilon_{ij}(\mathbf{x})$, and stress $T_{ij}(\mathbf{x})$. If the configuration is altered by an infinitesimal amount $\delta \mathbf{u}(\mathbf{x})$, the strain tensor correspondingly changes by

$$
\begin{aligned}
\delta \epsilon_{ij} &\equiv \frac{1}{2}\left[\frac{\partial}{\partial x_j}(u_i + \delta u_i) + \frac{\partial}{\partial x_i}(u_j + \delta u_j) \right] - \frac{1}{2}\left(\frac{\partial u_i}{\partial x_j} + \frac{\partial u_j}{\partial x_i} \right) \\
&= \frac{1}{2}\left(\frac{\partial}{\partial x_j} \delta u_i + \frac{\partial}{\partial x_i} \delta u_j \right) \qquad (63.44)
\end{aligned}
$$

In this way, the basic relation (63.43) becomes

$$\delta W = \sum_{ij=1}^{3} \int_A dA_j \, T_{ij} \, \delta u_i - \sum_{ij=1}^{3} \int_V d^3x \, T_{ij} \, \delta \epsilon_{ij} \qquad (63.45)$$

expressed in terms of surface and volume contributions. Note that only the symmetric components of the displacement derivatives ϵ_{ij} do work in (63.43) (not the antisymmetric components O_{ij}), confirming the discussion following Eqs. (63.8).

For simplicity, we shall treat only the special case of a bulk medium with no stress at infinity. The volume V can then be taken as the whole sample, and the surface term vanishes to give

$$\delta W = - \sum_{ij=1}^{3} \int d^3x \; T_{ij} \, \delta\epsilon_{ij} \qquad (63.46)$$

To clarify this general relation, we recall the familiar case of uniform hydrostatic pressure where the stress tensor takes the form $T_{ij} = p\delta_{ij}$ [see Eq. (63.27)]. Equation (63.46) then becomes

$$\delta W = -p \int d^3x \sum_{i=1}^{3} \delta\epsilon_{ii} \qquad (63.47)$$

Since $\text{Tr} \, \delta\underline{\epsilon} = \delta(\text{Tr} \, \underline{\epsilon})$ is the pressure-induced fractional change in the volume element $d^3x = dV$ [recall Eq. (63.17)], we recover the expected result that the work done on the total system is

$$\delta W = -p \, \delta V \qquad (63.48)$$

where δV is the change in the total volume.

In general, an infinitesimal deformation $d\epsilon_{ij}$ of the medium† will be accompanied by a reversible transfer of heat $T \, d\mathscr{S}$ per unit volume. The first and second laws of thermodynamics show that the internal energy density $\mathscr{V}$ increases by an amount

$$d\mathscr{V} = T \, d\mathscr{S} - \sum_{ij=1}^{3} T_{ij} \, d\epsilon_{ij} \qquad (63.49)$$

where we have used Eq. (63.46). If the process is isentropic and reversible $(d\mathscr{S} = 0)$, it follows that the internal energy density $\mathscr{V}$ acts like an elastic potential energy since it obeys the differential relation

$$T_{ij} = -\left(\frac{\partial\mathscr{V}}{\partial\epsilon_{ij}}\right) \qquad (63.50)$$

This equation is independent of any particular stress-strain relation. It is analogous to the elementary expression $\mathbf{F} = -\boldsymbol{\nabla} U$ for a particle in an external potential U.

These general expressions can be simplified in the present case of an elastic continuum that satisfies Hooke's law. Suppose that an initially unstrained medium is deformed adiabatically to a final configuration characterized by a strain ϵ_{ij}. Since the system is conservative, $\mathscr{V}$ is independent of the particular process $i \to f$ and we can choose a particularly simple one in which ϵ increases linearly from i to f. At any intermediate step of the process, the local strain has the value $\eta\epsilon_{ij}$, where η is a dimensionless parameter $(0 \le \eta \le 1)$. Correspondingly, the stress tensor becomes ηT_{ij}, and the work done in increasing the strain by the amount $d\epsilon_{ij} = \epsilon_{ij} \, d\eta$ is just

$$dW = - \sum_{ij=1}^{3} \int d^3x \; T_{ij}\epsilon_{ij}\eta \, d\eta \qquad (63.51)$$

† Although δu_i and $\delta\epsilon_{ij}$ really represent ordinary differentials, the use of a special symbol clarifies the manipulations involved in deriving Eq. (63.45). We now revert to the familiar notation.

The work W done in reaching the final configuration follows by integrating from $\eta = 0$ to $\eta = 1$. This stored energy is just the total elastic potential energy $\int d^3x \, \mathscr{V}$, where

$$\mathscr{V} = -\frac{1}{2} \sum_{ij=1}^{3} T_{ij}\epsilon_{ij} \tag{63.52}$$

is the potential energy associated with a local strain $\underline{\epsilon}$. Evidently, $\mathscr{V}$ is a quadratic form in $\underline{\epsilon}$; it can be expressed either in terms of Lamé coefficients [see Eq. (63.21)]

$$\mathscr{V} = \tfrac{1}{2}\lambda(\text{Tr }\underline{\epsilon})^2 + \mu \sum_{ij=1}^{3} \epsilon_{ij}^2 \tag{63.53a}$$

or, more physically, with Eq. (63.23)

$$\mathscr{V} = \tfrac{1}{2}K(\text{Tr }\underline{\epsilon})^2 + \mu \sum_{ij=1}^{3} (\epsilon_{ij} - \tfrac{1}{3}\delta_{ij} \text{ Tr }\underline{\epsilon})^2 \tag{63.53b}$$

Since the moduli K and μ are positive, we see that $\mathscr{V}$ is positive for any $\underline{\epsilon}$. If Eq. (63.53a) is rewritten as

$$\mathscr{V} = \sum_{ijkl=1}^{3} (\tfrac{1}{2}\lambda\delta_{ij}\delta_{kl} + \mu\delta_{ik}\delta_{jl})\epsilon_{ij}\epsilon_{kl} \tag{63.54}$$

differentiation with respect to the particular element ϵ_{mn} yields

$$\frac{\partial\mathscr{V}}{\partial\epsilon_{mn}} = \sum_{ijkl=1}^{3} [(\tfrac{1}{2}\lambda\delta_{ij}\delta_{kl} + \mu\delta_{ik}\delta_{jl})(\delta_{im}\delta_{jn}\epsilon_{kl} + \epsilon_{ij}\delta_{km}\delta_{ln})]$$

$$= \lambda\delta_{mn} \text{ Tr }\underline{\epsilon} + 2\mu\epsilon_{mn} \tag{63.55}$$

This expression is just $-T_{mn}$ [see Eq. (63.21a)], confirming the general thermodynamic relation Eq. (63.50).

64 DYNAMICAL BEHAVIOR

Throughout our study of mechanical systems, we have used two distinct approaches for obtaining the dynamical equations of motion. One is the direct application of Newton's law of motion, as exemplified, for example, in our treatment of a point mass in a central potential (Sec. 3) and hydrodynamics (Secs. 48 and 49). The other relies on either d'Alembert's principle or Hamilton's principle to derive Lagrange's equations, expressed in the appropriate generalized coordinates, as seen in numerous examples. For a conservative system, the choice is basically one of convenience, and we have obtained the general string equations (25.13) and (38.9) by both techniques. Sometimes, however, one approach can be considerably more intricate, as seen in the lagrangian formulation of nonviscous irrotational hydrodynamics in Sec. 48. Also, the presence of dissipation precludes use of the simple lagrangian method, necessitating a direct approach, as in Sec. 60 for a viscous fluid.

Equation of Motion

In the present case of an elastic continuum, either method is applicable. Each has its advantages, and we shall consider them in turn. As noted in the previous section, the general dynamical equation (60.9) is a direct expression of Newton's second law. Thus we need merely insert the elastic stress tensor (63.21) or (63.23) to obtain the general equation of motion. We observe that the velocity $\mathbf{v}(\mathbf{x}, t)$ at a point $\mathbf{x}$ is just $\partial \mathbf{u}(\mathbf{x}, t)/\partial t$ and that $\partial(\rho \mathbf{v})/\partial t \approx \rho\, \partial^2 \mathbf{u}/\partial t^2$ because the strain-induced change of density is higher order in the small quantities. We thus treat ρ *as a constant* throughout the remainder of this discussion. The dynamical equation of an elastic medium in the presence of external forces $\rho \mathbf{f}$ thus follows as

$$\rho \frac{\partial^2 u_i}{\partial t^2} = -\sum_{j=1}^{3} \frac{\partial}{\partial x_j} T_{ij} + \rho f_i$$

$$= \mu \sum_{j=1}^{3} \frac{\partial^2 u_i}{\partial x_j^2} + (K + \tfrac{1}{3}\mu) \sum_{j=1}^{3} \frac{\partial^2 u_j}{\partial x_i\, \partial x_j} + \rho f_i \tag{64.1}$$

where Eq. (63.23*b*) has been used in the second line. It should be compared with Eq. (60.24) for a viscous fluid. Although both equations involve the acceleration $\rho\, \partial \mathbf{v}/\partial t$, the viscous stress (60.23) depends on the *time rate of change of the strain*, introducing second spatial derivatives of $\mathbf{v}$, in contrast to the more familiar second spatial derivatives of $\mathbf{u}$ seen in Eq. (64.1). This elastic equation of motion is often rewritten in vector form

$$\rho \frac{\partial^2 \mathbf{u}}{\partial t^2} = \mu\, \nabla^2 \mathbf{u} + (K + \tfrac{1}{3}\mu)\nabla(\nabla \cdot \mathbf{u}) + \rho \mathbf{f} \tag{64.2}$$

To interpret this relation, we shall separate the general vector field $\mathbf{u}$ into its *longitudinal* and *transverse* components,

$$\mathbf{u} = \mathbf{u}_l + \mathbf{u}_t \tag{64.3}$$

where
$$\nabla \times \mathbf{u}_l = 0 \qquad \nabla \cdot \mathbf{u}_t = 0 \tag{64.4}$$

Note that $\mathbf{u}_t$ is a pure shear distortion with no change in volume [compare Eq. (63.17)], whereas $\mathbf{u}_l$ involves compression and rarefaction. Such a separation can always be accomplished, as we now demonstrate. Consider an arbitrary vector field $\mathbf{u}(\mathbf{x}, t)$. Assume that the field is sufficiently well localized in space to ensure that, at *any given time*, the spatial dependence can be expressed through a Fourier transform

$$\mathbf{u}(\mathbf{x}, t) = (2\pi)^{-3} \int e^{i\mathbf{k} \cdot \mathbf{x}}\, \boldsymbol{\phi}(\mathbf{k}, t)\, d^3 k \tag{64.5a}$$

The Fourier amplitude is evidently given by

$$\boldsymbol{\phi}(\mathbf{k}, t) = \int e^{-i\mathbf{k} \cdot \mathbf{x}}\, \mathbf{u}(\mathbf{x}, t)\, d^3 x \tag{64.5b}$$

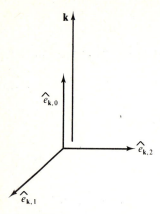

Figure 64.1 Complete orthonormal set of unit vectors for each Fourier component **k**.

For each **k**, we introduce a complete, orthonormal set of unit vectors as indicated in Fig. 64.1. Here

$$\hat{e}_{\mathbf{k}, 0} \equiv \frac{\mathbf{k}}{|k|} = \hat{k} \qquad (64.6)$$

lies along the direction **k**, and $\hat{e}_{\mathbf{k}, 1}$, $\hat{e}_{\mathbf{k}, 2}$ are orthogonal to this direction. The Fourier amplitude can be expanded in this complete basis according to

$$\boldsymbol{\phi}(\mathbf{k}, t) = \sum_{s=0}^{2} \hat{e}_{\mathbf{k}, s} a(\mathbf{k}, s; t) \qquad (64.7)$$

and we can use Eq. (64.5a) to write

$$\mathbf{u}(\mathbf{x}, t) = \mathbf{u}_l(\mathbf{x}, t) + \mathbf{u}_t(\mathbf{x}, t) \qquad (64.8)$$

where $\qquad \mathbf{u}_l(\mathbf{x}, t) \equiv (2\pi)^{-3} \int \hat{e}_{\mathbf{k}, 0} a(\mathbf{k}, 0; t) e^{i\mathbf{k} \cdot \mathbf{x}} d^3 k \qquad (64.9a)$

$$\mathbf{u}_t(\mathbf{x}, t) \equiv (2\pi)^{-3} \int \sum_{s=1}^{2} \hat{e}_{\mathbf{k}, s} a(\mathbf{k}, s; t) e^{i\mathbf{k} \cdot \mathbf{x}} d^3 k \qquad (64.9b)$$

If it is assumed that the derivatives can be taken under the integral, direct differentiation shows that

$$\nabla \times \mathbf{u}_l = 0 \qquad (64.10a)$$

$$\nabla \cdot \mathbf{u}_t = 0 \qquad (64.10b)$$

as was to be proved. Since **u**(**x**, t) has a *unique* Fourier transform, this prescription generates a unique decomposition into $\mathbf{u}_l$ and $\mathbf{u}_t$. In particular, if **u**(**x**, t) vanishes for all (**x**, t)

$$\mathbf{u}(\mathbf{x}, t) = \mathbf{u}_l(\mathbf{x}, t) + \mathbf{u}_t(\mathbf{x}, t) = 0 \qquad (64.11a)$$

we can readily prove that $\mathbf{u}_l$ and $\mathbf{u}_t$ must separately vanish

$$\mathbf{u}_l(\mathbf{x}, t) = 0 \qquad \mathbf{u}_t(\mathbf{x}, t) = 0 \qquad (64.11b)$$

This result is established by noting that the Fourier amplitude ϕ in Eq. (64.5b) vanishes identically because of Eq. (64.11a). Thus each of the Fourier coefficients also vanishes

$$a(\mathbf{k}, s; t) \equiv \boldsymbol{\phi}(\mathbf{k}, t) \cdot \hat{e}_{\mathbf{k}, s} = 0 \qquad s = 0, 1, 2 \tag{64.12}$$

and Eqs. (64.11b) then follow from Eqs. (64.9).

We now return to our discussion of the dynamical behavior of an elastic medium. Substitution of Eqs. (64.3) and (64.4) into Eq. (64.2) and use of the vector identity

$$\nabla^2 \mathbf{u} = -\nabla \times (\nabla \times \mathbf{u}) + \nabla(\nabla \cdot \mathbf{u}) \tag{64.13}$$

yields the equation of motion

$$\rho \frac{\partial^2 \mathbf{u}_l}{\partial t^2} + \rho \frac{\partial^2 \mathbf{u}_t}{\partial t^2} = (K + \tfrac{4}{3}\mu) \nabla(\nabla \cdot \mathbf{u}_l) - \mu \nabla \times (\nabla \times \mathbf{u}_t) + \rho(\mathbf{f}_l + \mathbf{f}_t) \tag{64.14}$$

where $\mathbf{f}_l$ and $\mathbf{f}_t$ are the appropriate parts of $\mathbf{f}$. Equations (64.11) imply that we may separate the longitudinal and transverse components of this relation. Thus we have

$$\rho \frac{\partial^2 \mathbf{u}_l}{\partial t^2} = (K + \tfrac{4}{3}\mu)\nabla(\nabla \cdot \mathbf{u}_l) + \rho \mathbf{f}_l \tag{64.15a}$$

or, equivalently, using Eq. (64.13) again

$$\rho \frac{\partial^2 \mathbf{u}_l}{\partial t^2} - (K + \tfrac{4}{3}\mu) \nabla^2 \mathbf{u}_l = \rho \mathbf{f}_l \tag{64.15b}$$

In the absence of external forces, the longitudinal displacement therefore satisfies a simple homogeneous wave equation

$$\frac{1}{c_l^2} \frac{\partial^2 \mathbf{u}_l}{\partial t^2} = \nabla^2 \mathbf{u}_l \tag{64.16}$$

where

$$c_l = \left(\frac{K + \tfrac{4}{3}\mu}{\rho} \right)^{1/2} \tag{64.17}$$

is the propagation speed for a longitudinal disturbance. Similarly, the transverse part of Eq. (64.14) together with Eq. (64.13) yields the analogous equation

$$\rho \frac{\partial^2 \mathbf{u}_t}{\partial t^2} - \mu \nabla^2 \mathbf{u}_t = \rho \mathbf{f}_t \tag{64.18}$$

In the absence of external forces, it becomes

$$\frac{1}{c_t^2} \frac{\partial^2 \mathbf{u}_t}{\partial t^2} = \nabla^2 \mathbf{u}_t \tag{64.19}$$

where

$$c_t = \left(\frac{\mu}{\rho} \right)^{1/2} \tag{64.20}$$

is the corresponding propagation speed for a transverse disturbance. Note that $c_l > c_t$, implying that an elastic medium offers greater resistance to a longitudinal disturbance than to a transverse one.

In addition to these field equations, the displacement field $\mathbf{u}$ must satisfy appropriate (vector) boundary conditions.

1. At a fixed surface, we naturally require

$$\mathbf{u} = 0 \qquad \text{on fixed surface} \qquad (64.21a)$$

2. At a free surface with outward normal $\hat{n}$, the force vanishes to yield the condition

$$\sum_{j=1}^{3} T_{ij}\hat{n}_j = 0 \qquad \text{on free surface} \qquad (64.21b)$$

3. At an interface between two elastic media 1 and 2, the net force must vanish to avoid an infinite acceleration. If $\hat{n}^{(1)}$ and $\hat{n}^{(2)}$ are the outward normals from the two media, then the relation $\hat{n}^{(1)} = -\hat{n}^{(2)}$ implies the condition

$$\sum_{j=1}^{3} (T_{ij}^{(1)} - T_{ij}^{(2)})n_j^{(2)} = 0 \qquad (64.21c)$$

Thus the normal projection of the stress tensor is continuous [compare Eq. (60.31b)]. Note again that Eqs. (64.21b) and (64.21c) are *vector* relations on the force at the boundary.

The dynamics of an elastic continuum also follows from Hamilton's principle, expressed in terms of the obvious lagrangian density

$$\mathscr{L} = \mathscr{T} - \mathscr{V} \qquad (64.22)$$

where

$$\mathscr{T} = \tfrac{1}{2}\rho \sum_{i=1}^{3} \left(\frac{\partial u_i}{\partial t}\right)^2 \qquad (64.23)$$

and $\mathscr{V}$ is the elastic potential energy from (63.53). Since $\mathbf{u}$ is now the generalized coordinate in Lagrange's equation, we must rewrite $\mathscr{V}$ with Eq. (63.9). It is slightly simpler to use the form (63.54), and we therefore have

$$\mathscr{V} = \frac{1}{4} \sum_{klmn=1}^{3} (\tfrac{1}{2}\lambda\delta_{kl}\delta_{mn} + \mu\delta_{km}\delta_{ln})\left(\frac{\partial u_k}{\partial x_l} + \frac{\partial u_l}{\partial x_k}\right)\left(\frac{\partial u_m}{\partial x_n} + \frac{\partial u_n}{\partial x_m}\right) \qquad (64.24)$$

Familiar manipulations now lead to Lagrange's equations for the ith component of the vector displacement $u_i(\mathbf{x}, t)$

$$\frac{\partial}{\partial t}\frac{\partial \mathscr{L}}{\partial(\partial u_i/\partial t)} + \sum_{j=1}^{3}\frac{\partial}{\partial x_j}\frac{\partial \mathscr{L}}{\partial(\partial u_i/\partial x_j)} - \frac{\partial \mathscr{L}}{\partial u_i} = 0 \qquad (64.25)$$

Here, the last term vanishes because $\mathscr{L}$ is independent of $\mathbf{u}$ in the absence of external forces. Moreover, the form of Eq. (64.24) facilitates evaluating the partial derivative of $\mathscr{L}$ with respect to $\partial u_i/\partial x_j$. An analysis similar to that for Eq. (63.55) leads to the final equation

$$\rho \frac{\partial^2 u_i}{\partial t^2} - \sum_{j=1}^{3} \left[\mu \frac{\partial^2 u_i}{\partial x_j^2} + (\lambda + \mu) \frac{\partial^2 u_j}{\partial x_i \, \partial x_j} \right] = 0 \tag{64.26}$$

Comparison with Eq. (63.24) immediately reproduces the previous dynamical equation (64.1) with $\mathbf{f} = 0$.

In addition to its formal advantages, the lagrangian description also allows a compact derivation of the energy-flux vector $\mathbf{S}$ for an elastic medium. The discussion in Sec. 45 applies with only slight changes required by the vector character of $\mathbf{u}$. Thus we immediately infer from Eqs. (64.22) to (64.24) that

$$\mathscr{H} = \mathscr{T} + \mathscr{V} \tag{64.27}$$

is the total energy density and from Eq. (45.11) that

$$S_i = \sum_{j=1}^{3} \frac{\partial \mathscr{L}}{\partial(\partial u_j/\partial x_i)} \frac{\partial u_j}{\partial t} \tag{64.28}$$

is the corresponding instantaneous energy-flux vector. By construction, these quantities satisfy the conservation law (45.10)

$$\frac{\partial \mathscr{H}}{\partial t} + \mathbf{V} \cdot \mathbf{S} = 0 \tag{64.29}$$

in the absence of applied forces. Use of Eqs. (64.22) to (64.24) gives the explicit form of the energy flux for an arbitrary small-amplitude elastic disturbance

$$S_i = -\lambda(\mathbf{V} \cdot \mathbf{u}) \frac{\partial u_i}{\partial t} - \mu \sum_{j=1}^{3} \left(\frac{\partial u_i}{\partial x_j} \frac{\partial u_j}{\partial t} + \frac{\partial u_j}{\partial x_i} \frac{\partial u_j}{\partial t} \right) \tag{64.30a}$$

Comparison with Eq. (63.21) gives the alternative form

$$S_i = \sum_{j=1}^{3} T_{ij} \frac{\partial u_j}{\partial t} \tag{64.30b}$$

Elastic Waves in an Unbounded Medium

For a general configuration, the solution of Eq. (64.1) for the elastic displacement $\mathbf{u}$ is extremely difficult, in large part because of the boundary conditions. For this reason, we shall here consider only an infinite region without boundaries, where the normal-mode solutions are plane waves

$$\mathbf{u}(\mathbf{x}, t) = \mathbf{u}^0 e^{i(\mathbf{k} \cdot \mathbf{x} - \omega t)} \tag{64.31}$$

Here $\mathbf{u}^0$ is the normal-mode amplitude, and the plane wave propagates along the direction $\hat{k}$ with wave number k. Substitution into Eq. (64.2) (with $\mathbf{f} = 0$) yields the equation

$$\rho\omega^2 \mathbf{u}^0 - \mu k^2 \mathbf{u}^0 - (K + \tfrac{1}{3}\mu)\mathbf{k}(\mathbf{k} \cdot \mathbf{u}^0) = 0 \tag{64.32}$$

For a given $\mathbf{k}$, this vector equation represents three coupled homogeneous equations for the eigenvectors (here, the components of $\mathbf{u}^0$). A solution exists only for the particular eigenfrequencies obtained from the secular equation

$$\det \left| (\rho\omega^2 - \mu k^2)\delta_{ij} - (K + \tfrac{1}{3}\mu)k_i k_j \right| = 0 \tag{64.33}$$

exactly as in Sec. 22. In the present case, however, the isotropy of the medium simplifies the procedure, and we can classify the solutions as longitudinal ($\mathbf{u}^0 \parallel \mathbf{k}$) and transverse ($\mathbf{u}^0 \perp \mathbf{k}$), exactly as for the waves in a viscous fluid (Sec. 62). Thus we separate $\mathbf{u}^0$ into its longitudinal and transverse parts

$$\mathbf{u}^0 = \mathbf{u}_l^0 + \mathbf{u}_t^0 \tag{64.34}$$

where

$$\mathbf{u}_l^0 \equiv \hat{k}(\hat{k} \cdot \mathbf{u}^0) \tag{64.35a}$$

$$\mathbf{u}_t^0 \equiv \mathbf{u}^0 - \hat{k}(\hat{k} \cdot \mathbf{u}^0) \tag{64.35b}$$

Note that the corresponding plane-wave solutions automatically satisfy the general conditions in (64.4), with $\mathbf{u}_l^0$ polarized along $\hat{k}$ and $\mathbf{u}_t^0$ having two independent orthogonal polarizations in the plane perpendicular to $\hat{k}$, as in Eqs. (64.9). Substitution into (64.32) readily gives the longitudinal and transverse dispersion relations

$$\omega_l^2 = c_l^2 k^2 \qquad \omega_t^2 = c_t^2 k^2 \tag{64.36}$$

where c_l and c_t are the longitudinal and transverse propagation speeds (64.17) and (64.20). The transverse wave is a pure shear disturbance with c_t determined by the shear modulus μ. The longitudinal wave is more complicated, and c_l cannot be expressed solely in terms of Young's modulus E because of the transverse stresses between adjacent regions.

It is instructive to evaluate the elastic energy flux $\mathbf{S}$ for the plane wave (64.31). Taking the time average of Eq. (64.30a) according to Eqs. (45.45) and (45.46), we find

$$\langle \mathbf{S} \rangle = \tfrac{1}{2}\omega\mu\mathbf{k}\,|\mathbf{u}^0|^2 + \tfrac{1}{2}\omega(\lambda + \mu)\,\mathrm{Re}\,[(\mathbf{k} \cdot \mathbf{u}^0)\mathbf{u}^{0*}] \tag{64.37}$$

where ω and $\mathbf{u}^0$ can refer to either the longitudinal or transverse normal modes. In the former case ($\omega = \omega_l$ and $\mathbf{k} \cdot \mathbf{u}_l^0 = k\mathbf{u}_l^0$), the corresponding time-average energy flux becomes

$$\langle \mathbf{S}_l \rangle = \tfrac{1}{2}\omega_l\mathbf{k}(\lambda + 2\mu)\,|\mathbf{u}_l^0|^2 = \tfrac{1}{2}\omega_l\mathbf{k}\rho c_l^2\,|\mathbf{u}_l^0|^2 \tag{64.38a}$$

where Eqs. (63.24) and (64.17) have been used to identify c_l. Similarly, for the transverse modes ($\omega = \omega_t$ and $\mathbf{k} \cdot \mathbf{u}_t^0 = 0$), we have

$$\langle \mathbf{S}_t \rangle = \tfrac{1}{2}\omega_t\mathbf{k}\mu\,|\mathbf{u}_t^0|^2 = \tfrac{1}{2}\omega_t\mathbf{k}\rho c_t^2\,|\mathbf{u}_t^0|^2 \tag{64.38b}$$

As expected, the time-average energy flux lies along $\hat{k}$, with a magnitude quadratic in the amplitude $|\mathbf{u}^0|^2$ [compare Eq. (45.33)]. These expressions are important in studying the reflection and transmission of elastic waves.

In comparing a discrete linear array of point masses with a continuous string (Sec. 24), the exact dispersion relation (24.58) simplified greatly in the limit that the wavelength was long compared with the interparticle spacing (see also Prob. 4.16 for a two-dimensional array). A similar situation occurs in the present three-dimensional situation of atoms arranged in a crystalline array. Assuming that these atoms interact through two-body potentials, we can define a total potential energy of the form (2.21), which must be minimum for static equilibrium. The analysis of small oscillations about this equilibrium configuration then follows as in Chap. 4. For a bulk medium, plane waves again provide the appropriate normal modes, and the complete dispersion relation is obtained by a three-dimensional generalization of the techniques used in Eqs. (24.40) to (24.43). As in that case, the exact dispersion relation for three-dimensional normal modes in a regular crystal simplifies in the long-wavelength limit to that obtained from the theory of elastic waves in such a crystal. In this limit, the microscopic approach therefore relates the phenomenological elastic constants to the actual interatomic potentials. Unfortunately, the details become quite intricate, and the analysis cannot be given here.†

The vector character of the elastic displacement **u** introduces several new features that do not occur in a string or membrane. We have already mentioned the three polarization states for bulk plane waves, where these different normal modes are uncoupled. More interesting is the behavior near boundaries, where this separation often becomes impossible. In particular, a *longitudinal* wave incident on a planar free surface gives rise to a reflected *transverse* wave (polarized in the plane of incidence) as well as a reflected *longitudinal* wave. The longitudinal waves obey the usual law of reflection, but the transverse wave propagates in a different direction. A similar situation occurs for an incident transverse wave polarized in the plane of incidence (see Prob. 13.2). Another qualitatively new phenomenon for an elastic continuum in a half space is the existence of waves that propagate *along* the free surface with an amplitude that attenuates exponentially into the bulk [compare Eq. (55.7) for a surface wave on a deep fluid]. This characteristic elastic disturbance is known as a *Rayleigh wave* (see Prob. 13.8). It propagates more slowly than either the bulk transverse or longitudinal waves and plays an important role in the observation of earthquakes, because the surface energy decreases only inversely with distance instead of the bulk inverse-square behavior.‡

PROBLEMS

13.1 An elastic medium with a free surface fills the half space $z \leq 0$. A transverse wave has an angle of incidence θ, that is, $\cos \theta = \hat{k}_t \cdot \hat{z}$, and $\mathbf{u}^0$ polarized perpendicular to the plane of incidence. Show that there is a single reflected wave with the same polarization at an angle of reflection that satisfies the usual optical law.

13.2 Repeat Prob. 13.1 for an incident transverse wave with $\mathbf{u}^0$ polarized in the plane of incidence.

(a) Show that this incident wave gives rise to a transverse reflected wave polarized in the plane of incidence and a longitudinal reflected wave. Verify that the transverse waves satisfy the usual optical

† See, for example, Ashcroft and Mermin [3], chap. 22.
‡ Stacey, [4] sec. 6.1; Bullen [5]; Ewing and Press [6].

law of reflection and that the angle of reflection θ', that is, $\cos \theta' = -\hat{k}_t \cdot \hat{z}$, of the longitudinal reflected wave satisfies $c_l \sin \theta = c_t \sin \theta'$, where θ is the angle of incidence.

(b) If u^i, u^t, and u^l are the amplitudes of the three waves in part (a), use the boundary condition $T_{xz} = T_{yz} = T_{zz} = 0$ at the free surface to obtain the ratios

$$\frac{u^l}{u^i} = \frac{c_t^2 \sin 2\theta \sin 2\theta' - c_l^2 \cos^2 2\theta}{c_t^2 \sin 2\theta \sin 2\theta' + c_l^2 \cos^2 2\theta}$$

$$\frac{u^t}{u^i} = -\frac{2c_t c_l \sin 2\theta \cos 2\theta}{c_t^2 \sin 2\theta \sin 2\theta' + c_l^2 \cos^2 2\theta}$$

13.3 (a) Consider a propagating elastic wave of the form $\mathbf{u}(\mathbf{x}, t) = \operatorname{Re} \mathbf{u}(\mathbf{x})e^{-i\omega t}$, where $\mathbf{u}(\mathbf{x}) = \mathbf{u}_l^0 e^{i\mathbf{k}_l \cdot \mathbf{x}} + \mathbf{u}_t^0 e^{i\mathbf{k}_t \cdot \mathbf{x}}$ with $k_l = \omega/c_l$ and $k_t = \omega/c_t$. Evaluate the energy flux vector with Eq. (64.30) and show that its time average separates into longitudinal and transverse contributions *when averaged over a spatial region containing many wavelengths*.

(b) Apply this result to Prob. 13.2. Prove that the corresponding reflection coefficients for each of the reflected waves are $R_t = |u^t/u^i|^2$ and $R_l = (c_l \cos \theta'/c_t \cos \theta)|u^l/u^i|^2$. Verify that $R_t + R_l = 1$.

13.4 A sample of an isotropic elastic continuum has the form of a rectangular parallelepiped with dimensions a by b by c and $a < b < c$. Let $\hat{n}$ and $\hat{t}^{(s)}$ with $s = 1$, 2 be unit vectors normal and transverse to the boundaries. Use Eq. (63.45) to show that the initial energy content of the medium is conserved for each of the following sets of boundary conditions:

(a) $\mathbf{u} = 0$ fixed surface
(b) $\sum_{j=1}^{3} T_{ij} \hat{n}_j = 0$ free surface
(c) $\mathbf{u} \cdot \hat{n} = 0$ $\sum_{ij=1}^{3} \hat{t}_i^{(s)} T_{ij} \hat{n}_j = 0$
(d) $\mathbf{u} \cdot \hat{t}^{(s)} = 0$ $\sum_{ij=1}^{3} \hat{n}_i T_{ij} \hat{n}_j = 0$

[Are there physical situations under which (c) or (d) would hold?]

13.5 With regard to the geometry and boundary conditions in Prob. 13.4, show that a separated solution analogous to that in Sec. 47 for a rectangular membrane *fails* for the boundary conditions (a) and (b) but *succeeds* for (c) and (d). In the latter situation, determine the eigenfrequencies and eigenfunctions. Discuss the first few low-lying modes.

13.6 The normal modes for Prob. 13.4 have the form $\mathbf{u}(\mathbf{x}, t) = \operatorname{Re} \mathbf{u}(\mathbf{x})e^{-i\omega t}$. Show that $\mathbf{u}(\mathbf{x})$ can be obtained from the variational functional

$$\omega^2[\mathbf{u}] = \frac{\int d^3x\, 2\mathcal{V}(\mathbf{x})}{\rho \int d^3x\, \mathbf{u}(\mathbf{x})^2}$$

where $\mathcal{V}(\mathbf{x})$ is the potential energy density given in Eq. (64.24) and $\mathbf{u}$ satisfies any of the boundary conditions (a) to (d).

13.7 Apply the result of Prob. 13.6 for fixed boundary conditions ($\mathbf{u} = 0$), using the trial function $\mathbf{u}(\mathbf{x}) = \mathbf{u}^0 \sin (\pi x/a) \sin (\pi y/b) \sin (\pi z/c)$ with $\mathbf{u}^0$ a constant vector. Discuss.

13.8 An elastic medium with a free surface fills the half space $z \leq 0$. An elastic surface wave propagates along the surface in the $\hat{x}$ direction.

(a) Show that the longitudinal part of the disturbance can be written in the form $\mathbf{u}_l(x, z, t) = \operatorname{Re} \nabla[\Phi(z)e^{i(kx - \omega t)}]$ and determine $\Phi(z)$ by solving Eq. (64.16).

(b) Show that the transverse part of the disturbance can be written in the form $\mathbf{u}_t(x, z, t) = \operatorname{Re} \nabla \times [\hat{y}\Psi(z)e^{i(kx - \omega t)}]$ and determine $\Psi(z)$ by solving Eq. (64.19).

(c) Apply the boundary condition $T_{xz} = T_{zz} = 0$ to the total displacement and obtain the dispersion relation for these *Rayleigh waves*

$$\left(1 - \frac{c^2}{2c_t^2}\right)^2 = \left(1 - \frac{c^2}{c_t^2}\right)^{1/2}\left(1 - \frac{c^2}{c_l^2}\right)^{1/2}$$

where $c = \omega/k$ is the propagation speed. Find c^2 numerically for the typical value $c_l/c_t = \sqrt{3}$ corresponding to $\sigma = \frac{1}{4}$.

13.9 A long elastic cylinder of radius a with a free surface executes axial oscillations of the form $\mathbf{u}(\mathbf{x}, t) = \text{Re } \hat{z}w(r)e^{-i\omega t}$, where $\mathbf{x}$ is expressed in cylindrical polar coordinates (r, ϕ, z). Explain why this motion is transverse. Use Probs. B.4 and B.8 to show that the boundary condition at $r = a$ becomes $T_{rz} = -\mu \, dw/dr|_a = 0$. Verify that $w(r) \propto J_0(\omega r/c_t)$ and that the eigenvalue equation becomes $\omega_n a/c_t = \alpha_{1,n}$, where $\alpha_{1,n}$ is the nth zero of $J_1(x)$.

13.10 Repeat Prob. 13.9 for radial oscillations with $\mathbf{u}(\mathbf{x}, t) = \hat{r}u(r)e^{-i\omega t}$. Explain why this motion is longitudinal. Show that the boundary condition at $r = a$ becomes

$$T_{rr} = -\lambda(u' + r^{-1}u) - 2\mu u' = 0$$

where $u' = du/dr$. Verify that $u(r) \propto J_1(\omega r/c_l)$ and apply the boundary condition to obtain the eigenvalue condition $\frac{1}{2}xJ_0(x) = J_1(x)(c_t/c_l)^2$, where $x = \omega a/c_l$. Sketch and discuss the solution of the eigenvalue condition and the first few normal modes.

13.11 Repeat Prob. 13.9 for torsional oscillations of the form $\mathbf{u}(\mathbf{x}, t) = \text{Re } \hat{\phi}v(r)e^{i(kz - \omega t)}$. Show that the boundary condition at $r = a$ becomes $T_{r\phi} = -\mu r \, d(v/r)/dr = 0$ (compare Prob. 12.11). Verify that $v(r) \propto J_1(pr)$, where $p^2 = (\omega/c_t)^2 - k^2$, and that the allowed frequencies are given by $\omega_n^2 = c_t^2(k^2 + \alpha_{2,n}^2/a^2)$. Prove that $v(r) = r$ is also a solution; discuss the corresponding frequency and torsional motion.

13.12 An elastic sphere of radius a with a free surface executes radial oscillations of the form $\mathbf{u}(\mathbf{x}, t) = \text{Re } \hat{r}u(r)e^{-i\omega t}$, with r the radial spherical polar coordinate.

(a) Explain why this motion is longitudinal. Use Probs. B.4 and B.8 to show that the boundary condition at $r = a$ reduces to $T_{rr} = -\lambda(u' + 2r^{-1}u) - 2\mu u' = 0$, where the prime denotes differentiation with respect to r.

(b) Solve the dynamical equation to show that $u(r) \propto j_1(\omega r/c_l)$.

(c) Apply the boundary condition at $r = a$ to obtain the eigenvalue equation $xj_0(x) = 4j_1(x)c_t^2/c_l^2$ and its equivalent form $x \cot x = 1 - \frac{1}{4}(xc_l/c_t)^2$, where $x = \omega a/c_l$. Sketch and discuss the graphical solution and the first few normal modes. Compare with Prob. 13.10.

(d) Treat the earth as a steel sphere and estimate the period of the lowest radial mode.

SELECTED ADDITIONAL READINGS

1. G. Joos, "Theoretical Physics," 3d ed. with I. M. Freeman, Hafner, New York, 1958.
2. L. D. Landau and E. M. Lifshitz, "Theory of Elasticity," Addison-Wesley, Reading, Mass., 1959.
3. N. W. Ashcroft and N. D. Mermin, "Solid State Physics," Holt, New York, 1976.
4. F. D. Stacey, " Physics of the Earth," 2d ed., Wiley, New York, 1977.
5. K. E. Bullen, in S. Flügge (ed.), "Handbuch der Physik," Springer-Verlag, Heidelberg, 1956, vol. 47, pp. 75–118.
6. W. M. Ewing and F. Press, in S. Flügge (ed.), "Handbuch der Physik," Springer-Verlag, Heidelberg, 1956, vol. 47, pp. 119–139.

THEORY OF FUNCTIONS

This appendix summarizes those results from the theory of functions of a complex variable which are necessary for our treatment of continuum mechanics.†

A1 COMPLEX VARIABLES

A complex variable is a number pair (x, y) defined by the relation

$$z = x + iy \qquad (A1.1)$$

This quantity has the properties of a two-dimensional vector running from the origin to the point (x, y) (see Fig. A1.1). The geometric length $(x^2 + y^2)^{1/2}$ is readily expressed as $|z| \equiv (|z|^2)^{1/2}$, where

$$|z|^2 \equiv zz^* = x^2 + y^2 \qquad (A1.2)$$

is the square of the *modulus* $|z|$ of the complex number z and z^* is the complex conjugate of z. Two complex numbers can be *added* by adding the real and imaginary parts

$$z_1 + z_2 = x_1 + x_2 + i(y_1 + y_2) \qquad (A1.3)$$

which is the usual rule for adding cartesian components of vectors.

Figure A1.1 also provides another representation of z. Letting $\rho = |z|$ be the length and ϕ the polar angle, we write

$$z = \rho(\cos \phi + i \sin \phi) = \rho e^{i\phi} \qquad (A1.4)$$

---

† Among the numerous references on complex variables, we have found the following particularly useful: Carrier, Krook, and Pearson [1]; Morse and Feshbach [2], chap. 4; Titchmarsh [3]; and Whittaker and Watson [4].

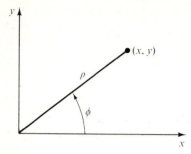

Figure A1.1 A representation of the complex number $z = x + iy$.

This result implies that two complex numbers can be multiplied according to the relation

$$z_1 z_2 = \rho_1 \rho_2 e^{i(\phi_1 + \phi_2)} \qquad \text{(A1.5)}$$

It is important to note here that the moduli are multiplied and the phases are added. Similarly, two complex numbers can be divided according to the relation

$$\frac{z_1}{z_2} = \frac{\rho_1}{\rho_2} e^{i(\phi_1 - \phi_2)} \qquad \text{(A1.6)}$$

A2 FUNCTIONS OF A COMPLEX VARIABLE

A function of a complex variable assigns to each complex number z another complex number $f(z)$ according to the relation

$$f(z) = \operatorname{Re} f + i \operatorname{Im} f \equiv u(x, y) + iv(x, y) \qquad \text{(A2.1)}$$

Here f has been explicitly separated into real and imaginary parts, $u(x, y)$ and $v(x, y)$, which are functions of the real and imaginary parts of z defined in Eq. (A1.1). The definition of length in Eq. (A1.2) allows us to introduce the concept of *continuity*. A function of a complex variable is called continuous at z_0 if, for arbitrarily small ϵ, the condition

$$|f(z) - f(z_0)| < \epsilon \qquad \text{(A2.2)}$$

holds for all z sufficiently close to the point z_0

$$|z - z_0| < \delta \qquad \text{(A2.3)}$$

These relations are equivalent to the continuity of two functions of two real variables; i.e.,

$$|u(x, y) - u(x_0, y_0)| < \frac{\epsilon}{\sqrt{2}} \qquad \begin{array}{l} \text{if } |x - x_0| < \delta/\sqrt{2} \\ \text{and } |y - y_0| < \delta/\sqrt{2} \end{array}$$

$$|v(x, y) - v(x_0, y_0)| < \frac{\epsilon}{\sqrt{2}} \qquad \begin{array}{l} \text{if } |x - x_0| < \delta/\sqrt{2} \\ \text{and } |y - y_0| < \delta/\sqrt{2} \end{array} \qquad \text{(A2.4)}$$

This equivalence is readily demonstrated by noting that the squared length of a complex vector as defined in Eq. (A1.2) is the sum of the squares of the lengths of the real and imaginary parts of the vector.

We shall be particularly interested in the special group of functions of the form (A2.1) known as *analytic*. This fundamental concept is defined in the following way:

If $f(z)$ is continuous, single-valued, and has a unique derivative

$$\lim_{z \to z_0} \frac{f(z) - f(z_0)}{z - z_0} \equiv f'(z_0) \qquad (A2.5)$$

independent of the direction with which $z \to z_0$, then $f(z)$ is analytic at z_0.

The situation is illustrated in Fig. A2.1. To appreciate the significance of *analyticity*, we shall investigate the quantity

$$\frac{f(z) - f(z_0)}{z - z_0} = \frac{u(x, y) + iv(x, y) - [u(x_0, y_0) + iv(x_0, y_0)]}{(x - x_0) + i(y - y_0)} \qquad (A2.6)$$

By assumption, the functions $u(x, y)$ and $v(x, y)$ are continuous and differentiable. Therefore, the right-hand side of Eq. (A2.6) can be written approximately

$$\frac{f(z) - f(z_0)}{z - z_0} \approx \frac{(x - x_0)\dfrac{\partial u}{\partial x} + (y - y_0)\dfrac{\partial u}{\partial y} + i\left[(x - x_0)\dfrac{\partial v}{\partial x} + (y - y_0)\dfrac{\partial v}{\partial y}\right]}{(x - x_0) + i(y - y_0)} \qquad (A2.7)$$

which becomes exact in the limit $z \to z_0$. Since the derivative is to be *unique*, the result must be independent of how the limit $z \to z_0$ is taken. For definiteness, consider the two different ways illustrated in Fig. A2.1. If the limit is taken along the line $x = x_0$ (that is, along the vertical direction), Eq. (A2.7) leads to

$$f'(z_0) = \frac{\partial v}{\partial y} - i\frac{\partial u}{\partial y} \qquad (A2.8)$$

If, however, the limit is taken along the line $y = y_0$ (that is, along the horizontal direction), Eq. (A2.7) now gives

$$f'(z_0) = \frac{\partial u}{\partial x} + i\frac{\partial v}{\partial x} \qquad (A2.9)$$

Uniqueness of the derivative requires that Eq. (A2.8) and (A2.9) have identical real and imaginary parts, which provides the *Cauchy-Riemann equations*

$$\frac{\partial u}{\partial x} = \frac{\partial v}{\partial y} \qquad (A2.10a)$$

$$\frac{\partial v}{\partial x} = -\frac{\partial u}{\partial y} \qquad (A2.10b)$$

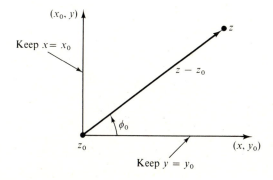

Figure A2.1 Different directions for the vector $z - z_0$ used in defining the derivative and analyticity.

They are clearly necessary conditions if the function $f(z)$ is to be analytic and have a unique derivative at z_0. They are also sufficient if u and v are continuous and continuously differentiable (which we henceforth assume to be the case), for then Eq. (A2.7) holds in a small region about z_0. Use of Eqs. (A2.10) gives [see Eqs. (A2.8) and (A2.9)]

$$\frac{f(z)-f(z_0)}{z-z_0} \approx \frac{(x-x_0)+i(y-y_0)}{(x-x_0)+i(y-y_0)}\left(\frac{\partial u}{\partial x}+i\frac{\partial v}{\partial x}\right)=f'(z_0) \qquad (A2.11)$$

which is evidently independent of the path chosen for the limit $z \to z_0$.

The following examples help to fix some of these ideas.

Example 1 Consider the function

$$f(z) = \text{Re } z = x \qquad (A2.12)$$

This is a well-defined function of a complex variable, but it is *not analytic* because

$$\frac{\partial u}{\partial x} = 1 \qquad \frac{\partial v}{\partial y} = 0 \qquad (A2.13)$$

and the Cauchy-Riemann condition (A2.10a) is violated.

Example 2 Consider the function

$$f(z) = z^n \qquad (A2.14)$$

for n a positive integer. We can compute the derivative from the expression

$$\frac{f(z+h)-f(z)}{h} = \frac{(z+h)^n - z^n}{h} \qquad (A2.15)$$

where h is an arbitrarily small complex number. Use of the binomial theorem on the right-hand side of Eq. (A2.15) shows that

$$\frac{(z+h)^n - z^n}{h} = nz^{n-1} + O(h) \qquad (A2.16)$$

The correction term in this expression is of order h, allowing us to take the limit $h \to 0$, and it follows that

$$f'(z) = nz^{n-1} \qquad (A2.17)$$

independent of the direction of the vector h. Since the polynomial defined in Eq. (A2.14) is continuous and single-valued for integral n [this follows immediately from Eq. (A1.4)], we conclude that $f(z)$ defined through Eq. (A2.14) is analytic.

Example 3 The exponential function is defined by the following infinite series

$$f(z) = e^z \equiv 1 + z + \frac{z^2}{2!} + \frac{z^3}{3!} + \cdots \qquad (A2.18)$$

It is readily established that this series converges absolutely and uniformly in any finite region of the complex plane. We again compute the derivative by writing

$$\frac{e^{z+h}-e^z}{h} = e^z \frac{e^h - 1}{h} \qquad (A2.19a)$$

$$= e^z + O(h) \qquad (A2.19b)$$

where Eq. (A2.19a) follows from the fundamental property of the multiplication of exponentials and the correction in Eq. (A2.19b) is again of order h. The function $f(z)$ defined in Eq. (A2.18) thus has a unique derivative given by

$$f'(z) = e^z \tag{A2.20}$$

Furthermore, this function is continuous and single-valued, so that the exponential function e^z is analytic.

Example 4 Consider the function

$$f(z) = \ln z \tag{A2.21}$$

where $\ln z$ is defined by the relation

$$e^{\ln z} = z \tag{A2.22}$$

Equation (A2.22) can be differentiated by using Eq. (A2.20)

$$\frac{dz}{d \ln z} = e^{\ln z} = z \tag{A2.23}$$

and the reciprocal of this relationship then leads to

$$f'(z) = \frac{1}{z} \tag{A2.24}$$

Thus the function defined in Eq. (A2.21) has a derivative for any finite value of z. Note, however, that Eq. (A1.4) and the fundamental property of the logarithm (A2.22) together imply that

$$\ln z = \ln \rho e^{i\phi} = \ln \rho + \ln e^{i\phi} = \ln \rho + i\phi \tag{A2.25}$$

If the phase ϕ of the complex number z is increased by 2π, then z returns to its original value. It is evident from Eq. (A2.25) that the real part of $\ln z$ also returns to its original value but the imaginary part of the logarithm increases by 2π. Thus, *the function defined in Eq. (A2.21) is not single-valued.* We can make it single-valued, however, by restricting ourselves to a region in the complex plane that does not allow us to circle the origin. One possibility is to imagine cutting the plane along the negative real axis into the origin, as illustrated in Fig. A2.2. The function defined in Eq. (A2.21) is then analytic in this *cut plane*, and the phase of z is confined to the region $-\pi < \phi \leq \pi$. The function $\ln z$ is said to have a *branch point* at the origin.

Example 5 Consider the function

$$f(z) = z^\alpha \tag{A2.26}$$

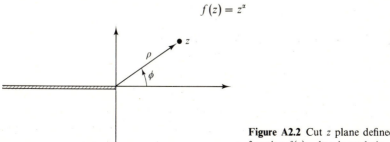

Figure A2.2 Cut z plane defined so that the function $f(z) = \ln z$ is analytic.

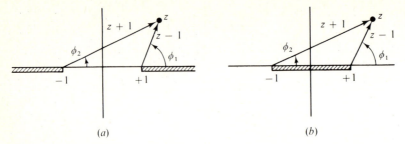

Figure A2.3 Two distinct cut structures that make the function $(z^2 - 1)^{1/2}$ analytic.

where α can now be complex. The arguments leading to Eq. (A2.17) again show that $f'(z)$ exists for $z \neq 0$. Like the logarithm considered previously, however, z^α is not in general single-valued. Use of Eq. (A1.4) shows that

$$z^\alpha = \rho^\alpha e^{i\alpha\phi} \qquad (A2.27a)$$

which does not return to its original value when ϕ increases by 2π unless α is an integer. Thus the function defined by Eq. (A2.26) typically has a branch point at $z = 0$ and can only be analytic in a cut plane, as in Fig. A2.2. More generally, $(z - z_0)^\alpha$ has a branch point at z_0, and it can be rendered single-valued by cutting the plane from z_0 to (say) $-\infty$. In that case, $(z - z_0)^\alpha$ becomes

$$|z - z_0|^\alpha e^{i\alpha\phi_0} = \exp\left(\alpha \ln |z - z_0|\right) e^{i\alpha\phi_0} \qquad (A2.27b)$$

where ϕ_0 is the polar angle measured with z_0 as the origin (compare Fig. A2.1).

To illustrate one common situation, we consider the function

$$f(z) = (z^2 - 1)^{1/2} \qquad (A2.28)$$

which has branch points at $z = \pm 1$. Although $f(z)$ is not single-valued on any path that encircles only one branch point, it *is* single-valued if both are enclosed. Thus in addition to drawing branch cuts from 1 to ∞ and from -1 to $-\infty$ (Fig. A2.3a) it is also permissible in this case merely to draw a cut between the two branch points (Fig. A2.3b) since Eq. (A2.28) can be rewritten (see Fig. A2.3) as

$$(z^2 - 1)^{1/2} = |z - 1|^{1/2} |z + 1|^{1/2} \exp\left[i\tfrac{1}{2}(\phi_1 + \phi_2)\right] \qquad (A2.29)$$

It is straightforward to evaluate the function at any particular point in the complex z plane. In Fig. A2.3a, for example, $f(0) = i$, whereas $f(z)$ approaches $\pm x$ for $x = \mathrm{Re}\, z \to \infty$ and $y \to \pm 0$ (that is, y approaches the right-hand branch cut from either above or below).

A3 COMPLEX INTEGRATION

Assume that $f(z)$ is continuous along a curve C in the complex z plane, as illustrated in Fig. A3.1, and divide the curve into p equal segments. Then the integral of the function $f(z)$ along the curve C is defined by the relation

$$\int_C f(z)\, dz \equiv \lim_{p \to \infty} \sum_{i=1}^{p} f(\bar{z}_i)(z_{i+1} - z_i) \qquad (A3.1)$$

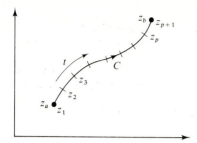

Figure A3.1 Curve C used in defining complex integration.

where $\bar{z}_i$ lies between z_{i+1} and z_i on the curve and we require that max $|z_{i+1} - z_i| \rightarrow 0$. Assume that the curve C has a continuously turning tangent (the arguments are immediately extended to any curve that can be divided into a finite number of such segments), and let it be parametrized by the representation

$$x = x(t) \quad \text{and} \quad y = y(t) \quad \text{for } 0 \leq t \leq 1 \tag{A3.2}$$

where the functions are continuous and differentiable. Equation (A3.1) can then be rewritten as

$$\int_C f(z)\, dz = \lim_{p \to \infty} \sum_{i=1}^{p} \{u[x(\bar{t}_i), y(\bar{t}_i)] + iv[x(\bar{t}_i), y(\bar{t}_i)]\}$$

$$\times \left| \frac{x_{i+1} - x_i}{t_{i+1} - t_i} + i\frac{y_{i+1} - y_i}{t_{i+1} - t_i} \right| (t_{i+1} - t_i) \tag{A3.3}$$

In this way, the problem of computing the complex integral defined in Eq. (A3.1) has been reduced to that of computing four real integrals. We now appeal to a fundamental theorem from real analysis that *every continuous function has a Riemann integral*. Thus all four limits in Eq. (A3.3) exist; this result

$$\int_C f(z)\, dz = \int_0^1 \{u[x(t), y(t)] + iv[x(t), y(t)]\} \left(\frac{dx}{dt} + i\frac{dy}{dt}\right) dt \tag{A3.4}$$

immediately defines the integral of any continuous function of a complex variable along a smooth curve or segment of a smooth curve in terms of real integrals. It is important to note that there is nothing mysterious about this expression. It is simply the standard result obtained by dividing up the curve, multiplying the function evaluated in the interval by the length of the interval, adding the contributions together, and then letting the length of all the intervals go to zero.

To clarify this procedure, we give a few simple examples. Consider the circle C_0 of radius ρ illustrated in Fig. A3.2. The angle ϕ now serves as the real parameter characterizing the curve

$$z = \rho e^{i\phi} \tag{A3.5}$$

and the differential of Eq. (A3.5) can be written

$$dz = i\, d\phi\, \rho e^{i\phi} \tag{A3.6}$$

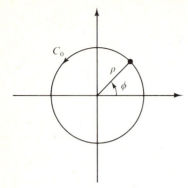

Figure A3.2 Integration around a circle.

corresponding to the last factor and differential in Eq. (A3.4). It immediately follows that the contour integral of z^n vanishes for n equal to any positive integer or zero

$$\oint_{C_0} z^n \, dz = i\rho^{n+1} \int_0^{2\pi} e^{i(n+1)\phi} \, d\phi = 0 \qquad (A3.7)$$

Similarly, it follows that

$$\oint_{C_0} \frac{dz}{z^n} = i\rho^{1-n} \int_0^{2\pi} e^{i(1-n)\phi} \, d\phi = 2\pi i \delta_{n,1} \qquad (A3.8)$$

Here the integral vanishes for n equal to any positive integer or zero *except* for the particular value $n = 1$, when the angular integral in Eq. (A3.8) is just 2π. The final result is independent of the radius of the circle, as indicated in the last equality in Eq. (A3.8). Note that Eqs. (A3.7) and (A3.8) require nothing about the *analyticity* of the function being integrated since they merely represent the total additive contribution obtained while moving along the curve.

Figure A3.1 illustrates two important properties of complex integration defined through Eqs. (A3.1) and (A3.3).

(i) The integral changes sign if the direction of integration is reversed.

$$\int_{z_a}^{z_b} f(z) \, dz = -\int_{z_b}^{z_a} f(z) \, dz \qquad (A3.9)$$

(ii) The modulus of the integral can be bounded in the following manner

$$\left| \int_{z_a}^{z_b} f(z) \, dz \right| \le ML \qquad \begin{aligned} M &= \max |f(z)| \text{ on } C \\ L &= \text{length of } C \end{aligned} \qquad (A3.10)$$

This follows immediately from Eq. (A3.1).

A4 CAUCHY'S THEOREM

The basic theorem of functional analysis, due to Cauchy, states that

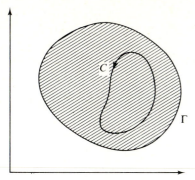

Figure A4.1 A simply connected region of analyticity Γ and a simple closed curve C lying inside Γ.

If $f(z)$ is analytic in a simply connected region Γ and C is a simple closed contour inside Γ, then

$$\oint_C f(z)\,dz = 0 \tag{A4.1}$$

The configuration is shown in Fig. A4.1. We proceed to prove this important result, making use of the three fundamental defining properties of an analytic function; i.e., if $f(z)$ is analytic, then $f(z)$ is continuous, single-valued, and has a unique derivative $f'(z)$.

The region of interest in Γ contains C and its interior. Divide this region with a mesh, as illustrated in Fig. A4.2. The interior of C evidently contains two types of small regions:

1. There are regular squares with four sides, each a straight line. These regions will be denoted by R_n.
2. There are irregular squares, part of whose boundary is defined by a piece of the curve C. Denote these regions by I_n.

The integral of the left-hand side of Eq. (A4.1) can now be written identically as

$$\oint_C f(z)\,dz = \sum \oint_{R_n} f(z)\,dz + \sum \oint_{I_n} f(z)\,dz \tag{A4.2}$$

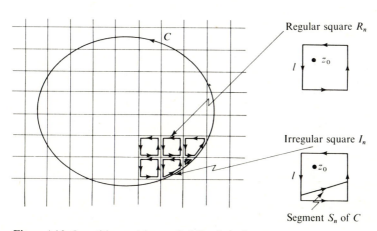

Figure A4.2 Quantities used in proof of Cauchy's theorem.

where the integrals run along the boundaries of all the regular and irregular squares. This result follows because Eq. (A3.9) shows that the contributions to the integral running in opposite directions along a common side cancel. Thus the only remaining contribution on the right-hand side of Eq. (A4.2) is that around the curve C (see Fig. A4.2), which is just the required result.

Suppose now that the mesh is very fine and the regions are very small. Take any one region, as enlarged in Fig. A4.2, and let z_0 be a point inside this region. Consider the following difference

$$\frac{f(z) - f(z_0)}{z - z_0} - f'(z_0) \equiv g(z) \tag{A4.3}$$

Since $f(z)$ is analytic, it follows from Eq. (A2.5) that

$$|g(z)| < \epsilon \tag{A4.4}$$

for all points in the interior and on the boundary of the region provided only that the square is small enough. Thus, for each small region interior to C we can solve Eq. (A4.3) for $f(z)$ and write

$$\oint f(z)\, dz = \oint f(z_0)\, dz + \oint f'(z_0)(z - z_0)\, dz + \oint g(z)(z - z_0)\, dz \tag{A4.5}$$

This relation is an identity.

We now note the following results

$$\oint dz = 0$$

$$\text{for each } R_n \text{ and } I_n \tag{A4.6}$$

$$\oint z\, dz = \oint \tfrac{1}{2}\, dz^2 = 0$$

which hold because *the integrands in these particular integrals are all perfect differentials of functions that are themselves single-valued* throughout the region. In Eq. (A3.7), these integrals were already shown to vanish for a circular contour, but the present argument demonstrates the results for any closed contour. From Eqs. (A4.6) we immediately conclude that the *first two integrals* on the right-hand side of Eq. (A4.5) vanish identically. The remaining integral on the right-hand side of Eq. (A4.5) can be bounded for each small region by making use of the result in Eq. (A3.10). Thus for the regular squares we evidently have

$$\left| \oint_{R_n} (z - z_0) g(z)\, dz \right| < (\epsilon \sqrt{2}\, l)(4l) \tag{A4.7}$$

The first factor in parentheses on the right-hand side is the maximum modulus of the integrand along the contour, as is evident from Fig. A4.2 and Eq. (A4.4), and the second factor in parentheses is just the total length of the contour. For the irregular squares (see Fig. A4.2) one has

$$\left| \oint_{I_n} (z - z_0) g(z)\, dz \right| < (\epsilon \sqrt{2}\, l)(4l + S_n) \tag{A4.8}$$

Here the first factor in parentheses is again the maximum modulus of the integrand. The second factor in parentheses is the length of the perimeter of the encompassing square *plus* the length of that piece of the curve C passing through the region; this total evidently exceeds the length of the actual contour for the small region I_n and hence the inequality in Eq. (A4.8). Equations (A4.2) to (A4.8) can be combined to yield the following bound on the left-hand side of Eq. (A4.2):

$$\left| \oint_C f(z)\, dz \right| < 4\sqrt{2}\, \epsilon \left(\sum_{R_n} l^2 + \sum_{I_n} l^2 \right) + \sqrt{2}\, \epsilon l \left(\sum_{I_n} S_n \right) \tag{A4.9}$$

Now observe that the first factor in parentheses on the right-hand side of Eq. (A4.9) is simply the total *area* of the squares either containing the curve C or inside of the curve C in Fig. A4.2. This total area remains bounded as the mesh is made finer and finer. The second factor in parentheses in Eq. (A4.9) is just the total *length* of the curve C. Since Eq. (A4.4) shows that the quantity ϵ on the right-hand side of Eq. (A4.9) can be made arbitrarily small by making the mesh fine enough, it follows that the total right-hand side of Eq. (A4.9) can also be made arbitrarily small.

To complete the proof, we notice that the left-hand side of Eq. (A4.9) is simply the modulus of some complex number which is independent of ϵ. Since the right-hand side of Eq. (A4.9) can be made arbitrarily small by taking ϵ as small as desired, we are forced to conclude that the left-hand side of Eq. (A4.9) must in fact vanish. If the modulus of a complex number vanishes, that complex number must be zero, which thus proves Cauchy's theorem

$$\oint_C f(z)\, dz = 0 \tag{A4.10}$$

A little thought will show that one aspect of the proof has been treated too cavalierly. Equation (A4.9) assumed that a single ϵ can be used in the inequality (A4.4) for *all* the regions under consideration. *Can we in fact choose ϵ uniformly for all the squares?* To answer this question, fix the quantity ϵ. Now either this ϵ suffices for the mesh we have chosen or there are some squares for which the inequality in Eq. (A4.4) does not hold. Subdivide these "bad" squares into smaller squares by using a finer and finer mesh. Then one of the following two possibilities must be true:

1. Either the inequality (A4.4) holds after a finite number of subdivisions of the mesh
2. Or there is at least one limit point inside the "bad square" where the inequality (A4.4) is violated and $|g(z)| > \epsilon$

But this second possibility contradicts the hypothesis that $f'(z_0)$ exists at all points in the region Γ, for Eqs. (A2.5) and (A4.3) imply that

$$|g(z)| < \epsilon \qquad \text{if } |z - z_0| < \delta \tag{A4.11}$$

We therefore conclude that the process must terminate after a finite number of steps, allowing us to choose ϵ uniformly for all the squares.

Cauchy's theorem has many important applications. Suppose that $f(z)$ is *analytic in the annulus* between the two curves C_1 and C_2 illustrated in Fig. A4.3 but assume nothing about $f(z)$ in the region *interior* to the curve C_1. Apply Cauchy's theorem to the contour C shown in Fig. A4.3, whose interior is a simply connected region. Since $f(z)$ is analytic, and hence single-valued, in the annulus, the contributions of the pieces *across the annulus cancel* identically because they run in opposite directions [see Eq. (A3.9)]. Thus in this case Cauchy's theorem takes the form

$$\oint_{C_2} f(z)\, dz + \oint_{C_1} f(z)\, dz = 0 \tag{A4.12}$$

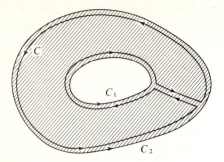

Figure A4.3 Contour C, with simply connected interior, used in application of Cauchy's theorem.

Here it is important to note that the first integral runs in the positive (counterclockwise) direction and the second integral runs in the negative (clockwise) direction. The direction of the second integral can be reversed by changing its sign, however, and Eq. (A4.12) then assumes the equivalent form

$$\oint_{C_1} f(z)\, dz = \oint_{C_2} f(z)\, dz \tag{A4.13}$$

where both integrals now run in the positive (counterclockwise) direction. This remarkable result permits us to deform an integration contour through any region where the integrand is analytic.

More generally, suppose that $f(z)$ is analytic in the region between the curve C and the curves $C_1, C_2, C_3, \ldots$ shown in Fig. A4.4. Repeated application of the idea illustrated in Fig. A4.3 produces a single master contour $\bar{C}$ with a simply connected interior, running counterclockwise along the curve C and clockwise around the curves $C_1, C_2, C_3, \ldots$. Cauchy's theorem now leads to the remarkable relation

$$\oint_C f(z)\, dz = \oint_{C_1} f(z)\, dz + \oint_{C_2} f(z)\, dz + \oint_{C_3} f(z)\, dz + \cdots \tag{A4.14}$$

which will play a central role in our subsequent analysis.

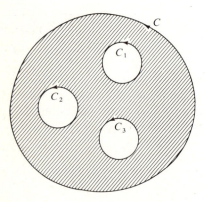

Figure A4.4 A multiply connected region of analyticity.

A5 CAUCHY'S INTEGRAL

We can also use Cauchy's theorem to derive an integral representation of an analytic function, known as Cauchy's integral. Suppose that $f(z)$ is analytic in the region Γ and C is a simple closed curve in Γ (see Fig. A5.1). Let z be a point inside the curve C, and consider the integral

$$\oint_C \frac{f(\zeta)\, d\zeta}{\zeta - z} = \oint_{C_0} \frac{f(\zeta)\, d\zeta}{\zeta - z} \tag{A5.1}$$

Here C_0 is a small circle of radius ρ surrounding the point z (see Fig. A5.1), and the equality in Eq. (A5.1) follows from Eq. (A4.13) because the integrand is an analytic function of ζ in the annulus lying between the curves C and C_0. The integral on the right-hand side of Eq. (A5.1) can be written identically as

$$\oint_{C_0} \frac{f(\zeta)\, d\zeta}{\zeta - z} \equiv f(z) \oint_{C_0} \frac{d\zeta}{\zeta - z} + \oint_{C_0} \frac{f(\zeta) - f(z)}{\zeta - z}\, d\zeta \tag{A5.2}$$

The parametrization

$$\zeta - z \equiv \rho e^{i\phi} \tag{A5.3}$$

from Fig. A3.2 allows us to evaluate the first term on the right-hand side of Eq. (A5.2) explicitly. Using the differential

$$d\zeta = i\, d\phi\, \rho e^{i\phi} \tag{A5.4}$$

we immediately obtain

$$\oint_{C_0} \frac{d\zeta}{\zeta - z} = 2\pi i \tag{A5.5}$$

just as in Eq. (A3.8). We next observe that, for small enough ρ, the numerator in the second term of Eq. (A5.2) can be made arbitrarily small

$$|f(\zeta) - f(z)| < \epsilon \tag{A5.6}$$

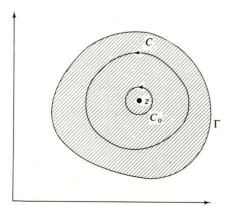

Figure A5.1 Quantities used in definition of Cauchy's integral.

because the function $f(z)$ is continuous. Thus the remaining integral around the circular contour C_0 in Figs. A3.2 and A5.1 can be bounded with Eq. (A3.10)

$$\left| \oint_{C_0} \frac{f(\zeta) - f(z)}{\zeta - z} \, d\zeta \right| < \frac{\epsilon}{\rho} 2\pi\rho \tag{A5.7}$$

Equations (A5.1), (A5.2), (A5.5), and (A5.7) together imply that

$$\left| \oint_C \frac{f(\zeta) \, d\zeta}{\zeta - z} - 2\pi i f(z) \right| < 2\pi\epsilon \tag{A5.8}$$

As in obtaining (A4.10), we notice that the left-hand side of Eq. (A5.8) is simply the modulus of some complex number whereas the right-hand side can be made arbitrarily small by choosing ρ small enough [Eq. (A5.6)]. Hence the left-hand side of Eq. (A5.8) must in fact vanish identically, and we therefore obtain *Cauchy's integral*

$$f(z) = \frac{1}{2\pi i} \oint_C \frac{f(\zeta) \, d\zeta}{\zeta - z} \tag{A5.9}$$

which holds for any function analytic inside and on C.

Cauchy's integral has several remarkable consequences.

1. The z dependence of the function $f(z)$ now appears *explicitly* in the denominator on the right-hand side of Eq. (A5.9). Note that we merely assumed that the function $f(z)$ is analytic inside the region Γ and that C is any closed curve inside that region (see Fig. A5.1). Thus Cauchy's integral provides an explicit representation of the z dependence for any analytic function.
2. Furthermore, Cauchy's integral only requires the value of $f(z)$ on the curve C. We conclude that knowledge of an analytic function $f(z)$ *on* a closed curve C (Fig. A5.1) fixes $f(z)$ *everywhere inside* the curve C.
3. Assume that the point z is strictly inside the curve C, so that the following inequality holds

$$|\zeta - z| > h > 0 \tag{A5.10}$$

and the denominator in Eq. (A5.9) can never vanish. *One can then differentiate the finite Cauchy integral any number of times simply by differentiating under the integral sign*, and the nth derivative of the function $f(z)$ assumes the remarkable form

$$f^{(n)}(z) = \frac{n!}{2\pi i} \oint_C \frac{f(\zeta) \, d\zeta}{(\zeta - z)^{n+1}} \tag{A5.11}$$

All that has been assumed here is *analyticity* of $f(z)$, namely, that its *first* derivative $f'(z)$ exists in the region Γ. But Eq. (A5.11) then demonstrates that every derivative of $f(z)$ exists by providing an explicit representation for each of these derivatives. Furthermore, if $f(z)$ is analytic, Eq. (A5.11) shows that $f'(z)$ is also *analytic* because it possesses a derivative, and this argument extends to all derivatives of $f(z)$. Thus the assumption that $f(z)$ is analytic (and hence possesses *one* derivative) suffices to establish that all its derivatives are themselves analytic.

Morera's Theorem

These same results can be used to demonstrate Morera's theorem, which is the converse of Cauchy's theorem. Consider a simply connected region Γ as illustrated in Fig. A4.1.

Given $f(z)$ continuous and single-valued and

$$\oint_C f(z)\, dz = 0 \tag{A5.12}$$

for any closed curve in Γ; then $f(z)$ is analytic in Γ.

PROOF Equation (A5.12) implies that

$$\int_{C_1} f(z)\, dz = \int_{C_2} f(z)\, dz \tag{A5.13}$$

where C_1 and C_2 are any two paths leading from the point z_0 to the point z in Γ. As a result, the integral from z_0 to z is *independent of the path* and depends only on the points z_0 and z, allowing us to define the function

$$F(z) \equiv \int_{z_0}^{z} f(\zeta)\, d\zeta \tag{A5.14}$$

From the basic definition of the complex integral, it follows that this function has a derivative given by

$$\frac{dF(z)}{dz} = f(z) \tag{A5.15}$$

We conclude that $F(z)$ is analytic. The relation (A5.11) then shows that its derivative $f(z)$ in Eq. (A5.15) is also analytic, which is the desired result.

A6 UNIFORMLY CONVERGENT SERIES

The following results for uniformly convergent series of functions of a complex variable are immediately established just as with series of functions of a real variable:

(i) A uniformly convergent series of continuous functions is continuous (A6.1)
(ii) A uniformly convergent series of continuous functions can be integrated term
  by term (A6.2)

  Differentiating an infinite series is a more difficult matter that depends in general on the properties of the differentiated series. We shall use Cauchy's integral together with the basic relations (A6.1) and (A6.2) to prove the following result, which will allow us to differentiate an infinite series of analytic functions term by term (see Fig. A4.1).

Given $f_n(z)$ analytic in Γ with C a simple closed curve in Γ and

$$\sum_{n=1}^{\infty} f_n(z) = f(z) \tag{A6.3}$$

a uniformly convergent series for z on C; then:
  (i) The series in Eq. (A6.3) converges to an analytic function everywhere
    inside C (A6.4)
  (ii) The series in Eq. (A6.3) can be differentiated term by term inside C (A6.5)

To establish this result, consider the integral

$$I(z) \equiv \frac{1}{2\pi i} \oint_C \frac{f(\zeta) \, d\zeta}{\zeta - z} \tag{A6.6}$$

where z is assumed to be strictly inside the curve C. Since the series is uniformly convergent along C, Eq. (A6.2) allows us to interchange summation and integration

$$I(z) = \frac{1}{2\pi i} \oint_C \frac{\sum\limits_{n=1}^{\infty} f_n(\zeta) \, d\zeta}{\zeta - z} = \sum_{n=1}^{\infty} \frac{1}{2\pi i} \oint_C \frac{f_n(\zeta) \, d\zeta}{\zeta - z} \tag{A6.7}$$

Each term of the series $f_n(z)$ is analytic and satisfies the conditions for Cauchy's integral (A5.9). Thus each term on the right-hand side of Eq. (A6.7) is simply $f_n(z)$, and we conclude that

$$I(z) = \frac{1}{2\pi i} \oint_C \frac{f(\zeta) \, d\zeta}{\zeta - z} = \sum_{n=1}^{\infty} f_n(z) \tag{A6.8}$$

This establishes the convergence of the series on the right-hand side of Eq. (A6.8) for all z inside C because $I(z)$ is bounded there. Furthermore, the integral (A6.6) represents an analytic function since the z dependence is explicit, and we can differentiate it any number of times, just as we did in obtaining the result in Eq. (A5.11). This observation establishes (A6.4) above. To prove the relation (A6.5), consider the derivative of the left-hand side of Eq. (A6.6)

$$I'(z) = \frac{1}{2\pi i} \oint_C \frac{f(\zeta) \, d\zeta}{(\zeta - z)^2} \tag{A6.9}$$

The argument leading from Eq. (A6.6) to Eq. (A6.8) now shows that

$$I'(z) = \frac{1}{2\pi i} \oint_C \frac{f(\zeta) \, d\zeta}{(\zeta - z)^2} = \sum_{n=1}^{\infty} f'_n(z) \tag{A6.10}$$

Thus the derivative of the analytic function is the sum of the derivatives, and Eq. (A6.5) is proved.

Power Series

The power series is a particular type of infinite series that plays a central role in the theory of analytic functions. It is defined according to the relation

$$f(z) = \sum_{n=0}^{\infty} a_n z^n \tag{A6.11}$$

where the a_n are constants. A power series has the following remarkable property (see Fig. A6.1):

> If a power series converges for some z_0, then it converges uniformly inside a circle

$$\frac{|z|}{|z_0|} \le r < 1 \tag{A6.12}$$

PROOF Since the series converges, the modulus of each additional term in the series must

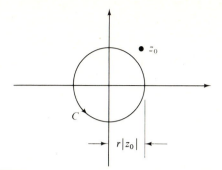

Figure A6.1 Circle of convergence of a power series. Here $r < 1$.

eventually approach zero, which follows from Cauchy's criterion for convergence of a series. Sufficiently far out in this series, we must have

$$|a_n z_0^n| < 1 \qquad \text{for } n \geq N \tag{A6.13}$$

Let R_N be the remainder after N terms of the series (A6.11). For any z in the region indicated in Fig. A6.1, R_N can be bounded by using the inequality (A6.13)

$$|R_N| \equiv \left| \sum_{n=N}^{\infty} a_n z^n \right| \leq \sum_{n=N}^{\infty} |a_n z_0^n| \left| \frac{z}{z_0} \right|^n < \sum_{n=N}^{\infty} r^n \tag{A6.14}$$

The last quantity in Eq. (A6.14) is the geometric series, which converges for any r satisfying the inequality (A6.12). Thus the comparison test establishes the convergence of the series (A6.11). Furthermore, the convergence is uniform because the geometric series being used for comparison is independent of z for any z satisfying the condition (A6.12). In particular, the power series (A6.11) converges uniformly along the curve C in Fig. A6.1. As a result, Eqs. (A6.3) to (A6.5) immediately imply that the power series converges to an analytic function everywhere inside the curve C and can be differentiated term by term any number of times inside the curve C. The power series can also be integrated term by term inside of C, as indicated in Eq. (A6.2).

The result (A6.12) is very powerful. It states that if a power series about the origin converges at any point z_0, then it converges uniformly to an analytic function in any disk extending out to that point. This disk is called the *circle of convergence* of the power series, and its radius is called the *radius of convergence*. It follows from these arguments that the circle of convergence of a power series extends out to the first singularity of $f(z)$ in Eq. (A6.11).

A7 TAYLOR'S THEOREM

Taylor's theorem asserts that "every analytic function can be expanded in a power series." More precisely, with reference to Fig. A7.1, the theorem can be stated as follows:

Given $f(z)$ analytic in a circle about z_0, and z any point inside the circle, then $f(z)$ has the expansion

$$f(z) = f(z_0) + (z - z_0)f'(z_0) + \cdots + \frac{(z - z_0)^n}{n!} f^{(n)}(z_0) + \cdots \tag{A7.1a}$$

$$f(z) = \sum_{n=0}^{\infty} a_n (z - z_0)^n \tag{A7.1b}$$

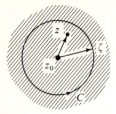

Figure A7.1 Taylor-series expansion about the point z_0.

Equation (A7.1b) evidently represents a power series centered about the point z_0.

Taylor's theorem can be proved from Cauchy's integral

$$f(z) = \frac{1}{2\pi i} \oint_C \frac{f(\zeta)\, d\zeta}{\zeta - z} \tag{A7.2}$$

where C is a circle centered on z_0 enclosing the point z (Fig. A7.1). Since the inequality

$$\frac{|z - z_0|}{|\zeta - z_0|} < 1 \tag{A7.3}$$

holds in this case, the denominator in Cauchy's integral has the expansion

$$\frac{1}{\zeta - z} \equiv \frac{1}{\zeta - z_0 - (z - z_0)} = \frac{1}{\zeta - z_0}\left[1 + \frac{z - z_0}{\zeta - z_0} + \left(\frac{z - z_0}{\zeta - z_0}\right)^2 + \cdots\right] \tag{A7.4}$$

The inequality (A7.3) ensures that this series converges uniformly by comparison with the geometric series. Thus the series (A7.4) can be inserted into the integral (A7.2) and integrated term by term to give

$$f(z) = \sum_{n=0}^{\infty} (z - z_0)^n \left[\frac{1}{2\pi i} \oint_C \frac{f(\zeta)\, d\zeta}{(\zeta - z_0)^{n+1}}\right] \tag{A7.5}$$

The analyticity of $f(z)$ inside and on C allows us to invoke (A5.11), and we therefore find

$$f(z) = \sum_{n=0}^{\infty} (z - z_0)^n \frac{f^{(n)}(z_0)}{n!} \tag{A7.6}$$

which demonstrates (A7.1) and Taylor's theorem.

A8 LAURENT SERIES

Taylor's theorem provides a representation of a function that is analytic throughout a circle. Suppose, however, that a function is analytic except at an isolated point z_0, as illustrated in Fig. A8.1. In this case, Cauchy's integral again provides a valuable representation of the function, known as the *Laurent series*. More precisely,

If $f(z)$ is analytic in an annulus centered about z_0 and z is a point in the annulus, then $f(z)$ has the expansion

$$f(z) = \sum_{n=0}^{\infty} a_n(z - z_0)^n + \frac{b_1}{z - z_0} + \frac{b_2}{(z - z_0)^2} + \cdots \tag{A8.1}$$

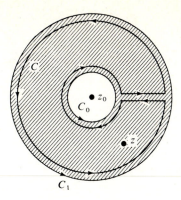

Figure A8.1 Laurent-series expansion about the point z_0.

PROOF Consider the contour C which surrounds the simply connected region illustrated in Fig. A8.1. Cauchy's integral states that

$$f(z) = \frac{1}{2\pi i} \oint_C \frac{f(\zeta)\, d\zeta}{\zeta - z} \tag{A8.2}$$

and the contributions *across* the annulus cancel because the contours run in opposite directions. Consequently, this relation can be rewritten as

$$f(z) = \frac{1}{2\pi i} \oint_{C_1} \frac{f(\zeta)\, d\zeta}{\zeta - z} - \frac{1}{2\pi i} \oint_{C_0} \frac{f(\zeta)\, d\zeta}{\zeta - z} \tag{A8.3}$$

and we treat the two terms separately. When ζ is on the exterior contour C_1, the following inequality holds

$$\left| \frac{z - z_0}{\zeta - z_0} \right| < 1 \qquad \text{for } \zeta \text{ on } C_1 \tag{A8.4}$$

because z lies inside the annulus. Thus the denominator in the first term on the right-hand side of Eq. (A8.3) can be expanded in the power series (A7.4), and term-by-term integration yields precisely the series in Eq. (A7.5). The second term in (A8.3) differs only in that ζ is on the interior contour C_0, and the appropriate inequality takes the form

$$\left| \frac{\zeta - z_0}{z - z_0} \right| < 1 \qquad \text{for } \zeta \text{ on } C_0 \tag{A8.5}$$

For z in the annulus, the corresponding denominator can be expanded in a uniformly convergent power series according to

$$\frac{1}{z - \zeta} \equiv \frac{1}{z - z_0 - (\zeta - z_0)} = \frac{1}{z - z_0} \left[1 + \frac{\zeta - z_0}{z - z_0} + \left(\frac{\zeta - z_0}{z - z_0} \right)^2 + \cdots \right] \tag{A8.6}$$

and again integrated term by term. These observations imply that Eq. (A8.3) takes the form

$$f(z) = \sum_{n=0}^{\infty} (z - z_0)^n \frac{1}{2\pi i} \oint_{C_1} \frac{f(\zeta)\, d\zeta}{(\zeta - z_0)^{n+1}}$$

$$+ \sum_{n=1}^{\kappa} \frac{1}{(z - z_0)^n} \frac{1}{2\pi i} \oint_{C_0} (\zeta - z_0)^{n-1} f(\zeta)\, d\zeta \tag{A8.7}$$

which now provides explicit expressions for the coefficients in Eq. (A8.1).

The previous Taylor's series can be recovered by noting that Cauchy's theorem ensures that

$$\oint_{C_0} (\zeta - z_0)^{n-1} f(\zeta) \, d\zeta = 0 \tag{A8.8}$$

if $f(z)$ is in fact analytic everywhere inside of the disk including the point z_0. In addition, Cauchy's integral (A5.11) then shows that

$$\frac{1}{2\pi i} \oint_{C_1} \frac{f(\zeta) \, d\zeta}{(\zeta - z_0)^{n+1}} = \frac{f^{(n)}(z_0)}{n!} \tag{A8.9}$$

but it is essential to note that Eq. (A8.9) holds *only* under these conditions.

The Laurent series in Eqs. (A8.1) and (A8.7) serves to classify the possible isolated singularities of an analytic function $f(z)$ at the point z_0.

(i) If the series of inverse powers in Eq. (A8.7) has a *finite* upper limit κ and terminates after κ terms

$$f(z) = \sum_{n=0}^{\infty} a_n(z - z_0)^n + \frac{b_1}{z - z_0} + \frac{b_2}{(z - z_0)^2} + \cdots + \frac{b_\kappa}{(z - z_0)^\kappa} \tag{A8.10}$$

then the function $f(z)$ is said to have a pole of order κ at the point z_0.

(ii) If the series of inverse powers in Eq. (A8.7) does not terminate and κ is in fact infinite, the function $f(z)$ is said to have an essential singularity at the point z_0.

A9 THEORY OF RESIDUES

We have shown that if a function is analytic in a region Γ except at an isolated point z_0, it possesses a Laurent expansion

$$f(z) = \sum_{n=0}^{\infty} a_n(z - z_0)^n + \frac{b_1}{z - z_0} + \frac{b_2}{(z - z_0)^2} + \cdots \tag{A9.1}$$

valid throughout an annulus lying inside Γ (see Fig. A8.1). The coefficients in this expansion are given by

$$a_n = \frac{1}{2\pi i} \oint_{C_1} \frac{f(\zeta) \, d\zeta}{(\zeta - z_0)^{n+1}} \tag{A9.2a}$$

$$b_n = \frac{1}{2\pi i} \oint_{C_0} (\zeta - z_0)^{n-1} f(\zeta) \, d\zeta \tag{A9.2b}$$

where C_0 is a small circle about the singularity at z_0 and C_1 is a larger circle containing z. In the following discussion we assume that the function $f(z)$ is *meromorphic* (has only isolated poles of finite order).

The coefficient b_1 of the first inverse power $(z - z_0)^{-1}$ in the Laurent expansion about the point z_0 is called the *residue*. It plays a very special role, being essentially the integral of the function $f(z)$ around the small circle C_0 surrounding the singularity z_0

$$b_1 = \frac{1}{2\pi i} \oint_{C_0} f(\zeta) \, d\zeta \equiv R_0 \qquad \text{residue at } z_0 \tag{A9.3}$$

If, as assumed here, the function $f(z)$ is analytic inside and on a contour C except for isolated singularities at the points $z_1, z_2, \ldots, z_n$ inside the curve C (see Fig. A4.4), then Eq. (A4.14) shows that the integral of $f(z)$ around the curve C can be written as a *sum* of the integrals of $f(z)$ taken around small circles enclosing the singularities

$$\oint_C f(z)\, dz = \left[\oint_{C_1} + \cdots + \oint_{C_n} \right] f(z)\, dz \tag{A9.4}$$

Equation (A9.3) makes it evident that each of these integrals is simply $2\pi i$ times the *residue* of the function $f(z)$ at the appropriate singularity. Thus

$$\oint_C f(z)\, dz = 2\pi i (R_1 + R_2 + \cdots + R_n) \tag{A9.5a}$$

$$= 2\pi i \left(\sum \text{residues} \right) \tag{A9.5b}$$

These relations provide a very powerful tool for evaluating integrals of meromorphic functions over a closed contour. All we need do is examine the behavior of the function in the vicinity of each of the enclosed singularities, make a Laurent expansion about that singularity, and identify the corresponding coefficient b_1 in the expansion (A9.1), which is the residue of the function at that point. The total integral is then given by $2\pi i$ times the sum of the residues, as indicated in Eq. (A9.5).

As an example, suppose we wish to evaluate the definite integral

$$I_1 = \int_{-\infty}^{\infty} \frac{dx}{1 + x^2} = 2 \int_0^{\infty} \frac{dx}{1 + x^2} \tag{A9.6}$$

where the last form follows from the symmetry of the integrand. We first define the contour integral

$$I \equiv \oint_c \frac{dz}{1 + z^2} \tag{A9.7}$$

where the contour C is indicated in Fig. A9.1. Here the integrand is the same as that in Eq. (A9.6) along the real axis, and this choice of contour will allow us to evaluate the integral in (A9.6) between the indicated limits.

The integrand in Eq. (A9.7) is analytic except at the points $z = \pm i$, and only $z = i$ lies

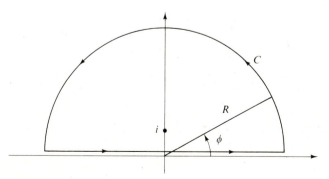

Figure A9.1 Contour C in Eq. (A9.7).

within our contour. To find the residue at this point, we write $z = i + \zeta$ and study the integrand for small $|\zeta|$

$$\frac{1}{z^2 + 1} = \frac{1}{(i + \zeta)^2 + 1} = \frac{1}{2i\zeta(1 - \frac{1}{2}i\zeta)} \tag{A9.8}$$

The binomial theorem immediately yields the desired Laurent expansion

$$\frac{1}{z^2 + 1} = \frac{1}{2i\zeta}[1 + \tfrac{1}{2}i\zeta + (\tfrac{1}{2}i\zeta)^2 + \cdots] \tag{A9.9}$$

and comparison with Eq. (A9.1) shows that the function $(z^2 + 1)^{-1}$ has a simple pole at $z = i$ with residue $(2i)^{-1}$. The general relation (A9.5) immediately gives

$$I = 2\pi i \frac{1}{2i} = \pi \tag{A9.10}$$

for the contour integral (A9.7).

This exact result holds for any contour C in Fig. A9.1 that encloses the point $z = i$ and hence for any value of the radius $R > 1$. We now separate the integral I into the explicit contributions coming from the straight line in Fig. A9.1 and from the large semicircle in the upper half plane, where

$$z = Re^{i\phi} \quad \text{and} \quad dz = i \, d\phi \, Re^{i\phi} \tag{A9.11}$$

In this way, we find

$$I = \int_{-R}^{R} \frac{dx}{1 + x^2} + \int_{0}^{\pi} \frac{i \, d\phi \, Re^{i\phi}}{1 + R^2 e^{2i\phi}} \equiv I_A + I_B \tag{A9.12}$$

In the limit $R \to \infty$, the first integral I_A is just the integral I_1 we are trying to evaluate in Eq. (A9.6)

$$I_A \xrightarrow[R\to\infty]{} \int_{-\infty}^{\infty} \frac{dx}{1 + x^2} = I_1 \tag{A9.13}$$

Furthermore, the second integral I_B in Eq. (A9.12) can be *bounded* in the following fashion

$$|I_B| \le \frac{R}{R^2 - 1}\pi \tag{A9.14}$$

Here the denominator has been replaced by its minimum modulus, and the final integral over ϕ gives the factor π. It is now evident that this contribution goes to zero in the limit $R \to \infty$

$$|I_B| \xrightarrow[R\to\infty]{} 0 \tag{A9.15}$$

Thus Eqs. (A9.10), (A9.12), (A9.13), and (A9.15) give the desired result that $I_1 = \pi$, or

$$\int_{0}^{\infty} \frac{dx}{1 + x^2} = \frac{\pi}{2} \tag{A9.16}$$

As a second, more complicated example, we evaluate the definite integral

$$I_1(\alpha) \equiv 2 \int_{0}^{\infty} \frac{ds \, s^{2\alpha - 1}}{1 + s^2} \qquad 0 < \text{Re } \alpha < 1 \tag{A9.17}$$

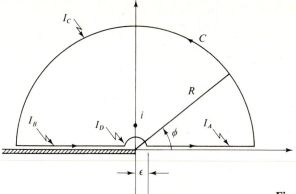

Figure A9.2 Contour C in Eq. (A9.18).

which converges only in the indicated interval. (This integral will play an important role in our subsequent discussion.) We first define the contour integral

$$I \equiv \oint_C \frac{dz \; z^{2\alpha-1}}{z^2+1} \tag{A9.18}$$

where the contour C illustrated in Fig. A9.2 differs from that in Fig. A9.1 only by a small indented semicircle in the upper half plane about the origin. The integrand now has a branch point at $z = 0$, as well as isolated singularities at $z = \pm i$, and a branch cut on the negative real axis makes the function single-valued, as in Figs. A2.2 and A2.3. To evaluate the residue at $z = i$ we first notice that i is now uniquely given in the cut plane by $e^{i\pi/2}$, and then write $z = e^{i\pi/2} + \zeta$. As in Eq. (A9.8) the integrand becomes

$$\frac{z^{2\alpha-1}}{z^2+1} = \frac{(e^{i\pi/2}+\zeta)^{2\alpha-1}}{2i\zeta+\zeta^2}$$

$$= \frac{e^{i\pi\alpha}e^{-i\pi/2}(1+e^{-i\pi/2}\zeta)^{2\alpha-1}}{2i\zeta(1-\tfrac{1}{2}i\zeta)} \tag{A9.19}$$

Use of the binomial theorem readily shows that this function has a simple pole at $\zeta = 0$ with residue $-\tfrac{1}{2}e^{i\pi\alpha}$; Eq. (A9.5) then gives the exact result

$$I = 2\pi i(-\tfrac{1}{2}e^{i\pi\alpha}) = -\pi i e^{i\pi\alpha} \tag{A9.20}$$

for the contour integral (A9.18).

This integral can also be evaluated by separating it into the pieces along the right and left real axis (I_A and I_B) and along the large and small semicircles (I_C and I_D). We consider them in turn. Just above the right half real axis, we have $z = \rho$ (see Fig. A1.1), so that

$$I_A = \int_\epsilon^R \frac{d\rho \; \rho^{2\alpha-1}}{\rho^2+1} \tag{A9.21a}$$

The situation is more complicated above the branch cut on the left half real axis, where $z = \rho e^{i\pi}$ and $dz = d\rho \; e^{i\pi}$

$$I_B = \int_R^\epsilon d\rho \; e^{i\pi} \frac{\rho^{2\alpha-1}e^{i\pi(2\alpha-1)}}{\rho^2 e^{2\pi i}+1}$$

Since $e^{2\pi i} = 1$ and $e^{2\pi i\alpha} = $ const, use of Eq. (A3.9) gives

$$I_B = -e^{2\pi i\alpha} \int_{\epsilon}^{R} \frac{d\rho\ \rho^{2\alpha-1}}{\rho^2 + 1} = -e^{2\pi i\alpha}I_A \tag{A9.21b}$$

In the limit $R \to \infty$, $\epsilon \to 0$, the sum of I_A and I_B reduces to

$$I_A + I_B \xrightarrow[\substack{R\to\infty \\ \epsilon\to 0}]{} (1 - e^{2\pi i\alpha}) \int_{0}^{\infty} \frac{d\rho\ \rho^{2\alpha-1}}{\rho^2 + 1} = \tfrac{1}{2}(1 - e^{2\pi i\alpha})I_1(\alpha) \tag{A9.22}$$

where I_1 is the desired integral (A9.17).

The remaining contributions I_C and I_D will now be proved to vanish in the limit $R \to \infty$ and $\epsilon \to 0$. The same argument as in (A9.14) yields the bound

$$|I_C| \le \frac{\pi\,|R^{2\alpha}|}{R^2 - 1} \tag{A9.23}$$

and the condition Re $\alpha < 1$ gives

$$|I_C| \xrightarrow[R\to\infty]{} 0 \tag{A9.24}$$

Similarly, the substitution $z = \epsilon e^{i\phi}$ shows that

$$I_D = \int_{\pi}^{0} \frac{i\,d\phi\ \epsilon^{2\alpha}e^{2i\alpha\phi}}{1 + \epsilon^2 e^{2i\phi}} \tag{A9.25a}$$

and we immediately obtain

$$|I_D| \le \pi\,|\epsilon^{2\alpha}|\,[1 + O(\epsilon^2)] \tag{A9.25b}$$

The condition $0 < $ Re α ensures that

$$|I_D| \xrightarrow[\epsilon\to 0]{} 0 \tag{A9.26}$$

Although the condition $0 < $ Re $\alpha < 1$ arose originally from the integral (A9.17), we see that it is also essential in bounding the contributions I_C and I_D. A combination of Eqs. (A9.20), (A9.22), (A9.24), and (A9.26) yields the final result

$$I_1(\alpha) = -\frac{2\pi i e^{i\pi\alpha}}{1 - e^{2\pi i\alpha}} = \frac{2\pi i}{e^{i\pi\alpha} - e^{-i\pi\alpha}} = \frac{\pi}{\sin \pi\alpha} \tag{A9.27}$$

for $I_1(\alpha)$ defined in Eq. (A9.17). Several other applications of contour integration to the evaluation of integrals are given in the problems and in Sec. 59.

A10 ZEROS OF AN ANALYTIC FUNCTION

If the function $f(z)$ is analytic in a circle about the point z_0, as indicated in Fig. A7.1, and $f(z_0) = 0$, it possesses a Taylor-series expansion

$$f(z) = \sum_{n=1}^{\infty} a_n(z - z_0)^n \tag{A10.1}$$

Either all the coefficients in this series vanish ($a_n = 0$) and the function is identically zero, or there is a first nonzero coefficient a_m such that the series takes the form

$$f(z) = a_m(z - z_0)^m + a_{m+1}(z - z_0)^{m+1} + \cdots \tag{A10.2}$$

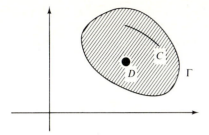

Figure A10.1 Behavior of an analytic function that vanishes along a continuous curve C or in some finite subregion D of Γ.

In the latter case, the leading power $(z - z_0)^m$ can be factored out to give

$$f(z) = (z - z_0)^m[a_m + a_{m+1}(z - z_0) + \cdots] \tag{A10.3a}$$

$$f(z) \equiv (z - z_0)^m \phi(z) \tag{A10.3b}$$

and the remaining power series in Eq. (A10.3a) converges everywhere inside the circle of convergence of the original power series (A10.1). As a result, its sum $\phi(z)$ is an analytic function everywhere within a slightly smaller circle, as was shown in Eq. (A6.4). Since $\phi(z)$ is analytic, it is also continuous. Observe that $\phi(z)$ is not zero at the point $z = z_0$, having the value

$$\phi(z_0) = a_m \neq 0 \tag{A10.4}$$

Therefore $\phi(z)$ must also differ from zero in a small region about z_0, since the statement of continuity requires that

$$|\phi(z) - \phi(z_0)| < \epsilon \quad \text{if } |z - z_0| < \delta \tag{A10.5}$$

Equivalently, a continuous function that is nonzero at a point must also be nonzero in a small region around that point, as indicated in Fig. A7.1. Thus this very simple argument gives the remarkable result that

Either the function $f(z)$ is identically zero in this circle or the only zero of $f(z)$ in this small region is that at $z = z_0$; hence the zeros of an analytic function are isolated points (A10.6)

Equation (A10.6) has the following corollary for the situations illustrated in Fig. A10.1:

If $f(z)$ is analytic in Γ and $f(z) = 0$ in any finite subregion D of Γ or along any curve C in Γ, then $f(z) = 0$ everywhere in Γ (A10.7)

To prove this result, we note that in any disk surrounding the zero of an analytic function either the function vanishes identically or the zero must be isolated. The generalization to an arbitrary simply connected region Γ will be given in the next section, where it is shown how to cover such a region with disks. Alternatively, the results in Eq. (A10.7) also follow by constructing a Taylor series centered on a point where $f(z)$ vanishes. The defining property of an analytic function is that the derivatives are independent of direction. In either of the situations in Fig. A10.1 the derivatives can be taken along a direction such that every derivative vanishes. Thus each coefficient in the Taylor series is zero, and the Taylor series vanishes identically throughout a disk whose radius of convergence extends to the first singularity of the function. In either form of proof, it still remains to extend the function obtained in a disk to the entire region Γ, which is discussed in the next section.

A11 ANALYTIC CONTINUATION

One of the most powerful and surprising results in the theory of analytic functions is the property of *analytic continuation*, defined in the following manner. Suppose that $f_1(z)$ is analytic in Γ_1 and $f_2(z)$ is analytic in Γ_2 and that

$$f_1(z) = f_2(z) \tag{A11.1}$$

in the intersection of the regions Γ_1 and Γ_2 (see Fig. A11.1). Then $f_2(z)$ is said to be the *analytic continuation* of $f_1(z)$ and vice versa, and each provides a representation of a single function $f(z)$.

As an example, consider the geometric series

$$f(z) = 1 + z + z^2 + z^3 + \cdots \qquad |z| < 1 \tag{A11.2}$$

This series converges inside the unit disk, as is immediately evident from comparison with Eq. (A6.14). Inside the disk, however, the series in Eq. (A11.2) is identical with the function

$$f(z) = \frac{1}{1 - z} \tag{A11.3}$$

since it is just the convergent power-series expansion. Unlike the series in (A11.2), however, the function in Eq. (A11.3) is *defined throughout the entire z plane*, being meromorphic with a simple pole at the point $z = 1$. Thus the function in (A11.3) provides an analytic continuation of the power series (A11.2) to the entire complex plane. As expected from our previous considerations, the first singularity of the function (A11.3) determines the radius of convergence ($= 1$) of the power series (A11.2).

The Taylor series can be used to define a *standard method of continuation*. Consider the situation illustrated in Fig. A11.1a, and suppose that an analytic function f is represented as a power series f_1 centered on the point z_1 with some nonzero circle of convergence. Choose another point z_2 within the circle of convergence of the first power series. Since the original

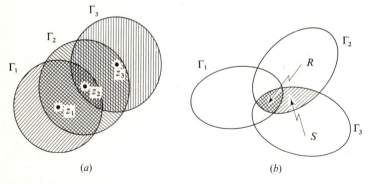

(a) (b)

Figure A11.1 Regions used in defining analytic continuation.

power series can be differentiated any number of times inside of its circle of convergence, we can compute all the derivatives of the function f at the point z_2. Equation (A7.1) then suffices to construct a new Taylor series f_2 centered about the point z_2. This new power series represents the function f within some new radius of convergence, which in general extends outside the original circle of convergence, as illustrated in Fig. A11.1a. By definition, the new function f_2 agrees with the original power-series representation f_1 of the function f in the region of overlap; thus we have constructed an analytic continuation of the function to a new region. This argument can then be repeated by choosing a point z_3 that lies outside the original region and inside the second region. In this manner, the function can be extended (perhaps to the entire complex plane). Note that in each case the radius of convergence of the power series will extend out to the first singularity of the function.

The preceding discussion merely defines a procedure for analytic continuation. We now demonstrate the important and amazing property that this process is *unique* for any simply connected region, as illustrated in Fig. A11.1b.

> Consider three regions Γ_1, Γ_2, and Γ_3 with a part R in common. Suppose that $f_1(z)$ is analytic in Γ_1, $f_2(z)$ in Γ_2, and $f_3(z)$ in Γ_3, that $f_2(z)$ provides an analytic continuation of $f_1(z)$ to the region Γ_2, and that $f_3(z)$ provides an analytic continuation of $f_1(z)$ to the region Γ_3 as defined in Eq. (A11.1). The statement of uniqueness is that $f_2(z) = f_3(z)$ in any further simply connected common region S (A11.4)

Thus *any* analytic continuation from the region R to the region S in Fig. A11.1b will give the same answer, provided that the union of the regions R and S in Fig. A11.1b is simply connected. Otherwise, by continuing in different directions about a *branch point* of a function, one can arrive at different results. The proof of uniqueness is as follows:

$$f_2(z) - f_3(z) \text{ is analytic in } R \oplus S \tag{A11.5}$$

This follows by assumption since $f_2(z)$ is assumed analytic in the region Γ_2 and $f_3(z)$ is analytic in Γ_3. Hence this difference is analytic in the intersection of the regions Γ_2 and Γ_3. Furthermore

$$f_2(z) - f_3(z) = 0 \qquad \text{in } R \tag{A11.6}$$

since both functions are assumed identical with $f_1(z)$ in the region R. Now the function $f_2(z) - f_3(z)$ is analytic in the region $R \oplus S$ but vanishes identically in the region R. We then invoke Eq. (A10.7) to conclude that

$$f_2(z) - f_3(z) = 0 \qquad \text{everywhere in } R \oplus S \tag{A11.7}$$

which is the statement of uniqueness.

Although the standard method of continuation works in principle for any function represented by a Taylor series, it often requires an infinite number of steps. Thus it is important to realize that other methods of continuation exist, and we here consider briefly the possibility of integral representations. One interesting example is the Laplace transform $\bar{f}(s)$ of a function $f(t)$ (see Sec. 59)

$$\bar{f}(s) = \int_0^\infty dt \, e^{-st} f(t) \qquad \text{Re } s > 0 \tag{A11.8}$$

If $t^{-N} f(t) \to 0$ as $t \to \infty$ for some real positive N, then $\bar{f}(s)$ is bounded and differentiable for real positive s. Thus the integral in Eq. (A11.8) can be taken to define an analytic function throughout the right half s plane, but it generally diverges for Re $s \le 0$. Suppose, however, that we can determine $\bar{f}(s)$ either independently or by doing the integral explicitly. The resulting representation typically remains analytic throughout a wider region of the s plane and therefore provides an analytic continuation into the left half plane. As an example of the former situation, where $f(t)$ is initially unknown, we recall the Laplace transform $s^{-1} \exp{(-s^{1/2}\xi)}$ from Eq. (59.15), which is analytic throughout the s plane with a branch cut on the negative real axis. An example of the latter possibility is $f(t) = 1$ when $\bar{f}(s) = s^{-1}$ is meromorphic with a simple pole at $s = 0$.

Among the many other examples of integral representations, we mention only the time Fourier transform (50.21), considered as a function of ω, and the integral in Eqs. (A9.17) and (A9.27)

$$I_1(\alpha) \equiv 2\int_0^\infty ds\, \frac{s^{2\alpha-1}}{s^2+1} = \frac{\pi}{\sin \pi\alpha} \qquad 0 < \text{Re } \alpha < 1 \qquad \text{(A11.9)}$$

Here the integral defines an analytic function of α in the indicated range. The right-hand side, however, is a meromorphic function of α throughout the entire α plane with simple poles at the integers. As expected, the singularities of the function $\pi/(\sin \pi\alpha)$ determine the convergence of the original representation. Note that $\pi/(\sin \pi\alpha)$ is periodic in α, which is in no way evident from the original integral representation. In a similar way, the coefficients in a Taylor series do not by themselves tell very much about the global properties of the full analytic function.

PROBLEMS

A.1 Consider the function $f(z) = [(1-z)(1+z^2)]^{1/2}$ with $f(0) = 1$. If z moves along a contour in the first quadrant from the origin to the point $z_0 = 2$, show that $f(z_0) = -i\sqrt{5}$.

A.2 Consider the following functions:

(a) $f(z) = [(1-z)(1+z^2)]^{1/2}$ (b) $f(z) = [(1-z)(1+z^2)]^{1/3}$

(c) $f(z) = \ln{(1-z^2)}$ (d) $f(z) = \ln \dfrac{1+z}{1-z}$

(e) $f(z) = \dfrac{\sin \sqrt{z}}{\sqrt{z}}$ (f) $f(z) = \dfrac{\cos \sqrt{z}}{\sqrt{z}}$

Where are the singularities in each case and what type of singularities are they? Consider the behavior of the function in circling each branch point, and consider all essentially different ways of drawing the branch lines in each case.

A.3 Determine the residues at the poles of the following functions:

(a) $z \csc z$ (b) $\dfrac{\cot \pi z}{(z-1)^2}$

(c) $\dfrac{1}{z(e^z - 1)}$ (d) $\dfrac{z^4}{(z^2+a^2)^2}$

A.4 Prove the following theorems, which we shall use frequently.

Theorem 1 Let $f(z, w)$ be a continuous function of the complex variables z and w where z ranges over a region Γ and w lies on a contour C. Let $f(z, w)$ be an analytic function of z in Γ, for every value of w on C. Then

$$F(z) = \int_C f(z, w)\, dw$$

is an analytic function of z in Γ and

$$F'(z) = \int_C \frac{\partial f}{\partial z}\, dw$$

and similarly for the higher derivatives.

Theorem 2 Let C be a smooth contour going to infinity. Suppose the conditions of the previous theorem are satisfied on any bounded part of C and that

$$\int_C f(z, w)\, dw$$

is uniformly convergent. Then the results of the previous theorem still hold.

A.5 This problem concerns *Jordan's lemma*.
(a) Consider the integral

$$I = \int_C dz\, f(z)e^{ikz} \qquad k > 0$$

where k is real and positive and C is a quarter circle of radius R in the first quadrant. Show that

$$|I| \le RM(R) \int_0^{\pi/2} e^{-kR \sin \phi}\, d\phi$$

where $M(R)$ is the maximum value of $|f(Re^{i\phi})|$ for $0 \le \phi \le \frac{1}{2}\pi$.
(b) Verify the inequality $2\phi/\pi \le \sin \phi$ for $0 \le \phi \le \frac{1}{2}\pi$ and hence show that $|I| \le \frac{1}{2}\pi k^{-1}M(R) \times (1 - e^{-kR})$. Thus obtain Jordan's lemma that $|I| \to 0$ as $R \to \infty$ if $M(R) \to 0$ in the same limit.
(c) Extend these results to a contour C which is a half circle in the upper half plane, as in Fig. A9.1. Show that $|I| \le \pi k^{-1}M(R)(1 - e^{-kR})$, where $M(R)$ is now the maximum value of $|f(Re^{i\phi})|$ for $0 \le \phi \le \pi$, and hence again draw the conclusion stated in (b).

A.6 Starting from

$$\oint_C \frac{e^{iz}}{z}\, dz = 0$$

where C is the contour in Fig. A9.2, show that

$$\int_0^\infty dx\, \frac{\sin x}{x} = \frac{\pi}{2}$$

(Be careful with all the contributions!)

A.7 Apply Jordan's lemma (Prob. A.5) to evaluate the definite integrals

$$\int_{-\infty}^\infty \frac{dx\, x \sin kx}{x^2 + a^2} = \pi e^{-ka} \qquad \text{and} \qquad \int_{-\infty}^\infty \frac{dx\, \cos kx}{x^2 + a^2} = \frac{\pi e^{-ka}}{a}$$

where a and k are real and positive. What happens for k negative?

A.8 A divergent integral of the form $\int_a^b dx\, f(x)(x - c)^{-1}$ with $a < c < b$ and $f(x)$ absolutely integrable is often replaced by its *principal value*

$$P \int_a^b dx\, \frac{f(x)}{x - c} \equiv \lim_{\epsilon \to 0} \left(\int_a^{c-\epsilon} + \int_{c+\epsilon}^b \right) dx\, \frac{f(x)}{x - c}$$

(a) Show that the principal value is well defined.

(b) Consider the complex integral

$$I(z) = \int_a^b dx \, \frac{f(x)}{x - z}$$

and prove that $I(z)$ is bounded for Im $z \neq 0$.

(c) If z approaches the real axis ($z \to c \pm i\eta$), use an appropriate contour deformation to verify that

$$I(c \pm i\eta) = P \int_a^b \frac{dx \, f(x)}{x - c} \pm i\pi f(c)$$

which is often expressed symbolically as $(x - c \mp i\eta)^{-1} = P(x - c)^{-1} \pm i\pi \, \delta(x - c)$.

A.9 Verify the following results with an appropriate contour integral:

(a) $\displaystyle\int_0^\infty dx \, \frac{x^{\alpha - 1}}{1 + x} = \frac{\pi}{\sin \pi\alpha}$ $\qquad 0 < \text{Re } \alpha < 1$

(b) $\displaystyle P \int_0^\infty dx \, \frac{x^{\alpha - 1}}{1 - x} = \pi \cot \pi\alpha$ $\qquad 0 < \text{Re } \alpha < 1$

(c) $\displaystyle\lim_{\epsilon \to 0} \int_{-\infty}^\infty \frac{dx}{2\pi i} \frac{e^{ixy}}{x - i\epsilon} = \theta(y) \equiv \begin{cases} 1 & y > 0 \\ 0 & y < 0 \end{cases}$

A.10 (a) Consider an integral of the form

$$I = \int_0^{2\pi} d\phi \, R(\cos \phi, \sin \phi)$$

where R is a rational function of its arguments. Show that the substitution $z = e^{i\phi}$ transforms I to a contour integral along the unit circle and discuss its evaluation with the residue theorem.

(b) Illustrate this technique for $R = (a + b \cos \phi)^{-1}$ with $a > b$.

A.11 Use an appropriate wedge-shaped contour to prove that $\int_0^\infty dx \, e^{\pm ix^2} = \frac{1}{2}\sqrt{\pi} \exp(\pm i\frac{1}{4}\pi)$ (these are related to the Fresnel integrals in optics).

A.12 The following techniques often allow us to sum a series in closed form.

(a) Show that the function $\pi \cot \pi z$ is meromorphic with simple poles and unit residues at the integers. If $f(n)$ is a rational function of n that vanishes at least as fast as n^{-2} for $n \to \infty$, explain how the contour integral $\int_C dz \, \pi \cot \pi z \, f(z)$ sums the series $\sum_n f(n)$, where C is a square contour with corners at $\pm(N + \frac{1}{2})$, $\pm(N + \frac{1}{2})$, $N \to \infty$, and n runs over all integers.

(b) Illustrate the method by deriving the result

$$\sum_{n=1}^\infty \frac{1}{n^2 + a^2} = \frac{\pi \coth \pi a}{2a} - \frac{1}{2a^2}$$

Hence obtain the relation $\sum_{n=1}^\infty n^{-2} = \frac{1}{6}\pi^2$ and a corresponding one for n^{-4}.

(c) What is the analogous result for the contour integral $\int_C dz \, \pi \csc \pi z \, f(z)$?

A.13 Evaluate the contour integral

$$\int_C \frac{dz}{z} \frac{1}{z^l - 1} \frac{1}{1 - p(z + z^{-1}) + p^2}$$

where C is a large circle of radius $R \to \infty$, l is an integer, and $0 < p < 1$. Hence sum the series

$$\sum_{n=1}^l \left(1 - 2p \cos \frac{2\pi n}{l} + p^2\right)^{-1} \qquad 0 < p < 1$$

SELECTED ADDITIONAL READINGS

1. G. F. Carrier, M. Krook, and C. E. Pearson, "Functions of a Complex Variable," McGraw-Hill, New York, 1966.
2. P. M. Morse and H. Feshbach, "Methods of Theoretical Physics," McGraw-Hill, New York, 1953.
3. E. C. Titchmarsh, "Theory of Functions," 2d ed., Oxford University Press, London, 1939.
4. E. T. Whittaker and G. N. Watson, "A Course of Modern Analysis," 4th ed., Cambridge University Press, London, 1969.

B

CURVILINEAR ORTHOGONAL COORDINATES†

Many developments in this text require the gradient, the divergence, and particularly the laplacian, which is the divergence of the gradient, in curvilinear coordinate systems. In this appendix, we develop a general procedure for defining these vector operations in an arbitrary curvilinear orthogonal coordinate system. A treatment of general tensors in such systems is discussed in the problems.

We start by considering a transformation from cartesian coordinates to a set of generalized coordinates (q_1, q_2, q_3). This transformation is specified by a set of analytic relations

$$x_i = x_i(q_1, q_2, q_3) \qquad i = 1, 2, 3 \tag{B.1a}$$

or their inverses

$$q_\lambda = q_\lambda(x_1, x_2, x_3) \qquad \lambda = 1, 2, 3 \tag{B.1b}$$

By this means a given point in space can be specified either by the triplet (x_1, x_2, x_3) or the triplet (q_1, q_2, q_3). We next introduce the three unit vectors $(\hat{q}_1, \hat{q}_2, \hat{q}_3)$, which are the normals to the surfaces $q_1 = $ const, $q_2 = $ const, and $q_3 = $ const, respectively, as illustrated in Fig. B.1. The present discussion is restricted to the case where these normals form an *orthogonal basis*; it will in general change from point to point in space.

To study this spatial variation, it is simplest to treat the change between two nearby points. For example, consider (q_1, q_2, q_3) and $(q_1 + dq_1, q_2, q_3)$ obtained from a small displacement of the coordinate q_1 with the coordinates q_2 and q_3 held fixed. The vector displacement joining these two points will be denoted $d\mathbf{l}_1$. For an infinitesimal change dq_1 the vector displacement $d\mathbf{l}_1$ will be proportional to dq_1; since q_2 and q_3 are constant, the displacement will be in the $\hat{q}_1$ direction, as indicated in Fig. B.1. Thus for

$$q_1 \rightarrow q_1 + dq_1 \tag{B.2}$$

† Morse and Feshbach [1], secs. 1.3, 1.4, and 5.1.

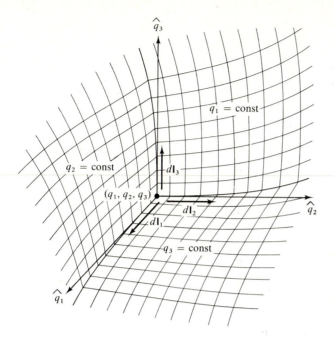

Figure B.1 Orthogonal curvilinear coordinates.

the vector displacement must have the form

$$dl_1 \equiv h_1(q_1, q_2, q_3)\hat{q}_1 \, dq_1 \tag{B.3}$$

where the constant of proportionality $h_1(q_1, q_2, q_3)$ can, in general, be obtained from a good figure and simple geometric considerations. (See Prob. B.1 for an alternative analytical approach.) Expressions similar to Eqs. (B.2) and (B.3) hold for the other two orthogonal directions. A general infinitesimal vector displacement dl can therefore be written as the sum of the three orthogonal components according to

$$dl = h_1\hat{q}_1 \, dq_1 + h_2\hat{q}_2 \, dq_2 + h_3\hat{q}_3 \, dq_3 \tag{B.4}$$

This quantity is known as the *line element*. The constants of proportionality h_1, h_2, and h_3 are all to be evaluated at the original point (q_1, q_2, q_3). Equation (B.4) is *exact to first order in the differentials* (dq_1, dq_2, dq_3) *for an arbitrary orthogonal curvilinear coordinate system.* We shall show that knowledge of the line element suffices to establish all the required vector relations.

Figure B.2 provides some familiar examples. For spherical coordinates, the appropriate unit vectors are shown in Fig. B.2a, and the line element becomes

$$dl = \hat{r} \, dr + \hat{\theta} \, r \, d\theta + \hat{\phi} \, r \sin \theta \, d\phi \tag{B.5}$$

In cylindrical coordinates (Fig. B.2b) the line element is

$$dl = \hat{\rho} \, d\rho + \hat{\phi} \, \rho \, d\phi + \hat{z} \, dz \tag{B.6}$$

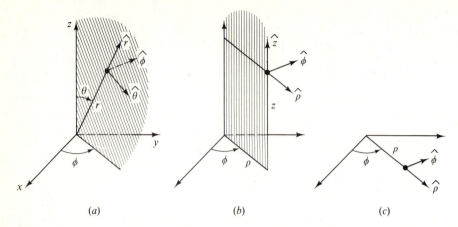

Figure B.2 (a) Spherical, (b) cylindrical, and (c) polar coordinate systems.

Polar coordinates in two dimensions represent a special case of cylindrical coordinates with the coordinate z fixed. This situation is illustrated in Fig. B.2c, and the line element takes the form

$$d\mathbf{l} = \hat{\rho}\, d\rho + \hat{\phi}\, \rho\, d\phi \tag{B.7}$$

Oblate spheroidal coordinates are discussed in Prob. B.2, and Morse and Feshbach [1] treat other curvilinear orthogonal coordinate systems in detail.

Gradient

Consider a scalar function Φ of the coordinates (q_1, q_2, q_3). Under a small vector displacement $d\mathbf{l}$, the change in the function $d\Phi$ is linear in $d\mathbf{l}$, and the coefficients define the gradient of this scalar function $\nabla\Phi$ according to the relation

$$d\Phi \equiv \nabla\Phi \cdot d\mathbf{l} \tag{B.8}$$

Alternatively, standard partial differentiation yields the result

$$d\Phi(q_1, q_2, q_3) = \frac{\partial\Phi}{\partial q_1}\, dq_1 + \frac{\partial\Phi}{\partial q_2}\, dq_2 + \frac{\partial\Phi}{\partial q_3}\, dq_3 \tag{B.9}$$

If the independent coordinates (q_1, q_2, q_3) are varied one at a time, a combination of Eqs. (B.9), (B.8), and (B.4) establishes the following general form of the gradient

$$\nabla\Phi(q_1, q_2, q_3) = \frac{1}{h_1}\frac{\partial\Phi}{\partial q_1}\,\hat{q}_1 + \frac{1}{h_2}\frac{\partial\Phi}{\partial q_2}\,\hat{q}_2 + \frac{1}{h_3}\frac{\partial\Phi}{\partial q_3}\,\hat{q}_3 \tag{B.10}$$

It is evident that Eqs. (B.4) and (B.10) reproduce Eqs. (B.8) and (B.9).

Divergence

Consider a vector field $\mathbf{a}$. Its divergence $\nabla \cdot \mathbf{a}$ in an arbitrary curvilinear orthogonal coordinate system can be defined by Gauss' theorem

$$\int_V (\nabla \cdot \mathbf{a})\, dV = \int_S \mathbf{a} \cdot d\mathbf{S} \tag{B.11}$$

We compute the right-hand side of Eq. (B.11) for an infinitesimal rectangular volume element associated with the point (q_1, q_2, q_3). This volume element has edges with lengths equal to the appropriate components of the line element in Eq. (B.4) and surfaces perpendicular to the unit vectors $(\hat{q}_1, \hat{q}_2, \hat{q}_3)$ as illustrated in Fig. B.3. An element of surface area dS has a magnitude equal to the differential surface area and a direction along the outward normal to that surface.

We first compute the contribution to the integral on the right-hand side of Eq. (B.11) from that part of the surface pointing in the direction $\hat{q}_1$. This quantity is simply the first component of the vector $a_1 \equiv \hat{q}_1 \cdot \mathbf{a}$ multiplied by the area of the surfaces that are perpendicular to the direction $\hat{q}_1$. Figure B.3 immediately gives the result

$$\int_{\substack{\text{surfaces} \perp \\ \text{to } \hat{q}_1}} \mathbf{a} \cdot d\mathbf{S} = a_1 h_2 h_3 \, dq_2 \, dq_3 \bigg|_{q_1 + dq_1} - a_1 h_2 h_3 \, dq_2 \, dq_3 \bigg|_{q_1} \qquad (B.12)$$

This expression is already of first order in $dq_2 \, dq_3$. Thus we can keep the variables (q_2, q_3) *fixed* when evaluating the right-hand side, since any variation of these quantities from their fixed values would produce contributions of higher order in the differentials (dq_2, dq_3). To leading order in the differentials (dq_1, dq_2, dq_3), the right-hand side of Eq. (B.12) therefore becomes

$$\text{rhs} = \frac{1}{h_1 h_2 h_3} \left(\frac{\partial}{\partial q_1} a_1 h_2 h_3 \right) (h_1 h_2 h_3 \, dq_1 \, dq_2 \, dq_3) \qquad (B.13)$$

Here the partial derivative indicates that this expression is evaluated for fixed q_2 and q_3. Since the final quantity in parentheses in Eq. (B.13) is just the differential volume element

$$dV = h_1 h_2 h_3 \, dq_1 \, dq_2 \, dq_3 \qquad (B.14)$$

the contribution of this part of the surface integral on the right-hand side of Eq. (B.11) has

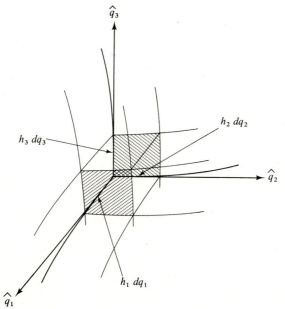

Figure B.3 Volume element used in defining the divergence.

now been put into the form that appears on the left-hand side of Eq. (B.11). The other additive contributions to the integral on the right-hand side of Eq. (B.11) come from the surfaces perpendicular to $\hat{q}_2$ and to $\hat{q}_3$; they are treated in precisely the same manner. Hence the divergence of **a** follows from Eqs. (B.11), (B.13), (B.14) and the corresponding additive contributions from the other sides:

$$\mathbf{V} \cdot \mathbf{a} = \frac{1}{h_1 h_2 h_3} \left(\frac{\partial}{\partial q_1} a_1 h_2 h_3 + \frac{\partial}{\partial q_2} h_1 a_2 h_3 + \frac{\partial}{\partial q_3} h_1 h_2 a_3 \right) \tag{B.15}$$

Laplacian

In curvilinear orthogonal coordinates, the laplacian of a scalar function is defined as the divergence of the gradient of that function

$$\nabla^2 \Phi \equiv \mathbf{V} \cdot (\nabla \Phi) \tag{B.16}$$

This quantity is immediately obtained from Eq. (B.15) by substituting for **a** the particular vector in Eq. (B.10); the result is

$$\nabla^2 \Phi = \frac{1}{h_1 h_2 h_3} \left[\frac{\partial}{\partial q_1} \left(\frac{h_2 h_3}{h_1} \frac{\partial \Phi}{\partial q_1} \right) + \frac{\partial}{\partial q_2} \left(\frac{h_1 h_3}{h_2} \frac{\partial \Phi}{\partial q_2} \right) + \frac{\partial}{\partial q_3} \left(\frac{h_1 h_2}{h_3} \frac{\partial \Phi}{\partial q_3} \right) \right] \tag{B.17}$$

Alternatively, Eq. (B.17) can be derived algebraically by starting with the expression for the laplacian in cartesian coordinates

$$\nabla^2 = \frac{\partial^2}{\partial x^2} + \frac{\partial^2}{\partial y^2} + \frac{\partial^2}{\partial z^2} \tag{B.18}$$

Use of the algebraic relations (B.1) and the rules for partial differentiation of implicit functions always provides a good, though generally extremely long, check of Eq. (B.17). Either procedure must, of course, give the same answer.

We proceed to give some examples.

Spherical Coordinates

Comparison of the line element (B.5) with the general form (B.4) immediately identifies

$$h_r = 1 \qquad h_\theta = r \qquad h_\phi = r \sin \theta \tag{B.19}$$

Thus the laplacian in spherical coordinates takes the form from (B.17)

$$\nabla^2 = \frac{1}{r^2 \sin \theta} \left[\frac{\partial}{\partial r} \left(r^2 \sin \theta \frac{\partial}{\partial r} \right) + \frac{\partial}{\partial \theta} \left(\frac{r \sin \theta}{r} \frac{\partial}{\partial \theta} \right) + \frac{\partial}{\partial \phi} \left(\frac{r}{r \sin \theta} \frac{\partial}{\partial \phi} \right) \right] \tag{B.20}$$

Since the other variables are held fixed when carrying out the indicated partial differentiations, this expression can be reduced to

$$\nabla^2 = \frac{1}{r^2} \frac{\partial}{\partial r} r^2 \frac{\partial}{\partial r} + \frac{1}{r^2 \sin \theta} \frac{\partial}{\partial \theta} \sin \theta \frac{\partial}{\partial \theta} + \frac{1}{r^2 \sin^2 \theta} \frac{\partial^2}{\partial \phi^2} \tag{B.21}$$

Cylindrical Coordinates

Comparison of the line element (B.6) with the general form (B.4) leads to the identification

$$h_\rho = 1 \qquad h_\phi = \rho \qquad h_z = 1 \tag{B.22}$$

The laplacian in Eq. (B.17) thus becomes

$$\nabla^2 = \frac{1}{\rho}\left[\frac{\partial}{\partial\rho}\left(\rho\frac{\partial}{\partial\rho}\right) + \frac{\partial}{\partial\phi}\left(\frac{1}{\rho}\frac{\partial}{\partial\phi}\right) + \frac{\partial}{\partial z}\left(\rho\frac{\partial}{\partial z}\right)\right] \tag{B.23}$$

Taking account of the variables held fixed in carrying out the indicated partial derivatives, we find

$$\nabla^2 = \frac{1}{\rho}\frac{\partial}{\partial\rho}\rho\frac{\partial}{\partial\rho} + \frac{1}{\rho^2}\frac{\partial^2}{\partial\phi^2} + \frac{\partial^2}{\partial z^2} \tag{B.24}$$

Polar Coordinates in Two Dimensions

As illustrated in Fig. B.2c, this situation can be obtained from cylindrical coordinates simply by keeping

$$z = \text{const} = 0 \tag{B.25}$$

Thus we obtain

$$\nabla^2 = \frac{1}{\rho}\frac{\partial}{\partial\rho}\rho\frac{\partial}{\partial\rho} + \frac{1}{\rho^2}\frac{\partial^2}{\partial\phi^2} \tag{B.26}$$

Specialization of the arguments in this appendix from three to two dimensions and the direct derivation of Eq. (B.26) in polar coordinates from the line element in Eq. (B.7) are left as an exercise (Prob. B.3).

The preceding analysis suffices for the scalar wave equation, which describes the dynamics of membranes and irrotational motion of nonviscous fluids, such as sound waves or surface waves. For more general phenomena, however, the vector character of the relevant field plays a crucial role, as seen in the dynamical equations of viscous fluids [Eq. (60.26)] or elastic continua [Eq. (64.2)]. These expressions are most simply evaluated in curvilinear coordinates through the use of the vector identities

$$(\mathbf{V}\cdot\nabla)\mathbf{V} = \nabla(\tfrac{1}{2}V^2) - \mathbf{V}\times(\nabla\times\mathbf{V}) \tag{B.27}$$

$$\nabla^2\mathbf{V} = \nabla(\nabla\cdot\mathbf{V}) - \nabla\times(\nabla\times\mathbf{V}) \tag{B.28}$$

which are readily proved in cartesian coordinates. In addition to the familiar operations grad and div, which have already been evaluated in Eqs. (B.10) and (B.15), the right-hand sides involve the curl. This quantity can be evaluated with Stokes' theorem to give (see Prob. B.4)

$$(\nabla\times\mathbf{V})_1 = \frac{1}{h_2 h_3}\left(\frac{\partial}{\partial q_2}h_3 V_3 - \frac{\partial}{\partial q_3}h_2 V_2\right) \tag{B.29}$$

with the remaining components found by cyclic permutation. Straightforward but tedious manipulations then allow us to express the vector field equations for viscous fluids or elastic continua in an arbitrary orthogonal curvilinear coordinate system. We may note that the three components of $\nabla^2\mathbf{V}$ generally differ from the quantities $\nabla^2 V_1$, $\nabla^2 V_2$, and $\nabla^2 V_3$ because of the spatial variation of the unit vectors $\hat{q}_1$, $\hat{q}_2$, and $\hat{q}_3$.

A complete study of a physical system frequently requires not only the solution of the dynamical equations but also the imposition of boundary conditions. For this reason, we also must be able to transform the symmetric tensor formed from the vector field $\mathbf{V}(\mathbf{x})$

$$S_{ij} = \frac{\partial V_i}{\partial x_j} + \frac{\partial V_j}{\partial x_i} \tag{B.30}$$

that appears in the viscous stress tensor [Eq. (60.17)] and in the elastic stress [Eq. (63.21)]. The corresponding curvilinear components are given in Prob. B.7.

Problems

B.1 An infinitesimal vector $d\mathbf{x}$ can be characterized either in a cartesian basis $\sum_i \hat{x}_i \, dx_i$ or in a curvilinear basis $\sum_\lambda h_\lambda \hat{q}_\lambda \, dq_\lambda$, where $x_i = x_i(q_1, q_2, q_3)$ determines the coordinate transformation.

(a) Use the invariance of the squared length $d\mathbf{x} \cdot d\mathbf{x}$ to show that

$$\sum_i \frac{\partial x_i}{\partial q_\lambda} \frac{\partial x_i}{\partial q_\mu} = \delta_{\lambda\mu} h_\lambda^2$$

This result provides an easy analytical means to find h_λ.

(b) Apply it to find h for cylindrical polar coordinates $x_1 = \rho \cos \phi$, $x_2 = \rho \sin \phi$, $x_3 = z$ and for spherical polar coordinates $x_1 = r \sin \theta \cos \phi$, $x_2 = r \sin \theta \sin \phi$, $x_3 = r \cos \theta$.

B.2 Consider *oblate spheroidal coordinates* defined by

$$x = a \cosh u \cos v \cos \phi$$

$$y = a \cosh u \cos v \sin \phi$$

$$z = a \sinh u \sin v$$

(a) Show that the coordinate surfaces are (i) oblate spheroids, $u = \text{const}$, $0 \leq u \leq \infty$; (ii) hyperboloids of one sheet, $v = \text{const}$, $-\pi/2 \leq v \leq \pi/2$; (iii) half planes through the z axis, $\phi = \text{const}$, $0 \leq \phi \leq 2\pi$; and that these surfaces generate an orthogonal curvilinear coordinate system with

$$h_u = h_v = a(\sinh^2 u + \sin^2 v)^{1/2}$$

$$h_\phi = a \cosh u \cos v$$

(b) Introducing $\xi = \sin v$ and $\zeta = \sinh u$, show that Laplace's equation for $\Phi(\xi, \zeta, \phi)$ becomes

$$\frac{\partial}{\partial \xi}\left[(1 - \xi^2)\frac{\partial \Phi}{\partial \xi}\right] + \frac{\partial}{\partial \zeta}\left[(1 + \zeta^2)\frac{\partial \Phi}{\partial \zeta}\right] + \frac{\xi^2 + \zeta^2}{(1 - \xi^2)(1 + \zeta^2)}\frac{\partial^2 \Phi}{\partial \phi^2} = 0$$

(c) Separate variables, and show that the solutions to this equation are

$$\begin{Bmatrix} P_n^m(\xi) \\ Q_n^m(\xi) \end{Bmatrix} \times \begin{Bmatrix} P_n^m(i\zeta) \\ Q_n^m(i\zeta) \end{Bmatrix} \times \begin{Bmatrix} \cos m\phi \\ \sin m\phi \end{Bmatrix}$$

(d) Show that the Helmholtz equation also separates in these coordinates.

B.3 Polar coordinates (ρ, ϕ) in two dimensions are defined by the equations $x = \rho \cos \phi$, $y = \rho \sin \phi$.

(a) Show that the coordinate surfaces $\phi = \text{const}$, $0 \leq \phi \leq 2\pi$ and $\rho = \text{const}$, $0 \leq \rho \leq \infty$ generate a two-dimensional orthogonal curvilinear coordinate system with $h_\rho = 1$, $h_\phi = \rho$.

(b) Use the general relations (B.10), (B.15), and (B.17) to write out the gradient, the divergence, and the laplacian in these coordinates.

(c) Show by direct algebraic transformation of Eq. (B.18) that the laplacian is that given in Eq. (B.26).

B.4 Stokes' theorem asserts that $\int_{\text{surface}} d\mathbf{S} \cdot (\nabla \times \mathbf{V}) = \oint_{\text{boundary}} d\mathbf{l} \cdot \mathbf{V}$, where the first integral is taken over an arbitrary surface and the second (line) integral is taken around the closed curve bounding that surface.

(a) Apply this formula to an infinitesimal rectangular area with sides $h_1 \, dq_1 \, \hat{q}_1$ and $h_2 \, dq_2 \, \hat{q}_2$ to obtain

$$(\nabla \times \mathbf{V})_3 = \frac{1}{h_1 h_2}\left(\frac{\partial}{\partial q_1} h_2 V_2 - \frac{\partial}{\partial q_2} h_1 V_1\right)$$

(b) In cylindrical polar coordinates, use this result and Eq. (B.15) to prove that

$$\nabla \cdot \mathbf{V} = \frac{1}{\rho} \frac{\partial}{\partial \rho} \rho V_\rho + \frac{1}{\rho} \frac{\partial V_\phi}{\partial \phi} + \frac{\partial V_z}{\partial z}$$

$$(\nabla \times \mathbf{V})_\rho = \frac{1}{\rho} \frac{\partial V_z}{\partial \phi} - \frac{\partial V_\phi}{\partial z}$$

$$(\nabla \times \mathbf{V})_\phi = \frac{\partial V_\rho}{\partial z} - \frac{\partial V_z}{\partial \rho}$$

$$(\nabla \times \mathbf{V})_z = \frac{1}{\rho} \frac{\partial}{\partial \rho} \rho V_\phi - \frac{1}{\rho} \frac{\partial V_\rho}{\partial \phi}$$

(c) In spherical polar coordinates, prove that

$$\nabla \cdot \mathbf{V} = \frac{1}{r^2} \frac{\partial}{\partial r} r^2 V_r + \frac{1}{r \sin \theta} \frac{\partial}{\partial \theta} (\sin \theta \, V_\theta) + \frac{1}{r \sin \theta} \frac{\partial V_\phi}{\partial \phi}$$

$$(\nabla \times \mathbf{V})_r = \frac{1}{r \sin \theta} \left[\frac{\partial}{\partial \theta} (\sin \theta \, V_\phi) - \frac{\partial V_\theta}{\partial \phi} \right]$$

$$(\nabla \times \mathbf{V})_\theta = \frac{1}{r \sin \theta} \frac{\partial V_r}{\partial \phi} - \frac{1}{r} \frac{\partial}{\partial r} r V_\phi$$

$$(\nabla \times \mathbf{V})_\phi = \frac{1}{r} \left(\frac{\partial}{\partial r} r V_\theta - \frac{\partial V_r}{\partial \theta} \right)$$

B.5 (a) A vector $\mathbf{V}$ can be defined by its components V_i on an arbitrary cartesian basis $\hat{x}_i$ through the familiar relation $\mathbf{V} = \sum_i V_i \hat{x}_i$. Show that its curvilinear components $V_\lambda \equiv \mathbf{V} \cdot \hat{q}_\lambda$ and its cartesian components satisfy the relations $V_\lambda = \sum_i (\hat{q}_\lambda \cdot \hat{x}_i) V_i$ and $V_i = \sum_\lambda (\hat{q}_\lambda \cdot \hat{x}_i) V_\lambda$.

(b) Similarly a tensor of arbitrary rank can be defined by its cartesian components $T_{ij \dots k}$ according to $T = \sum_{ij \dots k} T_{ij \dots k} \, \hat{x}_i \hat{x}_j \cdots \hat{x}_k$. Evidently, the components of T transform under spatial rotations like the appropriate direct product of vectors. Show that the curvilinear components of a second-rank tensor are given by

$$T_{\lambda \mu} = \sum_{ij} (\hat{q}_\lambda \cdot \hat{x}_i)(\hat{q}_\mu \cdot \hat{x}_j) T_{ij}$$

B.6 (a) Use the completeness of the orthonormal triad $\{\hat{x}_i\}$ to obtain the relation $\hat{q}_\lambda \cdot \hat{x}_i = (1/h_\lambda) \times (\partial x_i / \partial q_\lambda)$, where $x_i = x_i(q_1 \, q_2 , q_3)$.

(b) Assume that the transformation can be inverted to give $q_\lambda = q_\lambda(x_1, x_2, x_3)$. Use the completeness of the orthonormal triad $\{\hat{q}_\lambda\}$ to obtain the alternative relations $\hat{q}_\lambda \cdot \hat{x}_i = h_\lambda(\partial q_\lambda / \partial x_i)$.

B.7 Apply the results of Probs. B.5 and B.6 to the symmetric tensor $S_{ij} = \partial V_i / \partial x_j + \partial V_j / \partial x_i$ that appears in the stress tensor for viscous fluids and isotropic elastic continua.

(a) Justify the following sequence of steps

$$\frac{\partial V_i}{\partial x_j} = \sum_{\sigma \tau} \frac{\partial}{\partial q_\sigma} [V_\tau (\hat{q}_\tau \cdot \hat{x}_i)] \frac{\partial q_\sigma}{\partial x_j}$$

$$= \sum_{\sigma \tau} \left(\frac{\partial}{\partial q_\sigma} \frac{V_\tau}{h_\tau} \right) \frac{h_\tau}{h_\sigma} (\hat{q}_\tau \cdot \hat{x}_i)(\hat{q}_\sigma \cdot \hat{x}_j) + \sum_{\sigma \tau} \frac{V_\tau}{h_\tau} \frac{\partial^2 x_i}{\partial q_\tau \, \partial q_\sigma} \frac{1}{h_\sigma} (\hat{q}_\sigma \cdot \hat{x}_j)$$

$$= \sum_{\sigma \tau} \left(\frac{\partial}{\partial q_\sigma} \frac{V_\tau}{h_\tau} \right) \frac{h_\tau}{h_\sigma} (\hat{q}_\tau \cdot \hat{x}_i)(\hat{q}_\sigma \cdot \hat{x}_j)$$

$$+ \sum_{\sigma \tau} \frac{V_\tau}{h_\tau} \left[\frac{1}{h_\sigma} \frac{\partial h_\sigma}{\partial q_\tau} (\hat{q}_\sigma \cdot \hat{x}_i) + \frac{\partial}{\partial q_\tau} (\hat{q}_\sigma \cdot \hat{x}_i) \right] (\hat{q}_\sigma \cdot \hat{x}_j)$$

(b) Use the completeness of the orthonormal triad $\{\hat{x}_i\}$ and part (b) of Prob. B.5 to prove that

$$S_{\lambda\mu} = \frac{h_\lambda}{h_\mu}\frac{\partial}{\partial q_\mu}\frac{V_\lambda}{h_\lambda} + \frac{h_\mu}{h_\lambda}\frac{\partial}{\partial q_\lambda}\frac{V_\mu}{h_\mu} + \delta_{\lambda\mu}\frac{2}{h_\lambda}\sum_\rho\frac{V_\rho}{h_\rho}\frac{\partial}{\partial q_\rho}h_\lambda$$

B.8 (a) Use the general expressions from Probs. B.6 and B.7 to obtain the six independent components of $S_{\lambda\mu}$ in cylindrical polar coordinates (ρ, ϕ, z)

$$S_{\rho\rho} = 2\frac{\partial V_\rho}{\partial \rho}$$

$$S_{\phi\rho} = S_{\rho\phi} = \frac{1}{\rho}\frac{\partial V_\rho}{\partial \phi} + \frac{\partial V_\phi}{\partial \rho} - \frac{V_\phi}{\rho}$$

$$S_{\phi\phi} = \frac{2}{\rho}\frac{\partial V_\phi}{\partial \phi} + \frac{2V_\rho}{\rho}$$

$$S_{z\rho} = S_{\rho z} = \frac{\partial V_\rho}{\partial z} + \frac{\partial V_z}{\partial \rho}$$

$$S_{z\phi} = S_{\phi z} = \frac{\partial V_\phi}{\partial z} + \frac{1}{\rho}\frac{\partial V_z}{\partial \phi}$$

$$S_{zz} = 2\frac{\partial V_z}{\partial z}$$

Hence find the components of the viscous stress tensor for an incompressible fluid [Eq. (60.18)] and the elastic stress tensor [Eq. (63.21b)] in cylindrical polar coordinates.

(b) Repeat for spherical polar coordinates (r, θ, ϕ)

$$S_{rr} = 2\frac{\partial V_r}{\partial r}$$

$$S_{\theta r} = S_{r\theta} = \frac{1}{r}\frac{\partial V_r}{\partial \theta} + \frac{\partial V_\theta}{\partial r} - \frac{V_\theta}{r}$$

$$S_{\theta\theta} = \frac{2}{r}\frac{\partial V_\theta}{\partial \theta} + \frac{2V_r}{r}$$

$$S_{\phi r} = S_{r\phi} = \frac{1}{r\sin\theta}\frac{\partial V_r}{\partial \phi} + \frac{\partial V_\phi}{\partial r} - \frac{V_\phi}{r}$$

$$S_{\phi\theta} = S_{\theta\phi} = \frac{1}{r\sin\theta}\frac{\partial V_\theta}{\partial \phi} + \frac{1}{r}\frac{\partial V_\phi}{\partial \theta} - \frac{V_\phi\cot\theta}{r}$$

$$S_{\phi\phi} = \frac{2}{r\sin\theta}\frac{\partial V_\phi}{\partial \phi} + \frac{2V_r}{r} + \frac{2\cot\theta\, V_\theta}{r}$$

SELECTED ADDITIONAL READING

1. P. M. Morse and H. Feshbach, "Methods of Theoretical Physics," McGraw-Hill, New York, 1953.

SEPARATION OF VARIABLES

In this text, the standard approach for partial differential equations is to look for solutions in separated form

$$\psi(q_1, q_2, q_3) = \psi_1(q_1)\psi_2(q_2)\psi_3(q_3) \tag{C.1}$$

and then to construct the general solutions by superposition. Assuming a curvilinear orthogonal coordinate system (see Appendix B), we here discuss the separation of variables in the most common examples.

Normal Modes in Polar Coordinates (Two Dimensions)

The normal modes of two-dimensional systems in polar coordinates (see Fig. B.2c) satisfy the Helmholtz equation

$$(\nabla^2 + k^2)\psi(\rho, \phi) = 0 \tag{C.2}$$

where the laplacian in polar coordinates is given by Eq. (B.26). We seek a solution to Eq. (C.2) in separated form

$$\psi = R(\rho)\Phi(\phi) \tag{C.3}$$

Substitute this expression into Eq. (C.2) and then divide by ψ. The result is

$$\frac{1}{R}\left(\frac{1}{\rho}\frac{d}{d\rho}\,\rho\,\frac{dR}{d\rho}\right) + \frac{1}{\rho^2}\left[\frac{1}{\Phi}\frac{d^2\Phi}{d\phi^2}\right] + k^2 = 0 \tag{C.4}$$

The quantity in brackets in Eq. (C.4) depends only on ϕ, and the remaining terms in Eq. (C.4) depend only on ρ. By hypothesis, this equation must hold for all ϕ and all ρ, which requires that the factor in brackets be a constant, say $-m^2$. We then find

$$\frac{d^2\Phi}{d\phi^2} = -m^2\Phi \tag{C.5}$$

where m^2 is called the *separation constant*. To this point, it is completely arbitrary. Equation (C.5) is now an ordinary differential equation in one variable. It is already in the Sturm-Liouville form and has the simple general solution

$$\Phi = A \sin m\phi + B \cos m\phi \qquad (C.6a)$$

In any physical problem where the geometry allows the polar angle to increase by 2π and return to its original value (see Fig. B.2c), the function Φ must also return to its original value because any disturbance in classical physics must be a single-valued function of position. Expression (C.6a) will be single-valued only if m is an integer, and we therefore conclude that

$$m \text{ is an integer if } \Phi(\phi) \text{ is to be single-valued in the range } 0 \le \phi \le 2\pi \qquad (C.6b)$$

If Eq. (C.5) is substituted into Eq. (C.4), the remaining radial equation becomes

$$\frac{d^2R}{d\rho^2} + \frac{1}{\rho}\frac{dR}{d\rho} + \frac{k^2\rho^2 - m^2}{\rho^2} R = 0 \qquad (C.7)$$

This result can be rewritten in several ways. For some purposes the most useful is the Sturm-Liouville form

$$-\frac{d}{d\rho}\left(\rho \frac{dR}{d\rho}\right) + \frac{m^2}{\rho} R = k^2 \rho R \qquad (C.8)$$

which exhibits k^2 as the associated eigenvalue. Alternatively, the introduction of the dimensionless variable

$$z \equiv k\rho \qquad (C.9)$$

transforms Eq. (C.7) into *Bessel's equation*

$$z^2 \frac{d^2R}{dz^2} + z \frac{dR}{dz} + (z^2 - m^2)R = 0 \qquad (C.10)$$

Appendix D discusses the fundamental system of solutions to this equation.

Normal Modes in Spherical Coordinates (Three Dimensions)

The Helmholtz equation in spherical coordinates (see Fig. B.2a) takes the form

$$(\nabla^2 + k^2)\psi(r, \theta, \phi) = 0 \qquad (C.11)$$

where the laplacian in spherical coordinates is given by Eq. (B.21). We seek separated solutions of the form

$$\psi = R(r)P(\theta)\Phi(\phi) \qquad (C.12)$$

Substitution of this result in Eq. (C.11) and division by ψ yields

$$\frac{1}{R}\left(\frac{1}{r^2}\frac{d}{dr}r^2\frac{dR}{dr}\right) + \frac{1}{r^2}\left[\frac{1}{P}\frac{1}{\sin\theta}\frac{d}{d\theta}\sin\theta\frac{dP}{d\theta} + \frac{1}{\sin^2\theta}\left(\frac{1}{\Phi}\frac{d^2\Phi}{d\phi^2}\right)\right] + k^2 = 0 \qquad (C.13)$$

The final expression in parentheses depends only on ϕ, and the remaining terms in Eq. (C.13) depend only on r and θ. This relation can hold for all (r, θ, ϕ) only if the final expression in parentheses is, in fact, a constant, leading to an expression identical with Eq. (C.5) above

$$\frac{d^2\Phi}{d\phi^2} = -m^2\Phi \tag{C.14}$$

Again m must be an integer if the disturbance is to be a single-valued function of ϕ in the range $0 \le \phi \le 2\pi$.

If Eq. (C.14) is substituted into Eq. (C.13), the final quantity in brackets depends only on θ and the remainder of the equation depends only on r. The only way that this relation can hold for all r and θ is for the final quantity in brackets again to be a constant, which we denote by $-\lambda$

$$\frac{1}{\sin\theta}\frac{d}{d\theta}\left(\sin\theta\frac{dP}{d\theta}\right) - \frac{m^2}{\sin^2\theta}P = -\lambda P \tag{C.15}$$

At this point, the separation constant λ in the polar-angle equation is still completely arbitrary.

It is often preferable to rewrite (C.15) with the change of variables

$$x = \cos\theta \tag{C.16}$$

Use of the chain rule of differentiation

$$\frac{dP}{d\theta} = \frac{dP}{dx}\frac{dx}{d\theta} = -\sin\theta\frac{dP}{dx} \tag{C.17}$$

casts this equation in Sturm-Liouville form

$$-\frac{d}{dx}\left[(1-x^2)\frac{dP}{dx}\right] + \frac{m^2}{1-x^2}P = \lambda P \tag{C.18}$$

with the separation constant λ as the eigenvalue. If the values $\theta = 0$ and $\theta = \pi$ in Fig. B.2a are accessible regions in a given physical situation, the corresponding classical disturbance must be finite at these values of θ. This requirement of finite amplitude on the positive and negative z axis will be shown (Appendix D) to yield the following condition on the eigenvalue

$$\lambda = l(l+1) \qquad l = 0, 1, 2, 3, \ldots \tag{C.19}$$

where l must be an integer. Imposition of natural boundary conditions

$$(1-x^2)\frac{dP}{dx} = 0 \qquad x = \pm 1 \tag{C.20}$$

in the Sturm-Liouville problem (C.18) leads to the identical condition on the eigenvalues.

Equation (C.18) can also be written as

$$(1-x^2)\frac{d^2P}{dx^2} - 2x\frac{dP}{dx} + \left(\lambda - \frac{m^2}{1-x^2}\right)P = 0 \tag{C.21}$$

which is known as the *associated Legendre equation*. For $m = 0$, it is simply called *Legendre's equation*

$$(1-x^2)\frac{d^2P}{dx^2} - 2x\frac{dP}{dx} + \lambda P = 0 \tag{C.22}$$

A fundamental system of solutions to this equation and to Eq. (C.18) for integral m is constructed in Appendix D.

Assume that the eigenvalue condition (C.19) is satisfied. If Eq. (C.15) is substituted into Eq. (C.13), the resulting radial equation takes the form

$$r^2 \frac{d^2 R}{dr^2} + 2r \frac{dR}{dr} + [k^2 r^2 - l(l+1)]R = 0 \tag{C.23}$$

It is convenient to redefine the radial function as

$$R(r) \equiv \frac{f(r)}{r^{1/2}} \tag{C.24}$$

Substitution into Eq. (C.23) gives the equivalent form

$$r^2 \frac{d^2 f}{dr^2} + r \frac{df}{dr} + [k^2 r^2 - (l + \tfrac{1}{2})^2]f = 0 \tag{C.25}$$

which is just Bessel's equation (C.10) once again. The only difference is that here the index $l + \tfrac{1}{2}$ is now half an odd integer, whereas before the index m was an integer. The solutions (C.24) to the radial equation (C.23) are called *spherical Bessel functions*. A fundamental system of spherical Bessel functions will be discussed in Appendix D.

Normal Modes in Cylindrical Coordinates (Three Dimensions)

The scalar Helmholtz equation in cylindrical coordinates (see Fig. B.2b) is given by

$$(\nabla^2 + k^2)\psi(\rho, \phi, z) = 0 \tag{C.26}$$

where the laplacian in cylindrical coordinates is given by Eq. (B.24). We seek separated solutions of the form

$$\psi(\rho, \phi, z) = R(\rho)\Phi(\phi)Z(z) \tag{C.27}$$

Substitute this expression into Eq. (C.26) and divide by ψ

$$\frac{1}{R}\left(\frac{1}{\rho}\frac{d}{d\rho}\rho\frac{dR}{d\rho}\right) + \frac{1}{\rho^2}\left(\frac{1}{\Phi}\frac{d^2\Phi}{d\phi^2}\right) + \frac{1}{Z}\left(\frac{d^2Z}{dz^2}\right) + k^2 = 0 \tag{C.28}$$

It is evident that $Z(z)$ satisfies the simple equation

$$\frac{d^2 Z}{dz^2} = -k_z^2 Z \tag{C.29}$$

where the separation constant is denoted by $-k_z^2$. The general solution to this equation is

$$Z = Ae^{ik_z z} + Be^{-ik_z z} \tag{C.30}$$

Here k_z is completely arbitrary and must be chosen to meet the spatial boundary conditions in the z direction. The separated equations for the remaining radial and azimuthal functions in Eq. (C.27) then follow precisely as in the two-dimensional case, the only difference being the replacement

$$k^2 \to k^2 - k_z^2 \tag{C.31}$$

in Eq. (C.4) and Eqs. (C.7) to (C.9).

PROBLEMS

C.1 Consider a classical disturbance that satisfies the two-dimensional Helmholtz equation (C.2) in the interior of a pie-shaped region with radius R and wedge angle α and vanishes on the boundary. Separate variables and discuss the solution of the resulting polar-angle and radial equations.

C.2 Repeat Prob. C.1 for a three-dimensional disturbance that satisfies Eq. (C.11) in the interior of a sphere of radius R excluding conical regions of opening angle β along the positive and negative z axis.

D

INTEGRAL REPRESENTATIONS
AND SPECIAL FUNCTIONS

This appendix† uses the theory of complex variables (Appendix A) to define and study the fundamental solutions to the separated equations of mathematical physics that were derived in Appendix C. It is first necessary to discuss some properties of the Γ function, which serves both as an application of the techniques and as an introduction to our treatment of Bessel functions.

D1 THE Γ FUNCTION

The Γ function is defined by a definite integral

$$\Gamma(z) \equiv \int_0^\infty e^{-t}t^{z-1}\,dt \qquad \mathrm{Re}\,z > 0 \tag{D1.1}$$

which converges throughout the indicated region. Note that the variable z in the integrand appears in the exponent; the relevant factor can be rewritten as

$$t^{z-1} = (e^{\ln t})^{z-1} = e^{(z-1)\ln t} \tag{D1.2}$$

Consider any finite strip parallel to the y axis and lying on the right half z plane, as illustrated in Fig. D1.1

$$0 < b \le \mathrm{Re}\,z \le d \tag{D1.3}$$

---

† We have found the following references particularly useful: Arfken [1]; Carrier, Krook, and Pearson [2]; Lebedev [3]; Morse and Feshbach [4]; and Whittaker and Watson [5].

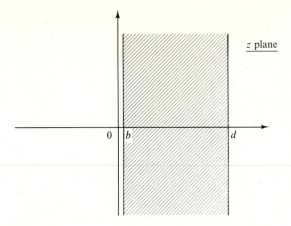

Figure D1.1 Strip used in definition of Γ function.

If this condition holds, we now show that the Γ function defined in Eq. (D1.1) can be differentiated an arbitrary number of times with respect to z

$$\frac{d^n}{dz^n}\,\Gamma(z) = \int_0^\infty (\ln t)^n e^{-t} e^{(z-1)\ln t}\, dt$$

$$= \int_0^\infty e^{-t}(\ln t)^n t^{z-1}\, dt \tag{D1.4}$$

Thus $\Gamma(z)$ is analytic in the strip (D1.3).

The proof consists in demonstrating that the integral in Eq. (D1.4) is uniformly convergent if the condition (D1.3) is satisfied. This result is established through the inequality

$$\left|\int_0^\infty\right| \leq \int_1^\infty (\ln t)^n e^{-t} e^{(d-1)\ln t}\, dt + \int_0^1 (-\ln t)^n e^{-t} t^{b-1}\, dt < \infty \tag{D1.5}$$

Here the integral has been broken into two contributions running from 1 to ∞ and from 0 to 1. The first contribution has been bounded by using the maximum of the modulus of the function in Eq. (D1.2), obtained by replacing z with its maximum real part d. The second integral, in which $\ln t$ is negative, is also bounded by replacing Eq. (D1.2) with its maximum modulus, obtained when the real part of z takes its *minimum* value b. Both integrals in Eq. (D1.5) converge because any number of powers of the logarithm leaves the convergence property unaltered. Thus Eq. (D1.5) places a bound on the integral in Eq. (D1.4) that is independent of z for z in the strip defined in Eq. (D1.3) (see Fig. D1.1). This proves that the integral in Eq. (D1.4) is uniformly convergent, and we are therefore permitted to differentiate under the integral in Eq. (D1.1) any number of times.† Since $\Gamma(z)$ defined in Eq. (D1.1) has any number of derivatives in the strip (D1.3), it is an *analytic function* of z in this strip. More generally, $\Gamma(z)$ is an analytic function of z for Re $z > 0$.

---

† The precise theorem is that one can differentiate under an integral provided that the resulting integral is uniformly convergent. The theorem is a direct analog of the corresponding theorem for series and is proved in Whittaker and Watson [5], sec 4.44.

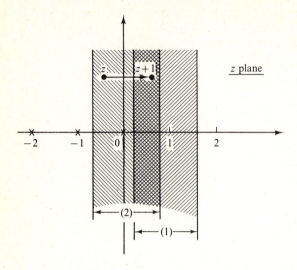

Figure D1.2 Analytic continuation of $\Gamma(z)$ to left half plane. Region 1 is specified by Re $z > 0$ and region 2 by Re $z > -1$.

The Γ function in Eq. (D1.1) has a simple value if z is replaced by a positive integer

$$\Gamma(n) = \int_0^\infty e^{-t}t^{n-1} \, dt = (n-1)! \qquad (D1.6)$$

As a result, it satisfies the functional relation $\Gamma(n+1) = n\Gamma(n)$. We now show that $\Gamma(z)$ obeys a similar relation by noticing that Eq. (D1.1) can be integrated by parts as long as Re $z > 0$. Introduce the following change of variables:

$$u = e^{-t} \qquad dv = t^{z-1} \, dt$$

$$du = -e^{-t} \, dt \qquad v = \frac{t^z}{z} \qquad (D1.7)$$

The integration by parts takes the form

$$\Gamma(z) = \left[\frac{e^{-t+z\ln t}}{z}\right]_0^\infty + \frac{1}{z}\int_0^\infty e^{-t}t^z \, dt \qquad (D1.8)$$

and the contributions from the endpoints vanish if Re $z > 0$. Since the remaining integral in Eq. (D1.8) is just $\Gamma(z+1)$, we immediately obtain the desired generalization

$$z\Gamma(z) = \Gamma(z+1) \qquad (D1.9)$$

The special case (D1.6) can be obtained from this general relation through iteration and the observation that

$$\Gamma(1) = 1 \qquad (D1.10)$$

Although the integral (D1.1) defines the Γ function only in the right half z plane, Eq. (D1.9) can be used to *analytically continue* the Γ function to the region Re $z \le 0$. The method for this procedure is illustrated in Fig. D1.2. In region 1, a vertical strip that lies in the right half plane where Re $z > 0$, $\Gamma(z)$ is defined by Eq. (D1.1) and satisfies the relation (D1.9). In region 2, a vertical strip part of which lies in the left half plane with Re $z > -1$ and part of which overlaps the original region 1, we *define*

$$\Gamma(z) \equiv \frac{\Gamma(z+1)}{z} \qquad (D1.11)$$

Since the numerator and denominator have a derivative everywhere in region 2 (see Fig. D1.2) and the denominator vanishes only at $z = 0$, we see that the function defined in Eq. (D1.11) is *analytic in region 2 except at the point $z = 0$, where it has a simple pole with unit residue*

$$\Gamma(z) \text{ has a simple pole at } z = 0 \qquad (D1.12)$$

Furthermore, in the intersection of the two regions, the function defined in region 2 by Eq. (D1.11) is identical with the function defined in region 1, which proves that Eq. (D1.11) provides a *unique analytic continuation* of the Γ function to the region 2 (Sec. A11). This argument can now be repeated; evidently the only singularity that ever appears in the Γ function is just that coming from the denominator in Eq. (D1.11). At the negative integers, repeated application of Eq. (D1.11) eventually relates the function to its value at the origin, where it has a simple pole. Thus we conclude that

The function $\Gamma(z)$ is analytic in the entire z plane except for simple poles at the values $z = -n$, $n = 0, 1, 2, 3, \ldots, \infty$, with residue $(-1)^n/n!$ (D1.13)

Among the many functional relations satisfied by the Γ function we here consider only

$$\Gamma(z)\Gamma(1 - z) = \frac{\pi}{\sin \pi z} \qquad (D1.14)$$

It is sufficient to prove this relation for real z lying between 0 and 1 (see Fig. D1.3) because both sides of Eq. (D1.14) are analytic except for poles at integral values of z. Thus the relation holds everywhere by analytic continuation [see Eq. (A10.7)] once it is proved to hold *on a finite section* of the real axis. First, let

$$t = u^2 \qquad (D1.15)$$

in Eq. (D1.1), which becomes

$$\Gamma(z) = 2 \int_0^\infty e^{-u^2} u^{2z-1} \, du \qquad (D1.16)$$

The left-hand side of Eq. (D1.14) can then be written

$$\Gamma(z)\Gamma(1 - z) = 4 \int_0^\infty du \int_0^\infty dv \; e^{-(u^2 + v^2)} u^{2z-1} v^{-(2z-1)} \qquad (D1.17)$$

This repeated integral runs over the first quadrant of the uv plane, and we shall transform (u, v) into polar coordinates

$$u \equiv \rho \cos \phi \qquad v \equiv \rho \sin \phi \qquad (D1.18)$$

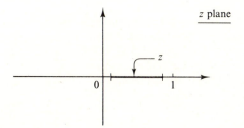

Figure D1.3 Region on real axis used in proof of Eq. (D1.14).

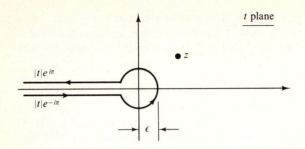

Figure D1.4 Contour used to define $[\Gamma(z)]^{-1}$ in Eq. (D1.23) and $J_\nu(z)$ in Eq. (D3.4).

In this way, Eq. (D1.17) takes the form

$$\Gamma(z)\Gamma(1-z) = 4 \int_0^\infty \rho \, d\rho \int_0^{\pi/2} d\phi \, e^{-\rho^2}\left(\frac{\cos \phi}{\sin \phi}\right)^{2z-1} \equiv I_1(z) \tag{D1.19}$$

The radial integral is now elementary, and the further change of variables

$$s \equiv \cot \phi \qquad \text{and} \qquad ds = -\csc^2 \phi \, d\phi \tag{D1.20}$$

reduces the integral to a simpler form

$$I_1(z) = 2 \int_0^\infty \frac{s^{2z}}{s} \frac{ds}{1+s^2} \tag{D1.21}$$

This integral, which is well defined for the region shown in Fig. D1.3, is precisely that considered in Eqs. (A9.17) and (A9.27). Recalling the expression

$$I_1(z) = \frac{\pi}{\sin \pi z} \tag{D1.22}$$

we immediately obtain the relation (D1.14).

One other integral representation of the Γ function obtained through the use of (D1.14) is particularly important. If C is the contour shown in Fig. D1.4, we claim that

$$I(z) \equiv \frac{1}{2\pi i} \int_C \frac{e^t}{t^z} \, dt = \frac{1}{\Gamma(z)} \tag{D1.23}$$

where the t plane is cut along the negative real axis. Since the integrand is exponentially small as $|t| \to \infty$ along the contour, this integral can be differentiated any number of times, showing that it is an analytic function of z for values lying outside the contour. In particular, let z lie on the real axis between 0 and 1 (see Fig. D1.3)

$$0 < z < 1 \tag{D1.24}$$

In this case, the contribution of the small circle about the origin (see Fig. D1.4) vanishes as its radius ϵ tends to zero. A simple power counting shows that the integral is proportional to ϵ^{1-z}, and the condition (D1.24) ensures that this quantity vanishes as $\epsilon \to 0$

$$\epsilon^{1-z} \to 0 \tag{D1.25}$$

Since $t = |t|e^{\pm i\pi} \equiv \tau e^{\pm i\pi}$ on the remaining contributions, we readily find

$$I(z) = \frac{1}{2\pi i}\left(\int_\infty^0 \frac{d\tau \, e^{-i\pi}e^{-\tau}}{\tau^z e^{-i\pi z}} + \int_0^\infty \frac{d\tau \, e^{i\pi}e^{-\tau}}{\tau^z e^{i\pi z}}\right)$$

$$= \frac{1}{2\pi i}\left(e^{i\pi z} - e^{-i\pi z}\right)\int_0^\infty d\tau \, e^{-\tau}\tau^{-z} \tag{D1.26}$$

The remaining integral is just $\Gamma(1 - z)$ in Eq. (D1.1), and we thus obtain

$$I(z) = \frac{\sin \pi z}{\pi} \Gamma(1 - z) \tag{D1.27}$$

Comparison with Eq. (D1.14) yields the result

$$I(z) = \frac{1}{\Gamma(z)} \tag{D1.28}$$

valid for $0 < z < 1$. Furthermore, $\Gamma(z)$ has no zeros, as seen from Eqs. (D1.13) and (D1.14). Thus, both sides of (D1.23) are analytic functions of z for z lying outside the contour in Fig. D1.4, they agree for z lying on the finite section of the real axis shown in Fig. D1.3, and our fundamental theorem (A10.7) shows that the two functions must be identical everywhere.

D2 LEGENDRE FUNCTIONS

In Appendix C we obtained Legendre's equation from the associated Legendre equation (C.21) by setting $m = 0$. It is convenient to write the separation constant as

$$\lambda \equiv \alpha(\alpha + 1) \tag{D2.1}$$

and Legendre's equation then takes the form [see Eq. (C.22)]

$$(1 - z^2)\frac{d^2u}{dz^2} - 2z\frac{du}{dz} + \alpha(\alpha + 1)u = 0 \tag{D2.2}$$

It has regular singular points at $z = \pm 1$ in the finite z plane.

Consider any two linearly independent solutions u_1 and u_2 of Eq. (D2.2). The linear independence merely requires that the wronskian of the two solutions be nonzero

$$W[u_1, u_2] \equiv u_1 u_2' - u_2 u_1' \neq 0 \tag{D2.3}$$

Since Legendre's equation is a second-order ordinary differential equation, its general solution can be expressed as a linear combination of *any* two linearly independent solutions

$$u = Au_1 + Bu_2 \tag{D2.4}$$

where A and B are constants. For this reason, any two linearly independent solutions are known as a *fundamental set* of solutions. Furthermore, the Sturm-Liouville form (C.18) of Legendre's equation and the analysis leading to Eq. (43.27) show that

$$W[u_1(z), u_2(z)] = \frac{C}{1 - z^2} \tag{D2.5}$$

where C is a constant that may depend on α *but not on z.*

We shall now solve Legendre's equation (D2.2) by constructing an integral representation. This approach has the great advantage of providing explicit expressions for a fundamental set of solutions *for all values of z in the complex plane and for all values of the separation constant α.* To construct the integral representation, we use the identity†

$$\left[(1 - z^2)\frac{d^2}{dz^2} - 2z\frac{d}{dz} + \alpha(\alpha + 1) \right] \frac{(t^2 - 1)^\alpha}{(t - z)^{\alpha + 1}} = \frac{d}{dt}\left[\frac{(\alpha + 1)(t^2 - 1)^{\alpha + 1}}{(t - z)^{\alpha + 2}} \right] \tag{D2.6}$$

† General rules for constructing identities of this type are discussed in Morse and Feshbach [4], sec. 5.3.

This algebraic identity is readily verified since the differentiation with respect to t on the right-hand side yields

$$\text{rhs} = \frac{(\alpha + 1)(t^2 - 1)^\alpha}{(t - z)^{\alpha + 3}}[2t(\alpha + 1)(t - z) - (\alpha + 2)(t^2 - 1)] \tag{D2.7}$$

and the differentiation with respect to z on the left-hand side yields

$$\text{lhs} = \frac{(\alpha + 1)(t^2 - 1)^\alpha}{(t - z)^{\alpha + 3}}[(1 - z^2)(\alpha + 2) - 2z(t - z) + \alpha(t - z)^2] \tag{D2.8}$$

A little algebra shows that these expressions are identical. Note that the differential operator with respect to z on the left-hand side of this identity is precisely the differential operator in Legendre's equation (D2.2), whereas the right-hand side of this expression is the total derivative with respect to a second variable t.

The identity (D2.6) immediately allows us to construct solutions to Eq. (D2.2) in the form

$$u_\alpha(z) \equiv \frac{1}{2\pi i} \oint_C \frac{(t^2 - 1)^\alpha}{(t - z)^{\alpha + 1}} \frac{dt}{2^\alpha} \tag{D2.9}$$

We now show that this integral is a solution to Legendre's equation for any contour in the t plane over which the integrand returns to its initial value. Apply the differential operator in Legendre's equation (D2.2) to this expression. If C is a finite contour, the differential operator can be taken under the integral sign to act on the integrand, where it produces the total derivative with respect to t appearing on the right-hand side of Eq. (D2.6). The contour integration in Eq. (D2.9) of this perfect differential simply gives the change in the quantity in brackets on the right-hand side of Eq. (D2.6) from the beginning to the end of the contour. Apart from the meromorphic factor $(t^2 - 1)(t - z)^{-1}$, this quantity is just the *integrand* in Eq. (D2.9), so that the contour integral of the perfect differential vanishes if this integrand has the same value at the ends of C. In particular, if the contour is closed (returns to its original position), and if the integrand in (D2.9) is single-valued on this contour, Eq. (D2.9) indeed satisfies Legendre's equation (D2.2).

In this way Eq. (D2.9) provides a *closed form* for the solution in the whole z plane and for any value of α, the separation constant in Legendre's equation, assuming only that the integrand is single-valued over the contour C. Moreover, we may use different contours to define different solutions. This freedom will allow us to generate a fundamental system of solutions to Legendre's equation.

$\alpha = l$ (an Integer)

Consider first for simplicity the case where α is a nonnegative integer l, which is the most common situation in our applications. The Legendre function $P_l(z)$ is now defined by Eq. (D2.9) with the contour C indicated in Fig. D2.1

$$P_l(z) \equiv \frac{1}{2\pi i} \oint_C \frac{(t^2 - 1)^l}{(t - z)^{l + 1}} \frac{dt}{2^l} \qquad l = 0, 1, 2, \dots \tag{D2.10}$$

Equation (D2.10) is known as *Schläfli's integral*. Since l is now an integer, the function $P_l(z)$ satisfies Legendre's equation because the integrand is meromorphic and hence single-valued on any closed contour. The only singularity enclosed in C is a pole of order $l + 1$ at the point $t = z$, and the contour integral can then be evaluated by the method of residues.

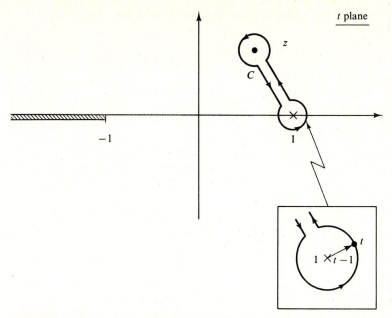

Figure D2.1 Contour C used to define Legendre functions in Schläfli's integral for $\alpha = l$ (an integer), for arbitrary α, and for all z.

The residue at the point $t = z$ is simply the coefficient of $(t - z)^l$ in the Taylor-series expansion of $2^{-l}(t^2 - 1)^l$ about the point $t = z$, which follows immediately from Eq. (A7.1). Thus the integral in Eq. (D2.10) reduces to

$$P_l(z) = \frac{1}{2^l l!} \left[\frac{d^l}{dt^l} (t^2 - 1)^l \right]_{t=z} = \frac{1}{2^l l!} \frac{d^l}{dz^l} (z^2 - 1)^l \qquad (D2.11)$$

The right-hand side of Eq. (D2.11) is a *polynomial* of order l with the symmetry property

$$P_l(-z) = (-1)^l P_l(z) \qquad l = 0, 1, 2, \ldots \qquad (D2.12)$$

These are known as *Legendre polynomials*, and Eq. (D2.11) is *Rodrigues' formula*. If $z = 1$, the integrand in Eq. (D2.10) simplifies greatly, having only a simple pole at $t = 1$ with unit residue. The method of residues shows that

$$P_l(1) = 1 \qquad (D2.13a)$$

and Eq. (D2.12) gives the value

$$P_l(-1) = (-1)^l \qquad (D2.13b)$$

Given any solution $P_\alpha(z)$ to Legendre's equation (D2.2) for arbitrary α, direct differentiation readily establishes that the function†

$$P_\alpha^m(z) \equiv (1 - z^2)^{m/2} \frac{d^m}{dz^m} P_\alpha(z) \qquad (D2.14)$$

† There are many phase conventions for these functions; we follow that of Morse and Feshbach [4], pp. 1325–1327.

is a solution of the *associated Legendre equation* (C.21)

$$(1 - z^2)\frac{d^2}{dz^2}P_\alpha^m - 2z\frac{d}{dz}P_\alpha^m + \left[\alpha(\alpha + 1) - \frac{m^2}{1 - z^2}\right]P_\alpha^m = 0 \qquad (D2.15)$$

for *m* a nonnegative integer

$$m = 0, 1, 2, \ldots \qquad (D2.16)$$

This last restriction ensures that the differentiation in Eq. (D2.14) will be well defined. In particular, if $\alpha = l$ (also a nonnegative integer), Rodrigues' formula (D2.11) for the Legendre polynomials serves to define the *associated Legendre polynomials*

$$P_l^m(z) \equiv (1 - z^2)^{m/2}\frac{d^m}{dz^m}P_l(z)$$

$$= \frac{1}{2^l l!}(1 - z^2)^{m/2}\frac{d^{l+m}}{dz^{l+m}}(z^2 - 1)^l \qquad (D2.17)$$

for integral values of $l \geq 0$ and $m \geq 0$. Since P_l is a polynomial of degree l, Eq. (D2.17) shows that

$$P_l^m(z) = 0 \qquad \text{for } m > l \qquad (D2.18)$$

Furthermore, the polynomials P_l^m are obviously bounded at $z = \pm 1$ so that they also satisfy the natural boundary conditions (C.20)

$$(1 - z^2)\frac{dP_l^m(z)}{dz} = 0 \qquad \text{at } z = \pm 1 \qquad (D2.19)$$

Thus they represent acceptable solutions for a physical problem whose domain includes $z = \pm 1$ (for example, the positive and negative polar axes in spherical coordinates with $z = \cos\theta$). Finally, we note that Eq. (D2.15) contains only m^2, so that the definition of P_l^m for negative *m* is simply a matter of convention.

We can now exhibit the angular part of the separated solutions to the Helmholtz equation (C.13) in spherical coordinates. For nonnegative integers *l* and *m*, the product of solutions to Eqs. (C.14) and (C.21) defines the *spherical harmonics*

$$Y_{lm}(\theta, \phi) \equiv (-1)^m\left[\frac{2l + 1}{4\pi}\frac{(l - m)!}{(l + m)!}\right]^{1/2}P_l^m(\cos\theta)e^{im\phi} \qquad m \geq 0 \qquad (D2.20)$$

Here, the normalization ensures that these functions will be orthonormal on the unit sphere

$$\int d\Omega\ Y_{lm}^*(\theta, \phi)Y_{l'm'}(\theta, \phi) = \delta_{ll'}\delta_{mm'} \qquad (D2.21a)$$

where the solid angle is defined according to (see Fig. B.2a)

$$d\Omega = \sin\theta\ d\theta\ d\phi \qquad (D2.21b)$$

Although the precise normalization factor requires a detailed application of Eq. (D2.17) (see Prob. D.2), the orthogonality is immediate because the functions $P_l^m(\cos\theta)$ and $e^{im\phi}$ are solutions to Sturm-Liouville problems (C.14) and (C.18). This observation also proves that the spherical harmonics defined in Eq. (D2.20) are a complete set of basis functions for any function that is single-valued in ϕ and satisfies the natural boundary conditions in θ, which is effectively the same as requiring that the function remain finite along the polar axis.

For negative integral values of m we *define* the spherical harmonics as

$$Y_{lm}^* \equiv (-1)^m Y_{l,\,-m} \tag{D2.22}$$

This choice is the common phase convention of quantum mechanics.†

Arbitrary α

The analysis of Eq. (D2.9) becomes more intricate if the separation constant α is an arbitrary complex number. In this case, the integrand in Eq. (D2.9) has branch points at $t = 1$, $t = -1$, and $t = z$ in the complex t plane. Nevertheless, the integrand is single-valued along the closed contour C illustrated in Fig. D2.1, which is easily proved by tracing the variable t around this contour. On the counterclockwise infinitesimal circle surrounding the point $t = 1$ (see insert in Fig. D2.1), the complex vector $t - 1$ running from 1 to the variable point t rotates through 2π in the counterclockwise (positive) sense (see Fig. A1.1). As a result,

$$\text{The phase of } (t - 1)^\alpha \text{ changes by } 2\pi\alpha \tag{D2.23a}$$

and the phase of all the other complex vectors in the integrand in Eq. (D2.9) remains unaltered. When we proceed counterclockwise around the point $t = z$ on an infinitesimal circle, the only change in the integrand of Eq. (D2.9) is that

$$\text{The phase of } (t - z)^{-\alpha-1} \text{ changes by } -2\pi(\alpha + 1) \tag{D2.23b}$$

It is now evident that the net change in the phase of the integrand along the contour C (see Fig. D2.1) is -2π, ensuring that the integrand in Eq. (D2.9) will indeed be single-valued around this contour. Thus again we can use Schläfli's integral (D2.10) with the contour C given in Fig. D2.1 to define the Legendre function $P_\alpha(z)$ for arbitrary α

$$P_\alpha(z) \equiv \frac{1}{2\pi i} \oint_C \frac{(t^2 - 1)^\alpha}{(t - z)^{\alpha+1}} \frac{dt}{2^\alpha} \tag{D2.24}$$

The preceding argument shows that C must not encircle the point $t = -1$ because the additional phase changes would destroy the single-valuedness. For this reason, if $\operatorname{Re} z < -1$, the contour C must be drawn wholly above or wholly below the point $t = -1$, according to the sign of $\operatorname{Im} z$. As a result, the Legendre function $P_\alpha(z)$ defined by Eq. (D2.24) and Fig. D2.1 has a cut along the negative real z axis from -1 to $-\infty$. In the cut z plane, it is clear from Schläfli's integral that $P_\alpha(z)$ can be differentiated with respect to z any number of times and is therefore analytic. At $z = 1$, the integrand simplifies considerably to give

$$P_\alpha(1) = \frac{1}{2\pi i} \oint_C dt \, \frac{(t + 1)^\alpha}{2^\alpha(t - 1)} \tag{D2.25a}$$

The only singularity in C is a simple pole at $t = 1$ with unit residue, and the method of residues immediately yields the value

$$P_\alpha(1) = 1 \qquad \text{arbitrary } \alpha \tag{D2.25b}$$

Legendre Functions of the Second Kind

Different choices of the contour in Eq. (D2.9) give different solutions to Legendre's equation. We now consider the contour D illustrated in Fig. D2.2 and claim that the integral

$$Q_\alpha(z) \equiv \frac{1}{4i \sin \pi\alpha} \oint_D \frac{(t^2 - 1)^\alpha}{(z - t)^{\alpha+1}} \frac{dt}{2^\alpha} \tag{D2.26}$$

† See, for example, Schiff [6], eqs. (14.12) and (14.16).

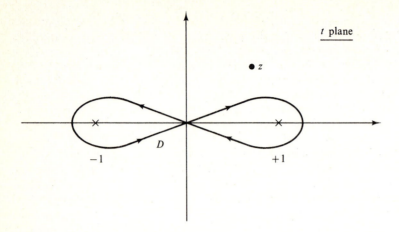

Figure D2.2 Contour D used to define Legendre functions of the second kind.

satisfies Legendre's equation [the difference in phase and overall normalization from Eq. (D2.9) is conventional]. As before, it is only necessary to demonstrate that the integrand in Eq. (D2.26) is in fact single-valued around the contour D indicated in Fig. D2.2. As we proceed around the point $t = 1$ on a small circle in the clockwise direction, the only phase change of the complex vectors in the integrand is that

$$\text{The phase of } (t - 1)^\alpha \text{ changes by } -2\pi\alpha \qquad \text{(D2.27)}$$

When we proceed around the point $t = -1$ in the opposite direction, the only phase change is that

$$\text{The phase of } (t + 1)^\alpha \text{ changes by } 2\pi\alpha \qquad \text{(D2.28)}$$

The net phase change of the integrand in Eq. (D2.26) is given by the sum of Eqs. (D2.27) and (D2.28); it clearly vanishes around the closed contour D. In this process, we must be certain to keep the point z outside the contour D or there will be additional phase changes that destroy the single-valuedness of the integrand. Thus the z plane must be cut on the real axis lying between -1 and $+1$. In addition, Eq. (D2.26) shows that $Q_\alpha(z)$ behaves like $z^{-\alpha-1}$ for $|z| \to \infty$. For nonintegral α, we see that $Q_\alpha(z)$ would not be single-valued as z moves around a large circular contour, and it is conventional to introduce an additional cut in the z plane from -1 to $-\infty$. Thus for arbitrary α, the function $Q_\alpha(z)$ is single-valued and analytic in the z plane cut along the real axis from 1 to $-\infty$.

These two solutions $P_\alpha(z)$ and $Q_\alpha(z)$ are linearly independent for any complex α because their wronskian can be proved to be $(1 - z^2)^{-1}$ (see Prob. D.3). They therefore provide a fundamental system of solutions to Legendre's equation (D2.2). In addition, the separation constant in Eq. (D2.1) is unchanged under the replacement

$$\alpha \to -\alpha - 1 \qquad \text{(D2.29)}$$

Thus all values of the separation constant will be included if

$$\text{Re } \alpha > -1 \qquad \text{(D2.30)}$$

We shall henceforth assume this to be the case.

In this case, the integral appearing in Eq. (D2.26) can be simplified by explicitly writing out the contribution of the four pieces shown in Fig. D2.3. The contributions from the small circles 3 and 4 surrounding the points $t = 1$ and $t = -1$ vanish in the limit that the radius ρ of these circles goes to zero. This conclusion follows because the entire dependence on ρ comes from the numerator in Eq. (D2.26). Collecting powers, we see that these contributions to the integral are proportional to $\rho^{1+\alpha}$, which vanishes as $\rho \to 0$ if Eq. (D2.30) holds. We define the phases of the integrand by assuming that

$$(t^2 - 1)^\alpha \text{ is real for real } \alpha \text{ and } t > 1 \qquad (D2.31)$$

On contour 1 in Fig. D2.3, the numerator in Eq. (D2.26) takes the form

$$(t^2 - 1)^\alpha = (1 + t)^\alpha (1 - t)^\alpha e^{-i\pi\alpha} \qquad (D2.32a)$$

because the vector $t - 1$ rotates in the negative sense to reach that part of the contour. Similarly, on part 2

$$(t^2 - 1)^\alpha = (1 + t)^\alpha (1 - t)^\alpha e^{-i\pi\alpha} e^{2\pi i\alpha} = (1 + t)^\alpha (1 - t)^\alpha e^{i\pi\alpha} \qquad (D2.32b)$$

Under the condition (D2.30), the integral in Eq. (D2.26) therefore takes the form

$$Q_\alpha(z) = \frac{1}{4i \sin \pi\alpha} \int_{-1}^{1} (e^{i\pi\alpha} - e^{-i\pi\alpha}) \frac{(1 - t^2)^\alpha}{2^\alpha (z - t)^{\alpha+1}} \, dt$$

which reduces to

$$Q_\alpha(z) = \frac{1}{2} \int_{-1}^{1} \frac{(1 - t^2)^\alpha}{2^\alpha (z - t)^{\alpha+1}} \, dt \qquad \text{Re } \alpha > -1 \qquad (D2.33)$$

For α an integer, the integrals appearing in Eq. (D2.33) can be done analytically. The first few Legendre functions of the second kind for integral α are given by

$$Q_0(z) = \tfrac{1}{2} \ln \frac{z + 1}{z - 1} \qquad (D2.34)$$

$$Q_1(z) = \tfrac{1}{2} z \ln \frac{z + 1}{z - 1} - 1 \qquad (D2.35)$$

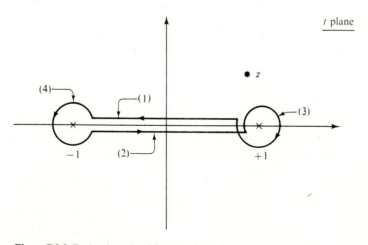

Figure D2.3 Evaluation of $Q_\alpha(z)$ in Eq. (D2.26).

As expected from our preceding discussion, these functions are single-valued in the z plane cut from -1 to $+1$ along the real axis. Thus they have a simpler analytic structure than $Q_\alpha(z)$ for general α. Just as in Eq. (D2.14), associated Legendre functions of the second kind can be defined according to

$$Q_l^m(z) = (-1)^m(z^2 - 1)^{m/2} \frac{d^m}{dz^m} Q_l(z) \qquad m \geq 0 \tag{D2.36}$$

These functions satisfy the associated Legendre equation for integral l and m; together with Eq. (D2.17), they define a fundamental system for that equation.

Arbitrary α

The functions $Q_\alpha(z)$ possess logarithmic singularities at the points $z = \pm 1$. For the integers $\alpha = 0$ and 1 this behavior is explicit in Eqs. (D2.34) and (D2.35), but the presence of these logarithmic singularities is more general. Indeed, *it is evident from the integral in Eq. (D2.33) that the function $Q_\alpha(z)$ has a logarithmic singularity at the points $z = \pm 1$ for any value of the separation constant α.* We proceed to make use of this important observation.

In addition to the symmetry property expressed in Eq. (D2.29), Legendre's equation (D2.2) is invariant under the substitution $z \to -z$. Thus $P_\alpha(-z)$ is also a solution. It is not, however, a wholly new solution, for it must be expressible as a linear combination of the fundamental system $P_\alpha(z)$ and $Q_\alpha(z)$. We shall use the integral representations (D2.24) and (D2.26) to prove the relation

$$P_\alpha(-z) = -\frac{2 \sin \pi\alpha}{\pi} Q_\alpha(z) + e^{\mp i\pi\alpha} P_\alpha(z) \qquad \text{Im } z \gtrless 0 \tag{D2.37}$$

It expresses the value of P_α at the reflected point $-z$ in terms of the values of the members of the fundamental system P_α and Q_α at z.

Suppose that Eq. (D2.37) is correct, and let $z \to 1$, where P_α is bounded [see Eq. (D2.25b)] and Q_α is logarithmically infinite. Equation (D2.37) then has the important corollary that $P_\alpha(z)$ *for general α has a logarithmic singularity at $z = -1$.* The only exception is if the coefficient of $Q_\alpha(z)$ vanishes identically

$$\sin \pi\alpha = 0 \tag{D2.38}$$

which requires

$$\alpha = l = 0, 1, 2, \ldots, \infty \tag{D2.39}$$

in the region $\alpha > -1$ [see Eq. (D2.30)]. Since most physical problems in spherical coordinates include the full range $-1 \leq x \equiv \cos \theta \leq 1$,

> The only nonsingular solutions to Legendre's equation and the associated Legendre's equation (with integral m) that include the polar axes $\theta = 0$ and π are $P_l(\cos \theta)$ and $Y_{lm}(\theta, \phi)$ for integral l (D2.40)

A logarithmic singularity also would violate the natural boundary condition (C.20) that $(1 - z^2)P_\alpha'(z)$ vanish at $z = -1$. Thus we conclude that the Legendre polynomials $P_l(z)$ for integral l are the only acceptable eigenfunctions of the Sturm-Liouville problem (C.18) and (C.20) and that the eigenvalues are just those in Eq. (C.19). Note that for the integral l given in Eq. (D2.39), the relation (D2.37) reproduces the previous relation (D2.12).

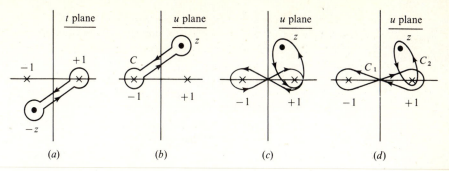

Figure D2.4 Contours used in proof of Eq. (D2.37).

We now prove the important relation (D2.37), which follows from the integral representations (D2.24) and (D2.26). Consider Schläfli's integral for $P_\alpha(-z)$ (Fig. D2.4a) and make the following change of variable in the integral:

$$t \to -u \tag{D2.41}$$

The points in Fig. D2.4a are reflected into the points in Fig. D2.4b, and Schläfli's integral takes the form

$$P_\alpha(-z) = -\frac{1}{2\pi i} \oint_C \frac{(u^2 - 1)^\alpha}{(z - u)^{\alpha + 1}} \frac{du}{2^\alpha} \tag{D2.42}$$

The contour C in Fig. D2.4b can be deformed into the shape shown in Fig. D2.4c without changing the value of the integral because the integrand in Eq. (D2.42) is analytic in the intervening region. We now note, however, that the contour in Fig. D2.4c is identical with the sum of the two contours C_1 and C_2 in Fig. D2.4d. Furthermore, the integrand in Eq. (D2.42) has already been proved to be single-valued around C_1 and C_2. Thus we have

$$P_\alpha(-z) = -\frac{1}{2\pi i} \left(\oint_{C_1} + \oint_{C_2} \right) \frac{(u^2 - 1)^\alpha}{(z - u)^{\alpha + 1}} \frac{du}{2^\alpha} \tag{D2.43}$$

Apart from a constant factor, the integral around the contour C_1 in Eq. (D2.43) is just $Q_\alpha(z)$ defined in Eq. (D2.26). The integral around C_2 differs from Schläfli's integral in Eq. (D2.24) by having $(z - u)^{-\alpha - 1}$ instead of the appropriate factor $(u - z)^{-\alpha - 1}$ in the integrand. Since α is arbitrary here, we must be careful in extracting the phase of this factor. The situation is illustrated in Fig. D2.5. We have already noted that $P_\alpha(z)$ and $Q_\alpha(z)$ are analytic in the z plane cut from -1 to $-\infty$ and from $+1$ to $-\infty$, respectively. As a result, the right-hand side of Eq. (D2.37) must be interpreted in the z plane cut from $+1$ to $-\infty$ (see Fig. D2.5), and the function $P_\alpha(-z)$ must have a similar character. Thus to go from the upper half z plane to the lower half z plane, we must cross the real axis *to the right* of the point $z = 1$, as illustrated in Fig. D2.6. If z lies in the upper half plane, it is obtained from the reflected point $-z$ as $z = e^{i\pi}(-z)$. The change of variables in Eq. (D2.41) must similarly be taken as $u = e^{i\pi}(-u)$ (see Fig. D2.6). Thus the denominator of the second integral in Eq. (D2.43) can be rewritten as

$$z - u = e^{i\pi}(u - z) \qquad \text{Im } z > 0 \tag{D2.44}$$

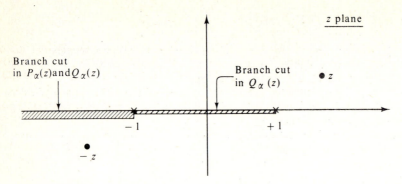

Figure D2.5 Interpretation of Eq. (D2.37).

If z lies in the lower half plane, it is obtained from the reflected point $(-z)$ as $z = e^{-i\pi}(-z)$ and

$$z - u = e^{-i\pi}(u - z) \qquad \text{Im } z < 0 \qquad (D2.45)$$

Thus the sign in the denominator of the integral over the contour C_2 in Eq. (D2.43) can be changed to give Schläfli's integral Eq. (D2.24) by taking either Eq. (D2.44) or (D2.45), as appropriate. Equation (D2.43) then reads

$$P_\alpha(-z) = -\frac{2 \sin \pi\alpha}{\pi} Q_\alpha(z) - \frac{P_\alpha(z)}{e^{\pm i\pi(\alpha+1)}} \qquad \text{Im } z \gtrless 0 \qquad (D2.46)$$

which is precisely the desired result in Eq. (D2.37). This equation can be interpreted as giving the phase of the function $P_\alpha(z)$ along the branch cut from -1 to $-\infty$ in terms of the functions $\{P_\alpha(z), Q_\alpha(z)\}$ along the real axis with $z > 1$. This phase evidently changes as one

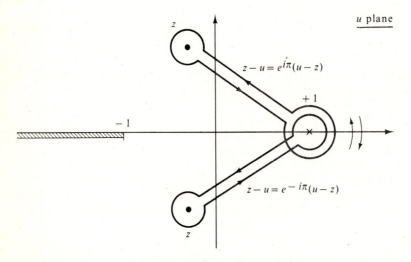

Figure D2.6 Determination of appropriate phase in Eq. (D2.37).

goes from the top to the bottom of the branch cut, passing the real axis to the right of the point $z = 1$. Note that this cut is unnecessary for $\alpha = l = 0, 1, 2, \ldots$.

We have already remarked that the Legendre functions typically occur as angular functions in spherical coordinates. If the polar axis is part of the allowed domain, the separation constant λ in Eq. (D2.1) must have the form $l(l + 1)$ and only the bounded Legendre polynomials are permissible. In other physical situations, however, both P_α and Q_α are acceptable. One example is the interior of a cone with apex angle θ_0, where the polar angle θ must satisfy $\theta < \theta_0$. Since $z = 1$ lies in the physical region, the function $Q_\alpha(z)$ is excluded owing to its logarithmic singularity. Thus only $P_\alpha(z)$ is allowed, but α need not be an integer because $z = -1$ is outside the domain of interest.

D3 BESSEL FUNCTIONS

For an arbitrary index v, Bessel's equation is given by [see Eq. (C.10)]

$$\frac{d^2u}{dz^2} + \frac{1}{z}\frac{du}{dz} + \left(1 - \frac{v^2}{z^2}\right)u = 0 \tag{D3.1}$$

This equation has a regular singular point at $z = 0$ in the finite z plane. We shall use the identity

$$\left[\frac{d^2}{dz^2} + \frac{1}{z}\frac{d}{dz} + \left(1 - \frac{v^2}{z^2}\right)\right]\left[\left(\frac{z}{2}\right)^v \frac{1}{t^{v+1}}\exp\left(t - \frac{z^2}{4t}\right)\right] = \frac{d}{dt}\left[\left(\frac{z}{2}\right)^v \frac{1}{t^{v+1}}\exp\left(t - \frac{z^2}{4t}\right)\right] \tag{D3.2}$$

to define an integral representation for the solutions to Bessel's equation. The identity is immediately established by noting that the differentiation on the right-hand side gives

$$\text{rhs} = \left(\frac{z}{2}\right)^v \frac{1}{t^{v+1}}\exp\left(t - \frac{z^2}{4t}\right)\left(-\frac{v+1}{t} + 1 + \frac{z^2}{4t^2}\right) \tag{D3.3}$$

and straightforward manipulations show that the left-hand side yields an identical result.

As in Sec. D2, the identity (D3.2) provides the following integral representation for solutions to Bessel's equation (D3.1) for arbitrary z and arbitrary v:

$$J_v(z) \equiv \frac{1}{2\pi i}\left(\frac{z}{2}\right)^v \oint_C \frac{1}{t^{v+1}}\exp\left(t - \frac{z^2}{4t}\right)dt \tag{D3.4}$$

provided only that the integrand returns to its initial value on executing the contour C. The validity of (D3.4) follows immediately by applying the differential operator in Bessel's equation (D3.1) to this expression. If the contour is finite, or if the integral is sufficiently convergent, the differentiation can be taken under the integral sign. The identity in Eq. (D3.2) then shows that the integrand becomes a perfect differential in t. The result is to evaluate the change in the integrand between the two ends of the contour.

We shall choose the contour C as indicated in Fig. D1.4, where z lies outside the contour and the variable t is real along the positive real axis. The variable t then takes the value indicated in Fig. D1.4 along those parts of the contour lying just above or below the negative t axis. At either end of the contour, we see that

$$t = -|t| \rightarrow -\infty \tag{D3.5}$$

so that the integral in Eq. (D3.4) is *exponentially* convergent as $t \rightarrow -\infty$. Although the contour in Eq. (D3.5) is closed only in the sense that it begins and ends at infinity, the integrand is single-valued along this contour since it starts and ends at zero. Thus Eq. (D3.4) indeed generates a solution to Bessel's equation.

If z remains outside the contour C indicated in Fig. D1.4, the integral in Eq. (D3.4) can be differentiated with respect to z under the integral sign an arbitrary number of times and the integral remains exponentially convergent. Thus all the derivatives with respect to the variable z exist, and the integral in Eq. (D3.4) defines an analytic function in the z plane cut from 0 to $-\infty$.

For many purposes, it is convenient to expand the factor containing z in the integrand with the familiar power series for the exponential

$$\exp\left(-\frac{z^2}{4t}\right) = \sum_{p=0}^{\infty} \frac{(-1)^p z^{2p}}{2^{2p} p!} \frac{1}{t^p} \tag{D3.6}$$

This series converges uniformly in t for t on the contour C in Fig. D1.4 because it can be bounded by the series in which t is replaced by its distance of closest approach to the origin. Such a series can be integrated term by term with respect to t when inserted in Eq. (D3.4). Furthermore, the required integral over t in Eq. (D3.4) is precisely that evaluated in Eq. (D1.23)

$$\frac{1}{2\pi i} \int_C \frac{e^t}{t^{v+p+1}} \, dt = \frac{1}{\Gamma(v+p+1)} \tag{D3.7}$$

where this relation was proved for arbitrary value of the separation constant v. A combination of Eqs. (D3.4), (D3.6), and (D3.7) gives the series expansion

$$J_v(z) = \left(\frac{z}{2}\right)^v \sum_{p=0}^{\infty} \frac{(-1)^p z^{2p}}{2^{2p} p! \, \Gamma(v+p+1)} \tag{D3.8}$$

This relation holds for arbitrary v and arbitrary z in the cut z plane. It is equivalent to the integral representation in Eq. (D3.4) and serves to define a class of solutions to Bessel's equation known as *Bessel functions of the first kind*.

Bessel's equation (D3.1) depends on only the parameter v^2. Thus, like the situation for Legendre's equation, the substitution $v \to -v$ generates a second solution $J_{-v}(z)$ to Bessel's equation. The analysis of Sec. 43 proves that their wronskian must have the form Cz^{-1}. To determine the constant C, we can use the leading terms for small $|z|$

$$J_v(z) \approx \left(\frac{z}{2}\right)^v \frac{1}{\Gamma(v+1)} \qquad J_{-v}(z) \approx \left(\frac{z}{2}\right)^{-v} \frac{1}{\Gamma(-v+1)} \tag{D3.9}$$

to find

$$W[J_v(z), J_{-v}(z)] = \frac{-2v}{z\Gamma(v+1)\Gamma(-v+1)} \tag{D3.10}$$

Equations (D1.9) and (D1.14) then yield the result

$$W[J_v(z), J_{-v}(z)] = -\frac{2 \sin v\pi}{\pi z} \tag{D3.11}$$

which differs from zero if v is not an integer. We therefore conclude that

If v is not an integer, then $J_v(z)$ and $J_{-v}(z)$ are two linearly independent solutions to Bessel's equation and form a fundamental system (D3.12)

This result has its most important application in the radial equation (C.25) obtained from separating the Helmholtz equation in spherical coordinates. In the most familiar case where m and l are both integers this equation is just Bessel's equation with an index

$$v = l + \tfrac{1}{2} \tag{D3.13}$$

The corresponding radial solutions (C.24) of Eq. (C.23) are known as *spherical Bessel functions*. The result (D3.12) immediately provides *a fundamental system of spherical Bessel functions*

$$j_l(z) \equiv \left(\frac{\pi}{2z}\right)^{1/2} J_{l+1/2}(z) \tag{D3.14}$$

$$n_l(z) \equiv (-1)^{l+1}\left(\frac{\pi}{2z}\right)^{1/2} J_{-l-1/2}(z) \tag{D3.15}$$

It is evident from condition (D3.13) that the spherical Bessel functions will have power-series representations, which can be calculated directly from the series in Eq. (D3.8). It turns out that these power series for j_l and n_l can be rewritten as simple combinations of sines and cosines multiplied by polynomials. Furthermore, Eqs. (D3.8) and (D3.14) show that the spherical Bessel function of the first kind $j_l(z)$ contains only positive integral powers of z; hence it is an *entire function* of z, that is, analytic throughout the finite z plane. For small $|z|$ an expansion readily yields

$$j_l(z) \xrightarrow[z \to 0]{} \frac{z^l}{(2l+1)!!} \tag{D3.16}$$

where $(2l+1)!! = 1 \cdot 3 \cdot 5 \cdots (2l+1)$. In contrast, the expansion of the spherical Bessel function of the second kind $n_l(z)$ contains inverse powers of z and has a singularity at the origin, where it becomes infinite. An expansion for $|z| \ll 1$ gives the leading term

$$n_l(z) \xrightarrow[z \to 0]{} -\frac{(2l-1)!!}{z^{l+1}} \tag{D3.17}$$

where $(-1)!! = 1$.

The situation for integral values of v is more complicated, for changing v to $-v$ in Eq. (D3.8) does not generate a second linearly independent solution. This result follows directly from Eq. (D3.11). Alternatively, a little manipulation with Eq. (D3.8) and the property of the Γ function in Eq. (D1.13) establishes the relation

$$J_m(z) = (-1)^m J_{-m}(z) \qquad m = 0, \pm 1, \pm 2, \ldots \tag{D3.18}$$

Thus $J_{-m}(z)$ is not a second linearly independent solution, and constructing a fundamental system for integral v requires a more detailed analysis. The conventional procedure is to define a *Neumann function* according to†

$$N_v(z) \equiv \frac{\cos \pi v \, J_v(z) - J_{-v}(z)}{\sin \pi v} \tag{D3.19}$$

A little manipulation with Eq. (D3.11) shows that

$$W[J_v(z), N_v(z)] = \frac{2}{\pi z} \tag{D3.20}$$

---

† This function is frequently denoted $Y_v(z)$, but we follow the common practice of physicists, which avoids confusion with the spherical harmonic Y_{lm}.

so that the set $\{J_\nu(z), N_\nu(z)\}$ forms a fundamental system for all ν. If one lets $\nu \to m$ (an integer) in Eq. (D3.19), the numerator vanishes by Eq. (D3.18) but the denominator also vanishes and the ratio (D3.19) remains finite. Thus we conclude that

> The functions $J_\nu(z)$ and $N_\nu(z)$ form a fundamental system for all values of the index ν (D3.21)

For integral ν, the Neumann functions $N_m(z)$ can be calculated from Eq. (D3.19) and either Eq. (D3.4) or Eq. (D3.8). They are quite complicated; for our purposes, the main property is that the functions $N_m(z)$ *are singular at the origin*

$$N_0(z) \xrightarrow[z \to 0]{} \frac{2}{\pi} \left[\ln \left(\tfrac{1}{2} z \right) + \gamma \right] \tag{D3.22}$$

$$N_m(z) \xrightarrow[z \to 0]{} -\frac{(m-1)!}{\pi} \left(\frac{z}{2} \right)^{-m} \qquad m = 1, 2, 3, \dots \tag{D3.23}$$

where $\gamma \approx 0.5772$ is Euler's constant. These relations can also be established directly by seeking a power-series solution to Eq. (D3.1) in the vicinity of the origin, although the series for $\nu = m =$ integer is not simple. The function $J_m(z)$, on the other hand, is well behaved [see Eq. (D3.18)]. It is again an entire function of z, with the behavior at the origin

$$J_m(z) \xrightarrow[z \to 0]{} \frac{1}{m!} \left(\frac{z}{2} \right)^m \qquad m = 0, 1, 2, \dots \tag{D3.24}$$

If the physical region in cylindrical coordinates or polar coordinates (two dimensions) includes the origin and allows $0 \le \phi \le 2\pi$, then the condition of finite amplitude eliminates the Neumann function as, for example, in the case of a vibrating circular drumhead. On the other hand, N_m must be retained if the origin is not part of the allowed physical region. This latter situation arises, for example, in the problem of tidal waves on an annular channel bounded by concentric circular cylinders. Similar remarks apply to spherical Bessel functions.

The integral representation (D3.4) allows us to determine the asymptotic form of $J_\nu(z)$ as $|z| \to \infty$ with the method of stationary phase (see Sec. 55). We first change variables to $u \equiv 2t/z$ to find

$$J_\nu(z) = \frac{1}{2\pi i} \int_{C'} \frac{du}{u^{\nu+1}} \exp \left[\tfrac{1}{2} z \left(u - \frac{1}{u} \right) \right] \tag{D3.25}$$

where C' can now be deformed through a region of analyticity of the integrand to assume the same shape in the u plane as that illustrated in Fig. D1.4. We furthermore take the circular part of the contour to run around the unit circle. On the straight sections just above and below the negative real axis, the variable u can be written as $e^{\sigma \pm i\pi}$, where σ runs from 0 to ∞; on the circular section, we write $u = e^{i\phi}$. A straightforward calculation gives the exact representation

$$J_\nu(z) = \frac{1}{2\pi} \int_{-\pi}^{\pi} d\phi \, e^{i(z \sin \phi - \nu\phi)} - \frac{1}{\pi} \sin \nu\pi \int_0^\infty d\sigma \, e^{-(z \sinh \sigma + \nu\sigma)} \tag{D3.26}$$

which generalizes Eq. (51.10) to arbitrary ν.

For simplicity, we assume that z is real and positive. For large z and arbitrary v, the second integral can be integrated by parts

$$\int_0^\infty d\sigma \, e^{-z \sinh \sigma} e^{v\sigma} = -\frac{1}{z} e^{-z \sinh \sigma} \frac{e^{v\sigma}}{\cosh \sigma} \Big|_0^\infty$$

$$+ \frac{1}{z} \int_0^\infty d\sigma \, e^{-z \sinh \sigma} \frac{d}{d\sigma} \frac{e^{v\sigma}}{\cosh \sigma}$$

$$= \frac{1}{z} + \frac{1}{z} \int_0^\infty d\sigma \, e^{-z \sinh \sigma} \frac{e^{v\sigma}}{\cosh \sigma} (v - \tanh \sigma) \qquad (D3.27)$$

which shows that this contribution is of order z^{-1} for $z \to \infty$. In contrast, the first term of Eq. (D3.26) will turn out to be of order $z^{-1/2}$, for it has precisely the form studied in Eq. (55.26) with the method of stationary phase. There are two stationary-phase points, at $\phi_0 = \pm\frac{1}{2}\pi$, and each one must be considered separately. Use of Eq. (55.37) immediately gives the desired result

$$J_v(z) \sim \frac{1}{z^{1/2}} \frac{\exp \left[i(z - \frac{1}{2}v\pi - \frac{1}{4}\pi)\right]}{(2\pi)^{1/2}}$$

$$+ \frac{1}{z^{1/2}} \frac{\exp \left[-i(z - \frac{1}{2}v\pi - \frac{1}{4}\pi)\right]}{(2\pi)^{1/2}} + O\left(\frac{1}{z}\right)$$

$$\sim \left(\frac{2}{\pi z}\right)^{1/2} \cos (z - \tfrac{1}{2}v\pi - \tfrac{1}{4}\pi) + O\left(\frac{1}{z}\right) \qquad (D3.28)$$

A more detailed analysis in fact shows that the next correction is of order $z^{-3/2}$ and that this asymptotic approximation holds throughout the z plane cut along the negative real axis.†

PROBLEMS

D.1 Show that

$$\int_0^1 t^{p-1}(1-t)^{q-1} \, dt = \frac{\Gamma(p)\Gamma(q)}{\Gamma(p+q)} \qquad \begin{matrix} |\text{Re } p > 0 \\ |\text{Re } q > 0 \end{matrix}$$

D.2 Use Eq. (D2.17) and integration by parts to verify the normalization factor for the spherical harmonic Y_{lm} defined in Eq. (D2.20).

D.3 For Re $\alpha > -1$, show that Eq. (D2.33) implies the asymptotic formula

$$Q_\alpha(z) \sim \frac{\Gamma(\tfrac{1}{2})\Gamma(\alpha+1)}{\Gamma(\alpha + \tfrac{3}{2})} \frac{1}{2^{\alpha+1} z^{\alpha+1}}$$

valid as $|z| \to \infty$.

(b) For $-1 < \text{Re } \alpha < 0$, and z real and greater than 1, show that Schläfli's integral (D2.24) can be rewritten

$$P_\alpha(z) = \frac{-\sin \pi\alpha}{2^\alpha \pi} \int_1^z \frac{dt(t^2 - 1)^\alpha}{(z - t)^{\alpha+1}}$$

† Whittaker and Watson [5], sec. 17.5.

(c) For $z \to \infty$, use part (b) to derive the asymptotic approximation

$$P_\alpha(z) \sim \frac{z^\alpha \; \Gamma(2\alpha + 1)}{2^\alpha \, [\Gamma(\alpha + 1)]^2} = \frac{z^\alpha 2^\alpha \Gamma(\alpha + \frac{1}{2})}{\Gamma(\frac{1}{2}) \Gamma(\alpha + 1)}$$

where the second form follows from the duplication formula for the Γ function. Use analytic continuation to argue that this formula holds for general values of α.

(d) Hence derive the wronskian

$$W[P_\alpha(z), Q_\alpha(z)] = (1 - z^2)^{-1}$$

D.4 Laplace's method of solving certain ordinary differential equations is to seek a solution in the form

$$u(x) = \int_C e^{xt} T(t) \, dt$$

where C is a suitably chosen contour.

(a) Apply this method to Bessel's equation of order zero

$$xu'' + u' + xu = 0$$

[*Hint:* Use the relation $xe^{xt} = d(e^{xt})/dt$ and an integration by parts.] Show that $T(t)$ satisfies the ordinary differential equation $(1 + t^2)T'(t) + tT(t) = 0$. Discuss the analytic structure of the resulting solution. What are the various ways of drawing the branch cuts?

(b) Choose an integration contour C such that $u(x)$ remains finite as $x \to 0$. Why is this solution a multiple of $J_0(x)$? Use the known result $J_0(0) = 1$ to determine the multiplicative constant, and hence find an integral representation for $J_0(x)$.

(c) Calculate the integral

$$\int_{-\infty}^{\infty} dx \, e^{-ikx} J_0(x)$$

by substituting your integral representation and thus evaluate the Fourier transform of $J_0(x)$.

D.5 (a) Evaluate the Fourier integral $(2\pi)^{-3} \int d^3k \; e^{i\mathbf{k} \cdot \mathbf{r}} (k^2 + \lambda^2)^{-1}$ in cylindrical polar coordinates. Take the limit $\lambda \to 0$ and compare with Eq. (50.12) to obtain

$$\int_0^{\infty} dk \, J_0(ka) e^{-kb} = (a^2 + b^2)^{-1/2}$$

where a and b are real and positive.

(b) Analytically continue this result to b near the positive imaginary axis and hence obtain the integral

$$\int_0^{\infty} dk \, J_0(ka) e^{-ikb} = \begin{cases} (a^2 - b^2)^{-1/2} & a > b \\ -i(b^2 - a^2)^{-1/2} & b > a \end{cases}$$

again valid for real positive a and b. Take the real and imaginary parts explicitly. Compare with part (c) of Prob. D.4.

D.6 Derive Poisson's integral representation for the spherical Bessel functions

$$j_l(z) = \frac{z^l}{2^{l+1} l!} \int_0^{\pi} \cos(z \cos \theta) \sin^{2l+1}\theta \, d\theta$$

by expanding and integrating term by term.

D.7 Use Rodrigues' formula for $P_l(\cos \theta)$ [Eq. (D2.11)], integration by parts, and Poisson's representation of $j_l(z)$ in Prob. D.6 to derive the plane-wave expansion

$$e^{i\mathbf{k} \cdot \mathbf{x}} = \sum_{l=0}^{\infty} (2l + 1)i^l \, j_l(kx) P_l(\hat{k} \cdot \hat{x})$$

SELECTED ADDITIONAL READINGS

1. G. Arfken, "Mathematical Methods for Physicists," 2d ed., Academic, New York, 1970.
2. G. F. Carrier, M. Krook, and C. E. Pearson, "Functions of a Complex Variable," McGraw-Hill, New York, 1966.
3. N. N. Lebedev, "Special Functions and Their Applications," Dover, New York, 1972.
4. P. M. Morse and H. Feshbach, "Methods of Theoretical Physics," McGraw-Hill, New York, 1953.
5. E. T. Whittaker and G. N. Watson, "A Course of Modern Analysis," 4th ed., Cambridge University Press, London, 1969.
6. L. I. Schiff, "Quantum Mechanics," 3d ed., McGraw-Hill, New York, 1968.

APPENDIX

E

SELECTED MATHEMATICAL FORMULAS†

E1 THE Γ FUNCTION

$$\Gamma(z) = \int_0^\infty dt \ e^{-t} t^{z-1} \qquad \text{Re } z > 0$$

$$\Gamma(z+1) = z\Gamma(z)$$

$$\Gamma(z)\Gamma(1-z) = \frac{\pi}{\sin \pi z}$$

$$\Gamma(2z) = \pi^{-1/2} 2^{2z-1} \Gamma(z)\Gamma(z+\tfrac{1}{2})$$

$$\ln \Gamma(z) \sim (z-\tfrac{1}{2}) \ln z - z + \tfrac{1}{2} \ln 2\pi + O(z^{-1}) \qquad \begin{aligned} &|z| \to \infty \\ &|\arg z| < \pi \end{aligned}$$

$$\Gamma(n+1) = n! \qquad \text{for } n = 1, 2, \ldots$$

$$\Gamma(1) = 1$$

$$\Gamma(\tfrac{1}{2}) = \sqrt{\pi}$$

$$\psi(z) \equiv \frac{d}{dz} \ln \Gamma(z) = \frac{1}{\Gamma(z)} \frac{d}{dz} \Gamma(z)$$

$$\psi(1) = \int_0^\infty dt \ e^{-t} \ln t = -\gamma \qquad \text{where } \gamma = 0.5772 \cdots \text{ is Euler's constant}$$

---

† A complete set of formulas and tables can be found in M. Abramowitz and I. A. Stegun (eds.), "Handbook of Mathematical Functions," NBS Appl. Math. Ser. 55, Washington, 1964.

$$\int_0^1 dx\, x^{\lambda-1}(1-x)^{\mu-1} = \frac{\Gamma(\lambda)\Gamma(\mu)}{\Gamma(\lambda+\mu)} \qquad \begin{matrix} \text{Re } \lambda > 0 \\ \text{Re } \mu > 0 \end{matrix}$$

$$= \int_0^\infty dx\, \frac{x^{\lambda-1}}{(1+x)^{\lambda+\mu}} = 2 \int_0^{\pi/2} d\theta\, \sin^{2\lambda-1}\theta\, \cos^{2\mu-1}\theta$$

E2 ERROR FUNCTION

$$\text{erf }(z) \equiv \frac{2}{\sqrt{\pi}} \int_0^z d\zeta\, e^{-\zeta^2}$$

$$\text{erf }(\infty) = 1$$

$$\text{erf }(z) \approx \frac{2}{\sqrt{\pi}}\left(z - \frac{z^3}{3} + \cdots\right) \qquad \text{for } |z| \ll 1$$

$$\text{erfc }(z) \equiv 1 - \text{erf }(z) = \frac{2}{\sqrt{\pi}} \int_z^\infty d\zeta\, e^{-\zeta^2}$$

$$\text{erfc }(z) \sim \frac{e^{-z^2}}{\sqrt{\pi}\, z}\,(1 + \cdots) \qquad \begin{matrix} |z| \to \infty \\ |\arg z| < \frac{3}{4}\pi \end{matrix}$$

E3 LEGENDRE FUNCTIONS

Recursion and General Relations [also for $Q_\alpha(z)$]

$$P'_{\alpha+1}(z) - (2\alpha+1)P_\alpha(z) - P'_{\alpha-1}(z) = 0$$

$$(\alpha+1)P_{\alpha+1}(z) - (2\alpha+1)zP_\alpha(z) + \alpha P_{\alpha-1}(z) = 0$$

$$W[P_\alpha(z), Q_\alpha(z)] = (1 - z^2)^{-1}$$

Addition Formulas

$$\frac{1}{(1-2hx+h^2)^{1/2}} = \sum_{l=0}^\infty h^l P_l(x) \qquad \text{for } |h| < 1$$

$$\frac{1}{|\mathbf{r} - \mathbf{r}'|} = \sum_{l=0}^\infty \frac{r_<^l}{r_>^{l+1}} P_l(\hat{r} \cdot \hat{r}')$$

$$P_l(\hat{r} \cdot \hat{r}') = \frac{4\pi}{2l+1} \sum_{m=-l}^l Y^*_{lm}(\theta', \phi') Y_{lm}(\theta, \phi)$$

$$e^{ikr\cos\theta} = \sum_{l=0}^\infty (2l+1)i^l j_l(kr) P_l(\cos\theta)$$

Explicit Forms for $l = 0, 1, 2, 3, \ldots, \infty$ and integral m

$$P_l(x) = \frac{1}{2^l l!} \frac{d^l}{dx^l} (x^2 - 1)^l$$

$$P_l^m(x) = (1 - x^2)^{m/2} \frac{d^m}{dx^m} P_l(x) \qquad m \geq 0$$

$$P_0(x) = 1$$

$$P_1(x) = x$$

$$P_2(x) = \tfrac{1}{2}(3x^2 - 1)$$

$$\int_{-1}^{1} dx \, P_l(x) P_{l'}(x) = \frac{2}{2l + 1} \delta_{ll'}$$

$$Q_0(z) = \tfrac{1}{2} \ln \frac{z + 1}{z - 1} \qquad \text{[see Eq. (D2.33)]}$$

$$Q_1(z) = \tfrac{1}{2} z \ln \frac{z + 1}{z - 1} - 1$$

$$Y_{lm}(\theta, \phi) = (-1)^m \left[\frac{2l + 1}{4\pi} \frac{(l - m)!}{(l + m)!} \right]^{1/2} P_l^m(\cos \theta) e^{im\phi} \qquad m \geq 0$$

$$Y_{l, -m} = (-1)^m Y_{lm}^*$$

$$\int d\Omega \, Y_{lm}^*(\theta, \phi) Y_{l'm'}(\theta, \phi) = \delta_{ll'} \delta_{mm'}$$

$$Y_{l0}(\theta, \phi) = \left(\frac{2l + 1}{4\pi} \right)^{1/2} P_l(\cos \theta)$$

$$Y_{00}(\theta, \phi) = (4\pi)^{-1/2}$$

$$Y_{10}(\theta, \phi) = (3/4\pi)^{1/2} \cos \theta$$

$$Y_{11}(\theta, \phi) = -(3/8\pi)^{1/2} \sin \theta \, e^{i\phi}$$

$$Y_{20}(\theta, \phi) = \tfrac{1}{2}(5/4\pi)^{1/2}(3 \cos^2 \theta - 1)$$

$$Y_{21}(\theta, \phi) = -(15/8\pi)^{1/2} \sin \theta \cos \theta \, e^{i\phi}$$

$$Y_{22}(\theta, \phi) = \tfrac{1}{4}(15/2\pi)^{1/2} \sin^2 \theta \, e^{2i\phi}$$

E4 CYLINDRICAL BESSEL FUNCTIONS

Recursion Relations [also for $N_\nu(z)$]

$$J_{\nu+1}(z) + 2J_\nu'(z) - J_{\nu-1}(z) = 0$$

$$J_{\nu+1}(z) - \frac{2\nu}{z} J_\nu(z) + J_{\nu-1}(z) = 0$$

$$\frac{d}{dz} z^\nu J_\nu(z) = z^\nu J_{\nu-1}(z)$$

$$\frac{d}{dz} z^{-\nu} J_\nu(z) = -z^{-\nu} J_{\nu+1}(z)$$

Series and Approximate Forms (*m* is a nonnegative integer)

$$J_\nu(z) = (\tfrac{1}{2}z)^\nu \sum_{p=0}^{\infty} \frac{(-1)^p}{p!\,\Gamma(p+\nu+1)} (\tfrac{1}{2}z)^{2p}$$

$$N_0(z) = \frac{2}{\pi} J_0(z) \ln (\tfrac{1}{2}z) - \frac{1}{\pi} \sum_{p=0}^{\infty} \frac{(-1)^p(\tfrac{1}{2}z)^{2p}}{(p!)^2} 2\psi(p+1)$$

$$H_\nu^{(1)}(z) = J_\nu(z) + iN_\nu(z)$$

$$J_\nu(z) \approx \frac{1}{\Gamma(\nu+1)} (\tfrac{1}{2}z)^\nu \qquad z \to 0$$

$$N_0(z) \approx \frac{2}{\pi}[\ln (\tfrac{1}{2}z) + \gamma] \qquad z \to 0$$

$$N_m(z) \approx -\frac{\Gamma(m)}{\pi}\left(\frac{2}{z}\right)^m \qquad \begin{array}{l} m \neq 0 \\ z \to 0 \end{array}$$

$$J_\nu(z) \sim \left(\frac{2}{\pi z}\right)^{1/2} \cos (z - \tfrac{1}{2}\nu\pi - \tfrac{1}{4}\pi) \qquad |z| \to \infty; \ |\arg z| < \pi$$

$$N_\nu(z) \sim \left(\frac{2}{\pi z}\right)^{1/2} \sin (z - \tfrac{1}{2}\nu\pi - \tfrac{1}{4}\pi) \qquad |z| \to \infty; \ |\arg z| < \pi$$

$$H_\nu^{(1)}(z) \sim \left(\frac{2}{\pi z}\right)^{1/2} \exp [i(z - \tfrac{1}{2}\nu\pi - \tfrac{1}{4}\pi)] \qquad |z| \to \infty; \ |\arg z| < \pi$$

$$e^{ikr \cos \phi} = \sum_{l=-\infty}^{\infty} i^l e^{il\phi} J_l(kr)$$

$$W[J_\nu(z), N_\nu(z)] = \frac{2}{\pi z}$$

$$I_\nu(z) = \exp (-\tfrac{1}{2}i\nu\pi)J_\nu(z \exp \tfrac{1}{2}i\pi) \qquad -\pi < \arg z \leq \tfrac{1}{2}\pi$$

$$K_\nu(z) = \tfrac{1}{2}\pi i \exp (\tfrac{1}{2}i\nu\pi)H_\nu^{(1)}(z \exp \tfrac{1}{2}i\pi) \qquad -\pi < \arg z \leq \tfrac{1}{2}\pi$$

$$I_\nu(z) \sim \frac{e^z}{(2\pi z)^{1/2}} \qquad |\arg z| < \tfrac{1}{2}\pi \qquad |z| \to \infty$$

$$K_\nu(z) \sim \left(\frac{\pi}{2z}\right)^{1/2} e^{-z} \qquad |\arg z| < \tfrac{3}{2}\pi \qquad |z| \to \infty$$

$$\int_0^1 x\,dx\, J_m(\alpha_{m,n} x)J_m(\alpha_{m,n'} x) = \delta_{nn'} \tfrac{1}{2}[J_{m\pm1}(\alpha_{m,n})]^2 \qquad \text{where } \alpha_{m,n} = n\text{th zero of } J_m(x)$$

$\alpha_{m,n} = $ nth zero of $J_m(x)$

$\alpha_{m,n}$	$n = 1$	$n = 2$	$n = 3$	$n = 4$
$m = 0$	2.4048	5.5201	8.6537	11.7915
$m = 1$	3.8317	7.0156	10.1735	13.3237
$m = 2$	5.1356	8.4172	11.6198	14.7960
$m = 3$	6.3802	9.7610	13.0152	16.2235

$\alpha'_{m,n} = $ nth root of $J'_m(x) = 0$

$\alpha'_{m,n}$	$n = 1$	$n = 2$	$n = 3$	$n = 4$
$m = 0$	0.0000	3.8317	7.0156	10.1735
$m = 1$	1.8412	5.3314	8.5363	11.7060
$m = 2$	3.0542	6.7061	9.9695	13.1704
$m = 3$	4.2012	8.0152	11.3459	14.5859

E5 SPHERICAL BESSEL FUNCTIONS

Specific Forms (l is a nonnegative integer)

$$j_l(z) = \left(\frac{\pi}{2z}\right)^{1/2} J_{l+1/2}(z) = (-z)^l \left(\frac{1}{z}\frac{d}{dz}\right)^l \frac{\sin z}{z}$$

$$n_l(z) = (-1)^{l+1} \left(\frac{\pi}{2z}\right)^{1/2} J_{-l-1/2}(z) = -(-z)^l \left(\frac{1}{z}\frac{d}{dz}\right)^l \frac{\cos z}{z}$$

$$h_l^{(1)}(z) = j_l(z) + in_l(z)$$

$$W[j_l(z), n_l(z)] = z^{-2}$$

$$j_0(z) = \frac{\sin z}{z} \qquad\qquad n_0(z) = -\frac{\cos z}{z}$$

$$j_1(z) = \frac{\sin z}{z^2} - \frac{\cos z}{z} \qquad n_1(z) = -\frac{\cos z}{z^2} - \frac{\sin z}{z}$$

$$h_0^{(1)}(z) = \frac{e^{iz}}{iz} \qquad\qquad h_1^{(1)}(z) = -\frac{e^{iz}}{z}\left(1 + \frac{i}{z}\right)$$

$$j_l(z) \approx \frac{z^l}{(2l+1)!!} \qquad z \to 0 \qquad (2l+1)!! = (2l+1)(2l-1)\cdots 3\cdot 1$$

$$n_l(z) \approx -\frac{(2l-1)!!}{z^{l+1}} \qquad z \to 0 \qquad (-1)!! \equiv 1$$

$$j_l(z) \sim z^{-1} \cos\left[z - \tfrac{1}{2}(l+1)\pi\right] \qquad |z| \to \infty;\ |\arg z| < \pi$$

$$n_l(z) \sim z^{-1} \sin\left[z - \tfrac{1}{2}(l+1)\pi\right] \qquad |z| \to \infty;\ |\arg z| < \pi$$

$$h_l^{(1)}(z) \sim z^{-1} \exp\left\{i[z - \tfrac{1}{2}(l+1)\pi]\right\} \qquad |z| \to \infty;\ |\arg z| < \pi$$

Recursion Relations [also for $n_l(z)$]

$j_{l+1}(z) - z^{-1}(2l + 1)j_l(z) + j_{l-1}(z) = 0$

$(l + 1)j_{l+1}(z) + (2l + 1)j_l'(z) - lj_{l-1}(z) = 0$

PHYSICAL CONSTANTS†

F1 ASTRONOMICAL CONSTANTS

Mass of sun	$M_s = 1.989 \times 10^{33}$ g
Mean sun-earth distance	$R_{se} = 1$ AU $= 1.4960 \times 10^{13}$ cm
Mass of earth	$M_e = 5.976 \times 10^{27}$ g
Equatorial radius of earth	$R_e = 6.37816 \times 10^8$ cm
Polar radius of earth	$R_p = 6.35678 \times 10^8$ cm
Angular velocity of earth's daily rotation	$\omega_e = 7.292 \times 10^{-5}$ rad s^{-1}
Gravitational acceleration at equator	$g_e = 978.03$ cm s^{-2}
Gravitational acceleration at pole	$g_p = 983.20$ cm s^{-2}
Angular velocity of earth's annual motion	$\omega_{se} = 1.991 \times 10^{-7}$ rad s^{-1}

F2 FUNDAMENTAL CONSTANTS

$G = 6.670 \times 10^{-8}$ dyn cm^2 g^{-2}

$c = 2.998 \times 10^{10}$ cm s^{-1}

$\hbar = \dfrac{h}{2\pi} = 1.055 \times 10^{-27}$ erg s

$e = 4.803 \times 10^{-10}$ esu

$k_B = 1.381 \times 10^{-16}$ erg K^{-1}

$R = 8.314 \times 10^7$ erg K^{-1} mol^{-1}

---

† Values from C. W. Allen, "Astrophysical Quantities," Athlone Press, London, 1973, chaps. 2, 6, 7, 9.

Avogadro number $N = 6.022 \times 10^{23} \text{ mol}^{-1}$

$$\alpha^{-1} = \left(\frac{e^2}{\hbar c}\right)^{-1} = 137.0$$

$$m_e = 9.110 \times 10^{-28} \text{ g}$$

$$m_p = 1.673 \times 10^{-24} \text{ g}$$

$$\frac{m_p}{m_e} = 1836$$

$$1 \text{ eV} = 1.602 \times 10^{-12} \text{ erg}$$

APPENDIX

G

BASIC TEXTS AND MONOGRAPHS

Coulson, C. A. (1958): "Waves," 7th ed., Oliver & Boyd, London.
Goldstein, H. (1950): "Classical Mechanics," Addison-Wesley, Reading, Mass.
Joos, G. (1958): "Theoretical Physics," 3d ed., with I. M. Freeman, Hafner, New York.
Konopinski, E. J. (1969): "Classical Descriptions of Motion," Freeman, San Francisco.
Lamb, H. (1945): "Hydrodynamics," 6th ed., Dover, New York.
Landau, L. D., and E. M. Lifshitz (1959): "Fluid Mechanics," Addison-Wesley, Reading, Mass.
Landau, L. D., and E. M. Lifshitz (1960): "Mechanics," Addison-Wesley, Reading, Mass.
Marion, J. B. (1970): "Classical Dynamics of Particles and Systems," 2d ed., Academic, New York.
Rayleigh, Lord (1945): "The Theory of Sound," vols. I and II, Dover, New York.
Sommerfeld, A. (1950): "Mechanics of Deformable Bodies," Academic, New York.
Sommerfeld, A. (1952): "Mechanics," Academic, New York.
Symon, K. R. (1971): "Mechanics," 3d ed., Addison-Wesley, Reading, Mass.
Whittaker, E. T. (1944): "A Treatise on the Analytical Dynamics of Particles and Rigid Bodies," Dover, New York.

INDEX

Page numbers in *italic* indicate problems.

Accelerating coordinate system, 37
Acceleration, 5, 37
  due to gravity, 40, *48,* 147
  finiteness of, 219
  neglect of vertical, 358
  of rocket, *28*
  (*See also* Newton's laws of motion, second law)
Acoustic disturbance, 317, *354*
Acoustic impedance, 307, *354*
Acoustic point source, *353*
Acoustic pressure, 307, 451
Acoustic radiation, 307, 320, 323, 324, 331, 334, 336,
  353, 354, 405
  (*See also* Sound wave)
Acoustic wave, 325
  in moving media, *351, 353*
  (*See also* Sound wave)
Action, 66, 173, 178, 184, 187
  and reaction, 2
  stationary, 67, 178
Action integral, 192, 196
Action variable, 192, 195, 196, 198, *205*
Additive separability, 188–190
Adiabatic equation of state, 340, 343
Affine connection, *85*
Airy disk, 331
Alfvén waves, *356*
Amplitude modulation, 107, 376–378, *404*
Analytic continuation, 418, 506, 528
  methods for, 506, 507
  uniqueness of, 507
Analytic function, 482, 496, 497, 500, 503, 505, 507,
  527
  derivative of, 494
Analyticity, 314, 428, 483, 488, 491, 498, *508*
  in annulus, 491
  of derivatives of analytic function, 494
  of time, Fourier transform, 251, 316

Angle variable, 193, 194
Angular differential operator, 333
Angular distribution of radiation, 324, *354*
Angular frequency, 146
  (*See also* Frequency)
Angular momentum, 3, 6, 10, 140, 142, 144, 153,
  436, 464
  conservation law of, 3, 7, 79, *455*
  internal, 8, 18
  Poisson brackets for, *205*
  quantization of, 199
  rate of change of, 143, 148, 149
  of rigid body, 136
Angular-momentum barrier, 12, 164
Angular velocity, 34–36, 135, 136, 138, 140, 144, 151,
  156, 159
  of earth, 39, 43, 46
  and Euler angles, 156
  rate of change of, 36
Annual penetration depth for thermal waves, 414
Annulus, 491, 498, 500
Antinodes, *351*
Apertures, 325–327, 329–331
Applied force, 70, 292
Area in phase space, 192
Areal velocity, 11
Associated Legendre equation, 523, 531, 534
Associated Legendre functions, 533*n.,* 534, 537
Associated Legendre polynomials, 534
Astronomical unit, 39
Asymmetric top, 153, *171*
Asymptote of hyperbola, 20
Asymptotic form, 338, *354, 405,* 427, 429
  of Bessel function, *405,* 544
  of Green's function, 326, 391
  of Legendre function, *545*
Atmospheric pressure, 358, 368, 381
Attenuation constant, 454